This book is the first to give a comprehensive overview of the important field of quantum chaos. It forms a statement of our present understanding of chaotic behaviour in a wide variety of quantum and semiclassical systems, and describes both experimental and theoretical investigations.

A general introduction sets out the main features of quantum chaos, highlighting the surprising fact that quantum mechanical systems can exhibit behaviour which is more regular and stable than in the classical limit. Thereafter, in an authoritative collection of new and previously published papers, prominent scientists put forward their particular interpretations of quantum chaos with reference to a broad range of interesting physical systems. Classical chaos and quantum localisation, atoms in strong fields, semiclassical approximations, level statistics and random matrix theory are all described in detail.

As yet, there is no universally accepted definition of quantum chaos. However, by dealing with such a wide range of topics from different branches of physics, this book provides a unique overview of this rapidly expanding field, and will therefore be of great interest to graduate students and researchers in many areas of physics and chemistry.

Quantum chaos

between order and disorder

DIETER WINTGEN
11 June 1957 – 16 August 1994

This book is dedicated to Dieter Wintgen, whose life was cut tragically short by a climbing accident in the Swiss Alps. Dieter made several key contributions to the development of the theory of quantum chaos and its physical applications. In particular, his work on the role of classical unstable orbits in the spectroscopy of hydrogen in strong external fields and the periodic orbit quantization of helium had much to do with making the whole field as vigorous as it is today. He was one of the foremost experts on quantum calculations, and performed some of the most accurate numerical evaluations of atomic spectra.

Dieter was an iconoclast. He always went around in sandals, chagrined older professors with the holes in his pants, was as tough a discussion partner as he was a soccer player, and worked with his students as though he himself were one of them. At the time of his unexpected death Dieter was in the midst of several exciting projects. He will be very much missed not only by his family, but also by all of us whose science he enriched and who looked forward to so much more to come from him.

Quantum chaos

between order and disorder

A selection of papers compiled and introduced by

GIULIO CASATI

Dipartimento di Fisica, Università di Milano

BORIS CHIRIKOV

Budker Institute of Nuclear Physics,
Novosibirsk, Russia

CAMBRIDGE
UNIVERSITY PRESS

CAMBRIDGE UNIVERSITY PRESS
Cambridge, New York, Melbourne, Madrid, Cape Town, Singapore, São Paulo

Cambridge University Press
The Edinburgh Building, Cambridge CB2 2RU, UK

Published in the United States of America by Cambridge University Press, New York

www.cambridge.org
Information on this title: www.cambridge.org/9780521432917

First published 1995
This digitally printed first paperback version 2006

A catalogue record for this publication is available from the British Library

Library of Congress Cataloguing in Publication data

Quantum chaos : between order and disorder : a paper selection
compiled and introduced by Giulio Casati, Boris Chirikov.
p. cm.
ISBN 0-521-43291-X
1. Quantum chaos. I. Casati, Giulio, 1942– , II. Chirikov, B.
V. (Boris Valerianovich)
QC174.17.C45Q36 1995
530.1′2–dc20 93-37135 CIP

ISBN-13 978-0-521-43291-7 hardback
ISBN-10 0-521-43291-X hardback

ISBN-13 978-0-521-03166-0 paperback
ISBN-10 0-521-03166-4 paperback

Contents

Preface

This volume presents a collection of basic papers, some already published others specially written for this volume, devoted to the study of a new phenomenon, the so-called quantum chaos. This problem arose from the by now, well known, classical dynamical chaos. However, unlike the latter, the study of quantum chaos is still in its early stages, attracting the ever growing interest of many physicists (but, unfortunately, of many fewer, as yet, mathematicians).

The original intention, of physicists at least, was mainly to understand the very important generic phenomenon of classical chaos from the viewpoint of the more deep and general quantum mechanics. At first sight it might seem that quantum chaos is simply a particular case of the general phenomenon of dynamical chaos in the well developed ergodic theory of dynamical systems; or it might be a trivial implication of the correspondence principle. Yet, Nature has turned out to be much more tricky, and more interesting!

As the present collection of papers clearly shows, there is no classical-like chaos at all in quantum mechanics. On the other hand since Nature, as is commonly accepted, obeys quantum mechanics, what is then the physical meaning of dynamical chaos? As a result of this surprising obstacle, the general situation in the study of quantum chaos, in the present state of research, might be characterized as some confusion and disorganization which is of course a typical situation in the early stages of a new field of scientific research. It greatly stimulates and encourages the active search for new approaches to the problem, quite often without any attempts to reconcile the different conceptions. The primary goal of this collection is just to help in the cure of such a disease of growth.

For the reader's convenience we have grouped the papers into four different main topics: a) "Classical chaos and quantum localization" which is the most extensively investigated subject; b) "Atoms in magnetic and microwave fields" which refers also to the laboratory experiments in quantum chaos; c) "Semi-classical approximations" which examines the transition between classical and quantum chaos; and d) "Level statistics and random matrix theory" which describes the relation with the well developed statistical theory of complex quantum systems.

The collection of papers is preceded by an introductory chapter in which we try to link different approaches to the problem of quantum chaos guided by our current personal understanding of this new field.

We thank Anna Auguadro for her valuable assistance in the preparation of this volume.

Novosibirsk – Como

GIULIO CASATI

BORIS CHIRIKOV

Acknowledgments

The following articles in this collection have already been published in the sources shown:

G. CASATI, B. V. CHIRIKOV, F. M. IZRAILEV and J. FORD
Stochastic behaviour of a quantum pendulum under a periodic perturbation
in Lecture Notes in Physics, **93**, p 334–52 (1979), reproduced by permission Springer-Verlag, Heidelberg

D. R. GREMPEL, R. E. PRANGE and S. E. FISHMAN
Quantum dynamics of a nonintegrable system
in Physical Review, A**29**, 1639–47 (1984), reproduced by permission of The American Physical Society, New York

R. BLÜMEL, S. FISHMAN and U. SMILANSKY
Excitation of molecular rotation by periodic microwave pulses. A testing ground for Anderson localization
in the Journal of Chemical Physics, **84**, 2604–14 (1986), reproduced by permission of The American Institute of Physics, New York

D. L. SHEPELYANSKY
Localization of diffusive excitation in multi-level systems
in Physica, **28D**, 103–14 (1987), reproduced by permission of North-Holland Physics Publishing, Amsterdam

F. HAAKE, M. KUS and R. SCHARF
Classical and quantum chaos for a kicked top
in Zeitschrift für Physik, B**65**, 381–95 (1987), reproduced by permission of Springer-Verlag, Heidelberg

L. E. REICHL and L. HAOMING
Self-similarity in quantum dynamics
in Physical Review, A**42**, 4543–61, (1990), reproduced by permission of The American Physical Society, New York

E. OTT, T. M. ANTONSEN JR and J. HANSON
Effect of noise on time-dependent quantum chaos
in Physical Review Letters, **53**, *2187–90, (1984), reproduced by permission of The American Physical Society, New York*

R. F. GRAHAM
Dynamical localization, dissipation and noise
in the Proceedings of the Varenna Summer School on Quantum Chaos, 1991, *reproduced by permission of the Italian Physical Society*

J. E. BAYFIELD, G. CASATI, I. GUARNERI and D. W. SOKOL
Localization of classically chaotic diffusion for hydrogen atoms in microwave fields
in Physical Review Letters, **63**, *364–7, (1989), reproduced by permission of The American Physical Society, New York*

R. V. JENSEN, M. M. SANDERS, M. SARACENO and B. SUNDARAM
Inhibition of quantum transport due to 'scars' of unstable periodic orbits
in Physical Review Letters, **63**, *2771–5 (1989), reproduced by permssion of The American Physical Society, New York*

C.-H. IU, G. R. WELCH, M. M. KASH, D. KLEPPNER, D. DELANDE and J. C. GAY
Diamagnetic Rydberg atom: confrontation of calculated and observed spectra
in Physical Review Letters, **66**, *145–8, (1991), reproduced by permission of The American Physical Society, New York*

D. WINTGEN, K. RICHTER and G. TANNER
The semiclassical helium atom
in Chaos, **2**, *19–32, (1992), reproduced by permission of The American Institue of Physics, New York*

M. V. BERRY
Semiclassical theory of spectral rigidity
in the Proceedings of the Royal Society, A**400**, *229–51 (1985), reproduced by permission of The Royal Society, London*

M. A. SEPULVEDA, S. TOMSOVIC and E. J. HELLER
Semiclassical propagation: how long can it last?
in Physical Review Letters, **69**, *402–5 (1992), reproduced by permission of The American Physical Society, New York*

N. L. BALAZS and A. VOROS
The quantized baker's transformation
in Annals of Physics, **190**, *1–31 (1989), reproduced by permission of Academic Press, New York*

M. SARACENO
Classical structures in the quantized baker transformation
in Annals of Physics, **199**, *37–60, (1989), reproduced by permission of Academic Press, New York*

O. BOHIGAS, M. J. GIANNONI and C. SCHMIT
Characterization of chaotic quantum spectra and universality of level fluctuation laws
in Physical Review Letters, **52**, *1–4 (1983), reproduced by permission of The American Physical Society, New York*

Introduction

The legacy of chaos in quantum mechanics

G. CASATI

Dipartimento di Fisica, Università di Milano
Via Castelnuovo – 22100 Como, Italy

B. CHIRIKOV

Budker Institute of Nuclear Physics
630090 Novosibisrk 90, Russia

The present collection of papers is devoted to a rather new and very controversial topic, the so-called "quantum chaos". Some researchers see nothing essentially new at all in this phenomenon (apart from a number of various examples and models), and they have good reason to believe so. Indeed, the problems in this field all belong to the traditional, "old-fashion", and rather "simple" quantum mechanics of finite-dimensional systems with a given interaction and no quantized fields.

Nevertheless, many, including ourselves, consider quantum chaos to be a new discovery, though in an old field, of a great importance for fundamental physics. To understand this, the phenomenon of quantum chaos should be put into its proper perspective in recent developments in physics. The central point of this perspective is the conception of dynamical chaos (also a rather new topic) in classical mechanics (for a good review see, e.g., Refs.[1-3]). Thus before discussing the current understanding of quantum chaos we need briefly to describe classical dynamical chaos.

1 Dynamical chaos in classical mechanics

The conception of dynamical chaos destroys the deterministic image of the classical physics and shows that typically the trajectories of the deterministic equations of motion are in some sense random and unpredictable. The mechanism for such a surprising property of classical mechanics is related to the most strong (exponential) local instability of motion. In what follows we restrict ourselves to Hamiltonian (non-dissipative) systems as more fundamental. The local instability is described by the linearized equations of motion:

$$\left. \begin{array}{l} \dot{\xi} = \left(\dfrac{\partial^2 H}{\partial p \partial q} \right)_r \xi \;\; + \left(\dfrac{\partial^2 H}{\partial p^2} \right)_r \eta \\[3mm] \dot{\eta} = - \left(\dfrac{\partial^2 H}{\partial q^2} \right)_r \xi \;\; - \left(\dfrac{\partial^2 H}{\partial q \partial p} \right)_r \eta \end{array} \right\} \tag{1}$$

where $H = H(q, p, t)$ is the Hamiltonian; q, p are N-dimensional vectors in the phase space and $\xi = dq$, $\eta = dp$ are N-dimensional vectors in the tangent space.

The coefficients of the linear equations (1) are taken on the reference trajectory and hence explicitly depend on time. This dependence is generally very complicated if, for example, the reference trajectory is chaotic.

The stability of the motion on the reference trajectory is characterized by the maximal Lyapunov exponent defined as the limit

$$\Lambda = \lim_{|t| \to \infty} \frac{1}{|t|} \ln d(t) \tag{2}$$

where $d^2 = \xi^2 + \eta^2$ is the length of the tangent vector and where $d(0) = 1$ is assumed.

The exponential instability of motion means positive maximal Lyapunov exponent $\Lambda > 0$. Notice that in Hamiltonian systems the instability does not depend on the direction of time and, hence, is reversible as is the chaotic motion.

The reason why the exponentially unstable motion is called chaotic is in that almost all trajectories of such a motion are unpredictable in the following sense: according to the Alekseev–Brudno theorem (see Ref.[4]) in the algorithmic theory of dynamical systems the information $I(t)$ associated with a segment of trajectory of length t is equal asymptotically to

$$\lim_{|t| \to \infty} \frac{I(t)}{|t|} = h = \sum \Lambda_+ \tag{3}$$

where $\sum \Lambda_+$ is the sum of all positive Lyapunov exponents and h is the so-called KS (after Kolmogorov–Sinai) entropy. This important result shows that in order to predict each new segment of a chaotic trajectory one needs an additional information proportional to the length of this segment and independent of the full previous length of trajectory. This means that this information cannot be extracted from observation of the previous motion, even an infinitely long one! If the instability is not exponential but, for example, only a power law, then the required information per unit time is inversely proportional to the full previous length of trajectory and, asymptotically, the prediction becomes possible. Needless to say, for a sufficiently short time interval the prediction is of course possible even for a chaotic system and can be characterized by the randomness parameter [5]

$$r = \frac{h|t|}{|\ln \mu|} \tag{4}$$

where μ is the accuracy of trajectory recording. The prediction is possible in the *finite* interval of "temporary determinism" ($r \lesssim 1$) while $r \gg 1$ corresponds to the *infinite* region of asymptotic randomness.

Notice that the information per unit time (3) does not depend on the accuracy of recording μ which determines the prediction time scale only. Recently a powerful technique has been developed which actually predicts a chaotic trajectory within this limit [6].

The important condition $h > 0$, which characterizes chaotic motion, is not invariant with respect to the change of time variable. (According to Ref.[3] the only invariant statistical property under change of time variable is ergodicity.) In our opinion the resolution of this difficulty is in that the proper characteristic of motion instability, important for dynamical chaos, should be taken with respect to the oscillation phases whose dynamics determines the nature of the motion. This implies that the proper time variable must vary proportionally with the phases.

The exponential instability implies a continuous spectrum of motion. (In a discrete spectrum the instability can be at most linear in time [57].) The continuous spectrum, in turn, implies correlation decay; this property, which is called mixing in ergodic theory, is the most important property of dynamical motion for the validity of the statistical description. The point is that mixing provides the statistical independence of different parts (sufficiently separated in time) of a dynamical trajectory. This is the main condition for the application of probability theory which allows the calculation of the statistical characteristics such as diffusion, relaxation, distribution functions etc. However, it is not clear as yet whether or not the strongest property of exponential instability is really required for a meaningful statistical description of dynamical motion.

In the modern theory of dynamical systems the latter property corresponds to one limiting case of dynamical motion which is called dynamical chaos (Fig. 1). The opposite limiting case is the so-called completely integrable motion which possesses the full set of N motion integrals (where N is the number of freedoms) in involution and in which, moreover, the action-angle variables can be introduced. The property of complete integrability is very delicate and non-typical as it is destroyed by an arbitrarily weak perturbation which converts a completely integrable system into a KAM-integrable system [7]. (Unlike integrable motion, chaotic motion is very robust: it is structurally stable and is not affected by weak perturbations [58].) The structure of the KAM motion is very intricate; the motion remains completely integrable for most initial conditions yet a single, connected, chaotic motion component (for $N > 2$) arises of exponentially small (with respect to perturbation) measure which is nevertheless everywhere dense. Interesting enough, the chaotic trajectories in this component approach arbitrarily close to any point on the energy surface, yet the motion is non ergodic due to the positive measure of the stable component. Ergodicity means that almost all trajectories not only approach every point on the energy surface, but cover it homogeneously that is the sojourn time of a trajectory in any region of phase space is proportional to its invariant measure. The property of ergodicity which has given the name to the whole ergodic theory turned out to be neither necessary nor sufficient for the statistical description. Indeed chaotic components of the motion may cover only a part of the energy surface (which is rather a typical case in dynamical systems); on the other hand for a meaningful statistical description a mixing property is needed which provides the relaxation to some statistical steady state.

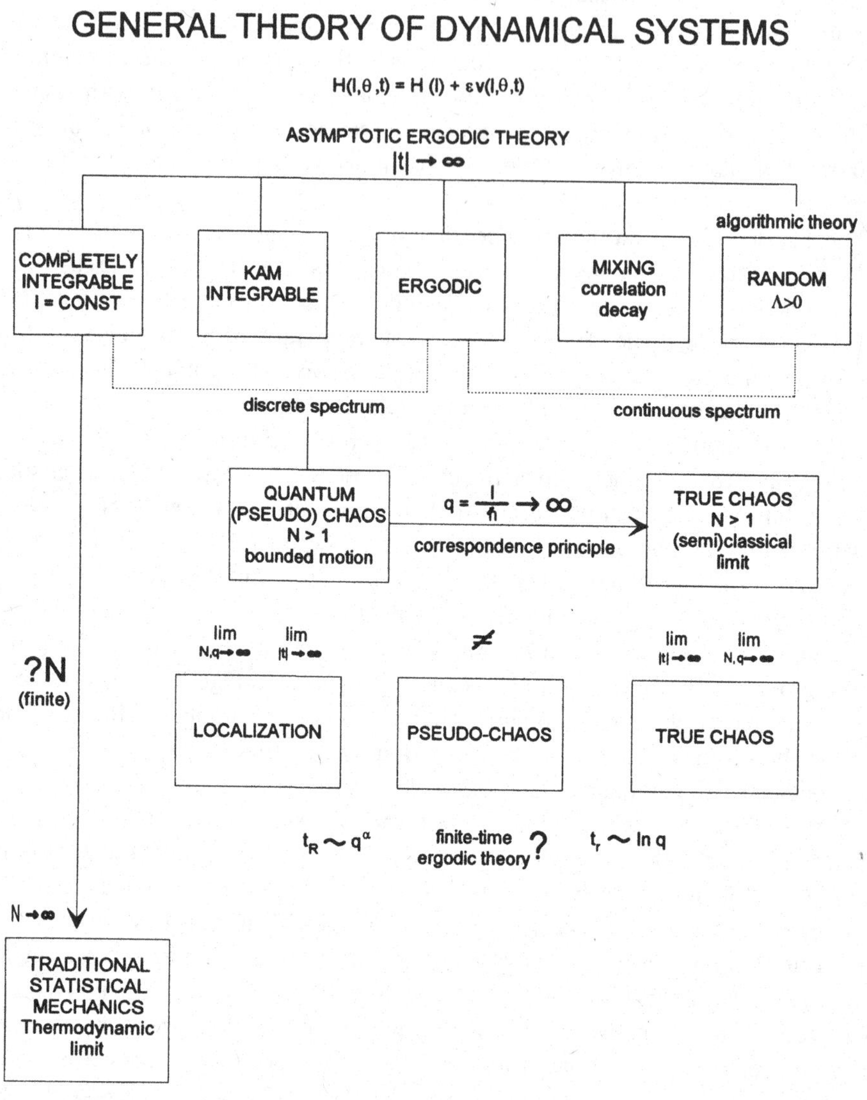

GENERAL THEORY OF DYNAMICAL SYSTEMS

$$H(I,\theta,t) = H(I) + \varepsilon v(I,\theta,t)$$

ASYMPTOTIC ERGODIC THEORY
$$|t| \to \infty$$

algorithmic theory

| COMPLETELY INTEGRABLE I = CONST | KAM INTEGRABLE | ERGODIC | MIXING correlation decay | RANDOM $\Lambda > 0$ |

discrete spectrum   continuous spectrum

| QUANTUM (PSEUDO) CHAOS $N > 1$ bounded motion | $q = \dfrac{I}{\hbar} \to \infty$ correspondence principle | TRUE CHAOS $N > 1$ (semi)classical limit |

$?N$
(finite)

$$\lim_{N,q \to \infty} \quad \lim_{|t| \to \infty} \qquad\qquad \neq \qquad\qquad \lim_{|t| \to \infty} \quad \lim_{N,q \to \infty}$$

| LOCALIZATION | PSEUDO-CHAOS | TRUE CHAOS |

$$t_R \sim q^\alpha \qquad \text{finite-time} \atop \text{ergodic theory} \, ? \qquad t_r \sim \ln q$$

$N \to \infty$

| TRADITIONAL STATISTICAL MECHANICS Thermodynamic limit |

Fig. 1. The place of quantum chaos in modern theories: action-angle variables I, θ; number of freedoms N; Lyapunov exponent Λ; quasiclassical parameter q; time scales t_R and t_r; Planck's constant $\hbar$. Two question marks indicate the problems in a new ergodic theory non-asymptotic in N and $|t|$ (see also section 10).

Instead in terms of trajectories an equivalent description of classical dynamics can be given by means of distribution functions which, if non-singular, represent not a single trajectory but a continuum of them. As is well known the distribution function obeys a linear Liouville equation. Thus the condition of non-linearity for dynamical chaos to occur is restricted to the description in terms of trajectories where the rôle of non-linearity is to bound the linearly unstable motion. In terms of distribution functions ergodicity means that the time-averaged distribution function is constant with respect to the invariant measure. Instead, the stronger property of mixing implies the approach, on average, of any initially smooth distribution to a steady-state. This process is called *statistical relaxation*. The relaxation is of course time-reversible as is the motion along a particular trajectory. However, unlike the latter, the evolution of the distribution function is non recurrent. Notice that according to the Poincaré theorem the trajectory recurs infinitely many times independent of the type of motion (regular or chaotic); the difference is that for regular motion the recurrence time is strictly bounded from above whereas in the latter case arbitrarily large recurrence times are possible. The time-reversibility of the distribution function is related to its very complicated structure which becomes more and more "scarred" as the relaxation proceeds. In the case of exponential instability of motion the spatial scale of the oscillations also decreases exponentially with time. It is in these fine spatial oscillations that the memory of the initial state is retained forever, which is only possible in a continuous phase space.

To get rid of this complicated structure, the distribution function must be "coarse-grained", that is, averaged on some domain. The evolution of the coarse-grained function is described by a kinetic equation, e.g., a diffusion equation. The coarse-grained function converges to a smooth steady state which is a constant with respect to the invariant measure.

In closing this section we would like to stress again the two crucial properties of classical mechanics necessary for dynamical chaos to occur: (i) a continuous spectrum of the motion, and (ii) a continuous phase space.

2 Quantum chaos vs the correspondence principle

The problem of quantum chaos arose from the attempts to understand the very peculiar phenomenon of classical dynamical chaos in terms of quantum mechanics. Preliminary investigations immediately unveiled a very deep difficulty related to the fact that the above mentioned conditions for classical chaos are violated in quantum mechanics. Indeed the energy and the frequency spectrum of any quantum motion, bounded in phase space, are always discrete. According to the existing theory of dynamical systems such motion corresponds to the limiting case of regular motion. The ultimate origin of this fundamental quantum property is discreteness of the phase space itself or, in modern mathematical language, a non-commutative geometry of the latter. This is the very basis of all quantum physics directly related to the fundamental uncertainty principle which implies a finite size of an elementary phase-space cell: $\Delta x \cdot \Delta p \gtrsim \hbar$ (per freedom).

The naive resolution of this difficulty would be the absence of any quantum chaos. For this reason it was even proposed to use the term "quantum chaology" [8] which essentially means the study of the absence of chaos in quantum mechanics. If the above conclusion were true, a sharp contradiction would arise with the correspondence principle which requires the transition from quantum to classical mechanics for all phenomena including the new one: dynamical chaos. Does this really mean a failure of the correspondence principle as some authors insist (see, e.g., Ref.[9]) ? If it were so quantum chaos would, indeed, be a great discovery since it would mean that classical mechanics is not the limiting case of quantum mechanics but a different separate theory. "Unfortunately", there exists a less radical (but also interesting and important) resolution of this difficulty which is discussed below.

A recent breakthrough in the understanding of quantum chaos has been achieved, particularly, due to a new philosophy which, either explicitly or implicitly, is generally accepted; namely the whole physical problem of quantum dynamics is considered as divided into two qualitatively different parts:

(i) proper quantum dynamics as described by a specific dynamical variable, the wavefunction $\psi(t)$; and

(ii) quantum measurement including the recording of the result and hence the collapse of the ψ function.

The first part is described by some deterministic equation, for example, the Schrödinger equation and naturally belongs to the general theory of dynamical systems. The problem is well posed and this allows for extensive studies. In the following, as well as in all papers of the present collection only the first part is discussed.

The second part still remains very vague to the extent that there is no common agreement even on the question whether this is a real physical problem or an ill-posed one so that the Copenhagen interpretation of (or convention in) quantum mechanics gives satisfactory answers to all the admissible questions. In any event there exists as yet no dynamical description of quantum measurement including the ψ-collapse.

The absence of classical-like chaos is true for the above mentioned first part of quantum dynamics only. Quantum measurement as far as the result is concerned, is fundamentally a random process. However, there are good reasons to believe that this randomness can be interpreted as a particular manifestation of dynamical chaos [10].

The separation of the first part of quantum dynamics, which is very natural from a mathematical viewpoint, was introduced and emphasized by Schrödinger who, however, certainly underestimated the importance of the second part in physics.

3 Characteristic time scales of quantum chaos

3.1 The models

One way to reconcile the discrete spectrum with the correspondence principle is to introduce some characteristic time scales of quantum motion. The main idea is that the distinction between discrete and continuous spectra is non ambiguous in the limit $t \to \infty$ only. This idea was suggested by our first numerical experiments in quantum chaos in which a very simple model was used (the so-called "kicked rotator")[11]. In the classical limit this model is described by the Hamiltonian

$$H = \frac{n^2}{2} + k \cdot \cos \vartheta \cdot \delta_T(t) \tag{5}$$

where (n, ϑ) are action-angle variables, k and T are the strength and period of the perturbation, and $\delta_T(t)$ is the δ-function of period T. The corresponding equations of motions are given by the so-called standard map (SM)

$$\left. \begin{array}{l} \bar{n} = n + k \sin \vartheta \\ \bar{\vartheta} = \vartheta + \bar{n} \, T \end{array} \right\} \tag{6}$$

The only parameter of this map is $K = kT$.

For $K < 1$ the motion is strictly bounded while for $K \gg 1$ it is ergodic, mixing and exponentially unstable with a Lyapunov exponent per step:

$$\Lambda \approx \ln \frac{K}{2} \tag{7}$$

This model can be considered either on the infinite cylinder (unbounded motion) or on a finite torus (bounded motion) of circumference

$$L = \frac{2\pi m}{T} \tag{8}$$

where m is an integer to avoid discontinuities.

The SM is very popular in studies of dynamical chaos both classical and quantal because of its apparent simplicity and intrinsic richness. (The nickname "kicked rotator" is related to a particular physical interpretation of this model as a rotator with angular momentum n driven by a series of periodic pulses, or "kicks".) Yet, it can also be considered also as the Poincare surface-of-section map for a conservative system of two freedoms. Particularly, the map on a torus (8) models the energy shell of a conservative system which is the quantum counterpart of the classical energy surface. What makes the SM almost universal is the local (in momentum) approximation it provides for a broad class of more complicated physical models.

One well studied example of such models is the photoeffect in Rydberg atoms.

The main features of this problem are described by the one-dimensional Hamiltonian [12] (see also Section.7)

$$H = -\frac{1}{2n^2} + \epsilon z\,(n,\,\vartheta)\,\cos\omega t \tag{9}$$

where ϵ and ω are the strength and the frequency of the linearly polarized electric field, n, ϑ are the action-angle variables and z the coordinate along the field direction.

If the field frequency exceeds the electron frequency the motion of system (9) is approximately described by a map over one Kepler period of the electron, the so-called Kepler map:

$$\overline{v} = v + k \cdot \sin\phi\,; \qquad \overline{\phi} = \phi + \frac{\pi}{\sqrt{2\omega}}(-\overline{v})^{-3/2} \tag{10}$$

where $v = E/\omega = -(2\omega n^2)^{-1}$, ϕ is the field phase at perihelion and k is perturbation parameter (in atomic units)

$$k \approx 2.58\,\frac{\varepsilon}{\omega^{5/3}} \tag{11}$$

Linearizing the second Eq.(10) in v reduces the Kepler map to the SM (6) with the same k, and parameter

$$T = 6\pi\omega^2 n^5 \tag{12}$$

provided $k \ll v$. Thus, the SM describes the dynamics locally in momentum. In this particular model the momentum v is proportional to the energy E as the conjugate phase ϕ is proportional to time.

Interestingly, the Kepler map can be derived from a simple expression for the electron free fall on the Coulomb center which, in reversed time, reads:

$$z = \left(\frac{3}{\sqrt{2}}\right)^{2/3} t^{2/3}$$

Then, the perturbation parameter is given by the integral

$$k = -2\varepsilon \int_0^\infty z(t) \cdot \cos\omega t\, dt = (48)^{1/6}\,\Gamma(2/3) \cdot \frac{\varepsilon}{\omega^{5/3}} \approx 2.58\frac{\varepsilon}{\omega^{5/3}} \tag{13}$$

Another, less known, example is the "kicked top", or the spin dynamics on a sphere [13] which is described by the Hamiltonian (cf. Eq.(5)):

$$H = \frac{s_z^2}{2} + \frac{k_0}{s} \cdot s_x \cdot \delta_T(t) = \frac{s_z^2}{2} + k_0\,\sqrt{1 - \frac{s_z^2}{s^2}}\,\cos\varphi \cdot \delta_T(t) \tag{14}$$

where s_z, s_x are the components of spin whose modulus squared $s^2 = s_z^2 + s_x^2 + s_y^2 = const$ is the motion integral, and φ is the azimuthal angle canonically conjugated to the momentum $s_z = s \cdot \cos\theta$ where θ is the polar angle. As is easily verified, the invariant measure $ds_x \cdot ds_y \cdot ds_z = s\,ds \cdot ds_z \cdot d\varphi$ is proportional to the area on the sphere.

If $k_0 \ll s$ the spin dynamics can be approximately described by the SM with the perturbation parameter

$$k(s_z) = k_0 \sqrt{1 - \frac{s_z^2}{s^2}} \qquad (15)$$

depending on action s_z. This approximation is valid if the change Δk of k in one kick is less than k. This is true on the whole sphere except a narrow 'polar' region

$$\frac{s - s_z}{s} \lesssim \left(\frac{k_0}{s}\right)^2 \ll 1 \qquad (16)$$

The restriction $s_z \ll s$ [14] is therefore not necessary.

An interesting version of this model [15] is the motion on a hyperboloid

$$s_x^2 + s_y^2 - s_z^2 = \pm s^2 \qquad (17)$$

In this case the SM perturbation parameter becomes

$$k(s_z) = k_0 \sqrt{\frac{s_z^2}{s^2} \pm 1} \qquad (18)$$

where the sign coincides with that of motion integral (17).

The local description via the SM allows the determination of the chaos border in all these models from the condition

$$K = T k(s_z) \approx 1 \qquad (19)$$

The statistical properties of the SM are described by a diffusion in action n with the rate

$$D_0 = \frac{<(\Delta n)^2>}{\tau} = \frac{k^2}{2} C(K) \qquad (20)$$

where the integer time $\tau = t/T$ is the number of map iterations, and the function $C(K)$ accounts for dynamical correlations. Particularly, $C(K) \to 1$ for $K \gg 1$, and $C(K) \to 0$ at the chaos border (19). The diffusion Green function is Gaussian

$$g(n, \tau) = \frac{\exp\left(-\frac{(n - n_0)^2}{2D_0\tau}\right)}{\sqrt{2\pi D_0 \tau}} \qquad (21)$$

For the bounded SM (on a torus) the diffusion leads to the statistical relaxation of any non-singular distribution function to the ergodic state:

$$f(n) \to \frac{1}{L} \qquad (22)$$

with a characteristic time $\tau_{cl} \sim L^2/D_0$.

The quantized standard map (QSM) on a cylinder, first introduced in Ref.[11], is described by the relation:

$$\bar{\psi} = \hat{U} \psi = \exp(-ik \cos \hat{\vartheta}) \cdot \exp(-i \frac{T}{2} \hat{n}^2) \psi \qquad (23)$$

where $\hat{U}$ is a unitary operator, $\hat{n} = -i\partial/\partial\vartheta$, and $\hbar = 1$. Notice that, while for the bounded classical SM on a torus (8) the only change is to take $n \bmod L$, for the QSM expression (23) becomes more complicated [16].

If $k \gg 1$ the perturbation couples approximately $2k$ unperturbed states per iteration of map (23). For $k \ll 1$ all the transitions are suppressed. Thus, $k \sim 1$ is a specific border of quantum stability due to the discrete spectrum and is independent of the behaviour in the classical limit [61,11]. This is also called the *perturbative localization border*.

For the spin model the unitary operator is obtained from the first expression of Hamiltonian (14):

$$\overline{\psi} = \exp(-i\frac{k_0}{s}\,\hat{s}_x) \cdot \exp(-i\frac{T}{2}\,\hat{s}_z^2)\,\psi \tag{24}$$

(The second expression in Eq.(14) is only a quasiclassical approximation for $s \gg 1$. The same is also true for model (17).) The transition to the classical limit corresponds to $k \to \infty$, $T \to 0$, while $K = kT = const$, $LT = const$ (for model (5)) and $k/s = const$, $s_z/s = const$ (for models (14) and (17)).

3.2 The relaxation time scale

In Fig. 2(a) we show an example of quantum 'diffusion' in the SM ($< n^2 > \sim \tau$). It is seen that during a finite time interval ($\tau \approx 200$) quantum diffusion is close to the classical diffusion (the straight line) in accordance with the correspondence principle. Moreover, during this time interval quantum diffusion follows many other details of classical diffusion as shown in Fig. 3. These are very satisfactory results; however, for longer times something breaks down and quantum diffusion, unlike classical diffusion, completely stops [11,17,18] (Fig. 2(b)).

The general explanation of this phenomenon is related to the fundamental uncertainty principle [19]. Indeed, the discrete spectrum cannot be resolved if

$$t \lesssim \rho_0 \sim t_R \tag{25}$$

where t_R is called the *relaxation time scale*, and ρ_0 is the energy (or quasienergy) level density for those eigenstates which are actually present in the initial state $\psi(0)$ and, hence, determine the system's dynamics. We call these *operative eigenstates*.

Generally, $\rho_0 \leq \rho$ where ρ is the total level density. The latter may even be infinite, like for the unbounded SM. On the torus the quasienergy density $\rho = L/(2\pi/T) = m$ which is, surprisingly, a classical quantity that does not change in the quasiclassical transition and which determines the upper bound of the relaxation time scale in continuous time t. This is because the physical time for the model under consideration is the number of map iterations $\tau = t/T$ in which the relaxation time scale

$$\tau_R = \frac{t_R}{T} \sim \frac{\rho_0}{T} \leq \frac{m}{T} \tag{26}$$

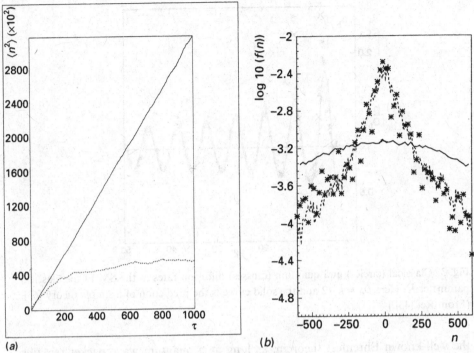

Fig. 2. Classical and quantum diffusion in the SM for $K = 5$, $k = 25$, $T = 0.2$. (a) Classical (solid curve) and quantum (dotted curve) unperturbed energy $< n^2(\tau) > = 2E$ as a function of time τ (number of map iterations). (b) Classical (solid curve) and quantum (dashed curve) probability distribution after time $\tau = 1000$. The stars give the Husimi distribution integrated over the angle ϑ.

grows indefinitely in the classical limit $T \rightarrow 0$, also in accordance with the correspondence principle.

For the kicked rotator the explicit estimate for this time scale has the remarkable form [19]

$$\tau_R \sim D_0 \qquad (27)$$

which relates the essentially quantum characteristic τ_R with the classical diffusion rate.

3.3 The random time scale

As discussed above the main peculiarity of quantum chaos is its restriction to a finite time interval: for this reason the term *quantum pseudo-chaos* is sometimes used to distinguish it from the "true" chaos in the classical limit. This is also true for a stronger chaotic property – the exponential instability. Indeed, according to

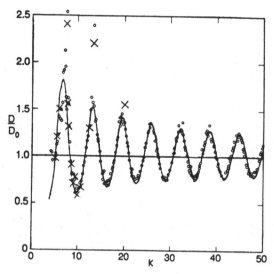

Fig. 3. Classical (circles) and quantum (crosses) diffusion rates in the SM vs the classical parameter K. Here $D_0 = k^2/2$ and the solid curve is the prediction of a simple theory [63]. (From Ref.[43].)

the well-known Ehrenfest theorem, as long as a quantum wave packet remains narrow it follows a beam of classical trajectories. During this time interval the wave-packet motion is as random as the classical trajectory. Particularly the packet is exponentially spreading with the classical rate h. However, the initial size of the quantum packet (unlike in the classical case) is restricted from below by the elementary cell of quantum phase space which is $\sim \hbar$ (equal 1 in our units). The final size for a bounded motion is proportional to some (large) quasiclassical parameter q which is of the order of the characteristic value of the action variable. Therefore, the full time for the exponential spreading of the packet is of the order

$$\tau_r \sim \frac{\ln q}{h} \tag{28}$$

This time scale has been introduced and explained in Ref.[20] (see also Ref.[69]). For the standard map there are two quantum parameters, k and $1/T$. If we consider the optimal, least-spreading, wave packet ($\Delta\vartheta_0 \sim (\Delta n_0)^{-1} \sim \sqrt{T}$) the latter estimate becomes [19]

$$\tau_r \sim \frac{|\ln T|}{\ln \dfrac{K}{2}} \tag{29}$$

This is another time scale, much shorter than τ_R (26), which we call the *random time scale*. It increases indefinitely as $T \to 0$, again in accordance with the correspondence principle.

In Fig. 4 an example of the evolution of an initially narrow wave packet is shown demonstrating both the random $((a)–(c))$ and the relaxation $((d)–(f))$ time scales compared with the classical evolution.

Even though the estimates (28, 29) appear very simple, almost trivial, some questions remain open. One is why only stretching of the packet shows up in Fig.4 without any substantial squeezing (see also Ref.[21]). Apparently, this is related to the particular phase-space density used (the Husimi distribution). Indeed, this density is obtained by the projection of an evolving quantum state on the coherent states whose width is fixed in *both* coordinate and momentum separately whereas the uncertainty principle restricts the product only. The problem is how long this product is going to remain unchanged (particularly minimal) which is the quantum counterpart of the phase-space volume conservation in classical mechanics. To the best of our knowledge, nobody has analysed this as yet. In our understanding, the Wigner function is much more suitable for the study of this problem.

The rate of "inflation" of the phase-space volume occupied by a quantum state can be estimated from the Liouville equation for the Wigner function [22]. In the particular case of the SM it can be written in the quasiclassical region as

$$\frac{dW}{dt} \approx -\frac{1}{24} \cdot \frac{\partial^3 H}{\partial \vartheta^3} \cdot \frac{\partial^3 W}{\partial n^3} \tag{30}$$

To obtain an estimate for the inflation we substitute in the rhs the "unperturbed" classical density $W_0(n, \vartheta, \tau)$ in the form of a Gaussian packet of rms dimensions A and $a \leq A$ with the minimal area $A \cdot a = 1/2$, stretching at angle $T \ll 1$ with the n-axis. Then, $\partial W/\partial n \sim W(T/a)$ while $\partial^3 H/\partial \vartheta^3 \sim k\,\delta_T(t)$, and we obtain from Eq.(30)

$$\frac{d}{d\tau} \ln W \sim k(AT)^3$$

Since $A = A_0 \cdot \exp(\Lambda\tau)$ where $\Lambda = \ln(K/2)$ is the Lyapunov exponent, and $A_0 = \Delta n_0 \sim 1/\sqrt{T}$ for the least spreading packet, we arrive at the estimate

$$\ln \frac{W}{W_0} \sim \frac{k(AT)^3}{\Lambda} \sim 1 \tag{31}$$

which determines the *inflation time scale* τ_{if} and the maximal packet length prior to a substantial inflation $\Delta\vartheta_{if} = (AT)_{if}$:

$$\tau_{if} \sim \frac{|\ln(TK^2/\Lambda^2)|}{6\Lambda}; \qquad \Delta\vartheta_{if} \sim \left(\frac{\Lambda}{k}\right)^{1/3} \gg \Delta\vartheta_0 \sim T^{1/2} \tag{32}$$

Another mechanism for the destruction of the unstable quantum packet is related to discrete (integer) values of action n. Apparently, it begins to work for $a \lesssim T$ when the continuous derivative $\dfrac{\partial W}{\partial n}$ breaks down. This determines the *destruction time scale*

$$\tau_d \sim \frac{|\ln T|}{2\Lambda}; \qquad \Delta\vartheta_d \sim 1 \tag{33}$$

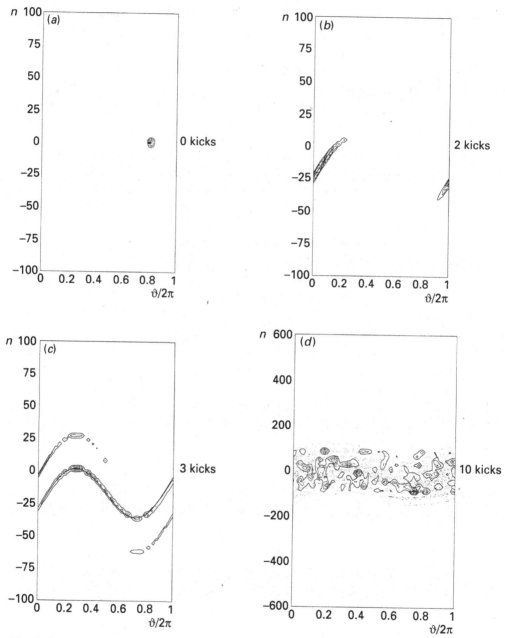

Fig. 4. A comparison between classical and quantum evolution in the kicked rotator for the same parameters as in Fig. 2. The initial quantum state is a coherent packet. As to the classical evolution we considered 2000 trajectories starting in the same area, of size $\hbar(=1)$, occupied by the initial quantum state (Fig. (*a*)). In the following figures ((*b*)–(*f*)), the dots represent the classical trajectories while the curves are level curves of the Husimi distribution. For each figure, taken at different times, we divide by 8 the maximum value of the Husimi distribution by 8 and then plot the seven level curves: (*b*) $\tau = 2$; (*c*) $\tau = 3$; (*d*) $\tau = 10$; (*e*) $\tau = 100$; (*f*) $\tau = 1000$. From this figure, as well as from fig. 2, the quantum localization phenomenon is clearly evident.

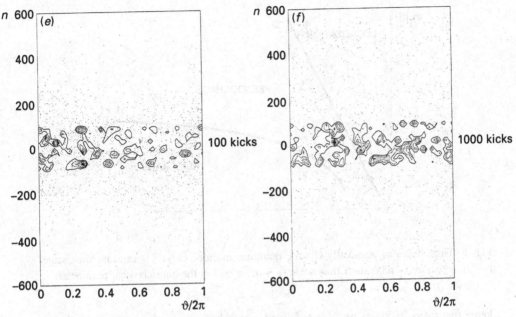

Fig. 4. —*Continued*, for figure caption see previous page

The scales τ_{if} and τ_d are comparable to each other and also to the well-known estimate (29). However, the critical values of the packet length, $\Delta\vartheta_{if}$ and $\Delta\vartheta_d$ are essentially different for $k \rightarrow \infty$. The numerical evidence seems to agree better with the destruction, rather than the inflation, mechanism but the latter may be hidden by the Husimi distribution.

We think that the concept of characteristic time scales of quantum dynamics is a satisfactory resolution of the apparent contradiction between the correspondence principle and the quantum transient (finite-time) pseudo-chaos. Some physicists, however, feel that such an explanation is, at least, ambiguous because it includes the two limits which do not commute:

$$\lim_{|t|\rightarrow\infty} \lim_{q\rightarrow\infty} \neq \lim_{q\rightarrow\infty} \lim_{|t|\rightarrow\infty}$$

While the first order leads to classical chaos, the second one results in an essentially quantum behaviour with no chaos at all. To resolve these doubts we note that in physics one does not need to take any limit at all, and, in principle, we can describe anything quantum-mechanically. If, nevertheless, we would like to make use of the much simpler classical mechanics (for practical purposes) then only one limit ($q \rightarrow \infty$) is quite sufficient as the physical time is certainly finite. Finally, even if it would be helpful for some reason (e.g., for mathematical convenience) formally to take the limit $|t| \rightarrow \infty$ this should be *conditional*; namely, one should

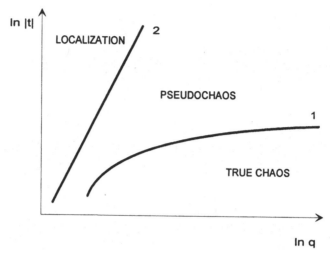

Fig. 5. Time scales of classically chaotic quantum motion: curve 1 - random time scale $t_r \sim \ln q$; curve 2 - relaxation time scale $t_R \sim q^{\alpha}$; $q \gg 1$ is the quasiclassical parameter.

keep the ratio $|t|/t_R(q)$ or $|t|/t_r(q)$ fixed. In other words the two above limits should be taken simultaneously. The general structure of quantum dynamics on the plane (q, t) is outlined in Fig. 5.

The limit $|t| \to \infty$ is related to the existing ergodic theory which is asymptotic in t. Meanwhile the new phenomenon of quantum chaos requires the modification of the theory to a finite time which is a difficult mathematical problem still to be solved. On the other hand, the practical importance of statistical laws even for a finite time interval is that they provide a relatively simple description of the *essential* behaviour for a very complicated dynamics.

In any event, if quantum mechanics is the universal theory, as is commonly accepted, then the phenomenon of the "true" (classical-like) dynamical chaos strictly speaking does not exist in nature. Nevertheless, the conception of "true" chaos is very important in the theory as the limiting pattern to compare with real quantum chaos.

3.4 Dynamical stability of quantum diffusion

Even though quantum diffusion and relaxation proceed on a fairly long time scale (25) they are very unusual and qualitatively different from their classical counterparts, namely they are dynamically stable. This was shown in several numerical experiments with time reversal [23]. Particularly, for the diffusive ionization of the Rydberg hydrogen atom in a microwave field (Fig.6) the electron velocity was reversed at $t = 60$ field periods, and the backward motion was observed. In the

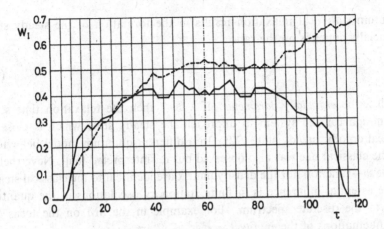

Fig. 6. Classical (dashed curve) and quantum (full curve) ionization probability as a function of time. Notice the perfect symmetry of the quantum curve about the time of reversal ($\tau = 60$ field periods). Both classical and quantal diffusive ionization lag in time which is a characteristic distinction from direct multiphoton ionization. (after Ref.[12]).

classical case the reversed motion reproduces the forward motion for a very short time only because, due to exponential instability, unavoidable computational errors immediately restore the diffusion. Unlike this, in the quantum system the "antidiffusion" goes back to the initial state with very high accuracy ($\sim 10^{-15}$) and only after passing the initial state does the quantum diffusion continue. The stability of quantun chaos over the relaxation time scale is comprehensible since the random time scale is much shorter. Yet, the accuracy of the reversal is surprising. Apparently, this is explained by the relatively large size of the quantum wave packet as compared to the unavoidable rounding-off errors (δ). In the SM, for example, the size of the least-spreading wave packet $\Delta\vartheta \sim \sqrt{T}$. On the other hand, any quantity in the computer must exceed the error δ. Since $T > \delta$, then $(\Delta\vartheta)^2/\delta^2 \sim (T/\delta)\delta^{-1} \gg 1$. This experiment clearly indicates that there is no appreciable instability in quantum chaotic motion. It is also an example which demonstrates that exponential instability is not necessary for a meaningful statistical description. An interesting version of the time-reversal experiment with controlled perturbation, much in excess of rounding-off errors, is described in Ref.[24].

The statistical relaxation to a steady state as described by a diffusion equation is typically exponential, yet this does not mean that the underlying dynamical motion is necessarily unstable as is sometimes assumed [25].

4 The quantum steady state

The quantum statistical relaxation results in the formation of the steady state which crucially depends on the ratio

$$\lambda^2 \sim \frac{\tau_R}{\tau_{cl}} \tag{34}$$

We call λ the *ergodicity parameter*. If $\lambda \gg 1$ then the relaxation time scale is long enough for the system to approach the steady state which is close to the classical ergodic steady state (22). (In a different quantum model behaviour close to the classical one was also observed but misinterpreted [65]). Nevertheless neither the ψ−function nor the density matrix are identical to the classical steady state. The essential difference is in finite stationary oscillations in the quantum case due to the discrete spectrum. For example in the SM on the torus the expected fluctuations of the energy $E = <n^2> /2$ are

$$\frac{\Delta E}{E_s} \sim \frac{1}{\sqrt{L}} \tag{35}$$

where $E_s = L^2/24$ is the average energy on the torus $(-L/2, L/2)$. This would imply that the ψ-function represents a finite ensemble of $\sim L$ systems even though formally it describes a single system. In other words the ψ-function plays an intermediate rôle between the trajectory and the distribution function in classical mechanics. Indeed, $|\psi|^2$ is not a constant like the classical (coarse-grained) distribution function in the steady state but its fluctuations are much smaller than those of a single classical trajectory which would be $\Delta E/E_s \sim 1$.

If $\lambda \ll 1$ the relaxation time scale is insufficient to reach the classical steady state and a qualitatively different quantum steady state is formed. For example, in the SM the steady state distribution in momentum is approximately exponential [26]

$$g_s(n) \sim \exp\left(-\frac{2|n-n_0|}{l_s}\right) \tag{36}$$

This is the well known phenomenon of quantum diffusion localization. Here l_s is called the *localization length*, and a sufficiently narrow initial state $g(n, 0) \sim \delta(n - n_0)$ is assumed. From the diffusion law $l_s^2 \sim D_0\tau_R$ hence from Eq.(27) $l_s \sim D_0$. Numerical experiments confirm this estimate and give the more accurate result [26]

$$l_s \approx D_0 \tag{37}$$

The distribution (36) is completely different from the classical ergodic state and we call it the *quantum steady state*. This state represents a finite ensemble of about l_s systems; hence, as with Eq.(35), we would expect the fluctuations to be of the order $1/\sqrt{l_s} \sim 1/k$. However, numerical experiments show that the fluctuations are much bigger:

$$\frac{\Delta E}{E_s} \sim k^{-0.6} \tag{38}$$

These are possibly related to the so-called Mott states (see below) [27,59].

Eq.(36) allows the definition of the ergodicity parameter (34) in a more precise way in terms of l_s:

$$\lambda = \frac{l_s}{L} \approx \frac{D_0}{L} \tag{39}$$

The first equality makes sense only for small λ, when the steady-state distribution is exponential (36). At larger $\lambda \gtrsim 1$ the approximate equality in Eq. (39) should be taken. The ratio D_0/L is an important parameter of this model and its relation to the ergodicity will be considered in Section 9.

Numerical experiments [19] show that all eigenfunctions are, on average, also exponentially localized:

$$\varphi_m(n) \sim \exp\left(-\frac{|m-n|}{l}\right) \pm \exp\left(-\frac{|m+n|}{l}\right) \tag{40}$$

where the signs correspond to symmetric and antisymmetric eigenfunctions, respectively, and the localization length is given by

$$l \approx \frac{l_s}{2} \approx \frac{D_0}{2} \tag{41}$$

The two exponential peaks in Eq.(40), separated by a distance of $2|m|$, are due to the exact parity conservation that is to the symmetry with respect to reflection $\vartheta \to -\vartheta$ or $n \to -n$ in the SM.

The difference between the two localization lengths, l and l_s, is due to very big fluctuations around the simple average law (40). This law actually holds asymptotically as $|m \pm n| \to \infty$. Indeed, the central part of each peak in Eq.(40) is not only substantially affected by big fluctuations but may have a very special shape in the case of the so-called Mott states [28] (see also Ref.[18]). These states appear as pairs of approximately symmetric and antisymmetric superposition of the two exponential peaks separated by a distance M (in n) for each of the peaks in Eq.(40). The main characteristic of Mott states is the dependence on M of the quasienergy splitting $\Delta \epsilon$ in the pair. According to numerical experiments [59] this dependence is approximately

$$\Delta\epsilon \approx \frac{0.3}{Tl_s}\left(1 + \frac{M}{l_s}\right)\exp\left(-\frac{M}{l_s}\right) \tag{42}$$

Even though there are only relatively few Mott states they essentially affect the asymptotic relaxation to the quantum steady state [18].

Another definition of the localization length l_H was introduced in Ref.[45]

using the entropy of eigenfunctions

$$l_H = \exp(H); \qquad H = -\sum_{n=1}^{L/2} |\varphi(n)|^2 \ln |\varphi(n)|^2 \qquad (43)$$

This definition is especially convenient in the intermediate case where the ergodicity parameter $\lambda \sim 1$ and the eigenfunctions are quite different from the exponential shape (40). The entropy localization length (43) is a particular case of the *Renyi participation ratio* ξ_q which can be defined as (cf. Ref.[30])

$$\frac{1}{\xi_q} = \left(\sum_n |\varphi(n)|^{2q} \right)^{\frac{1}{q-1}} = \langle |\varphi(n)|^{2(q-1)} \rangle^{\frac{1}{q-1}} \qquad (44)$$

where q is a continuous parameter. It is easily seen that $l_H = \xi_1$.

For a single exponential peak $\left(|\varphi(n)|^2 = \left(\frac{1}{l} \right) \exp \left(-2|n|/l \right) \right)$ we have:

$$\frac{l}{\xi_q} = \left(\frac{1}{q} \right)^{\frac{1}{q-1}} \rightarrow \begin{cases} q & q \ll 1 \\ 1/e & q = 1 \\ 1/2 & q = 2 \\ 1 & q \gg 1 \end{cases} \qquad (45)$$

All SM eigenstates are doubly degenerate. If this symmetry is broken by an additional perturbation the localization length, or better to say, the ratio l/D_0, generally rises [34] due to the increase of the number of operative eigenfunctions. However, this effect is generally different for different eigenstates [35].

The fluctuations of l_H for $\lambda \ll 1$, are fairly well described by the following empirical expression for the differential probability [60]

$$p(H) = \frac{1}{\cosh \left[\pi (H - \overline{H}) \right]} \qquad (46)$$

where $\overline{H}$ is the entropy averaged over all eigenstates. Therefore, unlike the fluctuations of the asymptotic localization length l, the fluctuations of l_H are quite big, namely, the rms $\Delta l_H / l_H \approx 1/2$. No explanation of the simple dependence (46) exists as yet. Notice that l is determined essentially by the tail of the eigenfunction, while l_H is mainly determined by the central part.

Numerical experiments [16] showed that the dimensionless localization length $\beta_{\overline{H}} = \exp \left(\overline{H} - H_e \right) = 2 l_{\overline{H}} (\gamma/L) (\gamma \approx 2)$ depends only on the dimensionless ergodicity parameter λ (39) and not separately on D_0 and L. Here $H_e = \ln(L/2\gamma)$ is the entropy of the ergodic state which is less than maximal $(\ln(L/2))$ because of fluctuations.

The explicit empirical dependence is given approximately by

$$\beta_{\overline{H}} = \begin{cases} \dfrac{\alpha\lambda}{1+\alpha\lambda} & \lambda \leq 0.5 \\[2mm] 1 - \dfrac{1}{\alpha\sqrt{\lambda}} & \lambda \geq 0.1 \end{cases} \qquad (47)$$

with $\alpha \approx 4$.

A partial explanation of the first scaling in Eq.(47) is related to the dependence of any ξ_q on the position of the exponential peak inside the interval $(0, L)$ due to the symmetry of eigenfunctions. Then, averaging over the different eigenvectors we obtain for $\lambda \lesssim 1$

$$\left\langle \frac{1}{\xi_q} \right\rangle \approx \frac{1}{\xi_q(\infty)} + \frac{C_q}{L} \tag{48}$$

where $\xi_q(\infty)$ is the localization length in the limit $L \to \infty$ and where the constant

$$C_q = \sum \left[\frac{1}{\xi_q} - \frac{1}{\xi_q(\infty)} \right] \tag{49}$$

depends on the shape of eigenfunctions, including fluctuations. Relation (48) is equivalent to the first scaling (47) with $C_1 = \gamma$ and $\gamma \xi_1(\infty) = \alpha D_0$. The second scaling (47) has still to be understood.

In the classical limit $\lambda \sim k^2/L = (K/LT)k \sim k \to \infty$ hence all eigenfunctions become ergodic in accordance with the Shnirelman theorem (see below).

An interesting microstructure of chaotic eigenfunctions was discovered by Heller (see Ref.[31]) and was termed "scars". These are enhancements of the density of eigenfunctions along classical periodic orbits in spite of the fact that all these orbits are unstable. A general theory of scars was developed in Refs.[29,32] using a very powerful method of Gutzwiller based on classical periodic orbits. In particular each such orbit determines the corresponding scar, the change of density being proportional to $\exp(-h_p T_p/2)$ where T_p is the orbit period, and $h_p \approx \sum \Lambda_+$ is the sum of positive Lyapunov exponents, the analogue of KS entropy for the unstable periodic orbit. Thus, appreciable scars appear only along the short-period orbits with $T_p \lesssim 1/h_p \approx 1/h$ where h is the classical KS entropy.

At first glance scars contradict the Shnirelman theorem [33] which states, loosely speaking, that classical ergodicity implies ergodicity for most eigenfunctions sufficiently far in the quasiclassical region (see also Refs.[54]). However, this is not the case because the spatial size of scars is minimal (of the order of the elementary quantum cell) while the Shnirelman theorem is of integral type. Indeed, according to Shnirelman the definition of an ergodic eigenfunction W_n (in Wigner's representation) is given by the expression

$$\int dp dq \, W_n(p, q) f(p, q) \xrightarrow[n \to \infty]{} \int dp dq \, g_\mu(p, q) f(p, q) \tag{50}$$

for any sufficiently smooth function f of phase space. Here

$$g_\mu = \delta \left(H(q, p) - E \right) \frac{dE}{dp dq} \tag{51}$$

is the microcanonical measure. The quantity $\rho(E) = dq dp/dE$ is the classical counterpart of the mean level density. The above definition of ergodicity is insensitive to the microstructure of the eigenfunction. Some eigenfunctions are known to be almost completely localized on a periodic orbit but the proportion of these rapidly decreases as $n \to \infty$ [55].

5 A phenomenological theory of quantum relaxation

In this section we derive a diffusion equation which describes the quantum relaxation process. As is known, in classical statistical mechanics the relaxation process is described by a diffusion equation. The same equation describes the quantum relaxation for $\lambda \gg 1$ when the final steady state is ergodic. However, in the case $\lambda \lesssim 1$ the quantum diffusion leads to a localized, non ergodic, steady state, and therefore it is necessary to modify the classical diffusion equation to include the phenomenon of localization. One way to do this is to use the complete Fokker–Planck equation with the so-called drift term [36]

$$\frac{\partial g}{\partial \tau} = \frac{1}{2} \frac{\partial}{\partial n} D(n) \frac{\partial g}{\partial n} - \frac{\partial}{\partial n} Bg \tag{52}$$

where

$$B = \frac{< \Delta n >}{\tau} - \frac{dD(n)}{dn} \tag{53}$$

In our problem this term describes the so-called backscattering, that is, the reflection of the ψ-wave propagating in n (see Section 6).

For sufficiently short times the diffusion is determined by the first term on the rhs of Eq.(52) and coincides with the classical diffusion. However, as time increases the backscattering eventually suppresses the diffusion and leads to a steady state (36). The general expression for the steady state $g_s(n)$ can be derived from Eq.(52) and is given by

$$\ln g_s = 2 \int \frac{B(n)dn}{D(n)} \tag{54}$$

In the case of homogeneous diffusion where $D = const$, the steady state Green function g_s is given by Eq.(36) with $l_s = D$. Hence, from Eq.(54) we obtain

$$B = \begin{cases} +1 & n < n_0 \\ -1 & n > n_0 \end{cases} \tag{55}$$

where n_0 is the initially excited state. It is remarkable that B turns out to be independent of the system's parameters and this may point to a quite general applicability of the method. Since the backscattering function $B(n)$ depends on the particular initial condition, the diffusion Eq.(52) describes the Green function only, which is initially $g(n, \tau) = \delta(n - n_0)$. Without loss of generality we can take $n_0 = 0$. Due to the symmetry with respect to $n = 0$ we can consider only the region $n > 0$. The diffusion equation then reads

$$\frac{\partial g}{\partial \tau} = \frac{D}{2} \frac{\partial^2 g}{\partial n^2} + \frac{\partial g}{\partial n} \qquad . \tag{56}$$

with boundary conditions

$$g(\infty, \tau) = 0$$

$$\frac{D}{2}\left(\frac{\partial g}{\partial n}\right)_{n=0} + (g)_{n=0} = 0$$

The solution of Eq.(56) can be found via Laplace transform and in the dimensionless variables $x = n/2D$, $s = \tau/2D$ reads:

$$g(x, s) = \frac{1}{\sqrt{\pi s}} \exp\left[-\frac{(x+s)^2}{s}\right] + \exp(-4x)\, \mathrm{erfc}\left(\frac{x-s}{\sqrt{s}}\right) \qquad (57)$$

where

$$\mathrm{erfc}(u) = \frac{2}{\sqrt{\pi}} \int_u^\infty \exp(-v^2)\, dv$$

Initially, all the probability $(\int_0^\infty g(x,s)\, dx = 1/2)$ is in the first term of (57) while asymptotically the whole probability goes in the second term, and for $s \to \infty$ the steady-state distribution is given by

$$g(x) = 2\exp(-4x) \qquad\qquad x > 0 \qquad (58)$$

Let us now compute the first two moments $<x>$ and $<x^2>$. The equations for them are

$$\left.\begin{aligned} \frac{d}{ds} <x> &= \frac{1}{2}g(0,s) - 1 \\ \frac{d}{ds} <x^2> &= \frac{1}{2} - 4<x> \end{aligned}\right\} \qquad (59)$$

which lead to the solution

$$<x> = \frac{1}{4} - \frac{1}{2}\mathrm{erfc}\left(\sqrt{s}\right)\left[\frac{1}{2} + s\right] + \frac{1}{2\sqrt{\pi}}\sqrt{s}\,\exp(-s)$$

$$<x^2> = \frac{E}{2D^2} = \frac{1}{8} + \frac{1}{2}\mathrm{erfc}\left(\sqrt{s}\right)\left[s + s^2 - \frac{1}{4}\right] - \frac{1}{2\sqrt{\pi}}\sqrt{s}\,\exp(-s)\left(\frac{1}{2} + s\right) \quad (60)$$

From this equation it follows that the relaxation to the steady state value $<x^2> \to 1/8$ is exponential. This is not in agreement with numerical computations where a power law relaxation was found for sufficiently large s [18]. To explain such a behaviour we need to take into account the explicit time dependence of $D(\tau) = D_0\, d(\tau)$ which follows from the discrete quantum spectrum. Notice that the ratio $D(\tau)/B(\tau)$ must be independent of time, at least asymptotically, to provide the exponential steady state (36). It follows that $B(\tau) = -d(\tau)$ with $d(\tau) \to 1$ as $\tau \to 0$.

As was remarked in Ref.[43] the diffusion rate must be proportional to the number of quasi-energy levels which are not yet resolved in time τ. This number decreases, for $\tau \geq \tau_R$ as $1/\tau$. We may therefore take

$$d(s) = \frac{\alpha}{\alpha + s}$$

where α is some unknown parameter ~ 1. The new solution of the diffusion equation (56), with time-dependent diffusion coefficient, remains the same "$g(x, \sigma)$ in (57)" in a new time variable

$$\sigma = \int d(s)\,ds \rightarrow \alpha \ln\left(1 + \frac{s}{\alpha}\right)$$

where now $x = n/2D_0$ and $s = \tau/2D_0$. Notice that the above logarithmic dependence is the only one which provides a power law relaxation in s for $g(x, \sigma(s))$.

Asymptotically, the relaxation goes as $s^{-\alpha}$ ($s \rightarrow \infty$). Comparison with numerical data [18,27] shows that $\alpha \approx 1$ which is also the result of a different theory [18].

From Eq.(60) we may then obtain the asymptotic relaxation law ($s \rightarrow \infty$)

$$\frac{E}{D_0^2} \approx \frac{1}{4} - \frac{1}{\sqrt{\pi}\,s\,(\ln s)^{3/2}}; \qquad \frac{dE}{ds} \approx \frac{D_0^2}{\sqrt{\pi}\,s^2\,(\ln s)^{3/2}} \qquad (61)$$

while for $s \rightarrow 0$ we have

$$\frac{E}{D_0^2} \approx s\left(1 - \frac{8}{3\sqrt{\pi}}\sqrt{s}\right)$$

in agreement with the classical diffusion.

Eq.(61) is close to but not identical with the result in Ref.[18] which, in our notations, reads:

$$\frac{dE}{ds} \approx c\,\frac{D_0^2 \ln s}{s^2} \qquad (62)$$

This expression seems to agree quite well with the numerical data in [18] with fitting parameter $c \approx .3$. On the other hand Eq.(60) was also confirmed by numerical data in Ref.[27] for the whole integration time interval. This problem therefore requires further analysis.

Another interesting characteristic of quantum relaxation is the so-called staying probability $g(0, \sigma)$ [30] which is given by

$$g(0, \sigma) = 2 + \frac{\exp(-\sigma)}{\sqrt{\pi\sigma}} - \mathrm{erfc}(\sqrt{\sigma}) \qquad (63)$$

and it has the same asymptotic behaviour (61) of the energy

$$g(0, s) \approx 2 + \frac{1}{2\sqrt{\pi}\,s\,(\ln s)^{3/2}} \qquad (64)$$

At this point it is necessary to stress that in the SM, where diffusion is homogeneous in n, quantum localization is a universal phenomenon. This is no longer the case when the diffusion rate depends on n. Consider, for example, the case

$$D(n) = D_0\,n^{2\alpha} \qquad (65)$$

Then, from Eq.(54) we obtain for the steady state distribution:

$$\ln g_s(n) = \begin{cases} -\dfrac{2n^{1-2\alpha}}{(1-2\alpha)D_0} & \alpha \neq \dfrac{1}{2} \\ -\dfrac{2}{D_0}\ln n & \alpha = \dfrac{1}{2} \end{cases} \qquad (66)$$

where we have assumed that $B = -1$ as before since it does not depend on the system's parameters. In agreement with previous results the critical value is $\alpha_c = (1/2)$. For $\alpha < \alpha_c$ the localization remains exponential (as was rigorously shown for a similar solid state model [70]) while for $\alpha > \alpha_c$ delocalization occurs because $g_s(n) \to const \neq 0$ as $n \to \infty$. In the critical case the steady state distribution is a power law

$$g_s \sim n^{-2/D_0} \qquad (67)$$

and the localization takes place for $D_0 < 2$ only, when $g_s(n)$ is normalizable.

Notice, however, that unlike the exponential localization, for a power law distribution the conditions of localization depend on the quantity one considers. The above condition $D_0 < 2$ refers to probability localization. However, the energy localization, that is, for the mean energy to be finite in the steady state, a stronger condition is required, namely $2/D_0 - 2 > 1$, or

$$D_0 < \frac{2}{3} \qquad (68)$$

which is in good agreement with numerical results, as is shown in Fig.7 [71].

One may also consider the SM (6) with parameter $k(n)$ depending on the dynamical variable n. In this case, however, the map is no longer canonical and in order to obtain the correct description of the global motion, the energy-time dynamical variables should be taken [43] as in the Kepler map (10). As is easily verified the expression for the steady state distribution (54) remain unchanged after appropriate rescaling of both D and B.

A more interesting example is the spin model (14) where the diffusion rate in s_z is given approximately by:

$$D(s_z) = \frac{k_0^2}{2}\left(1 - \frac{s_z^2}{s^2}\right) \qquad (69)$$

Consider, for instance, a steady state centred at $s_z = 0$. We have

$$g_s(s_z) \sim \left(\frac{s - |s_z|}{s + |s_z|}\right)^{1/\lambda} \qquad (70)$$

where $\lambda = l_0/s$ is the ergodicity parameter, and $l_0 = k_0^2/2$ is the localization length at $s_z \approx 0$. Contrary to the previous understanding [13,14] the diffusion in this model is localized, even for $\lambda \gtrsim 1$, provided $k_0 \ll s$. For a steady state centred at

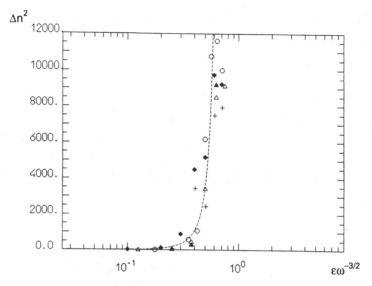

Fig. 7. The spread of the wave packet Δn^2 over the unperturbed states (second moment of the distribution in n), averaged in time from 100 to 200 field periods, vs the decimal logarithm of the variable $\varepsilon\,\omega^{-3/2}$ for $n_0 = 5$: filled diamonds: $\omega = 1.00$, $\varepsilon_0 = 0.2$; circles: $\omega = 2.00$, $\varepsilon_0 = 1.0$; triangles: $\omega = 2.52$, $\varepsilon_0 = 0.4$; crosses: $\omega = 4.60$, $\varepsilon_0 = 1.44$ and for $n_0 = 10$: filled triangles: $\omega = 2.52$, $\varepsilon_0 = 0.4$. The dashed line is drawn to guide the eye.

the "pole" $s_z = s$ we have

$$g_s(s_z) \sim \left(\frac{s + s_z}{s - s_z}\right)^{1/\lambda} \tag{71}$$

Unlike this, for the motion on hyperboloid (17) the diffusion is always delocalized since for $s_z \gg s$ the situation is the same as in model (65) with $\alpha = 1$.

Consider, finally, the diffusive photoeffect in Rydberg atoms. Instead of using the Kepler map (10) we may solve the problem in the principal quantum number n. Then, the diffusion rate [39] and the backscattering parameter are

$$D_n = \frac{a\,\varepsilon^2}{\omega^{7/3}}\,n^3; \qquad\qquad B(n) = B(v) \cdot \frac{dn}{dv} \cdot \frac{d\tau}{dt} = B(v) = \pm 1 \tag{72}$$

where t is measured in the number of field periods. From Eq.(54) we thus obtain

$$\ln g_s(n) = -\frac{2\,\omega^{7/3}}{a\,\varepsilon^2}\,|E - E_0| \tag{73}$$

that is an exponential localization in energy with the same localization length as from the Kepler map ($a \approx 3.3$) [12].

The above examples show that the phenomenological theory described in this section gives a reasonable description of the quantum relaxation process. In particular it would be interesting to check the predictions of Eqs. (70) and (71) numerically.

6 Quantum chaos and Anderson localization

The localization of eigenfunctions for the dynamical problem is similar to the celebrated Anderson localization in disordered solids [72,74]. For the latter problem, of particular interest are the so-called tight binding models which are lattice discretizations of the Schrödinger equation. These models play an important rôle in the investigation of transport properties of solids at low temperature, where the electron wave function becomes very sensitive to local impurities and imperfections of the crystal lattice.

The simplest well-known example is the Lloyd model which is described by the eigenvalue equation:

$$(Hu)_n = u_{n+1} + V_n u_n + u_{n-1} = E u_n \tag{74}$$

where the Hamiltonian H is a sum of a nearest-neighbour interaction term $(H_0 \psi)_n = \psi_{n+1} + \psi_{n-1}$ and a local term describing the interaction energy of the electron with the crystal site $(V\psi)_n = V_n \psi_n$. The boundary conditions are $u_0 = u_{N+1} = 0$ and the potential $\{V_n\}$ is a set of N independent random variables, with the same probability distribution:

$$P_w(V) = \frac{1}{\pi} \frac{W}{V^2 + W^2} \qquad W > 0 \tag{75}$$

It is mathematically proven that the above random model in the large N limit displays exponentially localized eigenfunctions, no matter how small the disorder W; the rate of decay is measured by the smallest Lyapounov exponent γ which may be evaluated by Thouless' formula [74] or by the transfer matrix method [75]. Although γ for a finite N depends on the realization of the disorder, in the limit $N \to \infty$ it converges to a non-random value, the inverse of which is known as *the localization length* ξ_∞.

For the Lloyd model, this quantity can be found analytically:

$$\xi_\infty^{-1} = \gamma = \text{Arcosh} \left[\frac{1}{4}\sqrt{(2+E)^2 + W^2} + \frac{1}{4}\sqrt{(2-E)^2 + W^2} \right] \tag{76}$$

The formal analogy between tight binding models and the quantum kicked rotator was discovered in Ref.[40]. It was shown that the equation for the quasienergies or for the Floquet operator

$$\hat{U}\psi = \exp\left(-ik\cos\hat{\vartheta}\right) \exp\left(-i\frac{T}{2}\hat{n}^2\right) \psi = \exp(-i\varepsilon T)\psi \tag{77}$$

can be written as an eigenvalue problem for a "tight binding model"

$$(W_0 - T_m)\bar{u}_m + \sum_r W_r \bar{u}_{m+r} = 0 \tag{78}$$

where

$$T_m = \tan\left(\frac{\varepsilon}{2}T - \frac{T}{4}m^2\right), \quad W_r = \frac{1}{2\pi}\int_0^{2\pi} d\vartheta\, \exp(i r\,\vartheta)\tan\left(\frac{k}{2}\cos\vartheta\right)$$

and $\bar{u}_r$ is the Fourier coefficient of the expansion of $\bar{\psi}(\vartheta) \simeq \frac{1}{2}[1 + \exp(i k\cos\vartheta)]\,\psi(\vartheta)$ in the momentum basis.

Eq.(78) is the eigenvalue equation for an electron in a crystal with site m, site energies T_m and hopping matrix elements W_r. This particular transformation must satisfy the bound $|k\cos\vartheta| < \pi$. This restriction can be avoided at the expense of a more complicated transformation as shown in Ref.[41].

For this particular example in the generic case in which T is an irrational multiple of 4π, the localization of the eigenfunction was proven by several numerical computations. This implies that Anderson localization is also possible in a non-random potential. It should be stressed that in the dynamical case no external random element is introduced since the perturbation is periodic and that localization is related here to quasienergy eigenfunctions and occurs in momentum instead of configuration space.

When $T/4\pi$ is rational it is apparent that the potential becomes periodic, the electron is described by Bloch waves and moves freely in the crystal. This situation corresponds to the so-called quantum resonance in the kicked rotator. Indeed let us observe that the quantum description is endowed with two different periods: the first is explicitly specified by the perturbation, the second is 4π and follows from the peculiarity of the free evolution of having a spectrum given by integers. Naively speaking, the free rotator has energy levels $E_n = n^2/2$ and the photon's energy is $2\pi/T$; the resonance condition is met whenever an integer number of photons matches the energy for a transition between unperturbed levels. This condition corresponds to rationality of the ratio $T/4\pi$ between the two periods.

The general *resonant case* $T/4\pi = p/q$ has been investigated by Izrailev and Shepelyansky [62]. In this case the search for quasienergies is reducible to the problem of diagonalizing a $q \times q$ unitary matrix. In fact, they obtained the following Floquet map depending on the parameter θ

$$\psi\left(\vartheta + \frac{2\pi}{q}m\right) = \sum_{n=0}^{q-1} S_{mn}^\theta\, \psi\left(\theta + \frac{2\pi}{q}n\right)$$

$$S_{rs}^\theta = \exp[-i k\cos(\theta + 2\pi r/q)]\frac{1}{q}\sum_{m=0}^{q-1}\exp\left[-2\pi i\left(\frac{p}{q}m^2 + m\frac{s-r}{q}\right)\right] \qquad (79)$$

The eigenvalues $\exp[i\lambda_j(\theta)]$ of the unitary matrix S^θ have a continuous dependence on the angle $\theta \in [0, 2\pi/q]$, and the spectrum of the resonant Floquet operator is therefore continuous with q bands.

The introduction of a second, incommensurate perturbing frequency in the kicked rotator produces a sharp increase in the localization length, which grows exponentially with the classical diffusion rate [79]. Again, this is in agreement with the theory of localization in two-dimensional disordered lattices [80].

Since new, incommensurate frequencies in the time-dependent problem introduce new dimensions in the extended phase space, a time-dependent model with three incommensurate frequencies is expected to correspond to a three-dimensional lattice problem, where a transition from localized to extended states occurs at some critical parameter value. Such a transition has indeed been observed [81] and by analysing it, some indications were obtained that may be helpful in clarifying the nature of the Anderson transition itself, which is still in discussion in solid-state physics. In this connection, we wish to emphasize that the one-dimensional character of our time-dependent model allows for a sharp reduction of the computation time needed to analyse the transition, so that recourse to scaling assumptions can be avoided [80].

Let us consider in general the following time-dependent Hamiltonian:

$$H = H_0 + V(\theta, t) \sum_{s=-\infty}^{+\infty} \delta(t - s) \tag{80}$$

The second term describes kicks occurring periodically in time with period one. The free evolution between kicks is given by the Hamiltonian H_0:

$$H_0|n\rangle = E_n|n\rangle, \qquad |n\rangle = \exp(in\theta)/(2\pi)^{1/2} \tag{81}$$

We assume the eigenvalues E_n to be random numbers uniformly distributed in $(0, 2\pi)$. Unlike for the usual kicked rotator, we also assume V to depend explicitly on time according to

$$V \equiv V(\theta, \theta_1 + \omega_1 t, \theta_2 + \omega_2 t) \tag{82}$$

where V is a periodic function of its three arguments to be specified later, and θ_1 and θ_2 are arbitrarily prescribed phases. We would like ω_1 and ω_2 to be incommensurate with each other and also with the frequency of the kicks. Therefore, we take $\omega_1 = 2\pi\lambda^{-1}$ and $\omega_2 = 2\pi\lambda^{-2}$, with $\lambda = 1.3247\cdots$ the real root of the cubic equation $\lambda^3 - \lambda - 1 = 0$. With this choice, ω_1 and ω_2 are a "most incommensurate" pair of numbers. Thus (80) describes the motion of a rotator subjected to periodic kicks, the strength of which is modulated in time by the frequencies ω_1 and ω_2.

The evolution of this rotator, from just after one kick to just after the next, is given by

$$\psi(\theta, t + 1) = \exp[-iV(\theta, t + 1)] \exp(-iH_0)\psi(\theta, t) \tag{83}$$

This formulation of the rotator dynamics is very convenient for numerical simulations because the time dependence of V is known explicitly. Nevertheless, in order to elucidate the connection of this time-dependent problem with a three-dimensional tight-binding model, we must resort to a different formulation as follows. First of all, we consider the phases θ_1 and θ_2 as new dynamical variables, with conjugate momenta n_1 and n_2. Then we consider the Hamiltonian

$$H' = H_0(\hat{n}) + \omega_1 \hat{n}_1 + \omega_2 \hat{n}_2 + V(\theta, \theta_1, \theta_2) \sum_{s=-\infty}^{+\infty} \delta(t-s) \qquad (84)$$

with $\hat{n}_{1,2} = -i\partial/\partial\theta_{1,2}$. Eq.(84) describes a quantum rotator with three freedoms $(\theta, \theta_1, \theta_2)$ subjected to periodic kicks, the strength of which is not explicitly time-dependent. The one-period propagator for this rotator is the unitary operator

$$\exp[-iV(\theta, \theta_1, \theta_2)] \exp\{-i[H_0(\hat{n}) + \omega_1 \hat{n}_1 + \omega_2 \hat{n}_2]\}$$

In order to show that the three-dimensional quantum model defined by (84) and the one-dimensional model defined by (80) and (82) are substantially equivalent, we rewrite the Schrödinger equation for the three-dimensional model

$$i\frac{d}{dt}\psi(\theta, \theta_1, \theta_2, t) = H'\psi(\theta, \theta_1, \theta_2, t)$$

in the interaction representation defined by

$$\psi(\theta, \theta_1, \theta_2, t) = \exp[-i(\omega_1 \hat{n}_1 + \omega_2 \hat{n}_2)t]\tilde{\psi}(\theta, \theta_1, \theta_2, t)$$

In this way we obtain

$$i\,d\tilde{\psi}/dt = {}^{\cdot}H_0\tilde{\psi} + V(\theta, \theta_1 + \omega_1 t, \theta_2 + \omega_2 t)\sum_{s=-\infty}^{+\infty}\delta(t-s)\tilde{\psi}$$

i.e., the Schrödinger equation for the evolution of the one-dimensional model.

Then, as shown above, the problem of determining the quasienergy eigenvalues and eigenvectors turns out to be formally equivalent to solving the equation

$$T_\mathbf{n}\, u_\mathbf{n} + \sum_{\mathbf{r}\neq 0} W_\mathbf{r} u_{\mathbf{n}+\mathbf{r}} = \epsilon u_\mathbf{n} \qquad (85)$$

where $\mathbf{n} \equiv (n, n_1, n_2)$ and $\mathbf{r}$ label sites in a three-dimensional lattice,

$$T_\mathbf{n} = -\tan\left[\frac{1}{2}\left(E_n + n_1\omega_1 + n_2\omega_2 + \lambda\right)\right]$$

Here λ is the quasi-energy, $W_\mathbf{r}$ are the coefficients of a threefold Fourier expansion of $\tan\left[\frac{1}{2}V(\theta, \theta_1, \theta_2)\right]$, and $\epsilon = -W_0$.

We now chose

$$V(\theta, \theta_1, \theta_2) = -2\tan^{-1}\left[2k(\cos\theta + \cos\theta_1 + \cos\theta_2)\right]$$

so that (85) becomes (compare with Eq.(78))

$$T_\mathbf{n} u_\mathbf{n} + k\sum_{\mathbf{r}}{}' u_\mathbf{r} = 0 \qquad (86)$$

where the sum $\sum'$ includes only the nearest neighbours to $\mathbf{n}$. The tight-binding model (86) with the potential $T_\mathbf{n}$ is in a sense equivalent to the original rotator problem. The quasienergy eigenfunctions of the rotator will be localized or extended over the unperturbed eigenstates of H_0 depending on whether the tight-binding model has localized or extended eigenstates; in the localized case, the localization length will be the same. Since the dynamics of the rotator is determined by the nature of its quasienergy eigenstates, any change from localized to extended states that may take place in the tightbinding model (86), as the coupling parameter k is increased, will be mirrored by a simultaneous change in the rotator dynamics, from a localized recurrent behaviour to an unbounded spreading over the unperturbed base. As we mentioned above, the latter type of transition can be numerically detected with less effort than directly tackling the three-dimensional tight-binding model.

The model was investigated by numerical simulation of the quantum dynamics defined by (80) with phases $\theta_1 = \theta_2 = 0$. A transition between two different types of motion was observed around a value $k_{cr} \approx 0.47$, with localization occurring for $k < k_{cr}$ and unbounded diffusion taking place for $k > k_{cr}$.

The dependence of the diffusion rate $D = \langle (n - n_0)^2 \rangle / t$ (in the delocalized regime) and of the inverse localization length $\gamma = l^{-1}$ (in the localized regime) on the perturbation parameter k is shown in Fig. 8. The dependences of D and γ near k_{cr} are consistent with power laws, $D \sim D_0 (k - k)_{cr}^s$, $\gamma \sim \gamma_0 (k - k_{cr})^v$ with $D_0 \sim 2.5$, $k_{cr} \approx 0.46$, and $s \approx 1.25$. An analogous fit for the dependence of γ gave $\gamma_0 \sim 3.5$, $k_{cr} \approx 0.469$, and $v \approx 1.5$. Thus the two fittings give very close values of k_{cr} consistent with renormalization theory predictions [78].

The fruitful analogy between Anderson localization and the dynamical problem is extensively used to share various methods and some results in both fields [76,77]. One of the important implications of this analogy is that a random potential is only a sufficient but not a necessary condition for Anderson localization. Hence a new problem arises: the localization in regular (but, of course, not periodic) solids.

One should bear in mind however that this analogy is restricted to the eigenfunctions only. For example the time evolution of a given initial state may be completely different in both problems. In particular, in one-dimensional disordered solids there is no diffusion stage [42] so that Anderson localization is the localization of the free spreading of the initial state. This is immediately seen from the same uncertainty principle which gives the following density of the operative eigenstates

$$\rho_0 \sim \frac{l\,dp}{dE} \sim \frac{l}{u} \tag{87}$$

This is just the localization (relaxation) time scale which is always of the order of the time interval for a free spreading of the initial wave packet at a characteristic velocity u. Thus, only backscattering remains, and its picture is expecially simple, being the interference of scattered waves from different parts of the potential.

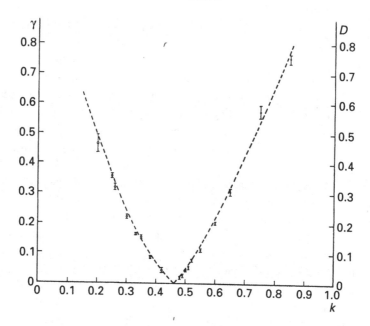

Fig. 8. Diffusion rate D (dots) and inverse localization length $\gamma = 1/l$ (circles) as a function of perturbation parameter k. Error bars were obtained from statistics over ten different realizations of the random spectrum. The dotted lines result from a three parameters least-squares fit of numerical data.

The reason why this general very complicated interference always gives an average backward flow is related to the fact that in a random (or sufficiently irregular) potential there is always a resonant harmonic which provides the complete reflection of the incoming wave. From this explanation it may seem that for localization one needs potentials with continuous spectra. However, this would only be true for an infinitely small amplitude of potential variation. For a finite amplitude the Schrödinger equation in periodic potential is the Mathieu equation with finite instability zones which almost overlap for a sufficiently large amplitude. However, in a periodic potential there are no solutions which would decay in both directions and hence localization is impossible. One needs at least two different periods in the potential to construct a solution decaying in both senses. This would not be universal localization because the instability zones overlap only for sufficiently strong amplitudes. Our conjecture is that the same is true for any discrete spectrum of the potential. Another plausible conjecture is that a continuous spectrum of the potential is a necessary condition for universal localization. However, it may not be sufficient as the interesting example in [73] demonstrates.

7 Experimental observation of localization: the hydrogen atom in a microwave field

One of the most significant cases where classical and quantum chaos confronted each other was in the explanation of an experiment on hydrogen atoms, first performed in 1974 by Bayfield and Koch [82]. Single atoms prepared in very elongated states with a high principal quantum number ($n_0 \approx 63 - 69$) were injected into a microwave cavity and the ionization rate was measured. The microwave frequency was 9.9 GHz, corresponding to a photon energy well below the ionization energy of level 66 and even lower than the transition from state 66 to 67. Much surprise therefore followed the discovery that a very efficient ionization occurred when the electric field intensity exceeded a threshold value of about 20 V per cm (for $n_0 = 66$), much lower than the static Stark value. More surprisingly, a numerical simulation by Leopold and Percival [83] showed that classical mechanics could reproduce the experimental data quite well. The subsequent analysis [39,85], still in classical terms, explained the threshold intensities as critical values for the onset of chaotic diffusion in action space. A condition for the occurrence of full chaotic diffusion is that the microwave frequency is greater than the frequency of the electron's motion, namely $\omega n_0^3 > 1$. However, the hydrogen atom is a quantum object. The quantum mechanical evolution was investigated in the one-dimensional approximation [86] and for $\omega n_0^3 > 1$ it was predicted an ionization threshold higher than the classical value, to overcome the occurrence of quantum localization. This effect vanishes when approaching the main resonant region $\omega n_0^3 \approx 1$, and this may explain why classical mechanics works so well at lower values of ωn_0^3.

The classical Hamiltonian for a one-dimensional hydrogen atom interacting with a time-periodic microwave field in the dipole approximation is

$$H(x', p', t') = \frac{p'^2}{2m} - \frac{e^2}{z'} + eE_0 z' \cos(\omega' t') \qquad z' \geq 0$$

One goes to natural atomic units by setting $z' = za_0$, where $a_0 = \hbar^2/me^2$ is the Bohr radius, and $t' = tT_0$, where $T_0 = \hbar^3/me^2$. Defining the rescaled parameters $\epsilon = E_0 a_0^2/e$ and $\omega = \omega' T_0$, the dimensionless Hamiltonian is (cf. Eq.(9))

$$H(z, p, t,) = \frac{p^2}{2} - \frac{1}{z} + \epsilon z \cos(\omega t) \qquad (88)$$

the energy being measured in units of e^2/a_0. The unperturbed Hamiltonian describes both bounded (with negative energy) and unbounded motions; since we are interested in exploring the dynamics that precedes ionization, we are confined to negative energies, and accordingly introduce action-angle variables (n, θ) thus obtaining Eq.(9).

In the time-independent formalism, the Hamiltonian (9) is replaced by

$$\mathscr{H}(n, \theta, v, \phi) = -\frac{1}{2n^2} + \omega v + \epsilon z(n, \theta) \sin \phi \qquad (89)$$

where $\phi = \omega t$ and v is the canonical momentum. Their variations in the auxiliary time η are given by $d\phi/d\eta = \omega$, implying $\eta = t + const$ and $dv/d\eta = -\epsilon z(I, \theta)\cos\phi$. Since $\mathcal{H}$ is a constant of motion, to zero order in ϵ one may view the variation of v as the number of photons exchanged by the atom with the field. We now want to write an approximate map for the canonical variables (v, ϕ). The equations are first integrated over one period of the unperturbed orbit, to first order in ϵ:

$$\left.\begin{array}{l} \bar{v} = v + k\sin\phi \\[1mm] \bar{\phi} = \phi + 2\pi\omega n^3 \end{array}\right\} \tag{90}$$

where $k \approx 2.6\,\epsilon/\omega^{5/3}$ (see eqs.(11) and (13)).

To remove the unwanted n-dependence, one may set from Eq.(89) $n \approx (2\omega v - 4\mathcal{H})^{-1/2}$ and replace v with $\bar{v}$. The resulting canonical map is Kepler's map.

$$\left.\begin{array}{l} \bar{v} = v + k\sin\phi \\[1mm] \bar{\phi} = \phi + 2\pi\omega(2\omega\bar{v} - 2\mathcal{H})^{-3/2} \end{array}\right\} \tag{91}$$

which coincides with Eq.(10) after setting the motion integral $\mathcal{H} = 0$. A linearization around the initial value $v_0 = v(n_0)$ and a phase shift, yield once again the SM:

$$\left.\begin{array}{l} \bar{v} = v + k\sin\phi \\[1mm] \bar{\phi} = \phi + T\bar{v} \end{array}\right\} \tag{92}$$

with $T = 6\pi\omega^2 n_0^5$. The actual value of v_0 is arbitrary, and reflects the arbitrariness in choosing the zero energy in (89). We decide to take $v_0 = 0$, corresponding to the initial value n_0. The relation between the action and the number of photons is therefore

$$v\omega = \frac{1}{2n_0^2} - \frac{1}{2n^2}$$

In spite of the various simplifications introduced so far, the map (92) still gives a good description of the mean behaviour of the system and allows some conclusions.

The condition $kT \approx 1$ gives the threshold for transition to classical chaos leading to fast ionization.

$$\epsilon_c \approx \frac{1}{50\,n_0^5\,\omega^{1/3}} \tag{93}$$

In the corresponding quantum model we expect, similarly to the kicked rotator, localization of the wave function in v. After the relaxation time τ_R the system reaches the steady state (cf. Eq.(36))

$$g(v) \approx \frac{1}{l_s}\exp\left(-2\frac{|v|}{l_s}\right) \tag{94}$$

The predicted localization length in the number of photons is

$$l_s \approx \frac{k^2}{2} \approx 3.3\epsilon^2\omega^{-10/3}$$

and equals the diffusion coefficient in ν-space. If l_s is greater than the number of photons required to ionize the atom $l_s \geq (2\omega n_0^2)^{-1}$, then localization cannot prevent ionization. This takes place for field intensities greater than the *quantum delocalization border*

$$\epsilon_q \approx 0.4\omega^{7/6}n_0^{-1} \tag{95}$$

Unexpected though these predictions may have been at their first appearance [12,86,87], they were confirmed by recent experimental results on the microwave ionization of hydrogen atoms [89, 90]. It was found that experimental and numerical data agree fairly well with localization theory and at the same time appreciably deviate from classical predictions. The experiments described in [90] were designed precisely for the purpose of checking localization theory; as a matter of fact, special care was taken in order that numerical computations could simulate as closely as possible the experimental conditions. Therefore, they provide experimental evidence of the quantum suppression of classically chaotic diffusion due to the localization phenomenon.

In Fig. 9 a comparison of the theory with the experimental data obtained in [89] is presented. The circles represent the experimentally observed threshold values of the microwave peak-field intensity for 10% ionization. Here the microwave frequency $\omega/2\pi = 36.02$ GHz, $\epsilon_0 = \epsilon n_0^4$ is the rescaled peak field intensity, $\omega_0 = \omega n_0^3$ is the rescaled microwave frequency and n_0 is the principal quantum number of the initially excited state value. Both in the experiment and in the quantum numerical computations the ionization probability is defined as the total probability above a cutoff level n_c. The dotted curve is the classical chaos border and the dashed curve in Fig. 9 is the theoretical prediction of localization theory for the 10% threshold value. Unlike the previous case of Ref. [90], the numerical data here were obtained from the numerical simulations of the "quantum Kepler map". In such simulations the interaction time, including the switching on and off of the microwave field, was chosen to be the same as in actual experiments.

The agreement between experimental and numerical data is the more remarkable, in that the quantum Kepler map is only a crude approximation for the actual quantum dynamics. In particular, from Fig. 9 it is seen that when the principal quantum number n_0 of the initial state is increased the data follow the predictions of localization theory.

Though the numerical model was one-dimensional, in actual experiments the initially excited state corresponds to a microcanonical distribution over the shell with a given principal quantum number. The classical counterpart for this would be a microcanonical ensemble of orbits. Nevertheless, the experimental data agree fairly well with the predictions of the one-dimensional quantum Kepler

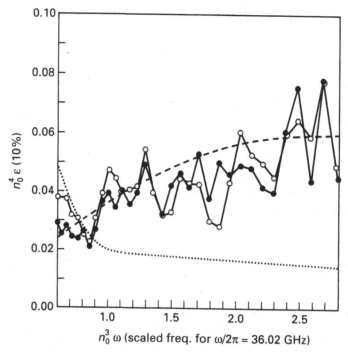

Fig. 9. Scaled 10% threshold fields from experimental results (taken from Fig. 2a, Ref.[89]), (circles), and from numerical integration of the quantum Kepler map (full circles). Curves have been drawn to guide the eye. The dashed line is the quantum theoretical prediction according to localization theory. The dotted curve is the classical chaos border (from Ref.[77]).

map. The reason for this agreement was found in Ref. [12]: due to the existence of an approximate integral of the motion, the main contribution to excitation turns out to be given by orbits which are extended along the direction of the (linearly polarized) external field. For such orbits, the use of the one-dimensional model is fully justified (see, e.g., Fig. 18b in Ref. [12]). Moreover, by using the Kepler map formulation, it has been theoretically and numerically shown [88] that the introduction of a second incommensurate frequency leads to a significant decrease of the threshold border for ionization. It would be interesting to have an experimental confirmation of this prediction.

In closing this section we would like to mention a new feature of this problem which is currently under investigation. Namely it has been shown recently [84] that, at field amplitudes much larger than the classical chaos border, the classical ionization probability *decreases* with *increasing* field intensity. This effect should be observable in laboratory experiments.

8 Fractal spectrum and anomalous diffusion

For motion bounded in phase space the quantum spectrum is always discrete as for example in the kicked rotator on a torus where the spectrum consists of L lines. For unbounded motion the spectrum has, generally, a very complicated structure, even for such a simple model as the SM on the cylinder. This problem has as yet no rigorous solution. In the QSM the spectrum is known to be continuous if the parameter $T/4\pi = r/q$ is any rational number other than $\frac{1}{2}$. Notice that such a spectrum does not mean any chaotic motion but corresponds to a peculiar process which is called *quantum resonance* since the quantity $T/4\pi$ is the ratio between the unperturbed and driving frequencies [11, 62]. In quantum resonance (which has no classical counterpart) the momentum n grows, on average, proportionally with time and hence the energy $E \sim \tau^2$. Using the analogy with the solid state problem described above quantum resonance corresponds to the free motion of the electron (quantum particle) in an exactly periodic potential (the so-called Bloch states).

Quantum resonance is a peculiarity of the kicked rotator model, which is periodic in n. Even though the resonant values of $T/4\pi$ have zero measure they are everywhere dense and this leads to difficulties in the analysis of the *typical* dynamics of this model when $T/4\pi$ is irrational. The resolution of this difficulty is in that the rate of the resonant motion (E/τ^2) rapidly decreases as the denominator q increases. This is approximately described by the semiempirical expression

$$< n^2 > \approx D\tau^2 \exp\left(-\frac{2q}{D}\right) \tag{96}$$

which is valid for $q \gtrsim D$.

A detuning from resonance $\epsilon(q) = |\,T/4\pi - r/q\,|$ would stop the growth after a time $\tau(\epsilon)$ which, on account of Eq.(23), can be assumed to be

$$\epsilon\tau n^2 = v \sim 1 \tag{97}$$

Consider now the irrational

$$\frac{T}{4\pi} = (m_1, m_2 \cdots m_i, \cdots) \equiv \cfrac{1}{m_i + \cfrac{1}{m_2 + \cdots}} \tag{98}$$

in the continued-fraction representation, and its convergents r_i/q_i

$$r_i/q_i = (m_1, \cdots m_i) \to \frac{T}{4\pi} \quad ; \qquad q_{i+1} = m_{i+1}\,q_i + q_{i-1}$$

Our aim is to identify the set of irrationals $T/4\pi$ for which a growth law of the form

$$< n^2 > = G\tau^\delta \tag{99}$$

occurs for any $G > 0$ and δ in the interval $(0, 2)$. To this end we substitute (99) in (97) and after eliminating τ with Eq.(96) we have

$$\epsilon(q_i) = \frac{v}{G} \left(\frac{D}{G}\right)^{\frac{1+\delta}{2-\delta}} \exp\left(\frac{-2q_i}{D}\frac{1+\delta}{2-\delta}\right) \tag{100}$$

On the other hand from the continued-fraction representation of $T/4\pi$ we have

$$\epsilon(q_i) \approx \frac{c}{q_i \, q_{i+1}} \tag{101}$$

with $c \sim 1$. Then, from (100) and (101) we obtain the relation which allows the construction of the irrational values of $T/4\pi$ giving the desired growth rate (99) and therefore the continuous spectrum

$$m_{i+1} \approx \frac{q_{i+1}}{q_i} \approx \frac{c\,G}{v\,q_i^2} \left(\frac{G}{D}\right)^{\frac{1+\delta}{2-\delta}} \exp\left(\frac{2q_i}{D}\frac{1+\delta}{2-\delta}\right) \tag{102}$$

In the whole interval $0 < \delta < 2$ the motion is unbounded and the spectrum is singular continuous with a fractal structure. The maximal allowed value $\delta = 2$ corresponds to the resonant values of $T/4\pi$ for which the spectrum is absolutely continuous. We would like to attract the reader's attention to the fact that Eq.(102) leads (for $\delta > 0$) to the irrationals that are approximated by rationals to exponential accuracy (the so-called Liouville, or transcendental, numbers) in agreement with known rigorous results [44] which prove the existence of the irrationals $T/4\pi$ leading to a continuous spectrum and unbounded energy growth. However, from the above discussion a new conclusion can be drawn, namely, that even inside those transcendental values of $T/4\pi$ there are infinitely many such ones which lead to localization. They correspond to $\delta = 0$ with finite G. This implies localization for typical irrationals. From Eq.(102) with $\delta = 0$ we can derive the asymptotic condition for the set of irrationals leading to localization:

$$m_i < \exp\left(\frac{q_i}{D}\right) \tag{103}$$

Therefore in the case of a quantum kicked rotator, dynamical localization takes place for almost all irrational values of the ratio between the unperturbed and external frequencies. Correspondingly the spectrum has a pure point character. The band structure obtained by approaching the irrational value via a sequence of rational approximants is characterized by bandwidths which shrink exponentially with respect to the number of bands and this excludes the possibility of self-similarity.

However, fractal features of the spectrum with a corresponding rich variety of dynamical behaviour have been shown to appear in an interesting model, the so-called kicked Harper model which is obtained by quantizing the following area-preserving map [94]

$$\left.\begin{array}{l} p_{n+1} = p_n + K \sin x_n \\ x_{n+1} = x_n - L \sin p_n \end{array}\right\} \qquad (104)$$

In particular this model describes the motion of electrons in two-dimensional lattices in the presence of a magnetic field and can lead to a better understanding of the physics of such systems.

The classical map (104) is known to be chaotic for $KL \gtrsim 1$. As in the case of the kicked rotator, the quantum motion is governed by the one-period evolution operator

$$\hat{U} = \exp\left[-i\left(\frac{L}{\hbar}\right)\cos(\hbar\hat{n})\right]\exp\left[-i\left(\frac{K}{\hbar}\right)\cos x\right] \qquad (105)$$

where $\hat{n} = -i\dfrac{d}{dx}$.

Quantum motion has been studied for a strongly irrational value $\hbar = 2\pi/(m + \rho_{GM})$ with $\rho_{GM} = (\sqrt{5}+1)/2$. Approaching this value via a sequence of rational convergents $\{p_n/q_n\}$ (by successive truncations of the continued fraction expansion of $\hbar/2\pi$) one obtains, at each convergent, a band spectrum with Bloch eigenfunctions. Analysis of the spectrum shows that regular scaling behaviour coexists with irregular scaling [91]. Moreover, the dynamical behaviour shows anomalous diffusion $< \Delta n^2(t) >\sim t^\alpha$ where the exponent $0 \leq \alpha \leq 2$ depends on the parameters K and L (Fig. 10(a)) [92]. This anomalous diffusion builds up at sufficiently large times while initially the diffusion is always close to the classical one ($\alpha = 1$) in accordance with correspondence principle (see Fig. 10(b)).

In conclusion the structure of the quasienergy spectrum and the quantum motion of classically chaotic systems appear to be much more rich than previously expected. While for the KR problem this complex structure is confined to a set of zero measure, for the kicked Harper model it appears to be the generic case. In spite of the progress made in the last 15 years the above results indicate that we are still far even from a qualitatively clear understanding of the quantum motion of classically chaotic systems and, moreover, a rigorous mathematical analysis appears to be more and more difficult.

There are also examples of the "true" chaos in quantum mechanics which have the strongest statistical property – the exponential instability of motion for the infinite time interval. One particular model is the following: consider a classical dynamical system on a N-dimensional torus specified by the equations:

$$\dot{\vartheta}_i = \omega_i(\vartheta_k) \qquad i, k, = 1, 2, \cdots N \qquad (106)$$

If $N \geq 3$ classical chaos is possible with exponentially unstable solutions of the linearized equations.

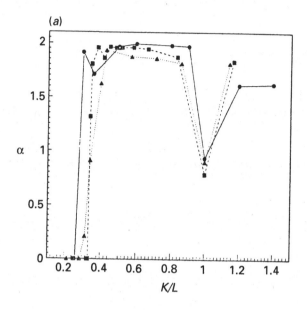

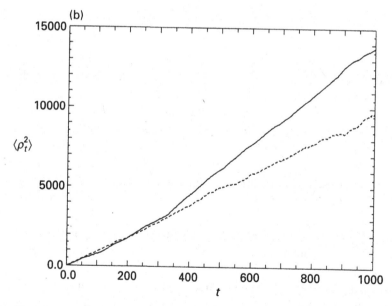

Fig. 10. (a) Dynamical exponent α vs K/L for some fixed values of L. Circles are for $L = 5$ and $1.3 < K < 7$, squares for $L = 6$ and $1.5 < K < 7$, triangles for $L = 7$ and $1.5 < K < 8$. Notice the minimum corresponding to the critical line ($K = L$). Each α was determined by examining a time series up to 3×10^4 kicks. (b) Example of anomalous quantum diffusion (solid line) for $K = L = 5$ and $\hbar = 2\pi/(18 + q_{GM})$, compared with classical diffusion (dotted line) for the same $K = L = 5$. Notice that the quantum diffusion is close to the classical up to a finite time (which increases as $\hbar \to 0$).

One particular example due to Arnold is

$$\left.\begin{aligned}
\vartheta_1 &= \cos \vartheta_2 + \sin \vartheta_3 \\
\vartheta_2 &= \cos \vartheta_3 + \sin \vartheta_1 \\
\vartheta_3 &= \cos \vartheta_1 + \sin \vartheta_2
\end{aligned}\right\} \tag{107}$$

Consider the Hamiltonian

$$H(n, \vartheta) = \sum_{1^k}^{3} n_k \, \omega_k \, (\vartheta) \tag{108}$$

The equations of motion are

$$\left.\begin{aligned}
\vartheta_i &= \omega_i \, (\vartheta) \\
\dot{n}_i &= -\sum_k n_k \frac{\partial \, \omega_k}{\partial \, \vartheta_i}
\end{aligned}\right\} \tag{109}$$

The equations for the momenta coincide (apart from a time-reversal) with the linearized equations of system (106). Therefore if system (106) is chaotic and time-reversible (like example (107)), then the momenta in (109) grow exponentially with time.

Consider now the quantized version of system (108)

$$\hat{H} = \frac{1}{2} \sum_{1^k}^{s} (\omega_k \, \hat{n}_k + \hat{n}_k \, \omega_k); \qquad \hat{n}_k = -i \frac{\partial}{\partial \, \vartheta_k} \tag{110}$$

From the Schrödinger equation we obtain, for the quantum probability density $f(\vartheta, t) = |\psi \, (\vartheta, t)|^2$:

$$\frac{\partial f}{\partial t} + \sum_k \frac{\partial}{\partial \, \vartheta_k} (f \, \omega_k) = 0 \tag{111}$$

This equation exactly coincides with the continuity equation of the classical system (106) and therefore, the quantum probability evolves in time exactly like the probability of the classical chaotic motion.

Even though the above, and other similar examples, are quite exotic, they are very useful for understanding the nature of quantum chaos. A common feature of these examples is that, in order to have the "true" quantum chaos, the momenta must grow exponentially with time. Indeed, as we stated in the introduction, the exact (not coarse-grained) classical distribution function does not approach a homogeneous distribution even in the case of chaotic motion. On the contrary, it becomes more and more scarred due to the local instability of motion and, correspondingly, the wave numbers of the Fourier harmonics grow exponentially. In quantum mechanics this corresponds to an exponential growth of momenta. Notice, however, that even in these exotic examples the true quantum chaos is restricted to configurational space only. (For further discussion of such examples see Ref.[56].)

9 Quantum chaos and the random matrix theory

The complete solution of the dynamical quantum problem is given by the diagonalization of the Hamiltonian to find the energy (or quasienergy) eigenvalues and eigenfunctions. The evolution of any quantity can be expressed as a sum over these eigenfunctions. For example the energy time dependence is

$$E(t) = \sum_{mm'} c_m c_{m'}^* E_{mm'} \exp[i(\omega_m - \omega_{m'})t] \tag{112}$$

where $E_{mm'}$ are the matrix elements and the initial state is $\psi(n, 0) = \sum_m c_m \varphi_m(n)$. For chaotic motion the dependence is generally very complicated but the statistical properties of the evolution can be related to the statistics of eigenfunctions $\varphi_m(n)$ (and hence of the matrix elements $E_{mm'}$) and of eigenvalues ω_m.

There exists a well-developed random matrix theory (RMT) which describes some average properties of a typical quantum system with a given symmetry of the Hamiltonian. At the beginning the object of this theory was assumed to be a very complicated, particularly many dimensional, quantum system as a representative of a certain statistical ensemble. After understanding the phenomenon of dynamical chaos it became clear that the number of freedoms of the system is irrelevant. Instead, the number of quantum states or the quasi-classical parameter is of importance provided the dynamical chaos is in the classical limit.

Until recently ergodicity of eigenfunctions was assumed in this theory which means that in the expansion

$$\varphi_m = \sum_j^N a_{mj} u_j \tag{113}$$

where the u_j form a physically significant, for example unperturbed, basis all probabilities

$$< |a_{mj}|^2 > = \frac{1}{N} \tag{114}$$

are equal on average, and N is the size of the matrix. Under this condition the distribution of neighbouring level spacings is given approximately by the Wigner–Dyson law

$$p(s) \approx A\, s^\beta \exp(-B\, s^2) \tag{115}$$

where A, B are obtained from normalization and the condition $< s >= 1$. The repulsion parameter β takes three values only (1, 2, or 4) depending on system's symmetry.

A new problem here is the impact of localization on the level statistics. This was first addressed in Ref.[45], and a new class of semiempirical spacing distributions was discovered

$$p(s) \approx A\, s^\beta \exp\left[-\frac{\pi^2}{16}\beta s^2 - \left(B - \frac{\pi\beta}{4}\right) s\right] . \tag{116}$$

where β is now a continuous parameter in the whole interval $(0, 4)$. This distribution is also called the *intermediate statistics* in contrast to the *limiting statistics* (115) for the ergodic eigenfunctions. This intermediate statistics should be distinguished from that of Berry–Robnik [66] which describes the lack of ergodicity in the classical limit.

The repulsion parameter β was shown to be closely related to the entropy localization length [16], namely

$$\beta \approx \beta_{\overline{H}} = \beta_e \exp\left[\overline{H} - H_e(\beta_e)\right] = \frac{\gamma(\beta_e)}{N} \, l_{\overline{H}} \tag{117}$$

where $\beta_e = 1, 2, 4$ is the repulsion parameter for ergodic eigenstates with limiting statistics (115) and average entropy H_e. The reason for this simple relation remains an open question (see also below).

Unlike the statistical RMT, the intermediate statistics was introduced and studied in dynamical systems like the SM. The statistical counterpart of the latter is the band random matrix (BRM) theory recently developed [48] which was a revival of Wigner's old work [67]. In this theory an ensemble of matrices is considered whose non-zero elements occupy some band of width $2b$ around the main diagonal.

Indeed the one-period evolution of the kicked rotator, in the angular momentum representation, is given by a unitary $N \times N$ matrix. This matrix has a band structure of width $2k$. The ergodicity parameter which describes the statistical properties in the regime of full classical chaos is the ratio k^2/N. This similarity suggests that the appropriate ergodicity parameter in BRM theory is [46]

$$\lambda_r = r \frac{b^2}{N} \tag{118}$$

where r is some numerical factor. This was recently confirmed analytically in Ref.[53]. It turns out [46] that the scaling $\beta_{\overline{H}}(\lambda_r)$ is similar but not identical to that for the dynamical problem (47). In fact, in the region of small λ_r up to $\lambda_r \approx 3$, the dependence is the same with $r \approx 1.5$. The second region ($\beta_{\overline{H}} \approx 1$) is apparently different but it has not yet been studied in detail. Notice, that the origin of the difference can be attributed not so much to the distinction between "random" and "deterministic" matrix elements as to the different boundary conditions for a square matrix and a torus. The scaling in the first region was recently theoretically confirmed in Ref.[50]. However, no explanation for the observed deviations from this scaling at large λ_r was given.

A similar scaling behaviour was also found for tridiagonal symmetric matrices describing the Anderson and Lloyd models in disordered solids [49]. Unlike previous models, no deviations from the first scaling were observed, and the scaling was presented in a 'model-independent' form (cf. Eq.(48)):

$$\frac{1}{\xi_1(N, W)} = \frac{1}{\xi_1(\infty, W)} + \frac{1}{\xi_1(N, 0)} \tag{119}$$

where W is the disorder parameter. However, the scaling may depend on the imposed boundary conditions (in the latter two models zero boundary conditions were chosen).

Notice that for $\lambda_r \ll 1$ the matrix of eigenfunctions (113) is also a band matrix with a_{ij} smoothly decreasing off the diagonal, and with a much larger effective width $\sim b^2$.

Until recently homogeneous BRMs were studied which do not describe the global structure of conservative systems. Indeed, the level density ρ of such matrices grows indefinitely as $N \to \infty$. Clearly, N is an irrelevant (technical) parameter for a conservative system with its energy shells of a finite width ΔE. Instead, inhomogeneous BRM were introduced [52] with increasing diagonal elements (also considered by Wigner [67])

$$\langle H_{mn} \rangle = \frac{m \cdot \delta_{mn}}{\rho}; \qquad\qquad \langle H_{mn}^2 \rangle = \sigma^2, \; m \neq n \qquad (120)$$

In such a model the localization length is bounded from above by the energy shell width [51–53]

$$l \leq l_\perp \approx 5.3 \, \sigma \rho \sqrt{b} \approx \rho \Delta E \qquad (121)$$

After the introduction of the parameter

$$\lambda = \frac{l(0)}{l_\perp} \approx \frac{1.2 \, b^2}{l_\perp} \qquad (122)$$

the scaling is described by the relation (cf. Eq.(47))

$$\beta_s(\lambda) = \frac{l(\lambda)}{l_\perp} = \lambda s(\lambda) \qquad (123)$$

The scaling function $s(\lambda)$ was found numerically in Ref.[51] and has the following properties [52,53,64]: $s(0) = 1$, $\lambda s(\lambda) \to 1$ as $\lambda \to \infty$. In the latter case the localization length reaches the maximum, $l_\perp$, which corresponds to the ergodic eigenfunctions (within an energy shell). For this reason we call λ the ergodicity parameter, similar to the parameter (39) for an SM on a torus. Nevertheless, the eigenfunctions remain globally localized. We call this the *transverse localization* (across the energy shell) as contrasted to the *longitudinal localization* (along the shell) for $l < l_\perp$.

As was shown in Ref.[52] the parameter λ determines the level statistics. Moreover, using numerical data from this paper we have found [51] that the level repulsion parameter $\beta \approx \beta_s(\lambda)$ is again surprisingly close to the scaling function (Fig.11) (cf. Eq.(117)) in spite of the fact that the scaling itself is qualitatively different. Namely:

$$\beta_s(\lambda) \approx 1 - \exp(-\lambda) \qquad (124)$$

in the whole range of available data ($0 < \lambda < 4$). Remarkably, this relation between statistical (β) and localization (β_s) characteristics of eigenfunctions has

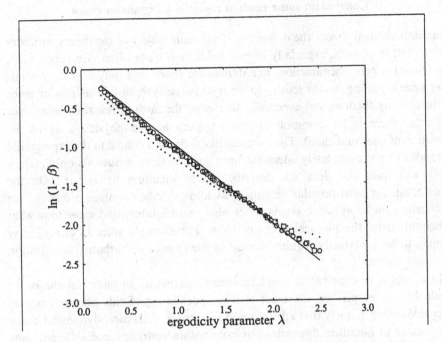

Fig. 11. The dependence of Brody's (points) and Izrailev's (circles) level repulsion parameters as well as of the localization parameter (squares) on the ergodicity parameter λ. The straight line is exponential scaling (124) (after Ref.[51]).

actually no fitting parameter. Yet, the scaling (124) may depend on the model via a particular structure of the energy shell, especially, of its borders.

Notice that the localization lengths ξ_q introduced in (44) depend on q and hence the surprisingly accurate relation $\beta \approx \beta_s$ holds for one localization length only namely the entropy localization length.

To conclude this section, we would like to attract the attention to a very interesting and less known theorem due to Shnirelman [37] (for the proof of both his theorems see Ref.[38]). It is related to KAM integrability in the classical limit which is intermediate between complete integrability with corresponding independent quantum levels (see Eq.(116), $\beta = 0$) and quantum chaos with level repulsion ($\beta \neq 0$). The classical KAM structure is highly intricate as its chaotic part, being of exponentially small measure, is everywhere dense. In quantum mechanics the beautiful Shnirelman theorem, which does not even need translation, asserts:

$$\forall N \exists C_N > 0, \quad \forall n > 1 \quad min(\lambda_{n+1} - \lambda_n, \lambda_n - \lambda_{n-1}) < C_N n^{-N} \qquad (125)$$

where λ_n^2 are the energy eigenvalues. Thus, asymptotically as $n \to \infty$, a half of level spacings are exponentially small. A striking difference from both the complete integrability and quantum chaos!

10 Conclusion: some random remarks on quantum chaos

In a mathematical theory the definition of the main object of the theory procedes the results; in physics, expecially in new fields, it is quite often vice versa. First, one studies a new phenomenon, like dynamical chaos and only at a later stage, after understanding it sufficiently, do we try to classify it, to find its proper place in the existing theories and eventually to choose the most reasonable definition.

So far there is no common agreement, even among physicists, as to the definition of quantum chaos. The classical-like definition related to the exponential instability is not completely adequate since such a chaos is possible only in very exotic examples and does not describe typical quantum behaviour. On the other hand, the most popular definition, as some specific quantum properties for classically chaotic systems, seems to us also unacceptable (and even somewhat unhelpful) from the physical point of view. For example such a "chaos" may happen to be a perfectly regular motion in the case of perturbative localization discussed above.

In attempts to construct a more reasonable definition of quantum chaos, we would like to emphasize the most striking peculiarity of this phenomenon as discussed above, namely that all statistical properties of classical dynamical chaos are present in quantum dynamics but only within restricted and different time scales. Thus we think that the best definition of quantum chaos is: "*finite-time* dynamical chaos". In other words, this new phenomenon reveals an intrinsic complexity and richness of the motion with discrete spectrum which has long been considered as the most simple and regular. This is also true for any classical linear waves but the linearity of quantum equations is not an approximation as in classical physics but a fundamental and universal physical property.

The practical importance of statistical laws even for a finite time interval is in that they provide a relatively simple description of the *essential* behaviour for a very complicated dynamics. The existing ergodic theory of dynamical systems which is asymptotic in time seems to be inadequate to describe this new phenomenon properly. With the latter definition of quantum chaos we feel that a new ergodic theory is required which could analyse the finite time statistical properties of dynamical systems. Of course this is much more difficult than the asymptotic relations in the existing theory but we believe that otherwise it would be impossible adequately to describe the new and important phenomenon of quantum chaos.

Since quantum mechanics is commonly accepted as a universal theory, the phenomenon of classical dynamical chaos strictly speaking does not exist in nature. Nevertheless it is very important in the theory as the limiting pattern to compare with real quantum dynamics. It is instructive to compare the brand new phenomenon of quantum chaos with the old mechanism for statistical laws in the thermodynamic limit $N \rightarrow \infty$ which is the standard approach in traditional statistical physics both classical and quantum. Thus, for infinitely dimensional quantum systems true chaos is possible and also non-physical. When we speak about the absence of true chaos in quantum mechanics, we mean finite, and even few-dimensional, systems.

It turns out that both mechanisms are very similar since for any finite N in the latter or q in the former, the dynamics is formally regular and in particular is characterized by a discrete spectrum. The main difference is in the nature of the large parameter N or q. The similarity comes from the fact that if any of these parameters is large, the motion is controlled by a large number of frequencies which makes it very complicated. The study of quantum chaos helps us to understand better the old mechanism for chaos in many-dimensional systems; particularly we conjecture the existence of characteristic time scales similar to those in quantum systems.

The direct relation between these two seemingly different mechanisms of chaos can be traced back in some specific dynamical models. One interesting example is the non-linear Schrödinger equation [47] (for another example see Ref.[69]). From a physical point of view this describes the motion of a quantum system interacting with many other freedoms whose state is expressed via the ψ function of the system itself (the so-called mean field approximation). This approximation becomes exact in the limit $N \rightarrow \infty$ which is a particular case of the thermodynamic limit. Therefore the mechanism for chaos in this system is the old one. On the other hand the non-linear Schrödinger equation has generally exponentially unstable solutions hence the mechanism of chaos here is the new one. Thus for this particular model both mechanisms describe the same physical process. We would like to emphasize that the true chaos present in these apparently few-dimensional models actually refers to an infinite-dimensional system.

Also, we would like to make a few comments on the problem of quantum measurement. Studies of quantum chaos suggest that it may have a close relation to this problem. First the measurement device is a macroscopic system for which the classical description is a very good approximation. In such a system true chaos with exponential instability is quite possible. The chaos in the classical measuring device is not only possible but unavoidable since the measuring system has to be, a highly unstable system; indeed, a microscopic intervention here produces a macroscopic effect. The importance of chaos for the quantum measurement is that it destroys the coherence of the initial pure quantum state to be measured converting it into an incoherent mixture. In the existing theories of quantum measurement this is described as the effect of external noise [93]. Chaos theory allows us to get rid of the unsatisfactory effect of the external noise and to develop a purely dynamical theory for the loss of quantum coherence (see also Ref.[68]). Unfortunately this is not yet the whole story. Indeed, besides the loss of coherence the most important effect of quantum measurement is the redistribution of probabilities $|\psi|^2$ according to the result of the measurement, the famous ψ-collapse, which remains to be explained. It seems that any dynamical explanation of the ψ-collapse requires some changes in the existing quantum mechanics and this is the main difficulty both technical and philosophical.

In conclusion we would like to emphasize the importance of the so-called numerical experiments (the computer simulation of motion equations) in this as well as in many other fields of research. As a matter of fact, most information about non-trivial dynamics both classical and quantal has been gained in precisely this way, not in laboratory experiments. With all their obvious drawbacks and limitations numerical experiments have very important advantage as they provide complete information about the system under study. In quantum mechanics this advantage becomes crucial because in the laboratory one cannot observe (measure) the quantum system without causing a radical change of its dynamics. Nevertheless, we believe that laboratory experiments remain vitally important because the basis of numerical experiments – the fundamental equations of physics – may (and eventually will) be found to be incomplete or even incorrect. No matter how negligible the probability, in view of the thousands of experiments already done, any new possibility like quantum chaos should be used carefully to check the fundamental equations in the laboratory again and again.

REFERENCES

[1]   A. Lichtenberg and M. Lieberman, *Regular and Stochastic Motion*, Springer, Berlin (1983); G.M. Zaslavsky, *Chaos in Dynamic Systems*, Harwood (1985).

[2]   V.I. Arnold and A. Avez, *Ergodic Problems of Classical Mechanics*, Benjamin, New York (1968).

[3]   I. Kornfeld, S. Fomin and Ya. Sinai, *Ergodic Theory*, Springer, Berlin (1982).

[4]   V.M. Alekseev and M.V. Yakobson, Phys. Reports **75**, 287 (1981).

[5]   B.V. Chirikov, in *Proc. 2d Intern. Seminar on Group Theory Methods in Physics*, Harwood (1985), Vol. 1, p. 553.

[6]   J. Theiler, S. Eubank, A. Longtin, B. Galdrikian and J.D. Farmer, Physica D **58**, 77 (1992).

[7]   B.V. Chirikov and V.V. Vecheslavov in *Analysis etc.*, Eds. P. Rabinowitz and E. Zehnder, Academic Press, New York (1990), p. 219.

[8]   M. Berry, Proc.Roy.Soc.London A **413**, 183 (1987).

[9]   J. Ford, G. Mantica and G. Ristow, Physica D **50**, 493 (1991).

[10] *Quantum Measurement and Chaos*, Eds. E. Pike and S. Sarkar, Plenum, New York (1987); *Quantum Chaos – Quantum Measurement*, Eds. P. Cvitanovic, I. Percival and A. Wirzba, Kluwer (1992).

[11] G. Casati, B.V. Chirikov, J. Ford and F.M. Izrailev, Lecture Notes in Physics **93**, 334 (1979), this volume.

[12] G. Casati et al, Phys. Reports **154**, 77 (1987); G. Casati, I. Guarneri and D.L. Shepelyansky, IEEE J. of Quantum Electr. **24**, 1420 (1988).

[13] F. Haake, *Quantum Signatures of Chaos*, Springer, Berlin (1991).

[14] F. Haake and D.L. Shepelyansky, Europhys.Lett. **5**, 671 (1988).

[15] Wei-Min Zhang and Da Hsuan Feng, Geometry on Quantum Non-Integrability (1992) (unpublished).

[16] F.M. Izrailev, Phys. Reports **196**, 299 (1990).

[17] G. Casati, J. Ford, I. Guarneri and F. Vivaldi, Phys. Rev. A **34**, 1413 (1986).

[18] D. Cohen, ibid. **44**, 2292 (1991).

[19] B.V. Chirikov, F.M. Izrailev and D.L. Shepelyansky, Sov. Sci. Rev. C **2**, 209 (1981); Physica D **33**, 77 (1988).

[20] G.P. Berman and G.M. Zaslavsky, Physica A **91**, 450 (1978).

[21] M. Toda and K. Ikeda, Phys. Lett. A **124**, 165 (1987); A. Bishop et al, Phys. Rev. B **39**, 12423 (1989).

[22] E. Wigner, Phys. Rev. **40**, 749 (1932); V.I. Tatarsky, Usp. Fiz. Nauk **139**, 587 (1983).

[23] D.L. Shepelyansky, Physica D **8**, 208 (1983); G. Casati et al, Phys. Rev. Lett. **56**, 2437 (1986); T. Dittrich and R. Graham, Ann.Phys. **200**, 363 (1990).

[24] K. Ikeda, this volume.

[25] A. Peres, in: *Quantum Chaos*, Eds. H. Cerdeira, R. Ramaswamy, M. Gutzwiller and G. Casati, World Scientific, Singapore (1991).

[26] B.V. Chirikov, D.L. Shepelyansky, Radiofizika **29**, 1041 (1986).

[27] G. Fusina, thesis, Physics Dept. Milan University (1992); G. Casati, B.V. Chirikov and G. Fusina, *Relaxation to the Quantum Steady State in the Kicked Rotator Model* (in preparation).

[28] N. Mott, Phil. Mag. **22**, 7 (1970).

[29] M. Berry, Proc. Roy. Soc. Lond. A **423**, 219 (1989).

[30] T. Dittrich and U. Smilansky, this volume.

[31] E. Heller, Phys. Rev. Lett. **53**, 1515 (1984).

[32] E.B. Bogomolny, Physica D **31**, 169 (1988).

[33] A.I. Shnirelman, Usp. Mat. Nauk **29**, #6, 181 (1974).

[34] R. Blümel and U. Smilansky, Phys. Rev. Lett. **69**, 217 (1992).

[35] F.M. Izrailev, private communication (1992).

[36] B.V. Chirikov, CHAOS **1**, 95 (1991).

[37] A.I. Shnirelman, Usp. Mat. Nauk **30** #4, 265 (1975).

[38] A.I. Shnirelman, On the *Asymptotic Properties of Eigenfunctions in the Regions of Chaotic Motion*, addendum in: V.F.Lasutkin, *The KAM Theory and Asymptotics of Spectrum of Elliptic Operators*, Springer, Berlin (1991).

[39] R. Jensen, Phys. Rev. Lett. **49**, 1365 (1982); Phys.Rev. A **30**, 386 (1984).

[40] R. Prange, D. Grempel and S. Fishman, this volume.

[41] D.L. Shepelyansky, Physica D **28**, 103 (1987).

[42] E.P. Nakhmedov et al, Zh. Eksp. Teor. Fiz. **92**, 2133 (1987).

[43] B.V. Chirikov, in: Proc. Les Houches Summer School on *Chaos and Quantum Physics*, Elsevier, North Holland, Amsterdam (1991).

[44] G. Casati and I. Guarneri, Comm. Math. Phys. **95**, 121 (1984).

[45] F.M. Izrailev, Phys. Lett. A **134**, 13 (1988); J. Phys. A **22**, 865 (1989).

[46] G.Casati et al, Phys. Rev. Lett. **64**, 1851 (1990); J. Phys. A **24**, 4755 (1991).

[47] F. Benvenuto et al, Phys. Rev. A **44**, R3423 (1991); G. Jona-Lasinio et al, Phys. Rev. Lett. **68**, 2269 (1992); D.L. Shepelyansky: Phys. Rev. Lett. **70**, 1787 (1993).

[48] T. Seligman et al, Phys.Rev.Lett. **53**, 215 (1985); M. Feingold et al, Phys. Rev. A **39**, 6507 (1989).

[49] G. Casati et al., J.Phys.: Condens. Matter **4**, 149 (1992).

[50] Ya.V. Fyodorov and A.D. Mirlin, Phys. Rev. Lett. **69**, 1093 (1992).

[51] G. Casati, B.V. Chirikov, I. Guarneri and F.M. Izrailev, Phys. Rev. E. **48**, R1613 (1993).

[52] M. Wilkinson, M. Feingold and D. Leitner, J. Phys. A **24**, 175 (1991); Phys. Rev. Lett. **66**, 986 (1991).

[53] Ya.V. Fyodorov and A.D. Mirlin, Phys. Rev. Lett. **67**, 2405 (1991).

[54] M. Berry, J. Phys. A **10**, 2083 (1977); A. Voros, Lecture Notes in Physics **93**, Springer, Berlin (1979), p. 326.

[55] P. O'Connor and E. Heller, Phys. Rev. Lett. **61**, 2288 (1988); P. Leboeuf et al, Ann. Phys. **208**, 333 (1991).

[56] S. Weigert, Z. Phys. B **80**, 3 (1990); M. Berry, *True Quantum Chaos? An Instructive Example*, Proc. Yukawa Symposium, 1990; F. Benatti et al, Lett. Math. Phys. **21**, 157 (1991); H. Narnhofer, J. Math. Phys. **33**, 1502 (1992).

[57] M. Born, Z. Phys. **153**, 372 (1958); G. Casati, B.V. Chirikov and J. Ford, Phys. Lett. A **77**, 91 (1980).

[58] D.V. Anosov, Dokl. Akad. Nauk SSSR **145**, 707 (1962).

[59] G. Casati, B.V. Chirikov, G. Fusina and F.M. Izrailev, *The Structure of Mott's States* (in preparation).

[60] B.V. Chirikov and F.M. Izrailev, *Statistics of Chaotic Quantum States: Localization and Entropy* (in preparation).

[61] E.V. Shuryak, Zh. Eksp. Teor. Fiz. **71**, 2039 (1976).

[62] F.M. Izrailev and D.L. Shepelyansky, Teor. Mat. Fiz. **43**, 417 (1980).

[63] A. Rechester et al, Phys. Rev. A **23**, 2664 (1981).

[64] A. Gioletta, M. Feingold, F.M. Izrailev and L. Molinari, Phys. Rev. Lett. **70**, 2936 (1993).

[65] N. Ben-Tal et al, Phys. Rev. A **46**, #3 (1992).

[66] M. Berry and M. Robnik, J. Phys. A **17**, 2413 (1984).

[67] E. Wigner, Ann.Math. **62**, 548 (1955); **65**, 203 (1957).

[68] L. Bonci et al, Phys. Rev. Lett. **67**, 2593 (1991).

[69] G.P. Berman and G.M. Zaslavsky, this volume.

[70] F. Delyon, B. Simon and B. Souillard, Ann.Inst.; H.Poincaré **42**, 283 (1985).

[71] F. Benvenuto, G. Casati, I. Guarneri and D.L. Shepelyansky, Z. Phys. B. **84**, 159 (1991).

[72] P.W. Anderson, Phys. Rev. **109**, 1492 (1958).

[73] D. Dunlap et al., Phys. Rev. Lett. **65**, 88 (1990).

[74] D.J. Thouless, J. Phys. C: Solid State Phys **5**, 77 (1972).

[75] H. Schmidt, Phys. Rev. **105**, 425 (1957).

[76] G. Casati and L. Molinari, Prog. Th. Physics suppl. **98**, 287 (1989).

[77] G. Casati, I. Guarneri and D.L. Shepelyansky, Physica A **163**, 205 (1990).

[78] P.A. Lee and T.V. Ramakhrishnan, Rev. Mod. Phys. **57**, 287 (1985).

[79] D.L. Shepelyansky, Physica D **8**, 208 (1983).

[80] E. Abrahams, P.W. Anderson. D.C. Licciardello and T.V. Ramakrishnan, Phys. Rev. Lett. **42**, 673 (1979).

[81] G. Casati, I. Guarneri and D.L. Shepelyansky, Phys. Rev. Lett. **62**, 345 (1989).

[82] J.E. Bayfield and P.M. Koch, Phys. Rev. Lett. **33**, 258 (1974).

[83] J.C. Leopold and I.C. Percival, Phys. Rev. Lett. **41**, 944 (1978).

[84] F. Benvenuto, G. Casati and D.L. Shepelyansky, Phys. Rev. A **47**, R786 (1993).

[85] N.B. Delone, B.P. Krainov and D.L. Shepelyansky, Sov. Phys. Usp. **26**, 551 (1983).

[86] G. Casati, B.V. Chirikov and D.L. Shepelyansky, Phys. Rev. Lett. **53**, 2525 (1984).

[87] G. Casati and I. Guarneri, Comments Atomic Molecular Physics, **25**, 185 (1991).

[88] G. Casati, I. Guarneri and D.L. Shepelyansky, Chaos, Solitons and Fractals **1**, 131 (1991).

[89] E.J. Galvez, B.E. Sauer, L. Moorman, P.M. Koch and D. Richards, Phys. Rev. Lett. **61**, 2011 (1988).

[90] J.E. Bayfield, G. Casati, I. Guarneri and D.W. Sokol, Phys. Rev. Lett. **63**, 364 (1989) (this volume).

[91] R. Artuso, G. Casati and D.L. Shepelyansky, Phys. Rev. Lett. **68**, 3826 (1992).

[92] R. Artuso, F. Borgonovi, I. Guarneri, L. Rebuzzini, G. Casati, Phys. Rev. Lett. **69**, 3302 (1992).

[93] J.A. Wheeler, W.H. Zurek Eds., Quantum Theory and Measurement, Princeton Univ. Press, Princeton (1983).

[94] P. Leboeuf, J. Kurchan, M. Feingold and D.P. Arovas, Phys. Rev. Lett. **65**, 3076 (1990); R. Lima and D.L. Shepelyansky, Phys. Rev. Lett. **67**, 1377 (1991); T. Geisel, R. Ketzmerick and G. Petschel, Phys. Rev. Lett. **66**, 1651 (1991); **67**, 3635 (1991).

PART ONE

Classical chaos and quantum localization

STOCHASTIC BEHAVIOR OF A QUANTUM PENDULUM
UNDER A PERIODIC PERTURBATION

G. Casati

Istituto di Fisica, Via Celoria 16, Milano, Italy

and

B. V. Chirikov and F. M. Izraelev

Institute of Nuclear Physics, 630090 Novosibirsk 90, U.S.S.R.

and

Joseph Ford

School of Physics, Georgia Institute of Technology
Atlanta, Georgia 30332, U.S.A.

ABSTRACT

This paper discusses a numerical technique for computing the quantum solutions of a driven pendulum governed by the Hamiltonian

$$H = (p_\theta^2/2m\ell^2) - [m\ell^2\omega_o^2\cos\theta]\,\delta_p(t/T),$$

where p_θ is angular momentum, θ is angular displacement, m is pendulum mass, ℓ is pendulum length, $\omega_o^2 = g/\ell$ is the small displacement natural frequency, and where $\delta_p(t/T)$ is a periodic delta function of period T. The virtue of this rather singular Hamiltonian system is that both its classical and quantum equations of motion can be reduced to mappings which can be iterated numerically and that, under suitable circumstances, the motion for this system can be wildly chaotic. Indeed, the classical version of this model is known to exhibit certain types of stochastic behavior, and we here seek to verify that similar behavior occurs in the quantum description. In particular, we present evidence that the quantum motion can yield a linear (diffusive-like) growth of average pendulum energy with time and an angular momentum probability distribution which is a time-dependent Gaussian just as does the classical motion. However, there are several surprising distinctions between the classical and quantum motions which are discussed herein.

Since we have not yet developed a completely adequate explanation for all these distinctions, this paper should be regarded as a progress report describing work on a highly interesting and numerically solvable model.

I. INTRODUCTION

Noteworthy recent progress[1-3] has been made in illustrating the truly stochastic behavior which can occur for the strictly deterministic systems of classical mechanics. Indeed, quite simple classical Hamiltonian systems can exhibit precise phase space trajectories so chaotic in their phase space wanderings that slightly imperfect observation cannot distinguish this deterministic motion from completely stochastic motion. Even though some of these model systems have only a few degrees of freedom, they nonetheless illustrate a generic type of wild classical mechanical trajectory behavior which is of great significance in the study of dynamical stability and of statistical mechanics; moreover, the nature of such wild behavior is now being studied for more general systems of widespread physical interest. Thus, even at this early stage, it becomes highly desirable to establish the effect introduced by quantum mechanics on this classical mechanical stochastic behavior. For even more than in classical mechanics, most of the exact work in quantum mechanics has been devoted to integrable (solvable) systems, with the remaining more difficult problems being left either to generally divergent perturbation theory or to quantum statistical mechanics whose foundations are less well understood than those of classical statistical mechanics. As a consequence, there is not only a dearth of quantum models exhibiting stochastic behavior, there is also some ambiguity even concerning the criteria for and the definition of quantum stochastic behavior.

Thus a number of recent papers have appeared, using a variety of techniques, which seek to develop viable criteria for and a definition of quantum stochastic behavior. To establish their criterion, Pukhov, et. al.,[4] generalize the notion of local instability (exponential separation) of initially close orbits valid for classical stochastic systems. In particular, they investigate the time growth of the change in the wave function $\delta\psi$ due to a small external perturbation added to the original Hamiltonian, and they establish criteria sufficient to insure an exponential growth in $\delta\psi$. Along somewhat related lines, Percival and Pomphrey[5] treat quantum mechanically a Hamiltonian system known to exhibit a classical transition from near-integrable to

stochastic behavior. They show that the quantum energy levels for
this Hamiltonian become highly sensitive to very small added external
perturbations only above the classical stochastic threshold, and they
argue that their study thus reveals a transition to quantum stochastic
behavior. Both of these studies investigate quantum stochasticity as
revealed through small but explicit outside perturbations.

The works of Nordholm and Rice[6,7] and of Shuryak[8] are more
directly concerned with the case of isolated Hamiltonian systems for
which the Hamiltonian splits naturally into dominant terms describing
independent degrees of freedom which are coupled by small nonlinear
interaction terms. Nordholm and Rice[6] also present a brief but clear
review of the older literature on quantum ergodic theory for isolated
systems, pointing out that this problem is still very much an open one.
Both Nordholm and Rice as well as Shuryak regard their isolated quantum
systems as behaving stochastically when the unperturbed eigenstates of
the dominant, independent modes are strongly coupled by the weak interaction in the sense that the expansion of the exact quantum state is a
sum of many unperturbed, independent mode states. The approach of
Nordholm and Rice is, in principle, exact but in practice is forced to
rely heavily on numerical (computer) computations whereas the work of
Shuryak is strictly analytic but only approximate since it is based on
a generalization of the order of magnitude resonance-overlap estimates
of Chirikov.[3] For further details on work in this area, the reader is
referred to the papers listed as Reference 9. Additional discussions
of quantum stochasticity appear, of course, in several of the compan-
ion papers in this volume.

But strangely enough, to our knowledge, no quantum investigation
has previously been made for the simplest possible Hamiltonian system
known to exhibit chaotic trajectories, namely a periodically driven
(i.e., time-dependent) one degree of freedom Hamiltonian system.
Perhaps this is because such driven systems are, in general, no easier
to solve than the conservative two degree of freedom systems whose
quantum behavior has been previously studied. However, there is one
notable exception to this general rule, and it is this exception which
we seek to exploit in this paper. In particular, both the classical
and quantum equations of motion for pendulum Hamiltonians of the type

$$H = (p_\theta/2m\ell^2) + V(\theta)\, \delta_p(t/T) \,, \tag{1}$$

where θ is angular displacement, p_θ is angular momentum, m is pendulum
mass, ℓ is pendulum length, $V(\theta)$ is angular potential energy, and

$\delta_p(t/T)$ is a periodic delta function of period T, can be reduced to mappings and solved numerically as we shall show. Indeed, the classical motion for Hamiltonian (1) using $V(\theta) = -mg\ell\cos\theta = -m\ell^2\omega_o^2\cos\theta$ has already been extensively investigated,[3] and it is for this reason that we chose to begin our quantum studies using this particular model.

In Section II, we derive the classical mapping equations for our model and discuss the nature of its solutions. In Section III, we derive and discuss the quantum mapping equations. Section IV presents our numerical results for the quantum problem and these results are then discussed in Section V. Brief concluding remarks appear in Section VI. This paper represents a progress report describing calculations on an exceptional type of driven quantum system which can be solved, at least numerically. It is our hope that future studies on this or related model systems may lead to a broader understanding of quantum chaotic behavior.

II. DISCUSSION OF THE CLASSICAL MODEL

The specific Hamiltonian we choose to study is given by

$$H = (p_\theta^2/2m\ell^2) - [m\ell^2\omega_o^2\cos\theta] \, \delta_p(t/T) \, , \qquad (2)$$

which is merely Hamiltonian (1) specialized to $V(\theta) = -mg\ell\cos\theta = -m\ell^2\omega_o^2\cos\theta$, where obviously $\omega_o^2 = g/\ell$. Here the periodic delta function, which may be expressed as

$$\delta_p(t/T) = \sum_{j=-\infty}^{\infty} \delta[j-(t/T)] = 1+2 \sum_{n=1}^{\infty} \cos(2n\pi t/T) \, , \qquad (3)$$

"turns on" the gravitational potential for a brief instant during each period T. The classical equations of motion for Hamiltonian (2) are

$$\dot{p}_\theta = -(m\ell^2\omega_o^2\sin\theta) \, \delta_p(t/T) \qquad (4a)$$

$$\dot{\theta} = p_\theta/m\ell^2 \qquad (4b)$$

where the dot notation indicates time derivative. Letting θ_n and $(p_\theta)_n$ be the values of θ and p_θ just before the nth delta function "kick", we may integrate Eq. (4a,b) to obtain the mapping equations

$$(p_\theta)_{n+1} = (p_\theta)_n - m\ell^2\omega_o^2T\sin\theta_n \, , \qquad (5a)$$

$$\theta_{n+1} = \theta_n + (p_\theta)_{n+1} T/m\ell^2 \quad . \tag{5b}$$

As an aside, let us observe that, when T tends to zero allowing us to replace T by dt, Eq. (5a,b) becomes

$$dp_\theta = - (m\ell^2\omega_o^2\sin\theta) \, dt \tag{6a}$$

$$d\theta = (p_\theta/m\ell^2) \, dt \quad , \tag{6b}$$

which are precisely the equations of motion for the conservative, gravitational pendulum Hamiltonian

$$H = (p_\theta^2/2m\ell^2) - mg\ell\cos\theta \tag{7}$$

recalling that $\omega_o^2 = g/\ell$. Thus as the time T between delta function "kicks" tends to zero, the gravitational potential is "turned on" continuously and the motion generated by Hamiltonian (2) becomes identical with that generated by the integrable[1] Hamiltonian (7). As T increases away from zero, the orbits of Hamiltonian (2) increasingly deviate from those of the integrable pendulum, eventually exhibiting chaotic or stochastic behavior. Consequently, one may adopt either of two viewpoints regarding Hamiltonian (2). As written above in Eq. (2), it obviously describes a free rotator perturbed by delta function "kicks". The remarks of this paragraph show that Hamiltonian (2) may also be regarded as describing a gravitational pendulum perturbed by a periodic driving force. This latter viewpoint becomes clearer if we rewrite Hamiltonian (2), using Eq. (3), as

$$H = [(p_\theta/2m\ell^2) - mg\ell\cos\theta] - 2mg\ell\cos\theta \sum_{n=1}^{\infty} \cos(2n\pi t/T) \tag{8}$$

In particular, it is the viewpoint represented by Hamiltonian (8) that led to the title of this paper.

Returning to Eq. (5a,b), let us now define the dimensionless angular momentum P_n via the equation $P_n = (p_\theta)_n T/m\ell^2$, and then let us write Eq. (5a,b) in the so-called standard form[3]

$$P_{n+1} = P_n - K\sin\theta_n \tag{9a}$$

$$\theta_{n+1} = \theta_n + P_{n+1} \quad , \tag{9b}$$

where $K = (\omega_o T)^2$ is the only remaining mapping parameter. As mentioned earlier when K tends to zero, this mapping becomes precisely integrable with all orbits lying on (or forming) simple invariant curves of the mapping. When K is small, the KAM theorem[2] insures that the mapping is near-integrable, meaning here that most mapping orbits continue to lie on simple invariant curves. Numerical computations[3,10] show that at least some of these simple invariant curves persist as K approaches unity. Moreover, for the K region 0<K<1, the momentum variation is bounded with $|\Delta P| \approx K^{1/2}$. As K reaches and exceeds unity, all the previously existing simple invariant curves completely disappear and most mapping orbits become chaotic point sets. For K>>1, numerical evidence shows that the P-motion is characterized by a simple, random walk diffusion equation having the form

$$\overline{P^2} \approx (K^2/2)t \ , \qquad\qquad (10)$$

where $\overline{P^2}$ is the average of the squared angular momentum at integer time t = n measured in units of the period T and where initially P is taken to be zero. The average here is over many orbits having distinct θ_o initial conditions or over many sections of the same orbit (normalizing P to zero at the beginning of each segment). In order to indicate the source of Eq. (10), let us note that from Eq. (9a) we may obtain

$$(P_n - P_o)^2 = K^2 \sum_{j,k=o}^{(n-1)} (\sin\theta_j \sin\theta_k) \ . \qquad\qquad (11)$$

Averaging Eq. (11) over θ_j and θ_k, taking both to be uniformly distributed random variables, yields Eq. (10); this procedure is validated by numerical iteration of chaotic orbits. But in addition to Eq. (10), the empirically observed angular momentum distribution itself has the time-dependent Gaussian form

$$f(P) = [K(\pi t)^{1/2}]^{-1} \exp [-P^2/K^2 t] \ , \qquad\qquad (12)$$

as would be expected from the central limit theorem provided we regard P_n in the equation $P_n = P_o - K \sum_{j=o}^{(n-1)} \sin\theta_j$ as being a sum of random variables. It is precisely this stochastic momentum (or energy) diffusion we shall seek to empirically observe in the quantum description of this model to which we now turn.

III. DISCUSSION OF THE QUANTUM MODEL

We now seek to solve Schrodinger's equation

$$H\psi(\theta,t) = i\,\hbar\,\partial\psi(\theta,t)/\partial t \tag{13}$$

for the system governed by Hamiltonian (2). In this section, we regard the system as a free rotator perturbed by delta function "kicks" and we expand the wave function $\psi(\theta,t)$ in terms of the free rotator eigenfunctions $(2\pi)^{-1}e^{in\theta}$. In particular, we write

$$\psi(\theta,t) = (2\pi)^{-1}\sum_{n=-\infty}^{\infty} A_n(t)e^{in\theta} . \tag{14}$$

Over any period T between delta function "kicks", the $A_n(t)$ evolve according to

$$A_n(t+T) = A_n(t)e^{-E_nT/\hbar} = A_n(t)e^{-in^2\tau/2} , \tag{15}$$

where

$$E_n = n^2\hbar^2/2m\ell^2 \tag{16}$$

are the free rotator energy eigenvalues and where

$$\tau = \hbar\,T/m\ell^2 . \tag{17}$$

During the infinitesimal time interval of a "kick", we may write Eq. (13) as

$$i\,\hbar\partial\psi/\partial t = -m\ell^2\omega_o^2\cos\theta\,\delta_p(t/T)\,\psi . \tag{18}$$

Integrating Eq. (18) over the infinitesimal interval $(t+T)$ to $(t+T^+)$, we find

$$\psi(\theta,t+T^+) = \psi(\theta,t+T)\,e^{ik\cos\theta} , \tag{19}$$

where

$$k = (m\ell^2\omega_o^2T)/\hbar . \tag{20}$$

Now expanding both sides of Eq. (19) in free rotator eigenfunctions yields

$$\sum_{n=-\infty}^{\infty} A_n(t+T^+) e^{in\theta} = \sum_{r,s=-\infty}^{\infty} A_r(t+T) b_s(k) e^{i(r+s)\theta}, \qquad (21)$$

where we have used the expansion

$$e^{ik\cos\theta} = \sum_{s=-\infty}^{\infty} b_s(k) e^{is\theta} \qquad (22)$$

with $b_s(k) = i^s J_s(k) = b_{-s}(k)$ and with $J_s(k)$ being the ordinary Bessel function of the first kind. Because of the orthogonality of the $e^{in\theta}$, we may use Eq. (21) to establish that

$$A_n(t+T^+) = \sum_{r=-\infty}^{\infty} A_r(t+T) b_{n-r}(k) \quad . \qquad (23)$$

Finally using Eq. (15), we obtain the quantum mapping

$$A_n(t+T^+) = \sum_{r=-\infty}^{\infty} A_r(t) b_{n-r}(k) e^{-ir^2\tau/2} \qquad (24)$$

giving the A_n at time $(t+T^+)$ in terms of the A_n at time t. Using Eq. (24) and Eq. (14), we could obtain $\psi(\theta, t+T^+)$ in terms of $\psi(\theta, t)$, but in the calculations of interest here, the A_n momentum-representation is more useful. Before turning to these calculations however, let us briefly discuss the quantum mapping of Eq. (24).

First, let us note that the mapping of Eq. (24) can actually be numerically iterated to obtain the time development of a quantum solution provided only one or a few of the initial $A_n(0)$ are non-zero. Even though the sum on r is infinite, the $b_s(k) = i^s J_s(k)$ coefficients become negligible[11] outside the range $s \approx 2k$. Thus for reasonable sized k-values starting from only one (or a few) initially non-zero $A_n(0)$, we may iterate through many periods T before the increasing number of non-zero $A_n(t)$ exceeds the practical limitations of a large computer. Moreover, accuracy can be monitored via the normalization condition $\Sigma |A_n|^2 = 1$. Next, let us observe that the classical mapping depends on the single parameter $K = (\omega_0 T)^2$. Thus one might, at first glance, expect the quantum mapping to depend only on the product $(k\tau)$ since $K = (k\tau)$ as one may immediately verify from Eq. (17,20). Indeed Ehrenfest's theorem makes it reasonable to suppose that the product $K = k\tau$ is particularly significant since it would control the time evolution of the center of a wave packet; however, the additional quantum spreading of the packet might not depend on this product alone. Not only does the numerical evidence presented later indicate that the quantum mapping does indeed depend separately on k and τ, but examination of Eq. (24) itself shows that this must be the case. First, independent of the value of k, Eq. (24) is invariant to the replacement of τ by (τ+4π); thus contrary to the classical case, the quantum motion places an upper bound on τ (or T), since in Eq. (24) there is no loss of generality in restricting τ to the interval $0 \leq \tau \leq 4\pi$. This quantum anomaly arises because the wave function for the unperturbed free rotator is periodic in τ independent of initial conditions. Also in Eq. (24) when k<1, only one or a very few $b_s(k)$ coefficients will be appreciably different from zero. For both these reasons, one would expect the classical and quantum motion to differ greatly when K is large due to a small k and a large τ. This is only one of a variety of perhaps interrelated classical-quantum distinctions. Finally since from Eq. (19) it is clear that k is the parameter which controls the amount of energy absorbed by the rotator due to the driving delta function "kicks", it may be worthwhile to use Eq. (20) and write Hamiltonian (2) as

$$H = (p_\theta/2m\ell^2) - (\hbar k T^{-1}\cos\theta)\, \delta_p(t/T) , \qquad (25)$$

thus revealing k as an explicit coupling parameter.

IV. NUMERICAL RESULTS FOR THE QUANTUM MODEL

For the quantum model discussed in Sec. III, we have numerically iterated Eq. (24) to obtain the time evolved $\{A_n(t)\}$ starting from various initial $\{A_n(0)\}$ sets. For each computer run, starting with a definite $\{A_n(0)\}$ set, we calculated at each iteration the probability distribution

$$\rho(n) = |A_n|^2 , \qquad (26)$$

the average energy (in units of $\hbar^2/m\ell^2$)

$$\langle E \rangle = \sum_n (n^2/2)\, \rho(n) , \qquad (27)$$

and the average angular momentum (in units of $\hbar$)

$$\langle p_\theta \rangle = \sum_n n\rho(n) \quad . \tag{28}$$

In all runs after fixing k and τ, we chose only one $A_n(0)$ or a few
(~ 10) adjacent $A_n(0)$ to be non-zero. Surprisingly, the computed final
state $\rho(n)$ appeared to be independent of the precise initial state, a
point to which we return later. Typical results for four runs are
discussed below. Each of these four runs was started with only the
ground state free rotator $A_0(0)$ being non-zero.

In order to investigate the extent to which the numerically computed quantum distribution $\rho(n)$ of Eq. (26) mimics the classical stochastic distribution of Eq. (12), let us write Eq. (12) in terms of the
quantum variables. Using the definitions of τ, P, and K and the fact
that $p_\theta = n\hbar$, we may write the quantum version of Eq. (12) as

$$f(n) = [k(\pi t)^{1/2}]^{-1} \exp(-n^2/k^2 t) \tag{29}$$

where t is integer time measured in multiples of the "kick" period T.
Note in Eq. (29) the dependence on k rather than K ($=k\tau$). If the
quantum system indeed mimics the classical motion for $K = k\tau \gg 1$, then
we would expect the $\rho(n)$ of Eq. (26) to equal the $f(n)$ of Eq. (29),
since loosely speaking the quantum solution is an "automatic" average
over many classical orbits. For ease of graphical comparison, let us
introduce the normalized variables $X = n^2/k^2 t$, $f_N(n) = k(\pi t)^{1/2} f(n)$,
and $\rho_N(n) = k(\pi t)^{1/2} \rho(n)$; in these variables we need only determine the
validity of the simple equation

$$\rho_N(n) = f_N(n) = e^{-X} \quad . \tag{30}$$

For k = 40 and τ = 1/8, we plot the numerically determined ($\ell n \rho_N$)
versus X at t = 25 in Fig. 1. The straight line $\ell n(e^{-X}) = \ell n f_N(n)$ is
graphed for comparison. One notes here that the comparison is rather
good indicating that the quantum motion indeed appears to be stochastic here. It must be noted in Fig. 1 that each plotted point for
$\ell n \rho_N(n)$ actually represents an average value of this quantity taken
over ten adjacent energy levels; this averaging reduces but does not
eliminate fluctuations. In Fig. 2 we present a graph of $\rho_N(n)$ and
$f_N(n)$ for the parameter values k = 10 and τ = 1/2. In Fig. 2, contrary
to Fig. 1, one notes that the quantum system is not behaving stochastically despite the fact that the classical value of K is five for

both cases and, classically, stochasticity would be expected for both. This is yet another of the several

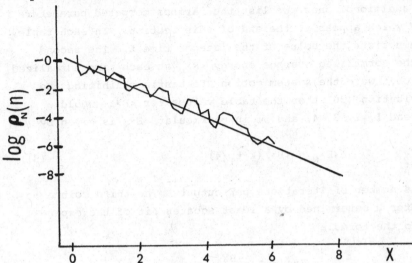

Fig. 1. A plot of the logarithm of the quantum probability distribution $\rho(n)$ versus the normalized variable $X = n^2/k^2t$ at time $t = 25$ for $k = 40$ and $\tau = 1/8$ ($K = 5$). The straight line is a plot of $\ln(e^{-X})$. Were the quantum motion stochastic, these two curves should be identical; the fact that they are quite close indicates a great similarity between the classical and quantum motion for this case.

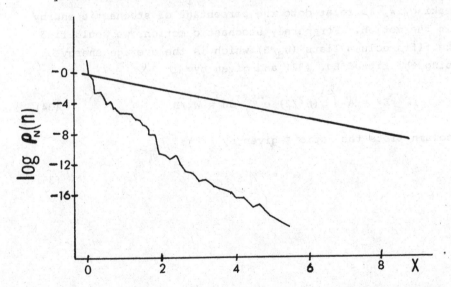

Fig. 2. A plot of the same variables as in Fig. 1 but for $k = 10$ and $\tau = 1/2$. Here the curves are quite distinct indicating a lack of stochasticity in the quantum motion even though $K = 5$ as in Fig. 1. We do not yet understand this quantum anomaly.

quantum anomalies which we shall discuss in the next section.

In order to summarize all the computed data for the above two runs plus two additional ones, we list the various computed parameters in Tables I-IV which appear at the end of this section. In each table, the first column lists the value of the integer time t. The second column lists the normalized average energy $<E>_N$ at each time normalized in such a way that were the system motion stochastic exhibiting the Gaussian distribution (30) then the table values for $<E>_N$ would sequentially read 1, 2, 3, 4, and 5; in particular $<E>_N$ is given by

$$<E>_N = <E>/(k^2 t_1 /4) , \qquad (31)$$

where t_1 is the number of iterations per output. The third column lists a parameter B determined by a least squares fit of the experimental data to the formula

$$\rho_N(n) = A e^{-BX} . \qquad (32)$$

If the motion were purely stochastic, B would equal unity. The fourth column lists a parameter W_d given by

$$W_d = [A/k(\pi t)^{1/2}] \int_{-\infty}^{\infty} \exp(-Bn^2/k^2 t) \, dn = A/B^{1/2} ; \qquad (33)$$

loosely speaking, W_d is related to the percentage of stochastic energy diffusion in the motion. For purely stochastic motion, we would find $W_d = 1$. The fifth column lists (W_d/B) which is the average energy computed using the fitted Eq. (32) and given by

$$<E> = A \int_{-\infty}^{\infty} (n^2/2) e^{-BX} \, dn = W_d/B . \qquad (34)$$

The sixth column lists the ratio ξ given by

$$\xi = <E> / <E>_d , \qquad (35)$$

TABLE I. A listing of stochastic energy absorbtion parameters as a function of time for one computer solution of the driven quantum pendulum. The parameters are defined in the text. Here $k = 40$, $\tau = 1/8$, and $K = k\tau = 5$. The number of $b_s(k)$ values used in Eq. (24) was 101.

t	$<E>_N$	B	W_d	W_d/B	ξ	R_0	$|a_0|^2$
5	1.55	3.67	331	90.3	1.55	0.38	0.0024
10	2.79	2.49	92.6	37.3	1.39	1.36	0.0061
15	3.77	0.76	0.90	1.18	1.26	1.01	0.0037
20	5.01	0.75	0.87	1.16	1.25	1.67	0.0053
25	5.66	0.83	0.87	1.05	1.13	1.43	0.0040

TABLE II. A listing of stochastic energy absorbtion parameters as a function of time for one computer solution of the driven quantum pendulum. Here $k = 40$, $\tau = 1/40$, and $K = k\tau = 1$. The number of N of $b_s(k)$ values used in Eq. (24) was 101.

t	$<E>_N$	B	W_d	W_d/B	ξ	R_0	$|a_0|^2$
30	0.089	29.1	0.28	0.0096	0.0056	16.50	0.043
60	0.100	53.2	0.27	0.0051	0.0029	6.70	0.012
90	0.068	76.4	0.24	0.0031	0.0018	9.48	0.014
120	0.077	94.2	0.14	0.0015	-0.0005	4.89	0.0063
150	0.048	116.0	0.15	0.0013	-0.0002	25.4	0.028

where $<E>$ is given by Eq. (27) and

$$<E>_d = (k^2/4)t \qquad (36)$$

is essentially Eq. (10) expressed in terms of the quantum parameters or, alternatively, it is $<n^2/2>$ computed using Eq. (29). If the motion were completely stochastic, the fifth and sixth columns of each table would be identical. The seventh column lists the ratio $R_0 = |A_0|^2 k(\pi t)^{1/2}$ of the actual probability of the zeroth energy level to that expected from the Gaussian distribution of Eq. (29). The last column lists

$|A_o|^2$ itself. The number of Bessel functions (or $b_s(k)$) needed in Eq. (22) to accurately compute each run is listed in the table captions. Finally, in all four runs at least 1,000 A_n-values ($-500 \leqslant n \leqslant 500$) were computed at each iteration and, as mentioned earlier, accuracy was checked by verifying normalization of the A_n-sum.

    We now turn to a discussion of these experimental results.

V.   DISCUSSION OF NUMERICAL RESULTS

    We have presented results for four runs selected to illustrate typical behavior in a modest variety of k and τ ranges. The parameter values k = 40>>1 and τ = 1/8<<1 (corresponding to a classical K = 5) used in the computations yielding Fig. 1 and Table I are those for which one would expect quantum stochastic behavior, since the corresponding classical system is certainly highly stochastic for this case. A survey of Fig. 1 and Table I reveal that these expectations are verified reasonably well. However, it must be emphasized that our calculations yield a solution only over a finite time interval, that the computed ρ(n) distribution contains large fluctuations, and that the various parameters in Table I are only crude indices which do deviate from their expected values. Pending further study, our results here must be regarded as providing only an indication of

TABLE III.   A listing of stochastic energy absorbtion parameters as a function of time for one computer solution of the driven quantum pendulum. Here k = 1, τ = 5, K = kτ = 5. The number N of $b_S(k)$ values used in Eq. (24) was 23.

t	$\langle E \rangle_N$	B	W_d	W_d/B	ξ	R_o	$|a_o|^2$
150	0.0108	13.1	0.023	1.78×10^{-3}	3.18×10^{-4}	12.6	0.58
300	0.0092	29.5	0.029	9.83×10^{-4}	-7.50×10^{-6}	27.9	0.91
450	0.0079	40.0	0.015	3.70×10^{-4}	-4.60×10^{-5}	22.9	0.61
600	0.0057	55.4	0.015	2.80×10^{-4}	-2.50×10^{-5}	38.2	0.88
750	0.0079	66.2	0.020	2.96×10^{-4}	1.90×10^{-6}	35.9	0.74

TABLE IV. A listing of stochastic energy absorbtion parameters as a
function of time for one computer solution of the driven
quantum pendulum. Here k = 10, τ = 1/2, and K = kτ = 5.
The number of N of $b_S(k)$ values used in Eq. (24) was 41.

t	$\langle E \rangle_N$	B	W_d	W_d/B	ξ	R_o	$\|a_o\|^2$
90	0.27	1.12	0.055	0.049	0.27	1.24	0.073
180	0.27	1.71	0.066	0.039	0.13	4.44	0.019
270	0.32	2.43	0.077	0.032	0.10	4.55	0.016
360	0.26	2.93	0.066	0.023	0.066	4.74	0.014
450	0.30	3.75	0.070	0.024	0.061	4.86	0.013

quantum stochastic behavior for the above parameter values.

The parameter values k = 40 and τ = 1/40 (classical K = 1) used
for Table II are those which classically lie on the border of stochasticity, and the data presented in Table II clearly indicates that the
quantum behavior for this case is also non-stochastic. Table III
which presents results for k = 1 and τ = 5 (classical K = 5) indicates
that the quantum motion is non-stochastic even though the corresponding
classical motion is stochastic. However, as mentioned earlier, this
difference between the quantum and classical case is understandable
and to be expected. For small k values, the impulses or "kicks" do
not give rise to many $e^{in\theta}$ "harmonics" in Eq. (22), and thus energy
cannot be absorbed into many energy levels as is the case for k>>1
where 2k "harmonics" are involved. This dependence of quantum stochasticity on k and not just the product (kτ) as in the classical case
receives support from the calculations resulting in Table III, but
more work will be required to establish the approximate k-value
determining the border of quantum stochasticity.

The results presented in Table IV and Fig. 2 which involve k = 10
and τ = 1/2 constitute a true puzzle, since here k>>1 and the corresponding classical K = 5 just as for Table I and Fig. 1. Thus one might
have expected stochastic behavior rather than the stable, non-stochastic data which actually appears. It is possible that the stochastic
border occurs for higher k-values than previously anticipated, or it
is possible that one here is observing a totally new and unexpected

effect; only further study can reveal the appropriate and correct
alternative. In this regard, let us mention that there are further
unique quantum effects. For example regardless of k-value or initial
$\psi(\theta,0)$, the free rotation mapping of Eq. (15) is the identity map
when $\tau = 4\pi$; for this τ-value after an integer number t of "kicks",
we have from Eq. (19) that

$$\psi \cdot (t) = \psi(0) \exp[itk\cos\theta] \qquad (37)$$

which yields the average energy growth given by

$$\langle E \rangle = -(\hbar^2/2m\ell^2) \int d\theta \psi^*(\partial^2/\partial\theta^2)\psi \sim t^2 \qquad (38)$$

that is proportional to t^2, corresponding to resonant rather than
diffusive energy absorption. Moreover, when $\tau = 2\pi$, one may use
Eq. (24) and a well-known Bessel function identity[11] to rigorously
prove that $A_n(t+2T^+) = A_n(t)$ for all n; that is, the full driven quantum solution is strictly periodic independent of $\psi(\theta,0)$ or the value
of k. Finally, resonant energy growth proportional to t^2 has also
been observed for $\tau = 4\pi/m$, where m = 3, 4, 8, and 32, although the
τ-widths of these resonances become increasingly narrow as m increases.
Apparently, these peculiar resonant (or anti-resonant at $\tau = 2\pi$)
effects are due to the strictly periodic nature of the unperturbed free
rotator wave function; while all unperturbed classical rotator orbits
are periodic, there is no period common to all solutions as occurs in
the quantum case.

Subsequent to the Como Conference, we continued the above Fig.
1-run for k = 40 and $\tau = 1/8$ and discovered that the diffusive quantum
energy absorption obeys Eq. (29) up to a time t_B (break-time) after
which diffusive energy absorption appears to continue but at a much
slower rate. Empirically, we find t_B proportional to k in sequences
of runs for which $K = (k\tau)$ is held fixed. This result further
"explains" the lack of any diffusive energy absorption in the data of
Table III where k = 1 and $\tau = 5$ which is classically stochastic. A
very crude, but possible explanation for the appearance of this
break-time t_B may lie in the uncertainty relationship $\Delta E \Delta t \sim \hbar$.
Note in Eq. (17) and Eq. (18) that the classical limit $\hbar \to 0$ is equivalent to $k \to \infty$ and $\tau \to 0$. Thus for large k and small τ ($\hbar$ fixed), we
might expect the quantum behavior to mimic the classical at least for
a time interval Δt during which the discrete nature of the free
rotator energy level spectrum is insignificant. From Eq. (15), one

notes that τ is a measure of the "effective" energy level spacing. Thus taking $\Delta E \sim \tau$ and $\Delta t \sim t_B$ in $\Delta E \Delta t \sim \hbar$, we have $t_B \sim \tau^{-1} \sim k$ for fixed $K(=k\tau)$, where t_B is the time required for the quantum system to "notice" that its energy levels are discrete. Further numerical study will be required to verify this possible explanation.

In closing this section, let us mention that we have sought to verify a sensitive dependence of final quantum state upon initial state similar to that implied by the exponential separation[3] of initially close classical orbits. Holding $K = k\tau$ fixed, we have tried k-values from $k = 1$ to $k = 100$ using various pairs of initially close $\{A_n(0)\}$ initial states only to find that the final state probability distribution $\rho(n) = A_n^* A_n$ was identical for each member of a pair to within numerical error. Even starting from non-close initial states such as $A_n = A_0 \delta_{no}$ and $A_n = e^{-n^2}$ yielded the same final $A_n^* A_n$ distribution. Each of these initial quantum states (for large k and small τ) appeared to be approaching the unique final state probability distribution given by Eq. (29), at least for times $t < t_B$. Moreover, Eq. (29) is being approached from each of these definite initial states without the need for a time or an ensemble average. These rather startling results tempt one to speculate that the unique final state probability distribution for an isolated (chaotic) quantum system might be the microcanonical distribution, however premature such a speculation might be. Regardless of such speculations however, our present computations reveal no sensitive dependence of final state upon initial state, indeed they indicate a surprising lack of such dependence. Certainly in the classical limit of sufficiently large k and sufficiently small $\tau(K = k\tau \gg 1)$, initially close wave packets must exponentially separate, but apparently even $k = 100$ and $\tau = 0.05$ does not lie in the classical parameter range.

VI. CONCLUDING REMARKS

This progress report has been presented in order to expose an example of a whole category of classically chaotic Hamiltonian systems whose exact quantum behavior can be investigated, at least numerically. It is our belief that future studies of the type Hamiltonian models revealed here can provide substantial information regarding the nature of chaotic behavior in deterministic quantum systems. Certainly the calculations for the specific pendulum Hamiltonian system considered herein provide at least an initial indication of the surprises and the possibly significant results which may await future investigators.

Regardless of final outcome, it now appears that a new doorway to quantum chaos may have opened and this progress report is an invitation for others to join us in crossing over its threshold.

In closing, we wish to express our profound appreciation to Ya. G. Sinai, E. V. Shuryak, G. M. Prosperi, G. M. Zaslavsky, D. Shepelyansky, and F. Vivaldi for many enlightening discussions regarding these problems. During the period of final editing for this paper, we received a specially prepared, handwritten preprint describing some splendid related work from Michael Berry, N. L. Balazs, M. Tabor, and A. Voros concerning "kicked" free particle systems. We have enormous admiration for the willingness of these authors to share their independent discoveries with us prior to publication.

REFERENCES

1. V. I. Arnold and A. Avez, Ergodic Problems of Classical Mechanics (W. A. Benjamin, Inc., New York, 1968); J. Moser, Stable and Random Motions in Dynamical Systems (Princeton Univ. Press, Princeton, 1973); Z. Nitecki, Differentiable Dynamics (MIT Press, Cambridge, 1971).

2. G. M. Zaslavsky, Statistical Irreversibility in Nonlinear Systems (Nauka, Moskva, 1970, in Russian); J. Ford in Fundamental Problems in Statistical Mechanics, III, Edited by E. G. D. Cohen (North-Holland, Amsterdam, 1975).

3. B. V. Chirikov, "A Universal Instability of Many-Dimensional Oscillator Systems," Physics Reports (to appear 1979).

4. N. M. Pukhov and D. S. Chernasvsky, Teor. i. Matem. Fiz. $\underline{7}$, 219 (1971).

5. I. C. Percival, J. Phys. $\underline{B6}$, 1229 (1973); J. Phys. $\underline{A7}$, 794 (1974); N. Pomphrey, J. Phys. $\underline{B7}$, 1909 (1974); I. C. Percival and N. Pomphrey, Molecular Phys. $\underline{31}$, 97 (1976).

6. K. S. J. Nordholm and S. A. Rice, J. Chem. Phys. $\underline{61}$, 203 (1974).

7. K. S. J. Nordholm and S. A. Rice, J. Chem. Phys. $\underline{61}$, 768 (1974).

8. E. V. Shuryak, Zh. Eksp. Teor. Fiz. $\underline{71}$, 2039 (1976).

9. G. M. Zaslavsky and N. N. Filonenko, Soviet Phys. JETP $\underline{38}$, 317 (1974). K. Hepp and E. H. Lieb in Lecture Notes in Physics, V. $\underline{38}$, Edited by J. Moser (Springer-Verlag, New York, 1974). A. Connes and E. Stormer, Acta Mathematica $\underline{134}$, 289 (1975).

10. John M. Greene, "A Method for Determining a Stochastic Transition," Preprint, Plasma Physics Laboratory, Princeton, New Jersey.

11. M. Abramowitz and I. A. Stegun, Handbook of Mathematical Functions
     (Dover Publications, Inc., New York, 1972), p. 363 and 385.

Quantum dynamics of a nonintegrable system

D. R. Grempel* and R. E. Prange†

Department of Physics and Center for Theoretical Physics, University of Maryland, College Park, Maryland 20742

Shmuel Fishman

Department of Physics, Israel Institute of Technology (Technion), 32000 Haifa, Israel
(Received 11 July 1983)

The quantum motion of a periodically kicked rotator is shown to be related to Anderson's problem of motion of a quantum particle in a one-dimensional lattice in the presence of a static-random potential. Classically, the first problem is nonintegrable and, for certain values of the parameters, exhibits chaos and diffusion in phase space; in the second problem, diffusion takes place in configuration space. Quantum phase interference, however, is known to suppress diffusion in Anderson's problem and to produce quasiperiodic motion. By establishing a mapping between the two systems we show that a similar effect determines the dynamics of the quantum rotator. As a result its wave functions are localized in phase space and their time evolution is quasiperiodic. This result explains the quantum recurrences and boundedness of the energy found in recent numerical work.

I. INTRODUCTION

In recent years the study of the effects of quantization on the properties of classically nonintegrable systems has attracted increasing attention.[1] The understanding of the nature of quantum behavior of these systems is not only of fundamental importance but it is also a problem of experimental relevance in fields as diverse as photochemistry,[2] electron dynamics in microstructures,[3] and other contexts. Understanding the relation between quantum problems and their classical limit may also shed light on the zero wavelength approximation to other wave equations, for example the eikonal approximation to the magnetohydrodynamic equations that are of great interest in plasma physics.[4] In the context of applications to photochemistry various model systems described by time-independent Hamiltonians have been recently studied.[5-7] Some of the evidence from these studies suggests that changes in the quantum behavior may take place when the corresponding classical system undergoes a transition to chaos. It is still unclear, however, whether these changes are only quantitative or if there is a sharp qualitative change in the nature of the long-time behavior as is the case in the corresponding classical system. Considerable effort[8-11] has also been made in order to develop semiclassical quantization rules for systems that are chaotic in their classical limit. These calculations involve many mathematical subtleties and their physical conclusions are still unclear to us.

As with the classical case, much insight into the properties of quantum nonlinear systems can be obtained from the study of simple maps. These are recursion relations defining the coordinates and momenta of the system at discrete time steps. The most studied area-preserving map is perhaps the Chirikov or standard map.[12,13] This map can be generated from a Hamiltonian that describes a planar rotator kicked at regular time intervals with a position-dependent force. It depends classically on a single parameter K, the dimensionless strength of the kick.

For each value of K the motion is chaotic or periodic depending on the initial conditions. For small K the chaotic regions are isolated and are separated by Kolmogorov-Arnol'd-Moser (KAM) trajectories[12,13] and consequently the motion is bounded. For $K = K_c \simeq 0.97164$ the last of these trajectories disappears[13] and for $K > K_c$ diffusion in p space takes place, namely, $p^2 \propto n$, for large n.[14]

Starting with the early work of Casati et al.[14] the quantized version of this map has been studied by several authors.[14-17] These investigations showed drastic differences between the classical and quantum motions. In particular, Chirikov et al.[15] found that (except in the special case of the quantum resonances[16]) the energy remains bounded and does not increase with time even for $K > K_c$. Later, Hogg and Huberman[17] showed numerically that the energy is quasiperiodic in time. This is in sharp contrast with the behavior in the classical case as described above.

Surprising as it is, this situation is quite reminiscent of the one encountered when studying a seemingly unrelated problem in solid-state physics, that of finding the motion of wave packets of electrons in a random lattice. The physics of this system is, by now, well understood. It has been known for a long time[18-20] that, in one dimension, all the eigenfunctions of a one-electron random Hamiltonian are exponentially localized in space and, consequently, are normalizable. As a result the electronic diffusion coefficient and the electron mobility vanish at zero temperature. The energy spectrum is pure point. It is, however, dense, so that in the thermodynamic limit the total density of states is continuous. Nevertheless, the local density of states, i.e., the density of states weighted by the wave function, is discrete. As a result, an electron initially localized around a point in the lattice undergoes quasiperiodic motion. These are pure quantum effects caused by destructive interference: If the electron were classical it would diffuse through the lattice.

It appears therefore natural to ask whether quantum interference effects can also destroy classical diffusion in *momentum* space and produce quasiperiodic motion. In this paper we will present evidence that the answer to this question is affirmative. We show that each of the two problems, the quantum rotator in a periodic time-dependent external field and a quantum particle moving in a static aperiodic potential, can be mapped into the other. By use of this mapping we show that the quantum dynamical system is, in effect, localized in angular momentum space and can therefore only reach a limited number of momentum states in the course of its time evolution. This in turn implies quasiperiodicity and thus boundedness and recurrence of the energy in time.[14-17] This is of course very different from the recurrence one would obtain ignoring these interference effects and considering the motion as an unrestricted random walk in momentum space.[21]

The organization of the rest of this paper is as follows. In Sec. II we introduce the quantum dynamical model studied in this work. The connection between it and the problem of electronic conduction in one-dimensional lattices is derived in Sec. III. In Sec. IV we exploit this connection to discuss the dynamics of the quantum system and present the results of numerical calculations. This is followed by a summary and concluding remarks in Sec. V. A preliminary account of part of this work was reported earlier.[22]

II. THE MODEL

In this section the specific quantum dynamical model to be used in the rest of the paper will be defined. We study the quantum motion of a system defined by the Hamiltonian

$$H = -\frac{\hbar^2}{2I}\frac{\partial^2}{\partial\theta^2} + \hat{k}V(\theta)\Delta(t) . \tag{2.1}$$

This is the Hamiltonian of a planar rotator with moment of inertia I driven by a time-dependent potential assumed to be factorizable into angle- and time-dependent parts. We shall assume that V is a periodic function of θ with period 2π and Δ has period t_0 (and dimensions of inverse time). Choosing t_0 as the unit of time the Hamiltonian depends on two dimensionless parameters $\tau = \hbar t_0/I$ and $k = \hat{k}/\hbar$. In these units the time-dependent Schrödinger equation is

$$i\frac{\partial}{\partial t}\psi(\theta,t) = \left[-\frac{1}{2}\tau\frac{\partial^2}{\partial\theta^2} + kV(\theta)\Delta(t) \right]\psi(\theta,t) . \tag{2.2}$$

One can readily generalize this class of Hamiltonians by replacing the kinetic energy $-\frac{1}{2}\tau\partial^2/\partial\theta^2$, by a general function $K(p)$ where $p = -i\partial/\partial\theta$ and deal with the operators

$$H = K(p) + kV(\theta)\Delta(t) . \tag{2.3}$$

A simple case is that in which $\Delta(t)$ is a sequence of δ-function kicks

$$\Delta(t) = \sum_{n=-\infty}^{\infty} \delta(t-n) . \tag{2.4}$$

The reason that (2.4) is a particularly simple choice is that, between kicks, the time evolution is that of free rota-

tion. At the kicks, the force is so strong that the kinetic energy is unimportant and Eq. (2.2) is readily integrated. The time evolution of the wave function between kicks is most easily expressed in the angular momentum representation. We call the wave function in this representation $\psi_n(t)$, where $p\psi_n = n\psi_n$. The relation of the angular momentum representation to the angle representation, $\psi(\theta,t)$ is $\psi(\theta,t) = \sum e^{in\theta}\psi_n(t)$. Since the time dependence between kicks is trivial, it is only necessary to consider times infinitesimally before or after the tth kick. We thus regard t as an integer and use the subscript $\mp$ to mean before or after the kick. We also absorb k into the definition of V.

The free propagation is thus represented by

$$\psi_n^-(t+1) = e^{-iK(n)}\psi_n^+(t) . \tag{2.5}$$

Direct integration of the Schrödinger equation over a kick gives

$$\psi^+(\theta,t) = e^{-iV(\theta)}\psi^-(\theta,t) . \tag{2.6}$$

Equations (2.5) and (2.6) are easily combined to obtain

$$\psi_m^+(t+1) = \sum_{n=-\infty}^{\infty} J_{m-n}\exp(-in^2\tau/2)\psi_n^+(t) , \tag{2.7}$$

where

$$J_{m-n} = (2\pi)^{-1}\int_0^{2\pi} d\theta\, e^{i(m-n)\theta}e^{-iV(\theta)} . \tag{2.8}$$

Note that the matrix J depends only on the difference $m-n$. This feature generally is lost for more general free Hamiltonians than $K(p)$. The recursion relations (2.7) can be solved numerically for various initial conditions. In order that this be feasible, it is necessary that J_r become suitably small for large $|r|$. This in turn requires that $V(\theta)$ be sufficiently smooth and small. The notation of Eq. (2.8) was chosen because J is a generalization of the Bessel function. Indeed, in the best studied case,[14-17] $V = k\cos\theta$, it is essentially just a Bessel function of the first kind.

III. MAPPING OF THE DYNAMICAL MODEL

In this section we will show that the problem defined by (2.7) is equivalent to a tight binding model for electronic conduction known as the Anderson model. Since the Hamiltonian is periodic in time, namely, $H(t) = H(t+1)$, we can classify the solutions of (2.7) according to the way the wave function transforms under translations in time. This leads to the introduction of a new quantum number, the quasienergy,[23] which is the only good quantum number in this problem. The states of fixed quasienergy ω have the form

$$\psi_\omega(\theta,t) = e^{-i\omega t}u_\omega(\theta,t) , \tag{3.1}$$

where $u_\omega(\theta,t) = u_\omega(\theta,t+1)$, and we have momentarily reverted to continuous time. Notice that this is an analog of the well known Bloch-Floquet theorem familiar in the case where potential is periodic in space. States of different quasienergies are orthogonal.[23] It is also believed that they form a complete set, but we know of no rigorous proof. In what follows we will expand arbitrary functions in quasienergy states. Due to the periodicity of the quasienergy states in time it is sufficient to study these states just before or after the kick, which we denote as before by $u^{\mp}$. We suppress, except when necessary, the dependence of the wave functions on ω. Again we use both the angle and angular momentum representations. Substitution of Eq. (3.1) into Eqs. (2.6) and (2.7) with t integer gives

$$u_m^- = e^{i\omega} e^{iK(m)} u_m^+ , \tag{3.2}$$

$$u^+(\theta) = e^{-iV(\theta)} u^-(\theta) , \tag{3.3}$$

and

$$u_m^+ = e^{i\omega} \sum J_{m-n} e^{iK(n)} u_n^+ . \tag{3.4}$$

The transformation we wish to make uses an alternative representation of the operator V in terms of an Hermitian operator W, namely, we define

$$e^{-iV(\theta)} = \frac{1 + iW(\theta)}{1 - iW(\theta)} \tag{3.5}$$

or equivalently

$$W(\theta) \equiv -\tan[V(\theta)/2] . \tag{3.6}$$

Defining

$$\bar{u}(\theta) = [u^+(\theta) + u^-(\theta)]/2 \tag{3.7}$$

we reformulate (3.3) as

$$\frac{u^+(\theta)}{1 + iW(\theta)} = \bar{u}(\theta) = \frac{u^-(\theta)}{1 - iW(\theta)} . \tag{3.8}$$

Substitution of (3.8) in (3.4) yields the equations for the Fourier components

$$u_m + i \sum W_{m-r} u_r = \left[u_m - i \sum W_{m-r} u_r \right] e^{-iE_m} , \tag{3.9}$$

where u_m (written without the bar for convenience) is the angular momentum representation of $\bar{u}(\theta)$. Here W_m is the Fourier transform of $W(\theta)$, and $E_m = \omega - K(m)$. Finally, (3.9) can be rewritten in the form

$$T_m u_m + \sum_{r\,(\neq 0)} W_r u_{m+r} = E u_m \tag{3.10}$$

with

$$T_m = \tan\left[\frac{E_m}{2} \right] = \tan\left[\frac{\omega - K(m)}{2} \right] \tag{3.11}$$

and

$$E = -W_0 . \tag{3.12}$$

Equation (3.10) describes a one-dimensional tight-binding model, with hopping W_r to the rth neighbor and "diagonal" potential T_m. This equation establishes therefore the correspondence between the quantum dynamical problem and the solid-state problem with the angular momentum in the quantum problem corresponding to the lattice sites in the solid-state problem. We will refer to (3.4) as the quantum dynamical problem or the rotator problem while (3.10) we will refer to as the tight-binding problem or the lattice problem.

In the tight-binding problem Eq. (3.10) is an eigenvalue equation for the energy E. In contrast, in the dynamical problem, E is a fixed parameter determined by the perturbation V [cf. (3.12) and (3.6)]. The quasienergy, on the other hand, which in the lattice problem simply determines which of several more or less equivalent sequences T_m is used, is the eigenvalue in the rotator problem. However, the relation between them is simple. Indeed, by applying the Feynman-Hellman theorem to Eq. (3.10) we find that $dE/d\omega = \sum \langle \sec^2 E_m \rangle /2$ is positive implying that E is a monotonic function of ω. Second, eigenvalues $E_\nu(\omega)$ belonging to distinct states ν will not become degenerate as ω is varied because the low symmetry of the system will, in general, prevent level crossing. Thus the functions $E_\nu(\omega)$ giving the lattice eigenvalues for fixed ω can in principle be inverted to give $\omega_\nu(E)$, the rotator quasienergies for fixed E, i.e., for fixed $V(\theta)$, and vice versa. The wave functions are, of course, essentially the same in the two pictures. The character of the spectrum is the same, as well.

The properties of the solutions of the lattice problem depend on the nature of the sequence T_m. Three cases are of great interest. The simplest of them occurs if T_m is periodic in m. Then the corresponding eigenstates are Bloch states. These states are extended and unnormalizable giving rise to electronic propagation, thus diffusion and conductivity in solids. In the dynamical problem this case corresponds to the quantum resonances.[16] Next in complexity is the case in which T_m is periodic, but with a period incommensurate with the (unit) lattice period. We discuss this case in detail elsewhere.[24] Finally, if $\{T_m\}$ constitutes a random sequence, each of its elements being chosen independently from a given fixed distribution, Anderson's model of localization in a one-dimensional random potential is obtained.

The properties of the solutions in the random case follow from rigorous results on the asymptotic behavior of products of random matrices.[25] It is known[18-20] that all the eigenstates of (3.10) are localized around some lattice site and decay exponentially away from that site with a characteristic length $\gamma^{-1}(E)$ which is solely determined by the probability distribution of the potential. Eigenstates with nearly identical energies are, however, generally localized around centers which are far apart. Conversely, two eigenstates localized at centers close compared with γ^{-1} will typically have eigenenergies separated by a finite energy spacing which is of order $\gamma\langle |W_r| \rangle$. The local density of states (i.e., as weighted by the square of the wave function) is discrete and, at a given site n, it consists of about γ^{-1} δ-function peaks at certain energies ϵ_n. The existence of a discrete local spectrum implies quasiperiodicity of the motion and absence of diffusion.

One can certainly arrive at the Anderson model by choosing for $\{K_m\}$ a random sequence. However, there is yet another possibility that, to our knowledge, has not been studied before in any detail and is of importance to understand the dynamics of the quantum rotator, namely, the one in which the sequence $\{K_m\}$ is *pseudorandom*, i.e., $\{K_m\}$ has some but not all the properties of a truly random sequence. This is of relevance for the problem at hand because, as will be seen shortly, the sequence

$K_m = \alpha m^2$ (mod 1) that corresponds to the rotator problem is, indeed, pseudorandom.

As an example of a pseudorandom sequence consider the case of $\{K_m = \alpha 2^m$ (mod 1)$\}$ where α is a given irrational number. This sequence reflects the random properties of the sequence of zeros and ones that occurs in the binary representation of α. It can be easily shown that, if α is irrational, the elements of $\{K_m\}$ are uniformly distributed in [0,1]. They are not independent, however, as the correlation function $\langle K_m K_{m+r} \rangle \cong 2^{-r}$ in this case. The existence of correlations of this type is a main difference between random and pseudorandom sequences. Little is known about the effect of these correlations on localization because the theorems mentioned above apply only to the case of independent random variables. While independence is certainly a sufficient condition for localization in one dimension, it is unlikely that it is necessary. For example, it is obvious on physical grounds that the fundamental results should still remain valid in the presence of weak, short-range correlations between the elements of $\{T_m\}$. The existence of localized states in a related quasiperiodic potential[24] can be taken as an indication that the conditions as stated in the relevant theorems[25] are too restrictive. These considerations are important because the numerical evidence that we present in Sec. III indicates that the sequence $\{T_m\}$ is "random enough" to localize all the solutions of Eq. (3.10).

The only rigorous result known to us which bears on this question has been provided recently by Bellissard[26] who shows that if W_r is of short range and small enough, and K_m is of the form $\tau(m+k)^2$, then for all k $(0 < k < 1)$ in a certain (Cantor) set C of measure approaching unity as W approaches zero, the spectrum of (3.10) is pure point, and the wave functions are exponentially localized. The theorem says nothing for larger W's or for k not in C, in particular for our case, $k = 0$. Bellissard conjectures that the spectrum in this case is singular continuous, or, at least, has a singular continuous component. It is almost surely true, however, that there is no absolutely continuous part of the spectrum. Although the mathematical problems posed are interesting and it is important that these issues be cleared up we expect, from a physical point of view, that it will be next to impossible to detect, in this case, any singular continuous part of the spectrum if finite size, finite time, finite temperature, dissipation, small random deviations from the m^2 law, etc., are taken into account. The main reason for this belief is that we always find numerically that the wave function drops off by 20 or 30 powers of e with no sign of comeback at large distances from the ostensible center. A singular continuous spectrum would imply that the apparent δ-function spikes in the local density of states (see Sec. IV) are really almost infinitely degenerate with the largest splittings of relative order e^{-30} or less. It must be a good approximation to neglect this if any other source of broadening is present. The numerical results may be an artifact of roundoff error, but this error probably mimics crudely the finite temperature effects mentioned above. In other words, we believe that the computer version of the model is more in accord with possible real systems than the idealized model is. Further, Bellissard's results provide evidence for this

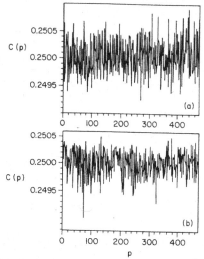

FIG. 1. Power spectrum of (a) the sequence $K_n = \sqrt{5}n^2$ (mod 1) and (b) a sequence of random numbers with uniform distribution in [0,1].

view, since with the finite smearing effects taken into account, the physical results surely cannot depend sensitively on whether k is zero or some small number in the set C.

IV. THE PSEUDO ANDERSON MODEL

We start this section by examining some statistical properties of $\{K_m = \tau m^2$ (mod 1)$\}$ for irrational τ. This sequence is ergodic with uniform distribution in the interval [0,1]. This is a rigorous result of a theorem by Weyl.[27] We have checked numerically that, for finite sequences of N elements, the deviations from uniformity follow the large-number behavior of random sequences, i.e., the fluctuation ΔN in the number of elements of the sequence that fall on any interval of fixed length contained in [0,1] is $\Delta N \cong \sqrt{N}$. The pair-correlation function is defined as

$$C(r) = \frac{1}{N} \sum_{m=1}^{N} K_m K_{m+r} . \tag{4.1}$$

Its Fourier transform,

$$C(p) = \sum_{r=1}^{N} e^{-i(2\pi r/N)p} C(r) , \tag{4.2}$$

is the power spectrum of the sequence. In Fig. 1(a) we plot $C(p)$ for a sequence of about 10^5 elements with $\tau = \sqrt{5}$. For comparison we show in Fig. 1(b) a plot of the Fourier transform of Eq. (4.1) with K_m replaced by "random" numbers as generated by a standard computer algorithm The general structure of the two plots is quite similar, the most important feature being the absence of correlations at any particular length scale. The statistical prediction for random numbers is that $C(p)$ is, on the average, independent of p with fluctuations of size $\sqrt{N}$. We observe this behavior in both the random and pseudoran-

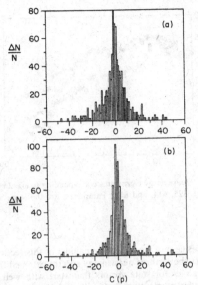

FIG. 2. Distribution of pair correlations for the potential $T_m \doteq \tan(x_n)$ with (a) $x_n = \sqrt{5n^2}$ and (b) $\{x_n\}$ a sequence of random numbers with uniform distribution in $[0,1]$.

dom sequences. Notice, however, that the peaks in the pseudorandom case are more numerous than those of the random one which shows that there are more correlations in the former case.

The statistical properties of $\{T_m\}$ follow from those of $\{K_m\}$. In particular, since $\{K_m\}$ is uniformly distributed, then $\{T_m\}$ will follow the Cauchy or Lorentzian distribution:

$$P(T_m) = \frac{1}{\pi} \frac{1}{1 + T_m^2} . \tag{4.3}$$

The correlation function for the potential can be computed by an expression like Eq. (4.1). In Fig. 2(a) we plot the distribution of the values taken by the correlation function for the same sample as used before. Figure 2(b) shows the same function in the case in which the argument of the tangent is a random number. In one sense both plots are similar: the correlations are of statistical nature (as opposed to systematic) and the widths of the distributions follow the law of large numbers for large sequences. The distribution function is, however, wider in the pseudorandom case, which shows again the presence of larger correlations.

If one is now willing to make the assumption that this degree of randomness is sufficient to localize all the solutions of Eq. (3.10) then, by virtue of the correspondence between the two problems established in Sec. III, the following picture of the quasienergy eigenstates emerges. For a given potential $V(\theta)$ [fixed E and W_r in Eq. (3.10)], each of the solutions of Eq. (2.12) is localized around some value of the angular momentum. Away from the center of localization, the solutions decay exponentially with an exponent $\gamma(E)$ which is independent of both the

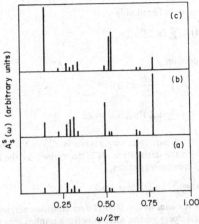

FIG. 3. Local quasienergy spectra as obtained from the time evolution of states with initial angular momenta (a) $l=46$, (b) $l=48$, and (c) $l=50$. The potential is the one shown in Eq. (4.8) and the parameters are $\kappa=2.8$, $\tau=4.867$.

values of the quasienergy and of τ and is uniquely determined by $V(\theta)$. In contrast, the center of localization depends upon the quasienergy. This dependence is not smooth, with states belonging to nearby quasienergies being centered, in general, around angular momenta that are far apart and states centered at nearby angular momenta having, in general, quite different quasienergies.

To see the effect of these properties on the time evolution of the quantum rotator, consider the expansion of the time-dependent wave function in terms of the quasienergy eigenstates. We have

$$\psi_n^+(t) = \sum_\nu C_\nu u_{n\nu}^+ e^{-i\omega_\nu t} . \tag{4.4}$$

The expansion coefficients are determined by the initial conditions. If, for simplicity, we start with an eigenstate of the angular momentum, $\psi_n^+(0) = \delta_{ns}$, the time-

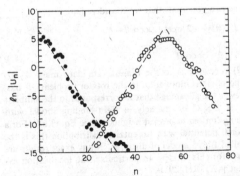

FIG. 4. Two quasienergy eigenstates for the same potential as in the previous plot. Quasienergies are $\omega = 2\pi j/2^{10}$ with $j=323$ (solid circles) and $j=621$ (open circles).

dependent wave function is

$$\psi_n^+(t) = \sum_\nu (u_{s\nu}^+)^* u_{n\nu}^+ e^{-i\omega_\nu t} \qquad (4.5)$$

and its spectrum,

$$A_n^s(\omega) \equiv \int_{-\infty}^{\infty} \frac{dt}{2\pi} \psi_n^+(t) e^{i\omega t}$$

$$= \sum_\nu (u_{s\nu}^+)^* u_{n\nu}^+ \delta(\omega - \omega_\nu) . \qquad (4.6)$$

A quantity of particular importance is $A_s^s(\omega)$, the projection of ψ on the initial state. It is determined by the local density of quasienergy states,

$$A_s^s(\omega) = \sum_\nu |u_{s\nu}^+|^2 \delta(\omega - \omega_\nu) . \qquad (4.7)$$

Although the sum in Eq. (4.7) is over all quasienergies the exponential decay of the wave functions implies that only a few states will effectively contribute to the sum: those centered on angular momenta within a distance of order γ^{-1} from the initial state. Since there is one quasienergy state for each angular momentum state, $A_s^s(\omega)$ will consist of about γ^{-1} peaks. The structure of Eq. (4.7) implies that $A_s^s(t)$ is an almost-periodic function.[28] Thus the state vector returns infinitely often to any given neighborhood of the initial state during the course of its evolution. Similar considerations show that, if the localization picture is correct, the expectation values of all observables are almost periodic functions. Notice that in these arguments no use is made of any properties of $V(\theta)$ other than that it produces a W_r of finite range.

To support these ideas we present results of numerical calculations. They consist of direct iteration of Eq. (2.7) for two different potentials and various initial conditions. The most efficient method to solve these equations that we found takes advantage of the translation invariance of the kernel in Eq. (2.8) and uses a forward-backward fast Fourier transform (FFT) technique. In this way one can easily include up to about 10^3 angular momentum states. Once a time series for the state vector is obtained, the FFT can be used again to calculate $A_s^s(\omega)$. From the latter, the spectrum and the eigenstates may be computed from Eqs. (4.6) and (4.7). In applications we took 2^{10} time steps.

The two potentials that we considered are

$$V_L(\theta) = -2 \arctan(\kappa \cos\theta - E) \qquad (4.8)$$

and

$$V_c(\theta) = k \cos\theta . \qquad (4.9)$$

$V_c(\theta)$ corresponds to the familiar standard map. $V_L(\theta)$ has no obvious motivation in the rotator problem. It has, however, the advantage that it corresponds to the simplest case of Eq. (3.10), namely, a tight-binding model with hopping limited to nearest neighbors [see Eq. (3.6)]. For a diagonal potential with Lorentzian distribution this is the so-called Lloyd model of disorder. Many exact results are known for this model. In particular, the localization exponent $\gamma(E)$ (Ref. 29) is

$$2\kappa \cosh\gamma(E) = [(E-\kappa)^2 + 1]^{1/2} + [(E+\kappa)^2 + 1]^{1/2} . \qquad (4.10)$$

Since this provides a quantitative test of our ideas we will

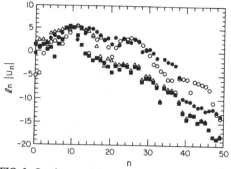

FIG. 5. Quasienergy eigenstates corresponding to $\omega = 2\pi j/2^{10}$ with $j = 50, 325, 614,$ and 810. Parameters are the same as in Fig. 3.

refer to this case in most of the remainder of this section.

Figure 3 shows three quasienergy spectra obtained from the time series for state vectors that start with well defined angular momentum $l = 46, 48,$ and 50, respectively. The parameters for this plot are $\kappa = 2.8$, $E = 0$, and $\tau = 4.867$. The main features to notice are that, as predicted above, (i) only a few peaks contribute effectively to the local spectrum, (ii) the same quasienergies appear in all three cases, and (iii) the amplitudes are very sensitive functions of the initial conditions. Although the initial states are closely spaced in angular momentum there are large differences in the weights of the peaks. The inverse of the localization exponent for this example is $\gamma^{-1}(E) \sim 1.5$. Thus quasienergy states that peak at more than a few units from a given one, l (say), make a very small contribution to the evolution of a state that starts at l. Conversely, from the localization of the peak of largest amplitude in the spectrum we see that states with very different quasienergy may be centered at nearby angular momenta.

Figure 4 shows two typical quasienergy eigenstates obtained as described above. These states peak at angular momenta $l = 0$ and 50 and rapidly decay away from their

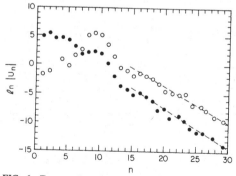

FIG. 6. Two nearby quasienergies with eigenfunctions centered at widely different angular momenta. $\omega = 2\pi j/2^{10}$ and $j = 323$ (solid circles) and $j = 325$ (open circles).

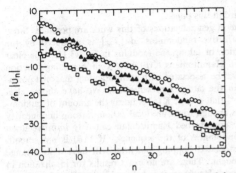

FIG. 7. A plot similar to Fig. 5 but with $\tau=0.1$. Other parameters are as before. $\omega=2\pi j/2^{10}$ with $j=298$, 741, and 858. Slope of these curves is the same as in Fig. 5.

centers. The decay is exponential as predicted. The dashed lines are the theoretical slopes calculated from Eq. (4.10). There is good agreement. Notice that the localization exponent is the same for both quasienergies. To further illustrate the independence of $\gamma(E)$ upon the quasienergy we plot in Fig. 5 four states all of which have appreciable weight at or near $l=10$. Their quasienergies are quite different but they all have the same long-distance behavior. The case of nearby quasienergies is shown in Fig. 6. We plot two eigenstates whose quasienergies differ by only 0.6%. Their centers are displaced by ten units of angular momentum. Thus, in general, the relation between quasienergy and center of localization is not smooth. This situation is familiar in the localization problem.

Results obtained for other values of τ are similar, provided it is not too small. Figure 7 shows three eigenstates corresponding to $\tau=0.1$, the rest of the parameters being fixed at their previous values. The states are localized with the same localization exponent that we found before. If τ is very small the short-distance behavior of the wave function may be considerably modified with respect to the one found for larger values. This is shown in Fig. 8 where we plot several states for $\tau=0.01$. The states are still lo-

calized and drop off rapidly as a function of distance but now we find relatively flat regions in between those in which the wave function decays exponentially. This can be understood by noticing that the origin of the pseudorandom behavior of $\{K_n\}$ lies in the discontinuities introduced by the operation of taking the modulum. If τ is small, however $\{K_n\}$ is smooth over distances of the order of the location of the first discontinuity. In the present example this occurs at $n_1\cong30$. Thus the phase of the wave function is coherent over about n_1 sites before interference effects set in. This coherence shows up as a plateau in a logarithmic plot of the amplitude of the wave function. Similarly, there are additional plateaus over scales corresponding to the distances between successive discontinuities. The phase of the wave function is coherent within each such region, but the correlation between the overall phases in different regions is rapidly lost. As n increases the distance between discontinuities becomes smaller and smaller and hence the width of the plateaus decreases until, at large distances (of order $\tau^{-1/2}$), they disappear completely. These features are clearly seen in Fig. 8.

The usual standard map corresponds to the potential of Eq. (4.9). The associated hopping potential may be calculated in closed form for $k<\pi$. For our present purposes it is sufficient to note its form for long distances:

$$W_r=\begin{cases}0 & \text{for even } r\\[2mm]\dfrac{2}{\pi}\exp\left[-r\ln\left|\dfrac{2\pi}{k}\right|\right] & \text{for odd } r\,.\end{cases}\qquad(4.11)$$

Thus the hopping potential is of finite range and we will find the same behavior as before. Since the localization exponent is unknown for this case we have no quantitative test of the theory. We have obtained numerical results for several values of k and τ. As an example we plot in Figs. 9 and 10 quasienergy spectra and wave functions computed from the time series for state vectors that start out at $l=0$ and 50, respectively. There is no qualitative

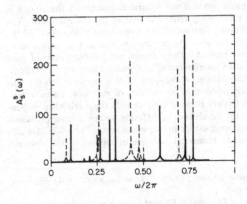

FIG. 9. Quasienergy spectra for the potential of the standard map, Eq. (4.9) for states starting with $l=0$ (dashed line) and $l=50$ (solid line). Parameters are $k=2.8$ and $\tau=4.867$.

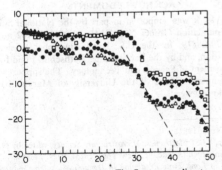

FIG. 8. A plot equivalent to Fig. 7 corresponding to $\tau=0.01$. Notice the presence of plateaus of decreasing extension in between regions of exponential decay.

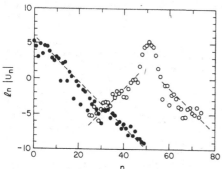

FIG. 10. Two quasienergy eigenstates for the same potential as in Fig. 9. Quasienergies are $\omega = 2\pi j/2^{10}$ with $j=511$ (solid circles) and $j=709$ (open circles).

difference between these and our previous results. The same is true for all other sets of parameters as well.

If $\tau/2\pi$ is a rational, p/q, the system is periodic, since $T_{m+q}=T_m$. However, if $q\gamma \gg 1$, the wave function will not realize that the potential is periodic until it has already become very small. The bands in this case can be estimated by the following procedure. Find approximate localized eigenstates, v_m^ν centered at site ν, $\nu=1,\ldots,q$, say, by replacing p/q by a nearby irrational. Then the wave function will be approximated by

$$u_m^{k\nu} = \sum_{-\infty}^{\infty} e^{irqk} v_{m-rq}^\nu .\qquad(4.12)$$

Here k is the continuous crystal momentum, $-\pi/q < k \le \pi/q$, and ν is the band index. The energy can be found by taking the expectation of H in this state. The bandwidth will be of order $e^{-\gamma q}$, the spacing between bands of order $1/q$.

V. CONCLUSIONS

We have presented evidence that the motion of the periodically pinged rotator is bounded and almost periodic in phase space if the natural frequency of the system is not rotationally related to that of the external field. The qualitative features of the motion are independent of the detailed form of the pinging potential as long as it is sufficiently smooth. These results are a consequence of pure quantum interference effects with no counterpart in classical mechanics. A mapping of this problem into the Anderson problem of localization of electronic states in random lattices allowed us to identify the mechanism responsible for the absence of diffusion and recurrence phenomena reported earlier in numerical work.[14-17] We have also presented numerical evidence in support of the ideas put

forward in this paper.

Several generalizations of this work are possible. Since the effective randomness in $\{T_m\}$ is produced by the operation of taking the modulus in Eq. (3.11), it is clear that all Hamiltonians $K(p)$ with K an increasing function of p will be associated with pseudorandom diagonal potentials in the lattice representation. We have shown elsewhere[24] that even if $K(p)$ is linear the amount of randomness is enough to ensure localization. Strong nonlinearity in the unperturbed Hamiltonian can only improve upon the randomness of the sequence. If $V(\theta)$ is not smooth the associated hopping potential will have a long-range component. There are no exact results for localization in this case. However, it is believed[30] that $W_r \cong 1/r$ [i.e., a discontinuous $V(\theta)$] constitutes the dividing line between the cases of localized and extended solutions. The simplest Hamiltonian that produces long-range hopping is the quantized version of "Arnol'd's Cat."[31] We have performed[32] numerical calculations of the type reported here for this potential. Our preliminary results indicate that, indeed, the states are extended in this case and there is diffusion in phase space.

It is of interest to consider the case in which the unperturbed spectrum has a continuous component, for this situation is closer to the one present in experimental systems. In these cases either there is a finite number of bound states or the spectrum has an accumulation point at the dissociation threshold. An example of the latter class was recently studied in the classical limit.[33] If one confines the attention to the bound states the same formal manipulations that we used in Sec. II of this paper can be performed to get a new type of equivalent lattice problem. This turns out to be that of the motion of a quantum particle in a semi-infinite chain with a diagonal potential that is nonrandom at long distances from the "surface." Under these circumstances one expects richer behavior than that found here because now, depending on the parameters, the states can be either extended or localized in the "surface region." The existence of extended states would make diffusion possible but quantum effects (i.e., the presence of localized states) will most likely introduce corrections to the classical results. We do not know at this stage whether these corrections are of quantitative or qualitative nature.

ACKNOWLEDGMENTS

This work was supported in part by the National Science Foundation Grants Nos. DMR-79-001172-A02 and DMR-79-08819, by the U.S. Israel Binational Science Foundation, and by the Bat Sheva de-Rothschild Fund for Advancement of Science and Technology. The support of the Computer Center of the University of Maryland is also acknowledged.

*Present address: Institute Laue-Langevin, 38042 Grenoble, France.
†Also at Institute for Physical Sciences and Technology, University of Maryland, College Park, MD.
[1]For a review see, e.g., G. Zaslavsky, Phys. Rep. **80**, 158 (1981).
[2]*Photoselective Chemisty*, edited by J. Jortner, R. D. Levine, and S. A. Rice, Vol. 47 of *Advances in Chemical Physics* (Wiley,

New York, 1981).
[3]See, e.g., *VLSI Electronics, Microstructure Science*, edited by N. G. Eimspruch (Academic, New York, 1981), Vols. 1 and 2.
[4]E. Ott, in *Long Time Prediction in Dynamical Systems*, edited by C. W. Horton, Jr., L. E. Reichl, and V. B. Szebehely (Wiley, New York, 1983), p. 281.
[5]See the article by S. A. Rice in Ref. 11.

[6]M. Bixon and J. Jortner (unpublished).

[7]G. Hose and H. S. Taylor, J. Chem. Phys. 76, 5356 (1982).

[8]M. V. Berry, Lecture Notes of the NORDITA School on Chaos, June 1982 (unpublished).

[9]M. V. Berry, N. L. Balazs, M. Tabor, and A. Voros. Ann. Phys. 122, 26 (1979).

[10]M. Tabor, Physica (Utrecht) 6D, 195 (1983).

[11]M. Gutzwiller, J. Math. Phys. 11, 1791 (1970); 12, 343 (1971); Phys. Rev. Lett. 45, 150 (1980).

[12]B. V. Chirikov, Phys. Rep. 52, 263 (1979).

[13]J. M. Greene, J. Math. Phys. 20, 1183 (1979).

[14]G. Casati, B. V. Chirikov, F. M. Izrailev, and J. Ford in *Stochastic Behavior in Classical and Quantum Hamiltonian Systems,* Vol. 93 of *Lecture Notes in Physics,* edited by G. Casati and J. Ford (Springer, Berlin, 1979).

[15]B. V. Chirikov, F. M. Izrailev, and D. L. Shepelyansky, Sov. Sci. Rev. Sec. C 2, 209 (1981).

[16]F. M. Izrailev and D. L. Shepelyansky, Teor. Mat. Fiz. 43, 417 (1980) [Theor. Math Phys. 43, 553 (1980)]; Dok. Akad. Nauk SSSR 249, 1103 (1979) [Sov. Phys—Dokl. 24, 996, 1979)].

[17]T. Hogg and B.A. Huberman, Phys. Rev. Lett. 48, 711 (1982); Phys. Rev. A 28, 22 (1983).

[18]P. W. Anderson, Phys. Rev. 109, 1492 (1958).

[19]P. W. Anderson, Rev. Mod. Phys. 50, 191 (1978).

[20]For a review, see, e.g., D. J. Thouless, in *Ill Condensed Matter, Proceedings of Les Houches Summer School,* edited by R. Balian, R. Maynard, and G. Toulouse (North-Holland, Amsterdam, 1979), Vol. 31.

[21]A. Peres, Phys. Rev. Lett. 49, 1118 (1982).

[22]S. Fishman, D. R. Grempel, and R. E. Prange, Phys. Rev. Lett. 49, 509 (1982).

[23]Ya. B. Zeldovich, Zh. Eksp. Teor. Fiz. 51, 1492 (1966) [Sov. Phys.—JETP 24, 1006 (1967)].

[24]D. R. Grempel, S. Fishman, and R. E. Prange, Phys. Rev. Lett. 49, 833 (1982); R. E. Prange, D. R. Grempel, and S. Fishman (unpublished).

[25]K. Ishii, Prog. Theor. Phys. Suppl. 53, 77 (1973).

[26]J. Bellissard (unpublished).

[27]I. P. Cornfeld, S. V. Forniu, and Ya. G. Sinai, *Ergodic Theory* (Springer, Berlin-Heidelberg-New York, 1982), Chap. 2.

[28]A. S. Besicovich, *Almost Periodic Functions* (Cambridge University, Cambridge, England, 1932).

[29]D. J. Thouless, J. Phys. C 5, 77 (1972).

[30]P. W. Anderson (private communication).

[31]J. H. Hannay and M. V. Berry, Physica (Utrecht) D1, 267 (1980).

[32]D. R. Grempel, S. Fishman, and R. E. Prange (unpublished).

[33]R. V. Jensen, Phys. Rev. Lett. 49, 1365 (1982).

Excitation of molecular rotation by periodic microwave pulses. A testing ground for Anderson localization

R. Blümel[a]
Department of Nuclear Physics, The Weizmann Institute of Science, Rehovot 76100, Israel

S. Fishman
Department of Physics, Technion, 32000 Haifa, Israel

U. Smilansky[b]
Department of Theoretical Physics, Oxford University, Oxford OX1 3NP, United Kingdom

(Received 22 March 1985; accepted 16 September 1985)

We study the excitation of molecular rotation by microwave pulses of duration σ which occur periodically with frequency ω. We analyze the molecular dynamics both classically and quantum mechanically and consider situations where the coupling of the field to the molecule is strong. In both approaches, the angular momentum transmitted to the molecule is confined to a finite band of width $\approx 1/\sigma$. But, while the classical dynamics displays chaotic features, the quantum treatment distinguishes clearly between two regimes. Resonance excitation occurs when ω is rationally related to the basic rotation frequency ω_0. Off resonance (ω/ω_0 irrational), the probability to transfer angular momentum to the molecule is small and the underlying mechanism for this effect is analogous to the Anderson model of localization in a one-dimensional random lattice with a finite number of sites. We show that the conditions required by our analysis can be achieved with, e.g., PbTe or CsI molecules and conventional field strengths and we propose this system as an experimental testing ground for the Anderson localization mechanism.

I. INTRODUCTION

Recent research work seems to indicate that a class of quantum systems which are chaotic in their classical limit ($\hbar = 0$) do not show any of the characteristic attributes of chaos if Planck's constant is kept finite. To this class belong, e.g., the 1 dim. Hamiltonian systems perturbed by an external driving force consisting of a train of periodic and sudden impulses ("δ kicks").[1] Due to the simple time structure of the external perturbation, the time evolution of periodically kicked systems is expressed in terms of a quantum mapping[2,3] which considerably facilitates the derivation of analytical and numerical properties. The most extensively studied system in this context is the planar rotor[4–8]

$$H_0 = \frac{\hbar^2 n^2}{2I} = \frac{-\hbar^2}{2I}\frac{\partial^2}{\partial\theta^2} \tag{1.1}$$

perturbed by a train of δ kicks

$$V(\theta,t) = k\cos(\theta)\sum_m \delta(t - mT). \tag{1.2}$$

The solution of the quantum problem indicates that the response of the rotor to the external field depends crucially on the ratio between the natural rotor frequency $\omega_0 = \hbar/2I$ and the driving frequency $\omega = 2\pi/T$. If this ratio is rational, the rotor absorbs energy due to a resonance mechanism and $E(t) \sim t^{2.5}$ If, on the other hand, ω/ω_0 is irrational, the wave function tends to localize near the original eigenstates of H_0 and the energy absorbed by the rotor is a quasiperiodic and

bounded function of time.[4] In both regimes, the quantum description differs markedly from its classical counterpart.

Our understanding of both aspects of the quantum kicked rotor is based on the spectral properties of the quasienergy operator.[9] In the resonance case (ω/ω_0 rational), the spectrum of the quasienergy operator is continuous and the eigenstates are extended. Off resonance, most probably one obtains a point spectrum and the states are localized. Although there is no rigorous proof of this statement it is supported by physical arguments as well as numerical evidence.[1,4,8] It was recently shown that the spectrum is continuous for some off resonance values of the parameters.[10] On the basis of a detailed analysis of a similar model,[11] we believe that the spectrum is continuous only for a set of measure zero, namely when ω/ω_0 is a Liouville (or rational) number. Therefore for any off resonance driving frequency ω, the probability for states to be localized is unity. The localization mechanism in this case is similar to the one which causes the electrons in a one-dimensional random lattice to localize in real space.

At first sight one may think that the behavior described above is peculiar to the kicked rotor, either because of the simple spectrum of H_0, or because of the singular nature of the δ kicks. However, some recent studies of other systems[12,13] show that the suppression of classical chaoticity due to quantum effects is of more general validity, even though the underlying mechanisms may differ from those responsible for the localization of the kicked rotor states. Together with the theoretical investigation of the problem, attempts have been made to propose actual experimental tests[14] where the predictions of the classical and the quantal theories could be compared vis a vis the experimental data. Until now, no such experiment was carried out.

[a] Present Address: Department of Theoretical Physics T30, Technical University Munich, 8046 Garching, FRG.
[b] On leave from the Weizmann Institute of Science, Rehovot, Israel.

In the present paper, we shall study the quantum dynamics of a diatomic molecule subject to a smooth radiofrequency electric field. We shall show that this system has much in common with the kicked planar rotor and, in particular, the same mechanism brings about the localization phenomenon off resonance and the enhanced resonance excitation. Thus, we are able to propose a system which is realistic and amenable to experimentation while showing the same peculiar effects displayed by the schematic δ-kicked rotor.

Such experiments may also be looked upon as demonstrations (in a finite basis) of the Anderson localization mechanism.[15] The system proposed here seems to be much less affected by the perturbations (e.g., electron–phonon interactions) which make the direct test of Anderson localization an almost impossible task.

The paper is organized in the following way: the general theoretical framework is presented in Sec. II and it is used to compare the dynamics of the well-known δ-kicked planar rotor with a periodically, but smoothly driven one. In Sec. III, we study the response of a diatomic molecule to an external (smooth) pulse. For large angular momenta and polarized external fields, the molecular dynamics approaches that of the planar rotor and we use the tools developed in Sec. II to discuss and compare the numerical results. Some experimental considerations and a summary of our findings will be given in Sec. IV.

II. CONTINUOUSLY DRIVEN PLANAR ROTOR

The quantum dynamics of a system which is driven by a time dependent force with period T is most conveniently discussed in terms of the time independent Hermitian (quasienergy) operator G, which is related to the propagator $U(t,t_0)$ by[16]

$$U(t,t_0) = P(t - t_0)\exp\left[-\frac{i}{\hbar} \cdot G \cdot (t - t_0) \right]. \quad (2.1)$$

Here, $P(t)$ is a time dependent unitary operator satisfying $P(t + T) = P(t)$ and $P(0) = 1$. As P is unity at times $t_N = t_0 + NT$, the time evolution operator $U(t_0 + T, t_0)$, which propagates the system over a full cycle of the external field, is diagonal in the quasienergy basis defined by

$$G|\chi_\alpha\rangle = \omega_\alpha |\chi_\alpha\rangle \quad (2.2)$$

and a wave function initially (for $t = t_0$) given by $|\psi_0\rangle$ will be mapped into

$$|\psi_N\rangle = \sum_\alpha e^{-(i/\hbar)\omega_\alpha NT} \langle \chi_\alpha|\psi_0\rangle |\chi_\alpha\rangle \quad (2.3)$$

after N periods of the external field. Thus, the time evolution of the wave function is entirely determined by the amplitudes $\langle \chi_\alpha|\psi_0\rangle$ and the quasienergies ω_α. In most cases, G cannot be written explicitly in terms of $U(t_0 + T, t_0)$ so that its spectral properties are extracted by diagonalizing the one cycle propagator $U(t_0 + T, t_0)$ directly

$$U(t_0 + T, t_0)|\chi_\alpha\rangle = e^{-(i/\hbar)\omega_\alpha T}|\chi_\alpha\rangle. \quad (2.4)$$

An alternative expression for $U(t_0 + T, t_0)$ is

$$U(t_0 + T, t_0) = e^{-(i/\hbar) H_0 T} \mathscr{T} \exp\left[-\frac{i}{\hbar} \int_{t_0}^{t_0 + T} \widetilde{V}(t,t_0)dt \right], \quad (2.5)$$

where H_0 is the Hamiltonian of the unperturbed system, $\mathscr{T}$ is the time ordering operator, and

$$\widetilde{V}(t,t_0) = \exp\left[\frac{i}{\hbar} H_0(t - t_0) \right] V(t) \exp\left[-\frac{i}{\hbar} H_0(t - t_0) \right] \quad (2.6)$$

is the external perturbation in the interaction representation. With a suitable Hermitian operator W, the propagator U can always be expressed as

$$U(t_0 + T, t_0) = \exp\left[-\frac{i}{\hbar} H_0 T \right] \frac{1 + iW}{1 - iW} \quad (2.7)$$

and Eq. (2.4) can be cast into the form

$$T_m^{(\alpha)} u_m^{(\alpha)} + \sum_n W_{mn} u_n^{(\alpha)} = 0, \quad (2.8)$$

where

$$\begin{aligned}
u_m^{(\alpha)} &= \langle m|(1 + iW)^{-1}|\chi_\alpha\rangle, \\
T_m^{(\alpha)} &= \tan\left[(1/2\hbar)(E_m - \omega_\alpha)T \right], \\
W_{mn} &= \langle m|W|n\rangle,
\end{aligned} \quad (2.9)$$

and $|m\rangle$ are the eigenstates of H_0 corresponding to the eigenenergies E_m.

For the kicked planar rotor, the hopping matrix element W_{nm} depends only on $|n - m|$ and Eq. (2.8) describes a one-dimensional tight binding model with the diagonal potential T_m. If $\{T_m\}$ is periodic in m, Eq. (2.8) describes a crystalline solid and the corresponding eigenstates are extended Bloch states. If $\{T_m\}$ is random, it is the one-dimensional Anderson model of localization, where all the states are exponentially localized, if the hopping falls off sufficiently fast with distance. Fishman, Grempel, and Prange[4] used the correspondence between the eigenvalue equation (2.4) and the tight binding model (2.8) in order to study the nature of the quasienergy states of the kicked rotor. They argued that since $E_m = \hbar^2 m^2/2I$, the sequence $\{T_m\}$ can be considered pseudorandom as long as $\tau \equiv \hbar T/I$ is not a rational multiple of π. Consequently, the quasienergy states of the kicked rotor are exponentially localized in momentum space (where the Hamiltonian of the unperturbed rotor is diagonal) by a mechanism similar to Anderson localization. This argument was substantiated by numerical evidence. For $\tau = \pi(p/q)$ where p and q are integers, the quasienergy states are extended, leading to the quantum resonances. We shall now investigate to what extent the mechanism of Anderson localization is preserved when the singular δ kicks in the driving potential of Eq. (1.2) are replaced by a properly chosen smooth pulse. The strict correspondence between the driven rotor [Eqs. (1.1) and (1.2)] and the tight binding model (2.8) holds only for periodic δ-function kicks. For a smooth pulse we will establish that over some range of momentum the evolution operator is similar to the one obtained for δ kicks and Anderson localization takes place in this range.

For this purpose, the Hamiltonian (1.1) and (1.2) is replaced by

$$H = -\frac{\hbar^2}{2I}\frac{\partial^2}{\partial\theta^2} + \hat{k}\cos(\theta)\Delta(t), \qquad (2.10)$$

where $\Delta(t)$ is the form factor of the force with dimension of inverse time. In terms of the dimensionless quantities τ and $k \equiv \hat{k}/\hbar$, the time dependent Schrödinger equation corresponding to the Hamiltonian (2.10) can be written as

$$i\frac{\partial}{\partial t}\psi(\theta,t) = \left[-\frac{1}{2}\tau\frac{\partial^2}{\partial\theta^2} + k\cos(\theta)\Delta(t)\right]\psi(\theta,t), \qquad (2.11)$$

where time is now measured in units of T. In these units, $\Delta(t)$ becomes a one-periodic function and the kicked rotor is recovered by choosing $\Delta(t) = \Sigma_n \Delta^{(\delta)}(t-n)$ and $\Delta^{(\delta)}(t) = \delta(t - 1/2)$, where for later convenience the kick was centered in the time interval [0,1]. The condition of zero pulse width can be relaxed in two obvious ways. Either we choose a periodic Gaussian pulse of width σ:

$$\Delta^{(\sigma)}(t) = \frac{1}{\sqrt{2\pi}\sigma}\exp\left[-\frac{(t-1/2)^2}{2\sigma^2}\right], \quad 0 < t < 1 \qquad (2.12)$$

or we truncate the Fourier series for a periodic δ function at some specified harmonic N:

$$\Delta^{(N)}(t) = 1 + 2\sum_{m=1}^{N}\cos[2m\pi(t - 1/2)]. \qquad (2.13)$$

A comparison of both pulse shapes is given in Fig. 1. It can be seen that the $\Delta^{(N=7)}(t)$ and $\Delta^{(\sigma=0.03)}(t)$ form factors are quite similar.

In order to predict the behavior of the continuously driven planar rotor, it is necessary to construct the one cycle propagator U. In the case of the kicked rotor the matrix elements U_{nm} of the propagator in the rotor basis can be expressed analytically in terms of Bessel functions of the first kind[7]

$$U_{nm} = \langle n|\exp\left[\frac{1}{2}i\tau\frac{\partial^2}{\partial\theta^2}\right]\exp[-ik\cos(\theta)]|m\rangle$$
$$= e^{-1/2\,i\tau n^2}(-i)^{n-m}J_{n-m}(k). \qquad (2.14)$$

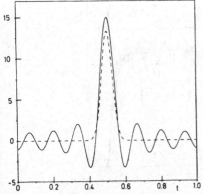

FIG. 1. Possible shapes of a microwave pulse. Broken line: Gaussian form factor $\Delta^{(\sigma=0.03)}(t)$ [see Eq. (2.12)]. Full line: truncated Fourier series for a periodic δ function $\Delta^{(N=7)}(t)$ [see Eq. (2.13)].

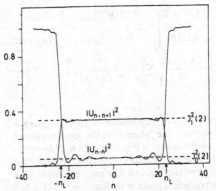

FIG. 2. Absolute square of diagonal and first off diagonal matrix elements of the one cycle propagator $U(T)$ for $\tau = 2$, $k = 2$, and $\Delta^{(N=7)}(t)$.

The matrix elements of appreciable magnitude lie in a band around the diagonal and the width of the band is of the order k. In order to calculate the one cycle propagator corresponding to a continuous perturbation, we expand the complete wave function $\psi(\theta,t)$ of the system in the set of unperturbed rotor states

$$|\psi(\theta,t)\rangle = \sum b_n(t)|n\rangle. \qquad (2.15)$$

Substituting this expansion into the time dependent Schrödinger equation (2.11), we end up with a set of coupled first order equations for the amplitudes $b_n(t)$:

$$ib_n(t) = \frac{1}{2}\tau n^2 b_n(t) + \frac{1}{2}k\Delta(t)[b_{n-1}(t) + b_{n+1}(t)]. \qquad (2.16)$$

The mth column of the propagator U is now constructed out of the vector of amplitudes $b_n(1)$ obtained by integrating Eq. (2.16) numerically over one complete cycle of the external force for initial conditions $b_n(0) = \delta_{nm}$.

Figure 2 shows the absolute squares of diagonal and first off diagonal elements of the propagator U in the case $\tau = 2$, $k = 2$, and $\Delta(t) = \Delta^{(N=7)}(t)$. The dependence of $|U_{nn}|^2$ and $|U_{n,n+1}|^2$ on n is characterized by an abrupt change which occurs at some n which will be denoted by n_L in the sequel. $|U_{nn}|^2$ and $|U_{n,n+1}|^2$ are almost constant for $|n| < n_L$ and, moreover, $|U_{nn}|^2 \simeq J_0^2(2)$, $|U_{n,n+1}|^2 \simeq J_1^2(2)$ in good agreement with the values of the corresponding kicked rotor matrix elements (2.14). Also, $|U_{n,n+s}|^2$ (which are not shown in the figure) behave according to Eq. (2.14). Beyond the limiting value n_L, $|U_{nn}|^2 \to 1$ and $|U_{n,n+s}|^2$ drop to zero very rapidly so that the states $|n\rangle$ with $|n| > n_L$ are almost decoupled from the $|n| < n_L$ states.

Although at least second order perturbation theory is required to understand the behavior of the diagonal matrix elements of the one cycle propagator U, first order perturbation theory is sufficient for a qualitative understanding of the first off diagonal matrix elements and the corresponding transition amplitudes W_{nm}:

$$U_{n,n\pm s}^{(1)}(T) = -2ie^{-(1/2)i\tau n^2}\,W_{n,n\pm s}^{(1)}, \quad s = 1,2,...,$$

$$W_{n,n\pm s}^{(1)} = (1/2\hbar)\int_{t_0}^{t_0+T}\widetilde{V}_{n,n\pm s}(t,t_0)dt = \tfrac{1}{4}k\delta_{s,1}\exp[\mp\tfrac{1}{2}i\tau(n\pm\tfrac{1}{2})]f^{(N)}[\tfrac{1}{2}\tau(n\pm\tfrac{1}{2})],$$

$$f^{(N)}(x) = \frac{\sin(x)}{x}\sum_{q=-N}^{+N}\frac{(-1)^q}{1-[q\pi/x]^2}; \quad f^{(N)}(x)\to 1 \text{ for } N\to\infty. \tag{2.17}$$

According to Eq. (2.17), only nearest neighbor hopping is possible in first order. The "plateau" in the absolute value of $U_{n,n+1}$ which was observed in Fig. 2 and which implies nearly constant nearest neighbor hopping probabilities $|W_{n,n+1}|^2$ in the plateau region is now readily explained as a delicate superposition of peaks in the function $f^{(N)}$. The last peak, i.e., the last resonance in $f^{(N)}[\tfrac{1}{2}\tau(n\pm\tfrac{1}{2})]$ appears at

$$n_L \simeq 2N\pi/\tau, \tag{2.18}$$

which for $\tau = 2$ and $N = 7$ predicts $n_L = 22$ (see Fig. 2). Beyond $|n| > n_L$ the hopping amplitudes become very small and states to the left or to the right of the plateau can neither be coupled to the plateau states, nor are they coupled amongst themselves. In the plateau region ($|n| < n_L$) on the other hand, the coupling between neighboring states of the unperturbed rotor is large. In this region the matrix elements of the evolution operator U are approximately equal to those of the kicked rotor given by Eq. (2.14). This clear distinction between the two regimes will, in turn, partition the quasienergy states into two sets: quasienergy states which have a large overlap with $|n| > n_L$ states will resemble the original state $|n\rangle$ because of the weak off diagonal coupling relative to the dominant diagonal matrix elements of H_0. These trivially localized states will be referred to as "perturbatively localized." States in the other set have a large overlap with $|n| < n_L$ states of H_0. Here, we expect Anderson localization to take place when τ is not a rational multiple of π. States localized on sites (in momentum space) that are separated from the edges of the plateau a distance larger than the localization length will be exponentially localized. States centered on sites closer to the edges will be localized even stronger due to "perturbative localization" beyond the edge of the plateau. This is analogous to Anderson localization in a finite solid.

If, on the other hand, τ is a rational multiple of π, we should observe *delocalization* of the quasienergy states over the whole plateau $|n| < n_L$ resulting in a rapid growth of the mean energy of the rotor, till the states with $|n| \approx n_L$ are populated.

In Fig. 3(a) we show the absolute squares of the expansion coefficients $|\langle n|\chi_\alpha\rangle|^2$ of some quasienergy states for the off resonance ($\tau = 2$) situation. The quasienergy states chosen are the ones which have a large overlap with the unperturbed $n = 0$ state of the rotor, and are therefore the important states if the rotor was initially in its ground state. All the quasienergy states in Fig. 3(a) are exponentially decaying in the plateau region, with an average fall-off rate, which is approximated by the dashed line in Fig. 3(a) and which corresponds to a localization length of roughly five states.

In the resonant case however, i.e., for $\tau = \tfrac{2}{3}\pi$, Fig. 3(b) shows that the quasienergy states are all maximally extended

over the plateau region and they fall to zero rapidly at the edges of the plateau.

Figure 4(a) shows the energy expectation value

$$E_N = \langle\psi_N|\tfrac{1}{2}\tau\frac{\partial^2}{\partial\theta^2}|\psi_N\rangle \tag{2.19}$$

as a function of time for the two values of τ. As expected, the rotor does not gain much energy in the localized case and since the ground state of the rotor can be reproduced with a relatively small number of quasienergy states, the recurrence time is very short. In the extended case, the rotor gains a tremendous amount of energy initially where $E\sim N^2$. But after a while, the system "realizes" that the extension of the delocalized quasienergy states is in fact finite and the resonant energy gain is replaced by the usual oscillatory pattern with quantum recurrences, although on an appreciably higher level than in the localized case.

All our results obtained so far do not crucially depend on the exact shape of the microwave pulse and it was checked that the Gaussian pulse in Fig. 1 gives essentially the same results as a pulse consisting of a coherent superposition of several harmonics of a basic driving frequency.

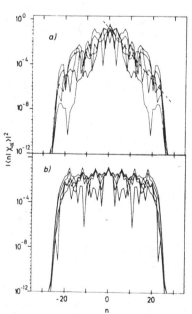

FIG. 3. Some quasienergy states characterized by a large overlap with the rotor ground state $|0\rangle$ for interaction strength $k = 2$ and (a) $\tau = 2$, (b) $\tau = 2\pi/3$.

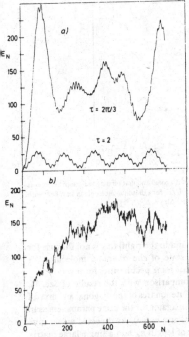

FIG. 4. Average energy of the rotor as a function of time for $k = 2$ and $\Delta(t) = \Delta^{(N-7)}(t)$. (a) Quantum mechanical calculations for the localized ($\tau = 2$) and extended ($\tau = 2\pi/3$) case. (b) Classical calculation ($\tau = 2$).

We conclude this chapter with a brief discussion of the classical description of the planar rotor.[6,17,18] If we measure energy in units of $\hbar/T$, the Hamiltonian is

$$H = \tfrac{1}{2}\tau n^2 + k\cos(\theta)\Delta^{(N)}(t), \qquad (2.20)$$

where the pulse was chosen to be of the form (2.13). The potential terms can be rewritten as

$$\cos(\theta)\Delta^{(N)}(t) = \sum_{m=-N}^{N} (-1)^m \cos[\theta - 2m\pi t] \quad (2.21)$$

and the pulse will be most effective if the phase $\psi_m(t) = \theta - 2m\pi t$ is stationary. From $\dot\psi_m(t) = 0$ and $\dot\theta = \partial H/\partial n = \tau n$, we obtain the resonance conditions

$$n_R^{(M)} = 2M\pi/\tau. \qquad (2.22)$$

Contrary to the kicked rotor, the sum in Eq. (2.21) is finite and the highest reachable resonance is $n_R^{(N)} = 2N\pi/\tau$. This estimate agrees remarkably well with the extension of the quantum mechanical plateau region (2.18). Quantum mechanically, all states with $|n| < n_L$ are strongly coupled, and if not for the Anderson localization, the excitation probability of the rotor initially concentrated in the ground state ($n = 0$) could spread over the rotor levels until it reaches the perturbatively localized region from whereon a further spread in n is impossible due to negligible coupling of higher n states. Classically we have a similar picture. Provided all resonances with $-N < M < N$ in Eq. (2.22) overlap, a trajectory started at $n = 0$ with some angle θ_0 could migrate from one resonance to another until it reaches the last resonance

($n = \pm n_R^{(N)}$) from whereon a further climbing in n is impossible due to the absence of higher resonances. Therefore, contrary to the kicked rotor, the dynamics of a pulsed rotor is confined in phase space and a classically stochastic strip can be compared to a quantum mechanical strong coupling region with exactly equal extension in angular momentum n. This classical confinement and quantal perturbation localization have a similar origin, namely, the limited number of frequencies of the driving force which are sufficiently close to the natural frequencies of the rotor.

In the vicinity of a resonance, the equations of motion for ψ_M and $\Delta n = (n - n_R^{(M)})$ are

$$\dot\psi_M(t) = \dot\theta - 2M\pi = \tau \cdot \Delta n \qquad (2.23)$$

and

$$\begin{aligned}(\dot{\Delta n})(t) &= -\frac{\partial H}{\partial\theta} = k\sum_{m=-N}^{+N}(-1)^m \sin[\psi_m(t)]\\ &\approx (-1)^M k\sin[\psi_M(t)]\end{aligned} \qquad (2.24)$$

which can be derived from the resonance Hamiltonian

$$H_R^{(M)} = \tfrac{1}{2}\tau(\Delta n)^2 + (-1)^M k\cos(\psi_M) \qquad (2.25)$$

which is that of a pendulum. The separatrix is the limit of bounded motion of the pendulum and corresponds to a total energy of $H_R^{(M)} = k$. In this case the maximal excursion in n is

$$\Delta n_{\max} = 2\sqrt{k/\tau}. \qquad (2.26)$$

The separation of the resonances (2.22) is $\Delta n_R = 2\pi/\tau$ and overlap is obtained for $2\,\Delta n_{\max} > \Delta n_R$ or

$$K = k\tau > \pi^2/4 \qquad (2.27)$$

which is just the overlap criterion for the kicked rotor and K is the dimensionless strength parameter of the standard mapping.[6]

For the parameters used in our numerical calculations ($\tau \approx 2$, $k = 2$), the overlap criterion (2.27) is fulfilled. Integrating the classical equations of motion for the Hamiltonian (2.20),

$$dn/dt = k\sin(\theta)\Delta(t),$$
$$\qquad (2.28)$$
$$d\theta/dt = \tau n,$$

we calculated the mean energy of the rotor as an ensemble average over N_c classical trajectories for initial conditions

$$n_j = 0; \quad \theta_j = \frac{2\pi}{N_c}(j - r_j); \quad j = 1,2,\dots,N_c, \qquad (2.29)$$

where r_j is a random variable equally distributed in the open interval (0,1).

Figure 4(b) shows the energy growth of the classical pulsed rotor for $\tau = 2$, $k = 2$, $N_c = 50$, $\Delta(t) = \Delta^{(N-7)}(t)$. After an initial diffusive energy gain, where the energy grows linearly in time, the rotor energy finally saturates around a mean energy which (due to the existence of some regular orbits of nonzero measure) is somewhat smaller than the value obtained by assuming equipartition of energy

$$\bar{E} = \frac{1}{2}\tau\,\frac{1}{n_R^{(N)}}\int_0^{n_R^{(N)}} n^2\, dn \approx 160. \qquad (2.30)$$

This behavior can be compared to the quantum mechanical calculations, where for $k = 2$ neither the case $\tau = 2$ nor the case $\tau = \frac{2}{3}\pi$ resembles the classical results. Instead, although in both cases states within the plateau region are strongly coupled, we observe localization of states and negligible energy growth if τ is not a rational multiple of π and delocalization over the plateau region and resonant energy growth if τ is rationally related to π.

From the above general discussion and the numerical examples, we can conclude that the pulsed rotor indeed displays the mechanisms of resonance absorption and Anderson localization in a finite basis. It differs essentially from the classical behavior which is chaotic and diffusive for frequencies, interaction strengths, and observation times which might be of interest for actual experiments with rotating molecules.

III. RF EXCITATION OF DIATOMIC MOLECULES

In the previous section we have shown that the basic conclusions of a study of the schematic planar rotor still hold if the singular $\delta(t)$ kicks are replaced by smooth pulses of finite width. The main question is now whether an actual experiment can be designed, which will eventually show the effects of Anderson localization in a finite basis. Therefore, in this section, we will study a diatomic molecule subject to a homogeneous time dependent electric field in terms of the Hamiltonian

$$H = H_0 + V(\theta,t) = (J^2/2I) + \mu E_0 T \cos(\theta)\Delta(t). \quad (3.1)$$

Here J is the angular momentum operator in three dimensions, I is the moment of inertia of the molecule about an axis perpendicular to the symmetry axis, and μ is the electric dipole moment. The external driving field with strength E_0 points in the z direction and varies in time according to the form factor $\Delta(t)$.

The equations of motion corresponding to the Hamiltonian (3.1) are again solved by expanding the full time dependent solution $\psi(\theta,t)$ into angular momentum eigenstates. Since the interaction matrix elements

$$\langle jm|\cos(\theta)|j'm'\rangle = \delta_{mm'}\left[C_j^{(m)}\delta_{f,j-1} + C_{j+1}^{(m)}\delta_{f,j+1}\right],$$

$$C_j^{(m)} = \sqrt{\frac{(j-m)(j+m)}{(2j-1)(2j+1)}} \quad (3.2)$$

are diagonal in m (the field is directed along the z axis), it is possible to label the time dependent expansion coefficients by the magnetic quantum number m and the following set of amplitude equations is obtained:

$$i\dot{b}_j^{(m)}(t) = \tfrac{1}{2}\tau j(j+1)b_j^{(m)}(t) + k\Delta(t)$$
$$\times \left[C_j^{(m)}b_{j-1}^{(m)}(t) + C_{j+1}^{(m)}b_{j+1}^{(m)}(t)\right], \quad (3.3)$$

where in analogy to the discussion in Sec. II the dimensionless variables $\tau = \hbar T/I$ and $k = \mu E_0 T/\hbar$ were introduced.

If initially the molecule is at rest ($j = m = 0$), the index m can be dropped and since $\langle jm|\cos\theta|j\pm 1,m\rangle$ quickly approaches $1/2$ as a function of j, we should expect that the dynamics of a diatomic molecule and a planar rotor essentially resemble each other [see Eq. (2.16)].

There are of course also important differences. Whereas in the case of the planar rotor every state has two nearest

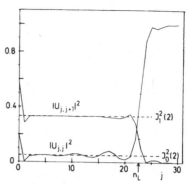

FIG. 5. Absolute squares of diagonal and first off diagonal matrix elements of the one cycle propagator $U(T)$ for a diatomic molecule in an rf field with parameters $\tau = 2$, $k = 2$, and $\Delta(t) = \Delta^{(N-7)}(t)$.

neighbors (to the left and to the right) this is not the case for the $j = 0$ rotational state of the diatomic molecule. We should therefore expect some peculiarities for low j.

To facilitate a comparison with the results of Sec. II, Fig. 5 shows the absolute squares of the diagonal and first off diagonal one cycle propagator for the same parameters as in Fig. 2 $\Delta(t) = \Delta^{(N-7)}(t)$ and $m = 0$. As expected we see now "one sided" plateaus of length n_L and a single large matrix element for $j = 0$. In Figs. 6(a) and 6(b) we display the absolute squares of the expansion coefficients $|\langle j|\chi_\alpha\rangle|^2$ of the quasienergy states which overlap appreciably with the

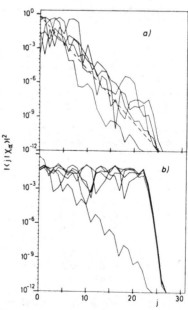

FIG. 6. Some quasienergy states characterized by a large overlap with the molecule ground state $|j = 0\rangle$ for interaction strength $k = 2$ and (a) $\tau = 2$, (b) $\tau = 2\pi/3$.

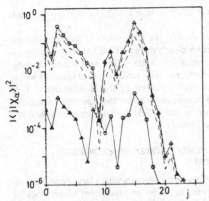

FIG. 7. Nearly degenerate quasienergy states. Full lines: quasienergy states localized at $|j=2\rangle$ and $|j=15\rangle$ for $\tau = 2 - 10^{-4}$. Dashed lines: the same states for $\tau = 2$.

$|j=0\rangle$ state, for off resonance $(\tau = 2)$ and resonance $(\tau = \frac{2}{3}\pi)$ conditions, respectively. We think that the large matrix element $U_{j=0, j=1}$ is responsible for exponentially localizing a single quasienergy state even in the resonance case, where all the quasienergy states are extended over the plateau region except for one.

The localization length in the nonresonant case is of the order of two sites. Compared to the width of the plateau, this decay length is an extremely small number and we can follow the states decaying over more than ten decades till they reach the edge of the plateau, where they will fall off even faster.

It is well known that the mechanism which brings about the exponential localization of quasienergy states does not always produce states which are concentrated at one site with amplitudes which decay exponentially with distance from the central site. Such an exception is illustrated in Fig. 7, which shows the quasienergy states with the third and fourth largest overlaps with the $|j=0\rangle$ state for $\tau = 2$. The

origin of this phenomenon is easily understood with the help of Fig. 8 where we show the dependence of the quasienergies of these states on τ in the interval $2 - 10^{-4} < \tau < 2 + 10^{-4}$. It is a typical example of avoided crossing of two levels and it so happens that at $\tau = 2$ the degeneracy is very near to its maximum. The corresponding eigenvectors change their identity in the usual way. Before the crossing occurs, the eigenvectors are localized on separate sites ($j = 2$ and $j = 15$) as shown in Fig. 7 for the value $\tau = 2 - 10^{-4}$. Only at the crossing the eigenvectors share with equal probabilities these two localized states and once τ reaches the value $2 + 10^{-4}$, the eigenvectors look again as the localized states with $\tau = 2 - 10^{-4}$. Avoided crossings of this type may occur, but with a low probability[19] and they may affect various observables as will be discussed below.

Figure 9 shows that for a frequency which is not rationally related to π, the molecule hardly gains energy. Nevertheless, we see a very slow parabolic growth of the mean energy due to the near degeneracy of two relatively important quasienergy states as discussed above. For $\tau = \frac{2}{3}\pi$, the mean energy of the molecule grows quadratically until it hits the edge of the plateau around cycle 30. From there on the energy oscillates around a mean value which—at least over 700 cycles—is well above the maximal mean energy in the nonresonant case.

The occupation probabilities $p_j(N) = |b_j(N)|^2$ of the jth rotor state after cycle N might serve as an experimental indication of localized or extended states. We found that in the nonresonant case, the occupation probabilities of states with an angular momentum between 10 and 22 was negligibly small ($< 10^{-3}$) for any N, $N = 1,2, \ldots ,700$ and decaying exponentially as a function of j. (The vicinity of the state $j = 15$ is an exception and its higher population is due to the accidental degeneracy of two quasienergy states as mentioned above.) In the resonant case, the excitation probabilities were of the order of 1% (see dashed lines in Fig. 10 for a snapshot after cycle 400).

Instead of considering the "snapshot" probability distribution $p_j(N)$, it is more instructive to study the distribution $\bar{p}_j$ which is the time average for the N different distributions $p_j(N)$. Starting from Eq. (2.3), it is easy to show that

$$\bar{p}_j = \sum_\alpha |\langle j|\chi_\alpha\rangle|^2 |\langle \chi_\alpha|\psi_0\rangle|^2 \tag{3.4}$$

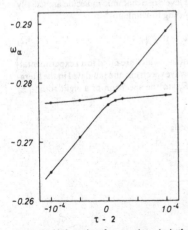

FIG. 8. Avoided crossing of two quasienergies in the vicinity of $\tau = 2$.

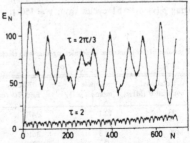

FIG. 9. Average energy of the molecule as a function of time in the localized and extended case.

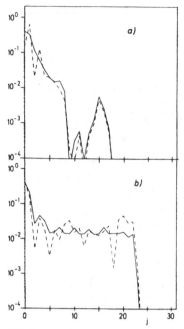

FIG. 10. Occupation probabilities of the angular momentum states of the molecule. Dashed line: snapshot after cycle 400. Full line: average of the occupation probabilities from cycle 150 to cycle 200 (a) off resonance, (b) on resonance.

provided that none of the quasifrequencies $(\omega_{\alpha'} - \omega_\alpha)/\hbar$ is a multiple of $2\pi/T$.

The $\bar{p}_j$ are very useful quantities, as they tell us how strongly a state $|j\rangle$ is coupled to the initial state $|\psi_0\rangle$ via the quasi energy states $|\chi_\alpha\rangle$. In the localized case, only a few quasienergy states will have an appreciable overlap with $|\psi_0\rangle$ and their overlaps with $|j\rangle$ will decay exponentially with j if we choose $|\psi_0\rangle$ to be the ground state of the molecule or at most a mixture of a finite number of low lying states. Therefore, we expect the $\bar{p}_j$ to decay exponentially with j. In the extended case, however, the quasienergy states $|\chi_\alpha\rangle$ have nearly equal overlap with all rotational states of the plateau region and we expect $\bar{p}_j$ to be reasonably large and constant throughout the plateau. This is verified in Figs. 10(a) and 10(b) where the full line shows the $\bar{p}_j$ obtained "experimentally" by averaging $p_j(\nu)$ over 51 cycles from $\nu = 150$ to

$\nu = 200$ in the localized $(\tau = 2)$ and extended $(\tau = \frac{3}{2}\pi)$ case, respectively.

Thus we see that the fingerprints of localized and extended states are clearly distinguished and related to experimentally measurable quantities. The only remaining question is now whether suitable molecules exist in nature whose dynamics are well described by the Hamiltonian (3.1) in some energy region, and whether the experimental requirements (frequency, power, field strength) are within the reach of modern technology.

IV. EXPERIMENTAL CONSIDERATIONS AND CONCLUSION

Only the heaviest molecules with a large permanent dipole moment are suited for an experiment which looks for the effects of Anderson localization in the dynamics of rf excited molecules. This is easily seen, if we investigate the potential term in the Hamiltonian (3.1) together with the form factor (2.13). The total electric field which acts on the molecule consists of a superposition of a constant field of magnitude E_0 and several harmonics of the basic driving frequency $\omega = 2\pi/T$ with a peak field strength of

$$E = 2E_0 = \frac{2\hbar k}{\mu T} = 4\left(\frac{\hbar^2}{2I}\right)\frac{I}{\mu}\left(\frac{k}{\tau}\right) = 4 \cdot \frac{B}{\mu} \cdot \frac{k}{\tau}, \quad (4.1)$$

where we introduced the rotational constant $B = \hbar^2/2I$ which is measured by microwave spectroscopy. Table I shows the field E for a few diatomic molecules togther with dipole moments and rotational constants. Among the molecules listed, CsI and PbTe seem to be the most promising candidates for actual experiments because they require the smallest field strengths.

Since E is inversely proportional to the moment of inertia and to the dipole moment of the molecule, one could as well try to perform the experiment with polyatomic linear molecules. The analysis of the experimental data, however, could be complicated because of additional modes of vibration ("bending modes"). Therefore, we feel that a clean experiment is only possible with a diatomic molecule such as CsI or PbTe.

The rotational levels of a diatomic molecule are usually parametrized according to Dunham[20]:

$$F_{\nu j} = \sum_{\alpha\beta} Y_{\alpha\beta}(\nu + \tfrac{1}{2})^\alpha j^\beta (j+1)^\beta, \quad (4.2)$$

where $Y_{\alpha\beta}$ are constants which are fitted to an experimentally measured microwave spectrum and tabulated in the literature. Comparing F_{0j} to the spectrum of a rigid rotor $[E_j$

TABLE I. Physical constants of some diatomic molecules together with electric field E [calculated according to Eq. (4.1)] for $\tau = k = 2$.

Molecule	Rotational constant B (MHz)	Dipole moment $\mu(D)$	Electric field E (V/cm)	References
CsI	708	12.1	470	2,21,31
KI	1825	11.1	1320	21
PbSe	1490	3.3	3610	32,33
PbTe	939	2.7	2780	23,33

$= B \cdot j(j + 1)]$ in the case of CsI[21,22] or PbTe[23] shows, that both molecules are excellent rotors and their free motion is well described by the H_0 part of the Hamiltonian (3.1).

Comparing the two molecules CsI and PbTe, we can say that CsI, due to its extremely large dipole moment (see Table I) demands much lower driving fields than PbTe. The constituent nuclei of PbTe on the other hand, do not possess any electric quadrupole moment and there are no additional complications because of hyperfine interactions in this case. The hyperfine structure of CsI rotational levels was measured for the transition $j = 2 \rightarrow 3$[24] and because of the symmetry in the electronic structure of the molecule (both atoms nearly reach a Xe configuration), the splitting of the unperturbed rotational levels was found to be very small. Moreover, the relative importance of level splitting becomes smaller for larger j values and the splitting is practically unresolvable for $j \approx 20$. Therefore, the hyperfine structure of the rotational levels of CsI does not necessarily rule out this molecule as a candidate for the proposed experiment.

Both molecules, CsI as well as PbTe, are relatively safe against vibrational excitation. The energy of the first vibrational level[25,26] is equal to the 71st and 82nd rotational levels, respectively, which is well above the $j \approx 22$ edge of the plateau.

We conclude that molecules appropriate for the Anderson localization experiment indeed exist in nature, which demand high, but not totally unrealistic driving fields.

A possible experimental setup is a molecular beam experiment, where supersonic jet techniques are used to cool the molecular rotation to approximately $T_{rot} \approx 1$ K[27]. The beam emerges with an average velocity $v_{av} \approx 10^5$ cm/s and the spread in the velocity amounts to less than 1% of v_{av}. The rf field is produced in a cavity with a typical dimension of 3.4 cm, so that the molecule experiences $1.5 \times 10^5 \pm 1.5 \times 10^3$ pulses during the time it traverses the cavity. The cavity is excited to its basic frequency (~ 4.5 GHz) and its first seven harmonics produce the pulse shape shown in Fig. 1. For CsI this corresponds to a repetition rate of 4.45 GHz, full width at half-maximum of 18 ps, and peak field $E_{max} = 3.5$ kV/cm. The change in the distribution of rotational states is monitored downstream and the experiment consists of comparing the resulting distributions for off resonance and on resonance conditions. The j distribution can be measured differentially by selective fluorescence methods, or integrally by measuring some moments of the j distribution such as the time averaged energy gain

$$\overline{\Delta E} = \lim_{N \to \infty} \frac{1}{N} \sum_{n=1}^{N} \frac{\mathrm{Tr}\, e^{-H_0/kT_{rot}}[(U^+)^n H_0 U^n - H_0]}{\mathrm{Tr}\, e^{-H_0/kT_{rot}}}$$

(4.3)

or the probability of occupying states with j higher than a given j_L:

$$\overline{\Delta p}_{j>j_L} = \lim_{N \to \infty} \frac{1}{N} \sum_{n=1}^{N} \frac{\mathrm{Tr}\, e^{-H_0/kT_{rot}}[(U^+)^n 1_{j_L} U^n - 1_{j_L}]}{\mathrm{Tr}\, e^{-H_0/kT_{rot}}},$$

where

$$1_{j_L} = \sum_{j>j_L} \sum_{|m|<j} |jm\rangle \langle jm|.$$

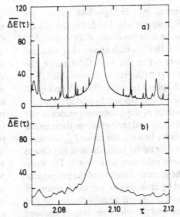

FIG. 11. Energy gain $\overline{\Delta E}(\tau)$ for CsI in the vicinity of the resonance at $\tau = 2\pi/3$. (a) $T_{rot} = 0$ K, (b) $T_{rot} = 1$ K.

The time average is necessary because due to the velocity spread the molecules will spend different time intervals in the cavity. This average is conveniently performed with the help of the time averaged probabilities $\overline{p}_j^{(m)}$ [see Eq. (3.4)].

For nonzero temperatures the above expressions assume a Boltzmann distribution for the initial distribution of rotational states. It is, however, well known[27] that due to the absence of collisions in the final stages of the cooling process, the rotational degrees of freedom do not reach thermal equilibrium and the Boltzmann distribution is an acceptable approximation only for very low angular momenta ($j \lesssim 7$ for $T_{rot} \lesssim 1$ K). But since our results do not dramatically depend on the details of the initial distribution and since for low rotational temperatures only low lying states are appreciably occupied, the Boltzmann distribution was chosen for simplicity.

In Figs. 11 and 12 we plot the moments $\overline{\Delta E}$ and $\overline{\Delta p}_{j>14}$ as a function of the period τ for CsI($B = 708$ MHz) in the

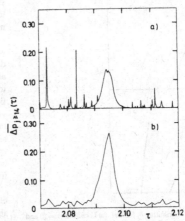

FIG. 12. The same as Fig. 11 for the high state occupation probability $\overline{\Delta p}_{j>14}(\tau)$.

vicinity of the resonance $\tau = 2\pi/3$ (~ 4.45 GHz) for $T_{rot} = 0$ and $T_{rot} = 1$ K, respectively. The full width at half-maximum of the resonance is of the order of $\Delta\tau = 4 \times 10^{-3}$ and $\Delta\omega/\omega = \Delta\tau/\tau = 2 \times 10^{-3}$, which shows that with the state of the art of generators in the 1–30 GHz range, the resonance could be scanned with very high resolution.

The "spikes" which appear sporadically both on and off resonance (for $T_{rot} = 0$ K) correspond to periods τ where accidental avoided crossings of the quasienergies occur (see Sec. III). At finite initial rotational temperatures these spikes are averaged out. This is so, because the finite T_{rot} amounts not only to a smearing of the initial j distribution, but also to the inclusion of magnetic substates with $m \neq 0$. The resonance conditions and the general structure of the propagator are not severely affected by changing m, but the positions of the avoided crossing spikes depend sensitively on the fine details of the interaction and therefore are largely reduced upon averaging over various initial values of j and m.

In Fig. 13 we compare the final probability distribution $\bar{p}_j$ for off and on resonance conditions, where the initial Boltzmann distribution is characterized in both instances by $T_{rot} = 1$ K. From a high resolution experiment of this type one could not only demonstrate the effect, but one could also obtain the effective localization length from the off resonance measurements. We also performed some tests at higher initial rotational temperature. While the effect of localized and delocalized situations on the energy gain or the high state occupation probability $\overline{\Delta p_{j>j_L}}$ could be demonstrated up to rotational temperatures $T_{rot} \approx 10$ K [$\overline{\Delta E}(\tau = 2\pi/3)/\overline{\Delta E}(\tau = 2) \approx 6$ for $T_{rot} = 10$ K], it is necessary to work at lower temperatures for an experimental estimate of the localization length. A temperature of ≈ 3 K should be a practical limit in this case. The prime factor which is assumed in the entire theoretical analysis is that *phase coherence*, which is responsible for localization or delocalization, is preserved during the entire interaction time. Ott *et al.*[28] analyzed the

quantum kicked rotor in which both the field strength k and the period τ were allowed to fluctuate at random around their mean values with root-mean-square amplitudes Δk and $\Delta\tau$, respectively. They find that small noise in the kick amplitudes k and the driving frequency τ might severely affect the long time behavior of the kicked rotor resulting in diffusion rather than localization or resonance excitation. Following their arguments and by studying the quantum kicked rotor for $k = 2$, $\tau \approx 2$, and several choices of Δk and $\Delta\tau$, we found out numerically that phase coherence can be guaranteed over $\approx 10^5$ cycles if the noise level is kept below $\Delta k < 10^{-3}$ and $\Delta\tau \leqslant 5 \times 10^{-5}$ which is easily achievable even with standard equipment. The sources of noise in the proposed experiment are electronic and thermal noise in the rf oscillator or cavity while effects due to Doppler shift can largely be suppressed if the beam traverses the cavity at right angles to the electric field.

Relaxation processes may alter the j distribution while the beam is transmitted from the source to the cavity or from the cavity to the detector. Neglecting collisions in the molecular beam, spontaneous or induced decay as well as induced absorption of photons are the prime source of this relaxation. The spontaneous lifetime[29] of a typical high j state ($j \approx 20$) is of the order of 10^4 s. If the experiment is conducted at room temperature, the black-body-radiation induces transitions which in the unfavorable case of low j transitions reduce the lifetime by a factor $2kT/h\nu \approx 10^4$. The resulting relaxation time which is of the order of 1 s is still exceedingly large compared to the travel time and therefore relaxation effects can be neglected.

In the analysis of an actual experiment one should also consider the effect of the inhomogeneous fields which the molecules encounter upon entering and leaving the cavity.[30] This effect corresponds to a redefinition of the initial and final angular momentum populations and can be incorporated in the formalism for any given stray field configuration.

Summarizing our results, we have shown that a close resemblance exists between the dynamics of the δ-kicked planar rotor and the behavior of diatomic molecules subjected to a train of smooth rf pulses. In both cases, Anderson localization takes place, where the rotational states of the molecule correspond to the lattice sites of the solid state model. In the case of the kicked planar rotor the ensemble of rotational states forms a 1 dim. infinite lattice on which conditions similar to those of the Anderson model are fulfilled and we are dealing with a global mapping of the planar rotor to the Anderson model. For real molecules, the infinite lattice becomes a semi-infinite one ($j > 0$) and complications beyond the Anderson model have to be expected for low angular momenta (e.g., some states are localized by the edge rather than by pseudorandomness). Moreover, if the singular δ kick is replaced by a smooth pulse, the rotational states can be classified into two categories—the "plateau states" for $j = 0,...,n_L$ and the perturbatively localized states for $j > n_L$. States with $j > n_L$ are localized because of a trivial effect—starting with probability 1 in such a state, the probability remains concentrated in this state simply because the transition probability to neighboring sites is very small which is due to a big mismatch of the natural rotational

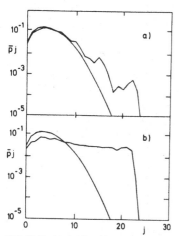

FIG. 13. Final probability distribution $\bar{p}_j$ for CsI. (a) Localized ($\tau = 2$) and (b) extended ($\tau = 2\pi/3$) case. The initial (Boltzmann) distribution, which corresponds to $T_{rot} = 1$ K, is also shown.

frequency of these states and the external driving frequencies. This effect is characteristic for the classical motion as well.

The plateau states on the other hand are strongly coupled and the probability for a transition from j to $j \pm 1$ after one cycle of the external field is as large as $\approx 1/3$ for the parameters studied in Secs. II and III. We could expect, therefore, that starting the molecule in its ground state, the occupation probabilities of the rotational states grow according to some sort of Markovian process and the molecule gains energy in a diffusive way. This behavior is indeed reflected in the classical calculation of Sec. II.

Quantum mechanically the picture is completely different. Since the plateau states are not coupled to the perturbatively localized states, they form an isolated subset of rotational states where Anderson localization takes place in the way it does in a finite disordered solid. Although this lattice consists of only a finite number of sites in the present case, the localization length of the quasienergy states turned out to be much smaller than the width of the plateau and the correspondence is indeed meaningful. For a generic driving frequency, the molecular dynamics is similar to the one of an electron in a one-dimensional dirty solid. Although the nearest neighbor hopping probabilities are large in the plateau region, we found that contrary to the classical results the molecule hardly gains energy. Driving frequencies for which τ is a rational multiple of π correspond to a periodic solid where the states are extended and therefore the molecule gains energy in a resonant way.

The correspondence with the Anderson model can be used in both directions. On the one hand, it allows for an interpretation of the dynamical behavior of rf molecular excitation with the help of a known model, on the other hand it can be used as a direct demonstration of the effects of Anderson localization in a system which might be more convenient experimentally than a solid state sample. The main advantage of molecular systems studied in this work is that they correspond to the solid state model at zero temperature, which is inaccessible to direct experiments. Changing the frequency of the driving force, the system changes from one corresponding to a disordered solid to one corresponding to a periodic solid—a change, which is much more difficult to perform for real solids.

ACKNOWLEDGMENTS

It is a pleasure to thank Professor M. Shapiro, Dr. B. J. Howard, and Professor J. Ford for helpful discussions and correspondence. One of us (R.B.) takes the opportunity to thank for a Minerva Grant which supported his stay at the Weizmann Institute where this work was initiated, and to thank for the hospitality of the Department of Theoretical Physics in Oxford, where the paper was completed.

This work was supported in part by the U.S.–Israel Binational Science Foundation, by the Bat-Sheva de Rothschild Fund for the Advancement of Science and Technology, and was carried out within the research program of the Einstein Center of Theoretical Physics at the Weizmann Institute of Science.

[1]T. Hogg and B. A. Huberman, Phys. Rev. Lett. **48**, 711 (1982); Phys. Rev. A **28**, 22 (1983).

[2]M. V. Berry, N. L. Balazs, M. Tabor, and A. Voros, Ann. Phys. **122**, 26 (1979).

[3]G. M. Zaslavsky, Phys. Rep. **80**, 158 (1981).

[4]S. Fishman, D. R. Grempel, and R. E. Prange, Phys. Rev. Lett. **49**, 509 (1982); D. R. Grempel, R. E. Prange, and S. Fishman, Phys. Rev. A **29**, 1639 (1984).

[5]F. M. Izrailev and D. L. Shepelyansky, Teor. Mat. Fiz. **43**, 417 (1980); Theor. Math. Phys. **43**, 553 (80); Sov. Phys.-Dokl. **24**, 996 (1979).

[6]B. V. Chirikov, Phys. Rep. **52**, 263 (1979).

[7]G. Casati, B. V. Chirikov, F. M. Izrailev, and J. Ford, in *Lecture Notes in Physics* (Springer, New York, 1979), Vol. 93, p. 334.

[8]B. V. Chirikov, F. M. Izrailev and D. L. Shepelyansky, Sov. Sci. Rev. Sec. C **2**, 209 (1981).

[9]Ya. B. Zeldovich, Sov. Phys JETP **24**, 1006 (1967).

[10]G. Casati and I. Guarneri, Commun. Math. Phys. **95**, 121 (1984).

[11]R. E. Prange, D. R. Grempel, and S. Fishman, Phys. Rev. B **29**, 6500 (1984); **28**, 7370 (1983).

[12]R. Blümel and U. Smilansky, Phys. Rev. Lett. **52**, 137 (1984); Phys. Rev. A **30**, 1040 (1984).

[13]G. Casati, B. V. Chirikov, and D. L. Shepelyansky, Phys. Rev. Lett. **53**, 2525 (1984).

[14]R. V. Jensen, Phys. Rev. Lett. **49**, 1365 (82); Phys. Rev. A **30**, 386 (1984).

[15]W. Anderson, Phys. Rev. **109**, 1492 (1958); Rev. Mod. Phys. **50**, 191 (1978); D. J. Thouless in *Proceedings of the Les Houches Summer School* (North–Holland, Amsterdam, 1979), Vol. 31; Phys. Rep. **13**, 94 (1974).

[16]F. Gesztesy and H. Mitter, J. Phys. A **14**, L79 (1981).

[17]G. M. Zaslavsky and B. V. Chirikov, Sov. Phys. Uspekhi **14**, 549 (1972).

[18]A. J. Lichtenberg and M. A. Lieberman, *Regular and Stochastic Motion* (Springer, New York, 1983).

[19]M. Feingold, S. Fishman, D. R. Grempel, and R. E. Prange, Phys. Rev. B **31**, 6852 (1985).

[20]J. L. Dunham, Phys. Rev. **41**, 721 (1932).

[21]A. Honig, M. Mandel, M. L. Stich, and C. H. Townes, Phys. Rev. **96**, 629 (1954).

[22]J. R. Rusk and W. Gordy, Phys. Rev. **127**, 817 (1962).

[23]E. Tiemann, J. Hoeft, and B. Schenk, Z. Naturforsch. Teil A **24**, 787 (1969).

[24]J. Hoeft, E. Tiemann, and T. Törring, Z. Naturforsch. Teil A **27**, 1017 (1972).

[25]R. F. Barrow and A. D. Caunt, Proc. R. Soc. London Ser. A **219**, 120 (1953).

[26]R. Lebargey, Proc. Phys. Soc. London **82**, 332 (1963).

[27]R. E. Smalley, D. H. Levy, and L. Wharton, J. Chem. Phys. **64**, 3266 (1976).

[28]E. Ott, T. M. Antonsen, and J. D. Hansen, Phys. Rev. Lett. **53**, 2187 (1984).

[29]C. H. Townes and A. L. Schawlow, *Microwave Spectroscopy* (McGraw–Hill, 1955), p. 336.

[30]J. G. Leopold and I. C. Percival, J. Phys. B **12**, 709 (1979).

[31]W. H. Rodebush, L. A. Murray and M. E. Bixler, J. Chem. Phys. **4**, 372 (1936).

[32]J. Hoeft and K. Manns, Z. Naturforsch. Teil. A **21**, 1884 (1966).

[33]J. Hoeft, F. J. Lovas, E. Tiemann, and T. Törring, Z. Naturforsch Teil A **25**, 539 (1970).

LOCALIZATION OF DIFFUSIVE EXCITATION IN MULTI-LEVEL SYSTEMS

D.L. SHEPELYANSKY

Institute of Nuclear Physics, 630090 Novosibirsk, USSR

Received 19 December 1986
Revised manuscript received 23 January 1987

The excitation of multi-level systems by a periodic field is considered in the regime of quasiclassical diffusion which takes place in the region of classical dynamical chaos. It is shown that quantum effects lead to a limitation of diffusion and to the localization of quasienergy eigenfunctions (QEE). The expression for the QEE localization length in terms of the classical diffusion rate ($l = D/2$) is obtained and the analogy between this phenomenon and the Anderson localization in solid-state problems is analyzed. The localization length for photon transitions in the energy spectrum is found.

1. Introduction

In recent years a number of experiments on the ionization of Rydberg (highly excited) atoms and dissociation of molecules by a strong monochromatic field have been carried out [1–5]. A characteristic peculiarity of such processes is the large number of absorbed photons $N_\phi \sim 100$ and the excitation of many unperturbed levels. Due to this the dynamics of excitation may be described in the first approximation by the classical equations of motion. Such an approach was used for molecules in ref. 6 and for Rydberg atoms in ref. 7. The process of excitation obeys the diffusion law. The appearance of diffusion in the absence of any random forces is connected with the chaotic dynamics of the corresponding classical system. The nature and the properties of such chaotic motion in classical mechanics is now well understood [8–10]. At the same time an investigation of simple models has shown that the dynamics of classically chaotic quantum systems has a number of peculiarities (see, e.g., refs. 9, 11 and 12). The most interesting one being the quantum diffusion limitation [11–16]. This limitation is due to the localization of quasienergy eigenfunctions (QEE),

(QEE decay exponentially with the serial number of unperturbed levels.)

We carried out an investigation of the QEE localization mainly on the examples provided by two models. The first one is the quantum rotator model which has been investigated in refs. 11–13, 18, 20, 27–29, 32–34. The second one is the Akulin–Dykhne model [17] which describes a general picture of the excitation of a system with irregular spectrum by a monochromatic field. It has been introduced as a model for the molecular excitation in a laser field. With the help of a simple estimate based on the uncertainty relation between frequency and time, which was first used in ref. 12, we obtain a simple expression for the QEE localization length in the rotator model (eqs. (7) and (13)). The generalization of this result and the estimate (7) is used to find the value of the length in the Akulin–Dykhne model (section 7).

We have checked and confirmed the theoretical results for the quantum rotator by a special numerical method. The advantage of this method consists in the fact that it allows to evaluate the value of the QEE localization length without the computation of the exact QEE. Indeed, we show that this evaluation can be reduced to a computation of the Lyapunov exponents (LEX) in some auxiliary classical Hamiltonian system. Due to the linearity of the Schrödinger equation, the equations of motion for this system are linear, with the coefficients explicitly depending on the serial number of the unperturbed level, which plays the role of a discrete "time". For one-dimensional systems the number of equations is determined by the number of unperturbed levels effectively coupled after one period of the field. This approach for the investigation of QEE was introduced in ref. 20.

There are two simple limiting cases for the obtained system of linear equations. The first one is the case in which the coefficients of the system periodically depend on discrete "time" (the serial number of the level). In that case the quasi-energy spectrum is continuous and the QEE are delocalized like the Bloch eigenfunctions in a perfect crystal. The second case corresponds to a random dependence of coefficients on "time". This situation is analogous to the quantum motion in a random potential. The analogy between these two physical problems has been established in ref. 18. In a one-dimensional random potential all eigenstates are localized, which is the well-known Anderson localization [19, 30]. This corresponds to the localization of all QEE [18]. In this approach the serial number of the unperturbed level plays the role of a spatial coordinate.

However, in spite of the usefulness of the analogy between the Anderson localization and the localization of dynamical chaos we need to stress two important differences between them. Firstly, the absence of randomness in the dynamical system, and secondly, the QEE localization occurs in a quite different class of systems than those considered in solid-state problems. In this paper we illustrate these differences mainly for the rotator model and the Akylin–Dykhne model. Another example is the diffusive photoeffect in a hydrogen atom (see refs. 14–16).

The contents of the paper is as follows. In section 2 we describe the rotator model and the LEX method, find the corresponding solid-state Hamiltonian, and give the estimate (7) for the

localization length l. In section 3 we obtain exact expressions for l in the dynamical Lloyd model and the quantum standard map, and compare them with the numerical results. The localization length for the steady-state distribution $l_s (\neq l)$ is obtained in section 4. The main results of sections 2–4 were briefly reported in ref. 20. In section 5 we discuss a simple variant of the Akulin–Dykhne model which can be reduced to the Anderson model. The conditions under which the excitation of systems with many degrees of freedom may be considered within the framework of one-dimensional localization are obtained in section 6. In section 7 we find the QEE length for the excitation of typical multi-level systems by a monochromatic field.

2. The quantum rotator model

In order to investigate the motion of quantum systems which are chaotic in the classical limit we chose the generalized model of a quantum rotator with the Hamiltonian

$$\hat{H} = H_0(\hat{n}) + V(\theta)\,\delta_T(t), \tag{1}$$

where $\hat{n} = -i\partial/\partial\theta$, $\delta_T(t)$ is the periodic delta-function with T the dimensionless period, θ the phase variable, $V(\theta)$ the external perturbation, $\hbar = 1$ [11–13, 18]. Here $H_0(n)$ determines the energies of unperturbed levels n. The dynamics of the corresponding classical system is determined by the equations of motion with the Hamiltonian (1), where n, θ are canonically conjugated action-phase variables. After integration over a period T, we obtain a map

$$\bar{n} = n - \frac{\partial V}{\partial\theta}, \quad \bar{\theta} = \theta + T\frac{\partial H_0(\bar{n})}{\partial\bar{n}}, \tag{2}$$

where $\bar{n}$ and $\bar{\theta}$ are the values of the variables n, θ after a period T. For strong perturbation the resonances overlap [8] and then the action grows beyond any limit according to the diffusion law: $\langle(\Delta n)^2\rangle = D\tau$, where τ is the number of periods.

Generally, the diffusion rate D is a complicated function of the system parameters. However, in the region of a strong chaos the phases $\theta(\tau)$ are random and independent. This allows one to use the quasilinear approximation in order to calculated D [10]. In this case the diffusion rate is equal to

$$D_{ql} = \frac{1}{2\pi} \int_0^{2\pi} \left(\frac{\partial V}{\partial \theta} \right)^2 d\theta. \tag{3}$$

The quasiclassical condition is satisfied if the number of levels excited in one period is large: $D \gg 1$, and if the dimensionless parameter $T \ll 1$ ($T \propto \hbar$) [9, 11–13].

The main part of our investigations was carried out for the quantum standard map

$$\hat{H} = \frac{\hat{n}^2}{2} + k \cos \theta \, \delta_T(t). \tag{4}$$

The motion of the corresponding classical system is described by the standard map [8–10]:

$$\bar{p} = p + K \sin \theta, \quad \bar{\theta} = \theta + \bar{p}, \tag{5}$$

where $p = Tn$, $K = kT$. For $K \le K_{cr} = 0.9716\ldots$ [24] the change of n is finite $|\Delta n| \lesssim \sqrt{k/T}$, but for $K > K_{cr} (\Delta n)^2$ grows according to the diffusion law with the rate $D = D_0(K)/T^2$, where $D_0(K)$ is the diffusion rate in p in the standard map (5). Within the chaotic component the dependence of the diffusion rate on K may be approximately described by the following expression [25, 13]:

$$D_0 = \begin{cases} \dfrac{K^2}{2} \left(1 + 2J_2(K) + 2J_2^2(K) \right), & K \ge 4.5, \\[2mm] 0.30(\Delta K)^3, & K < 4.5, \end{cases} \tag{6}$$

where $J_2(K)$ is a Bessel function, $\Delta K = K - K_{cr}$. For $K \ge 4.5$ the dependence of D_0 on K has the form of oscillations which decay as K grows. The limit value $D_0 = K^2/2$ corresponds to the quasilinear approximation (3), when the phases $\theta(\tau)$ in

(5) are random and independent. For $K \to K_{cr}$ we use in (6) the empirical formula which was obtained from numerical experiments in ref. 13. The value of the exponent $\eta \approx 3$ in the power law $D_0 \propto (\Delta K)^\eta$ is close to that given in ref. 26.

Numerical experiments [11–13, 27, 28] with the quantum standard map have shown that in the course of time $\langle n^2 \rangle$ stops growing. This means that the external field effectively excites only a finite number of unperturbed levels ($\Delta n \sim l$). An analogous result was obtained in ref. 29 for the dynamical Lloyd model with $V(\theta) = 2 \arctan(E - 2k \cos \theta)$, and $H_0 = \hat{n}^2/2$ which was introduced in ref. 18. It is natural to interpret this effect as the result of the QEE localization which is analogous to the Anderson localization in a one-dimensional lattice [18, 13, 20]. For the number of excited levels and the localization length of QEE, the following theoretical estimate was obtained in refs. 12 and 13:

$$l = \alpha D \sim \Delta n \sim \tau_D, \tag{7}$$

where α is an undetermined numerical constant. The derivation of (7) may be done in the following way. Let one unperturbed state contain l QEE with quasienergies ε_i. Since all these ε_i are distributed within the interval $[0, 2\pi]$, its average spacing is equal to $\Delta \varepsilon \sim 1/l$ (here we consider the case when the unperturbed levels are uniformly distributed in this interval). If initially we excite one unperturbed state then the diffusion will continue during the finite time τ_D until the discreteness of the QEE spectrum becomes effective. According to the uncertainty relation, $\tau_D \sim 1/\Delta \varepsilon \sim l$. After this time the number of diffusively excited levels will be equal to $\Delta n \sim (D\tau_D)^{1/2} \sim l$. From this relation we obtain eq. (7). The condition for its applicability is $D \gg 1$. In the case of a d-dimensional unperturbed system the number of excited levels $\Delta n \sim (D_1 \cdots D_d)^{1/2} \tau^{d/2}$ and hence in the absence of a degeneration of levels the condition for the quantum limitation of chaos takes the form

$$\tau_D \gtrsim (D_1 \cdots D_d)^{1/2} \tau_D^{d/2}. \tag{8}$$

For $d = 1$ we always have localization. For $d \geq 3$ the delocalization takes place if $D_1 \cdots D_d \geq 1$. This condition corresponds to the Anderson criterion [30] (a small random potential provides fast diffusion and large conductance). For $d = 2$ the estimate (8) gives the delocalization for $D_1 D_2 \gg 1$. However, a more rigorous consideration shows that in this case the localization always takes place but for $D_1 D_2 \gg 1$ its length is exponentially large: $\ln l \sim D_1 D_2$ (see, for example, ref. 31 and references therein). For any d it follows from (8) that the localization always takes place if each $D_i \ll 1$. This case corresponds to the quantum stability border [6].

Let us consider now the equation for the eigenfunction with quasienergy ε [18]:

$$u_n^- = e^{i(\varepsilon - TH_0(n))}u_n^+, \quad u^+(\theta) = e^{-iV(\theta)}u^-(\theta).$$

$$(9)$$

Here $u^{\mp}(\theta)$ are the values of the function u before and after a kick $\delta(t)$ and $u_n^{\pm}$ are the Fourier coefficients of $u^{\pm}(\theta)$. After simple transformations eq. (9) may be rewritten in the form

$$\hat{H}_{ss}u = \left\{ \cos\frac{\hat{V}}{2}\tan\left(\frac{\varepsilon}{2} - \frac{T}{2}\hat{H}_0\right)\cos\frac{\hat{V}}{2} - \frac{1}{2}\sin\hat{V} \right\}u$$

$$= 0, \qquad\qquad (10)$$

where $u = e^{\pm iV/2}u^{\pm}$. After the Fourier transformation $e^{-iV/2} = \sum_r W_r e^{i(r\theta + \varphi_r)}$ it is easy to see that the Hamiltonian H_{ss} corresponds to a one-dimensional lattice with interacting sites and energy $E = -\sum_r W_r^- W_{-r}^- \sin\varphi_r \cos\varphi_{-r}$. In such an approach the quasienergy ε determines the potential of interaction and the eigenvalue of energy E plays the role of a parameter. Moreover, the number of unperturbed levels n in the model (1) corresponds to a discrete spatial coordinate in the lattice.

Since all eigenfunctions in a one-dimensional random lattice are localized [19] it is natural to expect an exponential localization of QEE in (1). If $\cos(V/2) \neq 0$ we may introduce $\bar{u} = \cos(V/2)u$ and divide eq. (10) by $\cos(V/2)$. After that we

reduce the problem to the case with Hamiltonian $H_{ss} = \tan[\varepsilon/2 - (T/2)\hat{H}_0] - \tan(V/2)$. This procedure was implicitly used in ref. 18. However, it is necessary to stress that this approach leads to the appearance of a nonphysical singularity which does not allow for an analysis of the wide class of potentials with $|V(\theta)| \geq \pi$.

The form of eq. (10) is convenient for exploiting the analogy with the solid-state problems. However, for numerical experiments it is more convenient to rewrite eq. (9) and (10) as follows. We introduce $\bar{u} = e^{\pm iV/2}u^{\pm}/g$, where $g(\theta)$ is an arbitrary real function (we will consider the case when g and V are even functions of θ). Then we obtain from (9) the equation

$$\sum_r \bar{u}_{n+r}W_r \sin(\chi_n + \varphi_r) = 0, \qquad (11)$$

where

$$e^{-iV/2}g = \sum_r W_r e^{i(r\theta + \varphi_r)}, \quad \chi_n = (\varepsilon - TH_0(n))/2.$$

Let us assume that in (11) only W_r with $|r| \leq N$ differ from zero. Then a column of $2N$ known values of $\bar{u}_n$ determines the value of QEE for an arbitrary n. The recursive computation of $\bar{u}_n$ from (11) may be considered as the motion of a dynamical system in the discrete time n. The dynamics of the $2N$-component vector is determined by a transfer matrix M_n, the expression for which may be easily obtained from eq. (11). It may be shown that the product of matrices M_n can be transformed by a rotation into a simplistic matrix. Therefore, the dynamics in n is Hamiltonian and there are N positive and N negative Lyapunov exponents (LEX) $\gamma_i^+ = -\gamma_i^- \geq 0$ (see, for example, ref. 10). Asymptotically, the localization length of QEE is determined by the minimal positive LEX $\gamma_1 = 1/l$ [20–22]. The fact that γ_1 is different from zero leads to an exponential localization of QEE and to a discrete spectrum of quasienergies. The numerical method of computation for all LEX is described, for example, in ref. 10. This method allows not only to find all γ_i, but also to determine the dependence on n of the norm $\|u_n^{(i)}\| =$

$(\sum_{m=1}^{2N}|u_{n+m}^{(i)}|^2)^{1/2}$ of an eigenvector corresponding to a given exponent. In solid-state physics the LEX method is known as the transfer matrix method. This method is extensively used for the investigation of localization in two- and three-dimensional solid-state systems [21, 22]. For dynamical models it was first applied in ref. 20, and later in ref. 23.

3. Localization of QEE

Let us consider the rotator model with potential $V = 2\arctan(E - 2k\cos\theta)$, which was introduced in ref. 18. Taking $g = 1/\cos V/2$, we obtain $W_0 e^{i\varphi_0} = 1 - iE$, $W_{\pm 1}e^{i\varphi_{\pm 1}} = ik$ and $W_r = 0$ if $|r| > 1$. Eq. (11) now takes the form

$$E_n u_n + k u_{n+1} + k u_{n-1} = E u_n, \qquad (12)$$

where $E_n = \tan\chi_n$ and we dropped the bar on u_n. In the case when the phases χ_n are random and independent in the interval $[0, \pi]$ eq. (12) corresponds to the well-known Lloyd model [18, 19]. For this model the exact expression for the localization length is known (see, e.g., refs. 18–20), and for $l \gg 1$ it has the form $l = \sqrt{4k^2 - E^2}$. This may be used for the determination of the numerical factor in eq. (7) [20]. Due to the randomness of χ_n, the phases $\theta(\tau) \propto \partial\chi_n/\partial n$ in (2) are also random and independent, and $D = D_{ql}$. The calculation of the integral in (3) for $D_{ql} \gg 1$ and the comparison with l gives $\alpha = \frac{1}{2}$.

The same method of calculation for l can also be applied in a more complicated case when the interaction connects $2N$ sites and the potential has the form $V = 2\arctan(E - 2k\sum_{m=1}^{N}\cos m\theta)$. According to (7), $l = D_{ql}/2 \approx 2kN^2$, where the last equality takes place when $D_{ql} \gg N$. All values of D_{ql} were obtained by numerical calculation of the integral in (3). These values are in good agreement with the experimental values obtained by the LEX method (see fig. 3 in ref. 20).

Now we consider the dynamical Lloyd model with $H_0 = n^2/2$ ($N = 1$). Then the phases χ_n are not random. Moreover, for the rational values of $T/4\pi = p/q$, E_n in (12) becomes a periodic function of n, which corresponds to the case of an ideal crystal with delocalized eigenfunctions. In such a case of quantum resonance [32] the quasi-energy spectrum is continuous and consists of q zones. It was shown in ref. 33 that the continuous component exists also for the special irrational values of $T/4\pi$ which are very close to the rational numbers. However, we conjecture that for any irrational numbers $T/4\pi$ (so that $C_2 q^{-1-\varepsilon} < |qT/4\pi - p| < C_1 q^{-1}$ for any $\varepsilon > 0$), the measure of which on the interval $[0,1]$ is equal to one, an exponential localization will always take place with the same exponent as in the Lloyd model (12) with random phases χ_n.

This conjecture was confirmed by the numerical experiments based on the LEX method. They really showed that the localization length is the same as in the case of random phases χ_n. The parameters of the model were changed in the intervals $0.01 \leq k \leq 1000, 10^{-5} \leq T \leq 1, 0 \leq E \leq 2$. The value of γ was determined in the interval $1 \leq n \leq n_m = 10^5$ and it did not depend on ε and T. The relative accuracy of γ was equal to $(\Delta\gamma/\gamma)^2 = 2/\gamma n_m$ (see below).

The recursively obtained values of u_n may be considered to give a QEE in some interval of $n \to \infty$. For $T \ll 1$ the dependence of u_n on n is in steps of size $\Delta n \sim 1/Tn$. With the growth of n these steps decrease in size and for $Tn \sim 1$ they disappear. An estimate for Δn may be obtained from the condition $\Delta\chi_n = \frac{1}{2}Tn\Delta n \sim 1$. It is important to note that in the model under discussion the quasiclassical limit $(k \gg 1)$ always corresponds to the region of strong chaos where $l = D_{ql}/2$.

The LEX method has also been applied to the calculation of l in the quantum standard map (4). Taking $g = 1$ we obtain in (11) $W_r = J_r(k/2)$, $\varphi_r = -\pi r/2$, $\chi_n = (\varepsilon - Tn^2/2)/2$, where $J_r(k/2)$ is a Bessel function. Due to the fast decay of W_r for $|r| > k/2$ it is possible to use a finite number of sites $N \sim k/2$. The further increase of N is related only to exponents $\gamma \sim 1$. Another check consisted in verifying the relation $\gamma_i^+ + \gamma_i^- = 0$. For a large n interval of the order $\sim 10^5$ levels the sum of positive and negative exponents was small: $(\gamma_1^+ + \gamma_1^-)/\gamma_1^+ \sim 10^{-2}$.

The results of numerical experiments [20] show that in the quasiclassical region $T \leq 1$; $5 \leq k \leq 75$; $1.5 \leq K \leq 29$ ($T/4\pi$ is a typical irrational number) the localization length is satisfactorily described by the formula

$$l = D_0(K)/2T^2. \tag{13}$$

This relation holds not only in the region of strong stochasticity, $K \gtrsim 4.5$, where the measure of the stable component is negligibly small [8], but also in the region with $\Delta K = K - K_{cr} \ll 1$, where the diffusion rate is determined by a complicated critical structure [26] and the stable component covers approximately 50% of the whole phase space [8]. The numerical results obtained demonstrate a satisfactory agreement with the formula (13) in a range of four orders of magnitude for the diffusion rate (fig 1). However, it is important to note that the relation (13) holds only in the case when the localization length is larger than the number of interacting sites $2N \approx k$. In the opposite case $D_0/2T^2 \ll k$, the diffusion is too slow and does not lead to the increase of the localization length $l \sim k$. Thus the condition for the applicability of (13) is

$$k > k_{cr} = \frac{\kappa K^2}{D_0}$$

$$\approx \frac{3\kappa K^2}{(\Delta K)^3} \gg 1, \quad K > K_{cr}, \tag{14}$$

where $\kappa \approx 1.3$ is determined from numerical experiments. The inequality (14), first obtained in ref. 13, gives the condition for the so-called homogeneous (exponential) localization. In the opposite case, $k \ll k_{cr}$, the localization length $l \sim N \sim k$ is comparable with the period of the resonance structure $\Delta n = 2\pi/T$ and, like the diffusion rate, it is inhomogeneous too [13]. Examples of homogeneous and inhomogeneous localizations are

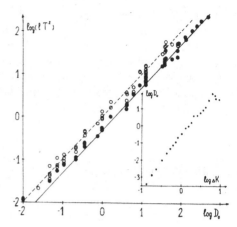

Fig. 1. The dependence of the localization length on the diffusion rate D_0 in the classical standard map (5). The circles represent the numerical data of ref. 13 for the values of l_s obtained from steady-state distributions (15). The dashed line corresponds to the average value $\langle \alpha_s \rangle = 1.04$. The points show the localization length obtained from QEE by the LEX method. The straight line shows the theoretical localization $l = D/2$. In the inset the numerical data from ref. 13 are shown, giving the dependence of D_0 on $\Delta K = K - K_{cr}$, $K_{cr} = 0.9716 \ldots$. Here and in fig. 5 the logarithm is decimal.

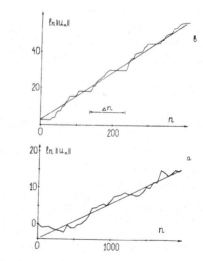

Fig. 2. The dependence on n of the averaged quantity $\|u_n^{(1)}\| = (\sum_{m=1}^{2N} |u_{n+m}^{(1)}|^2)^{1/2}$ in model (4): (a) the homogeneous localization at $k = 20$, $K = 5$; (b) the inhomogeneous localization at $k = 20$, $K = 1.3$. Periodic oscillations with $\Delta n = 2\pi/T$ are related to the resonance structure. The straight lines show experimental values of l obtained in the interval $n_m \approx 5 \times 10^4$.

shown in fig. 2. In the latter case, the size of the clearly observed steps is equal to $2\pi/T$. In contrast to the dynamical Lloyd model these steps persist for the arbitrarily large values of n as well. For example, the dependence of $\|u_n^{(1)}\|$ on n in fig. 2 is given in the region $n \approx 3 \times 10^4$. The reason for this difference is apparently related to the large number of interacting sites in the model (4). In the region of stability ($k \gg 1$, $K \ll 1$) the localization is almost homogeneous due to a small amount of nonlinear resonances. Numerically, the localization length in this case is comparable with the number of interacting sites: $l \approx k/4$.

The dependence of the diffusion rate D_0 on the classical parameter of chaos K (see (6)) leads to a significant change of the localization length with the parameter T even for a fixed value of k. An example of this effect is shown in fig. 3. According to (13), the localization length repeats all oscillations of D_0. These oscillations take place for $K \geq 4.5$ and are given by the formula (6) in which we may use the asymptotics of the Bessel function. The difference of D from the quasilinear value is related to the influence of the correlations: $D = D_{ql}[1 + 4\sum_{\tau=1}^{\infty} C(\tau)]$. Here the correlation function is $C(\tau) = \langle \sin\theta(\tau) \cdot \sin\theta(0) \rangle$ and the averaging is performed over the homogeneous steady-state distribution [25]. For $K \geq 5$, $C(\tau)$ decays very fast with τ and the main contribution to D is given by $C(2) = -\frac{1}{2}J_2(K)(C(1) = 0)$. The remaining correlations give only small corrections [25]:

$$C(4) = \tfrac{1}{2}J_2^2(K) \sim K^{-1}$$
$$C(3) = \tfrac{1}{2}\big[J_3^2(K) - J_1^2(K)\big] \sim K^{-2}.$$

The same method of computation for the diffusion rate may be also used in the quantum model with $T \gtrsim 1$, $k \gg 1$. In this case, a few first correlations $C_q(\tau) = \frac{1}{2}\langle 0|\sin\hat\theta(\tau)\sin\hat\theta(0) + \sin\hat\theta(0)\sin\hat\theta(\tau)|0\rangle$ give the quantum diffusion rate D_q using which we can determine the localization length and, in such a way, can take into account the influence of the residual correlations [28].

In order to calculate $C_q(\tau)$ it is convenient to write the operator $\sin\hat\theta(\tau)$ in a normal form with respect to the initial operators $\hat n(0)$ and $\hat\theta(0)$, as was done in ref. 34 (for example, all $\hat n(0)$ will be on the right and $\hat\theta(0)$ on the left). Using the expression for $\sin\hat\theta(\tau)$ obtained in ref. 34 and assuming that the distribution in the phase space is approximately homogeneous and, therefore,

$$\langle 0| e^{im_1\hat\theta(0)} e^{im_2\hat n(0)}|0\rangle = \delta_{m_1,0}\delta_{m_2,0},$$

we obtain that the first three correlations are the same as in the classical case, upon the substitution $K \to K_q = 2k\sin(T/2)$. The expression for $C_q(4)$ contains an additional term which is usually as small as $C(4)$:

$$C_q(4) = \tfrac{1}{2}\bigg\{ J_2^2(K_q) + \sum_{m \neq -2}\bigg[J_m(K_q)J_{m+4}(-K_q)$$
$$\times J_{2m+4}\Big(2k\sin\Big(\frac{T}{2}(m+2)\Big)\Big)$$
$$- J_m^2(K_q)J_{2m+2}\Big(2k\sin\Big(\frac{T}{2}(m+2)\Big)\Big)\bigg]\bigg\}.$$

Therefore, for $T \gtrsim 1$ the quantum diffusion rate is approximately given by the expression (6) with $K \Rightarrow K_q = 2k\sin T/2$. The numerical data for the localization length at $k = 30$, $0 < T < 2\pi$ demonstrate satisfactory agreement with the theoretical value $l = D_q/2$ (see fig. 3). In the region $T > 1$ the average ratio $\langle l/D_q \rangle \approx 0.6$ is slightly larger than $1/2$; this is apparently due to the not very small value of the ratio $k/l \approx 0.1$. From (6) it follows that the period of oscillations in $D(K)$ is equal to $\delta K_q \approx 2\pi$ (see fig. 3). Therefore, at fixed T the period in the dependence of l/D_{ql} on k is equal to $\delta k \approx \pi/\sin T/2$. Such oscillations for $T = 1, 2$ have been observed in ref. 23, with the period in satisfactory agreement with the above value.

Up to now we have discussed the properties of the minimal LEX which determine the asymptotic behavior of QEE. However, there is a spectrum of LEX γ_i^+ in the system (4). A typical spectrum of this type is shown in fig. 4 for both the stable and

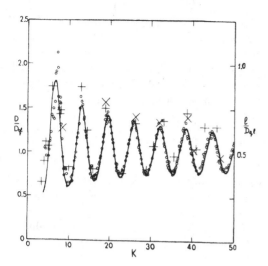

Fig. 3. The dependence of the localization length in the quantum standard map on the quantum parameter of chaos $K_q = 2k\sin T/2 \to K$. The circles and the curve are, respectively, the numerical data and the theory for the classical diffusion rate $D(K)$ from ref. 25 (+ for $0 < T < \pi$, × for $\pi < T < 2\pi$).

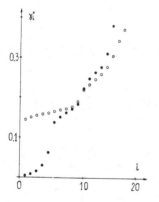

Fig. 4. The spectrum of the Lyapunov exponents in model (4). Quantum chaos: $k = 30$, $K = 5$ (points); stable motion: $k = 30$, $K = 0.003$ (circles).

chaotic regimes. Each γ_i is related to an eigenvector which is cross-orthogonal to all the other vectors. We conjecture that QEE is a linear superposition of all these vectors and that the probability of each vector is γ_i/N (γ_i appears in the normalization condition).

4. Steady-state distribution

If we initially excite one level then after some time the localization will lead to the steady-state distribution of the probabilities over unperturbed levels

$$\bar{f}(n) = \lim_{\tau \to \infty} \frac{1}{\tau} \int_0^\tau d\tau\, f(n, \tau).$$

The expression for this distribution function may be obtained by using the QEE with quasienergies ε_m. If initially only the $n = 0$ level was excited, then

$$\bar{f}(n) = \sum_m |\varphi_m(0)\varphi_m(n)|^2. \tag{15}$$

From this expression it is easy to see that $\bar{f}(n)$ is analogous to the density–density correlation function in solid-state problems [19]. The numerical experiments [13, 20] have shown that $\bar{f}(n) \propto \exp(-2|n|/l_s)$ and $l_s \approx D = 2l$ (see fig. 1). The cause of the difference between l and l_s is apparently related to fluctuations of the QEE $\varphi_m(n)$, as it was in solid-state problems. The QEE may be written in the form $\varphi_m(n) \propto \exp(-\gamma|n - m| + \xi_{mn})$, where ξ_{mn} describes fluctuations with zero average $\langle \xi_{mn} \rangle = 0$. However, the second moment may grow linearly with n: $\langle(\Delta\xi_{mn})^2\rangle = D_\xi \Delta n$, that leads to different values of $(1/n)\ln\langle|\varphi_m(n)|\rangle$ and $\langle(1/n)\ln|\varphi_m(n)|\rangle$. For Gaussian fluctuations we obtain

$$\langle|\varphi_0(n)|\rangle \sim \int_{\gamma n}^\infty \exp(-\gamma n + \xi)$$

$$\times \exp\left(-\frac{\xi^2}{2D_\xi n}\right) d\xi \sim e^{-\gamma_s n}.$$

Then $\gamma_s = 1/l_s$ and

$$\gamma_s = \gamma - \frac{D_\xi}{2}, \quad \gamma \geq D_\xi;$$

$$\gamma_s = \frac{\gamma^2}{2D_\xi}, \quad \gamma \leq D_\xi. \tag{16}$$

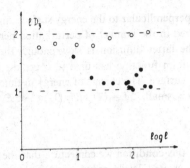

Fig. 5. The product of the diffusion rate D_ξ in QEE and the localization length for different l. The circles are for the dynamical Lloyd model ($T = 1$), and the points for the quantum standard map.

The value of D_ξ may be determined by the LEX method from $\ln\|u_n\|$. To this end, the whole interval n_m was divided into Δn_1 parts. The value of $(\Delta\xi_{mn})^2$ was computed for each part and then the average $\langle(\Delta\xi_{mn})^2\rangle$ of these values was determined. The value of γ was computed in the whole interval n_m. The results of numerical experiments have shown that $\langle(\Delta\xi_{mn})^2\rangle$ indeed grows linearly with n and have allowed to find D_ξ in the dynamical Lloyd model and in the quantum standard map. In the former case the whole interval $n_m \approx 5 \times 10^6$ was divided into $\Delta n_1 = 500$ and 1000 parts. The parameter k was changed in the interval 2–100 and no dependence on T was observed. The results obtained (fig. 5) show that $lD_\xi = 2$ for $l \gg 1$ and therefore $l_s = 4l$. This ratio is the same as the theoretical one in the case of a one-dimensional random potential when the localization length is much larger than the distance between sites [19, 35].

The situation is more complicated for the model (4). In this case we used $n_m \approx 10^6$, $\Delta n_1 = 200$–1000. The dependence of lD_ξ on l obtained from $\|u_n^{(1)}\|$ is shown in fig. 5 for the region of strong chaos ($5 \le K \le 10$; $5 \le k \le 20$; $T \le 1$). It is seen that for $l \gg 1$ the value of lD_ξ is close to one ($\langle lD_\xi \rangle = 1.14$). Therefore, $l_s = 2l$, according to (16), is in agreement with the results of ref. 13. The different values of the ratio l_s/l for these two

models are again due to the fact that in the Lloyd model there are only two interacting sites, while in the other this number of sites is $N \sim k/2 \gg 1$. A special check has shown that also in the model (4) with random phases ξ_n, $lD_\xi = 1$. However, the value of lD_ξ decreases with K for $K \le 1$, and for $K \ll 1$ and $k \gg 1$ reaches the limit $lD_\xi \approx 1/4$. No theory exists so far to explain this behavior.

5. The Akulin–Dykhne model

An interesting model of molecular excitation was introduced in ref. 17: in the vicinity of the energy ε_{nm} there is a zone of N nearby levels labelled by the index m, and a monochromatic field $\mathscr{E}$ produces dipole transitions with $\Delta n = \pm 1$, and the matrix element μ. We consider the case of a small zone width $W \ll \omega$ and a small perturbation $V = \mathscr{E}\mu \ll \omega$. Then one can neglect nonresonant terms and obtain the stationary equation for QEE [17]:

$$\varepsilon\psi_{nm} = \Delta_{nm}\psi_{nm} + \sum_{n',m'} V_{nm}^{n'm'}\psi_{n'm'}, \tag{17}$$

where $\Delta_{nm} = \varepsilon_{nm} - n\omega$.

Now we consider several different cases. If $N = 1$, $\Delta_{nm} = 0$, $V_n^{n'} = V$ then (17) corresponds to the case of a one-dimensional crystal. The quasienergy spectrum is continuous, $\varepsilon(p) = 2V\cos p$ (p is the quasimomentum), and the QEE are delocalized, thereby leading to an unlimited excitation ($\langle n^2 \rangle \propto \tau^2$). If Δ_n are randomly distributed in the interval $[-W/2, W/2]$ then (17) represents a one-dimensional Anderson model. All QEE are exponentially localized, and the excitation is limited. The quasienergies lie in the interval $[-W/2 - 2V, W/2 + 2V]$ and the localization length depends on ε. For $W \gg V$, the length $l \ll 1$. For a small disorder $W \ll V$ and $\varepsilon \sim W$, the length $l \approx 100(V/W)^2$ (see, for example, ref. 36).

For a zone with an unlimited number of levels, randomly distributed Δ_{nm} and

$$V_{nm}^{n'm'} = V\left[\delta_{n,\,n'-1}(\delta_{m,\,m'-1} + \delta_{m,\,m'+1})\right.$$
$$\left. + \delta_{n,\,n'+1}(\delta_{m,\,m'-1} + \delta_{m,\,m'+1})\right],$$

the model (17) corresponds to the two-dimensional Anderson model. According to refs. 22 and 31 all states in the model are localized and for $\varepsilon \lesssim W$ the localization length varies from $l \sim 1$ at $V/W \approx 0.07$ up to $l \sim 10^6$ at $V/W = 0.5$. Asymptotically, $\ln l_\infty \sim (V/W)^2$ and the QEE decay exponentially with n: $\psi_n \propto \exp(-|n|/l_\infty)$. For $1 \ll N \ll l_\infty$ the localization length is proportional to N: $l \sim N(V/W)^2$. In the case when all $\Delta_{nm} = 0$ and only the matrix element $V_{nm}^{n'm'}$ fluctuates, all QEE are also localized.

The typical variant of the Akulin–Dykhne model in which transitions from one to many sublevels of the neighboring zones are allowed is analyzed in section 7.

6. One-dimensional localization in many-dimensional systems

At a first glance it seems that the localization of chaos takes place only in one-dimensional systems (Section 2) and is therefore a rather special phenomenon. However, it may happen that under some appropriate conditions such a localization occurs also in multi-dimensional systems. Consider a conservative system with d degrees of freedom, the energy E and the density of levels $\rho(E)$. We assume that the motion of the system is chaotic and that the energy is the only motion integral. Then the motion on the energy surface is diffusive and may be characterized by a diffusion rate $D_\parallel$. If $D_\parallel$ is nearly independent of the direction then the time of spreading of a narrow quasi-classical distribution function throughout the surface is equal to $\tau_\parallel \sim E^2/D_\parallel$. If an external perturbation periodic in time is added, then also a diffusive growth of the energy will be observed with the rate $D_\perp = (\Delta E)^2/\Delta\tau$. This diffusion pro-

ceeds perpendicular to the energy surface, and τ is measured in the number of perturbation periods.

If the latter diffusion is slow enough then the distribution function has time to cover the whole energy surface. The change of energy in time $\tau_\parallel$ is relatively small, $\Delta E = (D_\perp E^2/D_\parallel)^{1/2} \ll E$, if

$$D_\perp \ll D_\parallel. \tag{18}$$

Under this condition we conjecture that the number of excited levels grows as if it was a one-dimensional diffusion: $\Delta n = \rho\Delta E \sim \rho\sqrt{D_\perp \tau}$. Here we assume that in the interval ΔE all levels are excited with approximately equal probabilities. This condition is satisfied if there are no special selection rules for the matrix elements. Besides that, for a high frequency $\omega \gg 1/\rho$ it is necessary to have $\mu\mathscr{E} \gtrsim \omega$, otherwise the excitation of non-resonant levels will be small. If these conditions are satisfied then the expression for the localization length can be obtained, as in the one-dimensional case, from eq. (7):

$$l = \tfrac{1}{2}\rho^2 D_\perp \sim \Delta n \sim \tau_D, \quad l_E = \frac{l}{\rho}, \tag{19}$$

where l_E is the localization length in the energy scale. The conditions for the applicability of this equation are a large number of absorbed photons $N_\phi = l_E/\omega \gg 1$, and the excited levels $\Delta n \sim l \gg 1$. For a monochromatic perturbation the diffusion goes on only for $\rho\omega \gtrsim 1$ and the first condition is decisive. Also, it is necessary to have $\tau_D \gg \tau_\parallel$, otherwise the localization will take place faster than the spreading of the wave function on the whole energy surface, and the value of l will be smaller than in (19).

The expression (19) holds in the case of inhomogeneous localization (when l_E depends on E) only if $l_E \sim \rho D_\perp \ll E$. Instead, for $l_E \gtrsim E$ the delocalization takes place. For example, if $\rho D_\perp = \rho_0 E^\beta$ then the delocalization occurs for $\beta \geq 1$ and $\rho_0 \gtrsim E^{1-\beta}$ (see also refs. 13 and 16).

Assuming that the above picture is true, we discuss the motion of a particle in a two-dimensional billiards under the influence of a periodic perturbation as an example of the described effect. Let the Hamiltonian of the system be

$$H = \frac{p_1^2}{2} + \frac{p_2^2}{2} + k \sin x_1 \sin\left(\frac{x_2}{\sigma}\right) \delta_T(t) \qquad (20)$$

and the wave function is equal to zero on the border of the billiard, $\sigma \approx 1$. In the case of a righ-angled billiards with $0 \le x_1 \le \pi$; $0 \le x_2 \le \pi\sigma$, the unperturbed spectrum is $E_{n_1 n_2} = (n_1^2 + n_2^2/\sigma^2)/2$. The average density of the levels is constant: $\rho \sim 1$. For $kT \gg 1$ the motion can be shown to be chaotic and diffusive excitation takes place in each degree of freedom: $(\Delta n_1)^2 \approx (\Delta n_2)^2 = D\tau$ with $D \sim k^2$. For $D \ll 1$ the localization length $l \ll 1$ (the quantum border, see section 2, ref. 6) and almost the whole probability is concentrated on the initial level. For $D \gg 1$ the localization length is exponentially large: $\ln l \sim D$ (see (8) and refs. 22 and 31). Now let the border be deformed inwards, which corresponds to the case of Sinai's billiards (see, for example, ref. 10). Then the dynamics would be chaotic even in the absence of the external perturbation. After one collision with a border, $\Delta p \sim p$ (the case of strong deformation). Taking into account that the time between collisions $\sim E^{-1/2}$ we obtain $D_{\parallel} \sim E^{5/2}T$. The diffusion rate is $D_{\perp} \sim Ek^2$. For high energies the conditions (18) and $\tau_{\parallel} \ll \tau_D$ are satisfied and we predict one-dimensional localization (19) with $l_E \sim Ek^2$. Due to the dependence of l_E on E the localization takes place only for $k \lesssim 1$. For $k \gtrsim 1$ the QEE are delocalized due to the increase in the diffusion rate. The significant difference from the integrable case (right-angled billiards) consists in the fact that even for a small perturbation ($k \ll 1$) the localization length may be large enough, $l \sim l_E \gg 1$ if $E \gg k^{-2}$.

7. Localization of photon transitions

In the previous section we calculated $D_{\perp}$ as the diffusion rate in the classical system. For a monochromatic field there is another way of calculating $D_{\perp}$. It is based on the expression for the probability of one-photon transition per unit time: $w = (\pi/2)|\mu(E, E + \omega)|^2 \mathscr{E}^2 \rho$. In such an approach the absorption and reabsorption of photons leads to diffusive excitation of the system. At $\omega\rho \gtrsim 1$, according to [14, 17, 37],

$$D_{\perp} = \frac{(\Delta E)^2}{\Delta \tau} = 2w\omega^2\frac{2\pi}{\omega} = 2\pi^2\mu^2\mathscr{E}^2\rho\omega. \qquad (21)$$

From (19) and (21) we obtain the localization length of QEE which is conveniently measured in terms of the number of absorbed photons:

$$l_\phi = \frac{l_E}{\omega} = \pi^2\mu^2\mathscr{E}^2\rho^2 = \pi D_\phi\rho, \qquad (22)$$

where $D_\phi = \pi\mu^2\mathscr{E}^2\rho$ is the diffusion rate in terms of the number of photons per unit time. This result does not depend on the field frequency and therefore it is natural to think that the assumed condition $\mu\mathscr{E} \gtrsim \omega$ may not be necessary at all for the derivation of (22). In order to confirm this conjecture let us consider the situation when $\rho^{-1} \ll \mu\mathscr{E} \ll \omega$, which corresponds to the typical variant of the Akulin–Dykhne model (see section 4 in ref. 17). In this case, an effective excitation takes place only for the levels close to resonance, lying in a zone of width $\Delta E \sim \mu^2\mathscr{E}^2\rho$ near the energies of one-photon transitions $n\omega$ (levels with $\Delta_{nm} \lesssim \mu^2\mathscr{E}^2\rho$ in (17)) [17]. According to (21) and [17], the number of absorbed photons (number of zones) grows diffusively with time: $N_\phi = \sqrt{\pi\mu^2\mathscr{E}^2\rho t}$. In each zone the number of excited levels is of the order of $\mu^2\mathscr{E}^2\rho^2$. Then the total number of excited levels is $N \sim \sqrt{\mu^2\mathscr{E}^2\rho t}\,\mu^2\mathscr{E}^2\rho^2$. Since the excited levels lie in narrow zones of width $\mu^2\mathscr{E}^2\rho \ll \omega$ near the energies $n\omega$, all quasienergies also lie in an interval of width $\sim \mu^2\mathscr{E}^2\rho$. Their average spacing is then $\Delta\varepsilon \sim \mu^2\mathscr{E}^2\rho/N$. In the same way as for (7), from the uncertainty relation, we obtain an estimate for the time after which a limitation of the diffusion occurs: $t_D \sim 1/\Delta\varepsilon \sim N_\phi\rho \sim \rho\sqrt{\mu^2\mathscr{E}^2\rho t_D}$ and for the localization length, $l_\phi \sim N_\phi \sim \mu^2\mathscr{E}^2\rho^2$. The obtained result is applicable in the quasiclassical region when $l_\phi \gg 1$. For $\mu \mathscr{E} \ll \omega$ the steady-state distribution looks like a chain of equally spaced (in energy) peaks with the maxima exponentially decaying. Since, as in the case of the quantum standard map, the number of interacting sites is large the localization length of the steady-state is equal to $l_{\phi s} = 2l_\phi$. It is important to note that the localization is homogeneous when $\mu \mathscr{E} \rho =$ constant. Therefore, quantum effects lead to the localization of photon transitions and to a limitation of the system excitation.

The estimates obtained for the localization length of QEE allow to investigate the excitation of different multi-level systems by a periodic field when the standard perturbation theory is inapplicable.

Acknowledgements

The author wishes to express his deep gratitude to B.V. Chirikov and I. Guarneri for attention to this work and valuable comments.

References

[1] J.E. Bayfield and P.M. Koch, Phys. Rev. Lett. 33 (1974) 258.

[2] J.E. Bayfield and L.A. Pinnaduwage, Phys. Rev. Lett. 54 (1985) 313.

[3] K.A.H. van Leeuven et al., Phys. Rev. Lett. 55 (1985) 2231.

[4] V.M. Akulin et al., Pis'ma Zh. Eksp. Teor. Fiz. 40 (1984) 432.

[5] A.V. Evseev, A.A. Puretzky and V.V. Tyakht, Zh. Eksp. Teor. Fiz. 88 (1985) 60.

[6] E.V. Shuryak, Zh. Eksp. Teor. Fiz. 71 (1976) 2039.

[7] J.G. Leopold and I.C. Percival, Phys. Rev. Lett. 41 (1978) 944.

[8] B.V. Chirikov, Phys. Rep. 52 (1979) 263.

[9] G.M. Zaslavsky, Stochasticity of Dynamical Systems (Nauka, Moscow, 1984).

[10] A.J. Lichtenberg and M.A. Lieberman, Regular and Stochastic Motion (Springer, Berlin, 1983).

[11] G. Casati, B.V. Chirikov, J. Ford and F.M. Izrailev, Lecture Notes in Physics (Springer) 93 (1979) 334.

[12] B.V. Chirikov, F.M. Izrailev and D.L. Shepelyansky, Sov. Sci. Rev. 2C (1981) 209.

[13] B.V. Chirikov and D.L. Shepelyansky, Radiofizika 29 (1986) 1041.

[14] N.B. Delone, V.P. Krainov and D.L. Shepelyansky, Usp. Fiz. Nauk 140 (1983) 335 (Sov. Phys. Uspekhy 26 (1983) 551).

[15] D.L. Shepelyansky, Proc. Int. Conf. Quantum Chaos, Como 1983 (Plenum, New York, 1985), p. 187.

[16] G. Casati, B.V. Chirikov and D.L. Shepelyansky, Phys. Rev. Lett. 53 (1984) 2525.

[17] V.M. Akylin and A.M. Dykhne, Zh. Eksp. Teor. Fiz. 73 (1977) 2098.

[18] S. Fishman, D.R. Grempel and R.E. Prange, Phys. Rev. A 29 (1984) 1639.

[19] I.M. Lifshitz, S.A. Gredeskul and L.A. Pastur, Introduction to the Theory of Disordered Systems (Nauka, Moscow, 1982 and Wiley, New York, to be published).

[20] D.L. Shepelyansky, Phys. Rev. Lett. 56 (1986) 677.

[21] J.L. Pichard and G.J. Sarma, J. Phys. C 14 (1981) L127.

[22] A. MacKinnon and B. Kramer, Phys. Rev. Lett. 47 (1981) 1546; Z. Phys. B 53 (1983) 1.

[23] R. Blümel, S. Fishman, M. Griniasti and U. Smilansky, Lecture Notes in Physics (Springer) 263 (1986) 212.

[24] J.M. Greene, J. Math. Phys. 20 (1979) 1183.

[25] A.B. Rechester and R.B. White, Phys. Rev. Lett. 44 (1986) 1586; A.B. Rechester, M.N. Rosenbluth and R.B. White, Phys. Rev. A 23 (1981) 2664.

[26] R.S. MacKay, J.D. Meiss and I.C. Percival, Physica 13D (1984) 55.

[27] T. Hogg and B.A. Huberman, Phys. Rev. Lett. 48 (1982) 711.

[28] D.L. Shepelyansky, Physica 8D (1983) 208.

[29] B. Dorizzi, B. Grammaticos and Y. Pomeau, J. Stat. Phys. 37 (1984) 93.

[30] P.W. Anderson, Phys. Rev. 109 (1958) 1492.

[31] D.E. Khmelnitskii, Pis'ma Zh. Eksp. Teor. Fiz. 32 (1980) 248.

[32] F.M. Izrailev and D.L. Shepelyansky, Teor. Mat. Fiz. 43 (1980) 417.

[33] G. Casati and I. Guarneri, Commun. Math. Phys. 95 (1984) 121.

[34] D.L. Shepelyansky, Teor. Mat. Fiz. 49 (1981) 117.

[35] J. Sak and B. Kramer, Phys. Rev. B 24 (1981) 1761.

[36] E. Roman and C. Wiecko, Z. Phys. B 62 (1986) 163.

[37] N.B. Delone, V.P. Krainov and B.A. Zon, Zh. Eksp. Teor. Fiz. 75 (1978) 445.

Classical and Quantum Chaos for a Kicked Top

F. Haake, M. Kuś*, and R. Scharf

Fachbereich Physik, Universität-Gesamthochschule Essen, Federal Republic of Germany

Received June 5, 1986

We discuss a top undergoing constant precession around a magnetic field and suffering a periodic sequence of impulsive nonlinear kicks. The squared angular momentum being a constant of the motion the quantum dynamics takes place in a finite dimensional Hilbert space. We find a distinction between regular and irregular behavior for times exceeding the quantum mechanical quasiperiod at which classical behavior, whether chaotic or regular, has died out in quantum means. The degree of level repulsion depends on whether or not the top is endowed with a generalized time reversal invariance.

1. Introduction

The quantum treatment of systems capable of chaotic motion in the classical limit is interesting for several reasons. First, there is the desire to see how the classical distinguishability between regular and chaotic motion gradually arises as the system is turned more and more classical by changing a suitable parameter. By a controlled increase of quantum mechanical time scales such as wave packet spreading times or inverse level spacings, for instance, one would like to study the growth of the life time of effectively chaotic evolution of suitable observables.

Perhaps an even greater incentive for quantum mechanical investigations lies in the question whether quantum chaos can be more than a mere transient mimicry of classical chaos. If so, we need intrinsically quantum mechanical criteria to distinguish regular an irregular behavior; the relation of such quantum criteria to the traditional ones for classical chaos would have to be clarified.

We shall here present a model system ideally suited to a study of the problems mentioned, a three dimensional angular momentum $\mathbf{J}$ moving such that its square is conserved. Quantum mechanically, we can identify the operator $\mathbf{J}^2$ with its eigenvalue $j(j+1)$ and thus have a Hilbert space with the finite dimensionality $(2j+1)$. As the quantum number j is increased the quantum behavior approaches the classical one. We choose the Hamiltonian so as to allow for classically chaotic behavior as $j \to \infty$. The simplest

such Hamiltonian accounts for a precession of $\mathbf{J}$ around a constant a constant external magnetic field as well as for a periodic train of impulsive nonlinear kicks. A stroboscopic description is then indicated, the basis ingredient being the unitary operator U which transports the state vector from kick to kick.

The perhaps most important feature of our model is the finite dimensionality of its Hilbert space. Previous studies of quantum billiards [1] and the kicked rotator [2] had to confront the intricacies of an infinite number of dimensions. As Casati et al. have recently emphasized [3], an equivalence of U to certain ensembles of random matrices is much more difficult to establish for an infinite then a finite number N of dimensions. Of course, actual calculations (of, say, eigenvalues or time dependent expectation values) for systems with infinite N always have to truncate the Hilbert space to some manageable finite size. There is the danger, then, of introducing a new or destroying some hidden symmetry. In fact, Izrailev [4, 5] has recently pointed out that the eigenvalues of U for the kicked rotator can be given drastically different statistical properties by different truncation schemes.

Our model can be endowed with various symmetries. Among those are discrete rotations and nonconventional time reversals. The classical analysis is greatly facilitated by these symmetries and yields a surprising wealth of analytic results for fixed points, periodic orbits, and the stability scenario. In the quantum case symmetries play an even greater role. Two variants of the model, one with and the other without time reversal invariance, belong to different universality classes with respect to the statistics of the eigenvalues of U.

* Permanent address: Institute of Theoretical Physics, Warsaw University, Hoza 69, PL-00-681 Warsaw, Poland

Due to the discreteness of the spectrum of our U all quantum expectation values behave quasiperiodically in time [6], the quasiperiod being of the order j (in units of the kick period). Since classical chaos can become manifest on a time scale $\sim \ln j$ (the time needed to amplify the minimum quantum uncertainty $\sim 1/\sqrt{j}$ of the orientation of $\mathbf{J}$ to a solid angle of order unity) rather modest values of j suffice to realize classical chaos as a transient and to observe the subsequent takeover of quasiperiodicity. It is most fascinating, however, to see quasiperiodicity on the time scale j to arise in two qualitatively different varieties.

Rather regularly shaped collapses and revivals of quantum means alternate with a (quasi)period $\sim j$ when all external parameters and the initial state are set such that the classical limit would yield regular trajectories. However, under the conditions of classical chaos quantum means display a seemingly erratic behavior even on the time scale j; recurrences to a close neighborhood of the initial means do occur on that scale but have no tendency towards constant temporal separation.

Much insight can be gained from a spectral synthesis of quantum means based on the eigenvalues and eigenvectors of U. We find a rather small number of modes to be excited under the conditions of classically regular motion while classical chaos always corresponds to a large fraction of all modes in action. Regularly alternating collapses and revivals are thus revealed as a quantum beat phenomenon while the erratic variety of quasiperiodicity corresponds to broad-band excitation.

Previous analyses of level statistics have mostly focused on autonomous Hamiltonian systems [1, 7, 8, 9]. In our case of a kicked system we have to discuss the eigenphases of the unitary operator U. Of special interest is the relative frequency of a spacing S of two neighboring ones among the $2j+1$ eigenphases. We expect and numerically confirm a Poisson distribution of S to correspond to classically regular motion [10]. To investigate the level statistics corresponding to classical chaos we extend previous theories for autonomous systems to kicked ones. We also discuss the influence of time reversal invariance on the level spacing distribution. On the basis of this discussion we predict linear level repulsion, $P(S) \sim S$ for $S \to 0$, for the variant of our model for which we have identified a time reversal invariance. After breaking that invariance by a slight modification of the dynamics and assuming that there is no unidentified hidden antiunitary symmetry we should expect quadratic level repulsion, $P(S) \sim S^2$ for $S \to 0$. These theoretical predictions are nicely confirmed by our numerical results.

Preliminary results of our work were reported in [11]. We should also refer the reader to independent work on a similar model by Frahm and Mikeska [12].

2. The Quantum Top

We imagine as system characterized by an angular momentum vector $\hbar \mathbf{J} = \hbar(J_x, J_y, J_z)$, $[J_i, J_j] = i \varepsilon_{ijk} J_k$. The dynamics of $\mathbf{J}$ is governed by the Hamiltonian

$$H(t) = +(\hbar p/\tau) J_y + (\hbar k/2j) J_z^2 \sum_{n=-\infty}^{+\infty} \delta(t - n\tau). \quad (2.1)$$

The first term in $H(t)$ describes a precession around the y axis with angular frequency p/τ while the second term accounts for a periodic sequence of kicks at a temporal distance τ. Each kick can be interpreted as an impulsive rotation around the z axis by an angle proportional to J_z, the proportionality factor involving a dimensionless coupling constant k/j.

The following investigation is most conveniently formulated with the help of the unitary time evolution operator

$$U = e^{-i(k/2j)J_z^2} e^{-ipJ_y} \quad (2.2)$$

which subjects the wave function to a precession around the y axis by an angle p and the subsequent kick. The powers U^n describe the time evolution on the sequence of discrete times $n\tau$, $n = 0, 1, 2, \ldots$. In the Schrödinger picture, an initial state $|0\rangle$ develops, within n units of time, into the state $|n\rangle = U^n |0\rangle$. In the Heisenberg picture the discrete time evolution generates a sequence of operators $J_{in} = U^{+n} J_i U^n$. The corresponding Heisenberg equations take the form of nonlinear operator recursion relations,

$$J_x' = \tfrac{1}{2}(J_x \cos p + J_z \sin p + i J_y) e^{i\frac{k}{j}(J_z \cos p - J_x \sin p + \frac{1}{2})}$$
$$+ \text{h.c.}$$
$$J_y' = \tfrac{1}{2i}(J_x \cos p + J_z \sin p + i J_y) e^{i\frac{k}{j}(J_z \cos p - J_x \sin p + \frac{1}{2})}$$
$$+ \text{h.c.} \quad (2.3)$$
$$J_z' = J_z \cos p - J_x \sin p.$$

We shall devote special attention to the precession angle $p = \pi/2$ for which the recursion relations (2.3) simplify considerably.

Evidently, the squared angular momentum is a conserved quantity,

$$[\mathbf{J}^2, H(t)] = 0,$$
$$[\mathbf{J}^2, U] = 0. \quad (2.4)$$

We can therefore restrict our discussion to the $(2j+1)$ dimensional Hilbert space spanned by the eigenvectors of J_z and $\mathbf{J}^2$,

$$J^2 |jm\rangle = j(j+1)|jm\rangle$$
$$J_z |jm\rangle = m |jm\rangle. \tag{2.5}$$

Initial states of special importance for our eventual goal of comparing quantum and classical dynamics are the directed angular momentum states [13, 14]

$$|\theta, \phi\rangle = (1 + \gamma\gamma^*)^{-j} e^{\gamma J_-} |j,j\rangle \equiv |\gamma\rangle,$$
$$\gamma = e^{i\phi} \tan\frac{\theta}{2}, \tag{2.6}$$
$$J_- = J_x - i J_y.$$

These states align, with minimum uncertainty, the vector $\mathbf{J}$ along a direction characterized by a polar angle θ and an azimutal angle ϕ,

$$\langle \theta\phi| J_z |\theta\phi\rangle = j \cos\theta$$
$$\langle \theta\phi| J_x \pm i J_y |\theta\phi\rangle = j e^{\pm i\phi} \sin\theta. \tag{2.7}$$

One such state is the basis vector $|jm\rangle$ with $m=j$. All other directed angular momentum states can be generated from the state $|jj\rangle$ by the unitary rotation operator

$$R(\theta, \phi) = \exp\{i\theta(J_x \sin\phi - J_y \cos\phi)\}. \tag{2.8}$$

The relative variance of $\mathbf{J}$ in a state $|\theta\phi\rangle$,

$$(1/j^2)\{\langle \theta\phi| \mathbf{J}^2 |\theta\phi\rangle - \langle \theta\phi| \mathbf{J} |\theta\phi\rangle^2\} = 1/j, \tag{2.9}$$

is the minimum one allowed by the angular momentum commutation relations, it evidently shrinks to zero as the quantum number j grows towards infinity, i.e. in the classical limit.

We shall characterize the quantum dynamics of our top by the means and the variances of the operators J_{in} with respect to the "coherent" initial states $|\theta\phi\rangle$.

Experimental realizations of the dynamics (2.2, 3) might be possible in different fields. One example are small magnetized solids; crystal anisotropies allowing for an easy plane of magnetization can be represented by a time independent term $\sim J_z^2$ in the Hamiltonian; a temporally periodic sequence of short magnetic field pulses oriented along the y axis would have to be imposed to generate dynamics described by (2.2, 3). Experiments performed by Waldner et al. [15] are of precisely this type but refer to truly macroscopic magnetizations (j effectively infinite) and thus pertain to the classical limit. It would be extremely interesting to redo this kind of investigation with small clusters of particles.

Another type of experiment is conceivable using Josephson junctions with a capacitance in parallel. In a crude approximation, the Cooper pairs of the two superconductors joined can be described so as to be created and annihilated by Bose operators a_i^+, a_i with $i=1, 2$. The net charge on the junction can then be represented by an operator $S_z = (a_2^+ a_2 - a_1^+ a_1)/2$ while operators $S_x = (a_2^+ a_1 + a_1^+ a_2)/2$, $S_y = (a_2^+ a_1 - a_1^+ a_2)/2i$ describe the tunneling of Cooper pairs through the junction. By a suitable temporal modulation of the capacitance an effective Hamiltonian of the form (2.1) might then be realized [16] since the operators S_i have angular momentum commutation relations among themselves.

3. Regular and Chaotic Classical Motion on the Sphere

Before facing the intricacies of the quantum dynamics it is well to study the classical behavior arising for $j \to \infty$. In order to perform the limit we may introduce the rescaled quantities $\mathbf{X} = \mathbf{J}/j$ which fulfill the commutation relations $[X_i, X_k] = (1/j)\, \varepsilon_{ikl} X_l$. As $j \to \infty$ the X_i obviously become c-number variables which lie on the unit sphere. Their stroboscopic time evolution is described by a map which follows from the quantum recursion relations (2.3) with $j \to \infty$,

$$\left.\begin{array}{c} X' \\ Y' \end{array}\right\} = \left\{\begin{array}{l} \mathrm{Re} \\ \mathrm{Im} \end{array}\right. (X\cos p + Z\sin p + iY)\, e^{ik(Z\cos p - X\sin p)}$$
$$Z' = -X\sin p + Z\cos p. \tag{3.1}$$

We shall sometimes use the shorthand $\mathbf{X}' = F(\mathbf{X})$ for the map (3.2) and correspondingly write $\mathbf{X} = F^{-1}(\mathbf{X}')$ for its inverse,

$$X = (X'\cos kZ' + Y'\sin kZ')\cos p - Z'\sin p$$
$$Y = -X'\sin kZ' + Y'\cos kZ' \tag{3.2}$$
$$Z = (X'\cos kZ' + Y'\sin kZ')\sin p + Z'\cos p.$$

The conservation law (2.4) entails $\mathbf{X}^2 = 1$ and thus makes our classical map a two dimensional one. We could obviously rewrite it as two recursion relations for a polar and an azimutal angle according to

$$X = \sin\theta \cos\phi, \ Y = \sin\theta \sin\phi, \ Z = \cos\theta. \tag{3.3}$$

These angles are the classical counterparts of the angles characterizing the orientation of the coherent states (2.6).

Two further symmetries of our map are of great help for the further analysis. Consider first the operations $\mathbf{X}' = T(\mathbf{X})$ and $\mathbf{X}' = \tilde{T}\mathbf{X}$ defined by

$$T\begin{pmatrix} X \\ Y \\ Z \end{pmatrix} = \begin{pmatrix} -X\cos p - Z\sin p \\ Y \\ -X\sin p + Z\cos p \end{pmatrix} \tag{3.4}$$

and

$$\tilde{T}\begin{pmatrix} X \\ Y \\ Z \end{pmatrix} = \begin{pmatrix} X\cos p + Z\sin p \\ Y \\ X\sin p - Z\cos p \end{pmatrix} \tag{3.5}$$

They are both involutions,

$$T^2 = \tilde{T}^2 = 1, \tag{3.6}$$

and have the determinants

$$\det T = \det \tilde{T} = -1. \tag{3.7}$$

It is easily verified that they both yield time reversal operations for our map F in the sense

$$TFT = \tilde{T}F\tilde{T} = F^{-1}. \tag{3.8}$$

Our top thus has (doubly reversible dynamics [17]. It is worth noticing, however, that neither T nor $\tilde{T}$ is the conventional time reversal $\mathbf{J} \to -\mathbf{J}$, the latter cannot yield an invariance since our dynamics involves precession around a magnetic field.

It follows immediately that the images under T and $\tilde{T}$ of n-periodic orbits of F, $F^n(\mathbf{X}) = \mathbf{X}$, are n-periodic orbits of F as well. It may, of course, happen that such an orbit coincides with this own image.

It is another important consequence of the time reversal invariance of F that we can decompose F into a product of two involutions

$$F = T(TF) \tag{3.9}$$

(and similarly with $\tilde{T}$) since $(TF)^2 = (FT)^2 = 1$, $(\tilde{T}F)^2 = (F\tilde{T})^2 = 1$. Such decompositions greatly facilitate the search for periodic orbits. As a first step towards their employment we introduce the "symmetry lines" for the various involutions, i.e. the set of all points $\mathbf{X}$ obeying

$$I(\mathbf{X}) = \mathbf{X} \quad \text{and} \quad \mathbf{X}^2 = 1 \tag{3.10}$$

for $I = T, \tilde{T}, FT, TF, \ldots$. For instance, the symmetry lines of T and $\tilde{T}$ (to be referred to as T line etc.) are the great circles on the unit sphere defined by $X \sin p - (\cos p - 1) Z = 0$ and $X \sin p - (\cos p + 1) Z = 0$, respectively. The other symmetry lines tend to have more complicated shapes for general values of the precession angle p. For $p = \pi/2$, however, we still have great circles with $Z = 0$ for $F\tilde{T}$ and with $X = 0$ for $\tilde{T}F$, while the intersections of the unit sphere with $X(\cos kZ + 1) + Y \sin kZ = 0$ and $Z(\cos kX + 1) + Y \sin kZ = 0$ are the symmetry lines for FT and TF, respectively.

As an illustration of the use of time reversal invariance we propose to consider a point $\mathbf{X}$ on, say, the T line. If the n-th iterate $F^n(\mathbf{X})$ also lies on the T line or on the symmetry line of either TF of FT, that point is periodic with period $2n$, $2n + 1$, and $(2n - 1)$, respectively [17, 18]. We may thus search for, say, period-$2n$ solutions of F by studying F^n (instead of F^{2n}).

It is also easily shown that any periodic orbit of F with even (odd) period which is invariant under any of the involutions in question has an even (odd) number of points on the corresponding symmetry line.

Beyond being symmetric under the time reversals T and $\tilde{T}$ our map F is invariant under rotations around the precession axis by π,

$$R_y F = F R_y \quad \text{with} \quad R_y \begin{pmatrix} X \\ Y \\ Z \end{pmatrix} = \begin{pmatrix} -X \\ Y \\ -Z \end{pmatrix}. \tag{3.11}$$

Actually, the three symmetries T, $\tilde{T}$, and R_y are not independent since $R_y T = T R_y = \tilde{T}$, $\tilde{T}T = T\tilde{T} = R_y$. It follows that the R_y image of every n-cycle of F is an n-cycle of F, too.

Specializing, for the remainder of this section, to $p = \pi/2$ we have the further symmetry

$$F R_x = R_x F R_y \quad \text{with} \quad R_x \begin{pmatrix} X \\ Y \\ Z \end{pmatrix} = \begin{pmatrix} X \\ -Y \\ -Z \end{pmatrix}. \tag{3.12}$$

This implies that the iterated map F^2 is invariant under rotations around the x-axis by π,

$$R_x F^2 = F^2 R_x. \tag{3.13}$$

The most important consequence of (3.12, 13) is that every $(2n + 1)$-cycle of F together with its R_y image is mapped, by R_x, into a $(4n + 2)$-cycle of F, provided the $(2n + 1)$-cycle in question is not symmetric under R_y.

Let us now look at the fixed points of F. Evidently, the poles

$$X = Z = 0, \quad Y = \pm 1 \tag{3.14}$$

with respect to the precession axis are invariant under F for any value of the kick strength k. Further, nontrivial fixed points appear for sufficiently large $|k|$. They are determined by

$$Z = -X, \quad Y = X \cot \frac{kX}{2}, \quad f(X) = 0$$

with $f(X) = \sin^2(kX/2)\{1 + \sin^2(kX/2)\}^{-1} - X^2$

and thus come in pairs, too. They all lie on the T line. One member of each pair can be generated from the other by R_y. It therefore suffices to consider only fixed points with $X > 0$. We may also confine ourselves to positive kick strengths since a change of sign in k is equivalent to changing the sign of Y for the fixed points.

A graphical analysis of (3.15) easily reveals that (i) the first nontrivial solution emerges for $k_0 = 2$ and (ii) additional pairs of nontrivial fixed points appear when k crosses the thresholds k_j, $j = 1, 2, 3, \ldots$ deter-

ed by $f(X)=0, f'(X)=0$, and $\mathbf{X}^2=1$. These thresh-
are easily found numerically.

To discuss stability we need to linearize the map
round the various fixed points. Due to angular
nentum conservation, $\mathbf{X}^2=1$, one of the three ei-
values of the linearized map is, for all fixed points,
al to unity and thus irrelevant for stability. The
·r two eigenvalues are equal to unity in modulus
ided the X and Y coordinates of the fixed point
uestion obey

$$+\cos kX - 1| < 2. \tag{3.16}$$

trivial fixed points are thus stable for $k<2$. The
trivial fixed point emerging for $k \geqslant k_0 = 2$ is stable
$k_0 \leqslant k \leqslant k_0' = \sqrt{2\pi}$. Of the pair of fixed points ap-
ing at $k > k_j$ the member with the smaller value
ζ is always unstable while the other member is
le for

$$k \leqslant k_j' = (2j+1)\sqrt{2\pi}, j=1,2,\ldots. \tag{3.17}$$

When the kick strength crosses one of the thresh-
k_j' the fixed point which looses its stability bifur-
s into the period-2 solution

$$
\begin{aligned}
&= X_2 = -Z_1 = -Z_2 = (2j+1)\pi/k \\
&- Y_2 = \{1 - 2(2j+1)^2 \pi^2/k^2\}^{1/2}.
\end{aligned} \tag{3.18}
$$

: the fixed point it originates from it lies on the
metry line of T. Once again linearizing we find
stability range

$$k < k_j'' = \sqrt{k_j'^2 + 4}. \tag{3.19}$$

ve increase k beyond k_j'' we generate a period-4
t and, for yet higher k, a whole cascade of period
olings with orbits symmetric under T.

Similar structures are revealed by investigating the
ie which is the R_y image of the T line. There are
od-2 solutions

$$= Z_1 = -X_2 = -Z_2, \qquad Y_2 = Y_1 = -X_1 \cot\frac{kX_1}{2} \tag{3.20}$$

X_1 again a root of the function $f(X)$ defined
.15). It is, of course, due to the invariance (3.12)
the operation R_x rotates each fixed point (3.15)
its R_y image to the location of the period-2 cycle
vice versa. The first of these cycles bifurcates from
'south" pole with respect to the precession axis
$k = k_0 = 2$ and pairs of cycles emerge at the
sholds k_j with $j=1, 2,\ldots$. Each cycle shares its
lity properties with the corresponding pair of
points. The cycle loosing stability at k_j bifurcates

into the period-4 orbit

$$
\begin{aligned}
X_1 &= Z_1 = -X_2 = -Z_2 = X_3 = Z_3 = -X_4 = -Z_4 \\
&= (2j+1)\,\pi/k, \\
Y_1 &= -Y_2 = Y_3 = -Y_4 = \{1 - 2(2j+1)^2\,\pi^2/k^2\}^{1/2}
\end{aligned} \tag{3.21}
$$

which remains stable in the range (3.19). Evidently,
it lies on the $\tilde{T}$ line and can be found by rotating
the 2-cycle (3.18) together with its R_y image around
the X-axis by π.

Due to the high symmetry of our model we can
easily find further periodic orbits. Of importance for
us will be the period-4 cycle

$$
\mathbf{X}_1 = -\mathbf{X}_3 = \begin{pmatrix} 0 \\ 0 \\ 1 \end{pmatrix}, \qquad \mathbf{X}_2 = -\mathbf{X}_4 = \begin{pmatrix} 1 \\ 0 \\ 0 \end{pmatrix}. \tag{3.22}
$$

Note that this orbit lies on the "equator" with respect
to the precession axis with $\mathbf{X}_1$ and $\mathbf{X}_3$ on the $\tilde{T}F$
line and $\mathbf{X}_2$ and $\mathbf{X}_4$ on the $F\tilde{T}$ line. It exists for arbitrary values of the kick strength but is stable only
for $(2\cos k + k \sin k)^2 < 4$.

Another sequence of 4-cycles

$$
\begin{aligned}
X_1 &= Z_1 = X_2 = -Z_2 = -X_3 = -Z_3 = -X_4 = Z_4, \\
Y_1 &= -Y_2 = Y_3 = -Y_4 = X_1 \tan\frac{kX_1}{2}, \\
\cos^2&(kX_1/2)/\{1 + \cos^2(kX_1/2)\} - X_1^2 = 0,
\end{aligned} \tag{3.23}
$$

has $\mathbf{X}_1$ and $\mathbf{X}_3$ on the $\tilde{T}$ line and $\mathbf{X}_2$ and $\mathbf{X}_4$ on the
T line. It is thus symmetric under both T and $\tilde{T}$. The
stability scenario is similar to the one of the fixed
points (3.15).

There are interesting 3-cycles

$$
\begin{aligned}
X_1 &= Z_1 = -Z_2 = -X_3, X_2 = Z_3 = 0, \\
Y_1 &= -X_1 \cot kX_1, Y_2 = Y_3 = -X_1/\sin kX_1, \\
\sin^2 &kX_1/\{1 + \sin^2 kX_1\} - X_1^2 = 0
\end{aligned} \tag{3.24}
$$

and

$$
\begin{aligned}
Z_1 &= Z_2 = -X_1 = -X_3 \\
Z_3 &= -X_2, Y_2 = Y_3 \\
X_2 &= -X_1 \cos kX_1 + Y_1 \sin kX_1 \\
Y_2 &= X_1 \sin kX_1 + Y_1 \cos kX_1
\end{aligned} \tag{3.25}
$$

where X_1, Y_1 are given by solutions of the equations:

$$
X_1\left[\operatorname{ctg}\left(\frac{k}{2}X_1\cos kX_1\right) - \sin kX_1\right] + Y_1\cos kX_1 = 0
$$

$$
2X_1^2 + Y_1^2 = 1.
$$

The 3-cycles (3.24) are symmetric under $\tilde{T}$ while the
ones given in (3.25) are T invariant. Both types arise
in pairs with each member the image of the other

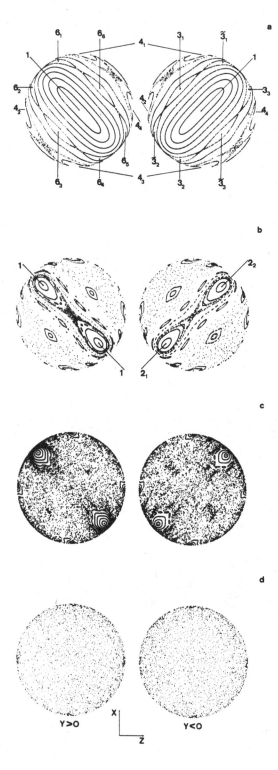

under R_y. Due to (3.12) each such pair yields, upon rotation around the X-axis by π, a 6-cycle.

All of the stable fixed points and periodic orbits discussed in the above show up in Fig. 1 which portraits trajectories generated by the map F for $p=\pi/2$ and various values of the kick strength k. Figure 1a pertains to $k=2$, a case where the fixed points at the poles (3.14) are still marginally stable. Most of the sphere is then covered by the stability islands around these fixed points, around the 4-cycle (3.22), and a pair of 3-cycles (3.25) together with their R_x image; narrow chaotic bands are visible near the first of the 4-cycles (3.23) and also near a pair of 3-cycles (3.24) and their R_x image which are unstable at $k=2$.

The somewhat richer structure arising for $k=2.5$ is displayed in Fig. 1 b. The precession poles have now become hyperbolic fixed points and thus have chaotic trajectories in their neighborhoods. A pair of stable fixed points (3.15) in the "northern" hemisphere ($Y>0$) and, as their R_x image, the period-2 orbit (3.20) in the "southern" hemisphere now have prominent stability islands, as does the 4-cycle (3.22). Quite conspicuous is a pair of 3-cycles (3.24) in the southern hemisphere and their northern correspondant by R_x, a 6-cycle. A southern pair of 5-cycles, one the R_y image of the other, and, at R_x symmetric locations, a northern 10-cycle are also discernible. Apart from further but much smaller islands of stability the rest of the sphere accommodates chaos.

Chaos has become much more predominant for $k=3$, as is shown in Fig. 1c. It is only the stability islands of the northern pair of fixed points (3.15) and the corresponding southern 2-cycle (3.20) and of the equatorial 4-cycle (3.22) that are easily detected numerically. Note that the points (3.15, 20) have approached the equator quite closely. For $k=k_0'=\pi\sqrt{2}$ they will have arrived at $Y=0$ and go unstable.

At $k=6$, the case described by Fig. 1d, only tiny islands of stability around the equatorial 4-cycle (3.22) remain visible. There may be other unresolved stability islands but clearly, chaos has expanded to near-global dominance.

We should emphasize once again that there are whole generations of solutions of the equations determining fixed points and other periodic orbits. Therefore, regular structures keep reappearing as k is in-

Fig. 1 a–d. Classical motion on the unit sphere. The left and right columns depict trajectories on the northern ($Y\geqslant0$) and the southern ($Y\leqslant0$) hemisphere, respectively. Fixed points are labeled by 1, points on n-cycles by n_i or $\tilde{n}_i$: **a**: $p=\dfrac{\pi}{2}$, $k=2$; **b**: $p=\dfrac{\pi}{2}$, $k=2.5$; **c**: $p=\dfrac{\pi}{2}$, $k=3$; **d**: $p=\dfrac{\pi}{2}$, $k=6$

creased. However, the corresponding stability islands on the sphere as well as the stability windows on the k axis shrink as we go through the sequence of generations.

4. Symmetries of the Quantum Top

There are quantum counterparts to all of the symmetries of the classical top. Their importance for the analysis of the quantum dynamics is even greater than in the classical case. For the sake of convenience of presentation we shall formulate them for integer values of j only. In that case the rotation operators

$$R_\nu = e^{\pm i \pi J_\nu} \tag{4.1}$$

are effectively hermitian.

Of special interest are the eigenvalues and eigenvectors of the unitary time evolution operator (2.2),

$$U|\phi\rangle = e^{i\phi}|\phi\rangle, \tag{4.2}$$

which we must construct, for a given j, as superpositions of the $2j+1$ states defined in (2.5). By observing the invariance of U under rotations around the y-axis by π,

$$[U, R_y] = 0, \tag{4.3}$$

the analogue of the classical symmetry (3.11), we find that the $2j+1$ eigenstates of U fall, for j even, into a group of $(j+1)$ states $|\phi_+\rangle$ which are even under R_y and another one of j states $|\phi_-\rangle$ which are odd,

$$U|\phi_\pm\rangle = e^{i\phi_\pm}|\phi_\pm\rangle, \quad R_y|\phi_\pm\rangle = \pm|\phi_\pm\rangle. \tag{4.4}$$

By again specializing to $p = \pi/2$ we obtain the quantum version of (3.12),

$$U R_x = R_x U R_y, \tag{4.5}$$

which entails the invariance of U^2 under R_x. With respect to the vectors $|\phi_\pm\rangle$ the identity (4.4) means

$$U R_x |\phi_\pm\rangle = \pm e^{i\phi_\pm} R_x |\phi_\pm\rangle. \tag{4.6}$$

An eigenphase ϕ_- is thus accompanied by $\phi_- + \pi$ as another one, both pertaining to states odd under R_y. On the other hand, the even states $|\phi_+\rangle$ and $P_x |\phi_+\rangle$ are either linearly dependent, or else the eigenvalue ϕ_+ is doubly degenerate.

For any value of the precession angle p we have, like in the classical case, two generalized time reversal symmetries which are represented by antiunitary operators. To construct them we first introduce an antiunitary conjugation operation as

$$K J_x K = J_x, K J_y K = -J_y, K J_z K = J_z, K^2 \\ = 1, K c |\psi\rangle = c^*|\psi\rangle^*. \tag{4.7}$$

where c is any complex c-number. With the help of two unitary operators,

$$S = e^{ipJ_y} e^{i\pi J_z}, \quad \tilde{S} = e^{-i\pi J_x} e^{-ipJ_y}, \tag{4.8}$$

we define

$$T = SK, \quad \tilde{T} = \tilde{S}K. \tag{4.9}$$

These operators obey the useful identities

$$[K, S] = [K, \tilde{S}] = 0, \quad S^2 = \tilde{S}^2 = T^2 = \tilde{T}^2 = 1. \tag{4.10}$$

Moreover, T and $\tilde{T}$ are revealed as time reversal operations by

$$T U T = \tilde{T} U \tilde{T} = U^{-1}, \tag{4.11}$$

a relation identical in appearance with (3.8). It is also easily shown using (4.6–9) that

$$\begin{aligned} T J_x T &= -J_x \cos p - J_z \sin p \\ T J_y T &= J_y \\ T J_z T &= -J_x \sin p + J_z \cos p. \end{aligned} \tag{4.12}$$

and

$$\begin{aligned} \tilde{T} J_x \tilde{T} &= J_x \cos p + J_z \sin p \\ \tilde{T} J_y \tilde{T} &= J_y \\ \tilde{T} J_z \tilde{T} &= J_x \sin p - J_z \cos p. \end{aligned} \tag{4.13}$$

Obviously, (4.12, 13) correspond to the classical transformations (3.4) and (3.5). Like the latter they are related to one another by the geometric symmetry R_y.

The consequences of time reversal invariance for the eigenvectors and eigenvalues of U are similar to the wellknown ones pertaining to time independent Hamiltonians. We shall discuss and use them in treating the statistics of the spacing of the eigenphases of U in Sect. 8.

5. Semiclassical Approximation

In studying the simplifications arising for large but finite values of j it is convenient to use the rescaled angular momentum operators

$$X_\pm = (J_x \pm i J_y)/j, \quad Z = J_z/j \tag{5.1}$$

which obey $[X_+, X_-] = 2Z/j, [Z, X_\pm] = \pm X_\pm/j$. For $p = \pi/2$ the stroboscopic Heisenberg equations (2.3) then take the form

$$\begin{aligned} X'_+ &= e^{\frac{ik}{2j}} [Z + \tfrac{1}{2}(X_+ - X_-)] e^{-ik(X_+ + X_-)} \\ Z' &= -\tfrac{1}{2}(X_+ + X_-). \end{aligned} \tag{5.2}$$

We now assume a coherent initial state (2.6) and look for the expectation value of the X_i after n kicks,

$$\sigma_n \equiv \langle \gamma | U^{-n} X_+ U^n | \gamma \rangle$$
$$\zeta_n = \langle \gamma | U^{-n} Z U^n | \gamma \rangle. \tag{5.3}$$

As follows from (2.6, 7) the sequence (5.3) begins with

$$\sigma = 2\gamma/(1 + \gamma\gamma^*)$$
$$\zeta = (1 - \gamma\gamma^*)/(1 + \gamma\gamma^*). \tag{5.4}$$

By using normal-ordering techniques and the commutation relations for the operators (5.1) it is easy to establish recursion relations for the expectation values (5.3) which generalize the classical map (3.1) by including first-order corrections with respect to $1/j$,

$$\begin{aligned}
\sigma_{n+1} &= [\zeta_n + \tfrac{1}{2}(\sigma_n - \sigma_n^*)] \, e^{-ik(\sigma_n + \sigma_{*n})} \\
&+ \tfrac{1}{2j} e^{-ik(\sigma_n + \sigma_{*n})} \{ ik[\zeta_n + \tfrac{1}{2}(\sigma_n - \sigma_n^*)] \\
&+ ik[D(\zeta_n, \sigma_n) + D(\zeta_n, \sigma_n^*) + \tfrac{1}{2} D(\sigma_n, \sigma_n) \\
&+ \tfrac{1}{2} D(\sigma_n, \sigma_n^*) - \tfrac{1}{2} D(\sigma_n^*, \sigma_n) - \tfrac{1}{2} D(\sigma_n^*, \sigma_n^*)] \\
&- \frac{k^2}{2} [\zeta_n + \tfrac{1}{2}(\sigma_n - \sigma_n^*)][D(\sigma_n, \sigma_n) \\
&+ D(\sigma_n, \sigma_n^*) + D(\sigma_n^*, \sigma_n) + D(\sigma_n^*, \sigma_n^*)]\}, \\
\zeta_{n+1} &= -\tfrac{1}{2}(\sigma_n + \sigma_n^*). \tag{5.5}
\end{aligned}$$

It is, of course, due to the linearity of the second one of the operator recursion relations (5.2) that no $1/j$ correction shows up in the second one of the relation (5.5). The $1/j$ correction in σ_{n+1}, on the other hand, involves the following bilinear differential form,

$$\begin{aligned}
D(A, B) &= -\sigma^2 \frac{\partial A}{\partial \sigma} \frac{\partial B}{\partial \sigma} - \sigma^{*2} \frac{\partial A}{\partial \sigma^*} \frac{\partial B}{\partial \sigma^*} - \sigma\sigma^* \frac{\partial A}{\partial \zeta} \frac{\partial B}{\partial \zeta} \\
&+ (1+\zeta)^2 \frac{\partial A}{\partial \sigma} \frac{\partial B}{\partial \sigma^*} + (1-\zeta)^2 \frac{\partial A}{\partial \sigma^*} \frac{\partial B}{\partial \sigma} \\
&+ (1-\zeta) \left[\sigma^* \frac{\partial A}{\partial \sigma^*} \frac{\partial B}{\partial \zeta} + \sigma \frac{\partial A}{\partial \zeta} \frac{\partial B}{\partial \sigma} \right] \\
&+ (1+\zeta) \left[\sigma \frac{\partial A}{\partial \sigma} \frac{\partial B}{\partial \zeta} + \sigma^* \frac{\partial A}{\partial \zeta} \frac{\partial B}{\partial \sigma^*} \right] \tag{5.6}
\end{aligned}$$

where A and B are functions of the initial expectation values σ, ζ and σ^*. Its formidable appearance notwithstanding the $1/j$ correction is not difficult to treat. Even though it represents a quantum effect we can work out its time dependence in zeroth order in $1/j$, i.e. by using classical dynamics; it thus behaves like an inhomogeneity in the first of the equations of motion (5.5).

The $1/j$ correction is bilinear in the elements of the matrix

$$M_n = \begin{pmatrix} \partial\sigma_n/\partial\sigma & \partial\sigma_n/\partial\zeta & \partial\sigma_n/\partial\sigma^* \\ \partial\zeta_n/\partial\sigma & \partial\zeta_n/\partial\zeta & \partial\zeta_n/\partial\sigma^* \\ \partial\sigma_n^*/\partial\sigma & \partial\sigma_n^*/\partial\zeta & \partial\sigma_n^*/\partial\sigma^* \end{pmatrix}. \tag{5.7}$$

We may use the chain rule of differentiation and represent the matrix M_n by a product of n matrices

$$M_n = \Delta_n \Delta_{n-1} \dots \Delta_1 \tag{5.8}$$

where each factor Δ_l has the same structure as M_n itself except that σ_l, ζ_l, σ_l^* are differentiated with respect to σ_{l-1}, ζ_{l-1}, σ_{l-1}^*. In fact, Δ_l is nothing else than the classical map F written for spherical rather than cartesian components of the angular momentum vector and linearized around σ_{l-1}, ζ_{l-1}, σ_{l-1}^*. All of the matrices Δ_l as well as their product M_n thus have determinants equal to unity.

We must now distinguish two radically different situations. The first one arises when the kick strength k and the orientation $\{\theta, \phi\}$ of the initial state are chosen such that the classical map generates a chaotic trajectory; regular classical motion characterizes the second situation.

For the chaotic case, the very definition of Lyapounov exponents [19] entails that the three eigenvalues of M_n approach the exponential growth $e^{\lambda_i n}$ for large times. Due to angular momentum conservation one of the Lyapounov exponents must vanish, $\lambda_1 = 0$. The other two must have equal magnitude but different signs since our top has Hamiltonian dynamics, $\lambda_2 = -\lambda_3 \equiv \lambda > 0$. It follows that the $1/j$ correction in (5.5) behaves like external noise with an intensity growing exponentially in time. The $1/j$ expansion used in deriving the semiclassical equations of motion is thus invalidated after a time

$$n_{\text{chaos}} \sim (1/\lambda) \ln j. \tag{5.9}$$

Turning now to the case of regular classical motion we are facing a matrix M_n whose element at worst grow in time like some power n^α, $\sigma > 0$. Correspondingly larger is the domain of validity of the semiclassical approximation (5.5),

$$n < n_{\text{reg}} \sim j^{1/\alpha}. \tag{5.10}$$

In neither the regular nor the chaotic case can the semiclassical approximation (5.5) be used on the especially interesting time scale $n \sim j$ on which, as we shall see below, the quasiperiodicity of the quantum motion becomes manifest. An asymptotic long-time treatment would require a large-j approximation of U^j rather than of U and is not available at present.

6. Regular Versus Erratic Quasiperiodicity

The $2j+1$ eigenphases ϕ_v of the unitary time evolution operator (2.2) can be thought of as lying in the interval $0 \leqslant \phi_v < 2\pi$. Their effective mean spacing is thus of the order π/j. Since the temporal resolution of a phase spacing $\Delta\phi$ requires times exceeding $1/\Delta\phi$

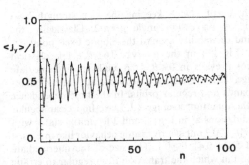

Fig. 2. Classical (crosses) and quantum (solid line) expectation value $\langle J_y \rangle_n$ for coherent initial state localized in classically regular region $\langle J_x \rangle_0 = -\langle J_z \rangle_0$, $\theta_y = 0.8$

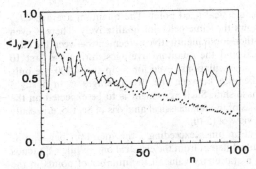

Fig. 4. As Fig. 2 but with coherent initial state localized in classically chaotic region at $\langle J_x \rangle_0 = -\langle J_z \rangle_0$, $\theta_y = 0.1$

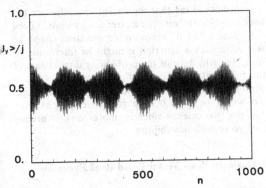

Fig. 3. Quantum mean as in Fig. 2 extended to large times to display the regular alternations of collapse and revivals

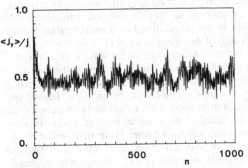

Fig. 5. Quantum mean as in Fig. 4 extended to large times to display the irregularity of recurrences

the quasiperiodicity of the quantum dynamics of our top becomes manifest for times of the order j. Classical chaos can therefore live in quantum expectation values as a transient only; it is definitely dead after a number of kicks of the order j.

Conversely, if we want to find out whether quantum chaos can be more than a mere transient mimicry of classical chaos we ought to study the quantum dynamics on a time scale of the order j, for large values of j. Such an investigation must be carried out numerically.

For the numerical calculation of time dependent expectation values we have employed two different strategies. One is to use the eigenstates $|jm\rangle$ of $\mathbf{J}^2$ and J_z to represent the operators U and $U^{-n} J_i U^n$ as $(2j+1)$ by $(2j+1)$ matrices. The alternative is to first diagonalize U and then spectrally synthesize time dependent expectation values by using the eigenvectors and eigenvalues of U. We have not been satisfied with our numerical results before both methods yielded agreement for $j = 100$ and times up to $n \approx 1,000$.

In Figs. 2–5 we depict the time dependence of the

expectation value of J_y for coherent initial states, $j = 100$, and $k = 3$. Figures 2 and 4 show the short-time behavior, $n \lesssim j$, and also contain classical ensemble averages $\langle Y \rangle$. The latter were evaluated from bundles of 1,000 classical trajectories originating from a cloud of initial points equal in location (θ, ϕ) and angular spread $(\sin\theta \, \Delta\theta \, \Delta\phi = 1/j)$ to the coherent state used in the quantum case.

For Fig. 2 we have chosen an initial state well within the classical stability island around the fixed point (3.15) with $X = -Z > 0$, $Y > 0$ (see Fig. 1c). The classical average (crosses) displays damped oscillations towards a stationary value $\langle Y \rangle_\infty \approx 0.5$. The oscillation corresponds to the classical orbiting of the regular trajectories around the stable fixed point; the oscillation period $\Delta n \approx 4$ compares favorably with the eigenvalues obtained by linearizing the map F around the fixed point. Actually, the nonlinearity of F attributes slightly different orbiting frequencies to the 1,000 trajectories in the classical ensemble. As the orbits gradually get out of phase with oneanother the classical average $\langle Y \rangle$ suffers a damping. The stationary value $\langle Y \rangle_\infty$ roughly equals the Y coordinate of the classical fixed point. The quantum average $\langle J_y \rangle$ shows the same behavior qualitatively. There is even rather good quantitative agreement between the classical and the quantum averages during the first 10 or so oscillations; even for n up to $j = 100$ the only difference between the two overages is in a slight phase shift. Such agreement is to be expected on the basis of the semiclassical analysis of Sect. 5, especially in view of (5.10).

For times exceeding j the quantum behavior is quite different from the classical one. While $\langle Y \rangle$ tends to a stationary value, if the number of points in the ensemble goes to infinity, the quantum average exhibits, as it must, quasiperiodicity. As shown in Fig. 3 the quasiperiodicity takes the form of a rather regular sequence of "collapses" and "revivals" [20]. The temporal separation of subsequent revivals, the quasiperiod, comes out as roughly 100 and that value nicely confirms the expectation based on the mean spacing $\Delta \phi \approx \pi / j$ of the eigenphases of U.

Figure 4 refers to an initial state localized within the classically chaotic region. In view of the semiclassical arguments presented in Sect. 5 it is not surprising to find significant differences between the classical and the quantum averages at rather early times, $n \approx 8$, already. Lateron, the classical average decays to zero. This decay signals a rather symmetric spread of the bundle of classical trajectories over the northern ($Y > 0$) and southern ($Y < 0$) hemispheres. Radically different is the quantum behavior for times $n > j$ (see Fig. 5). As in the regular case the quantum mean $\langle J_y \rangle$ keeps recurring to the neighborhood of its initial value [6, 21]. In striking contrast to the regular case the sequence of recurrences is seemingly erratic rather than having nearly equal spacings.

While the early-stage behavior of the quantum average in Figs. 4 and 5 can be interpreted as a reflection of classical chaos the erratic sequence of recurrences visible for times larger than j is a genuine quantum effect. It is quite interesting to see the quantum quasiperiodicity to manifest itself so drastically differently in Fig. 3 and Fig. 5. The difference certainly suggests that the distinction between regular and "chaotic" dynamics may not be an exclusive priviledge of classical mechanics.

In order to check whether the transition from orderly sequences of collapses and revivals to erratic recurrences can serve as at least a qualitative quantum criterion for chaos we have calculated the averages $\langle J_y \rangle_n$ for various coherent initial states. These states were chosen on the classical T line ($\langle J_x \rangle_0 = -\langle J_z \rangle_0$) with an angular separation $\theta_y = \langle (\hat{y}, \mathbf{J}_0) \rangle$ from the precession axis ranging from 0.1 (as in Figs. 4 and 5, close to the hyperbolic fixed point at the pole $Y = 1$.) to 1.1 (close to the classical elliptic fixed point

with $X > 0$), always keeping the kick strength $k = 3$ and the precession angle $p = \pi / 2$. Classically, the island of stability around the elliptic fixed point lies (see Fig. 1 b) in the interval $0.6 \lesssim \theta_y \lesssim 1.4$; classical chaos prevails in $0 \leqslant \theta_y \lesssim 0.6$ while the intermediate range, $0.5 \lesssim \theta_y \lesssim 0.6$, accomodates both narrow chaotic bands and regular orbits. In the time dependence of the quantum average $\langle J_y \rangle_n$ we find clear regular modulations as in Figs. 2 and 3 for initial states with $1.1 \gtrsim \theta_y \lesssim 0.7$ but clearly erratic behavior like in Figs. 4 and 5 for $0.4 \gtrsim \theta \gtrsim 0.1$. The size of the intermediate region in which the transition from regular to erratic recurrences takes place is in harmony with the classical transition range and the angular width of the initial coherent states, $\Delta \theta_y = 1 / \sqrt{j}$.

Needless to say that the qualitative difference between regular and erratic recurrences does not in itself constitute a "hard" criterion for quantum chaos. In searching such a criterion it might be interesting to study the width δn of the probability distribution of the temporal separations of recurrences of $\langle J_y \rangle$ to some close neighborhood of its initial value. The relative width $\delta n / j$ may behave quite differently at large j in the two cases. Evidently, more work is needed to explore such possibilities.

7. Quantum Beats Versus Broad-Band Excitation

Let us now turn to a spectral investigation of our quantum top based on the $2j + 1$ eigenvectors and eigenvalues of U. We have found the number of eigenvectors necessary for a satisfactory synthesis of $\langle J_y \rangle_n$ to vary considerably when the initial coherent state is moved from a region of classically regular motion to one of classical chaos. For a quantitative discussion of this phenomenon we may employ the minimum number $N_{\min}$ of eigenvectors of U necessary to exhaust the normalization of a coherent initial state to within, say, 1%. Figures 6a–c refer to coherent states on the classical T line and show $N_{\min}$ in its dependence on θ_y for $k = 2$, 3 and 6. It is quite interesting to compare the θ_y dependence of $N_{\min}$ to that of the classical Lyapounov exponent which is also displayed in Fig. 6. Roughly speaking, $N_{\min}$ is large when the Lyapounov exponent is.

The two quantities appear correlated most interestingly for $k = 3$. The rather flat minimum of $N_{\min}$, $N_{\min} \approx 8$, shows up close to the location of the classical elliptic fixed point, $\theta_y \approx 1.1$. As the neighboring chaotic regions are entered $N_{\min}$ grows by roughly an order of magnitude. The rather pronounced dips of $N_{\min}$ at the poles $\theta_y = 0$ and $\theta_y = \pi$ are due to a symmetry. For coherent states living close to those poles only eigenvectors which are even under R_y for j even are

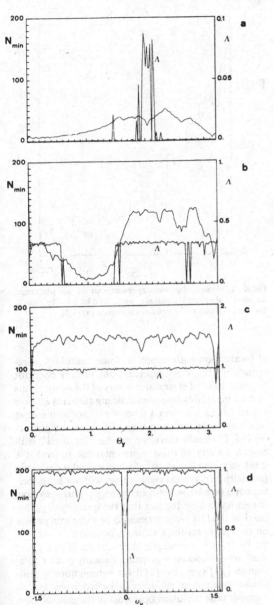

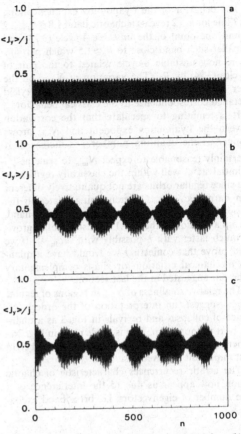

Fig. 7. Quantum mean $\langle J_y \rangle$ synthesized by the (a) four, (b) six, and (c) eight most important eigenvectors for $p = \frac{\pi}{2}$, $k = 3$ and the initial state localized in classically regular region at $\langle J_x \rangle_0 = -\langle J_z \rangle_0$, $\theta_y = 0.8$. These results should be compared with Fig. 3

Fig. 6a–d. Minimal number of eigenstates N_{min} exhausting 99% of the normalization of the initial coherent state localized at θ_y, ϕ_y, and the corresponding largest Lyapounov exponent Λ. **a** on the classical T line $\left(\phi_y = -\frac{\pi}{4}, \, 0 \leqslant \theta_y \leqslant \pi\right)$, $p = \frac{\pi}{2}$, $k = 2$; **b** on the classical T line $\left(\phi_y = -\frac{\pi}{4}, \, 0 \leqslant \theta_y \leqslant \pi\right)$, $p = \frac{\pi}{2}$, $k = 3$; **c** on the classical T line $\left(\phi_y = -\frac{\pi}{4}, \, 0 \leqslant \theta_y \leqslant \pi\right)$, $p = \frac{\pi}{2}$, $k = 6$; **d** on the equator $\left(\theta_y = \frac{\pi}{2}, \, -\frac{\pi}{2} \leqslant \phi_y \leqslant \frac{\pi}{2}\right)$, $p = \frac{\pi}{2}$, $k = 6$

appreciably populated while everywhere else along the T line even and odd eigenvectors tend to show up in approximately equal numbers. Clearly, it is the priviledge of coherent states located close to the poles defined by the y axis not to be displaced much by rotations around that axis by π and thus to have very small components along the eigenvectors of U odd under R_y.

For $k = 6$ our determination of the Lyapounov exponent does not reflect any regular motion anywhere along the T line. Correspondingly, N_{min} is large and rather uniform again except for the holes at the poles $\theta_y = 0, \pi$. On the other hand the dips corresponding to 4-cycle (3.22) are clearly visible when we move along the equator $\left(\theta_y = \frac{\pi}{2}\right)$ rather than along T line (Fig. 6d).

The behavior of the Lyapounov exponent along the T line for $k=2$ reveals a chaotic band $1.8 \lesssim \theta_y \leqslant 2.2$ around one point of the unstable 4-cycle (3.23) and a smaller such band close to $\theta_y \approx 1.5$ which pertains to the now unstable 6-cycle related to the pair of 3-cycles (3.24) by R_x. The variation of $N_{\min}$, on the other hand, is not sufficiently pronounced to yield a detailed correspondence to the classical behavior.

It is tempting to speculate that the correlation between the Lyapounov exponent and $N_{\min}$ grows stronger as the quantum number j is increased. It is certainly reasonable to expect $N_{\min}$ to scale as $\sqrt{j}$ for initial states well within the classically regular region since regular orbits are not qualitatively different from harmonic ones. Coherent initial states in the classically chaotic region should, on the other hand, pick up a fraction of the $2j+1$ eigenstates of U growing much faster with j, possibly with $N_{\min} \sim j$. If we could prove that conjecture we would have a quantum mechanical criterion for chaos complementary to the one suggested in the last section.

The relative smallness of $N_{\min}$ in regions of regular motion suggests an interpretation of the orderly sequence of collapses and revivals in Fig. 3 as a quantum beat phenomenon. This is made plain in Fig. 7a–c where $\langle J_y \rangle_n$ is synthesized by the four, six, and eight most important eigenvectors.

The erratic recurrences characteristic of chaotic motion now appear as due to the interference of a large number of eigenvectors, i.e. broadband excitation.

8. Level Repulsion

For kick strengths at which our top has regular classical trajectories over most of the sphere we can expect the eigenphases ϕ_n to have spacings with a Poisson distribution [10]. Figure 8 shows that expectation borne out nicely for $j=100$. Actually, to obtain a reasonably smooth level spacing distribution we had to superimpose the histograms pertaining to the 101 dimensional even subspace and the 100 dimensional odd subspace for five different kick strengths in the interval $0.1 \leqslant k \leqslant 0.3$ and $p=2$.

When we increase k to anywhere beyond 6 chaos dominates practically all of the classical sphere. The eigenphases ϕ_n must thus be expected to be equivalent in their statistical properties to the eigenvalues of "random" $(2j+1)$ by $(2j+1)$ matrices from an appropriate matrix ensemble. Among the ensembles which have been found to define different universality classes of level statistics we can confidently rule out the band-diagonal matrices whose eigenvectors show the effect

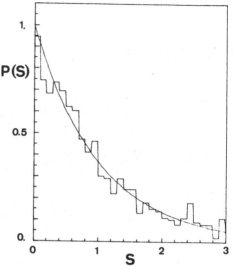

Fig. 8. Distribution of quasienergy spacings for the case when regular motion dominates classically, $p=2$, $0.1 \leqslant k \leqslant 0.3$. The smooth curve corresponds to the Poisson distribution $\exp(-S)$

of localization with respect to some "natural" representation [22]. If we represent our unitary operator U in the basis of eigenstates of any of the components of $\mathbf{J}$ we invariably find nonvanishing elements all over the matrix rather than a tendency of nonzero entries to cluster near the main diagonal. Typical eigenvectors of U should therefore not be "localized" with respect to any of these representations (provided k is set such one has global chaos classically). Speaking geometrically, typical eigenvectors will not have their supports on narrow solid angle ranges. This reasoning is also backed by the fact that the quantity $N_{\min}$ defined in the last section tends to be large everywhere on the sphere for the k values in question.

In order to investigate the statistics of our ϕ_n we shall now extend an argument originally given by Pechukas [23] (see also [24]) for autonomous Hamiltonian systems to periodically kicked ones. Let us consider time evolution operators of the general form

$$U(\lambda) = e^{i\lambda V} U_0, \qquad V = V^+ \tag{8.1}$$

and inquire about the dependence of the eigenvectors $|\phi_n(\lambda)\rangle$ and the eigenvalues $\phi(\lambda)$ on the parameter λ (For our kicked top λ can be taken to be the kick strength k while $V = -J_z^2/2j$). The eigenvectors and eigenvalues pertaining to λ and $\lambda + \delta\lambda$ can be related to one another by lowest-order perturbation theory with respect to $V\delta\lambda$. As $\delta\lambda \to 0$ we obtain the following set of differential equations

$$\frac{d}{d\lambda}\phi_n = V_{nn}$$

$$\frac{d}{d\lambda}V_{nm}$$

$$= \sum_{k(\neq n,m)} i\,V_{nk}V_{km}\left\{\frac{1}{e^{i(\phi_n-\phi_k)}-1}-\frac{1}{e^{-i(\phi_m-\phi_k)}-1}\right\}$$

$$+ i(1-\delta_{nm})\frac{V_{nm}(V_{mm}-V_{nn})}{e^{i(\phi_n-\phi_m)}-1}, \tag{8.2}$$

where

$$V_{nm} = \langle\phi_n(\lambda)|V|\phi_m(\lambda)\rangle.$$

We may interpret these differential equations as describing a flow in a phase space with ϕ_n and V_{nm} as generalized coordinates, and λ a "time". Alternatively, we may interpret them as showing how U moves about in a matrix ensemble the nature of which we want to find out. Having in mind that U tends to behave like a random matrix we assume the flow (8.2) to be ergodic. Instead of "time" averages over large λ intervals we can then employ ensemble averages to establish the statistics of the eigenvalues ϕ_n.

The phase space density $\rho(\{\phi_n\},\{V_{nm}\},\lambda)$ must obey the continuity equation

$$0 = \frac{\partial}{\partial\lambda}\rho + \mathrm{Div}(\mathbf{V}_\rho)$$

$$= \frac{d}{d\lambda}\rho(\{\phi_n(\lambda)\},\{V_{nm}(\lambda)\},\lambda) + \rho\,\mathrm{Div}\,\mathbf{V} \tag{8.3}$$

where the phase space divergence

$$\mathrm{Div}\,\mathbf{V} = \sum_n\frac{\partial\phi_{n'}}{\partial\phi_n} + \sum_{n,m}\frac{\partial V'_{nm}}{\partial V_{nm}} \tag{8.4}$$

is given by (8.2) as

$$\mathrm{Div}\,\mathbf{V} = \sum_{n\neq m} i(\phi'_n-\phi'_m)/\{e^{i(\phi_m-\phi_n)}-1\}$$

$$= -\frac{d}{d\lambda}\sum_{n\neq m}\ln|e^{i(\phi_n-\phi_m)}-1|. \tag{8.5}$$

The summation in (8.5) is over as many pairs n, m as there are independent real parameters in the set of off diagonal matrix elements V_{nm}. That number obviously depends of the dimension N of the Hilbert space the operators U and V live in (For our kicked top with even j we have $N=j+1$ and $N=j$ for the subspaces even and odd under R_y, respectively). If there is no restriction on V beyond Hermiticity, $V=V^+$, the number of independent pairs n, m is $N(N-1)$. Only half as many pairs are independent, however, if a generalized time reversal invariance holds as we shall show presently. Distinguishing the two cases by a parameter β which takes the values

1 and 2 for systems with and without such an invariance, respectively, we write the simplest solution of (8.3, 5) as

$$\rho \sim \prod_{n<m}|e^{i(\phi_n-\phi_m)}-1|^\beta. \tag{8.6}$$

If the flow (8.2) has no other constants of the "motion" the proportionality factor in (8.6) cannot depend on the phases ϕ_n and the matrix elements V_{nm} and is thus determined by normalization. The distribution ρ then yields the probability density of eigenphases in the ensemble of matrices within which $U(\lambda)$ moves ergodically. By suitably reducing [25] ρ we obtain the probability for two neighboring eigenphases to have the spacing S which reads, for $S\to 0$,

$$P(S)\,dS \sim S^\beta. \tag{8.7}$$

Kicked systems thus display, as do autonomous Hamiltonian ones, linear or quadratic level repulsion depending on whether or not they possess a suitable time reversal invariance.

To verify the asserted consequence of time reversal invariance we represent the generalized time reversal as in (4.9)

$$T = SK, \quad TUT = U^+. \tag{8.8}$$

where K is an antiunitary conjugation operation and S a unitary operator. We assume T not to depend on λ and to obey (4.10). With respect to any antiunitary operator T obeying $T^2=1$ it is possible to transform any complete set of orthogonal vectors into another one with T invariant basis vectors [26], $T|\psi_k\rangle = |\psi_k\rangle$. In such a basis U is represented by a symmetric matrix since $U_{kl}=\langle K\psi_k|KU\psi_l\rangle^* = \langle T\psi_k|TU\psi_l\rangle^* = \langle\psi_k|\;TUT\;|\psi_l\rangle^* = \langle U\psi_k|\psi_l\rangle^* = U_{lk}$. Moreover, the eigenvectors $|\phi_n\rangle$ of U have real components ϕ_n^k along those $|\psi_k\rangle$ and are thus T invariant themselves,

$$T|\phi_n\rangle = |\phi_n\rangle. \tag{8.9}$$

To prove this [25] we decompose the matrix U_{kl} as $U=U_1+iU_2$ with U_1 and U_2 both real and symmetric. The unitarity of U then yields $1=U^+U=U_1^2+U_2^2+i[U_1,U_2]$ and thus $[U_1,U_2]=0$. As any pair of commuting real symmetric matrices U_1 and U_2 can be diagonalized simultaneously by a real orthogonal matrix R the columns of which contain the components ϕ_n^k of the common eigenvectors of U_1, U_2, and U. The T invariance (8.9) is an immediate consequence of the reality of the ϕ_n^k, $T\sum_k\phi_n^k|\psi_k\rangle = \sum_k\phi_n^k$ $T|\psi_k\rangle = |\phi_n\rangle$.

We now differentiate in

$$Te^{i\lambda V}U_0 T = U_0^+e^{-i\lambda V} \tag{8.10}$$

with respect to λ, set $\lambda = 0$ and obtain the restriction imposed on the operator V by time reversal invariance,

$$TVT = U^+ VU. \tag{8.11}$$

By taking matrix elements with respect to eigenstates of U we derive the symmetry

$$e^{-i\phi_m} V_{mn} = V_{nm} e^{-i\phi_n}. \tag{8.12}$$

This symmetry could also have been obtained as a conservation law for the flow (8.2). We have now shown time reversal invariance to effectively halve the number of independent offdiagonal elements of V and thus to imply linear level repulsion.

Further insight into the role of time reversal can be obtained from a perspective established by Dyson [25]. Dyson's "orthogonal" ensemble of unitary symmetric N by N matrices is defined by its invariance under any transformation $S \to \tilde{W} S W$ where W is an arbitrary unitary matrix and $\tilde{W}$ its transpose. Any member U of this ensemble can thus be represented with the help of a suitable unitary matrix A as

$$U = A\tilde{A}, \qquad A^{-1} = A^+. \tag{8.13}$$

Assume such a decomposition to hold for a U of the form (8.1) with a particular value of λ. To find the condition under which U remains within the orthogonal ensemble when λ is varied we consider an infinitesimally neighboring one

$$U(\lambda + \delta\lambda) = (1 + iV\delta\lambda)\, U(\lambda) = (1 + i\, V\delta\lambda)\, A\tilde{A}$$
$$\equiv A(1 + iM\delta\lambda)\,\tilde{A}. \tag{8.14}$$

The last equation in (8.14) defines a matrix M as

$$M = A^{-1} VA \tag{8.15}$$

Evidently, then, $U(\lambda + \delta\lambda)$ is symmetric if and only if M is. Choosing the eigenvectors $|\phi_n\rangle$ of $U(\lambda)$ as a basis we have

$$A = \tilde{A} = \sum_n e^{i\phi_n/2} |\phi_n\rangle\langle\phi_n|$$

and

$$M_{nm} = \bar{e}^{i(\phi_n - \phi_m)/2}\, V_{nm}. \tag{8.16}$$

The symmetry of the matrix M, however, is just the identity (8.12) the compatibility of which with the evolution of $U(\lambda)$ in "time" is now obvious.

In order to compare the theoretically predicted level repulsion with numerical results for our top for $j = 100$ we had to superimpose several level spacing histograms. Figure 9 is based on $10 \times (101 + 100)$ eigenphases pertaining to ten different kick strengths in the interval $10.0 \leqslant k \leqslant 10.5$ and $p = 1.7$. We should note that we have chosen $p \neq \pi/2$ in order to avoid

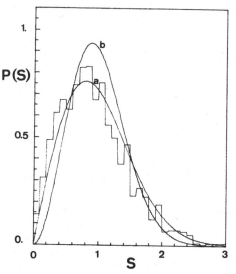

Fig. 9. Distribution of quasienergy spacings for the case when classically chaotic motion dominates; $p = 1.7$, $10.0 \leqslant k \leqslant 10.5$. The smooth curves correspond to (a) the Wigner distribution $\frac{1}{2}\pi S \exp\left(-\frac{\pi}{4}S^2\right)$ and (b) $\frac{32S^2}{\pi^2} \exp\left(-\frac{4}{\pi}S^2\right)$ which pertain to the cases of linear and quadratic repulsion, respectively

the symmetry (4.6) which would halve the number of independent ϕ_n in the odd subspace. Figure 9 quite convincingly reveals the expected linear level repulsion.

Let us finally turn to the most intriguing question as to how we have to modify the dynamics in order to achieve quadratic level repulsion [4, 9, 27]. As potential candidates we have studied the evolution operators

$$U(k', k, p) = e^{-i(k'/2j)J_x^2}\, e^{-i(k/2j)J_z^2}\, e^{-ipJ_y} \tag{8.17}$$

which differ from (2.2) by accounting for an additional nonlinear kick around the x axis. It is easy to see that in the special cases $p = \pi/2$ and $k' = k$ antiunitary generalized time reversal operators can again be constructed so that we must expect linear level repulsion. In the general case, however, we have not been able to identify any T invariance. Moreover, taking $U(0, k, p)$ as U_0 and $-J_x^2/2j$ as V in the sense of (8.1) we have verified that the flow (8.2) with $\lambda = k'$ is not compatible with the symmetry of the matrix (8.16). There is thus certainly no k' independent antiunitary operation T. While these findings do not prove that $U(k', k, p)$ typically lies within the so-called circular unitary ensemble we at least have no obvious reason to expect linear level repulsion. In fact, our numerical analysis for $j = 100$, $p = 1.7$, $\delta = 0.5$ and the same set

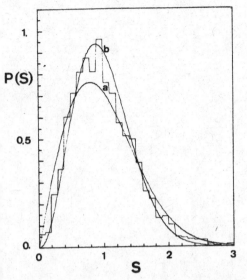

Fig. 10. Distribution of quasienergy spacings for the classically chaotic motion but with broken generalized time-reversal symmetry; $p = 1.7$, $10.0 \leqslant k \leqslant 10.5$, $k' = 0.5$. Smooth curves as in Fig. 9

6. Hogg, T., Huberman, B.A.: Phys. Rev. Lett. **48**, 711 (1982)
7. Bohigas, O., Giannoni, M.J., Schmit, C.. Phys. Rev. Lett. **52**, 1 (1984)
8. Haller, E., Köppel, H., Cederbaum, L.S.: Phys. Rev. Lett. **52**, 1165 (1984)
9. Seligman, T.H., Verbaarscot, J.J.M.: Phys. Lett. **108A**, 183 (1985)
10. Berry, M.V., Tabor, M.: Proc. R. Soc. London Ser. A **356**, 375 (1977)
11. Haake, F., Kuś, M., Mostowski, J., Scharf, R.: In: Coherence, cooperation and fluctuations. Haake, F., Narducci, L., Walls, D. (eds.), p. 220. Cambridge: Cambridge University Press 1986
12. Frahm, H., Mikeska, H.J.: Z. Phys. B – Condensed Matter **60**, 117 (1985)
13. Arecchi, F.T., Courtens, E., Gilmore, R., Thomas, H.: Phys. Rev. A **6**, 2211 (1972)
14. Glauber, R.J., Haake, F.: Phys. Rev. A **13**, 357 (1976)
15. Waldner, F., Barberis, D.R., Yamazaki, H. Phys. Rev. A **31**, 420 (1985)
16. Reynaud, S.: Private Communication
17. McKay, R.S.: Renormalization in area preserving maps. Ph.D. thesis, Princeton 1982
18. DeVogelaere, R.: In: Contributions to the theory of nonlinear oscillations. Lefschetz, S. (ed.), Vol. IV, p. 53. Princeton: Princeton University Press 1958
19. Lichtenberg, A.J., Lieberman, M.A.: Regular and stochastic motion. Berlin, Heidelberg, New York: Springer 1982
20. Eberly, J.H., Narozhny, N.B., Sanchez-Mondragon, J.J.: Phys. Rev. Lett. **44**, 1323 (1980)
21. Graham, R., Höhnerbach, M.: in: Proceedings of the Topical Meeting on Instabilities and Dynamics of Lasers. Rochester 1985 (to be published)
22. Feingold, M., Fishman, S., Grempel, D.R., Prange, R.E.: Phys. Rev. B **31**, 6852 (1985)
23. Pechukas, P.: Phys. Rev. Lett. **51**, 943 (1983)
24. Yukawa, T.: Phys. Rev. Lett. **54**, 1883 (1985)
25. Dyson, F.J.: J. Math. Phys. **3**, 140, 157, 166 (1962)
26. Porter, C.E.: In: Statistical theories of spectra. Porter, C.E. (ed.), p. 3. New York: Academic Press 1965
27. Robnik, M., Berry, M.V.: J. Phys. A **19**, 669 (1986)

of ten k values as in Fig. 9 suggests quadratic repulsion, as is shown in Fig. 10.

We gratefully acknowledge financial support of Marek Kuś by the Alexander von Humboldt-Stiftung and the Gesellschaft von Freunden und Förderern der Universität-Gesamthochschule Essen. We have benefitted from discussions with R. Graham, H. Frahm, S. Grossmann, H.J. Mikeska, S. Reynaud, and L. van Hemmen. During the early stages of this project we enjoyed the collaboration of J. Mostowski.

References

1. McDonald, S., Kaufman, A.N.: Phys. Rev. Lett. **42**, 1189 (1979)
2. Casati, G., Chirikov, B.V., Izrailev, F.M., Ford, J.: Stochastic behavior in classical and quantum hamiltonian systems. In: Lecture Notes in Physics. Casati, G., Ford, J. (eds.), Vol. 93, p. 334. Berlin, Heidelberg, New York: Springer 1979
3. Casati, G., Mantica, G., Guarneri, I.: In: Chaotic behavior in quantum systems. Casati, G. (ed.), p. 113. New York: Plenum Press 1985
4. Izrailev, F.M.: Quasi-energy level spacing distribution for quantum systems stochastic in the classical limit. (in Russian), Inst. of Nuclear Physics, Novosibirsk, Preprint No. 84–63 (1984)
5. Izrailev, F.M.: Phys. Rev. Lett. **56**, 541 (1986)

F. Haake
R. Scharf
Fachbereich Physik
Universität –
Gesamthochschule Essen
Postfach 10 37 64
D-4300 Essen 1
Federal Republic of Germany

M. Kuś
Institute of Theoretical Physics
Warsaw University
Hoza 69
PL-00-681 Warsaw
Poland

Self-similarity in quantum dynamics

L. E. Reichl and Li Haoming*

Center for Statistical Mechanics, University of Texas at Austin, Austin, Texas 78712

(Received 16 February 1990)

We have developed a renormalization transformation, based on the existence of higher-order nonlinear resonances in the double-resonance model, that gives good predictions for the extension of the wave function in that system due to nonlinear resonance overlap. The double-resonance model describes the qualitative behavior, in local regions of the Hilbert space, of many quantum systems with two degrees of freedom whose dynamics is described by a nonlinear Hamiltonian but a linear Schrödinger equation.

I. INTRODUCTION

The phase space of nonlinear nonintegrable classical conservative systems exhibits extremely complex structure consisting of regular Kolmogorov-Arnold-Moser (KAM) tori intermixed with chaos. In some regions of the phase space the structure is self-similar to all length scales and exhibits scaling behavior in space and time. KAM tori are the remnants of global conserved quantities. For many systems, when some parameter which characterizes the size of the nonlinearity is small, KAM tori dominate the phase space. However, as the non-linearity parameter is increased in size, nonlinear resonances in the system grow and overlap and destroy KAM tori lying between them. The existence of KAM tori can have a profound effect on the dynamics of a conservative system with two degrees of freedom because some KAM tori can divide the phase space into disjoint parts. When such a KAM torus is destroyed by nonlinear resonance overlap, the dynamics of the classical system may change dramatically.

The mechanism by which KAM surfaces are destroyed by nonlinear resonances has been studied extensively by Greene,[1] Shenker and Kadanoff,[2] MacKay,[3] and others. KAM tori have irrational winding numbers. Each irrational winding number can be represented uniquely by a continued fraction. Greene showed that associated with this continued fraction is a unique infinite sequence of nonlinear resonances which approximate the KAM torus. If the winding number of the KAM torus is represented by the continued fraction

$$w \equiv [a_0, a_1, a_2, \dots] = a_0 + \cfrac{1}{a_1 + \cfrac{1}{a_2 + \cdots}} \,,$$

then the resonances which approximate the KAM torus have periodic orbits with winding numbers

$$w_n \equiv [a_0, a_1, a_2, \dots, a_n, \infty]$$

with $\lim_{n \to \infty} w_n = w$. The resonances having periodic orbits with winding numbers w_0, w_1, w_2, etc. alternatively lie on opposite sides of the KAM torus. The width of each resonance depends on the nonlinearity parameter of the system. As this parameter is increased, resonances on an ever smaller scale grow until at a critical value of the nonlinearity parameter, all resonances in the sequence have grown large enough to punch holes in the KAM torus making it a Cantorus, a barrier with a Cantor set of holes through which phase-space trajectories can leak. The KAM torus breaks abruptly as a function of the non-linearity parameter. As the nonlinearity parameter increases further, the Cantorus gradually disappears.

The behavior of nonlinear resonances is easiest to study in nonlinear systems which are driven by a periodic external field. Because of the nonlinearity of the driven system, the external field induces infinite sets of nonlinear resonances in the phase space of the system. When the Hamiltonian is written in terms of the action-angle variables of the driven system, the induced resonances appear as traveling potential-energy waves in the phase space (cf. Appendixes A and B for examples). These are called the primary resonances of the system. The primary resonances interact to produce infinite families of higher-order resonances. There is considerable numerical evidence that the behavior of the phase space between any two primary resonances is largely determined by those two primary resonances and that the effects of primary resonances outside this region tend to average out. Therefore, in order to analyze the mechanism for destruction of KAM tori between two given primary resonances, it is often sufficient to consider a Hamiltonian composed of only those two resonances. It is this fact that underlies the renormalization procedure of Escande and Doveil.[4,5] They begin with a Hamiltonian which describes a system with two primary resonances. It consists of a particle moving in the presence of two cosine potential-energy waves, one of which is at rest and another which is traveling through phase space. This Hamiltonian, which they call the paradigm Hamiltonian, depends on only three parameters, the amplitudes of the two cosine waves and the relative wave number of the two waves. The renormalization proceeds as follows. They focus on a given KAM torus between the two primary resonances. By a sequence of canonical transformations they can select the two daughter resonances which bracket this KAM surface. They then write a paradigm Hamiltonian for these daughter resonances and neglect contributions coming from all other daughter resonances. In so doing they obtain a new paradigm Hamiltonian with two primary resonances and new amplitudes and wave number. By repeating the process they obtain a mapping of the amplitudes and wave number to ever smaller scales in the phase space. The properties of this map can be used to determine whether or not various KAM tori exist for a given initial paradigm Hamiltonian.

There is now considerable evidence that nonlinear resonances and KAM-like behavior exist in quantum-mechanical systems. Geisel, Radons, and Rubner[6] have studied the effect of KAM tori and Cantori, which are known to be present in the classical standard map, on the spread of probability in the quantum standard map. They find that KAM tori and Cantori act as barriers to the spread of probability and that probability decays exponentially across these barriers. Brown and Wyatt[7] have shown that a similar type of behavior exists for a driven oscillator model. The phenomenon of nonlinear resonance and resonance overlap in quantum dynamics has been studied extensively by Berman, Zaslavsky, and Kolovsky,[8] Lin and Reichl,[9,10] and Toda and Ikeda.[11] We now know that just as for classical systems, nonlinear resonance regions exist in the Hilbert space and for a small nonlinearity parameter remain isolated from one another. However, as the nonlinearity parameter is increased resonances can overlap leading to an extension of the wave function in the region of the influence of the overlapping resonances. In addition, resonance overlap leads to a change in the spectral statistics of states involved in the resonance overlap,[10] from a Poisson-like to a Wigner-like distribution. This is considered one of the manifestations of chaos in quantum systems. Recently, Reichl[12] has given numerical evidence that primary nonlinear resonances in Hilbert space generate higher-order resonances and that these play a role in facilitating resonance overlap in quantum systems just as they do in classical systems.

In this paper we shall show that renormalization techniques analogous to those of Escande and Doveil can be developed to describe the phenomenon of resonance overlap and extension of the wave function in quantum systems just as it can be for a classical system. However, rather than perform the renormalization on the Hamiltonian as is done classically, we shall perform the renormalization directly on the Schrödinger equation. We shall begin with a general double-resonance model which consists of two traveling cosine potential waves (two primary resonances). The renormalization procedure requires that the amplitude of one of the waves be larger than the other so that wave will dominate the system. We then rewrite the Schrödinger equation in terms of eigenstates of the system consisting of only a single large wave (this is an integrable system). By so doing, we come close to solving the problem. However, we find that in this new basis, the Schrödinger equation contains an infinite number of higher-order (daughter) resonances. We then focus on a given pair of daughters and write a double-resonance Schrödinger equation for them. This procedure can be repeated and gives a renormalization mapping which enables us to determine parameters of the

initial double-resonance model for which overlap occurs. Our results give much better estimates for resonance overlap than does the simple Chirikov estimate used until now for quantum systems.

We shall begin in Sec. II by writing a general double-resonance Schrödinger equation and then we will transform it into the form of a paradigm Schrödinger equation which is the basis of our renormalization transformation. In Sec III we obtain WKB solutions for energies and eigenstates of the single-resonance Schrödinger equation in the region outside the resonance and we show that the WKB energies agree fairly well with exact energies obtained numerically. These analytic expressions enable us to build the renormalization transformation. In Sec. IV we use the WKB solutions to write the double-resonance Schrödinger equation in terms of eigenstates of the single-resonance equation. This generates an infinite family of higher resonances. We show that the WKB amplitudes for the higher-order resonances agree fairly well with exact numerical expressions.

In Sec. V we write the renormalization mapping for this system. The renormalization map relates the relative wave number and amplitudes of a resonance pair at one level (scale) to those at a higher level (smaller scale). The renormalization mapping thus allows us to examine a sequence of resonance pairs on an ever smaller scale in the Hilbert space. Bounded quantum systems have a discrete spectrum so that in actuality we cannot go to infinitely small scale in such systems. However, the mappings have a stable manifold which separates regions in which resonance overlap at smaller scale does not occur (the stable side) from regions in which it does occur (the unstable side). In the unstable region, the amplitudes grow so rapidly as we go to small scale that the mapping still gives good predictions.

In Sec. VI we give numerical results showing that steps by which overlap of the two primary resonances occurs and we compare the renormalization predictions with observed results. Finally, in Sec. VII we make some concluding remarks.

II. THE PARADIGM SCHRÖDINGER EQUATION

The manifestations of chaos occur in quantum systems when nonlinear quantum resonances zones overlap in the unperturbed Hilbert space.[10] As we have shown in Ref. 12, resonance overlap is facilitated by the existence of higher-order nonlinear resonances that are generated by the interaction of primary resonances. In this paper we explore the structure of the network of nonlinear resonances generated by any given pair of primary resonances and develop a renormalization scheme to describe the overlap of any given sequence of successively higher-order resonance pairs.

Let us consider a typical pair of resonances found in the standard map or square-well system discussed in Appendixes A and B. The Schrödinger equation can be written

$$-i\frac{\partial\langle\theta|\Psi^{(0)}(t)\rangle}{\partial t}=-\frac{\partial^2\langle\theta|\Psi^{(0)}(t)\rangle}{\partial\theta^2}$$

$$+\{V_a(0)\cos[\mu_a(0)\theta-\omega_0 t]$$

$$+V_b(0)\cos[\mu_b(0)\theta\pm\omega_0 t]\}$$

$$\times\langle\theta|\Psi^{(0)}(t)\rangle \qquad (2.1)$$

where $V_a(0)$ and $V_b(0)$ are the amplitudes of the cosine waves, $\mu_a(0)$ and $\mu_b(0)$ are wave numbers of the cosine waves, ω_0 is a radial frequency associated with this time periodic system, and $|\Psi^{(0)}\rangle$ is the state of the system. The parameters $V_i(0)$ and ω_0 depend on Planck's constant $\hbar$ as $V_i(0)\sim\hbar^{-2}$ and $\omega_0\sim\hbar^{-1}$. The index 0 on the above

quantities indicates that we are at the zeroth level of the renormalization transformation. We shall assume that our system has periodic boundary conditions with period $2\pi N$ and therefore we require that $\langle\theta|\Psi^{(0)}(t)\rangle=\langle\theta+2\pi N|\Psi^{(0)}(t)\rangle$ and that $\mu_i(0)=M_i(0)/N$ $(i=a,b)$, where $M_i(0)$ are integers. We can also write Eq. (2.1) in terms of a traveling-wave basis. If we note that

$$\langle\theta|\Psi^{(0)}(t)\rangle=\sum_{k=-\infty}^{\infty}e^{ik\theta}\langle k|\Psi^{(0)}(t)\rangle , \qquad (2.2)$$

where k is a rational fraction, $k=n/N$, n is an integer, and the summation $\sum_{k=-\infty}^{\infty}$ is over all values in k. In terms of the states $\langle k|\Psi^{(0)}(t)\rangle$ Eq. (2.1) takes the form

$$-i\frac{\partial\langle k|\Psi^{(0)}(t)\rangle}{\partial t}=k^2\langle k|\Psi^{(0)}(t)\rangle+\frac{V_a(0)}{2}[e^{-i\omega_0 t}\langle k-\mu_a(0)|\Psi^{(0)}(t)\rangle+e^{i\omega_0 t}\langle k+\mu_a(0)|\Psi^{(0)}(t)\rangle]$$

$$+\frac{V_b(0)}{2}[e^{\pm i\omega_0 t}\langle k-\mu_b(0)|\Psi^{(0)}(t)\rangle+e^{\mp i\omega_0 t}\langle k+\mu_b(0)|\Psi^{(0)}(t)\rangle] . \qquad (2.3)$$

As we have shown in Refs. 9 and 12, the resonance due to cosine wave $\cos[\mu_i(0)\theta\pm\omega_0 t]$ occurs at $k=\bar{k}_i=\mp\omega_0/2\mu_i(0)$ where $i=a,b$, and has a half-width $\Delta k_i=[2V_i(0)]^{1/2}$.

We can transform to the rest frame of the cosine wave $\cos[\mu_a(0)\theta-\omega_0 t]$ via a unitary transformation

$$\hat{U}_{\mu_a(0)}(t)=\exp(-i\hat{P}_{\mu_a(0)}t) , \qquad (2.4)$$

where

$$\langle k'|\hat{P}_{\mu_a(0)}|k\rangle=(2k\bar{k}_a-\bar{k}_a^2)\delta_{k,k'} . \qquad (2.5)$$

We introduce a state $|\Phi^{(0)}(t)\rangle$ such that

$$|\Psi^{(0)}(t)\rangle=\hat{U}_{\mu_a(0)}(t)|\Phi^{(0)}(t)\rangle . \qquad (2.6)$$

The state $|\Phi^{(0)}(t)\rangle$ describes the behavior of the system in the rest frame of the cosine wave, $\cos[\mu_a(0)\theta-\omega_0 t]$. In terms of this state, the Schrödinger equation takes the form

$$-i\frac{\partial\langle k|\Phi^{(0)}(t)\rangle}{\partial t}=(k-\bar{k}_a)^2\langle k|\Phi^{(0)}(t)\rangle+\frac{V_a(0)}{2}[\langle k-\mu_a(0)|\Psi^{(0)}(t)\rangle+\langle k+\mu_a(0)|\Phi^{(0)}(t)\rangle]$$

$$+\frac{V_b(0)}{2}[e^{+i(v_0\pm1)\omega_0 t}\langle k-\mu_b(0)|\Phi^{(0)}(t)\rangle+e^{-i(v_0\pm1)\omega_0 t}\langle k+\mu_b(0)|\Phi^{(0)}(t)\rangle] \qquad (2.7)$$

where $v_0=\mu_b(0)/\mu_a(0)$ is the ratio of the wave numbers of the two cosine waves in Eq. (2.1). If we use the transformation in Eq. (2.2) to transform back to the angle picture, Eq. (2.7) takes the form

$$-i\frac{\partial\langle\theta|\Phi^{(0)}(t)\rangle}{\partial t}=\left[-i\frac{\partial}{\partial\theta}-\bar{k}_a\right]^2\langle\theta|\Phi^{(0)}(t)\rangle+\{V_a(0)\cos[\mu_a(0)\theta]+V_b(0)\cos[\mu_b(0)\theta+(v_0\pm1)\omega_0 t]\}\langle\theta|\Phi^{(0)}(t)\rangle .$$

$$(2.8)$$

Let us now rescale the angle so that $\Theta=\mu_a(0)\theta$. Then the stationary cosine oscillates once in period 2π and Eq. (2.8) takes the form

$$-i\frac{\partial\langle\Theta|\Phi^{(0)}(t)\rangle}{\partial t}=\left[-i\mu_a(0)\frac{\partial}{\partial\Theta}-\bar{k}_a\right]^2\langle\Theta|\Phi^{(0)}(t)\rangle+\{V_a(0)\cos(\Theta)+V_b(0)\cos[v_0\Theta+(v_0\pm1)\omega_0 t]\}\langle\Theta|\Phi^{(0)}(t)\rangle .$$

$$(2.9)$$

If we introduce a change of phase

$$\langle\Theta|\Phi^{(0)}(t)\rangle=\langle\Theta|\chi^{(0)}(t)\rangle e^{i\bar{k}_a\Theta/\mu_a(0)}\ ,$$

then Eq. (2.9) takes the form

$$-i\frac{\partial\langle\Theta|\chi^{(0)}(t)\rangle}{\partial t}=-[\mu_a(0)]^2\frac{\partial^2\langle\Theta|\chi^{(0)}(t)\rangle}{\partial\Theta^2}+\{V_a(0)\cos(\Theta)+V_b(0)\cos[\nu_0(\Theta+\bar{\omega}_0 t)]\}\langle\Theta|\chi^{(0)}(t)\rangle \tag{2.10}$$

where $\bar{\omega}_0=[(\nu_0\pm1)/\nu_0]\omega_0$. Equation (2.11) is the paradigm Schrödinger equation and is the starting point of our process of renormalization. In the subsequent sections, we shall always assume that $V_a(0)>V_b(0)$.

III. WKB EXPRESSIONS FOR PENDULUM STATES

In order to obtain an equation for higher-order resonances, we expand Eq. (2.10) in terms of eigenstates of Eq. (2.10) for the case when $V_b(0)=0$. When $V_b(0)=0$, Eq. (2.10) is integrable and is a form of Mathieu equation. If $V_a(0)\gg V_b(0)$ then we come close to solving Eq. (2.10) by doing this, or at least we are in a basis which more adequately describes the actual behavior of the system. However, as we shall see in Sec. IV, we also reveal the fact that this system contains an infinite number of higher-order nonlinear resonances.

Let us consider the quantum pendulum equation (a form of Mathieu equation)

$$E_l\phi_l(\theta)=-\frac{\partial^2\phi_l(\theta)}{\partial\theta^2}+V_a\cos(\mu_a\theta)\phi_l(\theta) \tag{3.1}$$

where $l=m/N\geq0$ (m an integer), $\phi_l(\theta)=\langle\theta|\phi_l\rangle$, and $\phi_l(\theta)$ satisfies the boundary condition $\phi_l(\theta)=\phi_l(\theta+2\pi N)$. Equation (3.1) has solutions of definite parity, $\phi_l^\alpha(\theta)$ ($\alpha=A,S$),

$$E_l^\alpha\phi_l^\alpha(\theta)=-\frac{\partial^2\phi_l^\alpha(\theta)}{\partial\theta^2}+V_a\cos(\mu_a\theta)\phi_l^\alpha(\theta)\ , \tag{3.2}$$

where the symmetric solutions $\phi_l^{(S)}(\theta)$ satisfy the conditions $\phi_l^{(S)}(\theta)=\phi_l^{(S)}(-\theta)$ and $\phi_{-l}^{(S)}(\theta)=\phi_l^{(S)}(\theta)$ and the antisymmetric solutions satisfy the conditions $\phi_l^{(A)}(\theta)=-\phi_l^{(A)}(-\theta)$ and $\phi_{-l}^{(A)}(\theta)=-\phi_l^{(A)}(\theta)$. Furthermore, $E_l^{(\alpha)}$ is an even function of l. In the limit $V_a\rightarrow0$, $\phi_l^{(S)}(\theta)\rightarrow(1/\sqrt{\pi N})\cos(l\theta)$ and $\phi_l^{(A)}(\theta)\rightarrow(i/\sqrt{\pi N})\sin(l\theta)$. Let us now introduce the following functions:

$$\phi_l^{(+)}(\theta)=\frac{1}{\sqrt{2}}[\phi_l^{(S)}(\theta)+\phi_l^{(A)}(\theta)]\ , \tag{3.3a}$$

$$\phi_l^{(-)}(\theta)=\frac{1}{\sqrt{2}}[\phi_l^{(S)}(\theta)-\phi_l^{(A)}(\theta)]\ , \tag{3.3b}$$

for $l>0$ with $\phi_0=\phi_0^{(S)}(\theta)$ and $\phi_0=\phi_0^{(A)}(\theta)=0$. These functions reduce to traveling waves in the limit $V_a\rightarrow0$.

and have the property that $\phi_{-l}^{(-)}(\theta)=\phi_l^{(+)}(\theta)$. From Eqs. (3.2) and (3.3) we obtain

$$E_l^{(+)}\phi_l^{(+)}(\theta)+E_l^{(-)}\phi_l^{(-)}(\theta)$$
$$=-\frac{\partial^2\phi_l^{(+)}(\theta)}{\partial\theta^2}+V_a\cos(\mu_a\theta)\phi_l^{(+)}(\theta) \tag{3.4}$$

where $E_l^{(+)}=\frac{1}{2}(E_l^{(+)}+E_l^{(+)})$ and $E_l^{(-)}=\frac{1}{2}(E_l^{(-)}-E_l^{(+)})$. Since $\phi_{-l}^{(-)}(\theta)=\phi_l^{(+)}(\theta)$, we can introduce a new function $\phi_p(\theta)=\phi_{|p|}^{(+)}(\theta)$ and $\phi_{-p}(\theta)=\phi_{|p|}^{(-)}(\theta)$ where p has the range $-\infty\leq p\leq\infty$. Then Eq. (3.4) takes the form

$$E_p^{(+)}\phi_p(\theta)+E_p^{(-)}\phi_{-p}(\theta)$$
$$=-\frac{\partial^2\phi_p(\theta)}{\partial\theta^2}+V_a\cos(\mu_a\theta)\phi_p(\theta)\ . \tag{3.5}$$

If we make the change of variables, $\Theta=\mu_a\theta$, Eq. (3.5) takes the form

$$E_p^{(+)}\phi_p(\Theta)+E_p^{(-)}\phi_{-p}(\Theta)$$
$$=-\mu_a^2\frac{\partial^2\phi_p(\Theta)}{\partial\Theta^2}+V_a\cos(\Theta)\phi_p(\Theta)\ . \tag{3.6}$$

Let us now obtain the WKB solutions to these equations. Assume a solution to Eq. (3.1) of the form $\phi_p(\theta)=e^{if_p(\theta)}$. Then

$$if_p''-f_p'^2+[E_p-V_a\cos(\mu_a\theta)]=0\ , \tag{3.7}$$

where $f_p'=df_p/d\theta$. In regions where $|f_p''|\ll1$, the solution to Eq. (3.1) can be written

$$\phi_p(\theta;\mu_a)\approx\psi_p(\theta;\mu_a)$$
$$=\frac{1}{[\varepsilon_p-V_a\cos(\mu_a\theta)]^{1/4}}$$
$$\times\exp\left[\frac{i}{\mu_a}\int_0^\theta[\varepsilon_p-V_a\cos(\mu_a\theta)]^{1/2}d\theta\right]\ , \tag{3.8}$$

with $E_p^{(S)}\approx\varepsilon_p$ and $E_p^{(A)}\approx\varepsilon_p$ so that in the WKB approximation $E_p^{(-)}=0$. It is easy to show that the condition of validity, $|f_p''|\ll1$, of the WKB solution implies that the WKB solution will only be valid when $\varepsilon_p\gg U_a$.

The integral appearing in Eq. (3.8) can be done explicitly. We find

$$\int_0^\theta d\theta[\varepsilon_p-V_a\cos(\mu_a\theta)]^{1/2}=\frac{1}{\mu_a}\int_0^{\mu_a\theta}d\Theta[\varepsilon_p-V_a\cos(\Theta)]^{1/2}$$
$$=\frac{2}{\mu_a}(\varepsilon_p+V_a)^{1/2}[E(\mu_a\theta/2+\pi/2,\kappa)+E(\pi/2,\kappa)]\ , \tag{3.9}$$

where $E(\Theta,\kappa)$ is the incomplete elliptic integral of the second kind, and κ is the modulus and is defined $\kappa^2 = 2V_a/(\varepsilon_p + V_a)$. The modulus κ in the region of validity of the WKB approximation will be small. Therefore we can expand the right-hand side of Eq. (3.9) in powers of κ. If we keep terms to second order in κ, we obtain

$$\int_0^\theta d\theta [\varepsilon_p - V_a\cos(\mu_a\theta)]^{1/2} = \frac{1}{\mu_a}\int_0^{\mu_a\theta} d\Theta[\varepsilon_p - V_a\cos(\Theta)]^{1/2}$$

$$\approx \frac{2}{\mu_a}(\varepsilon_k + V_a)^{1/2}\left[(1-\tfrac{1}{4}\kappa^2)\frac{\mu_a\theta}{2} - \tfrac{1}{8}\kappa^2\sin(\mu_a\theta)\right]. \tag{3.10}$$

The WKB values for the energy eigenvalues ε_p are obtained from the condition

$$\int_0^{2\pi N} d\theta[\varepsilon_p - V_a\cos(\mu_a\theta)]^{1/2} = 2\pi pN \tag{3.11}$$

where p is a rational fraction, $p = m/N$, and m is an integer. Using Eq. (3.6), the quantization condition becomes

$$(\varepsilon_p + V_a)^{1/2}(1-\tfrac{1}{4}\kappa^2) = p. \tag{3.12}$$

Thus the energy is given approximately by

$$\varepsilon_p \approx p^2\left[1 - \frac{1}{4}\frac{V_a^2}{p^4} + \frac{1}{4}\frac{V_a^3}{p^6} + \cdots\right] \tag{3.13}$$

and $\varepsilon_p \approx p^2$ when V_a/p^2 is small.

Let us now introduce the quantity

$$x_p = \frac{1}{4\mu_a}\kappa^2(\varepsilon_p + V_a)^{1/2}$$

$$= \frac{V_a}{2\mu_a(\varepsilon_p + V_a)^{1/2}} \approx \frac{V_a}{2p\mu_a}\left[1 + \frac{V_a^2}{8p^2\mu_a^2} + \cdots\right]. \tag{3.14}$$

Then the WKB wave function, which is a solution to Eq. (3.1), can be written

$$\psi_p(\theta) \approx \frac{1}{\sqrt{2\pi N}}e^{+i[p\theta - x_p\sin(\mu_a\theta)]}. \tag{3.15}$$

Equation (3.15) is a solution to Eq. (3.1) if we neglect terms of order V_a/p. These will always give contribu-

tions of order V_a/p^2 in the subsequent calculations. Equation (3.15) is valid for $p^2 \gg V_a$. The WKB wave function, Eq. (3.15), is normalized to one. A rather lengthy calculation shows that

$$\int_0^{2\pi N} d\theta[\psi_{p'}(\theta)]^*\psi_p(\theta) = \delta_{p'p}. \tag{3.16}$$

The WKB solution to Eq. (3.6) can be written

$$\psi_p(\Theta) \approx \frac{1}{(2\pi M_a)^{1/2}}e^{+i[(p/\mu_a)\Theta - x_p\sin(\Theta)]} \tag{3.17}$$

and satisfies the normalization condition

$$\int_0^{2\pi M_a} d\Theta[\psi_{p'}(\Theta)]^*\psi_p(\Theta) = \delta_{p',p} \tag{3.18}$$

where we have used the fact that $\mu_a N = M_a$. It is useful to note that the WKB wave functions can be expanded in a Fourier series

$$\psi_p(\Theta) = \frac{1}{(2\pi M_a)^{1/2}}\sum_{n=-\infty}^{\infty} J_n(x_p)e^{i(p/\mu_a - n)\Theta} \tag{3.19}$$

where $J_n(x)$ is the Bessel of integer order n.

It is of interest to compare the exact eigenvalues of Eq. (3.2) to the WKB approximation $\varepsilon_p \approx p^2$. In Fig. 1 we plot $E_p^{(+)}$ versus p and compare it to $\varepsilon_p \approx p^2$ for $N=1$ and $M_a = 1$ and in a range of values $V_a = 240, 480, 720, 960$. In Fig. 2, we plot $E_p^{(-)}$ versus p for the same values of N, M_a, and V_a. We see that $E_p^{(+)} \approx p^2$ when $E_p^{(-)} = 0$.

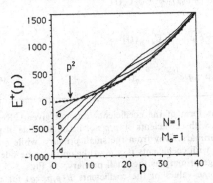

FIG. 1. Plot of $E^{(+)}(p)$ for $N=1$, $M_a=1$. Curves labeled a, b, c, and d correspond to $V_a = 240, 480, 720,$ and 960, respectively. The four curves were obtained numerically. Also shown is a plot of p^2 vs p. For p large enough, $E_p^{(+)} \approx p^2$.

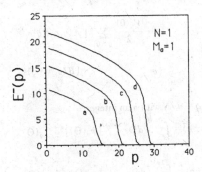

FIG. 2. Plot of $E^{(-)}(p)$ for $N=1$, $M_a=1$. Curves labeled a, b, c, and d correspond to $V_a = 240, 480, 720,$ and 960, respectively. These curves were obtained numerically.

IV. EXPANSION IN MATHIEU EIGENSTATES

Let us now return to the paradigm Schrödinger equation, Eq. (2.10), and expand it in terms of eigenstates of the Mathieu equation (3.2). We first expand the wave function $\langle \Theta | \chi^{(0)}(t) \rangle$ in a complete set of eigenstates,

$$\langle \Theta | \chi^{(0)}(t) \rangle = \sum_{l=0}^{\infty} [C_l^{\{S\}}(t)\phi_l^{\{S\}}(\theta) + C_l^{\{A\}}(t)\phi_l^{\{A\}}(\theta)] \quad (4.1)$$

where the functions $\phi_l^{\{S\}}(\Theta)$ and $\phi_l^{\{S\}}(\Theta)$ depend on parameters $\mu_a(0)$, N, and $V_a(0)$, and are defined similarly to $\phi_l^{\{S\}}(\Theta)$ and $\phi_l^{\{A\}}(\Theta)$. The functions $C_l^{\{S\}}(t)$ and $C_l^{\{A\}}(t)$ depend on parameters $\mu_a(0)$, N, v_0, $\overline{\omega}_0$, $V_a(0)$, and $V_b(0)$. If we substitute Eq. (4.1) into Eq. (2.10) and use the orthonormality of Mathieu eigenstates, we obtain

$$-i\frac{\partial C_l^{\{\alpha\}}(t)}{\partial t} = E_l^{\{\alpha\}}C_l^{\{\alpha\}}(t) + V_b(0)\sum_{l'=0}^{\infty}\int_0^{2\pi M_a}d\Theta[\phi_{l'}^{\{\alpha\}}(\Theta)]^*\cos[v_0(\Theta+\overline{\omega}_0 t)]\sum_{\alpha'=S,A}C_{l'}^{\{\alpha'\}}(t)\phi_{l'}^{\{\alpha'\}}(\Theta) , \quad (4.2)$$

where $\alpha = S, A$. Let us define

$$C_l^{\{+\}} = \frac{1}{\sqrt{2}}(C_l^{\{S\}} + C_l^{\{A\}}) \quad \text{and} \quad C_l^{\{+\}} = \frac{1}{\sqrt{2}}(C_l^{\{S\}} - C_l^{\{A\}}) . \quad (4.3)$$

If we now note that $C_l^{\{-\}} = C_{-l}^{\{+\}}$ and combine Eqs. (4.2) and (4.3) we can write (after considerable algebra)

$$-i\frac{\partial C_l^{\{+\}}(t)}{\partial t} = E_l^{\{+\}}C_l^{\{+\}}(t) + E_l^{\{-\}}C_l^{\{-\}}(t)$$

$$+ \frac{V_b(0)}{\sqrt{2}}\int_0^{2\pi M_a}d\Theta\cos[v_0(\Theta+\overline{\omega}_0 t)]\sum_{l'=0}^{\infty}[(\phi_l^{\{+\}})^*\phi_{l'}^{\{+\}}C_{l'}^{\{+\}} + (\phi_l^{\{+\}})^*\phi_{l'}^{\{-\}}C_{l'}^{\{-\}}] \quad (4.4)$$

and

$$-i\frac{\partial C_l^{\{-\}}(t)}{\partial t} = E_l^{\{+\}}C_l^{\{-\}}(t) + E_l^{\{-\}}C_l^{\{+\}}(t)$$

$$+ \frac{V_b(0)}{\sqrt{2}}\int_0^{2\pi M_a}d\Theta\cos[v_0(\Theta+\overline{\omega}_0 t)]\sum_{l'=0}^{\infty}[(\phi_l^{\{-\}})^*\phi_{l'}^{\{+\}}C_{l'}^{\{+\}} + (\phi_l^{\{-\}})^*\phi_{l'}^{\{-\}}C_{l'}^{\{-\}}] . \quad (4.5)$$

Let is now note that $C_l^{\{-\}} = C_{-l}^{\{+\}}$ and $\phi_l^{\{-\}} = \phi_{-l}^{\{+\}}$. Then we can write Eqs. (4.4) and (4.5) as a single equation. Let us introduce the function $\phi_p(\Theta) = \phi_{|p|}^{\{+\}}(\Theta)$ for $p > 0$ and $\phi_p(\Theta) = \phi_{-|p|}^{\{+\}}(\Theta)$ for $p < 0$. Similarly we define $C_p(t) = C_{|p|}^{\{+\}}(t)$ for $p > 0$ and $C_p(t) = C_{-|p|}^{\{+\}}(t)$ for $p < 0$. Then Eqs. (4.4) and (4.5) can be written

$$-i\frac{\partial C_p(t)}{\partial t} = E_p^{\{+\}}C_p(t) + E_p^{\{-\}}C_{-p}(t) + V_b(0)\int_0^{2\pi M_a}d\Theta\cos[v_0(\Theta+\overline{\omega}_0 t)]\sum_{p'=-\infty}^{\infty}[(\phi_p)^*\phi_{p'}C_{p'}] \quad (4.6)$$

where the index p has the range $-\infty \leq p \leq \infty$.

If we expand $\cos[v_0(\Theta+\overline{\omega}_0 t)]$ in exponentials, we can write Eq. (4.6) in the form

$$-i\frac{\partial C_p(t)}{\partial t} = E_p^{\{+\}}C_p(t) + E_p^{\{-\}}C_{-p}(t) + \frac{V_b(0)}{2}\int_0^{2\pi M_a}d\Theta\cos[v_0(\Theta+\overline{\omega}_0 t)](\phi_p)^*\phi_p C_p(t)$$

$$+ \frac{V_b(0)}{2}\sum_{q>0}\{[A(p,p+q)e^{iv_0\overline{\omega}_0 t}C_{p+q}(t) + A(p,p-q)e^{-iv_0\overline{\omega}_0 t}C_{p-q}(t)]$$

$$+ [B(p,p+q)e^{-iv_0\overline{\omega}_0 t}C_{p+q}(t) + B(p,p-q)e^{iv_0\overline{\omega}_0 t}C_{p-q}(t)]\} \quad (4.7)$$

where $q = n/N$ with n an integer,

$$A(p,p\pm q) = \int_0^{2\pi M_a}d\Theta e^{\pm iv_0\Theta}[\phi_p(\Theta)]^*\phi_{p\pm q}(\Theta) , \quad (4.8)$$

and

$$B(p,p\pm q) = \int_0^{2\pi M_a}d\Theta e^{\mp iv_0\Theta}[\phi_p(\Theta)]^*\phi_{p\pm q}(\Theta) . \quad (4.9)$$

The equation for $C_p(t)$ contains an infinite number of resonances. [Equation (4.7) and the coefficients $A(p,p\pm q)$ and $B(p,p\pm q)$ were also derived in Ref. 12 although an

analytic form of the coefficients was not given.] Resonances with coefficients $A(p,p\pm q)$ lie on the side of the large primary away from the small primary, while resonances with coefficients $B(p,p\pm q)$ lie on the same side of the large primary as the small primary. In Figs. 3–5 we plot some values of the coefficients $B(p,p-q)$ for the special cases $N=1$, $M_a=1$, and $v_0=\frac{1}{3}$, $\frac{3}{1}$, and $\frac{5}{1}$. [Note that $B(p,p-q) = B(p+q,p)$.] In Fig. 6 we plot $B(p,p-1)$ and $B(p,p-2)$ for $v_0=\frac{5}{1}$, $N=1$, and $V_a(0)=240$. Note that the amplitudes oscillate in the re-

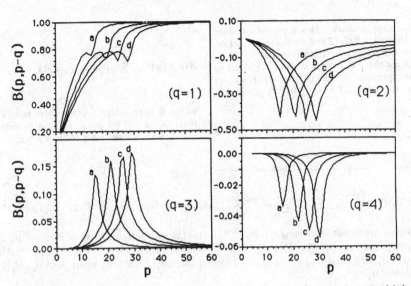

FIG. 3. Plot of amplitudes $B(p, p-q)$ for $v_0 = \frac{1}{\tau}$ and $N=1$ for the cases $q = 1$–4. In each figure the curves labeled a, b, c, and d correspond to $V_a = 240$, 480, 720, and 960, respectively.

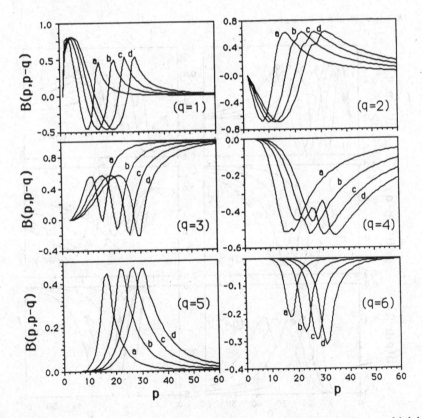

FIG. 4. Plot of amplitudes $B(p, p-q)$ for $v_0 = \frac{1}{\tau}$ and $N=1$ for the cases $q = 1$–6. In each figure the curves labeled a, b, c, and d correspond to $V_a = 240$, 480, 720, and 960, respectively.

gion inside the large primary. This a purely quantum effect. As we showed in Ref. 12 and we shall show below, the coefficients $A(p, p\pm q)$ are extremely small.

We can now use the WKB states derived in Sec. III to obtain explicit expressions for the coefficients $A(p, p\pm q)$ and $B(p, p\pm q)$ for values of $p > [V_a(0)]^{1/2}$. From Eq. (3.7) we find

$$A(p, p\pm q) = \sum_{K=-\infty}^{\infty} \delta_{qN, -M_b \pm K M_a} J_K(x_p - x_{p\pm q}) \quad (4.10)$$

and

$$B(p, p\pm q) = \sum_{K=-\infty}^{\infty} \delta_{qN, M_b \pm K M_a} J_K(x_p - x_{p\pm q}), \quad (4.11)$$

where K is an integer. Using these results and the fact that $E^{(-)}(p) = 0$ for $p > p_c$ (p_c can be determined from Fig. 2), we can rewrite Eq. (4.7) for $p > p_c$:

$$-i\frac{\partial C_p(t)}{\partial t} = p^2 C_p(t) + \frac{V_b(0)}{2} \sum_{K=-K^*}^{K^*} [U_K(p)e^{iv_0\bar{\omega}_0 t} C_{p-\mu_a(0)(v_0+K)}(t) + U_{-K}(p)e^{-iv_0\bar{\omega}_0 t} C_{p+\mu_a(0)(v_0+K)}(t)] + \cdots, \quad (4.12)$$

where the ellipsis represents remaining terms, K^* is the largest integer K such that $p - \mu_a(0)(K + v_0)x p_c$, $U_K(p) = J_K(x_p - x_{p-\mu_0(v_0+K)})$, and $U_{-K}(p) = J_{-K}(x_p - x_{p+\mu_0(v_0+K)})$. Note that $U_{\pm K}(p) = A(p, p\pm q)$ for $-\infty \le K \le -(\text{int}|M_b/M_a|)$ and $U_{\pm K}(p) = B(p, p\pm k)$ for $(\text{int}|M_b/M_a|) \le K \le \infty$, where $(\text{int}|M_b/M_a|)$ indicates the integer part of $|M_b/M_a| + 1$.

Let us now note that

$$x_p - x_{p\pm\mu_a(0)(K+v_0)} \approx \pm \frac{(K+v_0)V_a(0)}{2p^2}. \quad (4.13)$$

Thus for $p^2 \gg (K+v_0)V_a(0)$ and $K \neq 0$ we have

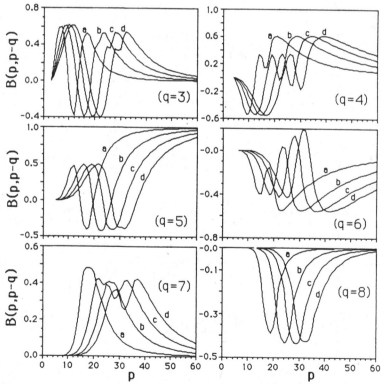

FIG. 5. Plot of amplitudes $B(p, p-q)$ for $v_0 = \frac{5}{1}$ and $N=1$ for the cases $q = 3$–8. In each figure the curves labeled a, b, c, and d correspond to $V_a = 240, 480, 720,$ and 960, respectively.

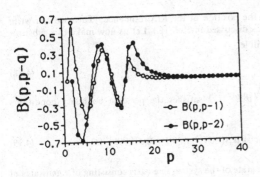

FIG. 6. Plot of amplitudes $B(p,p-1)$ and $B(p,p-2)$ for $v_0 = \frac{5}{1}$ and $N=1$ for the case $V_a = 240$.

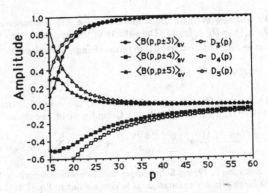

FIG. 8. A comparison of $\langle B(p,p\pm q)\rangle_{av}$ vs p and $D_{v_0+q-1}(p)$ vs p for $v_0 = \frac{3}{1}$, $N=1$, and $V_a = 240$, for the cases $q=3,4,5$.

$$J_{\pm K}(x_p - x_{p \mp \mu_a(0)(K+v_0)}) \approx \frac{(-1)^K}{2|K|\,|K|!}\left[\frac{(v_0+K)V_a(0)}{2p^2}\right]^{|K|} \equiv D_{K+v_0}(p) \,. \tag{4.14}$$

For $K=0$ we find

$$J_0(x_p - x_{p \pm v_0\mu_a(0)}) \approx 1 - \frac{1}{4}\left[\frac{(v_0)^2[V_a(0)]^2}{4p^4}\right] \equiv D_{v_0}(p) \,. \tag{4.15}$$

Using the above coefficients, we can write Eq. (4.12) in the form (for $p > p_c$)

$$-i\frac{\partial C_p^{(0)}(t)}{\partial t} = p^2 C_p^{(0)}(t) + \frac{V_b(0)}{2}\sum_{K=-K^*}^{K^*} D_{K+v_0}(p)[C_{p-\mu_a(0)(K+v_0)}^{(0)}(t)e^{iv_0\overline{\omega}_0 t} + C_{p+\mu_0(0)(K+v_0)}^{(0)}(t)e^{-i\mu_a(0)\overline{\omega}_0 t}] + \cdots . \tag{4.16}$$

In Figs. 7–9, we compare the average amplitude $\langle B(p,p\pm q)\rangle_{av} = \frac{1}{2}[B(p,p-q)+B(p,p+q)]$ for $v_0 = \frac{1}{1}$, $\frac{3}{1}$, and $\frac{5}{1}$, $N=1$, $M_a(0)=1$, with the approximate amplitudes D_K. We see that the WKB approximations are fairly good although they tend to overestimate the amplitude as we get too close to the edge of the large primary resonance. However, this is where we expect the WKB approximation to begin to break down.

In Eq. (4.16), the Kth resonance is located at

$$\overline{p}_K = -\frac{v_0\overline{\omega}_0}{2\mu_a(0)(K+v_0)} \,. \tag{4.17}$$

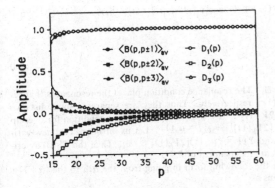

FIG. 7. A comparison of $\langle B(p,p\pm q)\rangle_{av}$ vs p and $D_{v_0+q-1}(p)$ vs p for $v_0 = \frac{1}{1}$, $N=1$, and $V_a = 240$, for the cases $q=1,2,3$.

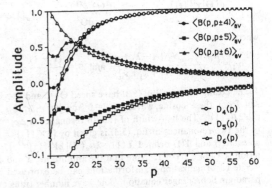

FIG. 9. A comparison of $\langle B(p,p\pm q)\rangle_{av}$ vs p and $D_{v_0+q-1}(p)$ vs p for $v_0 = \frac{5}{1}$, $N=1$, and $V_a = 240$, for the cases $q=4,5,6$.

We will now evaluate the amplitude of the Kth resonance at the position of the Kth resonance. Then we can write $D_{K+v_0}(p) \approx D_{K+v_0}(p_K)$. (The validity of this approximation was discussed in Ref. 12.) Let us now make this substitution in Eq. (4.16) and write it in terms of angle variables. We will let

$$\langle \theta | \Psi^{(1)}(t) \rangle = \sum_{p=-\infty}^{\infty} C_p^{(0)}(t) e^{ip\theta} . \tag{4.18}$$

Since $p = m/N$, where m is an integer, the wave function $\langle \theta | \Psi^{(1)}(t) \rangle$ satisfies the periodic boundary conditions $\langle \theta | \Psi^{(1)}(t) \rangle = \langle \theta + 2\pi N | \Psi^{(1)}(t) \rangle$. Using Eq. (4.18), we obtain

$$-i \frac{\partial \langle \theta | \Psi^{(1)}(t) \rangle}{\partial t} = -\frac{\partial^2 \langle \theta | \Psi^{(1)}(t) \rangle}{\partial \theta^2} + V_b(0) \sum_{K=-K^*}^{K^*} D_{K+v_0}(p_K) \cos[\mu_a(0)(K+v_0)\theta + v_0\bar{\omega}_0 t] \langle \theta | \Psi^{(1)}(t) \rangle + \cdots . \tag{4.19}$$

Equation (4.19) is a Schrödinger equation which describes the state of the system in a basis consisting of eigenstates of the large primary resonance. It is very similar to Eq. (2.1) in structure except that it contains an infinite number of resonances. We shall select two of these as the next step in our renormalization transformation.

V. RENORMALIZATION TRANSFORMATION

Having obtained the Schrödinger equation for secondary resonances, we will now select a pair of neighboring resonances and write the paradigm Schrödinger equation for this resonance pair. This will give us the renormalization mapping on the relative wave numbers and amplitudes of an arbitrary sequence of resonance pairs.

A. Renormalization mapping

Let us now select two neighboring resonance terms, $K = N_1$ and $K = N_1 + 1$ in Eq. (4.19). Then we obtain the following double-resonance Schrödinger equation:

$$-i \frac{\partial \langle \theta | \Psi^{(1)}(t) \rangle}{\partial t} = -\frac{\partial^2 \langle \theta | \Psi^{(1)}(t) \rangle}{\partial \theta^2} + V_b(0) D_{N_1+\lambda_1+v_0}(p_{N_1+\lambda_1}) \cos[\mu_a(0)(M_1+\lambda_1+v_0)\theta + v_0\bar{\omega}_0 t]$$

$$+ V_b(0) D_{N_1+1-\lambda_1+v_0}(p_{N_1+1-\lambda_1}) \cos[\mu_a(0)(M_1+1-\lambda_1+v_0)\theta + v_0\bar{\omega}_0 t] . \tag{5.1}$$

We have included a factor λ_1, where $\lambda_1 = 0$ or 1, in Eq. (5.1) so that at subsequent steps of the renormalization transformation we can choose either of the two resonances as a basis for building the WKB solutions. We can rewrite Eq. (5.1) in the form

$$-i \frac{\partial \langle \theta | \Psi^{(1)}(t) \rangle}{\partial t} = -\frac{\partial^2 \langle \theta | \Psi^{(1)}(t) \rangle}{\partial \theta^2} + \{V_a(1)\cos[\mu_a(1)\theta - \omega_1 t] + V_b(1)\cos[\mu_b(1)\theta - \omega_1 t]\} \langle \theta | \Psi_1(t) \rangle \tag{5.2}$$

where $V_a(1) = V_b(0) D_{N_1+\lambda_1+v_0}(p_{N_1+\lambda_1})$, $V_b(1) = V_b(0) D_{N_1+1-\lambda_1+v_0}(p_{N_1+1-\lambda_1})$, $\mu_a(1) = \mu_a(0)(N_1+\lambda_1+v_0)$, $\mu_b(1) = \mu_a(0)(N_1+1-\lambda_1+v_0)$, $\omega_1 = -v_0\bar{\omega}_0$.

Equation (5.2) now has the same form as Eq. (2.1) and we can use the same procedure as in Sec. II to write it in the form of a paradigm Schrödinger equation. We find

$$-i \frac{\partial \langle \Theta | \chi^{(1)}(t) \rangle}{\partial t} = -\mu_a(1)^2 \frac{\partial^2 \langle \Theta | \chi^{(1)}(t) \rangle}{\partial \Theta^2} + \{V_a(1)\cos(\Theta) + V_b(1)\cos[v_1(\Theta+\bar{\omega}_1 t)]\} \langle \Theta | \chi^{(1)}(t) \rangle \tag{5.3}$$

where

$$v_1 = \frac{\mu_b(1)}{\mu_a(1)} = \frac{N_1+1-\lambda_1+v_0}{N_1+\lambda_1+v_0} \quad \text{and} \quad \bar{\omega}_1 = \frac{(v_1-1)}{v_1} \omega_1 . \tag{5.4}$$

The cosine waves in Eq. (5.3) have speed $\dot{\Theta} = 0$ and $\dot{\Theta} = -\bar{\omega}_1$. The resonance condition places the resonances that result from these cosine waves at $\bar{p} = 0$ and $\bar{p} = -\bar{\omega}_1/2\mu_a(1)^2$, respectively. Thus they are separated by a distance $\bar{\omega}_1/2\mu_a(1)^2$. The half-width of the ith resonance is $\Delta p_i = [2V_i(1)]^{1/2}$ $(i = a, b)$.[3,4] The Chirikov condition for overlap of the two resonances in Eq. (5.3) is given by $[2V_a(1)]^{1/2} + [2V_b(1)]^{1/2} = \bar{\omega}_1/\mu_a(1)$. Let us introduce two new variables, $X(1)$ and $Y(1)$ defined $X(1) = 2\mu_a(1)[2V_a(1)]^{1/2}/\bar{\omega}_1$ and $Y(1) = 2\mu_a(1)[2V_b(1)]^{1/2}/\bar{\omega}_1$. Then the Chirikov criterion is $X(1) + Y(1) = 1$. Note that the amplitudes, X and Y, are independent of $\hbar$.

The renormalization transformations can be expressed in the following form in going from the αth to the $(\alpha+1)$st paradigm Schrödinger equation. The wave number transforms as

$$v_{\alpha+1} = \frac{N_{\alpha+1}+1-\lambda_{\alpha+1}+v_\alpha}{N_{\alpha+1}+\lambda_{\alpha+1}+v_\alpha} . \tag{5.5}$$

The frequency transforms as

$$\bar{\omega}_{\alpha+1} = -\frac{\nu_{\alpha+1}-1}{\nu_{\alpha+1}}\nu_\alpha\bar{\omega}_\alpha .$$ (5.6)

The mapping of $X(\alpha)$ and $Y(\alpha)$ from the αth level to the $\alpha+1$ level is given by

$$X^2(\alpha+1) = \frac{(-1)^{N_{\alpha+1}+\lambda_{\alpha+1}}}{2^{N_{\alpha+1}+\lambda_{\alpha+1}}(N_{\alpha+1}+\lambda_{\alpha+1})!}\left[\frac{(N_{\alpha+1}+1-\lambda_{\alpha+1}+\nu_\alpha)^2(N_{\alpha+1}+\lambda_{\alpha+1}+\nu_\alpha)^2}{\nu_\alpha^2}\right]$$

$$\times Y(\alpha)^2\left[\frac{(N_{\alpha+1}+\lambda_{\alpha+1}+\nu_\alpha)^3 X(\alpha)^2}{4\nu_\alpha^2}\right]^{N_{\alpha+1}+\lambda_{\alpha+1}} \quad \text{for } N_{\alpha+1}+\lambda_{\alpha+1}\neq 0 ,$$ (5.7a)

$$X^2(\alpha+1) = \left[\frac{(N_{\alpha+1}+1-\lambda_{\alpha+1}+\nu_\alpha)^2(N_{\alpha+1}+\lambda_{\alpha+1}+\nu_\alpha)^2}{\nu_\alpha^2}\right]Y(\alpha)^2\left[1-\left[\frac{\nu_\alpha^2 X^4(\alpha)}{64}\right]\right] \quad \text{for } N_{\alpha+1}+\lambda_{\alpha+1}=0 ,$$ (5.7b)

and

$$Y^2(\alpha+1) = \frac{(-1)^{N_{\alpha+1}+1-\lambda_{\alpha+1}}}{2^{N_{\alpha+1}+1-\lambda_{\alpha+1}}(N_{\alpha+1}+1-\lambda_{\alpha+1})!}\left[\frac{(N_{\alpha+1}+1-\lambda_{\alpha+1}+\nu_\alpha)^2(N_{\alpha+1}+\lambda_{\alpha+1}+\nu_\alpha)^2}{\nu_\alpha^2}\right]$$

$$\times Y(\alpha)^2\left[\frac{(N_{\alpha+1}+1-\lambda_{\alpha+1}+\nu_\alpha)^3 X(\alpha)^2}{4\nu_\alpha^2}\right]^{N_{\alpha+1}+1-\lambda_{\alpha+1}} \quad \text{for } N_{\alpha+1}+1-\lambda_{\alpha+1}\neq 0 ,$$ (5.8a)

$$Y^2(\alpha+1) = \left[\frac{(N_{\alpha+1}+1-\lambda_{\alpha+1}+\nu_\alpha)^2(N_{\alpha+1}+\lambda_{\alpha+1}+\nu_\alpha)^2}{\nu_\alpha^2}\right]Y(\alpha)^2\left[1-\left[\frac{\nu_\alpha^2 X^4(\alpha)}{64}\right]\right]$$

$$\text{for } N_{\alpha+1}+1-\lambda_{\alpha+1}=0 .$$ (5.8b)

The mappings in Eqs. (5.5)–(5.8) allow us to determine whether or not any given sequence of resonance pairs overlap as we go to small scale in the Hilbert space. A particular sequence is determined once we fix λ_α and N_α at each scale. If overlap has occurred between all sequences of higher-order resonance pairs between the two primary resonances, then we know that there is a continual path for probability to flow between the two primary resonances. The possibility of mappings such as those in Eqs. (5.5)–(5.8) has been proposed by Berman and Kolovsky,[13] however no explicit expressions were given.

B. Fixed points of wave-number mapping

The wave-number mapping, Eq. (5.5), can be studied independently of the amplitude mapping Eqs. (5.7) and (5.8) and determines the particular sequence of resonance pairs that is followed in the amplitude mappings. The relative wave number ν_α at the αth level of the renormalization transformation is determined by the sequence of values, $N_{\alpha'}$ and $\lambda_{\alpha'}$ $(\alpha'\leq\alpha)$, that precede it. Thus we can write ν_α schematically in the form

$$\nu_\alpha = \{N_\alpha,\lambda_\alpha;N_{\alpha-1},\lambda_{\alpha-1};\ldots;N_1,\lambda_1;\nu_0\}$$

For the special cases in which λ_α is fixed to be either 0 or 1 for all α, ν_α can be written as a continued fraction. Let us consider these two cases separately.

Case (i) ($\lambda_\alpha=0$ *for all* α). For this case the relative wave number can be written as the continued fraction

$$\nu_\alpha = \{N_\alpha,0;N_{\alpha-1},0;\ldots;N_1,0;\nu_0\} = 1+\cfrac{1}{N_\alpha+\nu_{\alpha-1}}$$

$$= [1,N_\alpha+1,N_{\alpha-1}+1,\ldots,N_2+1,N_1+\nu_0]$$

$$\equiv 1+\cfrac{1}{N_\alpha+1+\cfrac{1}{N_{\alpha-1}+1+\cdots\cfrac{1}{N_1+\nu_0}}} .$$ (5.9)

For the special case $N_\alpha=n$ for all α the mapping has fixed points $\bar{\nu}_n^{(0)}=\frac{1}{2}[1-n+(n^2+2n+5)^{1/2}]$, where $n\geq 0$. $\bar{\nu}_n^{(0)}$ can be expressed as a continued fraction

$$\bar{\nu}_n(\lambda=0) = [1,n+1,n+1,\ldots] .$$ (5.10)

For $n=0$ this is just the golden mean, $\nu_0^{(0)}=\gamma=[(1+\sqrt{5}/2)]$.

Case (ii) ($\lambda_\alpha=1$ *for all* α). For this case the relative wave number can be written as the continued fraction

$$\nu_\alpha = \{N_\alpha,1;N_{\alpha-1},1;\ldots;N_1,1;\nu_0\}$$

$$= \frac{N_\alpha+\nu_{\alpha-1}}{N_\alpha+\nu_{\alpha-1}+1}$$

$$\equiv [0,1,N_\alpha,1,N_{\alpha-1},1,\ldots,N_2,1,N_1+\nu_0] .$$

For the case $N_\alpha=n$ for all α, the mapping for the relative wave number takes the form

$$\nu_{\alpha+1} = \frac{\nu_\alpha+n}{\nu_\alpha+n+1} = \cfrac{1}{1+\cfrac{1}{n+\nu_\alpha}} .$$ (5.11)

This equation has fixed points at $\bar{v}_n^{(1)} = \frac{1}{2}[-n + (n^2 + 4n)^{1/2}]$ with $n \geq 1$. If we iterate Eq. (5.11), we find that $\bar{v}_n^{(1)}$ can be expressed as a continued fraction

$$\bar{v}_n^{(1)} = [0,1,n,1,n,\ldots] = \cfrac{1}{1 + \cfrac{1}{n + \cfrac{1}{1 + \cdots}}}. \quad (5.12)$$

For $n=1$ this is just the inverse golden mean, $\bar{v}_1^{(1)} = [(\sqrt{5}-1)/2]$. Continued fractions with the structure $[a_0, a_1, \ldots, a_n, 1, 1, \ldots]$ define "noble" resonance sequences. It appears that the noble resonance sequences in the quantum case are often the last to overlap as is true classically. This can be seen by studying the stable manifolds that result from the mappings Eqs. (5.7) and (5.8).

C. Stable manifolds

We have studied the mapping Eqs. (5.5)–(5.8) for three different choices of the primary resonances. We have taken $N=1$ and have considered $v_0 = \frac{1}{1}[\mu_a(0)=1, \mu_b(0)=1]$, $v_0 = \frac{1}{3}[\mu_a(0)=1, \mu_b(0)=3]$, and $v_0 = \frac{1}{5}[\mu_a(0)=1, \mu_b(0)=5]$. In terms of our original traveling-wave basis, the large primary resonance is centered at $\bar{k}_a = \omega_0/2\mu_a(0)$ while the small primary lies at $\bar{k}_b = \omega_0/2\mu_b(0)$. After the first step of the renormalization transformation, in which we expand the Schrödinger equation in terms of the eigenstates of the large primary resonance, we obtain an infinite number of secondary resonances whose relative wave numbers are given by

$$v_1 = \left\lceil \frac{N_1 + 1 + v_0}{N_1 + v_0} \right\rceil$$

where $-\infty \leq N_1 \leq \infty$. These secondary resonances are located in the Hilbert space of eigenstates of the large primary at

$$\bar{p}_{N_1} = -\frac{v_0 \bar{\omega}_0}{2\mu_a(0)(N_1 + v_0)}.$$

For $\bar{p}_{N_1} \gg [2V_a(0)]^{1/2}$, we can locate these resonances in the original Hilbert space at

$$\bar{k}_{N_1} \approx \bar{p}_{N_1} + \bar{k}_a$$

since far from the large primary resonance the eigenstates become approximately traveling waves. Let us now consider the three cases $v_0 = \frac{1}{1}$, $v_0 = \frac{1}{3}$, $v_0 = \frac{1}{5}$ separately.

Case (i) $(v_0 = \frac{1}{1})$. For the case $v_0 = \frac{1}{1}$ we must choose the plus sign in the cosine wave, $\cos[\mu_b(0)\theta \pm \omega_0 t]$, Eq. (2.1). Thus $\bar{\omega}_0 = [(v_0+1)/v_0]\omega_0 = 2\omega_0$. The mapping equations contain stable manifolds which separate values of $X(0)$ and $Y(0)$ for which a sequence resonance pairs overlap and for which they do not.

The two primary resonances (shaded) and two pairs of secondary resonances are shown schematically in Fig. 10 (the small primary becomes a secondary). If $N_1 = 0$ and $\lambda = 0$, then $v_1 = \frac{2}{1}$. The two resonances forming this resonance pair are located at $\bar{p}_0 = -\omega_0$ and $\bar{p}_1 = -\omega_0/2$ (or

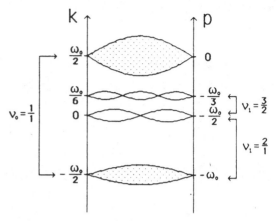

FIG. 10. A sketch of resonance pairs for $v_0 = \frac{1}{1}$ and $v_1 = \frac{2}{1}$ and $v_1 = \frac{3}{2}$. The relative spacings and wave numbers are shown accurately but the sizes are not to scale.

$\bar{k}_0 \approx -\omega_0/2$ and $\bar{k}_1 \approx 0$). The two resonances corresponding to $N_1 = 1$ and $\lambda = 0$ have relative wave number $v_1 = \frac{3}{2}$. They are located at $\bar{p}_1 = 0$ and $\bar{p}_2 = \omega_0/6$ (or $\bar{k}_0 \approx -\omega_0/2$ and $\bar{k}_1 \approx -\omega_0/3$). The mapping must be performed so that at each step, the larger of the two resonances is used to determine $X(\alpha)$ and the smaller determines $Y(\alpha)$.

We have searched for overlap of a large variety of resonance sequences. We find stable manifolds separating the stable region of values $(X(0), Y(0))$ for which no overlap occurs as we go to smaller scale [that is, $X(\alpha) \to 0$ and $Y(\alpha) \to 0$ as $\alpha \to \infty$], from the unstable region in which overlap does occur as we go to smaller scale [$X(\alpha) \to \infty$ and $Y(\alpha) \to \infty$ as $\alpha \to \infty$]. Once we cross the stable manifold from the stable to the unstable side, the growth in $X(\alpha)$ and $Y(\alpha)$ is rapid. We were interested in predicting the values $(X(0), Y(0))$ at which overlap of the resonance pair $v_1 = \frac{2}{1}$ occurs. This happens when all resonance sequences starting from the pair $v_1 = \frac{2}{1}$ have overlapped. We found that there was a significant range of values for which we could maintain $X(\alpha) > Y(\alpha)$ with $\lambda_\alpha = 0$ for all α. In Fig. 11, we show the stable manifolds corresponding to sequences (with $\lambda_\alpha = 0$ for all α) $v = [1,0,0,\ldots,0,0,\frac{1}{1}]$, $v = [1,1,1,\ldots,1,0,\frac{1}{1}]$, and $v = [1,2,2,\ldots,2,0,\frac{1}{1}]$ [these correspond to $v_1 = \frac{2}{1}$ and (a), (b), and (c), respectively, in Fig. 11]. We also studied overlap of the resonance pair $v_1 = \frac{3}{2}$. In Fig. 11 we show stable manifolds for sequences $v = [1,0,0,\ldots,0,1,\frac{1}{1}]$, $v = [1,1,1,\ldots,1,1,\frac{1}{1}]$, and $v = [1,2,2,\ldots,2,1,\frac{1}{1}]$ [these correspond to $v_1 = \frac{3}{2}$ and (d), (e), and (f), respectively, in Fig. 11]. It is interesting that these sequences each correspond to the last [for increasing $X(0)$ and $Y(0)$] overlapping sequence for some range of $X(0)$ and $Y(0)$. Note that as $X(0)$ and $Y(0)$ become roughly equal, the sequences $v = [1,0,0,\ldots,0,0,\frac{1}{1}]$ and $v = [1,0,0,\ldots,0,1,\frac{1}{1}]$ which are noble sequences are the last to overlap. In this case we approach the standard map. Also, note that overlap between resonance pair $v_1 = \frac{3}{2}$ occurs for smaller values of

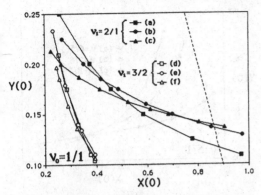

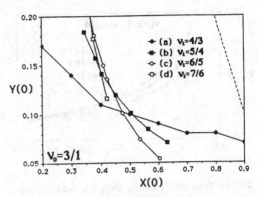

FIG. 11. Some stable manifolds for $\nu_1 = \frac{1}{1}$. Solid symbols indicate curves for $\nu_1 = \frac{2}{1}$ and (a) $n_\alpha = 0$ for $\alpha > 1$, (b) $n_\alpha = 1$ for $\alpha > 1$, and (c) $n_\alpha = 2$ for $\alpha > 1$. Open symbols indicate curevs for $\nu_1 = \frac{3}{2}$ and (d) $n_\alpha = 0$ for $\alpha > 1$, (e) $n_\alpha = 1$ for $\alpha > 1$, and (f) $n_\alpha = 2$ for $\alpha > 1$. The dashed line is the Chirikov prediction, $X(0) + Y(0) = 1$.

$X(0)$ and $Y(0)$ than does overlap of resonance pair $\nu_1 = \frac{2}{1}$.

Case (ii) ($\nu_0 = \frac{3}{1}$). For this case, $\bar{\omega}_0 = [(\nu_0 - 1)/\nu_0]\omega_0 = \frac{2}{3}\omega_0$. In Fig. 12, we have given a sketch of the two primary resonances (shaded) and two pairs of secondary resonances $\nu_1 = \frac{4}{3}(N_1 = 0)$ and $\nu_1 = \frac{5}{4}(N_1 = 1)$. Again, we show the relative wave numbers and positions of the resonances accurately but the width of the secondaries is exaggerated. For this case, the secondary pairs are much closer together than for the case $\nu_0 = \frac{1}{1}$ and lie further from the large primary. In Fig. 13, we show the stable manifolds for the last [as $X(0)$ and $Y(0)$ increase] resonance sequences to overlap for the resonance pairs $\nu_1 = \frac{4}{3}$, $\frac{5}{4}$, $\frac{6}{5}$, and $\frac{7}{6}$. For these four cases the resonance sequences were given by $\lambda_\alpha = 0$ for all α and $\nu = [1,0,0,\ldots,0,0,\frac{3}{1}]$, $\nu = [1,0,0,\ldots,0,1,\frac{3}{1}]$,

$\nu = [1,0,0,\ldots,0,2,\frac{3}{1}]$, and $\nu = [1,0,0,\ldots,0,3,\frac{3}{1}]$, respectively. Thus, the noble resonance sequences in these cases were the last to overlap.

In Figs. 14 and 15, we show additional stable manifolds for the cases $\nu_1 = \frac{4}{3}$ and $\nu_1 = \frac{5}{4}$. In Fig. 14, we show stable manifolds for the sequences (a) $\nu = [1,0,0,\ldots,0,0,\frac{3}{1}]$, (b) $\nu = [1,1,1,\ldots,1,0,\frac{3}{1}]$, and (c) $\nu = [1,2,2,\ldots,2,0,\frac{3}{1}]$ for the resonance pair $\nu_1 = \frac{4}{3}$. Clearly the noble resonance sequences are the last to overlap. In Eq. (5.6), we show the stable manifolds for the sequences (a) $\nu = [1,0,0,\ldots,0,1,\frac{3}{1}]$, (b) $\nu = [1,1,1,\ldots,1,1,\frac{3}{1}]$, and (c) $\nu = [1,2,2,\ldots,2,1,\frac{3}{1}]$ for the resonance pair $\nu_1 = \frac{5}{4}$. Again, the noble resonance sequences are the last to overlap over the range of $X(0)$ and $Y(0)$ for which our theory is valid.

Case (iii) ($\nu_0 = \frac{5}{1}$). For this case, $\bar{\omega}_0 = [(\nu_0 - 1)/\nu_0]\omega_0 = \frac{4}{5}\omega_0$. in Fig. 16, we show a sketch of the two primary resonances (shaded) and two pairs of secondary resonances $\nu_1 = \frac{6}{5}(N_1 = 0)$ and $\nu_1 = \frac{7}{6}(N_1 = 1)$. The relative wave numbers and positions of the resonances are given accurately but the width of the secondaries is exaggerated. For this case, the secondary pairs are still farther

FIG. 13. Some stable manifolds for $\nu_1 = \frac{3}{1}$. For all curves shown, $n_\alpha = 0$ for $\alpha > 1$. The dashed line is the Chirikov prediction, $X(0) + Y(0) = 1$.

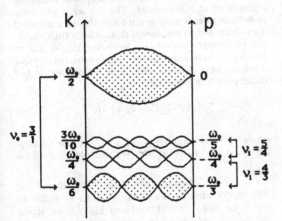

FIG. 12. A sketch of resonance pairs for $\nu_0 = \frac{3}{1}$ and $\nu_1 = \frac{4}{3}$ and $\nu_1 = \frac{5}{4}$. The relative spacings and wave numbers are shown accurately but the sizes are not to scale.

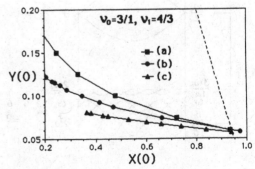

FIG. 14. Some stable manifolds for $\nu_0 = \frac{3}{1}$. and $\nu_1 = \frac{4}{3}$ and (a) $n_\alpha = 0$ for $\alpha > 1$, (b) $n_\alpha = 1$ for $\alpha > 1$, and (c) $n_\alpha = 2$ for $\alpha > 1$. The dashed line is the Chirikov prediction, $X(0) + Y(0) = 1$.

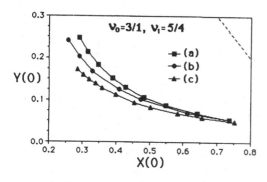

FIG. 15. Some stable manifolds for $v_0 = \frac{3}{1}$ and $v_1 = \frac{5}{4}$ and (a) $n_a = 0$ for $\alpha > 1$, (b) $n_a = 1$ for $\alpha > 1$, and (c) $n_a = 2$ for $\alpha > 1$. The dashed line is the Chirikov prediction, $X(0) + Y(0) = 1$.

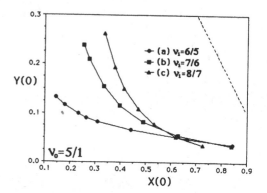

FIG. 17. Some stable manifolds for $v_0 = \frac{5}{1}$. For all curves shown, $n_a = 0$ for $\alpha > 1$. The dashed line is the Chirikov prediction, $X(0) + Y(0) = 1$.

away from the large primary than was the case for $v_0 = \frac{1}{1}$ or $v_0 = \frac{3}{1}$. The stable manifolds for the last resonance sequences to overlap for the resonance pairs $v_1 = \frac{6}{5}, \frac{7}{6}$, and $\frac{8}{7}$ are shown in Fig. 17. These stable manifolds correspond to $\lambda_a = 0$ for all α and (a) $v = [1, 0, 0, \ldots, 0, 0, \frac{5}{1}]$, (b) $v = [1, 0, 0, \ldots, 0, 1, \frac{5}{1}]$, and (c) $v = [1, 0, 0, \ldots, 0, 2, \frac{5}{1}]$. Other sequences we studied had stable manifolds lying at lower values of $X(0)$ and $Y(0)$ than did these resonance pairs.

VI. NUMERICAL DISCUSSION

There have been a number of studies which indicate that resonance overlap in quantum systems can lead to a fairly abrupt spreading of probability over the region of Hilbert space which is influenced by the resonances.[9,10] There is also evidence of KAM behavior.[6] That is, blockages to the spread of probability which in classical systems would be attributed to KAM surfaces. Probabili-

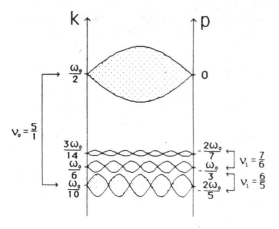

FIG. 16. A sketch of resonance pairs for $v_0 = \frac{5}{1}$ and $v_1 = \frac{6}{5}$ and $v_1 = \frac{7}{6}$. The relative spacings and wave numbers are shown accurately but the sizes are not to scale.

ty appears to decay exponentially across these barriers.

We have considered two cases, $\omega_0 = 120$ and $\omega_0 = 240$, of the double-resonance model with $N = 1$, $\mu_a(0) = 1$, and $\mu_b(0) = 3$ for both cases (this was also studied in Ref. 12), and we have compared the predictions of the renormalization transformation with the actual spread of probability. The effect of increasing the frequency ω_0 is to move the primary resonances farther apart and this also increases the number of states between the two primaries. Once we fix the frequency ω_0 the positions of the primaries are fixed. However, we can still adjust their size. In the following we will fix the size of the small primary and adjust the size of the large primary. We start with all the probability initially on the state at the center of the small primary and we then integrate the equations of motion, Eq. (2.2), for a long time (many periods $T = 2\pi/\omega_0$) to determine how far the probability can spread. We typically find that after a few periods the probability reaches its maximum extent and does not spread further although it may slosh around within the region in which it is confined. The manner in which the probability spreads gives us some idea about the nature of the barriers and/or resonances that lie in its path. In order to attempt to eliminate local fluctuations due to sloshing we have computed the spread after ten different times and have averaged over those ten different sets of values according to the equation

$$\bar{P}_k = \frac{1}{10} \sum_{j=1}^{5} \left[|\Psi_k^{(0)}(2j)|^2 + |\Psi_k^{(0)}(2.3j)|^2 \right] , \qquad (6.1)$$

where $\Psi_k^{(0)}(t)$ is the solution of Eq. (2.2) at time t assuming initial conditions $|\Psi_{\bar{k}_b}^{(0)}(0)|^2 = 1$ and $|\Psi_k^{(0)}(0)|^2 = 0$ for $k \neq \bar{k}_b$.

For the case when $\omega_0 = 120$ we fix $V_b(0) = 20$ and find $Y(0) = 0.158$, while for the case $\omega_0 = 240$ we fix $V_b(0) = 80$ and again find $Y(0) = 0.158$. From Fig. 13, we expect overlap of the resonance pair $v_1 = \frac{4}{3}$ to occur for $X(0) = 0.23$, while for resonance pair $v_1 = \frac{5}{4}$ we expect overlap to occur for $X(0) = 0.4$. We will now show how the probability spreads through these two regions and compare this spread with the predictions of the renormalization transformation. We will first consider the case when $\omega_0 = 120$.

Case I ($\omega_0 = 120$, $N = 1$, $\nu_0 = \frac{3}{7}$, $V_b(0) = 20$). For the case when $\omega_0 = 120$ the large primary resonance is located at $\bar{k}_a = 60$ and the small primary is located at $\bar{k}_b = 20$. Thus there are 40 quantum states separating the centers of these two resonances. In Ref. 12, we showed (and our current calculations support the fact) that the largest secondary resonances lie between the two primaries and this is the region we focus on here. We have probed the Hilbert space in the following way. We fix $V_b(0) = 20$ and choose values for $V_a(0)$ ranging between 30 and 190. We start all the probability at the center of the small primary. That is, we set $|\Psi_{20}^{(0)}(0)|^2 = 1$ and $|\Psi_k^{(0)}(0)|^2 = 0$ for $k \neq 20$ and we integrated Eq. (5.2) for many periods $T = 2\pi/\omega_0$ (a time much longer than necessary for the probability to reach its maximum spread). We then find how probability is spread over states, $\Psi_k^{(0)}(t)$. In Fig. 18, we show the average spread of probability, $\bar{P}_k$, for two cases, $V_a(0) = 20$ and $V_a(0) = 190$. Note the asymmetry in the spread of probability in the direction of the large primary.

In Fig. 19, we focus on the region $k = 30$ to $k = 44$ and show the extent of the spread for nine different values of $V_a(0)$ [with $V_b(0) = 20$]. The secondary resonance pairs $\nu_1 = \frac{1}{3}$ lie at $\bar{k}_0 = 20$ and $\bar{k}_1 = 30$. From Fig. 13 we predict overlap to occur when $X(0) = 0.23$ and $Y(0) = 0.158$. However, for this case, we do not expect the theory to work well. The half-width of the resonance at $\bar{k}_0 = 20$ is $\Delta k_b \approx [2V_b(0)]^{1/2} = 6.3$. Thus the resonance at $\bar{k}_1 = 30$ lies within three quantum states of it. We expect WKB to give too large a value for $X(0)$. In addition the stable manifold curve is almost flat and is very sensitive to any error in $X(0)$. For example $X(0) = 0.23$ we get $V_a(0) = 42$ while for $X(0) = 0.2$ we get $V_a(0) = 30$. At $V_a(0) = 30$ we see that overlap has occurred. Probability has spread between the resonance pair $\nu_1 = \frac{4}{3}$.

We expect better results for the resonance pair $\nu_1 = \frac{5}{4}$ which lies at $\bar{k}_1 = 30$ and $\bar{k}_2 = 36$. The stable manifold in Fig. 13 indicates that there is blockage between $k = 30$ and $k = 36$ until $X(0) \approx 0.4$ or $V_a(0) \approx 128$. In Fig. 19, we

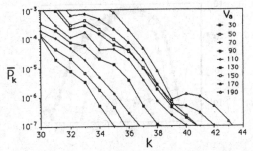

FIG. 19. The average probability, $\bar{P}_k = \frac{1}{10} \sum_{j=1}^{5} [|\Psi_k^{(0)}(2j)|^2 + |\Psi_k^{(0)}(2.3j)|^2]$ obtained by solving Eq. (2.3) numerically for $M_a(0) = 1$, $M_3(0) = 3$, $\omega_0 = 120$, $V_b(0) = 20$, and a variety of values of $V_a(0)$ as indicated in the figure. The initial conditions for all curves are $|\Psi_{20}^{(0)}(0)|^2 = 1$ and $|\Psi_k^{(0)}(0)|^2 = 0$ for $k \neq 20$. Note that the vertical axis is a log scale.

see exponential decay into this region until $V_a(0) \approx 90$. When $V_a(0) \approx 110$, the probability has spread into the region between resonance pairs $\nu_1 = \frac{1}{4}$ and $\nu_1 = \frac{6}{5}$ which, according to Fig. 13, overlap at about the same values of $X(0)$.

There is another interesting phenomenon at work here. We see that the probability for $V_a(0) = 130$ is pushed backward relative to where we expect it to be. This appears to be due to the fact that it is hitting the edges of the large resonance that is pushing into that region from the right. In Fig. 20 we show the edge of the large resonance (shaded area), whose position we have established using the estimate for the half-width, $\Delta k_a \approx [2V_a(0)]^{1/2}$. From Ref. 12, this gives fairly good predictions to within a few quantum states. For $V_a(0) = 110$ it has not yet reached $k = 44$ (rightmost extent of our figure). However, at $V_a(0) = 130$ it just begins to enter the figure and at

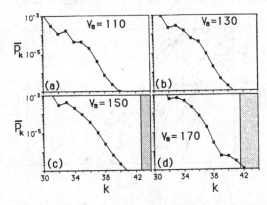

FIG. 20. Plot of curves from Fig. 19 indicating their position relative to the edge of the primary resonance. The large primary has a half-width of $[2V_a(0)]^{1/2}$ and for $\omega_0 = 120$ is centered at $k = 60$. The spread of probability appears to be blocked by the edge of the large primary [cf. (c) and (d)] before it finally penetrates into the interior of the large primary resonance region.

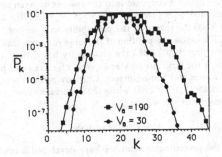

FIG. 18. A plot of $\bar{P}_k$ for $M_a(0) = 1$, $M_3(0) = 3$, $\omega_0 = 120$, $V_b(0) = 20$, and $V_a(0) = 30$ and 190 as indicated in the figure. The same initial conditions were used as in Eq. (6.1). Note that the vertical axis is a log scale.

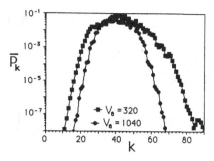

FIG. 21. A plot of $\bar{P}_k$ for $M_a(0)=1$, $M_3(0)=3$, $\omega_0=240$, $V_b(0)=80$, and $V_a(0)=320$ and 1040 as indicated in the figure. The same initial conditions were used as in Eq. (6.1). Note that the vertical axis is a log scale.

$V_a(0)=150$ it has blocked and pushed the probability back from its expected position. When $V_a(0)=170$, the probability has overcome KAM barriers at the edge of the large primary and has penetrated into it. At this point, we can say that overlap between the two primary resonances has occurred. Note that our stable manifolds predict that probability spreads between the two primary resonances for $V_a(0)\approx130$. We observe it occurring for $150\leq V_a(0)\leq170$. The Chirikov prediction gives $X(0)+Y(0)=1$ or $X(0)=0.843$ and $V_a(0)=570$ when $V_b(0)=20$. Thus the renormalization predictions are extremely good.

Case II ($\omega_0=240$, $N=1$, $\nu_0=\frac{3}{7}$, $V_b(0)=80$). For this case the large primary resonance is located at $\bar{k}_a=120$ and the small primary is located at $\bar{k}_b=40$ so that 80 quantum states separate the two primaries. We set $V_b(0)=80$ and allow $V_a(0)$ to range between $V_a(0)=320$ and $V_a(0)=1040$. As before, we start all probability at the center of the small primary and let the probability flow to the maximum extent allowed by the KAM barriers. In Fig. 21 we show $\bar{P}_n$ for two limiting cases,

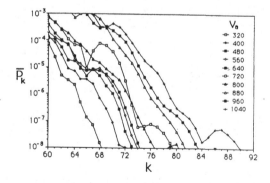

FIG. 22. The average probability, $\bar{P}_k=\frac{1}{10}\sum_{j=1}^{5}[\,|\Psi_k^{(0)}(2j)|^2+|\Psi_k^{(0)}(2.3j)|^2]$ obtained by solving Eq. (2.3) numerically for $M_a(0)=1$, $M_3(0)=3$, $\omega_0=240$, $V_b(0)=80$, and a variety of values of $V_a(0)$ as indicated in the figure. The initial conditions for all curves are $|\Psi_{40}^{(0)}(0)|^2=1$ and $|\Psi_k^{(0)}(0)|^2=0$ for $k\neq40$. Note that the vertical axis is a log scale.

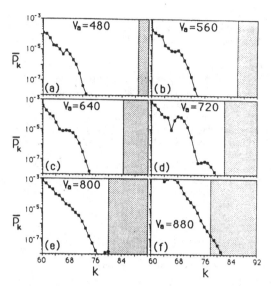

FIG. 23. Plot of curves from Fig. 22 indicating their position relative to the edge of the large primary resonance. The large primary has a half-width of $[2V_a(0)]^{1/2}$ and for $\omega_0=240$ is centered at $k=120$. The spread of probability appears to be blocked by the edge of the large primary [cf. (d) and (e)] before it finally penetrates into the interior of the large primary resonance region.

$V_a(0)=320$ and $V_a(0)=1040$. Again note the asymmetry. In Fig. 22, we show a sequence of values of $\bar{P}_n$ for values of $V_a(0)$ ranging from 320 to 1040. The resonance pair $\nu_1=\frac{4}{3}$ lies at $\bar{k}_0=40$ and $\bar{k}_1=60$. The half-width of the resonance at $\bar{k}_0=40$ is $\Delta k=12.6$ so the secondary lies further away than in the previous case. We predict overlap of this resonance pair at $X(0)=0.23$ or $V_a(0)=170$ [$X(0)=0.2$, $V_a(0)=128$]. So overlap has occurred by $V_a(0)=320$. The secondary resonance pair $\nu_1=\frac{5}{4}$ lies at $\bar{k}_1=60$ and $\bar{k}_2=72$, while secondary resonance pair $\nu_1=\frac{6}{5}$ lies at $\bar{k}_2=72$ and $\bar{k}_3=80$. We predict overlap when $X(0)=0.4$ or $V_a(0)=512$. From Fig. 22, we see that the probability has spread into the regions between resonance pairs $\nu_1=\frac{5}{4}$ and $\nu_1=\frac{6}{5}$ when $V_a(0)=720$. For $V_a(0)=800$ the spread of the probability has stopped. In Fig. 23 we show the position of the large primary (shaded region). For $V_a(0)\geq880$ the probability has entered the region of the large primary and overlap between two primary resonances had occurred. Chirikov predicts this to happen for $V_a(0)\approx2700$ while the renormalization predictions give $V_a(0)\approx720$.

VII. CONCLUSION

In the preceding sections we have developed a renormalization transformation which is based on the existence of higher-order nonlinear resonances in quantum-dynamical systems. It is interesting that the renormalization mapping itself depends on dimensionless variables and does not depend explicitly on Planck's constant $\hbar$. The mapping relates the wave numbers and amplitudes of the resonance zones on successively smaller scales in the Hilbert space, and gives fairly good predictions for the parameter values at which resonance overlap occurs in the Hilbert space. One might object that the renormalization mapping has no meaning as we go to scales where the resonance pairs have a size smaller than the spacing between quantum numbers. However, in practice, the growth of resonance amplitudes is so rapid as we cross the stable manifold into the unstable region that the stable manifold still appears to give good predictions in the quantum systems we have looked at.

We have not attempted in this paper to study the deviations of the predictions of our renormalization theory for quantum systems from those of classical renormalization theory. It is known from numerical experiments that it requires stronger external fields to remove KAM barriers in quantum systems than in the corresponding classical systems. There have been attempts to explain this based on semiclassical extensions of classical scaling theory.[14-19] Perhaps the simplest is that due to MacKay and Meiss[14,15] who simply require that a Cantorus ceases to be barrier to quantum wave functions when the flux (classical phase-space area) ΔW across the Cantorus satisfies the condition $\Delta W > \hbar$ rather than $\Delta W > 0$ as is true classically. This gives quite good qualitative predictions for the shifts from classical behavior observed in quantum systems.

Understanding when resonance overlap occurs in a given quantum system is important because it leads to extension of the wave function in that region and may have profound effects on the dynamics (such as ionization, if we are considering electron states in a molecule). We see that KAM behavior causes localization of the wave function in Hilbert space. However, this type of localization must be distinguished from Anderson localization[20] which occurs in regions of extreme resonance overlap (where KAM behavior has been destroyed).

It is important to note that *the systems we have considered exhibit highly nonlinear behavior* and in fact show many of the phenomena observed in nonlinear classical systems (except chaos if the spectrum is discrete) *even though the Schrödinger equation for this system is linear.* The type of behavior we have observed here is probably typical of most quantum systems which have two degrees of freedom and nonlinear Hamiltonians even though the Schrödinger equation is linear.

ACKNOWLEDGMENTS

L.E.R. wishes to thank the Institute for Nonlinear Science, University of California, San Diego; the University of Texas University Research Foundation; and the Welch Foundation of Texas, Grant No. F-1051 for partial support of this work. She also thanks M. Mikeska and J. Cornelius for their invaluable help with the I.N.L.S. computer system.

APPENDIX A: QUANTUM STANDARD MAP

In this appendix we write the quantum standard map in terms of the resonance picture. The standard map may be viewed as describing a rotor subject to repeated δ-function kicks occurring with period T and with an amplitude which depends on the angular position of the rotor. The Hamiltonian for the classical standard map can be written

$$H = \frac{J^2}{2I} + K\cos(\theta) \sum_{n=-\infty}^{\infty} \delta(t - nT) \qquad (A1)$$

where J is the angular momentum of the rotor, θ is its angular position, and I is its momentum of inertia. The parameter K is the strength of the kicks and T is their period. If we note the identity

$$\sum_{n=-\infty}^{\infty} \delta(t - nT) = \frac{2}{T}\sum_{k=1}^{\infty}\cos\left[\frac{2\pi kt}{T}\right] + \frac{1}{T} \qquad (A2)$$

and note that the angular momentum operator can be written $\hat{J} = -i\hbar\partial/\partial\theta$, then the quantum Hamiltonian can be written

$$\hat{H} = -\frac{\hbar^2}{2I}\frac{\partial^2}{\partial\theta^2} + \frac{K}{T}\cos(\theta)$$
$$+ \frac{K}{T}\sum_{k=1}^{\infty}\left[\cos\left[\theta - \frac{2\pi kt}{T}\right] + \cos\left[\theta + \frac{2\pi kt}{T}\right]\right].$$
$$(A3)$$

The Schrödinger equation for the δ-kicked rotor, in the angle picture, can be written

$$i\hbar\frac{\partial\Psi}{\partial t} = -\frac{\hbar^2}{2I}\frac{\partial^2\Psi}{\partial\theta^2} + \frac{K}{T}\sum_{k=-\infty}^{\infty}\left[\cos\left[\theta - \frac{2\pi kt}{T}\right]\right]\Psi$$
$$(A4)$$

where $\Psi = \Psi(\theta, t)$. In terms of the angular momentum quantum number n the Schrödinger equation takes the form

$$i\hbar\frac{\partial\Psi_n}{\partial t} = \frac{\hbar^2 n^2}{2I}\Psi_n + \frac{K}{2T}\sum_{k=-\infty}^{\infty}(e^{-i\omega kt}\Psi_{n-1} + e^{i\omega kt}\Psi_{n+1}).$$
$$(A5)$$

Equation (A5) can be written in dimensionless form. It becomes

$$i\frac{\partial\Psi_n}{\partial\tau} = n^2\Psi_n + \frac{\epsilon}{2}\sum_{k=-\infty}^{\infty}(e^{-i\omega k\tau}\Psi_{n-1} + e^{+i\omega k\tau}\Psi_{n+1})$$
$$(A6)$$

where $\tau = \hbar t/2I$, $\omega_0 = 2I\omega/\hbar$, and $\epsilon = 2KI/\hbar^2 T$. Thus, in terms of dimensionless quantities the primary resonance zones are located, in the Hilbert space of angular momentum states, at $n_k = \omega_0 k/2$ and have a half-width of $\Delta n_k = \sqrt{2\epsilon}$. In terms of the original units we have $n_k = \omega kI/\hbar$ and $\Delta n_k = 2(KI/\hbar^2 T)^{1/2}$. If comparison is made with the paper of Grempel, Prange, and Fishman[20] we find for the parameters used in Fig. 10 of that paper, $n_k = 1.3k$ and $\Delta n_k = 1.5$. Thus they are well within the regime of resonance overlap. This is reflected in the fact that their quasienergy eigenstate is spread over many states of the Hilbert space.

APPENDIX B: DRIVEN SQUARE WELL

Let us consider a particle in an infinite square-well potential [$V(x)=0$ for $0<x<2a$ and $V(x)=\infty$ otherwise] driven by a monochromatic external field. The Hamiltonian may be written

$$\hat{H}=\hat{H}_0-\epsilon(|\hat{x}|-a)\cos(\omega t)\ , \tag{B1}$$

where $\hat{H}_0$ is the Hamiltonian for the unperturbed particle in the infinite square-well potential, ϵ and ω are the external field amplitude and frequency, respectively, $\hat{x}$ is the particle position operator, and t is the time. The absolute value $|\hat{x}|$ appears in Eq. (2.1) because we will expand in traveling-wave states $|n\rangle$ (n is an integer) normalized on the interval $-2a$ to $2a$. In the position representation $\langle x|n\rangle=(1/\sqrt{4a})\exp(in\pi x/2a)$. In the traveling-wave basis, matrix elements of $\hat{H}_0$ are given by

$$\langle n'|\hat{H}_0|n\rangle=\frac{\hbar^2\pi^2n^2}{8ma^2}\delta_{n',n} \tag{B2}$$

where $\hbar$ is Planck's constant and m is the mass of the particle. Matrix elements of $\hat{x}$ are given by

$$\langle n'|\,|x|\,|n\rangle=\begin{cases}\dfrac{4a}{\pi^2(n'-n)^2} & \text{for }(n'-n)\text{ odd}\\[2mm]0 & \text{for }(n'-n)\text{ even}.\end{cases} \tag{B3}$$

The total Hamiltonian can be written

$$\hat{H}=\sum_{n=-\infty}^{\infty}\hbar\Omega n^2|n\rangle\langle n|$$
$$+\frac{4a\epsilon\cos(\omega t)}{\pi^2}\sum_{n=-\infty}^{\infty}\sum_{n'=-\infty}^{\infty}\frac{|n'\rangle\langle n|}{(n'-n)^2} \tag{B4}$$

($n-n'$ odd) where $\Omega=\hbar\pi^2/8ma^2$. It is useful to introduce a change of summation variables, $N=n'+n$ and $M=n'-n$. Then the Hamiltonian can be written

$$\hat{H}=\sum_{n=-\infty}^{\infty}\hbar\Omega n^2|n\rangle\langle n|+\frac{4a\epsilon\cos(\omega t)}{\pi^2}$$
$$\times\sum_{N=-\infty}^{\infty}\sum_{M=-\infty}^{\infty}\frac{1}{M^2}|\tfrac{1}{2}(N+M)\rangle\langle\tfrac{1}{2}(N-M)| \tag{B5}$$

(M and N odd) and the Schrödinger equation for this system can be written

$$i\hbar\frac{\partial\langle n|\Psi(t)\rangle}{\partial t}=\hbar\Omega n^2\langle n|\Psi(t)\rangle$$
$$+\frac{4a\epsilon\cos(\omega t)}{\pi^2}\sum_{M=-\infty}^{\infty}\frac{1}{M^2}\langle n-M|\Psi(t)\rangle \tag{B6}$$

(M odd) where $|\Psi(t)\rangle$ is the probability amplitude for the system at time t. Because the potential at the walls of the square well is infinite, the wave function must be zero at the walls. Thus $\langle 0|\Psi(t)\rangle=\langle 2a|\Psi(t)\rangle=0$ and $\langle n|\Psi(t)\rangle=-\langle -n|\Psi(t)\rangle$.

Let us now write the Schrödinger equation in terms of dimensionless quantities. Let $\tau=\Omega t$, $\omega_0=\omega/\Omega$, and $q=2a\epsilon/\hbar\Omega\pi^2$. We then find

$$i\frac{\partial\langle n|\Psi(\tau)\rangle}{\partial\tau}=n^2\langle n|\Psi(\tau)\rangle+q\sum_{\substack{M=-\infty\\M\ \text{odd}}}^{\infty}\frac{1}{M^2}\{[e^{-i\omega_0\tau}\langle n-M|\Psi(\tau)\rangle+e^{i\omega_0\tau}\langle n+M|\Psi(\tau)\rangle]\}\ . \tag{B7}$$

It is also useful to introduce angle variables $\phi=\pi x/2a$. We can transform to the angle picture by means of the transformation $\langle\phi|\Psi(\tau)\rangle=\sum_{n=-\infty}^{\infty}\langle\phi|n\rangle\langle n|\Psi(\tau)\rangle$, where $\langle\phi|n\rangle=e^{in\phi}/\sqrt{2\pi}$. The range of ϕ is $-\pi$ to π. The physical range (in the well) is 0 to π. Then the Schrödinger equation takes the form

$$i\frac{\partial\langle\phi|\Psi(\tau)\rangle}{\partial\tau}=-\frac{\partial^2\langle\phi|\Psi(\tau)\rangle}{\partial\phi^2}+2q\sum_{\substack{M=-\infty\\M\ \text{odd}}}^{\infty}\frac{1}{M^2}[\cos(M\psi-\omega_0\tau)]\langle\phi|\Psi(\tau)\rangle\ . \tag{B8}$$

Equation (B8) describes the behavior of a particle in the presence of an infinite number of traveling cosine potential wells, each traveling with different speed. Each traveling potential well gives rise to a nonlinear primary resonance zone in Hilbert space. In Refs. 8 and 11 it was shown that the primary resonance zone due to cosine potential $\cos(M\phi-\omega_0\tau)$ (where M can be positive or negative) is centered in Hilbert space at quantum number $n_M^{pr}=\omega_0/2M$ and has a half-width given approximately by $\Delta n_M^{pr}=2\sqrt{q/|M|}$.

---

*Permanent address: Department of Physics, Xuzhou Teacher's College, P.R. of China.

[1] J. Greene, J. Math. Phys. 20, 1183 (1979).
[2] S. J. Shenker and L. P. Kadanoff, J. Stat. Phys. 27, 631 (1982).
[3] R. S. MacKay, Physica 7D, 283 (1983).
[4] D. F. Escande and F. Doveil, J. Stat. Phys. 26, 257 (1981);

Phys. Lett. 83A, 307 (1981); Phys. Scr. T2, 126 (1982); D. F. Escande, Phys. Rep. 121, 165 (1985).
[5] L. E. Reichl and W. M. Zheng, in *Directions in Chaos*, edited by Hao Bai-lin (World Scientific, Singapore, 1987), Vol. I.
[6] T. Geisel, G. Radons, and J. Rubner, Phys. Rev. Lett. 57, 2883 (1986).

[7]R. C. Brown and R. E. Wyatt, Phys. Rev. Lett. **57**, 1 (1986).

[8]G. P. Berman and G. M. Zaslavsky, Phys. Lett. **61A**, 295 (1977); G. P. Berman, G. M. Zaslavsky, and A. R. Kolovsky, *ibid.* **87A**, 152 (1982); G. P. Berman and A. R. Kolovsky, *ibid.* **95A**, 15 (1983); G. P. Berman, G. M. Zaslavsky, and A. R. Kolovsky, Zh. Eksp. Teor. Fiz. **81**, 506 (1981) [Sov. Phys.—JETP **54**, 272 (1981)].

[9]L. E. Reichl and W. A. Lin, Phys. Rev. A **33**, 3598 (1986).

[10]W. A. Lin and L. E. Reichl, Phys. Rev. A **36**, 5099 (1987); **37**, 3972 (1988); **40**, 1055 (1989).

[11]M. Toda and K. Ikeda, J. Phys. A **20**, 3833 (1987).

[12]L. E. Reichl, Phys. Rev. A **39**, 4817 (1989).

[13]G. P. Berman and A. R. Kolovsky, Phys. Lett. A **125**, 188 (1987).

[14]R. S. MacKay and J. D. Meiss, Phys. Rev. A **37**, 4702 (1988).

[15]J. D. Meiss, Phys. Rev. Lett. **62**, 1576 (1989).

[16]D. R. Grempel, S. Fishman, and R. E. Prange, Phys. Rev. Lett. **53**, 1212 (1984).

[17]S. Fishman, D. R. Grempel, and R. E. Prange, Phys. Rev. A **36**, 289 (1987).

[18]G. Radons and R. E. Prange, Phys. Rev. Lett. **61**, 1691 (1988).

[19]R. V. Jensen, S. M. Susskind, and M. M. Sanders, Phys. Rev. Lett. **62**, 1472 (1989).

[20]D. R. Grempel, R. E. Prange, and S. Fishman, Phys. Rev. A **29**, 1639 (1984).

Time irreversibility of classically chaotic quantum dynamics

K. S. IKEDA

Faculty of Science and Engineering Ritsumeikan University,
Noji-cho 1916, Kusatsu 525, Japan

Abstract

The nature of classically chaotic quantum dynamics is studied by menas of a numerical time reversal experiment. It turns out that there is a fundamental quantum scale below which the system cannot exhibit classical irreversibility to external perturbations. On the other hand, a paradoxical phenomenon manifesting that quantum irreversibility mày exceed its classical counterpart is discovered. These features are explained in terms of semicalssical dynamical theory.

Introduction

Classically chaotic quantum systems in general can exhibit intrinsinc chaotic behaviour on a quite restricted time scale [1]. In particular localization phenomena provide direct evidence exhibiting the suppression of quantum ergodicity [2]. On the other hand, recent studies reveal that a many-dimensional system can mimic *some aspects* peculiar to chaotic dynamics quite well on an unexpectedly long time scale [3]. A remarkable feature that distinguishes chaotic motion from integrable motion is the sensitive dependence of dynamics on external perturbations. For classical dynamical systems such a sensitivity can be well defined by measuring how two nearby trajectories in the phase space separate in time. However, there has been proposed no systematic way to examine the sensitivity of quantum dynamics to external perturbation. In quantum dynamics, it is not possible to trace a well defined trajectory in phase space, because both quantum uncertainty and chaotic instability make a localized wavepacket spread suddenly over the phase space [4]. There is, however, a simple way in which we may test the quantum sensitivity quantitively. This is the time reversal experiment described below. In this chapter I describe several remarkable characteristics of quantal instability which have been clarified by the time reversal experiment. I emphasize here that we are concerned with the sensitivity of quantum dynamics within the time scale on which quantum dynamics apparently mimics it classical counterpart in the sense of ergodicity.

The time reversal experiment consists of the following three steps [3]: Starting with an initial quantum state $|in>$, we evolve the sytem with the normal (forward) time evolution rule described by the unitary transformation $\hat{U}(t)$. At time $t-T$ we apply a perturbation $\hat{P}$ which classically corresponds to shifting the position of the system in phase space. Finally, we evolve the system back according to the reversed (backward) time evolution rule. Thus the final state is $\hat{U}(-T)\hat{P}\hat{U}(T)\,|in>$. Without the perturbation the system will return to the initial state, but the shifted sytem can no longer return to the initial state if the system obeys unstable dynamics. By measuring how the irreversibility changes as the strength of P is varied, we can characterize the sensitivity of the systems to externally applied perturbation quantitively.

Time reversal experiment

As will be argued later, the phenomena studied here can be observed in general systems, but to be concrete we take specific systems as the sample systems. The model systems employed here are the single kicked rotor (SKR) described by the Hamiltonian

$$H^{(1)}(\hat{I}, \hat{\theta}, t) = \hat{I}^2/2 + K\ cos(\hat{\theta})\ f(t) \tag{1a}$$

and the coupled kicked rotor (CKR) with the Hamiltonian

$$H^{(2)}(\{\hat{I}_k\}, \{\hat{\theta}_k\},\ t) = H^{(1)}(\hat{I}_1, \hat{\theta}_1,\ t) + H^{(1)}(\hat{I}_2, \hat{\theta}_2,\ t) + \epsilon\ cos(\hat{\theta}_1 - \hat{\theta}_2)f(t), \tag{1b}$$

where $f(t) = \sum_n \delta(t-n)$ is a periodic impulsive force, and $\hat{\theta}_k$ and $\hat{I}_k$ are canonically conjugate operators corresponding to the angle and momentum (action) variables. The reason why I choose the kicked rotor systems as the sample systems is that their classical counterparts exhibit chaotic diffusion across momentum space, which enables us to quantify the irreversibility of the system in terms of the difference of the diffsuion between the forward and backward processes.

It is well known that a quantum SKR can mimic the classical chaotic diffusion only on a quite restricted time scale, much shorter than the characteristic time scale $T_s = D_{CL}/\hbar^2$, where D_{CL} is the classical diffusion constant [2]. However, this time scale can be made arbitrarily long by letting $\hbar \to 0$. Furthermore, T_s for the CKR is much longer than T_s for the SKR [3]. We are concerned with the instabilities which occur on a time scale much shorter than T_s.

The system starts with a momentum eigenstate $|in>$ with momentum I_{in}. Let us introduce the second order moment of momentum $M(t) =<\Psi(t)|(\hat{I}-I_{in})^2|\Psi(t)>$, where $\Psi(t) = \hat{U}^t|\,in>$ (for $t < T$) or $\Psi(t) = \hat{U}^{T-t}\hat{P}\hat{U}^T|\,in>$ (for $t > T$) and $\hat{U}$ is the single step unitary evolution operator $\mathscr{T}\ exp\{-\frac{i}{\hbar}\int_0^1 H^{(1),(2)}(s)\ ds\}$. Classical chaotic diffusion is characterized by the diffusion constant D_{CL} as $M(t) = D_{CL}t$. We quantify the irreversibility of the system by the difference between the moments computed before and after the shifting perturbation:

$$m_{irr}(\tau, \hat{P}) = M(T + \tau) - M(T - \tau)\ (\tau > 0). \tag{2}$$

We restrict the perturbation $\hat{P}$ to the one which shifts either the angle variable or the momentum alone, namely $\hat{P} = exp(i\xi\hat{\mathbf{I}}\Omega(\hat{\mathbf{I}})/\hbar)$ (θ-shifting) or $\hat{P} = exp(-i\xi\hat{\theta}\Omega(\hat{\theta})/\hbar)$ (I-shifting), where $\boldsymbol{\xi} = (\xi_1, \xi_2), \hat{\mathbf{I}} = (\hat{I}_1, \hat{I}_2)$ and $\hat{\boldsymbol{\theta}} = (\hat{\theta}_1, \hat{\theta}_2)$. Here the projection operator $\Omega(\hat{\mathbf{x}}) = \sum_{x\epsilon\Omega} |\mathbf{x}> < \mathbf{x}|(\hat{\mathbf{x}} = \hat{\mathbf{I}}$ or $\hat{\theta})$ controls the region Ω to which the perturbation is applied; the region Ω being defined in the x representation, for example, in such a way that $I_k^{(min)} < I_k < I_k^{(max)}$ $(k = 1, 2)$ in the case of $\hat{\mathbf{x}} = \hat{\mathbf{I}}$.

Global perturbation

First we consider the case where the shifting perturbation covers the whole phase space, i.e., $\hat{\Omega} = 1$. Without the perturbation, complete time reversibility $m_{irr}(\tau, \hat{P}) = 0$ is expected, but once the shifting perturbation is applied, chaotic dynamics forces the system to lose the memory of the state before $t = T$ and to restore again the stationary diffusion as $M(\tau + T) = D_{CL}(T + \tau - 2\tau_D)$ or $m_{irr}(\tau, \hat{P}) = 2D_{CL}(\tau - \tau_D)$, where $\tau_D = \tau_D(\xi)$ is the delay time beyond which the stationary diffusion is restored. Classical theory predicts that the delay time diverges logarithmically as

$$\tau_D(\xi) \sim log(\xi)/\lambda \tag{3}$$

if $\xi \rightarrow 0$ (λ is the Lyapunov exponent). It is measured directly by the deviation of $m_{irr}(\tau, \hat{P})$ from the one for sufficiently strong $\hat{P}$, that is, by $m_0(\tau, \hat{P}) = m_{irr}(\tau, \hat{P}_{\xi=strong\ enough}) - m_{irr}(\tau, \hat{P})$. Classical theory predicts $m_0(\tau, \hat{P}) = 2D_{CL}\tau_D(\xi) \sim D_{CL}\ log(\xi)/\lambda$. In Fig.1(a) we show a typical result of the θ-shifting time reversal experiment examined for the SKR $M(t)$ computed at various values of $\xi = \xi_n$ such as $\xi = \xi_0 \times 2^{n/2}$ $(n = 1, 2, 3 \ldots, \xi_0 \ll O(\hbar))$ are superimposed. After $t = T$ normal diffusion seems to be restored sooner or later. To demonstrate this more precisely, we plot $m_0(\tau, \hat{P})$ measured at two fixed τs$(\tau = T/2$ and $T)$ versus $log(\xi)$ in Fig.1(b), Note that the plots are on a straight line on the larger side of ξ, which is a classically expected behavior, i.e., $m_0 \sim log(\xi)/\lambda$. However, there is a threshold value of $\xi = \xi_{th}$, below which a remarkable deviation from classical behavior takes place. Extensive studies of SKR and CKR reveal that ξ_{th} is proportional to $\hbar$ as shown in Fig.2. This fact implies that classical exponential separation of nearby trajectories occurs only when the systems are separated by more than the quantum scale ξ_{th}. We, therefore, call this *quantized orbital instability*.

Then how can we explain the above threshold phenomenon? We introduce here a view point from semiclassical theory [5]: Let us consider the problem of finding the classical orbits starting with θ_{in} at the 0-th step and reaching θ_{fi} at the τth step (the orbit is not, of course, unique). We can associate a phase $\Phi(\theta_{in}, \theta_{fi})$ with each of the classical orbits connecting θ_{in} and θ_{fi} [6]. Indeed, the phase is the action computed along the orbit divided by $\hbar$. Then the criterion for the relative motion between the two systems started from slightly different positions θ_{in} and $\theta_{in} + \xi$ to behave in a classical manner is that the difference

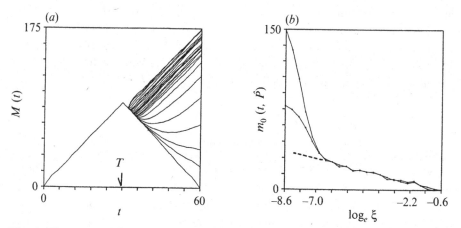

Fig. 1. Time reversal experiment. (a) $M(t)$ for various values of $\xi = \xi_n$. (b) $m_0(T,P)$ and $m_0(T/2,P)$ versus $log_e(\xi)$. The broken line indicates classical behavior. The SKR ($K = 9, \hbar = 2\pi \times 41/32768$) is used.

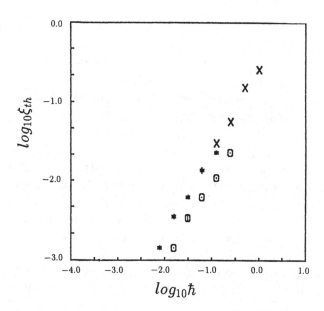

Fig. 2. Threshold perturbation strength versus Planck's constant for the CKR ($K_{1,2} = 5$, $\epsilon = 3$; $\times$), and for the SKR ($K = 5$; $\ast$, and $K = 21$; $\square$).

of the associated phases, i.e., $\Delta\Phi = \Phi(\theta_{in} + \xi, \theta_{fi}) - \Phi(\theta_{in}, \theta_{fi})$, is much larger than 2π. Semiclassical theory reveals that the difference of the phases is evaluated as [5]

$$\Delta\Phi = \Delta I(\tau)\xi/\hbar \tag{4}$$

Here $\Delta I(\tau)$ is the characteristic difference in momentum which is made during τ by two systems evolved by the classical backward process from the same θ_{fi} with the initial momentum difference less than 2π. In the case of the kicked rotors $\Delta I(\tau)$ increases in time as $\Delta I(\tau) = \sqrt{D_{CL}\tau}$, therefore the 'classicalization' of the motion is achieved beyond τ_{CL} which makes $\Delta\Phi > 2\pi$. Hence $\tau_{CL} \sim h^2/\xi^2 D_{CL}$ ($h = 2\pi\hbar$). From these considerations we come to the following conjecture: There is a threshold $\xi = \xi_{th}$, which is decided by the equality $\tau_{CL} = \tau_D(\xi_{th})$, or equivalently by the self-consistent uncertainty relation

$$\xi_{th}\Delta I(\tau_D(\xi_{th})) = h, \tag{5}$$

and (i) for $\xi \gg \xi_{th}$ the classical time irreversibility is observed and the past memory is lost beyond τ_D, but (ii) for $\xi \ll \xi_{th}$ the 'classicalization' and loss of memory occurs beyond τ_{CL}, which implies that the diffusion is restored after the divergently long time scale $T_{CL} \sim h^2/\xi^2 D_{CL}$. In eq.(5) $P(\xi) \equiv \Delta I(\tau_D(\xi))$ is the length scale which two nearby points can make before the memory of their initial relative positions is lost, and we call it the *maximum memory length*. (i) and (ii) are compared with the results of extensive numerical time reversal experiments, and they both agree quite well with experiment as depicted in Fig.3. Thus the quantum nature works so as to suppress the quantum instability by introducing a threshold length scale into the dynamics.

Local perturbation

However, the quantum nature does not necessarily suppress the irreversibility. Such a paradoxical feature is observed when the region Ω to which the perturbation is applied does not cover the whole phase space. In Fig.4, we show how the quantum irreversibility varies with the area of the perturbed region Ω. In this example we apply the θ-shifting perturbation only to the restricted region

$$\Omega = ((\theta,I)| - I_\Omega/2 + I_{in} \leqq I \leqq I_\Omega/2 + I_{in}, 0 < \theta \leqq 2\pi) \tag{6}$$

at $t = T$. According to pure classical theory, the classical irreversibility increases from $D_{CL}T$ to $2D_{CL}T$ as I increases from 0 to infinity. Quantum irreversibility is quite different from this. It starts from 0 and increases steeply as I_Ω exceeds a quantum scale $I_{th} \sim O(\hbar)$. The behavior for $I_\Omega < I_{th}$ is essentially the same as the threshold instability argued above and is not discussed further in detail [5]. The problem is the behavior observed above I_{th} : The quantum irreversibility first forms a sharp maximum just above I_{th}, and it decreases towards a value of almost *twice* the classical irreversiblity. With further increase of ξ the quantum

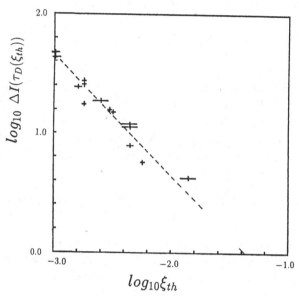

Fig. 3. Threshold perturbation strength versus maximum memory length at $\xi = \xi_{th}$ obtained at various values of K $(3 < K < 25)$. The SKR $(\hbar = 2\pi \times 81/32768)$ is used. The broken line indicates $\xi_{th} = 2.8\hbar/\Delta I(\tau_D(\xi_{th}))$.

irreversibility keeps the same level, and it is always larger than the classical irreversibility. What must be emphasized is that such a discrepancy between quantum and classical dynamics happens in the entirely classical regime $I_\Omega \gg I_{th}$ and $\xi \gg \xi_{th}$. Furthermore, it has been confirmed that such a characteristic persists even at very small $\hbar$, and, moreoever, it is observed even for the CKR, which has been thought of as a typical system being able to mimic classical chaotic diffusion [3].

We now present an intuitive picture explaining the above phenomenon, which will suggest that the anomalous excess is a quite generic phenomenon in classically chaotic quantum systems. Consider the classical counterpart of the time reversal experiment: Let U_{CL} be the single step mapping (the standard map in the case of SKR) describing the corresponding classical time evolution, and suppose Ω_i to be the set representing the initial conditions. As time elapses, the image $U_{CL}^t(\Omega_i)$ is stretched and folded due to chaotic dynamics to form a complex object in phase space. At $t = T$, the set of the points in $U_{CL}^T(\Omega_i) \cap \Omega$ is shifted by the perturbation and they all contribute to the irreversibility as the normal classical component of the irreversibility. Thus the excess part of the irreversibility should come from the remaining unperturbed part, i.e., $\Omega_{NP} = U_{CL}^T(\Omega_i) \cap \overline{(U_{CL}^T(\Omega_i) \cap \Omega)}$. At the τ step after the shifting perturbation, the time reversed classical dynamics makes the points in Ω_{NP} return to form a subset of $U_{CL}^{T-\tau}(\Omega_i)$. Thus the retrieved set at

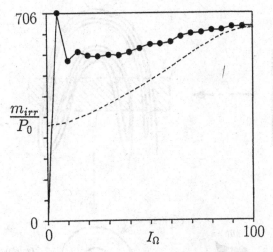

Fig. 4. Irreversibility versus the area of the perturbed region. The SKR ($K = 5$, $\hbar = 2\pi \times 21/32768$, $T = 30$ and $\xi = 0.1$) is used. Similar characteristics are obtained for the CKR as well.

$t = T + \tau$ is

$$U_{CL}^{-\tau}(\Omega_{NP}) = U_{CL}^{T-\tau}(\Omega_i) \bigcap \overline{(U_{CL}^{T-\tau}(\Omega_i) \bigcap U_{CL}^{-\tau}(\Omega))}, \tag{7}$$

namely $U_{CL}^{T-\tau}(\Omega_i)$ from which its intersection with $U_{CL}^{-\tau}(\Omega)$ is removed. The intersection $U_{CL}^{T-\tau}(\Omega_i) \cap U_{CL}^{-\tau}(\Omega)$ introduces a number of 'slits' along the set $U_{CL}^{T-\tau}(\Omega_i)$ and cut it into a number of parts. Considering the formation process of the set of slits, it can be identified with a *transient hyperbolic set*. The size of the 'slits' decreases exponentially in the direction tangent to the stretched set $U_{CL}^{T-t}(\Omega_i)$ due to the time reversed chaotic dynamics, and becomes too small to be compatible with the uncertainty principle. The presence of such very narrow slits in the classical set implies that the corresponding wavefunction is squeezed out at the 'slits' in the direction normal to the stretched set as illustrated in Fig.5 [5]. Since the part pushed out of the set $U_{CL}^{T-\tau}(\Omega_i)$ is out of the bundle of the trajectories going back towards Ω_i, they lose their memory at $t = T$ and eventually contribute to the irreversibility as the excess component. If we suppose that the squeezed 'slits' are pushed out far from their classical positions and that they all contribute to the irreversibility, then the irreversibility due to them reaches the same amount as that from the perturbed part. This is the origin of the excess factor '2'. This seems to provide an objection to the widely accepted idea that quantum nature suppresses classical chaotic instability. Thus the anomalous excess may be thought as direct evidence of the incompatibility of the formation of a hyperbolic set with the quantum uncertainly principle.

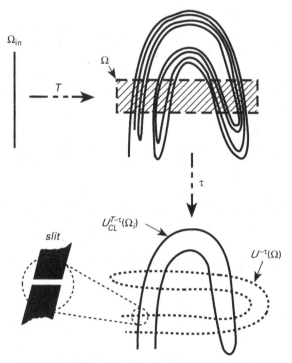

Fig. 5. Formation of squeezed 'slits' in time reversed dynamics. See the text.

Summary

In conclusion the time irreversibility inherent in classically chaotic quantum dynamics has been studied by means of the time reversal experiment, confining this to the time regime in which quantum dynamics mimics classical ergodicity precisely. A perturbation is applied just before the dynamics is reversed, and the irreversibility is quantified by the difference between the forward evolution process and the backward evolution process. The experimental results lead to the conclusion that the two nearby trajectories can lose the memory of their initial relative positions only when the initial separation is greater than a minimal quantum scale. This scale is determined by the 'maximum memory length' through the self-consistent uncertainly relation (eq. (5)). The presence of the threshold scale means that the quantum nature certainly suppresses the realization of classical irreversibility. However, a further study revealed that the quantum effect does not necessarily suppress the irreversibility. Indeed, when the perturbation is applied only to a part of the system, the quantum time irreversibility exceeds it classical counterpart by about two times at the maximum. Such an anomalous excess is explained in terms of the incompatibility of the hyperbolic set with the quantum uncertainty principle. Further studies are necessary in order to understand the more profound significance of the phenomena to the foundation of quantum statical mechanics.

This work is supported by the Grant-in-Aid for Scientific Research Ministry of Education, Science and Culture.

REFERENCES

[1]   M.V. Berry and N.L. Balazs, J. Phys. **A12** (1979) 625.

[2]   G. Casati, B.V. Chirikov, F.M. Izrailev and J. Ford; in *Stochastic Behavior in Classical and Quantum Hamiltonian Systems*, edited by G. Casati and J. Ford, Lecture Notes in Physics, vol 93, (Springer-Verlag, N.Y. 1979) pp. 334–352; B.V. Chirikov, F.M. Izrailev, and D.L. Shepelyansky; Sov. Sci. Rev. C2 (1981) 209.

[3]   S. Adachi, M. Toda and K. Ikeda, Phys. Rev. Lett. 61 (1988) 659; M. Toda, S. Adachi and K. Ikeda, Prog. Theor. Phys. Suppl. No.98 (1989) 323.

[4]   M. Toda and K. Ikeda, Phys. Lett. **A21** (1987) 165

[5]   For details, see K. Ikeda, 'Time reversal experiment for classically chaotic quantum systems' in preparation.

[6]   For semiclassical quantization of quantum maps, see K. Ikeda, Ann. Phys. **1** (1993) Appendix E, F.

Effect of Noise on Time-Dependent Quantum Chaos

E. Ott and T. M. Antonsen, Jr.

Laboratory for Plasma and Fusion Energy Studies, University of Maryland, College Park, Maryland 20742

and

J. D. Hanson

Institute for Fusion Studies, University of Texas, Austin, Texas 78712

(Received 12 June 1984)

The dynamics of a time-dependent quantum system can be qualitatively different from that of its classical counterpart when the latter is chaotic. It is shown that small noise can strongly alter this situation.

What is the nature of a quantum system whose classical counterpart exhibits chaotic dynamics? The subfield dealing with this question has been called *quantum chaos*. A striking result in quantum chaos has been obtained by Casati et al.[1] These authors considered a particular Hamiltonian and a potential representing periodic impulses kicking the system. If the strength of the kicks is large enough, then, in the classical description the motion is chaotic, and the momentum variable, p, behaves diffusively. That is, the average value of p^2 apparently increases linearly with time. Casati et al. considered numerically the quantum mechanical version of the same problem with $\hbar$ small. They found that for early times, the average value of p^2 increased linearly with time at roughly the classical diffusive rate, but that for long time this linear increase slowed and eventually appeared to cease. Thus, there was no numerically discernible diffusion in the quantum case.

The observed saturation of the growth of $\langle p^2 \rangle$ is understandable if the Schrödinger operator for this problem has an essentially discrete quasienergy level spectrum.[1-4] Recently, Fishman, Grempel, and

Prange[4] have presented strong arguments supporting the idea that the quasienergy spectra for systems of the type studied by Casati et al. are essentially discrete. These arguments are based on an analogy with Anderson localization of an electron in a solid with a random lattice. Futhermore, it has been pointed out that these results have implications for other physical systems[5-7] and experiments have been proposed. For example, the ionization of an atom by high-frequency electromagnetic waves and the interaction of electrons on the surface of a superconductor with an oscillating electric field have both been suggested[5,6] as systems for which the consequences of quantum localization in a classically chaotic system could be experimentally observed. A question then arises as to how sensitive the localization is to real effects not included in the model, e.g., finite bandwidth of the ionizing radiation, finite temperature, etc. In this Letter we crudely model such effects as noise. That is, we introduce a small random component into the quantum rotator equations[8] (see also Shepelyanski[3]). (Since our subsequent arguments are apparently not model dependent, we believe that they should be relevant to real physical experiments.) We find that the quantum interference leading to localization of p^2 is a delicate effect that is strongly affected by small noise. It is the goal of this paper to investigate the mechanisms by which small noise leads to diffusion, as well as the regimes of dependence of the quantum momentum diffusion coefficient on the noise and kicking strength.

We consider a Hamiltonian

$$H = \frac{P^2}{2I} + [\bar{\epsilon}R\cos\theta + \bar{\nu}\phi(\theta,t)] \sum_{n=-\infty}^{+\infty} \delta(t - nT),$$

(1)

where θ has period 2π, P is the angular momentum, I is the moment of inertia, $\bar{\epsilon}$ is the strength of a periodically applied (period T) horizontal impulsive force, R is the radius at which the force is applied, and the term $\phi(\theta,t)$ is a random function of time representing a noise component in the kicking with $\bar{\nu}$ a parameter governing the noise strength.

The classical problem corresponding to the Hamiltonian (1) yields the well-known *standard mapping*[9]

(including noise),

$$p_{n+1} = p_n + \epsilon\sin\theta_{n+1} - \nu\phi'_{n+1}(\theta_{n+1}),$$

$$\theta_{n+1} = \theta_n + p_n,$$

where $\phi_n(\theta) = \phi(\theta,nT)$, $\phi'_n = d\phi_n/d\theta$, (p_n,θ_n) denote the values of $(p(t),\theta(t))$ just after the nth kick (at $t = nT$), and $\epsilon = \bar{\epsilon}RT/I$, $p = PT/I$, and $\nu = \bar{\nu}T/I$. One possible choice for ϕ_n that we will use in all of our subsequent calculations is $\phi_n(\theta) = \sqrt{2}\Delta_n\cos(\theta + \alpha_n)$, where Δ_n is a Gaussian random variable $\langle\Delta_n\Delta_{n'}\rangle = \delta_{nn'}$, and α_n is random with a uniform distribution in $[0,2\pi]$. For the case where ϵ is large most initial conditions for the classical map generate orbits which are diffusive with a momentum diffusion coefficient given approximately by[9,10] $D_{cl} \simeq \epsilon^2/4 + \nu^2/2$. Thus, if $\nu^2 \ll \epsilon^2$ (which applies to all of our subsequent considerations), the noise has little effect on the value of D_{cl}.

Turning to the quantum problem, we impose periodic boundary conditions, $\psi(\theta,t) = \psi(\theta + 2\pi,t)$. Thus momenta are quantized at $p = l\hbar$ (l is an integer). Integrating Schrödinger's equation with Hamiltonian (1) through one time period,[1,11] setting $\psi_n = \psi(\theta,nT + 0^+)$, and normalizing $\hbar$ to I/T gives

$$\psi_{n+1}(\theta) = \exp[(i\nu/\hbar)\phi_{n+1}(\theta)]L[\psi_n(\theta)],$$

(2a)

$$L[\psi(\theta)] \equiv \sum_l \int_0^{2\pi} (d\theta'/2\pi)[-i\hbar l^2/2 + il(\theta - \theta') + i\epsilon\cos\theta/\hbar]\psi(\theta').$$

(2b)

In what follows we shall consider $\epsilon^2 \gg \nu^2$ and discuss the parameter dependence of D_q on ν, ϵ, and $\hbar$. We distinguish three regimes in terms of which we state our main results as follows: (a) $(\epsilon/\hbar)^2 \ll 1$ (large $\hbar$) for which we find $D_q \simeq \nu^2/2$; (b) $(\epsilon/\hbar)^2 \gg 1$ and $(\nu/\hbar)^2(\epsilon/\hbar)^2 \ll 1$ (moderate $\hbar$) for which we find $D_q \sim \nu^2(\epsilon/\hbar)^4$; and (c) $(\nu/\hbar)^2 \times (\epsilon/\hbar)^2 \gg 1$ (small $\hbar$) for which we find $D_q \simeq D_{cl}$.

Thus, from our result for regime (c), in the "classical limit" (i.e., $\hbar \to 0$) $D_q \to D_{cl}$ when $\nu > 0$ (see also Ref. 3). This is not so for $\nu = 0$, since then the quantum diffusion coefficient is apparently zero for any $\hbar > 0$ (hence, with $\nu = 0$, $\lim D_q = 0$ as $\hbar \to 0$). Thus we may say that noise, however small, restores the classical limit. Furthermore, we emphasize that $D_q \simeq D_{cl}$ can apply *even for very small noise* [i.e., $(\nu/\hbar)^2 \ll 1$] provided that we are

in the semiclassical regime.

Regimes (a) and (b) may be treated by random-phase-approximation perturbation theory considering the effect of finite noise ($\nu > 0$) as the perturbation. For $\nu = 0$, we assume that (2) has an essentially discrete quasienergy spectrum.[2,4] Thus $\psi_n(\theta)$ may be expanded as $\psi_n(\theta) = \sum A_m \exp(-i\omega_m n) \times u_m(\theta)$, where from Eq. (2) the $u_m(\theta)$ and $\exp(i\omega_m)$ are the eigenfunctions and eigenvalues of the unitary operator L, $L[u_m] = \exp(-i\omega_m)u_m$. Since $\nu/\hbar \ll 1$ for both regimes (a) and (b), the factor $\exp[i\nu\phi'_n(\theta)/\hbar] \simeq 1 + i\nu\phi'_n(\theta)/\hbar$ in Eq. (2), and, with the assumption that perturbation theory is valid, the probability per kick of a transition from u_m to $u_{m'}$ is $\alpha_{mm'} = (\nu/\hbar)^2|\langle u_{m'}|\phi'_n|u_m\rangle|^2_{ave}$ where the subscript "ave" indicates an average over the ensemble of random ϕ_n. With use of the transition probability $\alpha_{mm'}$, the diffusion coefficient is

$$D_q(m) = \frac{1}{2}(\nu/\hbar)^2\sum_{m'}|\langle u_{m'}|\phi'_n|u_m\rangle|^2_{ave}(p_{m'} - p_m)^2,$$

(3)

where p_m is the momentum expectation value for the state u_m. Note that, whenever Eq. (3) applies, D_q is proportional to ν^2.

We now consider regime (a). In this case the term $\exp[i(\epsilon/\hbar)\cos\theta]$ in L may be neglected to lowest order; thus the $u_m(x)$ are as in the freely rotating (unkicked) rotator, $u_m(x) \simeq (2\pi)^{-1/2}\exp(im\theta)$. For

$\phi_n = \sqrt{2}\Delta_n \cos(\theta + \alpha_n)$, we obtain $\alpha_{mm'} = (\nu/\hbar)^2$ $\times (\delta_{m,m'+1} + \delta_{m,m'-1})/2$. Since, in this approximation, u_m is an eigenfunction of the momentum operator corresponding to a momentum $p = m\hbar$, we obtain from (3) $D_q \simeq \nu^2/2$. This result is the same as the diffusion that one would obtain for the classical map with noise if ϵ were set equal to zero.

We now consider regimes (b) and (c). In these cases, $(\epsilon/\hbar)^2 \gg 1$, and the eigenvalue problem for $u_m(\theta)$ is not analytically solvable. Thus we shall only be able to obtain estimates for D_q. First, we note that, on the basis of Anderson localization, Fishman, Grempel, and Prange[4] have argued that, in the momentum representation, the eigenfunctions are exponentially localized about the "lattice points" $p = l\hbar$. The localization length in p (which we denote Δ) is large compared to $\hbar$. Furthermore, for $(\epsilon/\hbar)^2 \gg 1$, the momentum eigenfunctions,

$$\hat{u}_m(l) = (2\pi)^{-1} \int_0^{2\pi} \exp(-il\theta) u_m(\theta) d\theta,$$

are not smoothly varying on the lattice. That is, although *on average* there is a slow exponential decrease of $|\hat{u}_m(l)|$ with l away from the center of localization of $\hat{u}_m(l)$, there are also typically $\sim 100\%$ variations of $\hat{u}_m(l)$ on the lattice-spacing scale [i.e.; typically $|\hat{u}_m(l) - \hat{u}_m(l \pm 1)| \sim |\hat{u}_m(l)|$]. This results from the factor $\exp(-i\hbar l^2/2)$ in Eq. (2b) which for large l gives each $u_m(l)$ a nearly random phase.

We now obtain an estimate of Δ using the arguments of Chirikov, Izrailev, and Shepelyanski[2] We observe numerically, for the case with no noise, that $\langle p^2 \rangle$ increases with time initially at roughly the classical rate, but then turns over at some time $n \sim n_*$. This is interpreted as being due to the excitation of many Anderson-localized modes by the initial condition (which is localized near $p = 0$). Furthermore, those modes most strongly excited are those which are localized around momenta within Δ of $p = 0$. Hence the effective number of modes excited by an initial condition with $p = 0$ is of the order of $\Delta/\hbar$. Each mode has an associated eigenvalue $\exp(-i\omega_m)$. Thus the ω_m may be taken to lie in $[0, 2\pi]$. Since there are $\Delta/\hbar$ modes, the typical frequency spacing between modes is $\delta\omega \sim 2\pi/(\Delta/\hbar)$. For $n \lesssim 1/\delta\omega$, the system does not yet "know" that the quasienergy spectrum is discrete. Thus we expect that $\langle p^2 \rangle$ increases with time until $1/\delta\omega$, at which time the turnover in $\langle p^2 \rangle$ should occur. Thus $n_* \sim 1/\delta\omega \sim \Delta/\hbar$. In addition, at the turnover the characteristic spread in momentum will be the localization width of the modes, i.e., $\langle p^2 \rangle \sim \Delta^2$. Let n_d denote the time to classically diffuse the distance Δ, $n_d \sim \Delta^2/D_{cl} \sim \Delta^2/\epsilon^2$. Since the initial increase of $\langle p^2 \rangle$ is at the classical rate, we have $n_* \sim n_d$ or $\Delta/\hbar \sim \Delta^2/\epsilon^2$, which yields the result[2] $\Delta \sim \epsilon^2/\hbar$.

Before considering regime (b), we ask what is the limit of validity of perturbation theory, Eq. (3). Localization is dependent on the maintenance of phase coherence for the time it would take a wave packet to classically diffuse the distance Δ in p (e.g., see Thouless[12]). Thus, if noise destroys this phase coherence in the time n_d, then the localized modes will also be destroyed. With localization no longer operable we expect a return to the classical result $D_q \simeq D_{cl}$. To see how much noise is needed to do this, we recall that an eigenstate in the momentum representation has $\sim 100\%$ variations down to momentum separations of $\hbar$ (the lattice spacing). Thus, if the cumulative effect of the noise scatters p by an amount equal to $\hbar$, then the phases have been randomized. Noting that $\nu^2/2$ is the component of momentum diffusion due to the noise, the time n_c for the noise to scatter p by $\hbar$ is $n_c(\nu^2/2) \sim \hbar^2$ or $n_c \sim \hbar^2/\nu^2$. Thus, if $n_c < n_d$, or $(\nu/\hbar)^2(\epsilon/\hbar)^2 > 1$, then we expect that $D_q \simeq D_{cl}$. This defines the boundary between regimes (b) and (c).

To estimate D_q when $(\nu/\hbar)^2(\epsilon/\hbar)^2 < 1$ and $(\epsilon/\hbar)^2 > 1$ [i.e., regime (b)], we note that the phase coherence of the waves is maintained for a time n_c. Thus we expect transitions between localized modes on this time scale. Since transitions are appreciable only for modes within a localization length of each other, $D_q \sim \Delta^2/n_c$, or $D_q \sim \nu^2(\epsilon/\hbar)^4$.

The above arguments are similar to those of Thouless[12] who considered the effect of finite temperature on localization in a solid. Thus our numerical experiments testing the above arguments (described below) may also be viewed as a test of Thouless's heuristic treatment of the low-temperature conductivity of disordered solids. To our knowledge no other numerical experiments testing Thouless's arguments exist.

The estimate $D_q \sim \nu^2(\epsilon/\hbar)^4$ can also be obtained directly from (3) as follows:

$$u_m(\theta) = \sum_l \hat{u}_m(l)\exp(il\theta).$$

From the fact that the $\hat{u}_m$ are localized, there are effectively of the order of $\Delta/\hbar$ appreciable terms in the sum over l. Thus, with use of the $\hat{u}_m(l)$ representation, the quantity $\langle u_m|\phi_n'|u_m \rangle$, with $\phi_n = \sqrt{2}\Delta_n \cos(\theta + \alpha_n)$, will involve a sum over roughly $\Delta/\hbar$ appreciable terms. Since $\langle u_m|u_m \rangle = 1$, $|\hat{u}_m(l)|^2 \sim (\Delta/\hbar)^{-1}$. Now assuming that the $\hat{u}_m(l)$ are pseudorandom in l, we see that the sum involved in the calculation of $\langle u_{m'}|\phi_n'|u_m \rangle$ will be of the order of $(\Delta/\hbar)^{-1/2}$. Thus (3) yields $D_q \sim (\nu/$

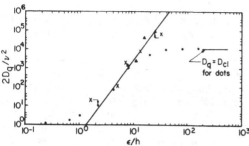

FIG. 1. $D_q/(\nu^2/2)$ vs $\epsilon/\hbar$ with $\nu = 0.0354$ in regime (b). Solid line corresponds to $D_q \propto (\epsilon/\hbar)^4$. Dots, $\epsilon = 5.0$, $\hbar$ varies; crosses, $\hbar = 5.0$, ϵ varies; triangles, $\epsilon = 55.26$, $\hbar$ varies. The iteration method is discussed by Hanson *et al.* (Ref. 11). For the dots, regime (a) corresponds to $\epsilon/\hbar \lesssim 0.8$, regime (b) to $2 \lesssim \epsilon/\hbar \lesssim 10$, and regime (c) to $\epsilon/\hbar \gtrsim 30$.

$\hbar)^2\Delta^2$ which again gives $D_q \sim \nu^2(\epsilon/\hbar)^4$.

As a test of these arguments, Fig. 1 shows numerical results obtained from long-time evolutions of Eq. (2). (Values of ϵ were chosen to avoid accelerator modes,[9] while values of $\hbar/4\pi$ are irrational to avoid quantum resonances.[13]) The dots show results for D_q versus $\epsilon/\hbar$ with $\epsilon = 5.0$, $\nu = 0.0354$, and $\hbar$ varying (horizontal axis). For $(\epsilon/\hbar)^2 \ll 1$ [regime (a)] there is good agreement with $D_q \simeq \nu^2/2$, and D_q apparently becomes asymptotic to D_{cl} for large $\epsilon/\hbar$ appropriate to regime (c). Figure 1 also shows other data (triangles and crosses) for regime (b). The triangles and dots have ν and ϵ fixed and $\hbar$ varying, while the crosses correspond to ν and $\hbar$ fixed and ϵ varying. The three sets of data fall close to each other and are consistent with an approximate proportionality of D_q to the fourth power of $\epsilon/\hbar$ in regime (b), as predicted theoretically (solid line in Fig. 1). In addition, we have obtained extensive data on the variation of D_q with ν (ϵ and $\hbar$ held fixed). Excellent agreement is found with the theoretically predicted proportionality to ν^2 in regimes (a) and (b) [cf. Eq. (3)].

In conclusion, the presence of small noise can greatly modify the behavior of a quantum mechanical system which is classically chaotic, particularly for systems in the semiclassical regime.

We thank S. Fishman, J. Ford, R. E. Prange, and R. Westervelt for fruitful discussions. This work was supported by the U.S. Department of Energy.

[1]G. Casati, B. V. Chirikov, F. M. Izrailev, and J. Ford, in *Stochastic Behavior in Classical and Quantum Hamiltonian Systems* (Springer, New York, 1979), p. 334.

[2]B. V. Chirikov, F. M. Izrailev, and D. L. Shepelyanski, Sov. Sci. Rev. C **2**, 209 (1981); T. Hogg and B. A. Huberman, Phys. Rev. Lett. **48**, 711 (1982).

[3]D. L. Shepelyanski, Physica (Utrecht) **8D**, 208 (1983).

[4]S. Fishman, D. R. Grempel, and R. E. Prange, Phys. Rev. Lett. **49**, 509 (1982). Actually the arguments of this reference do not imply a strictly discrete spectrum, and there *may be* continua present [G. Casati and I. Guaineri, Commun. Math. Phys. **95**, 121 (1984)]. That is why we have used the wording "essentially discrete." By this we mean that the evolution over a long but not too long time is as if the spectrum were discrete. See Prange, Grempel, and Fishman, in Proceedings of the Conference on Quantum Chaos, Como, Italy, June 1983 (to be published), for a heuristic discussion of why, for practical purposes, it may often be a good approximation to neglect the possible continuum aspects of the spectrum and approximate it as discrete. In our numerical experiments we see no evidence of continua, and, particularly for small $\hbar$, the presence of even tiny noise would seem to make the question moot [excluding low-order rational values of $\hbar$; cf. F. M. Izrailev and D. L. Shepelyanski, Theor. Math. Phys. **43**, 417 (1983)].

[5]R. V. Jensen, Phys. Rev. Lett. **49**, 1365 (1982); R. Blumel and U. Smilansky, Phys. Rev. Lett. **52**, 137 (1984).

[6]D. L. Shepelyanski, Institute of Nuclear Physics Report No. 83-61, 1983 (to be published).

[7]E. Ott, in *Long Time Prediction in Dynamics* (Wiley, New York, 1983), p. 281.

[8]Since we introduce randomness externally, our results do not address the question of whether, and to what extent, there is a quantum counterpart to *deterministic* chaos in classical systems. Indeed, for finite noise there is always diffusion [G. Casati and I. Guarneri, Phys. Rev. Lett. **50**, 640 (1983)].

[9]B. V. Chirikov, Phys. Rep. **52**, 265 (1979).

[10]A. B. Rechester and R. B. White, Phys. Rev. Lett. **44**, 1586 (1980).

[11]M. V. Berry, N. L. Balazs, M. Tabor, and A. Voros, Ann. Phys. **112**, 26 (1979); J. D. Hanson, E. Ott, and T. M. Antonsen, Phys. Rev. A **29**, 819 (1984).

[12]D. J. Thouless, Phys. Rev. Lett. **39**, 1167 (1977), and Solid State Commun. **34**, 683 (1980).

[13]Izrailev and Shepelyanski, Ref. 4. To avoid these resonances we use $\hbar = 4\pi r/(q + \gamma^{-1})$, where r and q are integers and γ is the golden mean.

Dynamical Localization, Dissipation and Noise

R. Graham

Fachbereich Physik, Universität Gesamthochschule Essen - D4300 Essen 1, B.R.D.

1. – Introduction.

Dynamical localization is a variant of Anderson localization appearing in quantum systems whose classical limit is chaotic. It is a novel quantum coherence effect—perhaps the most important new physical effect appearing in the field of quantum chaos. Like the familiar Anderson localization it is based on destructive interference of waves in random systems. What is new in dynamical localization is the fact that the randomness is not externally imposed, *e.g.*, by a random medium, but is produced dynamically by a simple and completely deterministic system. Here the parallel to chaos (*i.e.* stochasticity) in deterministic classical dynamical systems with few degrees of freedom is apparent.

Dynamical localization has been reviewed in depth in the lectures of FISH-MAN, SHEPELYANSKY and IZRAILEV. Another useful review has been given in [1]. In order to set the stage it is, therefore, enough to recall a few basic facts and to mention some physical examples where this effect appears.

The discussion will be restricted to Hamiltonian systems which are either autonomous with two degrees of freedom or externally driven periodically in time with one degree of freedom. In fact, extending phase space the latter case can be viewed just as a special case of the former [2]. Classically, under conditions of chaos, the two action variables, describing the system together with the canonically conjugate angles, will undergo a diffusion process. We shall always assume that the chaotic part of phase space is sufficiently large to neglect boundary effects on the diffusion. Taking a Poincaré surface of section keeping the total energy and one of the angles fixed, and using as coordinates the remaining angle θ and its action I, one then finds the diffusion law

$$(1.1) \qquad \langle (I(n) - I_\lambda)^2 \rangle = Dn,$$

where n is the iteration number of the map and $\langle ... \rangle$ denotes an average over the ensemble of initial conditions, for fixed initial value I_λ. Quantum-mechanically the action variable I is quantized, $I = \hbar l$, l integer. Dynamical localization is the phenomenon by which the eigenstates ϕ_λ of the Hamiltonian (of the extended Hamiltonian in the periodically driven case) for large $|l|$ decay exponentially in l:

$$(1.2) \qquad\qquad |\phi_\lambda| \sim \exp[-|l - l_\lambda|/\xi]$$

in a situation where the classical system obeys (1.1). Here ξ is the localization length of the state ϕ_λ. It is related to the classical diffusion constant D by

$$(1.3) \qquad\qquad \xi = \alpha \frac{D}{\hbar^2}.$$

We note that D in (1.1) may, in general, vary with I_λ and hence ξ may depend on l_λ. We shall always assume, for simplicity, that the variation of ξ_λ over a range of l_λ of size ξ_λ is negligible. Otherwise (1.1)-(1.3) would have to be generalized [3]. Let us mention a few examples:

   1) The kicked rotor (for references see Fishman's lecture):

The Hamiltonian in convenient units is

$$(1.4) \qquad\qquad H_S = \frac{p^2}{2} - K \cos q \cdot \partial^{(1)}(t),$$

where $\partial^{(1)}(t) = \sum_n \partial(t - n)$. For $K \gg 1$ and generic initial values p_λ avoiding accelerator modes [2] (1.1)-(1.3) hold with $I = p$, $\alpha = 1/2$ and

$$(1.5) \qquad\qquad D = \frac{1}{2} K^2(1 - J_2(K) + J_2^2(K) + O(K^{-2})),$$

where J_l is the ordinary Bessel function. Thus dynamical localization in p appears. (A mathematical proof of this statement is still lacking.)

   2) Rydberg atoms in microwave fields (see, *e.g.*, [4]):

The Hamiltonian for a 1-dimensional version of this system in atomic units (including $\hbar = 1$) is

$$(1.6) \qquad\qquad H_S = \frac{p^2}{2} - \frac{1}{x} + \varepsilon x \sin \omega t, \qquad\qquad x > 0,$$

and one obtains diffusion in the action variable $I = (1/2\pi) \oint p \, dx$ according to (1.1) with $n = \omega t/2\pi$ and the diffusion constant

$$(1.7) \qquad\qquad D(I_\lambda) = 2\xi_\lambda = 3.3 \frac{\varepsilon^2 I_\lambda^3}{\omega^{7/3}}$$

depending explicitly on I_λ.

3) Other periodically driven nonlinear oscillators:

These are described by Hamiltonians of the form

(1.8)
$$H_S = H_0(I) + \varepsilon x \sin \omega t.$$

E.g., for multiphoton absorption by molecular vibrations leading to excitation and dissociation $H_0(I)$ is the Morse Hamiltonian [5, 6]

(1.9)
$$H_0(I) = \begin{cases} I - \dfrac{I^2}{2}, & H_0 < 0, \\ 1 - I + \dfrac{1}{2}I^2, & H_0 > 0, \end{cases}$$

and one obtains the *I*-dependent diffusion constant

(1.10)
$$D(I_\lambda) = 2\hbar^2 \xi_\lambda = \frac{2\pi^2 \varepsilon^2}{\omega} \exp[-2\omega] \frac{1}{\left(1 - \dfrac{I_\lambda}{N_0}\right)^3}$$

and the associated localization length ξ_λ, where N_0 is the total number of bound vibrational states of the molecule and $\hbar$, in the employed units, is the ratio of the vibrational zero-point energy and the dissociation energy.

Another example is the driven pendulum realized in periodically driven Josephson junctions [7, 8] where H_0 is the free-pendulum Hamiltonian $H_0 = p^2/2 - k \cos x$ and where $D = \langle \Delta P^2 \rangle / (2\pi/\omega)$, with $P = p - (\varepsilon/\omega) \cos \omega t$, is obtained as

(1.11)
$$D = 2\hbar^2 \xi = \frac{k^2 \omega^2}{\varepsilon}.$$

The diffusion of P in the latter example arises from the crossing of the resonance $P = 0$ twice during each period of the driving field, and (1.11) holds if that crossing is fast, $\varepsilon/k \gg 1$.

4) Quantum optical examples:

The deflection of a laser-cooled beam of atoms crossing a standing-wave laser field in front of a vibrating mirror can be mapped on the periodically driven pendulum [9] and presents an example of dynamical localization in (chaotic) light scattering. Another quantum optical example is a high-*Q* single-mode cavity containing a medium with $\chi^{(3)}$ nonlinear susceptibility [10], kicked by a periodic train of light pulses. Its Hamiltonian is

(1.12) $\quad H_S = \dfrac{\hbar}{2} \chi^{(3)} (a^+ a)^2 +$

$$+ \frac{1}{2}\hbar\Omega_0 (a^+ \exp[i(\omega_0 - \omega)t] + a \exp[-i(\omega_0 - \omega)t]) \delta^{(1)}(t),$$

where a, a^+ describe the cavity mode in second quantization, ω is the frequency

of the driving field, and the time period of the kick has been normalized to 1.
The diffusion constant for the action $I = \hbar a^+ a$ obtained in this case is

$$(1.13) \qquad D_{I_\lambda} = 2\hbar^2\,\xi_\lambda = \frac{\hbar}{2}\Omega_0^2 I_\lambda\,.$$

In all the examples mentioned the dynamical localization, like any other quantum-mechanical coherence effect, is limited by dissipation and (or) noise, *i.e.* by any inelastic interaction of the localizing system with its environment. This inelastic interaction can be due to photons, as in the quantum optical examples or in Rydberg atoms, or due to phonons or electron-electron collisions etc. In the present lecture we wish to examine dynamical localization in the presence of such inelastic interactions with the environment. We shall proceed in three steps. First, for two typical working examples we briefly review three basic formalisms for the treatment of the interaction with the environment: the quantum Langevin equation in the Heisenberg picture[11, 12], the influence functional technique[13, 14] and the master equation[15, 16]. Then we shall consider in detail dynamical localization in the presence of dissipation for the kicked rotor[17-19]. Very similar results apply for the kicked $\chi^{(3)}$ cavity oscillator. The influence of classical external noise on the kicked rotor has been analysed in[20]. Finally we consider Rydberg atoms in a noisy waveguide, where the theory has been compared with experiment[21, 22].

2. – Dissipative quantum dynamics.

2˙1. *Model systems.* – The interaction of a system with Hamiltonian $H_S(t)$ interacting with the environment with Hamiltonian H_B is described by

$$(2.1) \qquad H = H_S(t) + H_B + H_{\text{int}}\,.$$

The environment is modelled by its normal modes, *i.e.* uncoupled harmonic oscillators:

$$(2.2) \qquad H_B = \sum_{k,\,\lambda} \hbar\omega_k\, b_{k\lambda}^+\, b_{k\lambda}$$

with the mode operators $b_{k\lambda}$, $b_{k\lambda}^+$ satisfying

$$(2.3) \qquad [b_{k\lambda},\, b_{k'\lambda'}^+] = \delta_{kk'}\,\delta_{\lambda\lambda'}\,,$$

where k, λ are a mode index and a polarization index, respectively. Among the many different models one may construct we shall consider two, exemplifying typical cases of low-temperature physics (*e.g.*, Josephson junctions) and quantum optics (*e.g.*, a nonlinear cavity), respectively.

1) Dipole coupling model for the kicked rotor.

Here we choose $H_S(t)$ as in (1.4) and

(2.4) $$H_{int} = \sum_k g_k \sum_\lambda (b_{k\lambda}^+ + b_{k\lambda}) f_\lambda(q),$$

where $\lambda = 1, 2$ and

(2.5) $$f_1(q) = \cos q, \qquad f_2(q) = \sin q.$$

The kicked rotor considered here has angular momentum p around the z-axis and a dipole moment d transverse to the z-axis with x, y components equal to $e\cos q$, $e\sin q$, respectively, where e is a charge. These components couple with equal coupling constant g_k to independent linearly polarized modes of the environment with polarization vectors e_λ. We note that we may write H_{int} as a scalar product $H_{int} = A \cdot d$ with $A = (1/e) \sum_{k,\lambda} g_k e_\lambda (b_{k\lambda}^+ + b_{k\lambda})$. Therefore, H is invariant under rotations and conserves angular momentum around the z-axis.

2) Quantum optical model.

We choose $H_S(t)$ as in (1.12) and

(2.6) $$H_{int} = \sum_{k,\lambda} g_{k\lambda}(a^+ \exp[i\omega_0 t] b_{k\lambda} + a \exp[-i\omega_0 t] b_{k\lambda}^+)$$

with the same properties of the mode operators $b_{k\lambda}$ as before. Here we made the rotating-wave approximation using the fact that ω_0 is a large optical frequency, hence only modes $b_{k\lambda}$ with frequencies ω_k near ω_0 will couple effectively, and terms oscillating with frequencies $\omega_0 + \omega_k$ will be extremely small on average.

The goal now is always to eliminate the modes of the environment to end up with a reduced dynamical description of the dissipative subsystem of interest. Three methods to achieve that goal are described next.

2'2. *Quantum Langevin equation.* – The reduced dynamics in this description is very easy to derive, and the central quantities describing dissipation appear very directly. On the other hand, except in cases where H_S describes free particles or linear systems or can at least be approximated by such systems, no methods of solution, not even numerical ones, are known. For a discussion of the quantum Langevin equation see [12]. We shall consider the dipole coupling model by this method.

The Heisenberg equations of motion for p, q, $b_{k\lambda}$ are easy to write down:

(2.7) $$\begin{cases} \ddot{q} = -K \sin q \, \delta^{(1)}(t) - \sum_{k\lambda} g_k \, f_\lambda'(q)(b_{k\lambda}^+ + b_{k\lambda}), \\ \dot{b}_{k\lambda} = -i\omega_k b_{k\lambda} - \dfrac{i}{\hbar} g_{k\lambda} f_\lambda(q). \end{cases}$$

The last equation may be formally solved:

$$(2.8) \quad b_{k\lambda}(t) = -\frac{ig_k}{\hbar} \int_{-\infty}^{t} d\tau \exp[-i\omega_k(t-\tau)+\varepsilon t] f_\lambda(q(\tau)) + b_{k\lambda}(-\infty)\exp[-i\omega_k t],$$

where $\varepsilon \to +0$ is required for convergence, and the result can be inserted in the first equation strictly preserving the ordering of the operators. After a partial integration and taking $\varepsilon \to 0$ we obtain

$$(2.9) \quad \ddot{q}(t) = -K\sin q(t)\,\delta^{(1)}(t) -$$

$$-\int_{-\infty}^{t} d\tau\,\gamma(t-\tau)\sum_\lambda f_\lambda'(q(t))\frac{d}{d\tau}f_\lambda(q(\tau)) + \sum_\lambda f_\lambda'(q(t))\xi_\lambda(t)$$

with the damping kernel

$$(2.10) \qquad \gamma(t) = \frac{2}{\hbar}\sum_k \frac{g_k^2}{\omega_k}\cos\omega_k t$$

and the Langevin operators

$$(2.11) \qquad \xi_\lambda(t) = -\sum_k g_k(b_{k\lambda}(-\infty)\exp[-i\omega_k t] + \text{h.c.}).$$

From the Heisenberg equation of motion for $f_\lambda(q)$ we obtain

$$(2.12) \qquad \frac{d}{dt}f_\lambda(q) = f_\lambda'(q)\dot{q} + \frac{\hbar}{2i}f_\lambda''(q).$$

The ξ_λ defined by (2.11) satisfy the fundamental commutation relation

$$(2.13) \qquad [\xi_\lambda(t),\,\xi_\lambda(t')] = i\hbar\delta_{\lambda\lambda'}\dot{\gamma}(t-t').$$

For a thermal environment with inverse temperature β the fluctuation dissipation relation also fixes their anticommutator

$$(2.14) \quad S_{\lambda\lambda'}(t,t') = \frac{1}{2}\langle\{\xi_\lambda(t),\,\xi_{\lambda'}(t')\}_+\rangle = \delta_{\lambda\lambda'}\int_0^\infty \frac{d\omega}{2\pi}\hat{\gamma}(\omega)\hbar\omega\,\text{ctgh}\frac{1}{2}\beta\hbar\omega\cos\omega(t-t'),$$

where $\hat{\gamma}(\omega)$ is the Fourier transform of $\gamma(t)$. (2.9) is difficult to handle, but it serves to bring out in the most direct manner the fundamental physical quantities $\gamma(t)$ and $S_{\lambda\lambda'}(t,t')$. We point out that the operator orderings in the second and third term of (2.9) are strictly correlated with each other.

Let us now use this formulation for a discussion of the Markov approximation for a thermal environment in the present model. It requires two limits:

1) Constant friction («Ohmic dissipation»):

$$(2.15) \qquad\qquad \hat{\gamma}(\omega) = 2\gamma, \qquad \gamma(t) = 2\gamma\delta(t).$$

In this limit we simply obtain $-\gamma\dot{q}(t)$ for the second term on the right-hand side of (2.9), using (2.12) and the explicit form of the $f_\lambda(q)$.

2) High temperature:

The Markov approximation also requires that $S_{\lambda\lambda'}(t, t')$ reduces to a δ-function of $t - t'$. This is the case, according to expression (2.14), only for time scales $|t - t'| \gg \hbar\beta/2\pi$ where (2.14) may be replaced by

$$(2.16) \qquad\qquad S_{\lambda\lambda'}(t - t') \to \frac{2\gamma}{\beta}\delta(t - t').$$

Thus, in order to validate the Markov approximation, we must have sufficiently high temperature. Non-Markovian effects will *always* appear for sufficiently low temperature. This result is typical for systems in low-temperature physics like Josephson junctions. Indeed, in the example mentioned in (1.11) the variable q of the present model would be identified with the Josephson phase. A very different result is obtained in quantum optical examples which we consider next.

2˙3. *Influence functional method.* – In this most powerful method for the treatment of dissipative quantum systems the real-time path-integral representation of the time-dependent density matrix of the total system is used as a starting point and the trivial Gaussian integration over the harmonic oscillators of the environment is then performed explicitly and exactly. For a discussion of the influence functional method see [14]. Non-Markovian effects can be treated by this method by evaluating the remaining nontrivial path integrals over the influence functional in the WKB approximation.

The discussion of the Markovian limit brings into play the same basic quantities appearing in the quantum Langevin equation, the expectation of (2.13) and (2.14), but the present method, of course, uses the Schrödinger picture. In the Markov limit the nontrivial path integrals are not evaluated directly but reduced, via the Feynman-Kac formula, to a partial differential equation for the density matrix of first order in time, the master equation of the reduced dissipative dynamics [23]. The same master equation can be obtained more easily and directly using second-order perturbation theory with respect to H_{int} in the von Neumann equation of the total system and also applying the Markov approximation. As we shall only be interested in the latter case, we follow this more direct approach. However, we mention that the derivation of the master equation via the influence functional method shows that, once the Markov approximation is made, the result obtained in second-order perturbation theory becomes exact, *i.e.* the truncation of the perturbation series after the second-order term is not an independent approximation.

2'4. *Master equation.* – In our discussion of the master equation we shall treat the quantum optical model of sect. 2.

During the kicks dissipation is completely negligible. Therefore, we only have to consider the time interval between two kicks. In the interaction representation with respect to H_B the von Neumann equation of the total system reads

$$(2.17) \qquad i\hbar\dot\rho(t) = [\tilde{H}(t), \rho(t)].$$

It can be formally integrated, assuming that $\rho(t)$ at $t \to -\infty$ factorizes $\rho(-\infty) = \rho_S(-\infty) \otimes \rho_B$, where $\rho_B = Z^{-1} \exp[-\beta H_B]$ is the canonical operator:

$$(2.18) \quad \rho(t) - \rho_S(-\infty) \otimes \rho_B + i \int\limits_{-\infty}^{t} dt' \, \frac{\chi^{(3)}}{2}[(a^+ a)^2, \rho(t')] =$$

$$= -\frac{i}{\hbar} \int\limits_{-\infty}^{t} dt' \, [\tilde{H}_{int}(t'), \rho(t')].$$

The perturbation series is generated by iterating this equation, taking the right-hand side equal to zero in zeroth order. The zeroth-order solution takes the form $\rho(t) = \rho_S(t) \otimes \rho_B$. We stop after the second order and take the trace over the Hilbert space of the environment in order to obtain $\rho_S(t) = \mathrm{Tr}_B \, \rho(t)$. We obtain

$$(2.19) \quad \rho_S(t) - \rho_S(-\infty) + i \int\limits_{-\infty}^{t} dt' \, \frac{\chi^{(3)}}{2}[(a^+ a)^2, \rho_S(t')] -$$

$$- \frac{\chi^{(3)}}{2\hbar} \int\limits_{-\infty}^{t} dt' \int\limits_{-\infty}^{t'} dt'' \, \mathrm{Tr}_B \left[(a^+ a)^2, [\tilde{H}_{int}(t''), \rho_S(t'') \otimes \rho_B]\right] =$$

$$= -\frac{i}{\hbar} \int\limits_{-\infty}^{t} dt' \, \mathrm{Tr}_B [\tilde{H}_{int}(t'), \rho_S(t') \otimes \rho_B] -$$

$$- \frac{1}{\hbar^2} \int\limits_{-\infty}^{t} dt' \int\limits_{-\infty}^{t'} dt'' \, \mathrm{Tr}_B [\tilde{H}_{int}(t'), [\tilde{H}_{int}(t''), \rho_S(t'') \otimes \rho_B]].$$

Then the trace can be carried out using $\mathrm{Tr}_B \, \rho_B \, b_{k\lambda} = 0$. It brings into appearance the functions (with $\mathrm{Tr}\,\rho_B\, b_{k\lambda}^{\dagger}\, b_{k\lambda} = \bar{n}_{k\lambda}$)

$$(2.20) \quad \begin{cases} J_{\pm}(t'-t'') = \sum\limits_{k\lambda} \dfrac{g_{k\lambda}^2}{\hbar^2} \exp\left[\pm i(\omega_0 - \omega_k)(t'-t'')\right], \\[2ex] S_{\pm}(t'-t'') = \sum\limits_{k\lambda} \dfrac{g_{k\lambda}^2}{\hbar^2} \bar{n}_{k\lambda} \exp\left[\pm i(\omega_0 - \omega_k)(t'-t'')\right] \end{cases}$$

replacing (2.10) and (2.14) in the present example, respectively. We now wish to pass to the Markov approximation, which consists in approximating

$$(2.21) \quad \begin{cases} J_{\pm}(t'-t'') \simeq 2(\kappa(\omega_0) \mp i\Delta(\omega_0))\,\delta(t'-t''), \\[2ex] S_{\pm}(t'-t'') \simeq 2(\kappa(\omega_0)\bar{n}(\omega_0) \mp i\Delta_s(\omega_0))\,\delta(t'-t''), \end{cases}$$

where

$$(2.22) \quad \begin{cases} \kappa(\omega) - i\Delta(\omega) = \displaystyle\int\limits_{-\infty}^{0} dt' \sum\limits_{k\lambda} \dfrac{g_{k\lambda}^2}{\hbar^2} \exp\left[-i(\omega - \omega_k)t'\right], \\[3ex] \kappa(\omega)\bar{n}(\omega) - i\Delta_s(\omega) = \displaystyle\int\limits_{-\infty}^{0} dt' \sum\limits_{k\lambda} \dfrac{g_{k\lambda}^2}{\hbar^2} \bar{n}_{k\lambda} \exp\left[-i(\omega - \omega_k)t'\right]. \end{cases}$$

The imaginary parts of these expressions are given by principal-part integrals

$$(2.23) \quad \begin{cases} \Delta(\omega) = \mathrm{P} \displaystyle\int\limits_{-\infty}^{+\infty} \dfrac{d\omega'}{\pi} \dfrac{\kappa(\omega')}{\omega - \omega'}, \\[3ex] \Delta_s(\omega) = \mathrm{P} \displaystyle\int\limits_{-\infty}^{+\infty} \dfrac{d\omega'}{\pi} \dfrac{\kappa(\omega')\bar{n}(\omega')}{\omega - \omega'} \end{cases}$$

and give rise to small frequency shifts which we shall neglect (or assume to be absorbed in the Hamiltonian H_S). The real parts of (2.22) define the damping constants

$$(2.24) \quad \begin{cases} \kappa(\omega) = \dfrac{\pi}{\hbar^2} \sum\limits_{k\lambda} g_{k\lambda}^2\, \delta(\omega_k - \omega), \\[2ex] \kappa(\omega)\bar{n}(\omega) = \dfrac{\pi}{\hbar^2} \sum\limits_{k\lambda} g_{k\lambda}^2\, \bar{n}_{k\lambda}\, \delta(\omega_k - \omega). \end{cases}$$

Clearly the transition from (2.20) to (2.21) is justified if the frequency dependence of $\kappa(\omega)$ and $\kappa(\omega)\bar{n}(\omega)$ is negligible in a frequency interval around ω_0 which may be tiny compared to ω_0 but must be sufficiently large so that its inverse is much smaller than the dynamical time scale of interest, which is here given by a kicking period. Thus in the present example, typical for quantum optics, the foundation of the Markov approximation is completely different from the first example, due to the appearance of the optical frequency ω_0. Sometimes this difference is not sufficiently appreciated and has given rise to confusion. The temperature β^{-1}, which only enters in $\bar{n}(\omega_0) = (\exp[\beta\hbar\omega_0] - 1)^{-1}$, may be taken to be effectively zero ($\beta\hbar\omega_0 \ll 1$) as long as $\beta^{-1} \ll \hbar\kappa$ without invalidating the Markov approximation, yielding $\bar{n}(\omega_0) \simeq 0$. Returning now to (2.19) with the approximation (2.21) and taking the time derivative, we obtain

$$(2.25) \qquad \dot{\rho}_S = -i\frac{\chi^{(3)}}{2}[(a^+ a)^2, \rho_S] + \kappa(1 + \bar{n})([a, \rho_S a^+] + [a\rho_S, a^+]) +$$

$$+ \kappa\bar{n}([a^+, \rho_S a] + [a^+ \rho_S, a]),$$

which is the desired master equation. (2.25) may also be turned into a master equation for the kicked rotor. To this purpose we rewrite it in the l-representation

$$(2.26) \qquad\qquad a^+ a|l\rangle = l|l\rangle, \qquad\qquad l \geq 0, \text{ integer},$$

and note that in this representation the unperturbed energy

$$(2.27) \qquad\qquad \frac{\chi^{(3)}}{2}(a^+ a)^2 |l\rangle = \frac{\chi^{(3)}}{2} l^2 |l\rangle$$

is the same as for the rotor, apart from a trivial scale factor, and apart from the fact that the integer l is restricted to $l \geq 0$, in the present case. Extending the master equation also to negative l (and removing the scale factor), we obtain for the rotor with $\langle l|\rho_S|l'\rangle = \rho_{l,l'}$

$$(2.28) \qquad \dot{\rho}_{l,l'} = -\frac{i\hbar}{2}(l^2 - l'^2)\rho_{l,l'} +$$

$$+ 2\kappa\left(\sqrt{(|l| + 1)(|l'| + 1)}\,\rho_{l + l/|l|, l' + l'/|l'|} - \frac{|l| + |l'|}{2}\rho_{l,l'}\right),$$

where we used the assumption $\bar{n} \simeq 0$, for simplicity.

In [17] this master equation was derived directly from a formal microscopic model. It is used in the following section to analyse the influence of dissipation on dynamical localization in the kicked-rotor model. We note, however, that the units of p, q (and consequently also $\hbar$) used in the present lecture differ by factors 2π (and for $\hbar$ by $(2\pi)^2$) from the units used in [17].

3. – Dynamical localization in the dissipative kicked-rotor model.

3˙1. *Quantum map.* – We solve the master equation (2.28) between the two kicks at the times $t = n$ and $T = n + 1$. Then the kick at time $t = n + 1$ is applied. The solution takes the form [17]

$$(3.1) \qquad \rho_{l', \, m'}(n+1) = \sum_{l, \, m} G(l' \, m' \, | \, lm) \, \rho_{l, \, m}(n)$$

with the following explicit expression for the kernel G (with $\lambda = \exp[-2\kappa]$):

$$(3.2) \qquad G(l' \, m' \, | \, lm) = \lambda^{(|l| + |m|)/2} \left\{ G_c(l' \, m' \, | \, lm) + \sum_{j \geqslant 1} \Theta_{l \cdot m} \binom{|l|}{j}^{1/2} \binom{|m|}{j}^{1/2} \cdot \right.$$

$$\left. \cdot \left(\frac{\lambda^{-1} - \exp[-i(|l| - |m|)]}{1 + i(|l| - |m|)/|\ln \lambda|} \right)^j G_c(l' \, m' \, | \, l - jl/|l|, \, m - jm/|m|) \right\},$$

where $\Theta_{l \cdot m} = 1/2 + lm/2|lm|$ and G_c is the unitary kernel describing the conservative map

$$(3.3) \qquad \begin{cases} G_c(l' \, m' \, | \, lm) = \langle l' \, | \, U \, | \, l \rangle (\langle m' \, | \, U \, | \, m \rangle)^*, \\ U = \exp\left[\dfrac{i}{\hbar} K \cos q\right] \exp\left[-\dfrac{i}{\hbar} p^2/2\right]. \end{cases}$$

The sum in (3.2) is taken over the number j of quanta of size $\hbar$ of the angular momentum p absorbed by the environment. Thus for fixed j quantum coherence is fully preserved, but between two terms of different j there is no quantum coherence left (the sum is over probabilities, not probability amplitudes).

3'2. *Semi-classical limit, quantum noise.* – The semi-classical limit of the quantum map is most easily discussed by introducing the Wigner phase space distribution

$$(3.4) \qquad W_l(q) = \mathrm{Tr}\left(\rho \sum_m \int\limits_{-1/2}^{1/2} \mathrm{d}x \exp\left[i \left[m(q - \hat{q}) + x \left(l - \frac{\hat{p}}{\hbar} \right) \right] \right] \right)$$

and then taking the saddle point approximation for $\hbar \to 0$ in the resulting kernel G. The result to leading (Gaussian) order for

$$(3.5) \qquad (K^2 \hbar)^{1/3} \ll (1 - \lambda)|p_n|$$

is mathematically equivalent to a c-number stochastic map [24]

$$(3.6) \qquad \begin{cases} p_{n+1} = \lambda p_n - K \sin q_{n+1} + \eta_n, \\ q_{n+1} = \left(q_n + \dfrac{1 - \lambda}{|\ln \lambda|} p_n + \zeta_n \right) (\mathrm{mod}\, 2\pi), \end{cases}$$

with the classical Gaussian noise defined by the correlation functions [17]

$$(3.7) \quad \begin{cases} \langle \eta_n \rangle = 0 = \langle \zeta_n \rangle, \\ \langle \eta_n \eta'_n \rangle = \hbar \delta_{nn'} |p_n| \lambda (1 - \lambda), \\ \langle \eta_n \zeta'_n \rangle = \hbar \delta_{nn'} |p_n| \lambda \dfrac{|\ln \lambda| - (1 - \lambda)}{|\ln \lambda|}, \\ \langle \zeta_n \zeta'_n \rangle = \hbar \delta_{nn'} \left(\dfrac{1 - \lambda}{4\lambda |p_n|} + \dfrac{\lambda}{1 - \lambda} \left(1 - \dfrac{1 - \lambda}{|\ln \lambda|} \right)^2 |p_n| \right). \end{cases}$$

The deterministic part of (3.6), after some rescaling, is known as the Zaslavski map.

The semi-classical reduction to a classical stochastic process with noise intensity $\sim \hbar$ of a quantum system with sufficiently strong dissipation (and no thermal noise, because we deliberately took $\bar{n} = 0$) is the theoretical foundation of the notion «quantum noise» frequently used in this context. Taking thermal noise into account ($\bar{n} \neq 0$) the form (3.6) subsists, but the noise intensities in (3.7) receive additional terms proportional to $\bar{n}$.

In fig. 1 we show the strange attractor with its invariant measure for the classical Zaslavski map with $K = 5$, $\lambda = 0.3$. In fig. 2a) the Wigner distribution

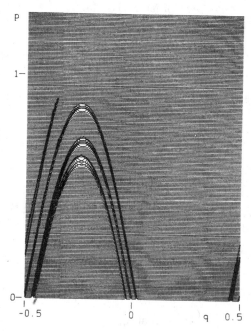

Fig. 1. – Strange attractor of the damped kicked rotor for $K = 5$, $\lambda = 0.3$ with its invariant measure in the phase plane. Here and in the following p and q are given in units of 2π (from [17]).

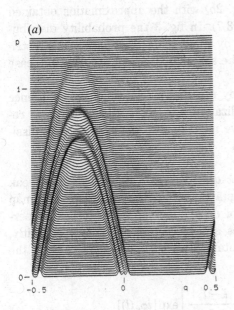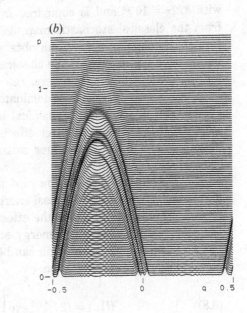

Fig. 2. – Wigner distribution (a)) and its semi-classical approximation (b)) in the steady state for the case of fig. 1 and $\hbar/2\pi = 0.01$ (from [17]).

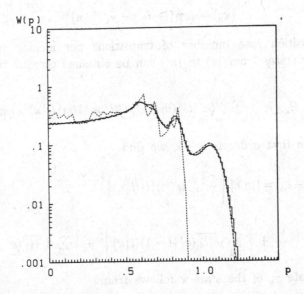

Fig. 3. – Angular-momentum distribution for the case of fig. 2; quantum (full line), semi-classical (dashed line), classical (dotted line) (from [17]).

obtained for the steady state of the quantum map is shown for the same case, with $\hbar/2\pi = 10^{-2}$ and is compared in fig. 2*b*) with the approximation obtained from the classical stochastic map (3.6), (3.7). In fig. 3 the probability distribution of the angular momentum obtained by integrating the phase space distribution over q is shown for the classical, the semi-classical and the quantum case for the same set of parameters.

These results are clearly dominated by quantum noise. Quantum coherence effects, and in particular dynamical localization, are not apparent in these results. In order to see the latter effects, we must turn to the case of weak dissipation, where (3.5) is no longer satisfied.

3̇3. *Dynamical localization and weak dissipation.* – For sufficiently weak dissipation the (localized) quasi-energy states $|\kappa\rangle$ of the nondissipative map form a useful basis in which the effects of dissipation can be discussed by perturbation theory. The quasi-energy states are, of course, not known explicitly, but a reasonable approximation can be obtained by assuming them to be of the form, in the $|l\rangle$-representation,

$$(3.8) \qquad \langle l|\kappa\rangle \simeq (2/\xi)^{1/2} \exp\left[-\frac{|l - l_\kappa|}{\xi}\right] \exp\left[i\varphi_\kappa(l)\right]$$

with a random phase $\varphi_\kappa(l)$.

Due to dissipation there occur transitions between the quasi-energy states which, therefore, decay in time exponentially:

$$(3.9) \qquad |\kappa_n\rangle \simeq \exp\left[(-i\omega_\kappa - \gamma_\kappa/2)n\right]|\kappa\rangle.$$

The transition rate (number of transitions per kicking period) between quasi-energy states from $|\kappa\rangle$ to $|\kappa'\rangle$ can be obtained directly from the master equation

$$(3.10) \qquad R_{\kappa\kappa'} = \sum_{l', m', l, m} \langle\kappa'|l'\rangle\langle m'|\kappa'\rangle G(l' m'|lm)\langle l|\kappa\rangle\langle\kappa|m\rangle.$$

Expanding to first order in $1 - \lambda$, we find

$$(3.11) \qquad R_{\kappa\kappa'} - \delta_{\kappa\kappa'} = |\ln\lambda|\left\{-\left|\sum_l \langle\kappa'|l\rangle|l|\langle l|\kappa\rangle\right|^2 +\right.$$

$$\left. + \left|\sum_{l>0} \sqrt{|l|}\langle\kappa'|l-1\rangle\langle l|\kappa\rangle\right|^2 + \left|\sum_{l<0} \sqrt{|l|}\langle\kappa'|l+1\rangle\langle l|\kappa\rangle\right|^2\right\}.$$

The decay rate γ_κ of the state κ follows from

$$(3.12) \qquad \gamma_\kappa = \sum_{\kappa' \neq \kappa} R_{\kappa\kappa'}$$

and the normalization condition

(3.13)
$$\sum_{\kappa'} R_{\kappa\kappa'} = 1$$

as

(3.14)
$$\gamma_\kappa = 1 - R_{\kappa\kappa}.$$

The same result can also be obtained directly from perturbation theory. Using the randomness of the phases $\varphi_\kappa(l)$ in (3.8) in order to evaluate (3.11) approximately, we obtain

(3.15)
$$\begin{cases} \gamma_\kappa \simeq |l_\kappa|(1-l)(1+O(1-\lambda, \xi^{-1})), \\ R_{\kappa, \kappa'} \simeq \dfrac{|l_\kappa| + |l_{\kappa'}'|}{\xi} \exp\left[-2|l_\kappa - l_{\kappa'}|/\xi\right](1-\lambda)(1+O(1-\lambda, \xi^{-1})). \end{cases}$$

Choosing the initial state $|l = 0\rangle$, we may replace in the prefactors of these expressions $|l_\kappa|$, $|l_{\kappa'}'|$ by ξ. We can define «weak dissipation» more precisely by the condition

(3.16)
$$\gamma_\kappa \cdot n^* = \gamma_\kappa \cdot \xi < 1,$$

where $n^* = \xi$ is the characteristic time (in kicking periods) which the system needs to establish dynamical localization. The relevant quantum numbers κ are determined by the initial state. From the transition rate $R_{\kappa\kappa'}$ a dissipation-induced diffusion process of the angular momentum $l = p/\hbar$ on time scales $n > n_c = \gamma_\kappa$ follows with the diffusion constant

(3.17)
$$\overline{D}_\kappa = \sum_{\kappa'} (l_\kappa - l_{\kappa'}')^2 R_{\kappa\kappa'} \simeq 2(1-\lambda)|l_\kappa|\xi^2.$$

In fig. 4 the result of a numerical simulation is shown (for $K = 10$, $\hbar/2\pi = 0.15((1+\sqrt{5})/2)$, $1 - \lambda = 0, 5 \cdot 10^{-6}, 10^{-4}, 10^{-3}$) and compared with the corresponding classical case without dissipation.

Dynamical localization for $n > n^*$ in the dissipationless case $\lambda = 1$ can be clearly seen, and its replacement, for times $n > n_c$, by a slow dissipation-induced diffusion process whose rate is proportional to $1 - \lambda$ as predicted by (3.17).

At the beginning the dissipation-induced increase of $\langle p^2 \rangle$ grows quadratically in n, as can be seen in the double logarithmic plot of a part of the same data in fig. 5.

This quadratic increase can be understood by the following scaling argument, which extends a well-known argument by CHIRIKOV *et al.* [25]: For the sake of this argument we first consider the case without dissipation. As is well known, the mean square $\langle \Delta p^2 \rangle = \langle (p_n - p_0)^2 \rangle$ then satisfies the scaling law

(3.18)
$$\langle \Delta p^2 \rangle = \hbar^2 \xi^2 f\left(\frac{n}{n^*}\right), \qquad\qquad n^* = \xi,$$

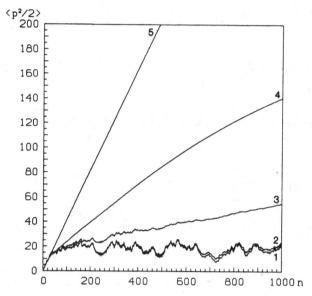

Fig. 4. – Mean kinetic energy of the kicked rotor as a function of time n (in kicking periods) for $K = 10$, $\hbar/2\pi = 0.075 (1 + \sqrt{5})$, $\lambda = 1$ (curve 1), $1 - \lambda = 5 \cdot 10^{-6}$ (curve 2), $1 - \lambda = 10^{-4}$ (curve 3), $1 - \lambda = 10^{-3}$ (curve 4), $\hbar = 0$, $\lambda = 1$ (curve 5) (from [17]).

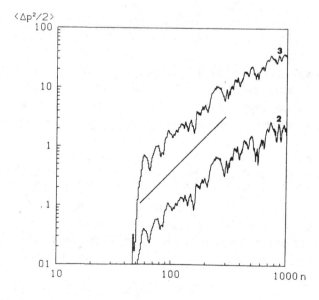

Fig. 5. – Double logarithmic plot of curves 2, 3 of fig. 4 (from [17]).

with a scaling function $f(x)$ with the properties

(3.19)
$$f(x) = \begin{cases} 1, & x \gg 1, \\ x/\alpha, & x \ll 1. \end{cases}$$

The saturation of $f(x)$ for $x \gg 1$ describes localization. Its reason is the gradual resolution, as a function of time, of the spacings between the discrete quasi-energy levels. Let $q(x)$, with $x = n/\xi$, be the fraction of quasi-energy level pairs which are spectrally resolved after the time n (in kicking periods) has elapsed. Due to the resolution of these quasi-energy level pairs the diffusion rate of p is reduced from D (the classical value) to $D(1 - q(x))$, with $D = \hbar^2 \xi/\alpha$, i.e.

(3.20)
$$\frac{\partial \langle \Delta p^2 \rangle}{\partial n} = D(1 - q(x)) = \hbar^2 \xi f'(x),$$

where the second equality follows from (3.18). It relates the scaling function $f(x)$ to $q(x)$. (By the uncertainty principle $q(x)$ also gives the quasi-energy level pair distribution function, $q(n/\xi)$ = probability for an ordered pair of quasi-energy levels to be separated by less than the distance $2\pi/n$.)

Now we return to the case with dissipation, where dissipation-induced transitions between resolved pairs of quasi-energy levels give rise to an excess diffusion.

We consider the fraction $dq(x)$ of level pairs resolved in the time interval $(x\xi, (x + dx)\xi)$. Its contribution to the excess diffusion for $n > x\xi$ is

(3.21)
$$\delta \langle \Delta p^2 \rangle = 2\hbar^2 (1 - \lambda) |l_\kappa| \xi^2 \, dq(x)(n - x\xi),$$

where we used (3.17). Thus the total excess diffusion up to time n is

(3.22)
$$\delta \langle \Delta p^2 \rangle = 2\hbar^2 (1 - \lambda) |l_\kappa| \xi^2 \int\limits_0^{q(n/\xi)} dq(x)(n - x\xi).$$

By partial integration and the use of (3.20) we obtain

(3.23)
$$\delta \langle \Delta p^2 \rangle = 2\hbar^2 (1 - \lambda) |l_\kappa| \xi^2 \left(\frac{n}{\xi} - \alpha f\left(\frac{n}{\xi} \right) \right).$$

It can be seen that the right-hand side is proportional to n^2 in the domain where $f(x) = x/\alpha + O(x^2)$, while for $n \gg \xi$ the linear diffusive increase of $\delta \langle \Delta p^2 \rangle$ takes over, which can also be seen in fig. 5.

Finally, we point out the dependence of the dissipation-induced diffusion rate on the centre of localization l_κ of the quasi-energy states. For $|l_\kappa| \to 0$ the diffusion rate $\overline{D}_\kappa$ and the decay rate γ_γ become arbitrarily small. The effect on the distribution of angular momentum can be seen in fig. 6, which is obtained for $n = 10^3$ time steps, starting with the initial state $|l = 0\rangle$, with the parameters $\hbar/2\pi = 0.15((1 + \sqrt{5})/2)$, $K = 10$, $1 - \lambda = 10^{-4}$. For comparison the corre-

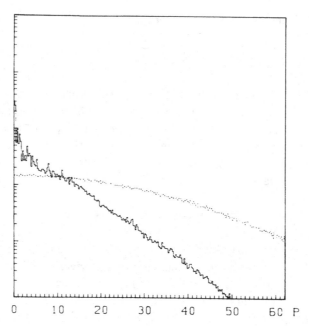

Fig. 6. – Probability distribution of the angular momentum after 10^3 kicks for $\hbar/2\pi =$
$= 0.075\,(1 + \sqrt{5})$, $K = 10$, $1 - \lambda = 10^{-4}$ compared with the classical result (dashed line)
(from [17]).

sponding classical result is also shown. It can be seen that for $p \ll \hbar\xi$ localization
persists, *i.e.* the quasi-energy states in that domain have not yet decayed. As
the time n increases, the size of this domain decreases inversely proportional to
n.

For times long compared to the classical relaxation time $n \gg (1 - \lambda)^{-1}$ the
quantum system approaches a unique steady state. It is an important feature of
dissipative systems that for $\hbar \to 0$ the quantum steady state approaches the
classical steady state (*e.g.*, in a description using the Wigner phase space distribution). Thus (barring exceptional cases) in dissipative systems the limits $t \to$
$\to \infty$ and $\hbar \to 0$ commute. If the dissipation becomes smaller and smaller, one has
to wait longer and longer before the steady state is reached. For vanishing dissipation a steady state is never reached and the limits $\hbar \to 0$ and $t \to \infty$ no
longer commute.

4. – Rydberg atoms in a noisy waveguide.

4'1. *Basic effects and ideas for an experiment.* – In the preceding section the
influence of the weak coupling to the environment on a quantum system exhibiting dynamical localization was studied for the example of the kicker rotor.
We summarize the main result, the distinction of four different dynamical
regimes:

1) Very short times: $n \ll n^* = \xi$:

Here the quantum system mimicks the classical system, *i.e.* chaotic diffusion, because by the uncertainty principle the discreteness of the quasi-energy levels is not yet resolved. There is no influence of the environment yet as the coupling is very weak.

2) Intermediate times $\xi \lesssim n \lesssim n_c = [(1 - \lambda)\xi]^{-1}$:

Here dynamical localization occurs. The environment has only a very small influence due to the weak coupling.

3) Long times $n_c \lesssim n \lesssim n_l = (1 - \lambda)^{-1}$:

Environment-induced diffusion occurs with a rate proportional to the square of the coupling constant.

4) Very long times $n_d < n$:

The steady state is reached (if the environment absorbs energy from the system, *i.e.* if there is dissipation).

In the present section we discuss how these results are compared with experiment [21, 22].

A sufficiently good realization of the kicked rotor is furnished by a microwave-driven Rydberg atom (cf. Shepelyanky's lecture). In that case the interaction with the environment is realized by the coupling to the electromagnetic field, either to its thermal and vacuum fluctuations, or to artificial and experimentally controlled noise. In order to have the latter possibility, the interaction region is realized in a waveguide, rather than a microwave cavity, as in other experiments. In order to have variable interaction times, which are the same for all atoms, a pulsed microwave field and an atomic beam with negligible dispersion during the interaction time are used. The experiment is run in a region of parameter space where ionization by the microwave field is negligible and the probability distribution over the Rydberg states after the interaction with the microwave field is measured. The actual experiment has been carried out (for circumstantial reasons only) with rubidium atoms for which a detailed theory of dynamical localization has not yet been worked out. However, a very reasonable qualitative comparison can be made with a simple 1-dimensional theory for hydrogen atoms, to which we turn next.

4'2. *Theory.* – As the theory closely follows the lines of sect. 2, 3 we can be brief. The Hamiltonian

$$(4.1) \qquad H(t) = H_S(t) + H_{int} + H_B$$

consists of the parts $H_S(t)$ given by (1.6), $H_{int} = \hbar \sum_k x(g_k b_k + g_k^* b_k^+)$ analogous to (2.4) and H_B as in (2.2) (where k, λ are combined to a single index k for brevity). The derivation of the master equation proceeds as in subsect. 2'4, with the important modification that now we use the representation provided by the

Floquet basis $|\phi_\alpha(t)\rangle$

(4.2)
$$U_S(t)|\phi_\alpha(0)\rangle = \exp[-i\mu_\alpha]|\phi_\alpha(t)\rangle,$$

where $U_S(t)$ is the time evolution operator generated by $H_S(t)$, which includes the microwave field. The interaction Hamiltonian, in interaction representation with respect to $H_S(t) + H_B$, can then be written as

(4.3)
$$\tilde{H}_{\text{int}}(t) = \hbar \sum_i g_i \sum_{\alpha\beta} \{|\phi_\alpha(0)\rangle\langle\phi_\beta(0)|b_i X^{(i)}_{\alpha\beta}(t) + \text{h.c.}\}$$

with

(4.4)
$$X^{(i)}_{\alpha\beta}(t) = \sum_k \frac{1 + \operatorname{sgn}\Omega_{\alpha\beta}(k)}{2} \exp[i(\Omega_{\alpha\beta}(k) - \omega_i)t]\overline{X}_{\alpha\beta}(k),$$

where we used the abbreviations

(4.5)
$$\begin{cases} \Omega_{\alpha\beta}(k) = \mu_\alpha - \mu_\beta + k\omega, \\[2mm] \overline{X}_{\alpha\beta}(k) = \dfrac{\omega}{2\pi} \int\limits_0^{2\pi/\omega} dt \exp[-ik\omega t]\langle\phi_\alpha(t)|x|\Phi_\beta(t)\rangle, \end{cases}$$

where ω is the frequency of the coherent microwave field. In (4.3) we introduced a rotating-wave approximation similar to (2.6), *i.e.* we neglected terms rotating with frequencies $\Omega_{\alpha\beta}(k) + \omega_i$.

The elimination of the reservoir now proceeds as in subsect 2.4, including the Markov assumption. In the present case the latter amounts to assuming that

(4.6)
$$\gamma^{(\lambda)}_{\alpha\beta}(k) = 2\pi\rho_\lambda |g_\lambda(\Omega_{\alpha\beta}(k) + \omega)|^2 |\overline{X}_{\alpha\beta}(k)|^2 \Theta(\Omega_{\alpha\beta}(k))$$

changes with ω only on scales $\Delta\omega$ large compared to $\gamma^{(\lambda)}_{\alpha\beta}(k)$. Here $\rho_\lambda(\omega)$ is the density of states of the modes of the waveguide (TM and TE) distinguished by the index λ and $g_\lambda(\omega)$ the corresponding coupling constant g_i. $\Theta(x)$ is the step function. The final master equation in the Floquet basis and the interaction picture then reads

(4.7)
$$\begin{cases} \dot{\tilde{\rho}}_{\alpha\alpha} = \sum_\mu (M_{\mu\alpha}\tilde{\rho}_{\mu\mu} - M_{\alpha\mu}\tilde{\rho}_{\alpha\alpha}), \\[2mm] \dot{\tilde{\rho}}_{\alpha\beta} = -\dfrac{1}{2}\tilde{\rho}_{\alpha\beta}\sum_\mu (M_{\alpha\mu} + M_{\beta\mu}), \end{cases}$$

where

(4.8)
$$M_{\alpha\beta} = \sum_{k,\lambda} \{\gamma^{(\lambda)}_{\alpha\beta}(k) + \overline{n}^{(\lambda)}(|\Omega_{\alpha\beta}(k)|) \cdot (\gamma^{(\lambda)}_{\alpha\beta}(k) + \gamma^{(\lambda)}_{\beta\alpha}(k))\}$$

with $\bar{n}^{(\lambda)}(\omega) = (\exp[\beta\hbar\omega] - 1)^{-1}$ for a thermal environment or

(4.9)
$$\bar{n}^{(\lambda)}(\omega) = \frac{2\pi}{\hbar\omega} R_\lambda(\omega)$$

for artificial noise with the spectral density $R_\lambda(\omega)$ of the noise power in the mode λ. The master equation (4.7) is easily solved:

(4.10)
$$\begin{cases} \rho_{\alpha\beta}(t) = \exp\left[-i(\mu_\alpha - \mu_\beta)t - \frac{t}{2}\sum_\gamma (M_{\alpha\gamma} + M_{\beta\gamma})\right]\rho_{\alpha\beta}(0), \\ \rho_{\alpha\alpha}(t) = \sum_\beta (\exp[\Lambda t])_{\alpha\beta}\rho_{\beta\beta}(0) \end{cases}$$

with

(4.11)
$$\Lambda_{\alpha\beta} = -M_{\beta\alpha} + \delta_{\alpha\beta}\sum_\mu M_{\alpha\mu}.$$

The localization of the distribution over the Rydberg states is quantified by the Shannon width W

(4.12)
$$\begin{cases} W = \exp[S], \\ S = -\sum_n P_n \ln P_n, \end{cases}$$

where P_n is the occupation probability of the Rydberg state with principal quantum number n.

A numerical computation using a basis of 12 atomic states ($n = 69, ..., 80$) has been performed[21] for a rectangular waveguide ($\approx 1 \times 2\,\text{cm}$), with $\varepsilon = 1.56 \cdot 10^{-9}$ a.u., $\omega = 1.6 \cdot 10^{-6}$ a.u. and initial state $n_0 = 71$. The result for $W(NT)$ as a function of $\ln N$ (N = number of microwave periods, T = temperature of the environment) is shown in fig. 7 for two different temperatures. The 4 dynamical regions mentioned in subsect. 4'1 and there labelled by 1) to 4) can be clearly seen in this figure. The coherence time n_c is here denoted by N^*. The dependence of N^* on T can be extracted and is shown in fig. 8 for $\varepsilon = 4\,\text{V/cm}$ (squares) and $\varepsilon = 8\,\text{V/cm}$ (circles) and compared with the theoretical prediction $N^*T = \text{const}$.

4'3. *Experiment.* – An experiment[21] has been performed with Rb atoms in a coherent microwave whose frequency could be varied in the range from 8 to 18 GHz with a pulse length between 20 ns and 5 µs. Artificial noise was used to realize the coupling to the environment. The details of the experiment and, in particular, the method to extract the Shannon width W from the data are described in[22]. In fig. 9 the Shannon width W is plotted as a function of the interaction time for various values of the noise power. Again the four dynamical

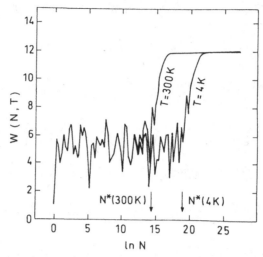

Fig. 7. – Shannon width of the P_n-function as a function of time for two different temperatures (from [21]).

regimes can be distinguished. In particular, a localized regime with $\xi \simeq W \simeq 14$ is followed by a noise-induced spreading of the wave packet. The time t^* after which the localization breaks up can be extracted from the data (fig. 10) and scales inversely proportional to the noise power, as predicted by theory. The main results of the experiment are, therefore, in qualitative agreement with the numerical results of the theoretical model based on 1-dimensional hydrogen atoms, and moreover with the qualitative results obtained for the kicked-rotor model. In particular, nontrivial dynamical localization with $\xi \simeq W \gg 1$ is found in the experimental data of fig. 9; also observed is the environment-induced

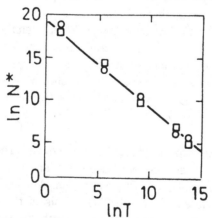

Fig. 8. – The relation between temperature and break time N^* for two different field strengths (squares: $F = 4\,\text{V/cm}$; circles: $F = 8\,\text{V/cm}$). The slope of the straight line reflects the relation $N^* \cdot T = \text{const}$ (from [21]).

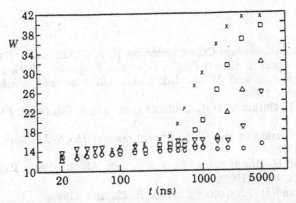

Fig. 9. – Shannon width of the P_n distribution extracted from the measurements as a function of time. Coherent microwave power 6.3 µW. Circles: coherent microwave only; additional noise power: crosses 12.6 µW, squares 6.3 µW, pyramids 3.2 µW, triangles 1.6 µW (from [22]).

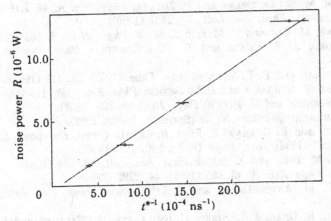

Fig. 10. – Relation between noise power and break time t^* extracted from fig. 9 (from [22]).

break-up after a time proportional to the square of the coupling strength or the noise intensity.

A more detailed quantitative analysis of the experimental data would, of course, be desirable. It would have to include the detailed level structure of Rb, possibly also the effects of the continuum and of the core electrons, and effects of the pulse shape. This remains a task for the future.

* * *

This work was supported by the Deutsche Forschungsgemeinschaft through the Sonderforschungsbereich 237 «Unordnung und große Fluktuationen».

REFERENCES

[1]   R. E. PRANGE: in *Quantum Chaos*, edited by H. A. CERDEIRA, R. RAMASWAMY, M. C. GUTZWILLER and G. CASATI (World Scientific, Singapore, 1991), p. 2.
[2]   A. J. LICHTENBERG and M. A. LIEBERMAN: *Regular and Stochastic Motion* (Springer, Berlin, 1983).
[3]   G. CASATI, B. V. CHIRIKOV, D. L. SHEPELYANSKY and I. GUARNERI: *Phys. Rep.*, 154, 77 (1987).
[4]   G. CASATI, I. GUARNERI and D. L. SHEPELYANSKY: *IEEE J. Quantum Electron.*, QE24, 1420 (1988).
[5]   R. GRAHAM and M. HÖHNERBACH: *Phys. Rev. Lett.*, 64, 637 (1990); *Phys. Rev. A*, 43, 3966 (1991); 45, 5078 (1992).
[6]   P. A. DANDO and D. RICHARDS: *J. Phys. B*, 23, 3179 (1990).
[7]   R. GRAHAM, M. SCHLAUTMANN and D. L. SHEPELYANSKY: *Phys. Rev. Lett.*, 67, 255 (1991).
[8]   R. GRAHAM and J. KEYMER: in *Quantum Chaos*, edited by H. A. CERDEIRA, R. RAMASWAMY, M. C. GUTZWILLER and G. CASATI (World Scientific, Singapore, 1991), p. 382; *Phys. Rev. A*, 44, 6281 (1991).
[9]   R. GRAHAM, M. SCHLAUTMANN and P. ZOLLER: *Phys. Rev. A*, 45, R19 (1992).
[10] J. R. KUKLINSKI: *Phys. Rev. Lett.*, 64, 2507 (1990).
[11] G. W. FORD, M. KAC and P. MAZUR: *J. Math. Phys. (N.Y.)*, 6, 504 (1965).
[12] G. W. FORD, J. T. LEWIS and R. F. O'CONNELL: *Phys. Rev. A*, 37, 4419 (1988).
[13] R. P. FEYNMAN and F. L. VERNON: *Ann. Phys. (N.Y.)*, 24, 118 (1963).
[14] H. GRABERT, P. SCHRAMM and G.-L. INGOLD: *Phys. Rep.*, 168, 115 (1988).
[15] R. K. WANGSNESS and F. BLOCH: *Phys. Rev.*, 89, 728 (1953).
[16] C. W. GARDINER: *Quantum Noise* (Springer, Berlin, 1992).
[17] T. DITTRICH and R. GRAHAM: *Z. Phys. B*, 62, 515 (1986); *Europhys. Lett.*, 4, 263 (1987); 7, 287 (1988); *Ann. Phys. (N.Y.)*, 200, 363 (1990).
[18] S. ADACHI, M. TODA and K. IKEDA: *Phys. Rev. Lett.*, 61, 655 (1988).
[19] D. COHEN: *Phys. Rev. A*, 43, 639 (1991); 44, 2292 (1991).
[20] E. OTT, T. M. ANTONSEN jr. and J. D. HANSON: *Phys. Rev. Lett.*, 23, 2187 (1984).
[21] R. BLÜMEL, R. GRAHAM, L. SIRKO, U. SMILANSKY, H. WALTHER and K. YAMADA: *Phys. Rev. Lett.*, 62, 341 (1989).
[22] R. BLÜMEL, A. BUCHLEITNER, R. GRAHAM, L. SIRKO, U. SMILANSKY and H. WALTHER: *Phys. Rev. A*, 44, 4521 (1990).
[23] R. GRAHAM: unpublished.
[24] R. GRAHAM: *Europhys. Lett.*, 3, 259 (1987).
[25] B. V. CHIRIKOV, F. M. IZRAILEV and D. L. SHEPELYANSKY: *Sov. Sci. Rev. C*, 2, 209 (1981).

Maximum entropy models and quantum transmission in disordered systems

J.-L. PICHARD AND M. SANQUER

Service de Physique de l'État Condensé.
Centre d'Études de Saclay, 91191 Gif sur Yvette Cedex, France.

Abstract

We consider the conductance g of disordered systems where the electronic quantum coherence extends over a large scale. In the first part, we show characteristic conductance fluctuations driven by a variation of the applied magnetic field B or of the Fermi energy E_F, which have been observed at very low temperature in a mesoscopic wire where the carrier density is controlled by a gate. Following the gate voltage, the wire is a conductor ($g \gg 1$) or an insulator ($g \ll 1$). The fluctuations of g have a normal distribution with a universal variance for conductors and a very large log-normal distribution for insulators. In a macroscopic insulator, the magnetoconductance is mostly governed by the field dependence of the localization length ξ. In the second part, we review a random matrix theory adapted to the transfer matrix. This macroscopic approach to quantum transmission allows us to describe in a unified and simple way the conductance fluctuations observed in conductors and insulators, and to predict new universal symmetry breaking effects on the variance of g and on ξ. This approach, based on symmetry considerations and on a maximum entropy criterion, gives the eigenvalue distribution of $\mathbf{t.t}^\dagger$ ($\mathbf{t}$ is the transmission matrix) in terms of a simple Coulomb gas analogy. In quasi-one dimension, the analogy is valid for conductors and insulators. Outside quasi-one dimension, we derive analytically the eigenvalue correlation functions of our maximum entropy model that we compare to their direct numerical evaluations from microscopic hamiltonians. For conductors, the actual correlations are correctly described on small intervals in all dimensions, but are slightly overestimated on larger intervals when the shape of the sample strongly differs from the quasi-one dimensional limit. For three-dimensional insulators, the eigenvalue spacing distribution is still close to the Wigner surmise, indicating the relevance of this model as far as the short range eigenvalue repulsion is concerned.

1 Introduction

Quantum interference effects in electron transport have motivated much work in the last three decades. In disordered systems, the interferences occur by definition between extremely complicated and chaotic trajectories. A typical behaviour is obtained after ensemble averaging a physical variable which has a more or less normal distribution. Sample to sample fluctuations can be observed around this appropriate mean.

When the disorder is weak, g in units of e^2/h is large and the system is a conductor. The underlying classical motions are random walks of step given by the elastic mean path l. Quantum mechanically these random walks interfer. The smaller is the Fermi wave vector k_F, the larger are these interferences and $(k_F.l)^{-1}$ is the relevant small parameter for perturbative expansion. The ensemble average conductance $< g >$ correctly describes experiments where thermal averaging processes are efficient. The so called weak localization corrections [1] to the Boltzman conductance, resulting on average from those interferences, are currently used for measuring the phase coherence length L_ϕ. This scale separates the fully coherent quantum regime on lengths $L \ll L_\phi$, which has to be described by a theory taking into account the interferences, from scales $L \gg L_\phi$ where the electronic motion becomes incoherent and where the classical Boltzmann theory applies.

For disordered insulators where $g \ll 1$, the finite size scaling analysis of long quasi-one-dimensional strips (extrapolated to dimension $d = 2$) and quasi-one-dimensional bars (extrapolated to $d = 3$), first introduced in localization theory by Pichard and Sarma [2,3] and extensively used later [4], is still the main method of investigation of the localized regime and of the mobility edges, as far as typical behaviours are concerned. Roughly speaking, the appropriate mean in this regime is $< \ln(g) >$ instead of $< g >$. In contrast to the diffusive regime, the relevant electronic paths which interfere are not clearly known. In the limit of extreme localization, one often assumes that only the shortest forward directed paths are relevant [5], but the role played by returning loops within the localization domain is certainly important [6].

For sample sizes L of the order of a few L_ϕ, experiments have shown the necessity of knowing not only the averages but also the whole distributions. Reproducible and observable conductance fluctuations, generated for instance by varying an external magnetic field, turned out to be one of the characteristics of these so-called **mesoscopic systems**. Large fluctuations were expected in the localized samples, but were not anticipated at all in good metallic samples [7]. Numerical and diagrammatic calculations by Stone [8], Altshuler [9] and Lee and Stone [10] have introduced the concept of "Universal Conductance Fluctuations" (UCF), since the variance of g in disordered metals is essentially of the order of one, independent of the average values of $< g >$. Much larger relative fluctuations were reported in insulators and explained in terms of resonant tunneling [11] (Azbel resonances).

For conductors, an essential explanation of these universal fluctuations has been given in terms of known general properties of random matrices, using either the Kubo–Greenwood formula for g and the hamiltonian matrix [12], or the Landauer formula and the transfer matrix M [13–16].

We first review the characteristic behaviour observed in a standard conductance study made on a mesoscopic Ga-As wire, in order to give an insight into the physics under consideration. Then we review the maximum entropy approach which have been developed for the transfer matrix description of quantum transport. Technicalities are avoided and we refer the reader to published works where detailed demonstrations can be found [17–20]. The ability of this random matrix theory to provide a unified and coherent explanation for the experimentally observed behaviour is underlined, both for disordered conductors and insulators, with a particular mention to symmetry breaking effects [21].

We compare the eigenvalue correlations of $t.t^\dagger$ yielded by our model to independent numerical calculations based on microscopic Anderson models, addressing the limitation of this theory for shapes of the sample far from the quasi-one-dimensional limit [22, 23], and when disorder induces transverse localization [24].

2 Experimental studies of quantum conductance in disordered systems

2.1 Conductance fluctuations in mesoscopic conductors and insulators

Progress in microfabrication allows us to study submicronic wires. When samples are cooled below $T = 4\,K$, their conductances exhibit chaotic but reproducible fluctuations when one varies the magnetic field, the Fermi energy or the disorder configuration. Three classes of wires can be considered: ballistic conductors without impurities, where the electrons can only suffer boundary scattering during their travel through the sample; disordered conductors, where the electrons are diffused by elastic scatterers and for which g on a scale $L_\phi = \sqrt{D\tau_{in}}$ (D is the elastic diffusion constant and τ_{in} the inelastic mean free path) is larger than e^2/h; and finally disordered insulators for which on the contrary the quantum coherent conductance is much less than e^2/h. Only the two last classes are considered here. Observation of conductance fluctuations have been widely reported either in conductors [7] or insulators [11]. In particular semiconductors offer the unique possibility of significantly varying the Fermi energy by capacitive technique in a given wire, and of illustrating the behaviour of the conductance continuously from the diffusive metallic to the disordered insulating regime. In figure 1, conductance fluctuations of a 5 micron wire etched from a MBE grown GaAs : Si layer doped at $10^{17}\,cm^{-3}$ are reported as function of the gate voltage V_G (curve A). The cross section of the wire is approximately $100\,nm \times 100\,nm$. V_G is applied between the wire and an aluminium film placed on the top of the sample (this techniques is widely used in field effect transistors). The number of electrons in the wire is proportional to the gate voltage, so that one can shift the Fermi level around the mobility edge which separates extended states from exponentially localized states. The static disorder seen by the electrons is mainly the topological disorder of silicon impurities. The spacing between impurities is of the order of 20 nm for our dopant concentration. In the diffusive regime, when the Fermi level is above the mobility edge, one observes typical UCF-fluctuations (figures 2(a) and 2(b)). Their main general properties are listed below: the distribution of $g(L_\phi)$ is gaussian with a variance which does not depend on the mean value. At $T = 0$ K, this variance is of order of unity (when g is measured in quantum units e^2/h), slightly dependent on the sample shape. The variance is halved when time reversal symmetry is broken by applying a magnetic flux quantum h/e through L_ϕ^2. One calls ergodicity the property that conductance fluctuations of similar amplitude can be induced by three different methods: by changing the magnetic field B, the disorder configuration or the Fermi wavelength k_F^{-1}. To decorrelate the quantum interference patterns, one needs typically a variation of B of order $B_C = (h/e)L_\phi^{-2}$, a change of the Fermi energy of order $E_c = hD/L_\phi^2$ (when the temperature is smaller than E_C), or to move on a scale of the order of k_F^{-1} [25] the locations of $(k_F.l)^{d-1}(L_\phi/l)^{d-2}$ impurities in each coherent box. For the sample presented in figure 1 in the diffusive regime, $L_\phi \approx 130$ nm and $E_c \approx 0.4$ K. Obviously the fluctuations are smaller than e^2/h since the sample is not entirely coherent ($L \gg L_\phi$), but consists in an incoherent series of $L/L_\phi \approx 40$ quantum coherent boxes. To include dissipation at finite temperature, one uses a very crude argument due to Thouless: electronic transport is totally coherent for scales smaller than L_ϕ and totally incoherent above L_ϕ. With this assumption one obtains for our sample a mean fluctuation given by $\delta g \simeq \frac{4}{15}^{1/2}(L_\phi/L)^{3/2} \simeq 2.7 \times 10^{-3}$ in units of e^2/h, in agreement with the observed fluctuations driven either by the gate voltage or by the applied magnetic field.

The transition to the insulating regime occurs when the conductance of each quantum coherent box becomes of order unity, corresponding to $g \simeq 1/40$ for the whole sample. This agrees with the activated behaviour which is characteristic of hopping transport between localized states and which is observed when $T < 4$ K and $g < 1/40$. In this insulating regime, very large fluctuations of the conductance are observed when one varies the gate voltage (or the applied magnetic field) such that a **logarithmic** scale for g is more adapted (figures 2(c) and 2(d)). The conductance distribution is log-normal and $\delta(\ln(g))$ shows a tendency to increase when the averaged conductance decreases, and apparently saturates to $\delta(\ln(g)) \simeq 1$ deep in the insulating regime. The vertical bar plotted on figures 2(c) and 2(d) corresponds to $\delta(\ln(g)) \simeq 0.6$, corresponding to a series of L/L_ϕ quantum coherent boxes and assuming that $var(\ln(g)) = - < \ln(g) >$ for each quantum coherent box. This relation is derived in section 2.1 and has been numerically checked in three dimensions (section 2.3). In insulators, L_ϕ is given by the Mott hopping length L_{Mott} which can be extracted from the temperature dependence of g. Very large fluctuations are also observed when the magnetic field is varied, but the field scale for these fluctuations becomes as large as a few teslas, such that only one or two fluctuations are observed between 0 and 5 T (for a given gate voltage). The large magnitude of this typical field has to be contrasted with the

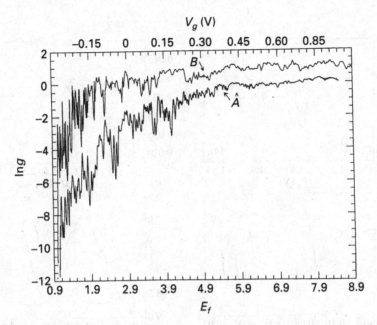

Fig. 1. (*A*) logarithm of *g* in units of e^2/h as a function of the gate voltage (upper coordinate) in the mesoscopic GaAs:Si wire at $T = 0.1\,\mathrm{K}$. *g* is multiplied by a factor $L/L_\phi \approx 40$ (L_ϕ estimated at $V_g \geq 1\,\mathrm{V}$). (*B*) Numerical simulation of ln(*g*) as a function of the Fermi energy (lower coordinate) calculated in a three-dimensional-quantum wire network (see section 2.3).

much smaller scale characterizing the conductors: i.e., the relevant interferences in insulators occur in much smaller areas than in conductors, of order ξ^2 or $\xi^{1/2}.L_{Mott}^{3/2}$ instead of L_ϕ^2.

Above $B = 5\,\mathrm{T}$, a flux quantum is applied inside an area corresponding to the Bohr radius L_B ($\simeq 10\,\mathrm{nm}$) of electronic states centred around the silicon impurities and a very large negative magnetoconductance is yielded by the shrinkage of the electronic orbitals. Interestingly, deeper in the insulating regime, a systematic study (not presented here) shows that the fluctuations in magnetic field become smaller than the fluctuations versus the Fermi energy. One concludes that ergodicity, which is more or less satisfied as far as one can see from figures 2(*c*) and 2(*d*), is broken for higher gate voltages (see also Ref. 26) when ξ becomes of the order of L_B.

2.2 Broken symmetries, magnetoconductance and localization lengths

In the insulating regime, the role of the quantum interferences has not been fully clarified, especially when one approaches the mobility edge where the

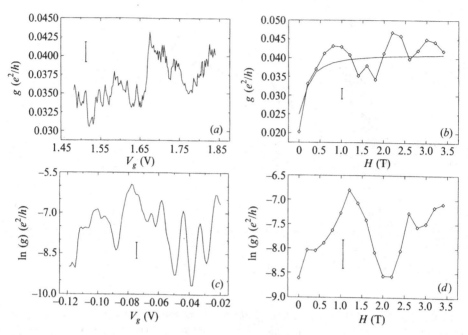

Fig. 2. (a) g as function of V_g close to the mobility edge but on the metallic side at T = 0.1 K. The vertical bar is the estimation of δg corresponding to the UCF-value with thermal averaging. (b) g as function of the transverse applied magnetic field in the same regime. The vertical bar is the same than in (a). The solid line is the one-dimensional weak-localization fit giving $L_\phi = 130$ nm. (c) $\ln(g)$ as function of V_g in the localized regime at T= 0.1 K. The vertical bar is an estimation of $\delta(\ln(g))$ assuming $var(\ln(g)) = - < \ln(g) >$ for each quantum coherent part. (d) $\ln(g)$ as function of the magnetic field in the same regime. The vertical bar is the same than in (c).

localization length is much longer than the distance between scatterers. Before understanding the effect of the phase coherence on the fluctuations in this regime, it is necessary to investigate the typical magnetoconductances which are observed in the macroscopic limit of large samples or after some averages over disorder.

In contrast to the localized regime, the averaged magnetoconductance is well understood in the diffusive regime, and we start with its description. In absence of a magnetic flux, the interferences between each couple of time reversal conjugated diffusion loops reinforce the backscattering probability. This leads to the well known weak localization phenomena [1]. When the magnetic field is applied, time reversal symmetry is removed and those interferences become random and average to zero, leading to a net positive magnetoconductance in the absence of spin-orbit scattering. An important point is that the effect of the interferences is reversed when there is a strong spin-orbit scattering: in the absence of a magnetic field, the time reversal conjugated loops decrease backscattering and the magneto-conductance becomes negative [1]. Therefore, the importance of the interferences between time reversal conjugated diffusion loops can be experimentally seen since the sign of the mean magnetoconductance is sensitive to the presence of impurities with large spin-orbit scattering amplitude (usually large Z atoms). This property can also be used in the insulating regime as shown below. As a consequence of quantum interferences for strong disorder, electrons are exponentially local-ized over a localization length ξ which is generally larger than the elastic mean free path l. At $T = 0\,\mathrm{K}$, the conductance decreases with the sample size L as $g_0.\exp(-L/\xi)$. At finite temperature, the phase coherence is typically preserved over the Mott hopping length L_{Mott} for which $\exp-[(L/\xi)+(E_L/k_BT)]$ is max-imum, $E_L \propto L^{-d}$ being the mean energy spacing for a box of size L. One gets $L_{Mott}/\xi \simeq (T_0/T)^{\frac{1}{d+1}}$ where T_0 is the level spacing at the scale ξ, $T_0 \simeq [n(E_F)\xi^d]^{-1}$ and $n(E_F)$ is the density of states at the Fermi level. Then, at finite temperature the conductance exhibits an activation law:

$$g(T) \simeq g_0.\exp\left(-\frac{L_{Mott}}{\xi}\right) = g_0.\exp\left[-(\frac{T_0}{T})^{\frac{1}{d+1}}\right] \tag{1}$$

when $T < T_0$.

The magnetoconductance of disordered insulators exhibiting this behaviour for different dimensionalities d and different strengths of the spin-orbit scattering has been studied [27, 28]. One can see in figure 3 that, despite a weak magnetic field dependence of g_0, the magnetoconductance at low temperature is controlled by the field dependence of ξ which appears in the exponential term of $g(T)$ and dominates the field dependence of the prefactor when $\xi > l$. This behaviour has been observed in an amorphous alloy of silicon and yttrium where a strong spin-orbit scattering is due to the yttrium atoms. One can see a large negative magnetoconductance, which corresponds to the field dependence of ξ given in the insert, assuming that $n(E_F)$ does not change with the applied magnetic field. This negative magnetoconductance corresponds to a decrease of ξ. The opposite situation is reported in Ref. 28, where indium oxide insulating samples have been studied at low temperature. The spin-orbit scattering is weak in indium oxides, and an increase of ξ is induced by the applied magnetic field. Except for the spin-orbit strength, these samples and our samples have similar parameters ($n(E_F)$ and T_0). Interestingly, the typical field scale B_ξ, where this change of the mean magnetoconductance occurs, is temperature independent and roughly agrees with the criterion $B_\xi \approx h/(e\xi^2)$ (figure 3). At least when $\xi > l$, these experimental results support the conclusion that ξ is almost halved in the presence of a strong spin-orbit scattering by a field $B > B_\xi$ (large negative magnetoconductance), while ξ is very approximately doubled in the absence of spin-orbit scattering (large positive magnetoconductance), indicating that time reversal conjugated loops remain important, at least inside the localization domains.

As ξ becomes of the order of l, i.e., deeper in the insulating regime, the mag-

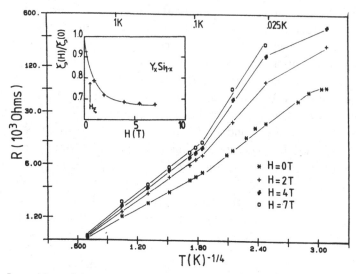

Fig. 3. Logarithm of the resistance as function of $T^{-1/4}$ for various magnetic fields in an insulating amorphous alloy of yttrium and silicon. Insert: Field dependence of ξ deduced from the magnetoconductance between $T = 4.2\,\mathrm{K}$ and $T = 0.08\,\mathrm{K}$.

netoconductance reduces due to a weaker field dependence of ξ, and the field dependence of g_0 dominates, leading eventually to a small positive magnetoconductance [27] at low fields even in the presence of a strong spin-orbit scattering. This probably indicates that the interferences between forward directed paths over L_{Mott} are much more sensitive to weak fields than the returning loops within the localization domain.

This short experimental survey illustrates typical fluctuations observed both in conductors and insulators and proves the relevance of time reversal symmetry (removed by an applied field) and of spin rotation symmetry (removed by spin-orbit scattering), both for the magnitude of the fluctuations and for the sign of the magnetoconductance.

3 Maximum entropy models for transmission

A random matrix theory using the multiplicative transfer matrix M has been developed in order to describe in a unified and simple way quantum electronic transport. This macroscopic and non-perturbative theory, based on symmetry considerations, leads us to propose statistical models of maximum information entropy, given certain physical constraints which are assumed to contain the essential physical features. One of these models, which could be expressed via a particularly simple Coulomb gas analogy, can qualitatively account for most of the previously described behaviour: the magnitude of the conductance fluctuations in disordered conductors and insulators, positive (negative) magnetoconductance in the absence (presence) of sufficient spin orbit scattering, the universal change of the variance of g and of the localization length yielded by symmetry breaking effects.... However, if one needs exact results instead of a qualitative understanding, as far as we can say from numerical studies and perturbative calculations, this model coincides with microscopic models only when the shape of the sample is close to the quasi-one-dimensional limit. But this approach also gives surprisingly good approximations for some distributions observed for instance in metallic and insulating cubes ($d = 3$).

The theoretical part of this paper is organized as follows. We present here a very brief summary of our theory, concentrating first on quasi-one-dimensional conductors and insulators. We then present very accurate quantitative studies of disordered conductors in two and three dimensions, showing the relevance and the limitation of this model when the shape of the sample is not quasi-one dimensional. In a third part, we study how some characteristic features of this approach survive in three dimensions for strong disorder (three-dimensional Anderson insulators). Our numerical results calculated from non-interacting electron models have conductance fluctuations, magnetic field and Fermi energy dependences extremely similar to the behaviour reported in the experimental part of this paper.

3.1 The theoretical model and quasi-one dimension

Consider a coherent quantum box of length L and of transverse width L_t, where carriers are incoherently injected by $N = (k_F.L_t)^{d-1}$ quantized channels, both from left and right sides. This box is a very complicated N-channel elastic scatterer which has a quasi-one-dimensional shape if $L >> L_t$. In this limit and for conductors, the average conductance and its fluctuations, calculated to leading order in power of $(k_F.l)^{-1}$, are dominated by the contribution of the diffusion mode of the zero transverse wave vector, and for insulators, the localization length is proportional to Nl. All the complex interference processes can be represented by a transfer matrix M, which relates at the Fermi energy the $2N$ quantized fluxes present on the right of the box to those present on the left side:

$$\begin{pmatrix} A_r \\ B_r \end{pmatrix} = M \begin{pmatrix} A_l \\ B_l \end{pmatrix} \tag{2}$$

A_l, B_l, A_r, B_r are the N-component vectors describing the flux amplitudes on the left and right sides respectively.

First of all, if one cuts the sample in many slices, M is not only a complicated random matrix, but results from the successive multiplication of many independent random matrices describing the transfer through the different slices. This multiplicative combination law, with corresponding laws of large numbers and the associated central limit theorem, is a crucial property for describing the distribution of M (or of a combination of M's) by appropriate maximum entropy models. Second, it is possible to extract from the transfer matrix a set of N real positive parameters $\{\lambda_a\}$, which are simply related to the conductance g via a standard two-probe Landauer formula. Denoting by t_{ab} the matrix element of the transmission matrix $\mathbf{t}$ between the channels a and b, one has

$$g = 2.\text{Tr}(\mathbf{t}.\mathbf{t}^\dagger) = 2.\sum_{a,b}^{N} |t_{ab}|^2 = 2.\sum_{a=1}^{N} \frac{1}{1 + \lambda_a}. \tag{3}$$

The $\{\lambda_a\}$ are then simply related to the eigenvalues of $\mathbf{t}.\mathbf{t}^\dagger$ and characterize the quantum transmission modes of the disordered sample. They can be defined as the N doubly degenerate eigenvalues of a matrix X:

$$X = \frac{(M^\dagger M) + (M^\dagger M)^{-1} - 2.\mathbf{1}}{4}, \tag{4}$$

or from a convenient parametrization [29] of M; $\mathbf{1}$ is the $2N * 2N$ identity matrix. As is clear from formula (3), we do not need to know the eigenvector statistics of X to have the statistics of g, but only the joint probability distribution $P(\lambda_1, \lambda_2, ..., \lambda_N)$.

The more usual method which allows us to associate a statistical ensemble to the matrices M or X representing the sample consists in beginning from microscopic hamiltonians (tight binding Anderson or Hofstadter models). Usually, a lattice is assumed and different statistical distributions of the diagonal and nearest neighbour off-diagonal elements are considered. This microscopic approach is adapted to diagrammatic calculations and to numerical studies. However, the huge amount of microscopic information contained in the huge hamiltonian matrix is probably not necessary to characterize the distribution of a much smaller matrix M or X, if a central limit theorem [30] applies. An alternative method has been proposed using maximum information entropy criteria. The necessary and sufficient information is supposed to be given by a certain choice of macroscopic constraints, while the remaining variables are assumed to be totally randomized. Since basic symmetry requirements are taken into account from the very beginning, this phenomenological approach is naturally designed for understanding transition between different symmetry classes [31] (Dyson threefold way). Applied to the matrix X, the method can be divided in two steps. One first defines from symmetry considerations the available matrix space and we calculate the measure $\mu(dX)$ of the infinitesimal volume element in this space. Since M belongs to a group, $\mu(dX)$ is unique and well defined. Then the statistical ensemble for X is completed by a density $\rho_X(X)$, giving for the probability $P(dX)$ to find the matrix X representing the sample inside a volume dX around X:

$$P(dX) = \rho_X(X).\mu(dX). \tag{5}$$

The symmetries of X result from Eq. (4) and from the symmetries of M (flux conservation and possibly time reversal symmetry). The corresponding measure

$\mu(dX)$ has been calculated in Refs 18 and 19, and expressed in terms of the N eigenvalues $\{\lambda_a\}$ and of complementary variables associated with the eigenvectors of X. One gets in $\mu(dX)$ a jacobian factor $J_\beta(\{\lambda_a\})$:

$$\mu(dX) \propto \prod_{a<b}^{N} |\lambda_a - \lambda_b|^\beta \equiv J_\beta(\{\lambda_a\}), \tag{6}$$

where β depends only on the symmetry of the system. Conductors with no magnetic field H and no spin-orbit interaction are described by $\beta = 1$ (Orthogonal Ensemble), with magnetic field by $\beta = 2$ (Unitary Ensemble), and for $H = 0$ and strong spin-orbit interaction by $\beta = 4$ (Symplectic Ensemble). $\mu(dX)$ contains only the elementary symmetries and the density $\rho_X(X)$ contains all the other relevant physical properties. An easy way to define a particular statistical model consists in assuming a given eigenvalue density $\rho(\lambda)$ and maximizing the information entropy $S(\rho_X)$

$$S[\rho_X(X)] = -\int \rho_X(X) \ln[\rho_X(X)]\mu(dX) \tag{7}$$

with $\rho(\lambda)$ as a constraint. This model necessarily gives the right average conductance and is characterized by a density

$$\rho_X(X) \propto \prod_{c=1}^{N} F(\lambda_c), \tag{8}$$

which does not depend on the eigenvectors of X. Using the standard method of statistical mechanics [32], one obtains $F(\lambda)$ in terms of $\rho(\lambda)$:

$$F(\lambda) \propto \exp - \int_0^\infty \rho(\lambda') \ln |\lambda - \lambda'|^\beta d\lambda'. \tag{9}$$

Integration over the eigenvectors of X is trivial in this particular model and yields for the joint probability distribution $P(\lambda_1, \lambda_2, ..., \lambda_N)$:

$$P(\lambda_1, \lambda_2, ..., \lambda_N) = C_N^\beta \prod_{a<b}^{N} |\lambda_a - \lambda_b|^\beta \prod_{c=1}^{N} F(\lambda_c) \tag{10}$$

C_N^β being a normalization constant. A convenient way [31] to re-express this distribution consists in introducing a fictitious hamiltonian of classical point charges, located in the complex plane on the positive part of the real axis at positions $\lambda_1, \lambda_2, ..., \lambda_N$, interacting with a two-dimensional logarithmic Coulomb interaction, and with a positive continuous jellium of density $\rho(\lambda')$ via the same logarithmic interaction:

$$H(\lambda_1, \lambda_2, ..., \lambda_N) = -\sum_{a<b}^{N} \ln |\lambda_a - \lambda_b| + \sum_{c=1}^{N} \int_0^\infty \rho(\lambda') \ln |\lambda_c - \lambda'| d\lambda' \tag{11}$$

The integral in Eq. (11) defines a single particle potential $V(\lambda_c)$. Then $P(\lambda_1, \lambda_2, ..., \lambda_N)$ is formally the Boltzmann weight of this fictitious Coulomb gas at a temperature $T = 1/\beta$. If the positive jellium of density $\rho(\lambda')$ is concentrated near the origin, the $\{\lambda_a\}$ will be small and g large (conductors). If the jellium is spread along the positive axis, the $\{\lambda_a\}$ will be large, and g small, eventually smaller than one (insulators).

The main assumption leading to this Coulomb gas model consists in neglecting possible relevant information contained in eigenvector statistics, which could modify the logarithmic eigenvalue interactions through the integration of $P(dX)$ over the variables associated with the eigenvectors. This approximation has clearly a limited validity for the hamiltonian matrix, since dramatic effects occur for the eigenvectors in the large disorder limit (Anderson localization). For a matrix describing transmission, we shall show that, if some eigenvalue–eigenvector correlations exist outside the quasi-one-dimensional limit, their consequences are much less dramatic than for the hamiltonian matrix.

The distribution (10) is similar to the eigenvalue distributions which are usual in random matrix theory [33] for the hamiltonian and scattering matrices, and allow us to use their powerful techniques. Let us note, however, two differences. The multiplicative character of M yields for $\rho(\lambda)$ a "pathological" generic behaviour. Instead of the $\{\lambda_a\}$ (characterized by the classic Coulomb interaction), it is also natural to use the logarithms $\{v_a\}$ of the eigenvalues of $M^\dagger.M$, since $\{(2L)^{-1}.v_a\}$ self-average [2, 3] for a fixed L_t in the large L-limit to the Lyapunov exponents of M. Since these exponents have a generic tendency to have a more or less uniform density, $\rho(\lambda)$ has a very long tail which differs from the eigenvalue densities generally considered for the hamiltonian or scattering matrices. One has

$$\lambda_a = \frac{\cosh(v_a) - 1}{2}.$$ (12)

The second new property of the eigenvalues of X is that they are by definition real positive and cannot cross the origin. For a problem of "interacting particles", this yields an "edge" contribution [22] to the correlation functions which does not exist in usual random matrix theories. This is particularly unfortunate since transport (transmission or reflection) is dominated by the contribution of the edges of the spectrum of X.

To have a complete theoretical description, we need to know the constraint $\rho(\lambda)$. In the quasi-one-dimensional limit, $\rho(\lambda)$ can be obtained from a diffusion equation derived by Mello, Pereyra and Kumar [17] for $P(\lambda_1, \lambda_2, ..., \lambda_N)$, giving the evolution of the $\{\lambda_a\}$ as a function of the sample length $S = L/l$

$$\frac{\partial P(\{\lambda_a\})}{\partial S} = \frac{2}{\beta N + 2 - \beta} \sum_{a=1}^{N} \frac{\partial}{\partial \lambda_a} \left[\lambda_a(1 + \lambda_a) J_\beta(\{\lambda_a\}) \frac{\partial}{\partial \lambda_a} \frac{P(\{\lambda_a\})}{J_\beta(\{\lambda_a\})} \right].$$ (13)

In the metallic regime ($g \gg 1$) one gets from (13) the evolution equations for the expectation value of pth moments of g, and one obtains [34] for the weak-localization corrections to the classical conductance and for the variance of g, at leading order in $< g >^{-1}$, the same results as those given by the microscopic diagrammatic calculations [9,10] for quasi-one-dimensional systems.

One can also show [35] also that a distribution $P(\lambda_1, \lambda_2, ..., \lambda_N)$ of the form (10) is compatible with the diffusion equation (13) in the large N-limit, if the variation of $\rho_S(\lambda)$ as a function of S satisfies:

$$\frac{\partial \rho_S(\lambda)}{\partial S} = 2 \frac{\partial}{\partial \lambda} \left[\lambda (1 + \lambda) \rho_S(\lambda) \mathcal{P} \int_0^\infty \frac{\rho_S(\lambda') d\lambda'}{\lambda - \lambda'} \right]. \tag{14}$$

In other words, if we consider a series of building blocks, a Coulomb gas model of the form (11) describing each block remains a good approximation for the series, the positive jellium of charges characterizing the Coulomb gas spreading out along the real positive axis according to Eq. (14) as one increases the number of blocks in the series. This implies that longitudinal localization does not limit the validity of this maximum entropy model, contrary to a similar description for the hamiltonian matrix. Taking the large L-limit, Eq. (14) can be simplified and gives for the variables $\{v_a\}$ a uniform density between 0 and $2S$.

Our model with such a density gives for the distribution of the $\{v_a\}$:

$$P_v(v_1, v_2, ..., v_N) \propto \exp{-[\beta H_v(v_1, v_2, ..., v_N)]}, \tag{15}$$

$$H_v = -\sum_{a<b}^{N} \ln[|\cosh(v_a) - \cosh(v_b)|] + \sum_{a=1}^{N} \{ -\frac{1}{\beta} \ln[2\sinh(v_a)] + \frac{Nl}{4L} v_a^2 \}. \tag{16}$$

Taking L_t fixed and $L \gg \xi$ where the localization length ξ is defined in this limit by $2L/v_1$, we note that

$$1 \ll v_1 \ll v_2 \ll ... \ll v_N \tag{17}$$

and we simplify the distibution (Eqs. (15), (16)):

$$P_v(v_1, v_2, ..., v_N) \propto \exp{-\left\{ \beta . \sum_{a=1}^{N} \left[-\left(a - 1 + \frac{1}{\beta} \right) v_a + \frac{Nl}{4L} . v_a^2 \right] \right\}}. \tag{18}$$

The two body interaction reduces to a one body potential in this quasi-one-dimensional localized limit due to a pathological density, not to a breakdown of the model. Minimizing the effective hamiltonian in (18), one obtains a "lattice" of charges at equilibrium positions v_a^o

$$v_a^o = \frac{2L}{Nl} . \left(a - 1 + \frac{1}{\beta} \right) \tag{19}$$

with independent gaussian fluctuations of variance:

$$< \delta^2(v_a) > = \frac{2L}{\beta N l}. \tag{20}$$

These results have major physical consequences. Since we have for quasi-one-dimensional insulators:

$$g \propto \exp\left(-\frac{2L}{\xi}\right) \equiv \exp(-\nu_1) \tag{21}$$

we get from (19)–(20) that the variance and the average of $\ln(g)$ are equal

$$\mathrm{var}[\ln(g)] = -<\ln(g)>= \frac{2L}{\beta N l}. \tag{22}$$

We note that β does not only appear in the variances, but also in the averages, implying **universal multiplication factors** for the localization lengths induced by symmetry breaking effects which change β

$$\xi(\beta) = \beta.\xi(\beta = 1), \tag{23}$$

while $\xi(\beta = 1) = N.l$ is the usual quasi-one-dimensional behaviour.

Numerical studies [21] of a disordered Anderson model confirm these predictions when time reversal symmetry is removed by an applied magnetic field with a characteristic cross-over magnetic field B_ξ given by $\xi^2(B_\xi).B_\xi = \Phi_o$. A quasi-halving of ξ seems at the root of the positive magnetoresistance observed in the yttrium–silicon alloy, according to formula (23) ($\beta = 4 \rightarrow \beta = 2$) and assuming that the removal of Kramer's degeneracy by the field is negligible in insulators. Arguments have been given by Bouchaud [36, 37] which agree with our result in quasi-one dimension, and similar symmetry breaking effects have also been confirmed in dynamical localization studies [38]. These changes of ξ were also obtained by Efetov and Larkin [39] from non-linear σ-models, ξ being defined in their case from the average conductance, and not from the distribution of ν_1. This result is proven in the quasi-one dimensional limit, and its extension outside this limit is still unclear. However, different experimental and numerical studies indicate that it could be qualitatively more general.

3.2 Validity of the model for disordered conductors outside quasi-one dimension

In the absence of analytical results for $\rho(\lambda)$ outside quasi-one dimension, we have to rely on numerical studies in order to know the physical constraint (9) of the maximum entropy model and to check its validity for different shapes of conductor (e.g. long strips, squares and cubes). We consider the usual Anderson–Hofstadter models where the diagonal elements are randomly distributed with a rectangular law of width W, and 0.02 flux quanta are applied through the lattice cell. Figure 4 gives the density for the variable ν calculated for three different shapes where W is adjusted in order to have identical average conductance for the three cases. We have taken $W = 0.85; 2; 5.5$ for $20 * 80$ strips (quasi-one-dimensional); $20 * 20$ squares (two-dimensional) and $6 * 6 * 6$ cubes (three-dimensional) respectively. The energy is equal to zero (band centre) and $N = 12$ in each case. The variances of the total transmission $T = \mathrm{Tr}.t.t^\dagger = g/2$ yielded by the microscopic models exhibit a

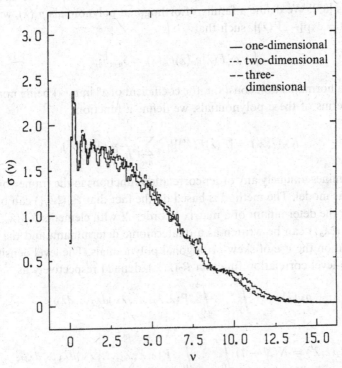

Fig. 4. Density $\sigma(v)$ calculated from tight binding Anderson–Hofstadter models with magnetic fields for different system shapes and identical $< g >$.

clear dimensionality dependence (0.15; 0.096; 0.065), in rough agreement with the perturbation theory results (0.148 (three-dimensional); 0.0925 (two-dimensional); 0.0666 (quasi-one-dimensional)). Looking at Figure 4, one can immediately guess that the dimensional dependence of the UCF-values will probably not result from the very weak dimensional dependence of $\sigma(v)$.

Instead of using relation (9), we prefer to evaluate directly the ensemble averaged quantity $< \sum_{a=1}^{N} \ln|\lambda - \lambda_a| >$ using the numerically obtained λ_a which should give a stationary state form for the single particle potential $V(\lambda)$. The results can be fitted for $0 < \lambda < 1000$ with a two parameter functional form given by

$$V(\lambda) = a \ln^2(1 + b\lambda) \tag{24}$$

for strips, squares and cubes. Our model consists in assuming that a parameter dependent single particle potential (24) together with a universal logarithmic interaction adequately approximates the joint probability distribution of the eigenvalues for a given symmetry. This model allows us to calculate exactly any n-point correlation function for any given N and for $\beta = 1, 2, 4$. The method is simpler for $\beta = 2$ (unitary case) and is based on the use of orthogonal polynomials [33]. We define a family of orthogonal polynomials $p_n(\lambda)$, with given weight $F(\lambda) = \exp[-2V(\lambda)]$, such that

$$\int_0^\infty d\lambda F(\lambda)p_n(\lambda)p_m(\lambda) = \delta_{n,m}h_n, \tag{25}$$

where h_n is a normalization constant, the coefficient of λ^n in $p_n(\lambda)$ being normalized to one. In terms of these polynomials, we define a function

$$K_N(\lambda, \lambda') = [F(\lambda)F(\lambda')]^{1/2}. \sum_{n=0}^{N-1} \frac{1}{h_n}p_n(\lambda)p_n(\lambda') \tag{26}$$

which determines uniquely any order correlation functions in the framework of the Coulomb gas model. The method is based on the fact that $P_N(\{\lambda_a\})$ can be writen for $\beta = 2$ as the determinant of a matrix of order N with elements $K_N(\lambda_a, \lambda_b)$. For $\beta = 1,4$, $P_N(\{\lambda_a\})$ can be written as a quaternionic determinant, and the method is then based on the use of skew-orthogonal polynomials. The level density $\rho_N(\lambda)$ and the two-level correlation function $R_2(\lambda, \lambda')$, defined respectively as

$$\rho_N(\lambda) = N \int_0^\infty \cdots \int_0^\infty P(\lambda, \lambda_2, \ldots, \lambda_N)d\lambda_2 \ldots d\lambda_N \tag{27}$$

and

$$R_2(\lambda, \lambda') = N(N-1) \int_0^\infty \cdots \int_0^\infty P(\lambda, \lambda', \lambda_3, \ldots, \lambda_N)d\lambda_3 \ldots d\lambda_N, \tag{28}$$

are given then by

$$\rho_N(\lambda) = K_N(\lambda, \lambda) \tag{29}$$

and

$$R_2(\lambda, \lambda') = K_N(\lambda, \lambda)K_N(\lambda', \lambda') - [K_N(\lambda, \lambda')]^2. \tag{30}$$

To determine the polynomials defined in Eq. (25) needed to evaluate $K_N(\lambda, \lambda')$, we use the fact that they satisfy a three term recursion relation for arbitrary $V(\lambda)$

$$\lambda p_n(\lambda) = p_{n+1}(\lambda) + S_n p_n(\lambda) + R_n p_{n-1}(\lambda). \tag{31}$$

The coefficients R_n are related to the normalization constants h_n by $h_{n+1} = R_{n+1}h_n$. In order to calculate the $p_n(\lambda)$ for any arbitrary potential and for a given finite N, we define the quantities

$$Q_{n,m} = \int_0^\infty \lambda^m F(\lambda)p_n(\lambda)d\lambda, \tag{32}$$

which satisfy

$$Q_{n,n} = h_n, \tag{33}$$

$$Q_{n,n+1} = h_n \sum_{k=0}^n S_k \tag{34}$$

and

$$Q_{n,m} = Q_{n-1,m+1} - S_{n-1}Q_{n-1,m} - R_{n-1}Q_{n-2,m}. \tag{35}$$

Therefore, the determination of the R_n and S_n necessary to calculate the polynomials of degree $n \leq N - 1$ requires only knowledge of the $2N + 1$ integrals

$$Q_{0,m} = \int_0^\infty \lambda^m F(\lambda)d\lambda \tag{36}$$

for $m = 0, \ldots, 2N$, which can be calculated either from (24) or using high precision cubic splines for $V(\lambda)$.

For the ensembles of cubes, squares and long strips in a magnetic field we solve numerically for the eigenvalues of the X-matrix assuming the Anderson–Hofstadter model and obtain the set $\{\lambda_a\}$ for each realization of randomness. We then proceed in two independent ways.

In one case we evaluate the ensemble averaged quantity $< \sum_{a=1}^N \ln|\lambda - \lambda_a| >$ that we fit with the form (24) or with cubic splines. We use these specific fits to calculate analytically the corresponding polynomials $p_n(\lambda)$ and $K_N(\lambda, \lambda')$ necessary to calculate the correlations in the Coulomb gas model.

In the second case we use the $\{\lambda_a\}$ directly to evaluate the same correlation function numerically without any reference to the potential or random matrix theory.

We first check that the density $K_N(\lambda, \lambda)$ is correctly given from the polynomials and we introduce a function

$$Y(\lambda, \lambda') = \frac{(K_N(\lambda, \lambda'))^2}{K_N(\lambda, \lambda)K_N(\lambda', \lambda')} \tag{37}$$

which determines with the density the two-point correlation function $R_2(\lambda, \lambda')$. Figure 5 shows this correlation function $Y(v, v')$ as the function of the variable v around values for v' choosen in the bulk of the spectrum and near the origin. The variable $v = \cosh^{-1}(2\lambda + 1)$ has been used for clarity in the figure. One can compare the correlation functions implied by the maximum entropy model and calculated from the polynomials obtained from the analytical fit (solid lines) or from the cubic splines (dashed lines) to the actual correlation function (circles) directly calculated from microscopic Anderson models. We just show the three-dimensional results, the two-dimensional results look similar. When v' is choosen in the bulk of the spectrum (Figure 5(a)), one can see that, at least in the metallic regime considered here, our maximum entropy model is not only exact in the quasi-one-dimensional limit, but is capable of giving a good description of the correlations in dimension two and three. When v' is taken in the vicinity of the origin (Figure 5(b)), we see also that the short range correlations are very well described by our theory, the agreement being better if the cubic spline approximations are used. We note that the actual correlation of our model is not the standard Gaussian Unitary one (GUE) [33], as seen in the dissymmetry of $Y(v, v')$ when v' is small (edge effect). We note, however, that the agreement is not good as far as the correlations between the edge ($v < 0.5$) and the bulk of the spectrum ($v > 0.5$) are concerned in the two- and three-dimensional samples.

A more precise check of the theory requires quantities integrated in some intervals of the spectrum. Consider a linear spectral statistics $F = \sum_{a=1}^{N} f(\lambda_a)$ whose the variance is given by

$$\text{var}(F) = \int_0^\infty K_N(\lambda, \lambda) f^2(\lambda) d\lambda - \int_0^\infty \int_0^\infty K_N^2(\lambda, \mu) f(\lambda) f(\lambda') d\lambda d\lambda'. \tag{38}$$

If one is interested by the total transmission T, $f(\lambda) = 1/(1 + \lambda)$, while the total reflection R corresponds to $f(\lambda) = \lambda/(1 + \lambda)$; g is given by $2.T = 2.(N - R)$. These two functions enhance the role of the correlations at the lower and upper edges of the spectrum, and must give identical variances, as required by current conservation. We have evaluated the averages and the variances of:

$$F(\lambda_{max}) = \sum_{\lambda_a < \lambda_{max}} f(\lambda_a), \tag{39}$$

both for transmission and reflection, where the summation is restricted to $\lambda_a < \lambda_{max}$. The averages are very accurately reproduced from the $p_n(\lambda)$ and the variances of $T(\lambda_{max})$ and $R(\lambda_{max})$, are presented in Figure 6 as a function of $v_{max} = \cosh^{-1}(2\lambda_{max} + 1)$ for strips close to a quasi-one-dimensional shape and three-dimensional cubes. We can see that the model reproduces better the correlations for the 20 * 80 strips than those for the 6 * 6 * 6 cubes when v_{max} is large enough. The model does not reproduce the dimensional dependence of the total variance, but gives for var(T) and var(R) values close to the quasi-one-dimensional UCF value 0.0666. We conclude that our model correctly describes the short range correlations for every shape, but overestimates outside quasi-one dimension the correlations between the edges and the bulk of the spectrum: i.e. between the best transmitting (reflecting) channels and the other channels. More detailed comparisons between our model and microscopic calculations can be found in Ref.23 , showing that the correlations are correctly given by our theory for a cube in a very large interval centred in the middle of the spectrum. If the interval is centred near the edges, the width of the window where the random matrix theory works has to be taken smaller. It is interesting to compare this conclusion for a matrix describing transmission to the case of the hamiltonian matrix, where Wigner–Dyson correlations are limited [12,40] to energy intervals smaller than E_c.

Our approach assumes than the eigenvectors of X are totally randomized, yielding for the eigenvalues a logarithmic interaction. It is probable that this represents the more correlated possible spectrum, appropriate only for quasi-one-dimensional shapes, and that, outside this limit, appropriate constraints for some eigenvectors could weaken the eigenvalue interaction within a maximum entropy approach. It is worth noting that neither the spacing distribution nor

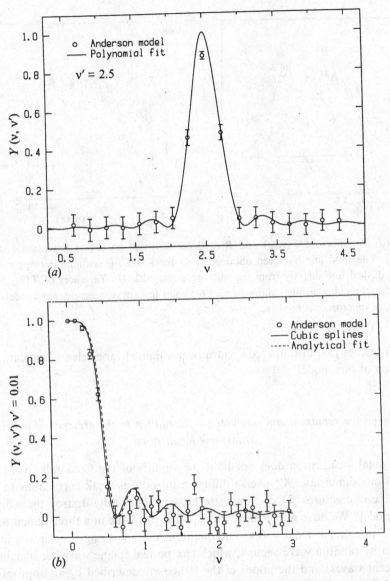

Fig. 5. (a) Correlation function $Y(v, v')$ in the bulk of the spectrum ($v' = 2.5$) as a function of v, showing identical behaviour for the maximum entropy model and a microscopic model in three dimensions. (b) Correlation function $Y(v, v')$ near the origin ($v' = 0.01$), showing identical behaviours for short range ($v < 0.5$), and an overestimation of the correlations by the theoretical model at longer range in three dimensions.

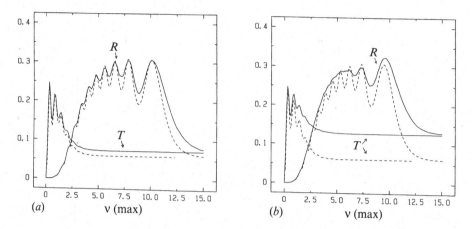

Fig. 6. (a) Variances of $T(v_{max})$ and $R(v_{max})$ for strips of quasi-one-dimensional shape (20 * 80). The solid line has been obtained from the $K(\lambda, \lambda')$ (maximum entropy model) and the dashed line directly from the microscopic model. (b) Variances of $T(v_{max})$ and $R(v_{max})$ for three-dimensional cubes (6 * 6 * 6): solid line (maximum entropy model) and dashed line (microscopic model).

the Δ_3-statistics [33] of unfolded spectra over a limited range clearly indicate this limitation of our model.

3.3 Eigenvalue repulsion and conductance fluctuation in the presence of a strong transverse localization

Longitudinal localization does not limit the validity of the Coulomb gas model in quasi-one dimension. Tranverse diffusion introduces weak corrections to the eigenvalue interactions. Does transverse localization totally destroy the validity of the model? We have numerically investigated this issue in a three-dimensional lattice of one-dimensional wires. The quantum motion along the wires is characterized by random wave vectors, which can be real (plane wave) or imaginary (evanescent waves) and the nodes of the lattice are described by an appropriate scattering matrix. Firstly, this microscopic model which does not contain electron interactions, remarkably reproduces the behaviour observed in real GaAs:Si samples, as a function of the Fermi energy (figure 1, curve (B)) and of the applied magnetic field. Secondly, for a fixed E_F, the localization length ξ varies above a critical strength W_c of the disorder parameter W as

$$\xi \propto (W - W_c) \qquad (40)$$

up to a strongly localized regime where ξ becomes of the order of the lattice spacing. This agrees with the scaling theory of localization with a critical exponent of order one, as experimentally observed in three dimensions, and indicates that the scaling regime is very broad around the critical point.

In contrast to quasi-one dimension, the v-levels are not widely separated for an insulating cube since N is not fixed, but varies proportionally to $L^2 = L_t^2$. For strong disorder, v_1 is large, but is followed by v_2, v_3 ... without large level spacings. A density effect cannot weaken the possible two body interactions. Our Coulomb gas model assumes that successive $\{v_a\}$ interact as:

$$\ln|\cosh(v_a) - \cosh(v_b)| \approx \ln|v_b - v_a| \tag{41}$$

if their spacing is small. This interaction must prevent small spacings (level repulsion) and the Wigner surmise should give an approximate fit for the distribution of their spacing measured in average spacing units. Figure 7 gives this spacing distribution for spacings between v levels in the bulk of the spectrum and very deep in the localized regime. One can see that level repulsion persists even in the presence of a very strong transverse localization. This distribution changes with the field B (Wigner surmises for $\beta = 1 \rightarrow \beta = 2$) when ξ is not too small and when B is larger than the cross-over field B_ξ. When ξ is close to the lattice spacing, the spacing distribution slightly deviates from the Wigner surmise (figure 7) and cannot be changed by B. This is not surprising since the applied flux in those tight binding models is a periodic Aharonov–Bohm flux which cannot affect loops smaller than the lattice cell. We have noticed that the B dependences of ξ and of $\delta \ln(g)$ are related, as is clear from our analysis in quasi-one dimension. Another relation that we have shown in section 2.1, but whose the validity seems more general, concerns the sample to sample conductance fluctuations (formula (22)). Figure 8 gives the result of a study of $8 * 8 * 8$ samples in the localized regime, where the variance of the sample to sample fluctuations of $\ln(g)$ is of the order of the average $< \ln(g) >$. The magnetoconductance fluctuations observed on a given sample have a similar magnitude to the sample to sample fluctuations for large variation $\Delta B \approx B_\xi$ if ξ is large enough, while they are weaker when ξ is small. All these features agree with the experimental data of Section 1.

In summary, this maximum entropy description adapted to the transfer matrix keeps some validity in the limit of extreme (transverse) localization, as far as the short range eigenvalue repulsion is concerned. This has to be contrasted with a similar approach for the hamiltonian matrix which breaks down with the localization of its eigenvectors. For X, localization first appears in the presence of exponentially large eigenvalues giving exponentially small conductances. The persistence of level repulsion for X must have a simple physical origin: electronic paths associated with neighbouring eigenvalues of $t.t^\dagger$ must continue to have a good spatial overlap in the presence of strong localization, possibly related to resonant localized states dominating transmission at the considered energy. In addition, this numerical work indicates that some of the behaviour derived in section 2.1 for quasi-one dimension still occurs in a genuine three-dimensional Anderson insulator. We conclude that this very simple Coulomb gas model

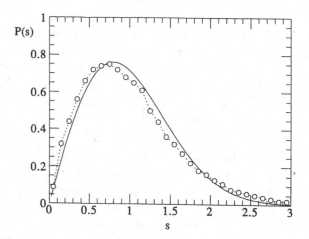

Fig. 7. Spacing distribution $P(v_7 - v_6/ < v_7 - v_6 >)$ calculated from $8 * 8 * 8$ insulating cubes (circles) deep in the localized regime ($< v_6 > \approx 20.5, < v_7 > \approx 21.4, < \ln(g) > \approx -10$). The continuous line is the random matrix prediction (Wigner surmise for $\beta = 1$).

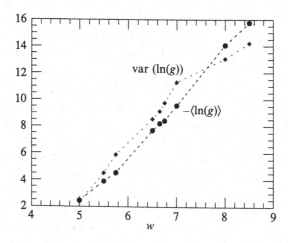

Fig. 8. Average and variance of $\ln(g)$ as a function of the disorder parameter W calculated on $8 * 8 * 8$ insulating cubes.

for the $\{\lambda_a\}$, which is appropriate for conductors and insulators in the quasi-one-dimensional limit, constitutes otherwise an approximation which must be improved to take into account tranverse diffusion and localization.

Acknowledgements

We are very grateful for the collaboration of P. Hernandez, F. Ladieu and D. Mailly for the experiments, and for collaboration of Y. Avishai, K. Muttalib and K. Slevin with whom the results of Sections 2.2 and 2.3 have been obtained. This work has been supported in part by the EEC Science project SCC-CT90-0020.

REFERENCES

[1] Altshuler B., Aronov A., Gershenson M. and Sharvin Y., 1987, *Soviet Scientific Rev., Section A.*, **9**, 225.

[2] Pichard J.-L. and Sarma G., 1981, *J. Phys. C* **14**, L127.

[3] Pichard J.-L. and Sarma G., 1981, *J. Phys. C* **14**, L617.

[4] Kramer B., Broderix K., MacKinnon A and Schreiber M., 1990, *Physica*, **A 167**, 163.

[5] Shklovskii B. and Spivak B., 1991, in *Hopping Transport in Solids*, Pollack M. and Shklovskii B. ed., Elsevier Science Publishers, Amsterdam.

[6] Feng S. and Pichard J.-L., 1991, *Phys. Rev. Lett.*, **67**, 753.

[7] Washburn S., 1991, *in Quantum Coherence in Mesoscopic Systems*, Kramer B. ed., NATO ASI Serie, Plenum Press, New York and ref. therein.

[8] Stone A. D., 1985, *Phys. Rev. Lett*, **54**, 2692.

[9] Altshuler B. L., 1985, *Sov. Phys. JETP Lett.* **41**, 648.

[10] Lee P. A. and Stone A. D., 1985, *Phys. Rev. Lett* **54**, 1622.

[11] Fowler A., Wainer J. and Webb R., 1991, in *Hopping Transport in Solids*, Pollack M. and Shklovskii B. ed., Elsevier Science Publishers.

[12] Altshuler B.L. and Shklovskii B.I., 1986, *Sov. Phys. JETP*, **64**, 127.

[13] Imry Y., 1986, *Europhys. Lett.*, **1**, 249.

[14] Pichard J.-L. and Sanquer M., 1990, *Physica*, **A 167**, 66.

[15] Pichard J.-L., 1991, in *Quantum Coherence in Mesoscopic Systems*, Kramer B. ed., NATO ASI Serie, Plenum Press, New York and refs. therein.

[16] Stone A., Mello P., Muttalib K. and Pichard J.-L., 1991, in *Mesoscopic Phenomena in Solids*, Altshuler B., Lee P.A and Webb R. ed., North Holland, Amsterdam.

[17] Mello P.A., Pereyra P. and Kumar N., 1988, *Ann. Phys.*, **181**, 290.

[18] Muttalib K., Pichard J.-L. and Stone A.D., 1987, *Phys. Rev. Lett.*, **59**, 2475.

[19] Zanon N. and Pichard J.-L., 1988 *J. Phys. France*, **49**, 907.

[20] Pichard J.-L., Zanon N., Imry Y. and Stone A.D., 1990, *J. Phys. France*, **1**, 1.

[21] Pichard J.-L., Sanquer M., Slevin K, and Debray P., 1990, *Phys. Rev. Lett.*, **65**, 1812.

[22] Slevin K, Pichard J.-L., and Mello P.A., 1991, *Europhys. Lett.*, **16**, 649.

[23] Slevin K, Pichard J.-L., and Muttalib K., 1993, *J. Phys. France* **3**, 1387.

[24] Avishai Y., Pichard J.-L., and Muttalib K., 1993, to appear in *J. Phys. France*.

[25] Mailly D. and Sanquer M., 1992, *J. Phys. France*, **2**, 357.

[26] Orlov, A., Savchenko A. and Koslov A., 1989, *Solid State Com.*, **72**, 743.

[27] Hernandez P. and Sanquer M., 1992, *Phys. Rev. Lett.*, **68**, 1402.

[28] Milliken F. and Ovadyahu Z., 1990, *Phys. Rev. Lett.*, **65**, 911.

[29] Mello P.A. and Pichard J.-L., 1991, *J. Phys. France*, **1**, 493.

[30] Mello P.A. and Shapiro B., 1988, *Phys. Rev.*, **B37**, 5860.

[31] Dyson F., 1962, *J. Mat. Phys.*, **3** 140.

[32] Wigner E.P., 1965, in *Statistical Theories of Spectra: Fluctuations.*, Porter C.E., ed. Academic Press, New York, 188.

[33] Mehta M.L., 1991, *Random Matrices*, Academic Press, New York, second edition.

[34] Mello P.A., 1988, *Phys. Rev. Lett.*, **60**, 1089.

[35] Mello P.A. and Pichard J.-L., 1989, *Phys. Rev. Rapid Com.*, **B40**, 5276.

[36] Bouchaud J.-P., 1991, *J. Phys. France*, **1**, 985.

[37] Bouchaud J.-P. and Sornette D., 1992, *Europhys. Lett.*, **17**, 721.

[38] Blumel B. and Smilansky U., 1992, *Phys. Rev. Lett.*, **69**, 217.

[39] Efetov K. and Larkin A., 1983, *Sov. Phys. J.E.T.P.*, **58**, 444.

[40] Argaman N., Imry Y. and Smilansky U., 1992, *Phys Rev.*, **B47**, 4440.

Solid-state "atoms" in intense oscillating fields

M. S. SHERWIN

Department of Physics,
Center for Nonlinear Science and Center for Free-Electron Laser Studies,
University of California, Santa Barbara, California 93106

Abstract

Modern semiconductor technology has enabled the fabrication of solid-state analogues of one-dimensional atoms. These are electrons confined in quantum wells by a graded band gap. Such structures typically have energy level spacings between 1 and 100 meV, and depths up to 300 meV. The development of a free-electron laser that is tuneable between 0.5 and 20 meV has now made possible the study of such solid-state atoms in oscillating electromagnetic fields with amplitudes sufficient to ionize them at frequencies much smaller than their binding energies. Thus experiments analogous to those carried out on atoms in strong electromagnetic fields can be performed (for example, ionization, harmonic generation). This chapter first introduces the physics of quantum wells, then discusses preliminary experimental results on ionization and harmonic generation from electrons in quantum wells, and finally describes the results of recent computer simulations. The chapter concludes by discussing a number of new issues in the interaction of light with matter which are raised in the study of solid-state atoms.

1 Introduction

Much of the theoretical work on quantum chaos[1] in periodically-driven systems[2] has been motivated by classic experiments on the microwave ionization in hydrogen.[3] In these experiments, Rydberg hydrogen atoms are driven by microwaves with photon energy $h\nu \ll$ ionization energy E_I of the Rydberg atom, and with electric field energies comparable to E_I. For $h\nu$ smaller than the separation between Rydberg levels (scaled frequency < 1), simple classical models predict remarkably well observed ionization thresholds.[4] Ionization is associated with the destruction of classical invariant tori and the onset of global chaotic diffusion. For scaled frequencies > 1, the ionization occurs at microwave fields higher than classically-predicted.[5] This suppression of classically-chaotic diffusion has been variously associated with dynamical localization[6] and "scarring".[7]

Experiments have also been performed on ground-state atoms in IR and visible electromagnetic fields with electric field energies comparable to E_I. These experiments report multiphoton ionization in tandem with the generation of odd harmonics up to 33 times the fundamental frequency.[8] Multiphoton ionization and multiple harmonic generation are also observed in quantum-mechanical computer simulations.[9,10] Harmonic radiation emitted by strongly-driven electrons is a useful probe of their dynamics en route to ionization.

Modern semiconductor technology has enabled the fabrication of solid-state analogues of one-dimensional atoms. Using molecular beam epitaxy, electrons in $Al_xGa_{1-x}As$ can be confined in "quantum wells" (QWs) parallel to the direction of epitaxial growth (z-axis), while remaining free perpendicular to z.[11,12] The confining potential $V(z)$ can be tailored to an arbitrary shape. QWs are typically 200–300 meV deep, with spacing between quantized subbands between several meV and several 100 meV for wells 1000s and 10s of Å wide, respectively. A QW with depth 300 meV and intersubband spacing of a few meV contains dozens of bound states, and dozens of far-infrared (FIR) photons can fit into such a well. The classical dynamics of single particles in wide QWs of many shapes are chaotic for experimentally-realizable oscillating electric fields that do not damage the sample.[13] Thus, wide QWs driven by intense FIR electromagnetic fields are a new system in which to search for quantum manifestations of chaos. The quantum dynamics of strongly-driven triangular[14] and square wells,[15,17], which are easily realizable shapes for QWs, have been modeled by various authors.

In addition to their fundamental interest, wide QWs populated with electrons are highly nonlinear materials, and thus have great promise as nonlinear devices (e.g., frequency multipliers and mixers) operating in the FIR. Considerations arising from the theory of quantum chaos will be crucial to engineering and understanding the performance of such devices. Even more exciting is the possibility that new devices may be engineered based on the principles of quantum chaos.

This chapter is aimed primarily at a reader who has some familiarity with quantum chaos in driven systems, but no familiarity with semiconductor physics or devices. The remainder of this chapter is organized as follows. Section 2 gives an introduction to the fabrication and physics of $Al_xGa_{1-x}As$ QWs. This section emphasizes the limits within which a description of a doped QW in an intense electromagnetic field can be described in terms of a single electron with no dissipation in a one-dimensional potential. Section 3 discusses the FIR lasers that are used to drive wide QWs at frequencies and intensities of interest. Section 4 describes the kinds of measurements that can be used to probe the dynamics of electrons in strongly-driven QWs. Experiments demonstrating linear absorption, ionization and harmonic generation in wide QWs are discussed. Section 5 describes preliminary attempts at modeling the dynamics of strongly-driven electrons in quantum wells. The chapter closes in Section 6 with a set of questions requiring further experimental and theoretical study.

2 $Al_xGa_{1-x}As$ quantum wells

The physics of $Al_xGa_{1-x}As$ QWs is an enormous subject on which thousands of papers are published each year. Excellent review articles and books exist, in particular one by Weisbuch and Vinter.[18] This section provides an introduction to those properties of $Al_xGa_{1-x}As$ QWs that are most relevant to experiments on quantum chaos. The interested reader is referred to references for greater depth on QW physics. This section begins in 2A by discussing the elementary excitations and energy scales of bulk GaAs and $Al_xGa_{1-x}As$. Section 2B discusses the epitaxial growth of heterostructures with graded band gaps, or "QWs." Section 2C discusses methods of doping QWs, in particular the technique of "modulation doping." Section 2D discusses the dynamics and relaxation mechanisms for carriers in doped QWs. Section 2E discusses the effects of electron-electron interaction. Finally, Section 2F illustrates important aspects of the physics of semiconductor quantum wells with a discussion of a modulation-doped $Al_xGa_{1-x}As$/GaAs heterojunction.

2.1 Bulk GaAs

GaAs is a semiconductor with a direct band gap of $E_g = 1.5\,eV$ at liquid helium temperatures. The effective mass of an electron in GaAs near $k = 0$ is $m* \approx m_e/15$, where m_e is the bare electronic mass. The elementary excitations of pure, bulk GaAs (which is an insulator at low temperatures) are, in order of decreasing energy:

(i) Excitons. Excitons are bound electron-hole pairs, which have an excitation energy $E_g - E_b$, where $E_b = 5$–7 meV is the binding energy of the exciton.

(ii) Phonons. GaAs is a polar semiconductor, and thus its optical phonons at 36 meV are strongly coupled to photons and electrons. GaAs of course has three acoustic phonon branches as well, but these do not couple directly to photons.

2.2 Graded $Al_xGa_{1-x}As$ heterostructures

In the alloy $Al_xGa_{1-x}As$, the band gap can be varied almost linearly from 1.5 eV for $x = 0$ to 2 eV for $x = 0.4$. The ratio $Q = \Delta E_c/\Delta E$ of the conduction band difference ΔE_c to the total band gap difference ΔE between GaAs and $Al_xGa_{1-x}As$ is still the subject of some debate. We shall assume the widely used value $Q = 0.6$ in this article, although recent measurements indicate that Q may be as high as 0.7.[19] Assuming $Q = 0.6$, the difference in conduction band minima between GaAs and $Al_{0.4}Ga_{0.6}As$ is approximately 300 meV. The devices of interest to us are generally made with $x_{max} \leq 0.4$, because for $x > 0.4$ the band gap of AlAs becomes indirect (e.g., the lowest minimum of the conduction band occurs at finite electron momentum).

The realization of high-quality epitaxial layers of $Al_xGa_{1-x}As$ grown on GaAs substrates is possible because the lattice constants of AlAs and GaAs differ by less than 0.1%. With current computer-controlled molecular beam epitaxy (MBE) technology, it is possible to make structures in which $x(z)$, and hence $\Delta E_c(z)$ have an almost arbitrary shape (z is the distance from the surface of the epitaxial layer). The grading of the band gap can be achieved in one of two ways.[11]

   (i) Continuous grading: the Al concentration x can be varied continuously as a function of time as the sample is being grown, resulting in the desired $\Delta E_c(z)$. This procedure is cumbersome, since the temperature of the Al furnace in the MBE machine, which controls x, must be changed continuously with time.

   (ii) Digital alloying: the temperature of the Al furnace is held constant at a value that would yield $Al_xGa_{1-x}As$ with a particular value of x. The Al beam is turned on and off by an automatic shutter to grow alternately (e.g., 25 Å) layers of $Al_xGa_{1-x}As$ and layers of GaAs. The average potential is then tailored by varying the duty cycle of the Al beam. If the well is much wider than the width of the $Al_{0.3}Ga_{0.7}As$ layers, an electron with kinetic energy near the bottom of the well will respond only to the average potential and not to the high-frequency potential variations.

2.3 Doping

Carriers can be added to pure semiconductor material by doping with electron donors (for electrons) or acceptors (for holes). We will consider only QWs populated with electrons in this chapter, since additional complications arise upon doping with holes.[20] The most common electron donor used in $Al_xGa_{1-x}As$ is Si. In bulk material with $x < 0.22$, the Si donor can be thought of as a hydrogenic impurity with a shallow binding energy of 6 meV. For $x > 0.22$, the Si donor levels become "deep," with a binding energy in excess of 100 meV.

The deep donors are a complex of an impurity with a configuration of Al, Ga and As nearest-neighbors that was until recently unknown,[21] and thus became known as "DX centers" (D for donor, X for the unknown local environment). DX centers dominate electrical transport in $Al_xGa_{1-x}As$ with $x > 0.22$.[22] The filling of a DX center is accompanied by a significant lattice distortion which gives rise to a capture barrier $V_c \approx 200$ meV. The DX center can be ionized by radiation with $hv > 0.8 eV$. At temperatures below 77 K the ionization of the DX center, and hence photoconductivity, persist almost indefinitely (days) because thermal energies are insufficient to overcome V_c. This remarkable phenomenon is known as "persistent photoconductivity." This phenomenon is important in experiments on ionization of QWs by FIR radiation, as described below.

There are two methods of adding carriers to a QW. Dopants can be placed directly in the well. This technique has the disadvantage that electrons scatter strongly from ionized impurities, and hence they have poor mobility (mobility

$\mu = e\pi/m*$, where e is the electronic charge, π is the momentum relaxation time, and $m*$ is the effective mass). QWs can be populated with carriers of high mobility by the technique of "modulation doping." Here, dopants are placed in the barriers just outside the QWs. If the bottom of the QW is below the energy of the filled donor, the donors are ionized and electrons populate the QW. Since the ionized donors are now far removed from the electrons, scattering is minimized and high mobilities can be achieved. World record mobilities of $1.1 \times 10^7 \, \text{cm}^2/\text{V-sec}$, corresponding to a mean free path of $100 \, \mu\text{m}$, have been achieved in $Al_xGa_{1-x}As$ heterojunctions. In wide modulation-doped QWs, typical mobilities are of order $2 \times 10^5 \, \text{cm}^2/\text{V-sec}$.

2.4 Dynamics

2.4.1 Intersubband transitions

The energy for electrons in a QW can be written as $E = E_n + h^2k_\perp^2/2m*$, E_n is the quantized energy in the growth ($z-$) direction, and $k \perp$ is the continuous wavevector for motion perpendicular to z. All $k_\perp$ states associated with a particular value of E_n are called a "subband." Intersubband transitions couple strongly to the electromagnetic field and have enormous oscillator strength compared with other excitations in solids.[23] Intersubband transitions have been extensively studied in narrow QWs (several tens of angstroms), where the transition between first and second subbands is of order 100 meV, and only one or two bound states fit into a well. Less work has been done on wide QWs. For a 500 Å square well with Al concentration $x = 0$ in the well and $x = 0.3$ in the barriers, the depth is approximately 225 meV, the bare transition energy between the first and second subband (not taking into account electron-electron interactions) is approximately 6 meV, and there are approximately ten bound states.

2.4.2 Relaxation times

In the absence of disorder, an electron in a QW sees a flat and featureless potential in the x- and y-directions, and there are thus three good quantum numbers: E_n, the quantized energy associated with the motion parallel to z, and k_y, the wavevectors in the x- and y-directions. However, some disorder is always present in the potential felt by an electron in a real QW the potential in the x- and y-directions has small, random bumps caused by residual impurities, the Coulomb field of the randomly-spaced remote donors in a modulation-doped QW, fluctuations in the width of the well, and alloy disorder in $Al_xGa_{1-x}As$. In the presence of disorder, the Schrödinger equation is no longer separable and the only quantum number that is strictly good is the total energy. However, if disorder is weak, an electron will maintain a definite value of the momentum for a time τ_m, the momentum or elastic relaxation time, and k_x and k_y remain approximately good quantum numbers for times shorter than τ_m. In wide QWs of good quality, typical mobilities are of order $200,000\,\mathrm{cm^2/V\text{-}sec}$,[24] corresponding to a scattering time of $\tau_m = 8\,\mathrm{psec}$.[25]

Excited electrons can relax their energy by emitting optical or acoustic phonons (decay by photon emission is negligibly slow). If an electron is in an excited state more than 36 meV above the ground state, it can decay in less than 1 psec by emitting 36 meV optical phonons.[26] If the electron has kinetic energy less than 36 meV, it can decay only by emitting acoustic phonons, which takes 100s of psec to 1 nsec.[27] Thus, for wide QWs with mobilities of order $10^5\,\mathrm{cm^2/V\text{-}}$ sec, the momentum relaxation time is roughly two orders of magnitude shorter than the energy relaxation time, and is the dominant source of broadening for intersubband absorption. A momentum relaxation time of 8 psec would naively imply a linewidth of $1\,\mathrm{cm^{-1}}$.[28] Such narrow lines are rarely observed, although intersubband absorptions near $\nu = 50\,\mathrm{cm^{-1}}$ with FWHM $< 2\,\mathrm{cm^{-1}}$ have been observed in a parabolic QW by P. Pinsukanjana et al.[29]

2.5 *Electron-electron interaction*

Important corrections to single-electron dynamics arise in a modulation-doped QW. The electrons that have fallen into the well feel the potential $V_0(z)$ due to the graded band gap, the electrostatic potential $V_{imp}(z)$ caused by the ionized impurities, and the potential $V_{ee}(z)$ due to the other electrons. The ionized impurities may be simply treated as a fixed sheet with a positive charge density eN_s, where e is the magnitude of the electronic charge and N_s is the two-dimensional sheet density. The electric field associated with the sheet of ionized impurities is $E_{imp} = 2\pi eN_s/\epsilon$ (in CGS units), where $\epsilon = 13$ is the dielectric constant of GaAs. The electrons in the well form a layer of charge with charge density $-eN_s$. However, since the electrons are mobile, the potential that they feel must be determined self-consistently. Calculations are usually performed in the local density approximation, with the Schrödinger equation

$$\frac{-\hbar^2}{2m*}\frac{\partial^2}{\partial z^2}\psi_i(z) + V_{tot}(z)\psi_i(z) = E_i\psi_i(z) \tag{1}$$

where

$$V_{tot}(z) = V_0(z) + V_{Hartree}(z) + V_{xc}(z) \tag{2}$$

Here, $V_0(z)$ is the potential due to the grading of the band gap. $V_{Hartree}$ is the self-consistent potential which takes into account the Coulomb interaction of a single electron with the electric field of all of the other electrons, and obeys Poisson's equation

$$\frac{d^2 V_{Hartree}(z)}{dz^2} = -\frac{4\pi e^2}{\epsilon}\sum_i N_i|\psi_i(z)|^2 \tag{3}$$

where e is the electronic charge, ϵ is the dielectric constant, and N_i is the sheet density in the ith subband. The constant electric field associated with the ionized impurities is lumped into $V_{Hartree}$. $V_{xc}(z)$ is the exchange-correlation potential, which takes into account the Pauli exclusion principle by subtracting off overcounting in the Hartree term. The reader is referred to Ref. 30 for a discussion of V_{xc}, for which there is not a universally-agreed upon parameterization.

Intersubband transitions are also shifted by many-body effects. Naively, one might expect electromagnetic radiation to be absorbed at the Bohr frequencies $(E_i - E_j)/h$, where E_i and E_j are the self-consistent energy levels that solve Eqs. (1)–(3). In fact, additional shifts occur because the electromagnetic radiation causes the charge in the well to oscillate. The so-called "plasma" and "excitonic" shifts are tied to changes in $V_{Hartree}$ and V_{xc} caused by the oscillating electromagnetic field. The plasma shift reflects the tendency of the electron gas density to shift to cancel out externally applied electric fields. The excitonic shift reflects the attraction between an electron excited to a higher subband and the hole it left behind. In a modulation-doped parabolic QW, it has been shown theoretically[31] and verified experimentally[32] that the plasma and excitonic shifts exactly cancel for arbitrary charge density in the well. Thus a parabolic QW absorbs electromagnetic radiation at the bare harmonic oscillator frequency, independent of the charge density in the well. This cancellation is very special, and does not hold for other well shapes.

2.6 Example: a modulation-doped heterojunction

Considerations important for determining the charge density and potential felt by electrons in real heterostructures are nicely illustrated by considering the simplest of all heterostructures, the modulation-doped $Al_xGa_{1-x}As$ heterojunction. Fig. 1(a) shows a compositional profile for a typical heterojunction. The substrate on which the epitaxial layers are grown is "semi-insulating" GaAs (not intentionally doped, Fermi level in the gap). On top of the substrate, a layer of nominally undoped GaAs (unintentional doping levels are typically between 10^{14} and 10^{15} cm^{-3}) is grown. Then a "smoothing superlattice" alternating thin layers of $Al_{0.35}Ga_{0.65}As$ – is grown to improve the quality of the overlayer by smoothing out inhomogeneities of the substrate. Then a thick layer of undoped GaAs is grown, followed by 250 Å of undoped $Al_{0.3}Ga_{0.7}As$, and 250 Å of $Al_{0.3}Ga_{0.7}As$ doped with Si. Finally, a 200 Å cap layer of GaAs is grown to prevent oxidation of the $Al_{0.3}Ga_{0.7}As$.

Fig. 1(b) shows the charge distribution and self-consistent potential calculated in the local density approximation for the compositional profile shown in Fig. 1(a). Qualitatively Fig. 1(b) can be understood as follows. The Fermi level is pinned near the middle of the energy gap at both the surface and in the substrate, providing boundary conditions for the solution of Poisson's equation. In thermal equilibrium, the Fermi energy is constant throughout the sample. Thus, if one imagines adding Si donors to the $Al_{0.3}Ga_{0.7}As$, the donated electrons first go to the

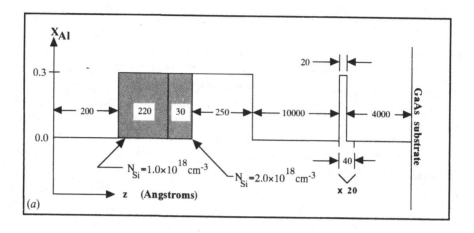

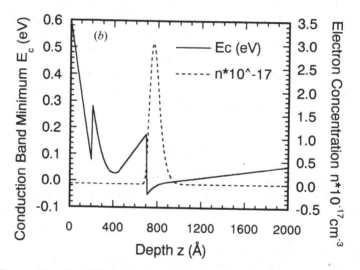

Fig. 1. (*a*) Compositional profile for a typical heterojunction. This particular structure was used in second-harmonic generation measurements described below. On top of the substrate, 4000 Å of GaAs are grown, followed by 20 periods of a 20 Å "smoothing superlattice" which improves the epitaxy. Then, 10,000 Å of clean GaAs are grown, followed by 500 Å of $Al_{0.3}Ga_{0.7}As$. The first 250 Å are undoped, the next 250 Å are doped with the Si concentrations indicated. The electrons donated by 30 Å of 2×10^{18} cm^{-3} are designed to go into a highly-mobile 2-DEG at the $Al_{0.3}Ga_{0.7}As$ interface, while the rest are designed to go to the surface. (*b*) Conduction band minimum and charge distribution calculated for the doping profile of (*a*) in the local density approximation. Off the graph to the right, the conduction band minimum increases until it reaches 0.9 eV at 20,000 Å. The calculation was performed by Keith Craig using a program written by Dr Greg Snider.

mid-gap states at the surface and the substrate, leaving behind ionized donors. The electric field due to this sheet of ionized donors "bends" the conduction band down. When enough Si donors are added (the 220 Å of $Al_{0.3}Ga_{0.7}As$ with $N_{Si} = 1.0 \times 10^{18}$ cm^{-3} are meant for this purpose), the conduction band at $z = 0$ crosses the Fermi level. At this point, electrons begin to accumulate in a two-dimensional electron gas (2-DEG) at the $Al_{0.3}Ga_{0.7}As$ interface (the 30 Å with $N_{Si} = 1.0 \times 10^{18}$ cm^{-3} are meant for this purpose). The electrons end up confined at the $Al_{0.3}Ga_{0.7}As$ interface by a built-in electric field E_{bi} that is entirely a result of electrostatic "band-bending." Second-harmonic generation experiments discussed later in this chapter were performed on the heterojunction depicted in Fig. 1.

A simplistic but useful model for a heterojunction is a gas of independent electrons confined by a triangular potential whose slope is determined by the sheet density N_S in the heterojunction and the background density of acceptors N_A on the GaAs side of the heterojunction.[30] For this heterojunction with the measured $N_S = 5 \times 10^{11}$ cm^{-2} and an assumed $N_A = 10^{15}$ cm^{-3}, the slope of the triangular well is 39 kV/cm and the first intersubband transition frequency is 290 cm^{-1}.

As demonstrated by Fig. 1, the potential felt by an electron in a AlGaAs/GaAs heterostructure well is not simply that introduced by gradations in the band gap. This fact will be emphasized again below, in data on the intersubband absorptions of a single square quantum well. In modeling the response of electrons in heterostructures to oscillating electric fields $E \parallel z$, care must be taken to understand the built-in electric fields caused by the pinning of the Fermi level at the surface and in the substrate.

3 Sources of intense FIR radiation

In analogy to experiments on atoms in a strong, low-frequency electromagnetic field, we would like to pump electrons in QWs with electromagnetic fields that have electric field energies comparable to the depth of the well ΔE_c but photon energies small compared to ΔE_c. Photon energies of 1–20 meV are ideal, since the ΔE_c is typically 200–300 meV. The University of California at Santa Barbara Free-Electron Laser currently produces radiation with photon energy continuously tuneable between 0.5 meV and 20 meV. Peak powers of 3 kW in pulses as long as 20 μsec are currently achieved. Peak powers of order 10 kW for 30 nsec have been achieved using a cavity dumping scheme. Pulses as short as 500 psec have been generated by dumping a small cavity external to the laser. Alternately, a molecular gas laser (MGL)–a transversely excited atmospheric (TEA) pressure CO_2 laser pumping Raman transitions in a molecular gas can produce pulses with peak powers of more than 100 kW and pulse widths of 50 nsec at a few discrete lines in the frequency range of interest. The MGL is unfortunately plagued by unrepeatable intensity fluctuations on a time scale of a few nsec.[33]

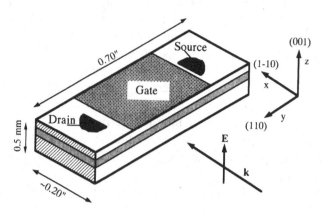

Fig. 2. Sample and contact geometry. The shaded layer in the middle of the sample represents the 2-DEG confined in the QW. Also shown are the crystallographic axes. From the Ph.D. thesis C. L. Felix.

4 Experiments

One of the beauties of the AlGaAs/GaAs QW system is the wide variety of experimental probes available. On a sample with a single QW, one can perform electrical transport, absorption, harmonic generation, and luminescence measurements. This section will briefly describe some of our experimental techniques, the samples we have studied, and results and significance of experiments on linear absorption (low FIR power) and ionization and harmonic generation (high FIR power).

4.1 Techniques

4.1.1 Sample preparation
Given a wafer containing a doped QW, samples are prepared as shown in Fig. 2. Ohmic contacts to the 2-DEG in the well are prepared by alloying strips of Au-Ge-Ni into the sample. Upon heating, the metal in the strips "spikes" down to contact the electrons in the 2-DEG, so that transport measurements can be performed. In the language used for transistors, these ohmic contacts are called the source and drain. A gold "gate" is then evaporated onto the surface of the sample in between the contacts. Unlike the contacts, the gate (ideally) does NOT "spike" down to the 2-DEG, but rather is separated from the 2-DEG by a Schottky barrier.

4.1.2 Controlling carrier concentration
If one applies a negative voltage to the gate then the negatively-charged electrons underneath the gate are repelled and forced out of the QW through the ohmic

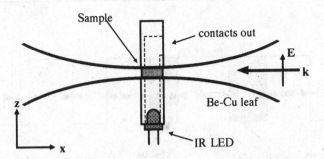

Fig. 3. Stripline couplier. In order to couple radiation in and out of the sample with $\vec{E} \parallel z$, the sample is used as a stripline. Radiation is coupled to the stripline using a Be-Cu horn coupler. The IR LED is used to ionize deep traps and populate the QW with electrons. From the Ph.D. thesis of C. L. Felix.

contacts. The presence of a gate that does not leak enables one to tune the number of electrons in a sample. If one applies a sufficiently large negative voltage to the gate ($-1\,$V to $-5\,$V for samples studied here), all of the electrons can be depleted from the QW. Thus, optical measurements can be normalized by comparing the results for a full and a depleted QW. The number of electrons in the QW at a given gate voltage can be deduced from the capacitance between the gate and the 2-DEG[34], or from magnetotransport measurements.[18]

For QWs with barrier Al concentrations in excess of 0.22, DX centers play an important role in charge transfer. Often, after cooling in the dark, a QW will contain almost no free electrons and its source-drain resistance will be many MΩ. Electrons can be added to a QW by using the persistent photoconductivity effect discussed earlier. The sample is illuminated with photons of energy greater than 0.8 eV – typically, we use a light-emitting diode (LED) mounted near the sample. These photons ionize DX centers, and a large fraction of the associated electrons go into the QW and stay there as long as the sample is kept cold (below 77 K). This is a less reproducible method of controlling the carrier concentration in a QW than applying a gate bias.

4.1.3 Coupling of FIR radiation with $\vec{E} \parallel z$

A number of techniques have been developed for coupling electromagnetic radiation to quantum wells with $\vec{E} \parallel z$. We have chosen the stripline technique, illustrated in Fig. 3. The sample is placed between two strips of metal. Radiation with $\vec{E} \parallel z$ is coupled into and out of the stripline via a flared horn. The advantages of this technique are that the component of $\vec{E} \parallel z$ is maximum, and the interaction length of the radiation with the sample is long

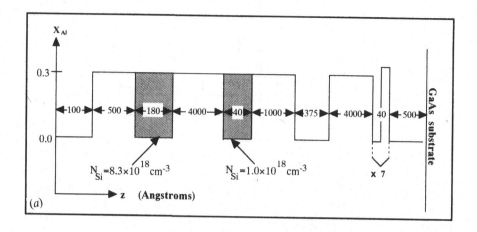

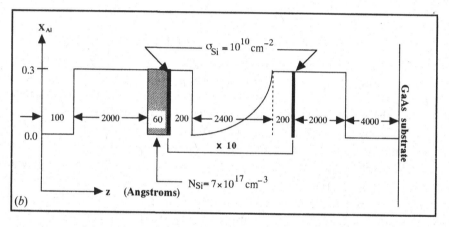

Fig. 4. (*a*) Compositional profile for the nominally square well on which intersubband absorption measurements were performed. (*b*) Compositional profile for the half-parabolic well sample, on which second-harmonic generation and ionization measurements were performed.

4.2 Samples

I will discuss experimental results on three sample: a nominally square well 375 Å wide, a $Al_{0.3}Ga_{0.7}As$ heterojunction, and a sample containing ten half-parabolic QWs. Compositional profiles of the three samples are given in Fig. 4. Mobilities and carrier concentrations for the three samples are given in Table 1.

Table 1. Carrier concentration and mobility for investigated samples. All values are quoted for cooling the sample below 10 K in the dark, then illuminating with an IR LED. For the heterojunction and half-parabolic wells, carrier concentration and mobility were determined from magnetotransport measurements (See Weisbuch and Vinter, Ref.[18]). For the half-parabolic well sample, the charge distribution among the ten wells is not known, but most of it was probably forced into the first well by built-in electric fields. For the square well, there was insufficient charge for a reliable magnetotransport measurement, so the carrier concentration was determined from a capacitance-voltage measurement, and the mobility is unknown.

Well	Carrier concentration	Mobility	Carriers when dark-cooled?
Square	3×10^{10} cm^{-2}	?	No
Heterojunction	5×10^{11} cm^{-2}	4.4×10^5 cm^2/V-sec	Yes
Half-parabolic	6×10^{11} cm^{-2}	2×10^5 cm^{-2}/V-sec	No

4.3 Results

4.3.1 Low FIR intensity: linear intersubband absorption

Fig. 5 shows a series of absorption spectra for the square well. These were recorded with a low-power source of FIR radiation, a Fourier transform interferometer (FTIR).[35]

Each spectrum is the ratio

$$-\frac{\Delta T}{T} = \frac{T(empty) - T(V)}{T(empty)}$$

where $T(empty)$ is the transmittance through the stripline with a gate voltage sufficient to deplete the well of all electrons (-1 V), and $T(V)$ is the transmittance of with a gate voltage V. Fig. 5(a) shows a clear absorption peak centered at 149 cm^{-1} with a FWHM of 21 cm^{-1}. The solid line is the best Gaussian fit to the absorption line, from which the peak and FWHM were extracted. Figs. 5(b), (c) and (d) show the peak shifting to lower frequency and decreasing in strength as the gate voltage is made more negative and charge is removed from the well. The fact that a clear absorption from a single QW can be observed even with a sheet density of only 3.7×10^9 cm^{-2} is a testament to the sensitivity of the stripline coupling technique.

The sample was designed to have its first intersubband absorption near 90 cm^{-1}. We suspect that the transition has been shifted to higher energy because of the built-in electric field caused by the Fermi level being pinned at mid gap in the substrate (the same linear potential that is solely responsible for confinement in the 2-DEG at the heterojunction of Fig. 1). The application of a negative gate bias would tend to pull the bottom of the well up, flattening the potential. This is consistent with the downward shift of the intersubband absorption frequency

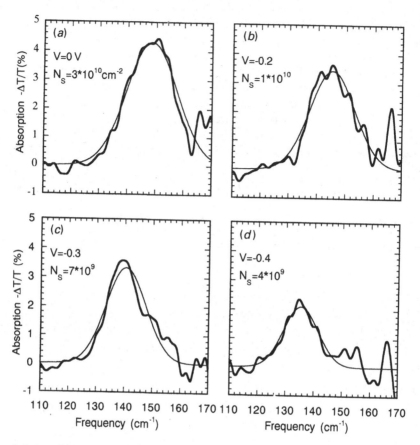

Fig. 5. Intersubband absorptions at different gate biases for the nominally square well shown in Fig. 4(a). Plotted is the differential absorption

$$-\frac{\Delta T}{T} = \frac{T(empty\_ T(V)}{T(empty)},$$

where $T(empty)$ is the transmission with $-1\,\mathrm{V}$ applied to the gate. Thin lines are the best Gaussian fits to the experimental data. $T = 7\,\mathrm{K}$. From the Ph.D. thesis of C. L. Felix.

with decreasing gate bias. The 11–14% width of this intersubband absorption is consistent with an estimated 6% variation in well thickness across the 5 mm width of this sample.[36]

These data have been chosen to give the reader a feeling for both the flexibility and the complexity of real QWs. One lesson is that, for the wide QWs used to achieve intersubband transition energies in the FIR frequency range, it is crucial to consider built-in electric fields in addition to the potential intentionally introduced by grading of the Al concentration x. For the half-parabolic wells, in the absence of built-in electric fields, intersubband absorptions in the half-parabolic wells would have occurred at

$$f = \frac{1}{2\pi}\sqrt{\frac{8\Delta}{W^2 m*}} = 50\,\mathrm{cm}^{-1},$$

where Δ is the depth of the well, W is the width of the half-parabola and $m*$ is the effective mass. No such intersubband absorptions were observed, and it is likely that built-in electric fields pushed them above $200\,\mathrm{cm}^{-1}$. Efforts are currently being made to minimize built-in electric fields and achieve a more ideal square well.

4.3.2 High FIR intensity

Measurements have been carried out at high FIR intensity on the heterojunction, parabolic and half-parabolic QW samples using the stripline coupler. Observed phenomena include:

(i) persistent ionization of parabolic QWs, half-parabolic QWs, and hetero-junctions by intense FIR radiation, and

(ii) generation of the second and third harmonics of the FIR radiation for half-parabolic QWs and heterojunctions.

Ionization measurements give an idea of the kinetic energy reached by electrons under intense irradiation. Measurements of the harmonic content of emitted radiation probe the Fourier components of the electronic motion under irradiation.

This section will discuss experimental results on ionization and second-harmonic generation of a modulation-doped heterojunction induced by intense FIR radiation. We have also observed bleaching of the intersubband absorption at high intensity in a parabolic QW,[37] but this will not be discussed here.

a. Ionization During experiments on FIR harmonic generation on the HPW sample, it was serendipitously discovered that intense FIR radiation with photon energy of a few meV and $\dot{E} \parallel z$ can persistently ionize a QW that is over 200 meV deep. Similar results have been obtained for the heterojunction and a parabolic QW.

Fig. 6 shows the ionization of the half-parabolic wells. The source-drain resistance of the HPW sample was greater than 20 MΩ when cooled in the dark, and dropped to 400 Ω after illumination with an IR LED. In Fig. 6(a), the source-drain resistance is plotted vs. time with the IR LED off. Each vertical jump in resistance coincides with a single laser pulse. Successive laser pulses drive the resistance from near 400 Ω to an asymptotic value near 650 Ω. In Fig. 6(b), the source-drain resistance is plotted vs. time with the IR LED on. Here, each FIR laser pulse drives the resistance up, but the resistance recovers between each FIR pulse as radiation from the LED repopulates the well with carriers.

We propose a two-step process to explain the results of Fig. 6.

(1) The FIR radiation first transfers sufficient kinetic energy to ionize the well (i.e., to send the electrons into the doped $Al_{0.3}Ga_{0.7}As$ barriers). For a FIR electric field of 10 kV/cm, the electric field energy drop across the width of the 2400 Å

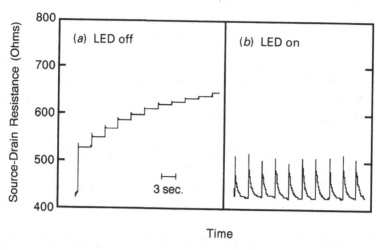

Fig. 6. Ionization of the half-parabolic well by intense FIR radiation. (*a*) Source-drain resistance R_{SD} vs. time with LED off. Each laser pulse causes a persistent jump in R_{SD}, indicating that electrons have been persistently removed from the well. (*b*) R_{SD} vs. time with LED on. The resistance recovers after each laser pulse, as the LED repopulates the QW with electrons. The FIR source was the molecular gas laser. FIR pump frequency = 43.3 cm^{-1}, FIR pulse length $\approx$ 5 mJ, repetition rate = 0.25 Hz, $T \approx 7$ K. Similar behavior occurred for FIR pump frequency = 29.5 cm^{-1}, and in the heterojunction sample. From the Ph.D. thesis of J. J. Plombon.

well is 240 meV, comparable to the depth of the well. Thus ionization appears likely. Unfortunately, the fluctuations of the molecular gas laser prevented a clear determination of an ionization threshold.

(2) The electrons are then captured by DX centers. We have proven that the second step can occur in experiments on a sample containing only a Si-doped, 2-μm thick Al$_{0.3}$Ga$_{0.7}$As epilayer, and no QW.[38] In the latter experiments, the sample was cooled in the dark and irradiated with an IR LED to ionize DX centers and induce persistent photoconductivity. The sample was then exposed to intense FIR radiation with photon energy of a few meV. The carrier concentration was reduced with each pulse of FIR radiation, indicating that intense FIR radiation can indeed induce capture of electrons by DX centers. Possible mechanisms are discussed in Ref. 38.

b. Harmonic generation   FIR harmonic generation studies were performed on the half-parabolic well sample and the heterojunction, again mounted in the stripline coupler (see Fig. 3).[39] Fig. 7 shows the second-harmonic power generated by the electrons in the heterojunction sample as a function of incident power. The sample was pumped by the molecular gas laser at $f = 29.5$ cm^{-1}, much lower than the estimated 290 cm^{-1} first intersubband transition. For powers below (above) 1 kW, the second-harmonic power depends quadratically (subquadratically) on

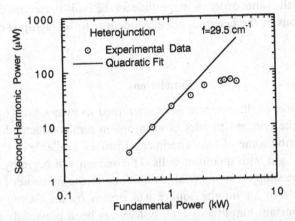

Fig. 7. Second-harmonic power vs. fundamental power generated by the heterojunction sample. The second-harmonic power is well fit by a quadratic for fundamental powers (fundamental electric fields) below 1 kW (10 kV/cm). Due to uncertainties in collection efficiency of the second harmonic, the estiamted uncertainty on the overall scale of the ordinate is ± 0.75. Source and parameters as for Fig. 6, except FIR frequency = 29.5 cm^{-1}. Data from the Ph.D. thesis of W. W. Bewley.

the fundamental power. The rms amplitude of the oscillating electric field felt by the sample at the highest FIR power level was approximately 20 kV/cm. Peak electric field amplitudes were larger, due to the complicated time structure of the molecular gas laser. Significant ionization occurred with each pulse at the highest powers. In order to maintain a constant electron density from pulse to pulse, the well was repopulated between each FIR laser pulse by flashing the IR LED.

Second-harmonic generation is usually treated theoretically in second-order time-dependent perturbation theory. The fundamental field E_ω with frequency ω generates a second-harmonic polarization $P_{2\omega} = \chi^{(2)}(\omega, 2\omega) E_\omega 2$. This second-order polarization is then used as a source for Maxwell's equations, from which second-harmonic power is calculated. From our data below 1 kW, we have calculated a surface nonlinear susceptibility $\chi^{(2)}s = (1.0 \pm 0.75) \times 10^{-8}$. Modeling the heterojunction as a triangular well with slope 39 kV/cm (see Section 2F), and ignoring electron-electron interactions, we calculate $\chi^{(2)}s = 8 \times 10^{-9}$, in agreement with our experimental results. The magnitude of $\chi^{(2)}s$ is enormous compared to bulk materials. If the QW were replaced in the stripline by an equivalent volume of bulk LiTaO$_3$, a highly nonlinear material in the FIR, between 1000 and 10,000 times less power would be generated (at low intensities where $\chi^{(2)}s$ is independent of FIR electric field amplitude).

It is necessary to go beyond perturbation theory to explain the subquadratic dependence of the second-harmonic power on fundamental power for powers larger than 1 kW (or rms fundamental electric fields larger than 10 kV/cm.) The breakdown of perturbation theory is not surprising since the incident electric field amplitudes are of the same order of magnitude as the built-in electric fields which confine the electrons in the heterojunction. Modeling in the high-power limit has not been done.

5 Simulations

The previous section of this chapter have attempted to give a brief but realistic view of some of the relevant physics of electrons in semiconductor heterostructures, and to describe some of the experimental probes available in the study of the interaction of light with quantum wells. This section will begin by discussing some results of the simplest non-trivial simulations of the dynamics of electrons in QWs – a particle in an infinite square will driven by an intense oscillating electric field of constant amplitude. This system has been previously studied by others,[15,16] although not with real experiments in mind. In classical simulations, the dynamics are shown to be chaotic at experimentally-accessible parameters. In quantum simulations, the quasienergy spectrum is computed as a function of electric field amplitude and frequency. For experimentally-accessible parameters, avoided crossings occur in the quasienergies associated with low-lying states. Possible experimental manifestations of these avoided crossings are discussed. Finally, the effects of some of the approximations inherent in our model calculations will be discussed.

5.1 Classical simulations

The Hamiltonian for a driven particle in a one-dimensional infinite square potential is

$$H(p, q, t) = \frac{p^2}{2} + V_0(q) - \kappa \sin(\tau) \tag{4}$$

where

$$V_0(q) = \begin{cases} 0, & |q| < 1 \\ \infty, & |q| > 1 \end{cases}$$

The only free parameter is κ, the unitless driving amplitude. The Hamiltonian (4) is general, noting the substitutions

$$\tau \equiv \omega t, \quad q \equiv \frac{x}{a}, \quad \kappa \equiv \frac{F}{m\omega^2 a^2}$$

Here, a is one half the width of the physical well, F is the magnitude of the driving force, m is the effective mass ($0.065\,m_e$) and ω is the frequency of the drive. The physical position and time are x and t, while q and τ are unitless. The appropriate unitless momentum p and kinetic energy U are

$$p = \frac{P}{m\omega a}, \quad U \equiv \frac{p^2}{2} = \frac{E}{m\omega^2 a^2}$$

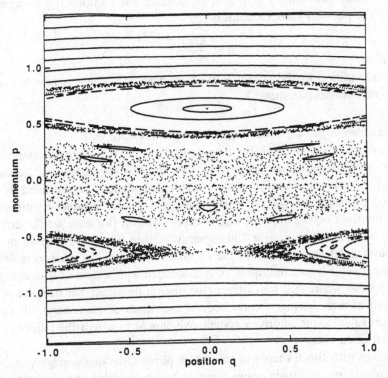

Fig. 8. The classical Poincaré section associated to the Hamiltonian (Eq. (4)). with scaled drive amplitude $\kappa = 0.04$. Trajectories are sampled at $\tau = 0$, mod. 2π. From B. Birnir et al., Phys. Rev. B **47**, 6795 (1993)

where P and E are the physical momentum and kinetic energy.

Fig. 8 shows the Poincaré section obtained by iterating Hamilton's equations for Hamiltonian 1 with $\kappa = 0.04$. For a 500 Å well and a frequency of $40\,\text{cm}^{-1}$, this corresponds to $F/\approx 270\,\text{V/cm}$, easily achievable in our experiments. Trajectories are sampled at $\tau = 0$, mod 2π. At low momenta, the dynamics are completely chaotic as all of the resonances of order higher than 3 have overlapped. The 1:1 resonance is clearly visible with its center near 0.7. Above the 1:1 resonance, the motion is mostly regular, and trajectories shown here are invariant tori that have not been destroyed by the nonlinearity.

5.2 Quantum simulations

The quantum dynamics of the infinite square well are computed using the Schrödinger equation

$$i\hbar\partial_\tau\psi(q,\,\tau) = \left[-\frac{\hbar^2}{2}\frac{\partial^2}{\partial q^2} - q\kappa\sin(\tau)\right]\psi(q,\,\tau) \qquad (5)$$

with the boundary conditions $\psi(|q| \geq 1, \tau) - 0$. Here, $\hbar$ is a unitless free parameter, related to the physical Planck's constant by

$$\hbar = \frac{\hbar_{phys}}{m\omega a^2}$$

Thus, in addition to the purely classical parameter κ, there is now a purely quantum mechanical parameter $\hbar$.

5.2.1 Quasienergies for a periodically-driven infinite square well

The Hamiltonian (4) is explicitly time-dependent, so there are no stationary states or energy eigenvalues. However, since the time-dependence is perfectly periodic, the temporal evolution of state vectors is completely determined by the one-period propagator $U_{2\pi}$. The eigenvalues of $U_{2\pi}$ are "Floquet multipliers," complex numbers on the unit circle $\mu_n = \exp(-2\pi i \epsilon_n / \hbar)$ (we have set the driving frequency $\omega = 1$) where ϵ_n is a "quasienergy." Since the time-evolution of the state vectors is determined only by the Floquet multipliers, the quasienergies are not unique. Each Floquet multiplier μ_n is associated with a class of quasienergies $\epsilon_{n,m} = \epsilon_{n,0} + m\hbar$, where m is any integer (the photon frequency has been set to 1). In the limit of zero driving, one member of each class of quasienergies is equal to the eigenvalue of the undriven system. We choose to assign the label $\epsilon_{n,\,0}$ to this special quasi-energy.

In analogy with Bloch theory in solid-state physics, the quasienergy spectrum can be divided into Brillouin zones, each of which contains one representative of each class of quasienergy. In general, for a periodically-forced particle in a one-dimensional potential, quasienergies will shift as the strength of the driving in increased. In the absence of any symmetries, for a periodically-driven one-dimensional potential, all quasienergies are coupled and may not cross. The Hamiltonian (4) is symmetric under the spatio-temporal parity operation $P(x, t)$ which takes $x- > -x$ and $t- > t + \pi$. Quasienergy states with opposite (the same) spatio-temporal parity may (not) cross.[16]

Fig. 9(a) plots several quasifrequencies (quasienergy/$\hbar$) in the first quasifrequency Brillouin zone as functions of the driving strength κ (and oscillating electric field E). The physical parameters of this simulation are: $\omega/2\pi = 40\,\text{cm}^{-1}$, $2a = 500\,\text{Å}$ and $m = 0.065\,m_e$, corresponding to $\hbar \approx 0.368$. Notable in the quasifrequency spectrum is an avoided crossing between $\epsilon_1, 0$ and $\epsilon_3, 4$ at a value of $\kappa = 0.78$ (14.0 Statvolts/cm). Parameters at which the avoided crossings of quasienergies occur are accessible to the UCSB free-electron laser.

5.2.2 What good are the quasienergies?

The quasienergies and Floquet states for a periodically-driven Hamiltonian are useful for many of the same reasons that the energies and eigenstates of an autonomous Hamiltonian are useful. For an autonomous Hamiltonian, time-dependent perturbation theory predicts that for weak perturbations, energy is

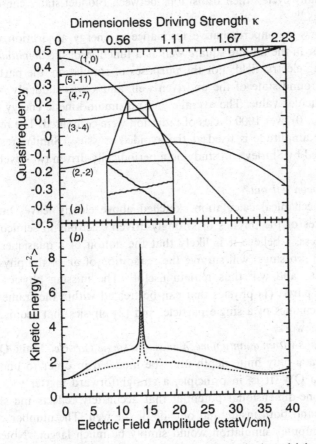

Fig. 9. (a) Quasifrequencies=quasienergy/$\hbar$ vs. dimensionless driving strength κ and electric field amplitude E. Quasifrequencies $(1, 0)$ and $(3, -4)$ show an avoided crossing at $\kappa = 0.78$ ($E = 14.0$ Statvolt/cm). (b) Peak (solid) and average (dotted) kinetic energy over 1000 cycles of drive. A nonperturbative resonance occurs at the avoided crossing. From Galdrikian et al., to be published.

absorbed or emitted when the photon energy is equal to a different between two energy eigenstates. For a time-periodic Hamiltonian, an analogous perturbation theory can be performed[40] which predicts that a weak perturbation (i.e., a weak laser which probes the response of a system driven periodically by a strong laser field) will be absorbed or amplified when the photon energy is equal to the difference between two quasienergies. Quasienergy states also obey an adiabatic principle: in the example given here, if κ is increased slowly, with the system beginning in an eigenstate of the undriven system, then it will always remain in the connected Floquet state. However, if κ is increased rapidly, or if the laser field is on for very many cycles, then transitions between Floquet states can occur at avoided crossings.[41]

Fig. 9(b) shows a nonperturbative resonance in energy absorption which is expected to occur in the infinite square well as a function of the *amplitude* of the intense oscillating electric field. For the purposes of simulation, the particle was initially in the ground state of the undriven well, and κ was suddenly increased from 0 to a particular value. The average and maximum kinetic energy ($< H$ at the instants $\tau \equiv_{2\pi} 0$) over 1000 cycles of radiation were computed. The resonance occurs when the amplitude is fixed at the avoided crossing. Similar resonances were first observed by Shirley[42] in studying a periodically-driven two-level system.

5.2.3 What has been left out?

The quantum mechanical calculation sketched above clearly leaves out many important features of the physics of strongly-driven electrons in semiconductor QWs. Nevertheless, I believe it is likely that the notion of a quasienergy (and hence of avoided crossings) will survive the readdition of all of the physics that has been left out, and will thus remain useful. The missing physics can be divided into two parts: (1) physics that can be treated within the framework of the quantum mechanics of a single particle, and (2) physics that requires going beyond that framework.

a) Considerations within quantum mechanics of a single particle. **Finite QW:** The fact that QWs have only finite depth must be modeled in order to understand ionization from a QW. It is, in principle, a straightforward matter to solve for a particle in a one-dimensional potential that accurately reflects the shape of the conduction band both inside and outside of the well. The number of states included in a computer simulation would simply be much larger. Note that it would not be particularly accurate to a simply replace the infinite square well by the finite square well solved in textbooks ($V(x) = -V_0$ for $|x| < a$, $V(x) = 0$ for $|x| > a$) since this would not accurately reflect band-bending outside the well (see Fig. 1(b)).

Three-dimensional effects, elastic scattering: Arguments parallel to those given for the width of intersubband absorptions (Section 2D.2 can be given for quasienergies. In the absence of disorder, the Schrödinger equation is separable for an electron in a three-dimensional QW driven by a periodic force in the z-direction, and k_x and k_y remain good quantum numbers for motion in the x- and y-directions. In the presence of weak disorder, an electron will maintain definite values of its momentum k_x and k_y and quasienergy for a time which we guess will be comparable to the momentum relaxation time which determines the width of intersubband absorptions. For much longer times, the electron will scatter between nearby quasienergy states with different values of k_x and k_y. If the system is near an avoided crossing of quasienergies which is comparable to the linewidth, one might conjecture that the system will elastically scatter between the nearby quasienergy states. In this case, nonperturbative resonances such as that shown in Fig. 9(*b*) would occur independent of how fast the excitation pulse were turned on, as long as the pulse were much longer than the elastic scattering time. For the UCSB-FEL, the pulse can be as long as 10 μsec, corresponding to several million cycles. This expected insensitivity to the shape of an excitation pulse is in stark contrast to the experiments on Rydberg atoms.[43,44]

b) Beyond the quantum mechanics of a single particle Real QWs contain between a few times 10^9 and 10^{12} electrons/cm^2. For densities between 10^{10} and a few 10^{11} cm^{-2}, the electrons come into thermal equilibrium with themselves on time scales of hundreds to tens of femtoseconds[45] – shorter than one cycle of 30 cm^{-1} FIR radiation (1 psec). Times for thermal equilibration with the lattice are much longer – of order 1 psec for electrons more energetic than a longitudinal optical (LO) phonon, and 100s of psec to 1 nsec for electrons less energetic than an LO phonon.[27] Thus, to model the physics of periodically-driven electrons in QWs electron-electron and electron-phonon interaction must be considered.

In the limits of low carrier density and low electron kinetic energy, calculations ignoring many-body and dissipative effects should serve as a qualitative guide to the real physics. For example, in a 500-Å wide, symmetrically modulation-doped square well with $N_s = 5 \times 10^{10}$ cm^{-2}, the self-consistent wavefunction and transition energy between ground and first-excited states differ by less than 10 and 5%, respectively, from the wavefunction and transition energy for a single electron. One hopes that single-particle quasienergies and Floquet states will be perturbed by a similar magnitude. In the nonperturbative resonance of Fig. 9(*b*), the electron's kinetic energy is approximately that of the third excited state of the undriven well, or 18 meV. This is well below the 36 meV LO phonon energy, and the inelastic scattering time should be hundreds of laser cycles, which is plenty of time to resolve the quasienergy spectrum.

6 Questions for further study

Since only preliminary experiments have been performed, and the theory necessary for their interpretation is still in its infancy, I cannot call the end of this chapter a conclusion. I close instead with a few fundamental questions.

Are the notions of a quasienergy and a Floquet state useful in a many-electron system? (One certainly hopes so.) For example, can one observe dramatic phenomena such as nonperturbative resonances associated with many-electron systems near avoided crossings of many-electron quasienergies? In a QW driven by a first, intense laser beam, will a second, weak laser be absorbed – or, more interestingly, amplified – when its photon energy coincides with a difference between quasienergies? If the notion of a quasienergy is useful, then some statistical mechanics based on quasienergies must be developed. Under steady-state excitation, what distribution governs the occupation probability of different quasienergies? These fundamental questions are not encountered in studies of real atoms in intense oscillating fields. They can only be answered with much further theoretical and experimental effort on the study of solid state atoms in intense oscillating fields.

Acknowledgments

The experimental data presented in this chapter were recorded and analyzed by W. W. Bewley, K. Craig, C. L. Felix, and J. J. Plombon. The samples were grown by M. Sundaram and P. F. Hopkins in A. C. Gossards experimental group. Classical and quantum computer simulations were performed by B. Galdrikian and B. Birnir. Martin Holthaus taught me about Floquet theory and quasienergies. For stimulating discussions, I wish to thank S. J. Allen, K. Ensslin, R. Grauer, E. Gwinn, W. Kohn, H. Metiu, A. Markelz, R. Prange, R. Scharf, B. Sundaram, and A. Wixforth.

This work was supported by NSF-DMR 8901651 (Sherwin, Craig and Galdrikian); the Alfred P. Sloan Foundation (Sherwin); AFOSR 88-0099 (Gossard, Sundaram and Hopkins); SDIO-ONR N00014-K-0110V (Bewley, Felix and Plombon); NSF DMS-9803012 (Birnir); the NSF Science and Technology Center for Quantized Electronic Structures, Grant no. DMR91-20007 (clean room access); and INCOR (Galdrikian).

REFERENCES

[1] By "quantum chaos," I mean the quantum mechanics of systems whose classical dynamics are chaotic

[2] For recent reviews of the theory of quantum chaos, see G. Casati and L. Molinari, Prog. Theor. Phys. **98**, 287 (1989); B. V. Chirikov, "Time-Dependent Quantum Systems," in Chaos and Quantum Physics, Proceedings of Les Houches, Session LII, 1989, M.-j. Giannoni, A. Voros and J. Zinn-Justin, eds., Elsevier Science Publishers, New York (1990)

[3] For recent reviews of experiments on microwave ionization of hydrogen, see P. M. Koch, E. J. Galvez K. A. H. van Leeuwen, B. E. Sauer, and L. Moorman, Physica Scripta T26, 51 (1989); D. Richards, B. E. Sauer, E. J. Glavez, J. G. Leopold, L. Moorman, K. A. H. van Leeuwen, P. M. Koch and R. V. Jensen, J. Phys. B 22, 1307 (1989)

[4] K. A. H. van Leeuwen, G. v. Oppen, S. Renwick, J. B. Bowlin, P. M. Koch R. V. Jensen, O. Rath, D. Richards and J. G. Leopold, Phys. Rev. Lett. **21**, 2231 (1985)

[5] E. J. Galvez, P. M. Koch, B. E. Sauer, L. Moorman and D. Richards, Phys. Rev. Lett. **61**, 2011 (1988)

[6] G. Casati, I. Guarneri and D. L. Shepelyansky, IEEE Journal of Quantum Electronics **24**, 1420 (1988)

[7] R. V. Jensen, M. M. Sanders, M. Saraceno and B. Sundaram, Phys. Rev. Lett. **63**, 2771 (1989)

[8] See, for example, M. Ferray, A. L'Huillier, X. F. Li, L. A. Lompré, G. Mainfray and C. Manus, J. Phys. B **21**, L31 (1988)

[9]   J. H. Eberly, Q. Su and J. Javanainen, Phys. Rev. Lett. **62**, 881 (1989)

[10] P. L. DeVries, J. Opt. Soc. Am. B **7**, 517 (1990), and other articles in this volume

[11] See, for example, A. C. Gossard, IEEE J. Quantum Electron. **QE-22** 1649 (1986)

[12] M. Sundaram, S. A. Chalmers, P. F. Hopkins and A. C. Gossard, Science **254**, 1326 (1991)

[13] B. Galdrikian, B. Birnir, M. S. Sherwin and R. Grauer, Phys. Rev. B **47**, 6795 (1993)

[14] F. Benvenuto et al., Z. Phys. B **84**, 159 (1990)

[15] W. A. Lin and L. Reichl, Phys. Rev. A **37**, (1988)

[16] H. P. Breuer, K. Dietz and M. Holthaus, Z. Phys. D8, 349 (1988)

[17] B. Galdrikian, B. Birnir and M. S. Sherwin, Phys. Lett. A, in Press

[18] C. Weisbuch and B. Vinter, Quantum Semiconductor Structures, Academic Press, Boston (1991)

[19] D. J. Wolford et al., J. Vac. Sci. and Tech. **B4**, 1043 (1986); K. Karrai et al., Phys. Rev. B **42**, 9732 (1990)

[20] There are two kinds of holes, so-called light and heavy holes, and their degeneracy is lifted in a quantum well.

[21] A recent theory (D. J. Chadi and K. J. Chang, Phys. Rev. B **39**, 10063 (1989)) appears to be quite successful in explaining the many baffling properties of the DX center.

[22] For a recent review on DX centers, see P. M. Mooney, J. Appl. Phys. **67**, R1 (1990)

[23] L. C. West and S. J. Eglash, Appl. Phys. Lett. **46**, 1156 (1985)

[24] P. F. Hopkins, A. J. Rimberg, E. G. Gwinn, R. M. Westervelt, M. Sundaram, and A. C. Gossard, Appl. Phys. Lett. **57**, 2823 (1990)

[25] Scattering time $\tau = \mu m * / e$.

[26] J. F. Ryan and M. C. Tatham in Hot Carriers in Semiconductor Nanostructures, edited by J. Shah, Academic Press, Boston (1992) p. 345

[27] D. Y. Oberli, D. R. Wake, M. V. Klein, J. Klem, T. Henderson and H. Morkoc, Phys. Rev. Lett. **59**, 696 (1987). The energy relaxation time due to acoustic phonon emission in wide quantum wells is not well established experimentally. It has been argued that the ≈ 1 nsec relaxation time measured by Oberli et al. was not the relaxation time of a single electron due to acoustic phonon emission, but rather was dominated by many-body effects. See Ref. 26

[28] The linewidth $\Delta v = 1/2\pi\tau_m$

[29] P. Pinsukanjana, E. Yuh, E. Gwinn, M. Sundaram, A. C. Gossard and J. Dobson, Phys. Rev. B **46**, 7284 (1992)

[30] F. Stern and S. das Sarma, Phys. Rev. B **34**, 840 (1984)

[31] L. Brey, N. F. Johnson and B. I. Halperin, Phys. Rev. B **40**, 647 (1989)

[32] K. Karrai, H. D. Drew, H. W. Lee and M. Shayegan, Phys. Rev. B **39**, 1426 (1989); A . Wixforth, M. Sundaram, J. H. English and A. C. Gossard, Proceedings of the XXth ICPS (Thessaloniki, Greece) World Scientific, Singapore (1990)

[33] C. T. Gross, J. Keiss, A. Mayer and F. Keilman, IEEE J. Quantum Electron. QE-23, 377 (1987)

[34] G. L. Miller, IEEE Trans. on Electron. Dev., ed-19, 1103 (1972); M. Sundaram and A. C. Gossard, "Capacitance-Voltage profiling through graded heterojunctions: theory and experiment," submitted

[35] C. L. Felix, Ph.D. thesis

[36] Samples are usually rotated in the MBE chamber as they are grown, in order to even out gradients in the flux from the ovens across the width of the well. The rotation stage was broken when this sample was grown, and unusually large gradients in well thickness were present

[37] M. S. Sherwin et al., in "Free-Electron Laser Spectroscopy in Biology, Medicine and Materials Science," Proceedings of the SPIE, vol. 1854, p. 36, SPIE, Bellingham, Washington (1993)

[38] J. J. Plombon, W. W. Bewley, C. L. Felix, M. S. Sherwin, P. Hopkins, M. Sundaram and S. C. Gossard, Appl. Phys. Lett. **60**, 1972 (1992)

[39] A detailed account of the second-harmonic generation studies can be found in W. W. Bewley, C. L. Felix, J. J. Plombon, M. S. Sherwin, P. F. Hopkins, M. Sundaram and A. C. Gossard, Phys. Rev. B **48**, 2376 (1993)

[40] H. Sambé, Phys. Rev. **A7**, 2203 (1973)

[41] H. P. Breuer and M. Holthaus, Z. Phys. D **11**, 1 (1989)

[42] J. H. Shirley, Phys. Rev. **138**, 979 (1965)

[43] M. C. Baruch and T. F. Gallagher, Phys. Rev. Lett. **68**, 3515 (1992)

[44] B. F. Sauer, M. R. W. Bellermann and P. M. Koch, Phys. Rev. Lett. 1633 (1992)

[45] W. H. Knox, in Hot Carriers in Semiconductor Nanostructures, edited by J. Shah, Academic Press, Boston (1992), p. 313

PART TWO

Atoms in strong fields

Localization of Classically Chaotic Diffusion for Hydrogen Atoms in Microwave Fields

J. E. Bayfield,[1] G. Casati,[2] I. Guarneri,[3] and D. W. Sokol[1]

[1]*Department of Physics and Astronomy, University of Pittsburgh, Pittsburgh, Pennsylvania 15260*
[2]*Dipartimento di Fisica, Universita di Milano, 20133 Milano, Italy*
[3]*Departimento di Fisica Nucleare e Teorica, Universita di Pavia, 27100 Pavia, Italy*

(Received 17 November 1988)

New experimental results are presented for short-pulse microwave ionization of highly excited hydrogen atoms. A comparison of these results with quantum numerical computations and analytical predictions provides for the first time experimentally grounded evidence of the localization phenomenon that leads to the suppression of the quantum version of the chaotic diffusion in action space occurring in the classical limit.

The onset of chaotic motion in externally driven classical systems is a key problem in nonlinear dynamics that is now well understood in its essential aspects. In many physically interesting cases, as the perturbation strength increases beyond some critical value, the system starts absorbing energy in a diffusive way.[1] The question whether "diffusive" excitation processes can take place also in externally driven quantum systems is then a very interesting one for the physics of atoms and molecules in external electromagnetic fields.[2] This is a deep question involving the nature and the validity of the quasiclassical approximation when the underlying classical dynamics is chaotic.

The question of "quantum diffusion" was first addressed in a simple model system, the kicked rotator.[3] It was chosen because the features of classical chaos in it were relatively clean and well understood. Besides that, the numerical simulation of its quantum dynamics could be easily accomplished.[3,4] The major indication of this model[3,4] was that quantum mechanics suppresses the classical chaotic diffusion in action via a destructive interference effect similar to that responsible for the Anderson localization well known in condensed matter physics.[5] However, unlike Anderson localization, this new type of localization does not require the introduction of random elements from the outside; to stress this essential difference, it was called "dynamic localization." It was then submitted that dynamic localization should not be considered an artifact of the peculiar kicked rotator model, but a general effect that would eventually stop any diffusive quantum excitation, unless the related growth in time of the number of populated excited states is fast enough.[6,7]

The microwave ionization of the hydrogen atom is an ideal testing ground for this localization theory. Since the first experiments,[8] increasingly accurate numerical simulations of a classical model for electron motion under the combined influence of a Coulomb field plus a monochromatic oscillating electric field have shown that the experimentally observed thresholds for the onset of ionization follow the classical thresholds for the onset of chaotic instability, when the ratio ω_0 of the microwave frequency ω to the unperturbed Kepler frequency is less than 1.[9] Most theoretical analyses up to recent times were carried out on a simplified one-dimensional model, which is adequate for experimental situations in which the atom is initially prepared in a state very extended along the direction of the field.[10] Numerical simulations of the quantum dynamics of this model in the region $\omega_0 < 1$ confirmed the agreement between classical and quantum thresholds.[7,11] On the other hand, in the region $\omega_0 > 1$, localization theory applied to the one-dimensional model predicted that, due to the localization phenomenon, the quantum threshold for ionization should rise above the classical one, following a "quantum delocalization" border.[6,7,12] These predictions were supported by extensive numerical simulations of the one-dimensional quantum dynamics.[6,7,12,13]

A formulation of the classical dynamics at $\omega_0 \gtrsim 1$ by means of an appropriate map, the so-called "Kepler map,"[12] made it possible to find a connection between the one-dimensional hydrogen atom model and the kicked rotator model, and thus provided even firmer grounds for the application of localization theory. Moreover, by constructing a similar Kepler map for a two-dimensional model it was shown that the classical excitation in this model develops along very similar lines as in the one-dimensional model, due to the existence of an approximate integral of the motion which practically decouples the two degrees of freedom. This circumstance enforced the prediction that a localization theory very similar to the one-dimensional one should apply in the two-dimensional case also. Therefore, this localization theory predicts that a localization phenomenon should be experimentally observable even for atoms initially prepared in nonstrictly one-dimensional states and with magnetic quantum number (with respect to the direction of the field) $m = 0$,[12] as is the case for the experiments described in this paper.

Presented here are new experimental results and the comparison of these results with numerical computations and analytical predictions. These reveal for the first time the localization phenomenon through an average nearly linear increase of the ionization threshold with increasing ω_0, with the rate of increase having the predicted value.

The use of fast atom beams for the study of microwave ionization of highly excited hydrogen atoms is a well established technique.[8,14-17] First, a nearly monoenergetic proton beam with particle kinetic energy typically near 23 keV is converted to a mixed-state neutral hydrogen atom beam by means of charge-transfer collisions. State-selective laser transitions in static electric fields are then used to select atoms in the beam with parabolic quantum numbers $n, n_1, m = 7, 0, 0$ and excite them to a state with numbers $n_0, 0, 0$.

In the present experiments the static field strength was reduced to $F_S = 0.87$ V/cm in the microwave field region, which would static-field ionize $n, 0, 0$ atoms only for n above 164. This field reduction was observed to produce a scrambling of hydrogen atom substates. The scrambling was due to electric field components before and after the microwave region that were not along the nominal electric field direction, vertical in the laboratory. As the apparatus had reflection symmetry through the vertical plane containing the atom beam axis, no scrambling in m should have occurred. A field-ionization scan of the substate-scrambled beam was precisely fitted with an almost uniform spread in n_1 between 0 and $\frac{3}{4} n_0$, taking $m = 0$. This "two-dimensional" situation is to be compared with that of "three-dimensional" experiments that utilize almost statistically weighted scrambling in both n_1 and m.[15,17]

As the laser-excited atoms passed through holes in the sidewalls of a TE_{10}-mode microwave interaction waveguide (WR-62), they were exposed to a microwave pulse with a nominal envelope $\sin[\pi t/(7.5 \text{ nsec})]$. The peak microwave field ranged from 2.5 to 3.8 V/cm at $n_0 = 98$ to 13 to 21 V/cm at $n_0 = 64$, depending upon ω, which was varied over the band 12.4 to 18 GHz. Machining the holes reduced the measured waveguide power transmission by only 4%. Hence leakage of microwaves out of the holes produced an additional long tail on the envelope of about 1% the peak field strength. Numerical calculations indicated that the added presence of this tail did not significantly change the predicted ionization probabilities.

The atomic beam leaving the microwave region was in a microwave-induced distribution of final atom quantum states that included the continuum. The beam then passed through a region of static field F_L which was adjustable and would ionize $n, 0, 0$ atoms with n above a corresponding adjustable value $\bar{n}$. It then entered a microwave ionization cavity field for converting all remaining excited atoms with n above 58 into protons again. The apparatus was designed to detect just these cavity-field-produced protons, and not protons produced, for instance, in the waveguide region. Therefore the ionization probability was defined experimentally as the reduction in cavity-field proton production rate induced by the

waveguide microwaves, and hence was the sum of waveguide-induced probabilities for transitions to final bound states above $\bar{n}$ and for final continuum states. The initial-state quantum number n_0 was varied between 64 and 98. The value of $\bar{n}$ was increased with n_0 so as to keep their ratio roughly equal to 1.5.

A search was carried out for any alteration of ionization probabilities arising from residual in-band microwave noise, with negative results. A tunable narrow-band filter for $\omega = 16-18$ GHz reduced the rms noise field strength from 4% to 0.06% and possibly increased sine-wave field strength values for 10% ionization probability by only (3 ± 3)%. Another check of the apparatus found that preionization of the excited atoms in a static field before the waveguide region reduced the proton detector signal down to the same background level to within 2% for the waveguide microwaves either on or off. In addition, the observation that $F_S = 0.87$ V/cm gave microwave ionization probabilities similar to those at still lower values of F_S was taken to mean that the static electric field in the wave guide was not playing a major role in the results now to be reported.

In Fig. 1 we present thresholds for 10% ionization obtained experimentally and by quantum numerical simulations. These thresholds are defined as the field strengths at which a 10% probability was observed above the cutoff values $\bar{n}$. The error bars shown on a few of the experimental data points are the final rms uncertainties for averages over several runs taken on different days. A further ± 10% uncertainty in the overall microwave field

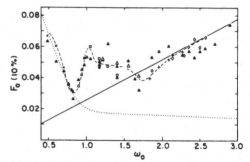

FIG. 1. A comparison at identical parameter values of experimental and quantum-mechanical values for the microwave field strength for 10% ionization probability, as a function of microwave frequency. The field and frequency are classically scaled, $\omega_0 = n_0^3 \omega$ and $F_0 = n_0^4 F$. Ionization includes excitation to states with n above $\bar{n}$. The theoretical points are shown as solid triangles. The dashed curve is one drawn through the entire experimental data set shown in Fig. 2. Values of $n_0, \bar{n}$ are ●, 64, 114; ×, 68, 114; ○, 71, 114; ■, 76, 114; ◻, 80, 120; △, 86, 130; +, 94, 130; ◇, 98, 130. Multiple theoretical values at the same ω_0 are for different compensating experimental choices of n_0 and ω. The dotted curve is the classical chaos border. The solid line is the quantum 10% threshold according to localization theory for the present experimental conditions.

strength is not included in the error bars. The numerical calculations were made in one dimension, but otherwise closely simulated the important experimental conditions and choices of parameters, including F_S and the pulse envelope. Two theoretical borders are also plotted in the figure. One of them (dotted curve) is the chaos border. (In the region $\omega_0 < 1$, the chaos border is extrapolated smoothly from the region $\omega_0 \approx 1$ to lower values of ω_0.) The other border (solid line) is the quantum theoretical prediction obtained for the present experimental situation by following the analytical procedure described in Ref. 13. This border is the field strength for a 10% probability flow into levels above the cutoff value, before the setting up of the final localized distribution. This quantum border is obtained by inserting the analytical expression for the localization length into the quantum equation for the exponential final-state distribution, integrating the latter above the cutoff value, and requiring the result to be a 10% probability. The result for the present experimental conditions is $F_0(10\%) = (0.23\omega^{1/6})\omega_0$, which exhibits only a weak $\omega^{1/6}$ deviation from classical scaling that amounts to about 8% over the present experimental range in ω.

The overall agreement between experimental and quantum numerical calculations seen in Fig. 1 is quite good, the overall rms deviation between these being 12%. For $\omega_0 > 1.5$ both sets of data rise with increasing ω_0, independent of particular values chosen for n_0 and ω. This almost linear rise is seen to agree in slope and location with the prediction of the analytical theory for localization. It also exhibits the predicted near-classical scaling. To be emphasized is that the analytical theory does not describe either quantum or classical resonance effects, as it is based upon only the smoothed behavior of stationary state distributions in quantum number or classical action. Also one should not expect the weak signs of narrow structure seen in the quantum numerical results to coincide with experiment, as in a static field; two-dimensional atom distributions have a different and more smooth averaged quantum energy-level structure than do one-dimensional atoms. Yet the high-frequency agreement between experiment, quantum calculations, and the analytical theory of localization is quite remarkable.

The experimental and quantum numerical results in the low-frequency region $\omega_0 \leq 0.8$ are consistent with theory, which predicts no localization for these frequencies. This region and the intermediate frequency region $0.8 \leq \omega_0 \leq 1.5$ are best discussed by including a comparison with classical numerical computations; see Fig. 2. For $\omega_0 \leq 0.8$, classical and quantum numerical values closely agree, and on average are 14% and 8% below experiment, respectively. Since the experimental uncertainty in microwave field strength is $\pm 10\%$, these deviations are not very significant. In the intermediate frequency region, both quantum numerical and experimental values are usually larger than classical ones, signs of a tendency for localization. The broad classical struc-

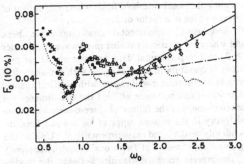

FIG. 2. Experimental data as in Fig. 1, compared with the results of one-dimensional model classical calculations (dotted line) and with one-dimensional model predictions for quantum suppression by cantori in phase space (dash-dotted curve). The solid line of Fig. 1 is also included. The dotted line was created by drawing through every point obtained for the parameter values used for the solid triangles of Fig. 1.

tures near $\omega_0 = 1$ and 2 are resonances, with only the former apparently playing a large role under present experimental conditions. For $\omega_0 > 0.8$, the classical results are well above the classical chaos border primarily because the microwave pulse time was less than the time for complete classical ionization $T_I \sim \omega_0^{7/3}/2F_0$.[7]

Also included in Fig. 2 is a smooth curve for the ionization threshold obtained from a model for quantum suppression arising from cantori in phase space just above the threshold for chaos.[18] This curve wave obtained from an analytical expression that was evaluated for the present experimental values of the parameters, with the choice of the typical microwave frequency 15 GHz. Although this curve very roughly follows the trend of the data between $\omega_0 = 1$ and 2, at higher frequencies it clearly departs from our experimental curve. Above $\omega_0 = 2$ the experimental data rise with ω_0 2.5 times faster than that predicted by the cantori model. In addition, numerical quantum calculations up to $\omega_0 = 4$ also show the more rapid rise.[13]

The present quantum calculations were based upon the one-dimensional model and employed numerical techniques previously described.[7,13] The procedure for dealing with the $n = 165$ static-field ionization threshold for the field in the waveguide region was as follows. As the microwave pulse rose towards its peak value, there was a time delay for a probability greater than 0.1% to arrive at the level $n = 165$, which typically was about 40 field oscillations. After a further time ΔT about equal to the Kepler period at $n = 165$, all population above $n = 164$ was removed from the system of coupled-state equations. Before doing this, it was checked that the removed population had not reached up to the $n = 365$ top of the hydrogenic basis set. This population removal procedure was contained for the rest of the microwave pulse. Calculations were carried out for various values of ΔT, with the finding that final threshold fields were independent of ΔT over a range of a factor of 2.

In conclusion, experimental conditions have been found where for classically scaled microwave frequencies $\omega_0 \gtrsim 2$, both experimental data and the numerical predictions of quantum mechanics agree with the analytical theory of localization and disagree with classical theory. The classical computations utilized the one-dimensional approximation, but the failure of three-dimensional classical theory in the present range of ω_0 has just been independently established experimentally.[19] Unlike the long-pulse experiments at twice our unscaled frequency ω,[19] the present results distinguish between the predictions of the localization and cantori models. A comparison of the two experimental results suggests that the cantori model's dependence of ω is not the correct one.

Two of us (J.E.B. and D.W.S.) thank the U.S. National Science Foundation for continued support of their experiments. S.Y. Luie and N. Adhikari participated in the taking of some of the experimental data.

---

[1] A. J. Lichtenberg and M. A. Liberman, *Regular and Stochastic Motion* (Springer-Verlag, New York, 1983).

[2] J. E. Bayfield, Comments At. Mol. Phys. **20**, 245 (1987).

[3] G. Casati, B. V. Chirikov, J. Ford, and F. M. Izrailev, in *Stochastic Behavior in Classical and Quantum Hamiltonian Systems,* edited by G. Casati and J. Ford (Springer-Verlag, Berlin, 1979).

[4] D. L. Shepelyansky, Physica (Amsterdam) **23D**, 103 (1987).

[5] S. Fishman, D. R. Grempel, and R. E. Prange, Phys. Rev. Lett. **49**, 49 (1982).

[6] G. Casati, B. V. Chirikov, and D. L. Shepelyansky, Phys. Rev. Lett. **53**, 2525 (1984).

[7] G. Casati, B. V. Chirikov, D. L. Shepelyansky, and I. Guarneri, Phys. Rep. **154**, 77–123 (1987), and references therein.

[8] J. E. Bayfield and P. M. Koch, Phys. Rev. Lett. **33**, 258 (1974).

[9] M. M. Sanders, R. V. Jensen, P. M. Koch, and K. A. H. van Leeuwen, Nucl. Phys. B (Proc. Suppl.) **2**, 578 (1987).

[10] J. E. Bayfield, in *Quantum Measurement and Chaos,* edited by E. R. Pike and Sarben Sarkar (Plenum, New York, 1987), pp. 1–33.

[11] R. Blumel and U. Smilansky, Phys. Rev. Lett. **58**, 2531 (1987).

[12] G. Casati, I. Guarneri, and D. L. Shepelyansky, IEEE J. Quantum Electron. **24**, 1420 (1988), and references therein.

[13] G. Brivio, G. Casati, I. Guarneri, and L. Perotti, Physica (Amsterdam) **33D**, 51 (1988).

[14] P. M. Koch, J. Phys. (Paris), Colloq. **43**, C2-187 (1982).

[15] K. A. H. van Leeuwen et al., Phys. Rev. Lett. **55**, 2231 (1985).

[16] J. E. Bayfield and L. A. Pinnaduwage, J. Phys. B **18**, L49 (1985).

[17] P. M. Koch et al., in *The Physics of Phase Space,* edited by Y. S. Kim and W. W. Zachary (Springer-Verlag, Berlin, 1987).

[18] R. S. MacKay and J. D. Meiss, Phys. Rev. A **37**, 4702 (1988); J. D. Meiss, Phys. Rev. Lett. **62**, 1576 (1989).

[19] E. J. Galvez, B. E. Sauer, L. Moorman, P. M. Koch, and D. Richards, Phys. Rev. Lett. **61**, 2011 (1988).

Inhibition of Quantum Transport Due to "Scars" of Unstable Periodic Orbits

R. V. Jensen and M. M. Sanders

Department of Applied Physics, Yale University, New Haven, Connecticut 06520

M. Saraceno

Departamento de Fisica, Comisión Nacional de Energia Atomica, 1429 Buenos Aires, Argentina

B. Sundaram

Los Alamos National Laboratory, MS J-569, Los Alamos, New Mexico 87545

(Received 18 August 1989)

A new quantum mechanism for the suppression of chaotic ionization of highly excited hydrogen atoms explains the appearance of anomalously stable states in the microwave ionization experiments of Koch *et al.* A novel phase-space representation of the perturbed wave functions reveals that the inhibition of quantum transport is due to the selective excitation of wave functions that are highly localized near unstable periodic orbits in the chaotic classical phase space. These "scarred" wave functions provide a new basis for the quantum description of a variety of classically chaotic systems.

Advances in the study of the chaotic behavior of strongly coupled and strongly perturbed nonlinear oscillators has lead many researchers to the question of how quantum mechanics modifies the chaotic classical dynamics in corresponding quantum systems like atoms or molecules in strong fields.[1] Until Heller's detailed studies of the stadium billiard,[2] highly excited states of chaotic systems were expected to resemble random functions in configuration space.[3] However, Heller found that typical eigenfunctions were often strongly peaked along the paths of unstable classical periodic orbits (PO).[3] In the past year these so-called "scars" of the PO have also been identified in several other systems that are classically chaotic,[4] where the peaking of the quantum probability near PO was found to be even more pronounced in the phase-space (PS) representation of the wave functions. These quantum PS distributions also revealed that the stable and unstable manifolds associated with the PO also play an important role in determining the regular structure of the eigenfunctions. Most recently, scars have also been found for highly excited hydrogen atoms in strong magnetic fields[5] where the scarred wave functions have a clear experimental signature in the regular modulations of the photoabsorption spectrum.[6]

The purpose of this Letter is to show that the wave functions of highly excited hydrogen atoms in strong microwave fields also exhibit surprisingly regular structure that is closely correlated with unstable PO and their associated stable and unstable manifolds. A novel PS representation of the wave functions, which is closely related to the Husimi or coarse-grained Wigner distribution,[7] is introduced, which clearly and simply reveals this underlying structure. Moreover, we show that these scarred wave functions are responsible for some of the large fluctuations in the threshold fields for the onset of ionization in the experimental measurements[8,9] shown in Fig. 1.

The classical description of the electron dynamics in the combined Coulomb and microwave field exhibits a sharp transition from regular behavior to chaos when the perturbing field exceeds a critical threshold which is typically 10%–20% of the binding field.[10] For low scaled frequencies, $n_0^3 \Omega < 1$, Fig. 1 shows that the predictions of the onset of chaotic ionization for a one-dimensional model of the experiment, described by the Hamiltonian (in a.u.)

$$H(x,p,t) = p^2/2 - 1/x + xF(t)\cos\Omega t \quad (x > 0) \quad (1)$$

[where n_0 is the princpal quantum number of the initial state, Ω is the microwave frequency, and $F(t)$ is the field

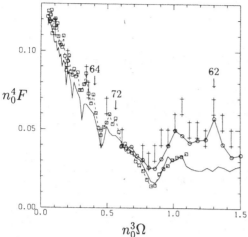

FIG. 1. The experimental measurements (Refs. 8 and 9) of the scaled threshold fields, $n_0^4 F$, for 10% ionization at 9.9 GHz with $n_0 = 32$ to 90 (squares) and 36 GHz with $n_0 = 45$ to 80 (circles) are compared with classical (Ref. 10) (solid curve) and quantum (Ref. 11) (crosses) predictions for a one-dimensional model of the experiment.

strength which slowly turns on and off as the atoms enter and exit from the cavity], are in remarkably good agreement with the experiment.[8] However, for $n_0^3 \Omega > 1$ the thresholds for the onset of ionization in the experiments[9] and the quantum calculations[11] gradually rise above the classical thresholds in rough agreement with the "localization" theory of Casati, Guarneri, and Shepelyansky.[12] In addition, the experimental measurements and quantum calculations exhibit large peaks near resonant frequencies $n_0^3 \Omega = \frac{4}{3}$, $\frac{3}{2}$, 2, $\frac{5}{2}$, and $\frac{8}{3}$.[9,11] At low scaled frequencies the prominent peaks at $n_0^3 \Omega = 1$, $\frac{1}{2}$, and $\frac{1}{3}$ in the classical and experimental ionization thresholds were previously shown[13] to be associated with regular, resonance island structures that persist in the classical PS and stabilize both the classical and the quantum dynamics. However, in Fig. 1 the striking peak near $n_0^3 \Omega = \frac{4}{3}$ in the experiment and the quantum calculations does not appear in the classical simulations.

Our quantum simulations[11] of the experiments use a one-dimensional model of the hydrogen atom in an oscillating electric field that provides a good description of the threshold fields for significant excitation above the cutoff $n_c \sim 95$ which is measured as ionization.[8,9] Insight into the relative stability of $n_0 = 62$ in the 36-GHz field ($n_0^3 \Omega = 1.3$) is provided by a detailed examination of the structure of the quasienergy states[14] (QES) (eigenstates of the one-period time evolution operator at full perturbation strength) that are excited at the end of the slow switch on of the microwave field. If the perturbation were switched on suddenly, the perturbed wave function would be determined by the overlap of the initial state

with each of the QES. For the parameters of the experiments, ten or more of these QES would be excited at a level of 1% or more. With the slow turn on this number is reduced. For example, five different QES are excited for $n_0 = 61$, with probabilities of 47.6%, 40.1%, 4.9%, 2.6%, and 1.1%. However, for $n_0 = 62$ only a single QES is excited with a probability of 97.8%.

If we examine the PS representation of these QES using the Husimi distributions described below, we find that many states exhibit regular structure that is reminiscent of the "scars" of unstable PO found in other chaotic systems. For example, Fig. 2 shows the Husimi distributions in action-angle space for three of the QES that are excited when the microwave perturbation is slowly applied to $n_0 = 61$. The first state in Fig. 2(a), which is excited with 40.1% probability, is strongly peaked near the unstable PO in the classical PS at $\theta = 0$ (mod 1) that is associated with the large resonance island centered at $n^3 \Omega = 1.0$. In addition, this QES exhibits distinct ridges along the classical stable and unstable manifolds which are embedded in the chaotic region surrounding the stable remnants of the nonlinear resonance. Figure 2(b) shows another state, excited with 47.6% probability, that is largely peaked on an annulus interior to the primary stable and unstable manifolds. [Several other states are excited with probabilities less than 1% (not shown) that are concentrated on annuli interior to the regular islands; however, these states are not scars since they are associated with the quantization of tori[3] surrounding the stable PO.] Because all of these QES are highly localized to the vicinity of the classical PS structure, they do not contribute to the ionization of $n_0 = 61$. Finally, Fig. 2(c) shows an example of a QES, which is excited with 2.6% probability, that is highly delocalized. Since this QES and those excited with 4.9% and 1.1% probability extend to highly excited states with $n > n_c$, the excitation of these states contributes significantly to the ionization of $n_0 = 61$.

In contrast, Fig. 3 displays the Husimi distribution for the single QES that is excited when the microwave perturbation is slowly applied to $n_0 = 62$. This QES is highly localized to the separatrix region near the stable and unstable manifolds associated with the primary classical resonance at $n^3 \Omega = 1.0$. The remarkable feature of this wave function is that it remains localized in a region of PS where the classical dynamics is unstable and chaotic. Because this scarred QES does not extend to states with $n > n_c$ and because no other delocalized states are excited with probability greater than 1%, a larger microwave field is required to ionize $n_0 = 62$ than 61.

The recognition that the quantum dynamics is dominated by the excitation of highly structured wave functions may provide an explanation for the remarkable experimental observation,[9] that the ionization threshold at $n_0^3 \Omega = 1.3$ and for several other relatively stable states does not appear to depend on the value n_c as it is varied from ≈ 95 to ≈ 160. If an "exponentially localized"

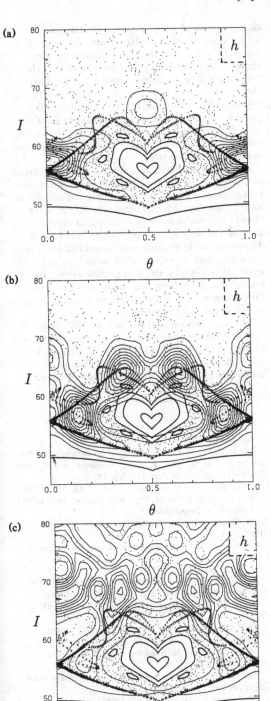

wave function[12] gradually extended over more and more excited states as the perturbing field is increased, then the threshold for 10% ionization would be a strong function of n_c. However, in the case of $n_0 = 62$ a highly localized QES is excited that does not spread significantly with increasing field. In this case ionization only occurs when the field is strong enough that delocalized states are also excited to an appreciable level. Consequently, the excitation of highly excited states above either value of n_c sets in abruptly.

To produce these figures we introduced a very simple and effective method for representing the wave functions in the classical action-angle space that should be useful in many other problems.[15] Since the dominant features of the classical dynamics are usually most apparent in the space of the unperturbed action-angle variables, it is desirable to have representation of the wave function in action-angle space. Unfortunately, since quantum mechanics is not invariant under canonical transformations,[16] the Wigner or Husimi distributions, which are usually defined in position-momentum space,[7] cannot be easily transformed from one space to the other.

These difficulties are circumvented using the following prescription. If the wave functions of the perturbed system, $\psi(x,t) = \sum_{n=0}^{\infty} a_n(t)\phi_n(x)$, are expanded in terms of the unperturbed eigenfunctions, $\phi_n(x) = \langle x \mid n \rangle$ with quantum numbers n that correspond to the unperturbed classical action variables, then a Husimi distribution of $\psi(x,t)$ in action-angle space can be generated by choosing appropriate functional forms for the distributions for the unperturbed eigenstates. In particular, the coherent states that define the Husimi distributions for the one-dimensional hydrogen atom may be chosen to be Poisson wave packets

$$|I,\theta\rangle = \sum_{n=0}^{\infty} [A_n(\alpha)I^{an}e^{-aI}]^{1/2}e^{i2\pi n\theta}|n\rangle, \qquad (2)$$

where α is a "squeezing" parameter that determines the relative width of the wave packet in θ vs I consistent with the uncertainty principle[17] and $A_n(\alpha) = a^{(an+1)}/2\pi\Gamma(an+1)$ is the normalization factor. Then the Husimi distribution of a single unperturbed eigenstate in action-angle space is

$$\rho_n(I,\theta) = |\langle n \mid I,\theta\rangle|^2 = A_n(\alpha)I^{an}e^{-aI},$$

which is peaked at an action $I = n$ with a width $\Delta I = [(an+1)/a^2]^{1/2}$ and is uniformly distributed in θ. With this definition, the Husimi distributions in action-

FIG. 2. The level contours (solid curves) of the Husimi distributions in action-angle (I,θ) space are superimposed on a Poincaré section (Refs. 10 and 13) (dots) of the classical dynamics for three of the QES that are excited for $n_0 = 61$ with $n_0^3\Omega = 1.24$ and $n_0^4F = 0.05$. The lines of arrows show the primary stable and unstable manifolds and their homoclinic oscillations. The relative size of Planck's constant, h, is indicated by the boxed area.

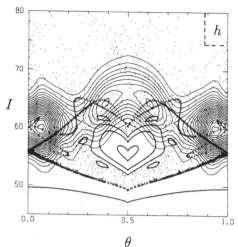

FIG. 3. Same as Fig. 2 for the single QES that is excited for $n_0 = 62$ with $n_0^3\Omega = 1.3$ and $n_0^4F = 0.05$.

angle space for the QES displayed in Figs. 2 and 3 are given by

$$\rho(I,\theta) = \left| \sum_{n=0}^{\infty} a_n \langle I, \theta | n \rangle \right|^2, \qquad (3)$$

where the coefficients a_n are supplied by the numerical solutions.[18]

Our analysis of these numerical "experiments" indicates that the stable peak in the ionization thresholds measured by Koch and co-workers[9] near $n_0^3\Omega = \frac{4}{3}$ is due to the excitation of a single QES that corresponds to a localized scar in phase space. To assess the broader significance of our results, we first studied the dependence of the ionization probability of $n_0 = 62$ to small changes in the microwave frequency and found that at a fixed field the excitation above $n_c = 95$ exhibits a broad minimum centered at $n_0^3\Omega = 1.3050$ that appears to extend smoothly from $n_0^3\Omega \simeq 1.28$ to 1.32. We also examined the dependence of the excitation of the single scarred QES to the shape and length of the microwave pulse and found little change when the turn-on, turn-off, and total duration was varied by a factor of 2. So the appearance of a relatively stable state in both the experiments and the quantum calculations near $n_0^3\Omega = 1.3$ is no accident.[19]

Secondly, we found that the sharp peaks in the ionization thresholds for $n_0 = 64$ and $n_0 = 72$ in the 9.9-GHz field,[8] that could not be explained by the classical theory, also appear to be caused by the excitation of single QES that are highly localized in the chaotic classical phase space. In fact, the Husimi distributions of many different QES for $n_0^3\Omega = 0.4$ to 2.8 and $n_0^4F = 0.05$ to 0.07 appeared to exhibit regular structure that was often

closely associated with PO and their associated manifolds. Coupled with similar results for other time-dependent and time-independent quantum systems,[4,5] it appears that scarred wave functions should be a ubiquitous feature of most quantum systems that are classically chaotic in atomic, molecular, solid-state, and nuclear physics and that these scars should have interesting, observable effects on the quantum dynamics as has already been demonstrated for hydrogen atoms in strong magnetic fields[5,6] and now for highly excited hydrogen atoms in intense microwave fields.

Finally, we note that, although a number of different theories have recently been proposed based on the wave-packed dynamics in the vicinity of unstable periodic orbits[3,20] and on the representation of the quantum Green's function in terms of semiclassical sums over the classical PO introduced by Gutzwiller,[21] the detailed understanding of the physical mechanisms that cause some wave functions to exhibit scars is still lacking. The empirical results presented here provide another illustration of this phenomena that should serve to stimulate further experiments and classical and quantum analyses of this simple, physical system.

This work was supported in part by NSF Grant No. PHY82-17853, supplemented by funds from NASA, and by NSF Grant No. PHY83-51418. In addition R.V.J. and M.S. are very grateful to the Institute for Theoretical Physics at Santa Barbara where this work began.

[1] B. Eckhardt, Phys. Rep. **163**, 205 (1988).

[2] E. J. Heller, Phys. Rev. Lett. **53**, 1515 (1984).

[3] M. V. Berry, in *Chaotic Behavior of Deterministic Systems*, edited by G. Iooss, R. H. G. Helleman, and R. Stora (North-Holland, Amsterdam, 1983), p. 172.

[4] R. L. Waterland et al., Phys. Rev. Lett. **61**, 2733 (1988); G. Radons and R. E. Prange, Phys. Rev. Lett. **61**, 1691 (1988); M. Saraceno (to be published).

[5] D. Wintgen and A. Hönig, Phys. Rev. Lett. **63**, 1467 (1989).

[6] D. Wintgen and H. Friedrich, in *Atomic Spectra and Collisions in External Fields 2*, edited by K. T. Taylor et al. (Plenum, New York, 1988).

[7] S.-J. Chang and K.-J. Shi, Phys. Rev. A **34**, 7 (1986); M. J. Stevens and B. Sundaram, Phys. Rev. A **39**, 2862 (1989).

[8] K. A. H. van Leeuwen et al., Phys. Rev. Lett. **55**, 2231 (1985).

[9] E. J. Galvez et al., Phys. Rev. Lett. **61**, 2011 (1988).

[10] R. V. Jensen, Phys. Rev. A **30**, 386 (1984).

[11] R. V. Jensen, S. M. Susskind, and M. M. Sanders, Phys. Rev. Lett. **62**, 1476 (1989).

[12] G. Casati, I. Guarneri, and D. L. Shepelyansky, IEEE J. Quantum Electron. **24**, 1240 (1988).

[13] R. V. Jensen, Phys. Scr. **35**, 668 (1987).

[14] J. N. Bardsley et al., Phys. Rev. Lett. **56**, 1007 (1986); R. Blümel and U. Smilansky, Z. Phys. D **6**, 83 (1987).

[15] P. Leboeuf and M. Saraceno (to be published).

[16] P. Carruthers and M. M. Nieto, Rev. Mod. Phys. **40**, 411

(1968).

[17]The phase-space distributions for the QES are not very sensitive to the choice of α as long as $1 \ll \alpha \ll n_0$. Figures 2 and 3 were generated with $\alpha = 10$ or 20.

[18]Since the equation of evolution for $\rho(\theta, I, t)$ reduces to the classical Liouville equation with corrections of order $\hbar$, this quantum distribution function is properly identified with the classical phase-space distribution in the limit $\hbar \rightarrow 0$.

[19]Although the quantum simulations of the experiment of J. E. Bayfield *et al.* [Phys. Rev. Lett. **63**, 364 (1989)] also exhibit a stable peak near $n_0^3 \Omega = 1.3$, this structure appears to be absent from the experimental measurements, possibly due to the presence of a small static field or noise in the waveguide.

[20]B. Eckhardt *et al.*, Phys. Rev. A **39**, 3776 (1989).

[21]M. C. Gutzwiller, J. Math. Phys. **12**, 343 (1971); E. B. Bogomolny, Physica (Amsterdam) **31D**, 169 (1988); M. V. Berry, Proc. R. Soc. (London) A **423**, 219 (1989); A. Ozorio de Almeida (to be published); M. Robnik (to be published).

Rubidium Rydberg atoms in strong fields

O. BENSON, G. RAITHEL and H. WALTHER

Sektion Physik der Universität München
and Max-Planck-Institut für Quantenoptik,
D-8046 Garching, Germany

1 Introduction

In this chapter experiments on rubidium Rydberg atoms in external fields are reviewed [1−8]. In the first part results on the interaction of microwave radiation with rubidium Rydberg atoms are described. The Rydberg atoms interact with a microwave pulse of a well-defined duration. When applying coherent microwave radiation with a frequency exceeding a certain value, the experimentally observed microwave field strength at which 10% of the atoms are ionized has to be much larger than the value found by classical trajectory calculations. The atoms are stabilized by the dynamical localization. An extremely weak microwave broadband noise which is added to the coherent wave at least partially destroys the localization. This phenomenon depends on the bandwidth of the added noise. In further experiments the dependence of the 10% ionization microwave field strength on the duration τ of the microwave pulse has been investigated. For increasing noise level the 10% ionization field strength $\epsilon_{0.1}$ develops from a behavior $\epsilon_{0.1} \sim \tau^{-1/4}$ to a $\epsilon_{0.1} \sim \tau^{-1/2}$ dependence, which is characteristic for a classical diffusion process.

The second part of the chapter deals with experiments in strong static crossed electric and magnetic fields. Below as well as above the ionization energy the observed quasi-Landau(QL)-resonances are explained using classical trajectories. The ionization behavior of the QL-resonances is clearly influenced by the spatial structure of the associated classical orbits. In further experiments the dependence of the observed ionization energy on the magnetic field was investigated in the presence of a fixed static electric field. The ionization energy plotted versus the magnetic field exhibits a characteristic double well structure which follows the classical scaling laws of the system: as the magnetic field is increased, the ionization energy first increases, passes a maximum value, then a minimum, and at very high magnetic field values it increases again. The different regions of the ionization behavior belong to qualitatively different domains of the classical dynamics. The ionization in the high magnetic field regime is caused by a quantum-mechanical coupling of the classically chaotic region close to the Coulomb center, and the outer configuration space volume, where in the classical treatment regular drift orbits exist.

2 Rubidium Rydberg atoms in microwave fields

The Hamiltonian of hydrogen in a microwave field with amplitude ϵ in atomic units is given by

$$H = \frac{p^2}{2} - \frac{1}{r} + \epsilon z \cos(\omega t). \tag{1}$$

It is convenient to introduce the scaled electric field $\epsilon_s = \epsilon W^{-2} = \epsilon n^{*4}$ which is the electric field in units of the Coulomb field acting on the Rydberg electron, and the scaled frequency $\omega_s = \omega W^{-3/2} = \omega n^{*3}$ measuring the microwave frequency in units of the Kepler frequency. The classical dynamics only depends on the scaled parameters ϵ_s and ω_s. For $\omega_s > 1$ the classical evolution of the electron between successive perihelion transitions can be approximated by the standard map which also describes the well-known kicked rotator [9, 10, 11]. The electron diffusively gains energy and ionizes if the scaled electric field exceeds

$$\epsilon_s > \frac{1}{49\omega_s^{1/3}}. \tag{2}$$

The quantum treatment of the kicked rotator [9, 12] reveals the localization phenomenon occurring for $\omega_s > 1$. As in the classical system, after switching on the microwave interaction, the quantum system starts to diffuse very rapidly on the energy scale. However, contrary to the classical system, after a time of the order of the inverse average quasi-energy level spacing the diffusion ceases, and further transitions to higher and lower energy levels are prohibited by destructive interference of the different (nonresonant) excitation channels starting from the already populated levels. After the initial diffusion phase which cannot be resolved in our experiment the atoms are frozen in an approximately stationary distribution $f(W)$ describing the probability of finding the system in an (undisturbed) state with energy W. The function $f(W)$ is centered at the energy of the initially excited level, and decreases exponentially with the number of microwave photons required to reach the particular energy W. The width of the distribution function in units of the photon energy, the so-called localization length l, is given by $l = 3.33\epsilon^2\omega^{-10/3}$ [9]. A common measure of the strength of the applied microwave radiation is the 10% ionization electric field strength: By integrating the energy distribution $f(W)$ over the continuum $W > 0$ and the atomic states coupled to the continuum by the experimental electric stray fields the 10% ionization scaled electric field strength $\epsilon_{s,0.1}$ is obtained [12]:

$$\epsilon_{s,0.1} = \frac{\omega_s^{7/6}}{\sqrt{8}\,n_0} \sqrt{1 - \frac{n_0^2}{n_c^2}}, \tag{3}$$

where n_0 is the initial state principal quantum number and n_c the principal quantum number above which the atoms are ionized in the apparatus by stray fields and other perturbing influences (n_c in our experiments was about 170). Due to the dynamical localization the value of $\epsilon_{s,0.1}$ is considerably larger than the classical value following from Eq. (2).

The different power-laws in ω_s (compare Eq. (2) and Eq. (3)) which are found by classical and quantum treatment are a challenge for an experiment: in the classical case the scaled ionization field strength ϵ_s for all ω_s decreases with ω_s. The quantum treatment exhibits the same behavior in the regime $\omega_s < 1$, whereas in the localization regime $\omega_s > 1$ the scaled ionization field strength should increase. Thus, the scaled 10% ionization field strength versus ω_s should exhibit a minimum which marks the lower frequency limit of the localization regime.

The experimental setup is shown in Fig. 1. A collimated rubidium atomic beam enters the apparatus from the right. One cycle of the pulsed experiment consists of the following steps:

- At the laser excitation point a few atoms per cycle are directly excited from the $5S$ ground state into a nP-state by a horizontally polarized UV laser pulse (duration about $6\,\mu s$) which is obtained by chopping an intracavity frequency doubled Rhodamine 6G dye laser with an electro-optical modulator.
- After a well-defined time, which is determined by the average velocity of the atoms and thus by the oven temperature, almost all Rydberg atoms are inside the waveguide. Now a short travelling microwave pulse (up to about $10\,\mu s$ in length) is sent into the waveguide which is terminated by $50\,\Omega$; thus the interaction time of the atoms with the microwave radiation is well defined. Since furthermore the microwave electric field strength does not depend on the position of the atoms within the waveguide, all atoms experience exactly the same microwave pulse. The orientation of the waveguide allows only $\Delta m = 0$ transitions.
- When the atoms have reached the field ionization zone a voltage ramp which is applied to the lower field electrode ionizes the atoms. The electron counts are registered as a function of the electric field in the ionization zone, thus providing information on the width of the population distribution for the atomic energy levels.

For the microwave ionization experiments it is actually not necessary to analyze the final state distribution function; by measuring the total number of detected Rydberg atoms N_M obtained with microwave interaction, and the number of counts N_0 with no microwave, the fraction of atoms ionized by the microwave is $P_{ion} = 1 - N_M/N_0$.

As in most complicated microwave networks the electric field strength of the microwave which actually acts on the atoms not only depends on the microwave power but also on the microwave frequency. For Rydberg transition frequencies starting from the initially excited states nP the transmission ratio of the microwave system has been calibrated by measuring the Rabi frequencies of two-photon transitions. In order to have a well-defined microwave field strength the microwave frequency was usually set to one of the calibrated frequencies. The scaled microwave frequency was scanned by varying the initial principal

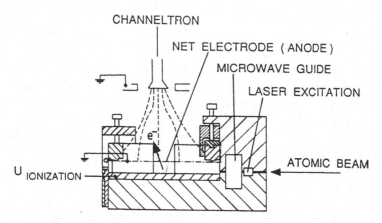

Fig. 1. Experimental setup. The figure shows the central part of the experiment where the atoms are excited, irradiated with microwaves and state-selectively field ionized.

quantum number (see the above mentioned scaling law). This is easily performed by changing the laser wavelength.

Another problem arises from the fact that in the experiments rubidium has been used, and not hydrogen. Since all the cited calculations have been performed for hydrogen, it is no surprise that the experimental results exhibit qualitative deviations from the theoretical results: Whereas for hydrogen theoretically as well as experimentally a minimum of the scaled 10% ionization electric field strength is found for $\omega_s \approx 1.0$ [9, 13, 14], in the rubidium experiments the corresponding minimum is found at $\omega_s \approx 0.3$. In hydrogen the microwave frequency is normalized by the Kepler frequency which is the transition frequency to the nearest upward neighboring level which can be reached by an allowed one-photon transition. In our experiment the rubidium Rydberg states interacting with the microwave radiation are P-states. Due to the large quantum defects for angular momenta $l < 3$ for those states the nearest allowed transition is the $P \rightarrow D$ transition the frequency of which is 0.3 times the Kepler frequency. Since the results for hydrogen suggest that the physical scaling frequency is the frequency of the nearest upward dipole allowed transition we normalize the microwave frequency by the PD transition frequency: $\omega_{s,pd} = \omega/\omega_{n_0 P \rightarrow (n_0-1)D}$. This empirical definition of the scaled frequency for rubidium gives qualitatively the same result as an averaging argument used in earlier work [2]. With the PD scaling the minimum of the scaled 10% ionization field strength is found for $\omega_{s,pd} \approx 1.0$. This is one of the reasons why the modified scaled frequency is useful for a qualitative comparison of our results with results obtained in experiments and calculations with hydrogen. However, it should be clearly distinguished from the notion of the scaling properties of the classical Hamiltonian which if they exist for rubidium are not at all obvious and for which there has been no rigorous investigation.

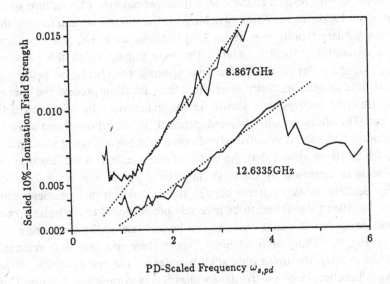

Fig. 2. The scaled 10% ionization electric field as a function of the PD-scaled microwave frequency $\omega_{s,pd}$ for two different microwave frequencies $\omega = 2\pi 8.867\,\text{GHz}$ and $\omega_{,} = 2\pi 12.6335\,\text{GHz}$. Both plots exhibit a minimum at $\omega_{s,pd} \approx 1$. For $\omega_{s,pd} > 1$ the curves can be well approximated by functions proportional to $\omega_{s,pd}^{7/6}/\sqrt{n_0^*}$ (see dotted lines). The deviation at the high frequency end of the 12.6335 GHz curve is discussed in the text.

Fig. 2 shows the experimentally observed values of $\epsilon_{s,0.1}$ with microwave frequencies 8.876 GHz and 12.6335 GHz, and pulse duration $\tau = 5\mu s$, as a function of the modified scaled microwave frequency $\omega_{s,pd}$, which is tuned via the inital state n_0. The value of n_0 was varied between 55 and about 97. Except for the high scaled frequency part of the 12.6335 GHz curve in the localization regime $\omega_{s,pd} > 1$ the experimental curves can be well approximated by $\epsilon_{s,0.1} = const. \cdot \omega_{s,pd}^{7/6}/\sqrt{n_0^*}$ (see also Eq. (3)). Since also in the localization regime the ionization probability actually weakly depends on the interaction time τ [15], the value of *const.* was considered as a free parameter here (see also below). The high frequency curve of Fig. 2 clearly drops for large $\omega_{s,pd}$. This is due to the cutoff number n_c (see Eq. (3)) which in our experiments was about 170, and which becomes important if the atoms enter the waveguide already in high principal quantum numbers n_0. In the mentioned part of the 12.6335 GHz plot obviously an increasing fraction of the atoms is excited by the microwave interaction into the range $\infty > n > n_c$, leading to the observed kink in Fig. 2.

Fig. 3 shows the destruction of the localization by additional broadband noise. The initial state $84P_{3/2}$ interacts with 12.059 GHz 6.3 mW coherent microwave for 20 ns and 1 μs. In two plots shown in Fig. 3 weak broadband noise was added. Without noise the field ionization patterns after 20 ns and 1 μs exhibit similar broadening with respect to the interaction-free pattern. This feature again demonstrates the localization: After switching on the microwave interaction the wave-function diffuses rapidly over about 12–15 atomic states [2], and the atoms end up in a dynamically localized state. The time during which the diffusion takes place is of the order of the inverse level spacing, thus being far below the experimental time resolution. Since after the initial diffusion process the energy distribution function evolves very slowly, in the experiment the observed field ionization patterns are essentially time-independent. If weak broadband noise is added, the initial behavior is not strongly affected. The evaluation of short pulse field ionization patterns shows that the width of the localized wave-function on the energy scale is increased by the noise only by a few atomic states (see for example Fig. 3, 20 ns plots). After a certain time t_B, which in agreement with theoretical calculations was found to be inversely proportional to the noise power [2], the noise destroys the localization, as can be seen clearly by comparing the 1 μs plots in Fig. 3. Thus, at fixed noise power there are two time regimes: for pulse width $\tau < t_B$ the noise only slightly enlarges the energy width of the localized wave-function. For $t > t_B$ the localization is completely destroyed by the noise. For fixed microwave pulse duration τ the situation is similar: Below a certain noise power the energy width of the localized wave-function is only slightly increased, whereas above that noise level the localization is destroyed.

In further experiments the influence of different types of noise on the scaled 10% ionization field strength has been investigated. Again the scaled frequency was scanned via the initial quantum number n_0. This implies that the noise power also has to be changed depending on n_0: in analogy to the scaling property of the coherent microwave field strength it is reasonable to introduce the scaled noise power $P_s = P n_0^{*8}$ with noise power P. Calculations of the diffusion constant describing the ionization of Rydberg atoms by microwave noise [16] also indicate that this definition of the scaled microwave power is physical. In the experiment microwave noise of constant scaled noise power P_s was added to the strong coherent microwave. The scaled 10% ionization field strength of the coherent microwave radiation was measured for different values of the noise bandwidth (see Fig. 4). The total noise powers in all the series displayed in Fig. 4 were equal. Thus the influence of noise in different frequency ranges on the microwave ionization behavior can be directly compared in Fig. 4.

First the influence of the broadband noise is considered. It can be noticed that in the lower part of the localization regime (here $1 < \omega_{s,pd} < 2$) the broadband noise lowers $\epsilon_{s,0.1}$ of the coherent microwave to a very low level, thus approaching the classical result which is $\sim \omega_s^{-1/3}$. It must be assumed that in this range of $\omega_{s,pd}$ the localization is completely destroyed by the additional noise. In the upper localization regime (here $\omega_{s,pd} > 2$) the scaled 10% ionization field strength with additional noise approaches the slope value of the result without noise. Obviously in this upper range of $\omega_{s,pd}$ the localization is not completely lifted by

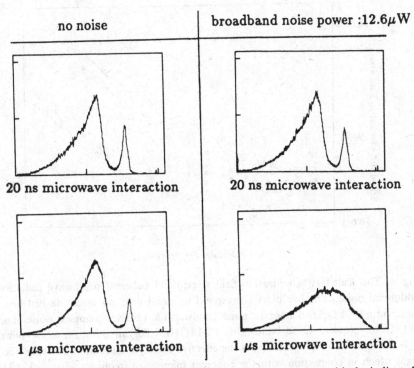

no noise | broadband noise power :12.6μW

20 ns microwave interaction | 20 ns microwave interaction

1 μs microwave interaction | 1 μs microwave interaction

Fig. 3. Field ionization patterns after interaction with microwave pulses with the indicated duration. The relative field ionization probability is plotted versus the voltage of the ionizing voltage ramp which ranges from 0 to 36.6 V (from right to left). In the right column a weak broadband noise was added to the coherent microwave. (For more details see Ref. [2].)

the noise. Qualitatively the observed behavior can be explained by assuming that the noise enhances the diffusion constant describing the short diffusion process after switching on the microwave pulse. In effect, the atoms are left in a broader final energy distribution, resulting in a lowered 10% ionization field strength. Since for $\omega_{s,pd} > 2$ the localization is not completely destroyed, the applied pulse duration of 5 μs is obviously below the time t_B after which the irradiated noise level would destroy the localization (see Fig. 3 and Ref. [2]). For $1 < \omega_{s,pd} < 2$ the localization is destroyed, indicating that for those parameters the applied pulse width exceeds the value of t_B. It can be concluded that at fixed scaled noise power the value of t_B increases with $\omega_{s,pd}$. This is a quite intuitive result since for large $\omega_{s,pd}$ the localization should be more resistive against disturbing influences.

Now the plots corresponding to bandwidth-reduced noise are considered (see Fig. 4). Theoretical calculations [16] show that, the larger the number of strong resonant transition frequencies which start from n_0P and which are within the

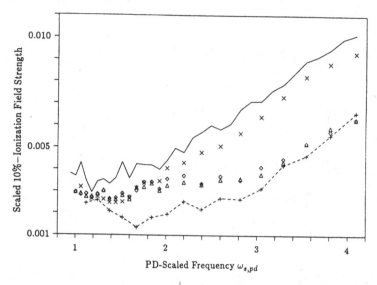

Fig. 4. The scaled 10% ionization field strength of coherent microwave radiation with additional noise. The five plots correspond to: solid line: no noise; dashed line: noise bandwidth 8–18 GHz; diamonds: noise bandwidth 8–12 GHz; triangles: noise bandwidth 8–15 GHz; crosses: noise bandwidth 12–18 GHz. For all plots the noise power was $P = 5.8 \times 10^{13} \mathrm{mW}/n_0^{*8}$, where n_0^* is the effective principal quantum number of the initial state which in connection with the coherent microwave frequency $\omega = 2\pi 12.6335\,\mathrm{GHz}$ defines the scaled frequency $\omega_{s,pd}$. The pulse width was $\tau = 5\,\mu s$.

noise bandwidth the more the noise enhances the diffusion constant. Thus in the localization regime $\omega_{s,pd} > 1$ the low frequency noise should have a greater effect than the high frequency noise. This behavior is also found in the experiment, where for $\omega_{s,pd} > 2$ the high frequency noise only slightly alters the ionization curve, whereas the low frequency noise almost causes the same dramatic effect as the broadband noise (see Fig. 4).

Finally the dependence of the 10% ionization field strength on the pulse width τ is briefly discussed. Usually one finds a power law $\epsilon_{0.1} \sim \tau^{\alpha}$ with α depending on the noise power. Since weak noise is always present in the experiment (thermal microwave photons), the case of weak noise is realized by irradiating only the coherent microwave. We found the value $\alpha \approx -0.25$ which can also be deduced from the diffusion constant which is found in Ref. [17] for the case of moderate noise level. When increasing the noise power up to about -10 dBm, the exponent α gradually shifts to $\alpha \leqslant -0.5$, as also follows from the strong noise level result in Ref. [17]. It should be also remarked that contrary to Eq. (3) even in the absolutely noise-free case the 10% ionization field strength should slightly depend on τ [15].

3 Rubidium Rydberg atoms in strong static crossed electric and magnetic fields

The spectra of hydrogen in strong magnetic fields are modulated by many types of quasi-Landau(QL)-resonances being associated with classical trajectories [18]. The modulations take their typical sinusoidal shape if the scaled energy $\omega = WB^{-2/3}$ enters the chaotic regime $\omega > -0.3$. QL-resonances are also found in the spectra of Rydberg atoms in crossed electric and magnetic fields. Assuming an external electric field $\mathbf{E} = -E\mathbf{e}_x$ and a magnetic field $\mathbf{B} = B\mathbf{e}_z$ the corresponding Hamiltonian is (neglecting the core potential):

$$H = \frac{p^2}{2} - \frac{1}{r} - Ex + \frac{B}{2} l_z + \frac{B^2}{8} (x^2 + y^2) . \tag{4}$$

The classical dynamics only depends on the scaled electric field $\epsilon = EB^{-4/3}$ and the already defined scaled energy $\omega = WB^{-2/3}$: with the scaled phase space variables $\mathbf{r}_s = \mathbf{r}B^{2/3}$ and $\mathbf{p}_s = \mathbf{p}B^{-1/3}$ one gets the B-independent scaled Hamiltonian

$$H_s = \frac{p_s^2}{2} - \frac{1}{r_s} - \epsilon x + \frac{1}{2} l_{z,s} + \frac{1}{8} (x_s^2 + y_s^2). \tag{5}$$

It turns out that the best way to detect QL-resonances is to Fourier-transform scaled spectra which are taken as a function of $B^{-1/3}$ at constant values of ϵ and ω. This requires one to change both external fields during the laser scan. The Fourier-transforms of the scaled spectra yield the scaled actions S_s of the QL-resonances and allow one to estimate the corresponding modulation intensities. The associated trajectories can be found with a computer. Below the ionization energy many QL-resonances have already been found and the corresponding classical orbits identified [5–7]. The modulation strengths which depend on the classical stability of the trajectories can be reproduced reasonably well by a formalism developed by Du and Delos [19]. The theoretical modulation strengths are determined by the ground state quantum numbers, the laser polarization and the classical stability behavior of the associated closed orbits. Fig. 5 shows two Fourier spectra which nicely demonstrate that the scaled spectra can be decomposed into well-distinguished QL-resonances. Fig. 5 also clearly shows the polarization dependence of the QL-resonance strengths: since according to Ref. [19] the modulation strength of symmetric orbits should be $\sim \cos^4\alpha$ the QL-resonances exhibit completely different intensities in the upper and lower spectra of Fig. 5. Modulations associated with orbits starting in x- or y-directions even occur only in one of the spectra shown in Fig. 5.

Performing spectroscopy in crossed fields above the ionization threshold is not straightforward since the photoelectrons excited by direct ionization of rubidium atoms by the UV radiation drift uncontrollably in the periphery of the fields. This problem was managed by using the arrangement shown in Fig. 6. The electrodes and the magnetic field essentially make up a Penning electron trap providing a region with homogeneous electric field and exhibiting a hole allowing

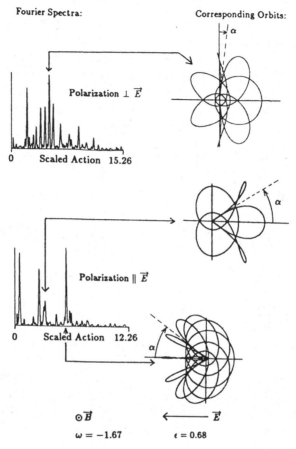

Fig. 5. QL-resonances for the indicated scaled parameters for two different laser polarizations. The squares of the Fourier amplitudes are plotted versus the scaled action σ which is the action S in units of h times $B^{1/3}$ with magnetic field B in atomic units. The three depicted classical orbits associated with strong resonances show that the angle between laser polarization and starting angle of the orbit α has to be small in order to obtain a large resonance strength.

the photoelectrons to escape. The hole is imaged onto a position sensitive electron detector in a way which allows one to determine whether the atom was photo-ionized in the excitation field region or field-ionized in the conventional field ionization region (see Fig. 6).

Fig. 7 shows Fourier spectra obtained above the ionization energy and laser polarization parallel to the electric field. As the excitation energy increases the observed modulation strengths decrease rapidly (this cannot be seen in Fig. 6 since the spectra have been normalized to the resonance with the highest intensity).

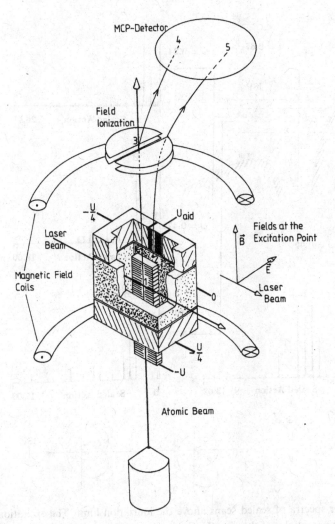

Fig. 6. Experimental setup for spectroscopy in crossed electric and magnetic fields above the ionization limit. The Rydberg atoms are excited at point 1 which is designed to provide a homogeneous electric field being unperturbed by the outer trap potentials. Electrons, which originate in Rydberg atoms ionized at location 3 in the usual field ionization region, arrive at position 4 on the detector. Electrons arriving at location 5 originate from excited atoms which were ionized in the trap region, i.e. immediately after excitation. The photoelectrons leave the trap through a 'hole' (2) generated by the potential U_{aid}.

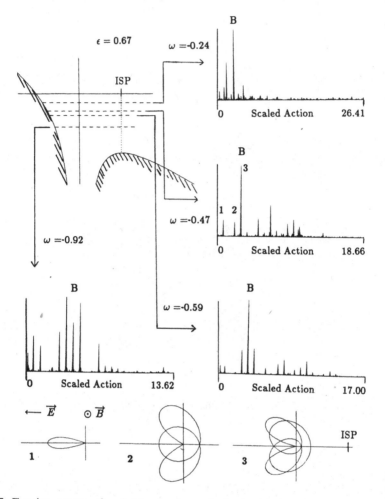

Fig. 7. Fourier spectra of scaled scans above the ionization limit. The excitation energies are shown in a potential plot showing the Coulomb-plus electric potential along the electric field direction. For $\omega = -0.47$ a few modulating trajectories are depicted.

This overall decrease reflects the decreasing stability of the modulating trajectories which is also found by classical trajectory calculations.

A dominating feature of the spectra is that there is a group of resonances labeled by B which shifts to smaller actions when the scaled energy ω is increased. This group is associated with trajectories which start approximately in the direction of the electric field and return from this direction (trajectory 3 in Fig. 7 is such an orbit). This type of orbit was found to have a dominating modulation strength over a wide parameter range [7]. According to the increasing influence of the external fields these orbits consist of a decreasing number of loops as the energy is increased. Consequently the major peak in the Fourier spectra of Fig. 6 shifts to the left with increasing ω.

Far above the ionization energy the trajectories associated to the QL-resonances tend to be oriented opposite to the classical field ionization saddle point (ISP), as shown for a few examples in Fig. 7. The reason is very simple: the energy is so far above the classical ionization energy $-2\sqrt{E}$ that a classical electron starting more or less in the direction of the ISP always disappears in the continuum. Nevertheless, the existing recurring orbits lead to QL-resonances which can be described by the formalism developed in Ref. [19]. It is interesting to note that this formalism applies even for QL-resonances in the continuum.

At the ionization energy it was observed that the resonance strength associated with individual QL-resonances shifts from the Fourier spectrum of the nonphotoionized atoms to the photoionized atoms at slightly different energies (Fig. 8). Thus the QL-resonances exhibit a resonance-specific ionization behavior. This can be understood in the following way: Similar to the oscillator strength the probability distribution of the excited states is enhanced along the modulating orbits according to the classical stability of the orbits [20]. It is expected that QL-resonances generated by orbits extending in direction of the ISP 'ionize' at lower energies than those generated by orbits extending in other directions. In Fig. 8 the associated orbits for three strong QL-resonances are plotted. The expected trend is clearly visible: the resonances corresponding to orbits which are not oriented in the direction of the ISP appear relatively strong in the Fourier spectrum of the stable atoms.

In recent experiments the ionization threshold of rubidium Rydberg atoms in crossed electric and magnetic fields has been carefully investigated. It turned out that the ionization energy as a function of the scaled electric field ϵ exhibits an interesting structure which exhibits classical scaling behavior. Fig. 9 shows a few results. In scaled variables the open and filled dots, corresponding to two series of experiments in completely different parameter ranges, follow approximately the same curve. From Fig. 9 and other measurements it can be concluded that the ionization energy is essentially determined by the (classical) scaled parameters. The most prominent features of the data displayed in Fig. 9 are the intermediate ionization energy maximum at $\epsilon \approx 0.2$, and a minimum at $\epsilon \approx 0.06$. In order to explain the unexpected low ionization energy for $\epsilon \lesssim 0.06$ it is reasonable to assume that beyond the ionization energy maximum in the range $0.2 > \epsilon > 0.06$ a new ionization channel occurs which lowers the ionization energy, and thus determines the ionization energy for $\epsilon < 0.06$. Since the ionization curve is determined by classical parameters it should be possible to understand this ionization process in a (semi)classical way.

The semiclassical explanation of the ionization process in the low ϵ regime will be the subject of the remaining part of this chapter. Since one deals with ionization, it is useful to investigate the classical dynamics close to the Coulomb center, where the electron starts, as well as in the vicinity of the ISP, which in

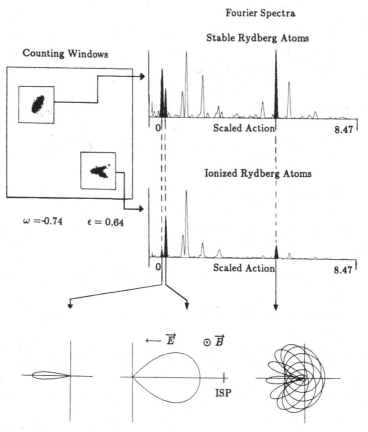

Fig. 8. Comparison between Fourier spectra at the ionization energy separately evaluated for ionizing Rydberg atoms and stable Rydberg atoms. The orbits which are shown correspond to the resonances differing most. 'ISP' indicates the position of the classical ionization saddle point.

the classical picture is passed by the electron during the ionization. Fig. 10 shows pairs of calculated trajectories at the ionization energy, one trajectory launched at the Coulomb center, the other lauched in the outer configuration space region. For $\epsilon > 0.2$ all trajectories appear regular. Since the magnetic field is relatively weak, the trajectories widely extend transverse to the magnetic field direction (ρ-direction). Since in particular the trajectories launched at the center cover almost the whole energetically allowed configuration space, according to the correspondence principle the excited states do so as well. The classical behavior also allows the conclusion that for $\epsilon > 0.2$ the ionization occurs via leakage of

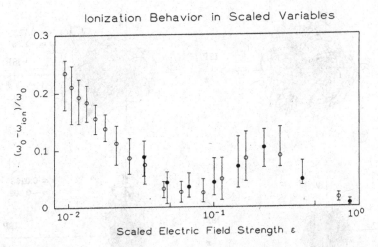

Fig. 9. Investigation of the ionization energy in crossed fields as a function of the scaled electric field. Two sets of measurements are shown: one at $E = 9000$ V/m (filled dots), one at $E = 3000$ V/m (open dots). In order to vary the scaled electric field $\epsilon = EB^{-4/3}$ the magnetic field B was changed. The observed ionization energy ω_{ion} relative to the classical ionization energy $\omega_0 = -2\sqrt{E}$ is displayed in units of ω_0. The error bars indicate the energies where 20% and 90% ionization was observed.

the excited wave-functions right away across the ISP into the free $\mathbf{E} \times \mathbf{B}$-space beyond the ISP.

The situation is completely different in the high magnetic field regime $\epsilon < 0.1$: Due to the increasing magnetic field the trajectories starting at the center are more and more restricted in the ρ-direction, the extension essentially determined by the diamagnetic potential in Eq. (5). As found in Ref. [21], the central trajectories are chaotic for $\omega \lesssim 0.3$. In the outer configuration space, however, there exist bound trajectories which are located on more or less cylindrical shells around the Coulomb center. On these shells the electrons perform a regular drift motion which has three typical time scales which are also found in the motion of an electron in a Penning trap. These are the fast cyclotron motion frequency $\omega_c = B$, an oscillation in the magnetic field direction with an intermediate frequency ω_z, and in the case of bound drift trajectories there is also a slow magnetron motion frequency ω_m.

It is straightforward to quantize this motion, leading to what can be called drift states. The drift trajectories in general show no or only a small spatial overlap with the space which is covered by trajectories launched at the Coulomb center. The drift orbits can only be found for $\epsilon \lesssim 0.1$, and only sufficiently far from the Coulomb center, since otherwise the adiabaticity condition is lost, i.e. the three mentioned time scales get too similar. Furthermore, since with decreasing

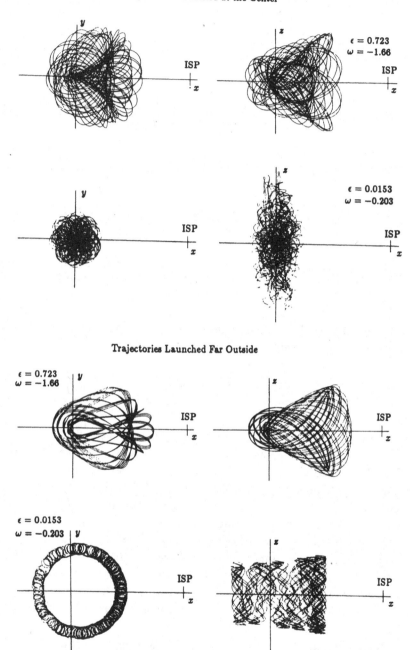

Fig. 10. Top and side view of a few trajectories calculated with the indicated parameters. In the upper part trajectories starting from the Coulomb center are displayed, the trajectories in the lower part start in the outer configuration space. The trajectory at the bottom is a bound drift trajectory.

magnetic field the dimension of the cyclotron motion increases, the electron more and more experiences the electric field inhomogeneity of the Coulomb potential, and the drift motion loses its regular character.

There are also drift trajectories which extend into the continuum. As found by numerical computations, the drift trajectories are most likely to ionize if the energy of the z-oscillation is as large as possible. One also finds that at a given energy the drift trajectories with maximum energy in the z-oscillation determine how deep a free electron may penetrate into the Coulomb potential. The free electrons which go around the Coulomb center approach the Coulomb center closest at the 'back side' (i.e. the opposite side to the ISP).

In the high magnetic field regime an electron moving on a trajectory launched at the center with the actual ionization energy usually exhibits spatial overlap with free drift trajectories. Since the free drift trajectories with maximum z-energy penetrate deepest into the Coulomb potential, it is no surprise that an appreciable overlap only exists with the maximum z-energy drift trajectories. The overlap region is localized around the point $(-x_0, 0, 0)$ in Fig. 11. Despite the mentioned overlap the electron started at the Coulomb center never leaves the inner configuration space region, or, in other words, the atom classically would not ionize at the observed ionization energy. This classical result can be explained in the following way: in the classical treatment it is found that the electron on the central trajectory, when it happens to cross the mentioned overlap region, has a kinetic momentum which is more or less perpendicular to the magnetic field. This is a natural consequence of the fact that in the symmetric gauge, which is adopted in Eq. (4), with decreasing scaled electric field the canonic angular momentum l_z in the magnetic field direction 'tends to be conserved', i.e. the probability distribution of measuring a certain value of l_z is more and more peaked at the fixed value determined by the polarization of the exciting laser. The kinetic momentum of the electron on the drift trajectory is essentially parallel to the magnetic field, when it crosses the overlap region. Therefore a large 'momentum kick' would be necessary to bring the electron from the inner trajectory in Fig. 11 to the free drift trajectory. In phase space this 'kick' would shift the electron from the central chaotic phase space region onto a torus describing a (regular) drift state. Since in classical mechanics this 'kick' never happens, the atom classically never ionizes at the actually observed ionization energy.

Thus, the question of why there is an ionization energy minimum in Fig. 9, is identical with the question of how the ionization in the high magnetic field regime is possible at all. In order to give a qualitative explanation we consider the mentioned overlap region quantum mechanically. Let us assume we have found approximate eigenstates in the vicinity of the Coulomb center by truncating the space well beyond the surface determining the space region which is covered by classical electrons starting at the center. According to the correspondence principle the extension of the excited states will also be essentially limited by this surface. Particularly, in the vicinity of the point $(-x_c, 0, 0)$ in Fig. 11 the wave-function

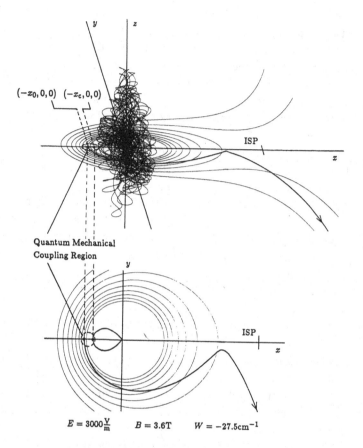

Fig. 11. Visualization of the ionization process. The region where the amplitude of the excited wave-function may be large is represented by the chaotic trajectory in the center. In the electric field direction (i.e. opposite to the ISP direction) the extension x_c of this region is found by means of the simple classical one-loop trajetory depicted in the lower plot. The ionization takes place via a quantum mechanical coupling with a free drift state. In the figure the projection of the drift motion of a free drift trajectory with maximum z-energy, averaged over a cyclotron period, is displayed. The coupling occurs in the region surrounded by the dashed circle. Due to energy transfer from the z-oscillation to potential energy in the electric potential the drifting electron leaves the Coulomb attraction zone. This energy transfer is reflected by the fact that the drift motion is not parallel to the indicated electric potential lines.

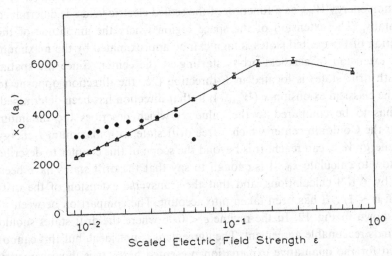

Fig. 12. Test of the ionization model discussed in the text. The triangles correspond to the values x_c following from the experimentally observed ionization energy (see also Fig. 11). The dots show the minimum distance between the Coulomb center and a free drift state with the observed ionization energy.

will be of the form $\sim \exp(iky + i\Phi(\mathbf{r}))$ with gauge-dependent $\Phi(\mathbf{r})$. The quantum mechanical drift states can be approximately found by a Born–Oppenheimer separation of the three already mentioned types of motion, yielding two (three) discrete quantum numbers for free (bound) drift states. The form of drift states with maximum z-energy would be essentially $\sim (\exp(ikz) \pm \exp(-ikz))\exp(i\Phi(\mathbf{r}))$. If there exists a nonvanishing overlap between a central state and a drift state with maximum z-energy, the coupling matrix element of H (Eq. (5)) would be very small due to the strongly $\mathbf{r}$-dependent phase difference between the drift state and the central state wave-functions in the overlap region. However, since the overlap region is limited, it would be nonzero. Therefore, a central state carries admixtures of those drift states with which it overlaps and which have approximately the same energy. Except in the case of accidental degeneracy the admixtures are small, which again can be seen as a consequence of the approximate conservation of l_z.

Provided that first order coupling determines the behavior, ionization of an inner state (which is initially excited) takes place if one of the drift states, which are in first order admixed to it, is a free drift state. As already mentioned, the maximum z-energy free drift states define how close a free electron may approach the Coulomb center. Thus, ionization should take place, when the excitation energy is high enough that at least the free maximum z-energy drift states start to overlap with the space region where the initially excited states have a large amplitude. In the latter case ionization may occur.

The outlined explanation of the ionization process can be tested as follows. First the ionization energy W_{ion} as a function of the scaled electric field ϵ is determined experimentally. The extension of the space region where the amplitude of the wave-function of the excited states is high can be approximated by the maximum transverse extension of classical orbits starting at the center. Since the spatial overlap with drift states is located in E-direction (i.e. the direction opposite to the ISP), the classical extension $x_c(W_{ion}(\epsilon))$ in that direction has been determined. Then x_c has to be compared to the value x_0 which describes the minimum distance to the Coulomb center which a free drift state with maximum z-energy and total energy W_{ion} can reach. It is beyond the scope of this chapter to describe in detail how to calculate x_0. It is enough to say that the drift states have been quantized by WKB calculations, and that the transverse extension of the drift state which is $\sim 1/\sqrt{B}$ has been taken into account. The comparison between x_c and x_0 is shown in Fig. 12. In the regime $\epsilon < 0.06$ where the drift states should exist one finds reasonable agreement. The agreement is not ideal, but this cannot be expected for the qualitative explanation presented here. It is, however, very likely that the presented ionization scheme determines the ionization energy of Rydberg atoms in strong crossed fields for $\epsilon < 0.06$.

REFERENCES

[1]   A. Buchleitner, L. Sirko, H. Walther, in *Adriatico Research Conference on Quantum Chaos*, Eds. H.A. Cerdeira, R. Ramaswamy, M.C. Gutzwiller, G. Casati, World Scientific, Singapore (1991), p.395

[2]   R. Blümel, A. Buchleitner, B. Graham, L. Sirko, U. Smilanski, H. Walther, Phys. Rev. A **44**, 4521 (1991)

[3]   A. Buchleitner, L. Sirko, H. Walther, Europhys. Lett. **16**, 35 (1991)

[4]   M. Arndt, A. Buchleitner, R.N. Mantegna, H. Walther, Phys. Rev. Lett. **67** (1991)

[5]   G. Raithel H. Walther, in *10. International Conference on Laser Spectroscopy*, Eds. M. Ducloy, E. Giacobino, G. Camy, World Scientific, Singapore (1992), p.437

[6]   G. Raithel, M. Fauth, H. Walther, in *Adriatico Research Conference on Quantum Chaos*, Eds. H.A. Cerdeira, R. Ramaswamy, M.C. Gutzwiller, G. Casati, World Scientific, Singapore (1991), p.409

[7]   G. Raithel, M. Fauth, H. Walther, Phys. Rev. A **44**, 1898 (1991)

[8]   G. Raithel, H. Walther, Phys. Rev. A, March 1994

[9]   G. Casati, I. Guarneri, D.L. Shepelyanski, IEEE Journal of Quantum Electronics **24**, 1420 (1988)

[10] B.V. Chirikov, 'Time Dependent Quantum Systems' in *Chaos and Quantum Physics*, Les Houches Lecture Notes (1989), eds A. Voros, M. Gianonni, O. Bohigas, North Holland, Amsterdam (1990)

[11] B.V. Chirikov, Phys. Rep. **52**, 263 (1979)

[12] G.Casati, I.Guarneri, D.L.Shepelyanski, Physica A **163**, 205 (1990)

[13] E.J. Galvez, B.E. Sauer, L.Moorman, P.M. Koch, D. Richards, Phys. Rev. Lett. **61**, 2011 (1988)

[14] J.E. Bayfield, G. Casati, I. Guarneri, D.W. Sokol, Phys. Rev. Lett. **63**, 364 (1989)

[15] G. Casati, B.V. Chirikov, D.L. Shepelyansky, I. Guarneri, Phys. Rep. **154**, 77 (1987)

[16] B. Meerson, Phys. Rev. Lett. **62**, 1615 (1989)

[17] E. Ott, T.M. Antonson, J.D. Hanson, Phys. Rev. Lett. **53**, 2187 (1984)

[18] J. Main, G. Wiebusch, A. Holle, K.H. Welge, Phys. Rev. Lett. **57**, 2789 (1986)

[19] M.L. Du, J.B. Delos, Phys. Rev. **A 38**, 1896 and 1913 (1988)

[20] E.B. Bogomolny, Physica **D 31**, 169 (1988)

[21] G. Wunner, U. Woelk, I. Zech, G. Zeller, T. Ertl, F. Geyer, W. Schweitzer, H. Ruder, Phys. Rev. Lett. **57**, 3261 (1986)

Diamagnetic Rydberg Atom: Confrontation of Calculated and Observed Spectra

Chun-ho Iu, George R. Welch, [a] Michael M. Kash, [b] and Daniel Kleppner

Research Laboratory of Electronics, Department of Physics, The George R. Harrison Spectroscopy Laboratory,
Massachusetts Institute of Technology, Cambridge, Massachusetts 02139

D. Delande and J. C. Gay

Laboratoire de Spectroscopie Hertzienne de l'Ecole Normale Superieure, 4, place Jussieu,
Tour 12-E1, 75252 Paris CEDEX 05, France
(Received 31 October 1990)

We present a detailed comparison of the observed and computed negative- and positive-energy spectrum of a Rydberg atom in a strong magnetic field. The study extends from -30 to $+30$ cm^{-1} at a field of 6 T. The experimental resolution is sufficiently high to provide well-resolved spectra over the entire range. The spectrum calculated for hydrogen is in remarkable agreement with the spectrum observed in lithium.

As described in the preceding Letter,[1] hereafter referred to as DBG, the hydrogen atom in a magnetic field has attracted unusual interest because it is among the simplest nonseparable systems that are physically realizable, because it is one of the small number of systems whose classical motion displays chaotic behavior in regimes where accurate quantum-mechanical calculations are possible,[2] and because it can be studied experimentally with high precision. The simplicity of this problem is deceiving, however, for carrying forward theory and experiment have both proven to be formidable undertakings. DBG describes a breakthrough in the problem of calculating the positive-energy spectrum at laboratory-sized magnetic fields. We report here the results of a comparison of calculated spectra with spectra observed experimentally by the MIT group who are the co-authors of this joint paper.

The most successful previous study of this kind was a comparison of the observed and computed spectrum for deuterium by Holle *et al.*[3] for energy in the range of -190 to -20 cm^{-1}. However, the experimental resolution was too low to achieve fully resolved spectra at the highest energies, and the computational method was limited to the negative-energy region. The work described here overcomes these limitations.

As described in DBG, the Hamiltonian of a hydrogen atom in a magnetic field along the z axis, in cylindrical coordinates, is (atomic units)

$$H = \frac{p^2}{2} - \frac{1}{(\rho^2+z^2)^{1/2}} + \frac{\gamma}{2}L_z + \frac{\gamma^2}{8}\rho^2. \qquad (1)$$

L_z, the z component of the angular momentum, is conserved. γ is the magnetic field in atomic units: $\gamma = B/B_c$, where $B_c = 2.35 \times 10^5$ T. A great deal is known about the negative-energy behavior of this system, but the positive-energy regime remains largely unexplored. It seems natural that the spectrum should tend to the Landau (i.e., free electron) limit of the Hamiltonian, perturbed by the Coulomb field, but understanding how this takes place constitutes a serious challenge because of the ionizing character of the Landau channels and the chaotic nature of the classical system.

The region chosen for study is at a magnetic field of approximately 6 T, and an energy range of -30 to $+30$ cm^{-1}. The experiments were carried out with $L_z = 0$ odd-parity states of lithium, rather than hydrogen, a point to which we shall return. The spectra were recorded by laser spectroscopy of an atomic beam in a superconducting solenoid. The spectral resolution was 1×10^{-3} cm^{-1}, the magnetic field was determined to an accuracy of 1×10^{-3} T, and the energy was determined to a relative accuracy of 1×10^{-3} cm^{-1} and an absolute accuracy of 5×10^{-3} cm^{-1}. The dynamical range of the spectra exceeds 10^4. Further details have been described previously.[4,5]

The spectra were computed as described in DBG, but using the nuclear mass of lithium-7. The size of the Sturmian basis was up to 90000. The positions and widths of the resonances are well converged with an uncertainty less than 1×10^{-3} cm^{-1}. The intensities of the lines are determined to within 1%. Typically, the calculation of a spectrum at $B = 6$ T over a 5-cm^{-1} interval requires 2000 s of Cray-2 CPU time.

An observed and a computed spectrum are shown in Fig. 1. Because the details of these spectra are too fine to display well, both spectra were convoluted with a Gaussian window with 0.05-cm^{-1} linewidth. Rather than superposing the spectra, which would lose all detail, we have shown them as "mirrors" of each other. The magnetic field was monitored periodically during the course of the measurements and found to drift up to 0.019 T. The effects of this on the calculated spectrum displayed in Fig. 1 were investigated, and found to be too small to be observable. It is to be emphasized that aside from an overall intensity scale factor, there are *no* adjustable parameters.

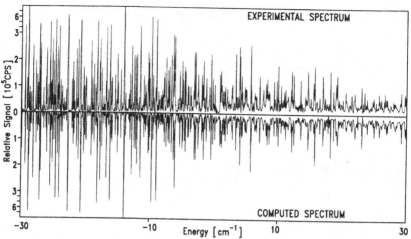

FIG. 1. Comparison of the experimental spectrum of the lithium atom in a magnetic field $B = 6.113$ T ($L_z = 0$, odd parity) with the theoretical spectrum of the hydrogen atom computed using the complex-coordinate method. To facilitate the comparison, the spectra were convoluted with a Gaussian window of 0.05-cm^{-1} linewidth and displayed as mirrors of each other. The scale, which is linear, decreases by a factor of 5 at 2×10^5 counts/s. Aside from the scale of the computed spectrum, there is no adjustable parameter.

The energies of the computed and measured spectra in Fig. 1 agree almost perfectly. There are, however, some differences in oscillator strengths. Possible reasons for this include perturbations due to the lithium ionic core, effects of the stray electric field (60 mV cm^{-1} in Fig. 1), experimental effects due to the finite step size of the laser sweep, and nonlinear response of the detector. Fourier transforms of these spectra reveal the expected short time periodicities associated with classical closed orbits[1,2] that have been previously observed.[6]

Because the spectra are too complex to allow detailed comparison on the scale of Fig. 1, we have separately displayed a 1-cm^{-1} interval in Fig. 2. The measured magnetic field was 6.1131 ± 0.001 T; the spectrum was computed at a field of 6.1143 T. (A spectrum computed at the measured field of 6.1131 T showed a small systematic discrepancy in the energy, $\sim 5 \times 10^{-3}$ cm^{-1}.) The calculated spectrum was convoluted with a Gaussian profile with the experimental linewidth, 1×10^{-3} cm^{-1}. The agreement of the calculated and measured spectra is remarkable in every respect except for some differences in oscillator strength, presumably arising from the effects discussed above. Save for the narrowest lines, which are difficult to resolve in the figure, the linewidths are in excellent agreement, as are the zeros of the excitation probability that arise from accidental destructive interference.[1] It is to be noted that this detailed agreement occurs well above the ionization limit.

The agreement of the computed and observed spectra gives great confidence in both methods. However, far more physical insight can be obtained from energy-level maps—plots of energy versus magnetic field—than from

individual spectra. In particular, such a map first revealed the existence of regular Rydberg progressions at positive energies,[5] an unexpected finding since it implies a separation of the ρ and z motions in a regime where

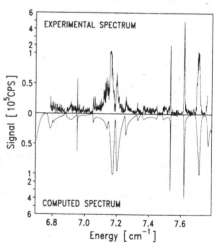

FIG. 2. Same as Fig. 1 for an energy interval of 1 cm^{-1}. Note that this spectrum lies above the ionization threshold. The compound spectrum was convoluted with a Gaussian window with 1×10^{-3}-cm^{-1} linewidth, equal to the experimental resolution. The scale decreases by a factor of 7.5 at 1×10^5 counts/s. The agreement is better than might be expected considering that the computation is for the hydrogen atom and the experiment is on the lithium atom.

they are expected to be strongly mixed. Consequently, an energy-level map was computed to compare with the MIT results. The maps, shown in Fig. 3, are in excellent agreement, each clearly revealing a Rydberg progression. Note that the existence of Rydberg progressions in hydrogen was heretofore unproven. This progression has been characterized within a framework of separated longitudinal and transverse motion characterized by quantum numbers n_z and n_ρ, respectively.[5] The progression in Fig. 3 corresponds to $n_\rho = 1$, $n_z \cong 115$. The intensities of the lines are not in complete agreement, possibly for the reasons discussed above, but the experimental features are well reproduced with the following exception.

Some of the experimental Rydberg lines reveal a fine structure of uniformly spaced oscillations with a period $\sim 6 \times 10^{-3}$ cm^{-1} (see inset in Fig. 3). One possible origin could be a Stark effect due to the stray electric field ~ 20 mV cm^{-1} (dominantly along the magnetic field axis). Such a field would destroy parity and modify the continuum as well as the Rydberg progression. In particular, the field could mix the $n_\rho = 1$ Rydberg progression with high-lying Rydberg states of $n_\rho = 0$.

Another puzzling feature is the relatively broad widths of some calculated lines, typically 10×10^{-3} cm^{-1}. These lines lie below the ionization threshold, ~ 3 cm^{-1}

(the first Landau threshold), where the spectrum is discrete. The artificial width in the computed spectrum in this energy region is due to the truncation of the basis. We have verified that this effect is small, though possibly a computation with a larger basis could reproduce the observed modulations. We do not see what physical phenomenon could induce oscillations with the observed period. Experimentally, the observed modulations appear to lie in an envelope whose width agrees well with the calculated width (see Fig. 3). This fact suggests that the linewidth is not a computational artifact and deserves physical explanation.

The detailed agreement between theory and experiment shown in these spectra is surprising if one considers that the experiments were carried out with lithium rather than hydrogen. However, the largest quantum defect of the odd-parity states is the p-state defect $\delta_p = 0.05$: The defects for the higher-angular-momentum states are essentially negligible. Because l is not a good quantum number in the magnetic field, the p state becomes mixed with many other levels, and the effect of its quantum defect is vastly diluted. Rough estimates suggest that this could lead to shifts of $\sim 4 \times 10^{-3}$ cm^{-1}, which would be marginally discernible. The stray electric field, $\sim 20-60$ mV cm^{-1}, might also be expected to perturb the experimental spectrum. Aside from the fine structure of the

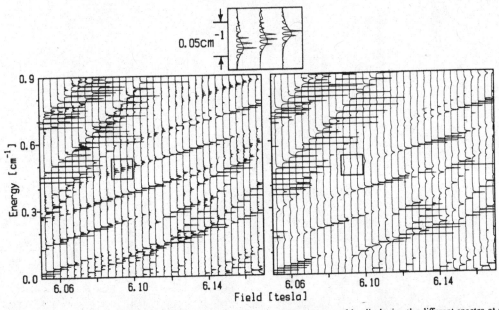

FIG. 3. Energy-level maps: experiment (left) and computed (right). These are created by displaying the different spectra at successively higher fields. Both spectra display a Rydberg progression of uniformly spaced levels converging to the second Landau threshold. The progressions are most easily seen by viewing close to the plane of the page from the left. The box above, an overlay of the two boxes inset in the spectra, displays the observed structure of fine oscillations and the corresponding calculated line width that are discussed in the text.

Rydberg levels described above, however, we have not been able to identify features that could be directly associated with an electric field.

By demonstrating the possibility of achieving high experimental and calculational precision for the positive-energy spectrum of the diamagnetic hydrogen atom, these results give hope of attaining new physical insight into the nature of nonseparable quantum systems. An important result is the confirmation of the experimental finding of Rydberg series in the spectra. Their occurrence suggests a resurgence of the Landau and Coulomb symmetries in a strongly mixed, classically chaotic, regime. Studies of the eigenfunctions, especially in the Rydberg windows, may give some clue as to how such a reorganization of symmetry arises.

The research at MIT was sponsored by the National Science Foundation Grant No. PHY 89-19381, and ONR Grant No. N00014-90-J-1322. We thank M. W. Courtney for assistance. CPU time on the Cray 2 computer was provided by the Conseil Scientifique du Centre de Calcul Vectoriel pour la Recherche. The Laboratoire de Spectroscopie Hertzienne de l'Ecole Normale Supérieure et de L'Université Pierre et Marie Curie is Unité Associée No. 18 du Centre National de la Recherche Scientifque.

(a)Present address: Department of Physics, Duke University, Durham, NC 27706.

(b)Present address: Lake Forest College, Lake Forest, IL 60045.

[1]D. Delande, A. Bommier, and J. C. Gay, preceding Letter, Phys. Rev. Lett. 66, 141 (1991).

[2]See for example, H. Friedrich and D. Wintgen, Phys. Rep. 183, 37 (1989); H. Hasegawa, M. Robnik, and G. Wunner, Prog. Theor. Phys. 98, 198 (1989); in *Atomic Spectra and Collisions in External Fields,* edited by M. H. Nayfeh, C. W. Clark, and K. T. Taylor (Plenum, New York, 1989), Vol. 2.

[3]A. Holle, G. Wiebusch, J. Main, K. H. Welge, G. Zeller, G. Wunner, T. Ertl, and H. Ruder, Z. Phys. D 5, 279 (1987).

[4]G. R. Welch, M. M. Kash, C. Iu, L. Hsu, and D. Kleppner, Phys. Rev. Lett. 62, 893 (1989).

[5]C. Iu, G. R. Welch, M. M. Kash, L. Hsu, and D. Kleppner, Phys. Rev. Lett. 63, 1133 (1989).

[6]A. Holle, G. Wiebusch, J. Main, B. Hager, H. Rottke, and K. H. Welge, Phys. Rev. Lett. 56, 2594 (1986); J. Main, G. Wiebusch, A. Holle, and K. H. Welge, Phys. Rev. Lett. 57, 2789 (1986); A. Holle, J. Main, G. Wiebusch, H. Rottke, and K. H. Welge, Phys. Rev. Lett. 61, 161 (1988).

Semiclassical approximation for the quantum states of a hydrogen atom in a magnetic field near the ionization limit

M. Yu. KUCHIEV

A.F. Ioffe Physical-Technical Institute,
194021 St. Petersburg, Russia

O. P. SUSHKOV

Budker Institute of Nuclear Physics,
630090 Novosibirsk, Russia

Abstract

Continuous and discrete spectrum states near the ionization limit are considered. The origin of the continuous spectrum narrow resonances is elucidated. We suggest the semiclassical equation for the states near the ionization limit. The way of analytical continuation of the equation for discrete spectrum states into the region of resonances in continuum is pointed out. It is shown that the wave function structure is intermediate between the regular structure and the chaotic one.

1 Introduction

In the works [1, 2, 3] on the photoexcitation of an Li atom in a magnetic field a number of continuous spectrum narrow resonances as well as discrete spectrum states were observed. The energies of all these states were comparable with the cyclotron frequency ω. The magnetic field was 6.1 T, and corresponding $\omega = 5.7\,\text{cm}^{-1}$. These are very high Rydberg excitations and therefore the spectrum of an Li atom probably does not differ substantially from that of an H atom. In recent years a number of theoretical works have been devoted to the investigation of an H atom in a magnetic field (see e.g. a review paper [4]). Progress has been achieved in numerical integration of the Schrödinger equation for the continuous spectrum [5, 6, 7]. In Ref. [8] a detailed comparison of experimental data with the results of numerical solution is carried out. There is good agreement after some averaging over the energy.

The analytical solution and explanation of the narrow resonances were suggested in our previous works [9, 10]. Our approach is somewhat similar to the semiclassical quantization suggested in Ref. [11]. The present chapter is based on Refs. [9] and [10]. We also formulate further problems which are still not resolved in the semiclassical quantization approach.

For the conditions of the experiments [1, 2, 3] the magnetic length is $a = 1/\sqrt{\omega} \approx 196$ (we use atomic units). The Coulomb turning point is given by

$r_Q \sim 1/E \sim 1/\omega = a^2 \approx 196^2$. These large distances can probably make the difference between the Li atom and the H atom not very prominent. However, this statement is not quite obvious because of a high density of the spectrum. It would be interesting to analyze the influence of an atomic core (cf. Ref. [7]). However, in the present work we restrict ourselves to the consideration of an H atom. The Hamiltonian of an H atom in a magnetic field $(B = B_z)$ is of the form

$$H_{tot} = \frac{\mathbf{p}^2}{2} - \frac{1}{\sqrt{z^2 + \rho^2}} + \frac{1}{8}\omega^2\rho^2 + \frac{1}{2}\omega L_z. \tag{1}$$

Here $\rho^2 = x^2 + y^2$. Let us introduce the reduced Hamiltonian

$$H = H_{tot} - \frac{1}{2}\omega L_z = \frac{\mathbf{p}^2}{2} - \frac{1}{\sqrt{z^2 + \rho^2}} + \frac{1}{8}\omega^2\rho^2. \tag{2}$$

The angular momentum L_z is an exact quantum number $(L_z = m)$. Therefore all the eigenstates of H are the eigenstates of H_{tot} with $\epsilon_{tot} = \epsilon + \frac{1}{2}\omega m$. Further we will consider the reduced Hamiltonian (2). Since $r_Q \gg a$ one can expect that the states are strongly stretched along the magnetic field. We restrict our consideration to these states only. Therefore we can expand (2) at small ρ^2/z^2:

$$H \approx \frac{1}{2}(p_z^2 + \mathbf{p}_\perp^2) - \frac{1}{z} + \frac{1}{8}(\omega^2 + \frac{4}{z^3})\rho^2. \tag{3}$$

At $z \gg \omega^{-2/3} = a^{4/3}$ the eigenstates of Hamiltonian (3) are given by

$$\psi_{n,m,\epsilon}(z,\rho) = A\frac{1}{\sqrt{p_z}}exp(\pm i\int_0^z p_z dz)\psi_{nm}(\rho)exp(im\phi), \tag{4}$$

where

$$\psi_{nm}(\rho) = \frac{1}{a^{1+|m|}|m|!}\left[\frac{(|m|+n)!}{2^{|m|}n!}\right]^{1/2} exp(-\rho^2/4a^2)\rho^{|m|}F(-n,|m|+1,\rho^2/2a^2) \tag{5}$$

is the wave function of a transverse motion in the magnetic field [12]. F is the confluent hypergeometric function, p_z is the longitudinal momentum: $p_z = \sqrt{2(1/z - \epsilon_{nm} + \epsilon)}$, $\epsilon_{nm} = \omega(n + \frac{|m|+1}{2})$.

The first impression is that the explanation of the continuous spectrum resonances is very simple: these are the one dimensional Coulomb levels which are built on the states (4): $\epsilon = \epsilon_{nm} - \delta\epsilon$. The value of $\delta\epsilon$ as well as the normalization constant A in the state (4) can be found using the Bohr–Sommerfeld quantization rule.

$$\delta\epsilon = \frac{1}{2n_Q^2}, \qquad A = \frac{1}{\sqrt{\pi n_Q^3}}. \tag{6}$$

Let us stress that the WKB phase as well as the normalization of the state (4) is determined by the region $z \gg a^{4/3}$ where the approximation (4) is applicable. A further scenario could look as follows. For example, the Coulomb state built on the Landau level $|n = 1, m = 0 >$ (but with the total energy $\epsilon > \epsilon_{IP} = \frac{1}{2}\omega$)

could acquire width due to the small mixing with the $|n = 0, m = 0 >$ state. The problem is that the mixing of the asymptotical states (4) which happens at $z \leq a^{4/3}$ is not small. The stationary state is a complicated combination of the basis states (4). The narrowness of the resonances is connected to the special structure of the mixing matrix.

2 Coulomb center scattering matrix

In the region $a^{4/3} \ll z \ll r_Q \sim n_Q^2$ the problem of the states (4) mixing can be reduced to the scattering problem. Let the wave incident on the origin from the right side $(z > 0)$

$$|n>_{in} = \frac{1}{\sqrt{p_z}} exp\left(-i\int_0^z p_z dz\right)\psi_{nm}(\rho)exp(im\phi). \tag{7}$$

It is reflected and we should calculate the Coulomb center scattering matrix $S^{(0)}$ to the states

$$|l>_{out} = \frac{1}{\sqrt{p_z}} exp\left(i\int_0^z p_z dz\right)\psi_{nl}(\rho)exp(im\phi). \tag{8}$$

It is shown below that the transmitted wave is small, therefore it is omitted in the basis of the $|>_{out}$ states. Due to this reason all the stationary states of the opposite parity (reflection $z \to -z$) are degenerate (cf. Ref.[13]).

It is convenient to use the operator formalism for calculation of $S^{(0)}$. The Heisenberg equations of motion corresponding to the Hamiltonian (2) are of the form

$$\ddot{z} \approx -\frac{1}{z^2}, \tag{9}$$

$$\ddot{x}_i \approx -\frac{1}{4}(\omega^2 + \frac{4}{z^3})x_i. \tag{10}$$

Here i=1,2; $x_1 = x$, $x_2 = y$. The solution of Eq.(9) is given by

$$z_0(t) = (9/2)^{1/3}|t|^{2/3}. \tag{11}$$

Solution (11) implies that z is not an operator, but the usual c-number. This result is connected with the fact that the motion along z at $z \gg \rho$ is always semiclassical. After the substitution of (11) into Eq.(10) one can find $x_i(t)$

$$\hat{x}_i(t) = \sqrt{|t|}\left[\hat{b}_i J_{1/6}(\omega|t|/2) + \hat{a}_i J_{-1/6}(\omega|t|/2)\right]. \tag{12}$$

Here $\hat{b}_i$ and $\hat{a}_i$ are operators which are defined by the initial conditions. J_ν is the Bessel function. At $z \gg a^{4/3}$ $(\omega|t| \gg 1)$

$$\hat{x}_i(t) \approx \frac{2}{\sqrt{\pi\omega}}\left[\hat{b}_i cos(\frac{\omega|t|}{2} - \frac{\pi}{3}) + \hat{a}_i cos(\frac{\omega|t|}{2} - \frac{\pi}{6})\right]. \tag{13}$$

It is evident from this equation that $b_i \sim d_i \sim \sqrt{n}$. Expanding the solution (12) at $z \ll a^{4/3}$ ($\omega|t| \ll 1$) we get

$$\hat{x}_i(t) = \hat{d}_i \frac{(2/9)^{1/6}}{\Gamma(5/6)} \left(\frac{\omega}{4}\right)^{-1/6} \sqrt{z} + \hat{b}_i \frac{(2/9)^{1/3}}{\Gamma(7/6)} \left(\frac{\omega}{4}\right)^{1/6} z. \qquad (14)$$

This expansion is valid at $na^{2/3} \ll z \ll a^{4/3}$. At $z \leq na^{2/3}$ the condition $\rho^2 \ll z^2$ is violated. To go through the region $z \leq na^{2/3}$ one should observe that at $z \ll a^{4/3}$ the magnetic field in the Hamiltonian (2) can be neglected, and therefore the problem is reduced to the pure Coulomb one. The expansion (14) describes the motion in a Coulomb field at a large distance from the center. Actually, even at $z \sim a^{2/3}$ due to (14)

$$x_i \sim d_i a^{2/3} + b_i a^{1/3} \gg 1. \qquad (15)$$

(The reader is reminded that we are using atomic units.) It is well known that the motion in a Coulomb field at a large distance from the center is always semiclassical (the distance is still less than the Coulomb turning point). Actually, Eq.(14) describes a parabola slightly rotated with respect to the z-axis. This is exactly the case of motion with zero energy in a Coulomb field. After orbiting around the nucleus an electron goes from one branch of the parabola to the other one. This corresponds to the reflection $\hat{d}_i \to -\hat{d}_i$. We would like to stress that the description of a motion in a magnetic field is not restricted by the semiclassic formulation. We consider the case of an arbitrary Landau level n. Nevertheless, due to the condition $a \gg 1$, the motion in a Coulomb field is always semiclassical. Thus, after a scattering on the center the trajectory is still described by solution (12) provided the substitution $\hat{d}_i \to -\hat{d}_i$ is carried out. Due to the complete reflection of the trajectories we omit the transmitted wave in the basis of $|>_{out}$ states for $S^{(0)}$ - matrix (Eq.(8)).

Let us fix the initial time $t_0 = -T$, so that $\omega T \gg 1$, $\omega T/2 - \pi/6 = 2\pi k + \pi/2$. Let at this moment $\hat{x}_i = \hat{x}_i(t_0)$, $\hat{p}_i = \hat{p}_i(t_0)$. Using Eq.(13) one can easily verify that at the moment $t_1 = +T$

$$\hat{x}_i(t_1) = \hat{x}_i(t_0), \qquad \hat{p}_i(t_1) = \hat{p}_i(t_0) - \sqrt{3}\omega\hat{x}_i(t_0). \qquad (16)$$

This is an exact quantum operator mapping.

Now we can calculate the scattering matrix $S^{(0)}$ for the asymptotic states (7),(8). By definition

$$< a_{out}|b_{in} >= S_{ab}. \qquad (17)$$

Therefore the matrix elements of any operator $\hat{A}$ before and after the scattering are related in the following way

$$\hat{A} \to S^{(0)+}\hat{A}S^{(0)}. \qquad (18)$$

Thus we should find the unitary operator which generates the mapping (16).

One can easily verify that it is of the form

$$S^{(0)} \sim exp\left(-i\frac{\sqrt{3}}{2}\omega\rho^2\right).$$ (19)

This is not yet the end of calculation. The times t_0 and t_1 were chosen rather arbitrarily. Therefore $S^{(0)}$ can be arbitrarily transformed to

$$S^{(0)} \rightarrow exp[-i(t_1' - t_1)H_0]S^{(0)}exp[i(t_0' - t_0)H_0],$$ (20)

where

$$H_0 = \frac{p_\rho^2}{2} + \frac{1}{2}\omega^2\rho^2.$$ (21)

In order to eliminate this ambiguity one should consider the dynamics in the magnetic field, see Eq.(13).

It is useful to formulate the same statement using the other language. The mappings (16) and (18) are relevant just to the transverse dynamics. However $|in>$- and $|out>$-states are the products of the transverse and longitudinal wave functions. At $a^{4/3} \ll z \ll r_Q$ one can arbitrarily transfer part of the phase dependence from the longitudinal wave function to the transverse one. For example, let us expand p_z in Eqs.(4),(7),(8):

$$\int_0^z p_z dz = \int_0^z \sqrt{2(1/z - \epsilon_{nm} + \epsilon)}dz \approx \int_0^z \sqrt{\frac{2}{z}}\left[1 + \frac{1}{2}z(\epsilon - \epsilon_{nm})\right]dz =$$

$$= \int_0^z \sqrt{\frac{2}{z}}dz + \frac{\sqrt{2}}{3}z^{3/2}(\epsilon - \epsilon_{nm}) = \int_0^z \sqrt{\frac{2}{z}}dz + (\epsilon - \epsilon_{nm})t.$$ (22)

Now one can transfer the $(\epsilon - \epsilon_{nm})t$ term to the phase of the transverse wave function.

The longitudinal motion is always semiclassical. Therefore the semiclassical calculation of $S^{(0)}$ presented in the appendix gives the most simple way to find the phase of $S^{(0)}$. Comparing Eqs.(19),(20) with Eq.(A17) we find

$$S^{(0)} = -exp\left(-i\frac{2\pi}{3\omega}H_0\right)exp\left(-i\frac{\sqrt{3}}{2}\omega\rho^2\right)exp\left(-i\frac{2\pi}{3\omega}H_0\right).$$ (23)

The standard calculation [12] using expression (5) gives $S^{(0)}$ in the Landau-states representation

$$S_{ln}^{(0)} = (-1)^{|m|}(-i)^{n+l}3^{\frac{n+l}{2}}2^{-n-l-|m|-1}\frac{1}{|m|!}\left[\frac{(n+|m|)!(l+|m|)!}{n!l!}\right]^{1/2}$$

$$\times F(-n,-l,|m|+1,-1/3).$$ (24)

F is a hypergeometric function. The only condition used in the derivation of Eqs.(23) and (24): $\rho^2/z^2 \ll 1$ when $z \ll a^{4/3}$. This condition is violated at $n, l, |m| \sim a^{2/3}$. Thus Eqs.(23) and (24) are valid at $n, l, |m| \ll a^{2/3}$.

3 The equation for the eigenstates of the discrete spectrum. Generalization of the Bohr–Sommerfeld quantization rule

The discrete spectrum state corresponds to an energy level below the ionization limit $\epsilon < \epsilon_{IP} = \frac{1}{2}\omega$. Let us write down the straightforward generalization of the Bohr–Sommerfeld quantization condition keeping in mind that the wave function is stretched along the z-axis. The matrix $S^{(0)}$ defines the boundary condition at small z. To solve the eigenvalue problem one should add the boundary condition at large z. It is convenient to write down this condition introducing the formal scattering matrix $S^{(\infty)}$ from the state (8) to the state (7) on the Coulomb turning point. Due to the standard semiclassical formulae

$$S_{ln}^{(\infty)}(\epsilon) = \delta_{ln} \cdot exp\left(2i\int_0^{z_n} p_z dz - i\frac{\pi}{2}\right) = \delta_{ln} \cdot exp\left(\frac{i\pi\sqrt{2}}{\sqrt{(\epsilon_{nm} - \epsilon)}} - i\frac{\pi}{2}\right). \tag{25}$$

Here $z_n = 1/(\epsilon_{nm} - \epsilon)$ is the Coulomb turning point. In operator form

$$S^{(\infty)}(\epsilon) = -iexp\left(\frac{i\pi\sqrt{2}}{\sqrt{(H_0 - \epsilon)}}\right). \tag{26}$$

Let us represent the wave function of a stationary state in the form $\Psi = \psi + \bar{\psi}$, where ψ is a combination of the states (7) incident on the origin and $\bar{\psi} = S^{(0)}\psi$ is a combination of the states (8). It is evident that the equation for the stationary states is of the form

$$S^{(\infty)}(\epsilon)S^{(0)}\psi = \psi. \tag{27}$$

The energy levels are given by the equation

$$|S^{(\infty)}(\epsilon)S^{(0)} - 1| = 0. \tag{28}$$

Eqs.(27),(28) generalize the Bohr–Sommerfeld quantization rule for the case in which we are interested. They are somewhat similar to the equation for semiclassical quantization derived in Ref.[11].

Let us discuss the general structure of the solution of Eqs.(27),(28). For simplicity we consider the case $m = 0$. First, let us note that at $\epsilon < \epsilon_0 = \epsilon_{IP}$ the matrix $S^{(\infty)}$ is the unitary one and therefore for any solution of Eq.(28) $Im\epsilon = 0$ at $Re\epsilon < \epsilon_0$. According to Eqs.(24),(A5) the matrix $S^{(0)}$ strongly mixes different Landau levels. Moreover, $S^{(\infty)}$ ((25),(26)) has a rather complicated dependence on ϵ. Therefore it is very natural to suppose that the solution of Eqs.(27),(28) is of the form

$$\psi = \sum_{n=0}^{N_{max}} \alpha_n |n>_{in}, \tag{29}$$

where all the α_n are of the same order of magnitude: $|\alpha_n| \sim \alpha$. What is N_{max}? To formulate the scattering problem (section 2) the condition $a^{4/3} \ll r_Q$ for the Coulomb turning point r_Q must be fulfilled. If the total energy $|\epsilon| \leq \omega$ the turning point for the state (4) built on the nth Landau level is $r_Q \sim 1/\omega n$. Therefore the above condition is violated at $n \sim a^{2/3}$. Thus $N_{max} \sim a^{2/3}$. We would like to stress that expression (24) for $S_{ln}^{(0)}$ is valid exactly for $l, n \ll a^{2/3} \sim N_{max}$. The solution (29) is written in terms of the unnormalized scattering states (7). In the basis of normalized states (4) it is as follows

$$\psi = \frac{1}{\sqrt{C}} \sum_{n=0}^{N_{max}} \beta_n \psi_{n,\epsilon}, \tag{30}$$

$$\beta_n = \frac{\alpha_n}{[\omega(n + 1/2) - \epsilon]^{3/4}}, \qquad C = \sum_{0}^{N_{max}} |\beta_n|^2.$$

We recall that $\epsilon < \frac{1}{2}\omega$. Thus the contribution of the higher Landau levels in the state (30) is suppressed. For $\frac{1}{2}\omega - \epsilon \ll \omega$ the state (30) is saturated by the zero Landau level. This is the case of regular dynamics. The spectrum of these states is Coulomb-like: $\frac{1}{2}\omega - \epsilon = 1/2n_Q^2$, $n_Q \gg a$.

For $|\epsilon| \gg \omega$ a lot of Landau levels ($\sim |\epsilon|/\omega$) are mixed in the wave function. This is a chaotic state in the sense of random matrix theory [14]. It is similar to the compound nucleus state. However, it is known that there are regions of regular dynamics even at $|\epsilon| \gg \omega$ [13, 15].

At $\frac{1}{2}\omega - \epsilon \sim \omega$ we have the intermediate case. For example the contributions of the different Landau states to the normalization decrease as $n^{-3/2}$. So the main contribution comes from the lowest levels. The situation for the value of $< \rho^2 >$ is opposite. Due to Eq.(30) it equals

$$< \psi|\rho^2|\psi > = \frac{1}{C} \left(\sum_{n=0}^{N_{max}} |\beta_n|^2 < n|\rho^2|n > + \sum_{n,l=n\pm2}^{N_{max}} \beta_n^* \beta_l < n|\rho^2|l > \right). \tag{31}$$

To estimate the mean value of $< \rho^2 >$ we should omit the interference term. If $\frac{1}{2}\omega - \epsilon$ is not very small we get from (31),(30)

$$<< \psi|\rho^2|\psi >> \sim 2a^2 \sum_{n=0}^{N_{max}} (n + 1/2)(n + 1)^{-3/2} \sim 4a^2 \sqrt{N_{max}} \sim 4a^{7/3}. \tag{32}$$

The double angle brackets indicate double averaging: over the quantum state and over the states. The estimation (32) agrees reasonably well with experimental data [1, 2, 3].

4 Continuous spectrum resonances.

Now we are going to show that with the correct definition of $S^{(\infty)}$ (25),(26), Eqs.(27),(28) are exact equations for the resonances in continuum. This means that their solutions give the positions and widths of the resonances. Let ϵ be in the region $\epsilon_1 > Re\epsilon > \epsilon_0$ (we consider the case $m = 0$). Then the components ψ_0 and $\bar{\psi}_0$ have no classical turning points. The standard condition [12] for the quasi-stationary level is the disappearance of the incident wave: $\psi_0 = 0$ at $z \to \infty$.

The $z \to \infty$ means $z \gg z_0 \sim 1/|\epsilon_0 - \epsilon|$. However, Eqs.(27),(28) are derived at $z \ll z_0$. To go to this region from $z \gg z_0$ one should consider the above barrier reflection of the wave going to infinity. The reflection happens at $z \sim z_0$. Due to the analytical dependence on ϵ it is evident that $S_{00}^{(\infty)}$ (25) is the correct amplitude of the above barrier reflection. The phase $\sqrt{(\epsilon_0 - \epsilon)}$ should be defined in such a way that $S_{00}^{(\infty)}$ is exponentially small: $\sqrt{(\epsilon_0 - \epsilon)} = -i\sqrt{(\epsilon - \epsilon_0)}$. Thus at $z \ll z_0$ we come back to Eqs.(27),(28). However, now the matrix $S^{(\infty)}(\epsilon)$ is not the unitary one and therefore the solutions of Eq.(28) have an imaginary part: $Im\,\epsilon = -\Gamma/2$, where Γ is the width of the level.

We would like to note that $S_{00}^{(\infty)}$ practically vanishes if ϵ is not very close to the ionization limit, i.e. $\epsilon - \epsilon_0 \sim \omega$. Actually

$$S_{00}^{(\infty)} \sim exp(-\pi\sqrt{2}/\sqrt{\omega}) = exp(-\pi\sqrt{2}a). \qquad (33)$$

Thus the exact analytical continuation of (25) is an academic exercise. In practical calculations one can set $S_{00}^{(\infty)} = 0$. A similar statement is also valid if many channels are open: $\epsilon > \epsilon_n > \epsilon_0$. In this case the matrix elements $S_{00}^{(\infty)}, ..., S_{nn}^{(\infty)}$ are exponentially small.

Thus at the appropriate continuation of $S^{(\infty)}$ Eqs.(27), (28) give the positions and widths of the resonances. For some conditions we can find the solution explicitly. Let ϵ once more be in the region $\epsilon_1 > Re\,\epsilon > \epsilon_0$. Let us switch off, for a moment, the coupling with the continuous spectrum. This means that in Eqs.(27),(28) instead of $S^{(\infty)} = 0$ we set $S^{(\infty)} = exp(i\gamma)$. γ is arbitrary. Then the states are stationary and the wave function has the same form as given by Eq.(30). However, only the components with $n \geq 1$ have the physical meaning. Similar to the discrete spectrum case at $\delta\epsilon = \frac{3}{2}\omega - \epsilon \ll \omega$ the state is saturated by the first Landau level (n=1), and the spectrum is Coulomb-like: $\delta\epsilon = 1/2n_Q^2$, $n_Q \gg a$. At $\delta\epsilon \sim \omega$ there are many Landau levels mixed in the state with an averaged weight $\sim n^{-3/2}$.

Now switch on the coupling with the continuous spectrum. It is quite obvious that for the levels with $\delta\epsilon \ll \omega$

$$\Gamma = v|S_{10}^{(0)}|^2 \approx 0.188\frac{\Omega}{2\pi} \qquad (34)$$

Here $v = 1/T = \Omega/2\pi$ is the frequency of classical motion; $\Omega = \Delta\epsilon = 1/n_Q^3$ is the splitting between the resonances. The value $|S_{10}^{(0)}|^2 \approx 0.188$ is calculated using Eq.(24). Thus

$$\frac{\Gamma}{\Delta\epsilon} = \frac{0.188}{2\pi}. \qquad (35)$$

For $\delta\epsilon \sim \omega$ we cannot calculate either ϵ or Γ analytically. However, it is obvious that as an estimation Eq.(35) is valid here as well. If $\delta\epsilon \sim 1/2n_Q^2 \sim \frac{1}{2}\omega$ then $n_Q \sim a$. For such a level and for the conditions of the experiments [1, 2, 3] Eq.(35) gives $\Gamma \sim 10^{-3}$ cm^{-1}. This agrees reasonably well with the experimental values.

Now let $(n_0 + \frac{1}{2})\omega > \epsilon > (n_0 - \frac{1}{2})\omega$. As above at $\delta\epsilon = (n_0 + \frac{1}{2})\omega - \epsilon \ll \omega$ there is a series of one dimensional Coulomb levels: $\delta\epsilon = 1/2n_Q^2$. At $n_0 \gg 1$ the width is equal to

$$\frac{\Gamma}{\Delta\epsilon} = \frac{1}{2\pi} \sum_{l < n_0} |S_{ln_0}^{(0)}|^2 = \frac{1}{12\pi}. \tag{36}$$

Here we have used the result obtained by D.Shepelyansky in the semiclassical formulation: $\sum_{l < n_0} |S_{ln_0}^{(0)}|^2 = \frac{1}{6}$, at $n_0 \gg 1$ [15]. In conclusion of this section we would like to emphasize that the narrowness of the resonances is due to the peculiarity of the $S^{(0)}$ matrix: the probability of downwards transition is relatively small.

5 Conclusion

In the present work we have suggested the semiclassical equation for the states of an H atom in a magnetic field near the ionization limit. This is a kind of generalization of the Bohr–Sommerfeld quantization rule. The Coulomb center scattering matrix is calculated analytically for $n \ll a^{2/3}$. It is shown that the wave function structure is intermediate between the regular structure and the chaotic one. The way of analytical continuation of the equation for discrete spectrum states into the region of resonances in continuum has been pointed out. The origin of the continuous spectrum narrow resonances has been elucidated. The widths of the Coulomb like resonances ($\delta\epsilon \ll \omega$) have been calculated analytically.

Below we would like to formulate some important problems which are still not solved in the approach suggested.

(i) The matrix equations (27),(28) have the dimension $\sim N_{max} \sim a^{2/3}$. This is ~ 30 for the conditions of the experiments [1, 2, 3]. This looks like a trivial problem for a computer. Nevertheless at the moment we cannot solve Eqs.(27),(28) explicitly and compare the calculated spectrum with experiment. The problem is the Coulomb center scattering matrix. Expression (24) is valid only for $l, n \ll N_{max}$. Thus we have to calculate the Coulomb center scattering matrix $S^{(0)}$ for $l, n \sim N_{max}$.

(ii) The Coulomb like levels discussed in the present chapter lie at $\delta\epsilon \ll \omega$. In Refs. [2] and [3] similar levels were observed at $\delta\epsilon \sim \omega$. The nature of these states is still unclear.

(iii) The slow oscillations in the photoabsorption spectra have long been known under the name of quasi-Landau resonances (see e.g. the review paper [4]). In nuclear physics terminology this is the modulation of the strength function due to the door-way states [14]. The standard interpretation of the door-way states in the single particle problem is as a closed classical trajectory (see e.g. the review paper [4]). Currently it is not clear how to match this approach to our approach. We think if we could understand how to match these two approaches this would help in the calculation of spread widths of the door-way states, as well as in the calculation of the dipole strength function.

Acknowledgments

We are grateful to B.V. Chirikov, for his interest in the present work.

APPENDIX

Here we calculate in the semiclassical approximation ($n \gg 1$) the scattering matrix $S^{(0)}$ between the states (7),(8). This approach is very useful for understanding the physical picture and the classical limit. We also use the semiclassical result to determine the phase of $S^{(0)}$ in the exact calculation presented in section 2.

In cylindrical coordinates the Hamiltonian (2) is as follows

$$H = \frac{p_\rho^2}{2} + \frac{m^2}{2\rho^2} + \frac{1}{8}\omega^2\rho^2 - \frac{1}{\sqrt{z^2+\rho^2}}. \tag{A1}$$

The wave function in the semiclassical approximation is of the form

$$\psi(\rho,z) = B(\rho,z) \cdot exp(i\sigma(\rho,z)) \cdot exp(im\phi), \tag{A2}$$

where σ is the classical action. At $t \to -\infty$ ψ coincides with the state (7) and

$$\sigma(\rho,z) = -\int_0^z p_z dz + \int_0^\rho p_\rho d\rho, \qquad B(\rho,z) = \frac{1}{\sqrt{p_z}}\frac{C}{\sqrt{p_\rho}}, \tag{A3}$$

$$p_z = \sqrt{2(1/z - \epsilon_{nm} + \epsilon)}, \qquad p_\rho = \sqrt{2\left(\epsilon_{nm} - \frac{m^2}{2\rho^2} - \frac{1}{8}\omega^2\rho^2\right)}.$$

Here $C = \sqrt{\omega/\pi}$ is the normalization constant for transverse motion, $\epsilon_{nm} = \omega(n + \frac{|m|+1}{2})$. To find the wave function ψ after the scattering we have to calculate the action along the classical trajectory. The trajectory is given by Eqs.(11),(12),(13),(14),(16). In this case $\tilde{x}_\perp$ and $\tilde{p}_\perp$ are the usual numbers not the operators. Due to the mapping (16) the scattering causes the variation of the transverse energy:

$$\epsilon_\perp = \epsilon_n = \frac{1}{2}\mathbf{p}_{\perp 0}^2 + \frac{1}{8}\omega^2\rho^2 \to \epsilon_l = \frac{1}{2}\mathbf{p}_{\perp 1}^2 + \frac{1}{8}\omega^2\rho^2 =$$

$$= \epsilon_n + \frac{3}{2}\omega^2\rho^2 \pm \sqrt{3}\omega\rho\sqrt{2\left(\epsilon_{nm} - \frac{m^2}{2\rho^2} - \frac{1}{8}\omega^2\rho^2\right)}. \tag{A4}$$

It is evident from this relation that the scattering strongly mixes the states. After a simple calculation one can find the limits within which ϵ_l lies. At small angular momentum m

$$1/\lambda \leq \epsilon_l/\epsilon_n \leq \lambda, \qquad \lambda = 7 + 4\sqrt{3} \approx 13.9. \tag{A5}$$

To find the correction to the solution $z_0(t)$ (11) we use the energy conservation law

$$\frac{dz}{dt} = \pm\sqrt{2(1/z - \epsilon_\perp + \epsilon)} \approx \pm\sqrt{2/z}(1 - \frac{1}{2}z\epsilon_\perp + \frac{1}{2}z\epsilon), \qquad (A6)$$

$$\epsilon_\perp = \frac{1}{2}(\dot{x}^2 + \dot{y}^2) + \frac{1}{8}(\omega^2 + \frac{4}{z_0^3})\rho^2.$$

The transverse energy should be calculated using Eqs.(11),(12). At $z \ll r_Q \sim a^2$ the correction to (11) is small: $\delta z \ll z$. Nevertheless the product $p_z \delta z$ is not small and therefore the correction is essential. At $\omega t \gg 1$ ($z \gg a^{4/3}$) $\epsilon_\perp$ is conserved and Eq.(A6) can easily be integrated explicitly:

$$z(t) \approx z_0(t)\left[1 + \frac{1}{5}(\epsilon - \epsilon_\perp)z_0(t) - -0.030(\epsilon - \epsilon_\perp)^2 z_0^2(t) + \ldots\right]. \qquad (A7)$$

To calculate the variation of the action along the classical trajectory,

$$\Delta\sigma = \int_{-T}^{T}(p_z^2 + p_\perp^2)dt, \qquad (A8)$$

we divide the trajectory into three intervals: $[-T, -\tau]$, $[-\tau, \tau]$, $[\tau, T]$. We recall that T corresponding to the mapping (16) obeys the conditions: $\omega T \gg 1$, $\omega T/2 - \pi/6 = 2\pi k + \pi/2$. Let us choose τ in such a way that $na^{2/3} \ll z(\tau) \ll a^{4/3}$. In the middle interval the magnetic field can be neglected and the pure Coulomb action is

$$\Delta\sigma[-\tau, \tau] = 4\sqrt{2z_\tau + p} \approx 4\sqrt{2z_\tau} + p\sqrt{2/z_\tau}, \qquad (A9)$$

$$z_\tau = z(\tau) \approx z(-\tau), \qquad p = \frac{d_x^2 + d_y^2}{(9\omega)^{1/3}\Gamma^2(5/6)}.$$

In the first and second intervals let us represent integrand in (A8) in the following form,

$$\mathbf{p}^2 dt = 2\left(\frac{1}{z} + \epsilon - \frac{1}{8}(\omega^2 + \frac{4}{z^3})\rho^2\right)dt$$

$$= \left(\frac{2}{z} + \epsilon - \epsilon_\perp\right)dt + \left(\epsilon + \epsilon_\perp - \frac{1}{4}(\omega^2 + \frac{4}{z^3})\rho^2\right)dt$$

$$\approx \sqrt{\frac{2}{z}}dz + \left(\epsilon + \frac{1}{2}(\dot{x}^2 + \dot{y}^2) - \frac{1}{8}(\omega^2 + \frac{4}{z_0^3})\rho^2\right)dt, \qquad (A10)$$

where we have used relations (A6). After the partial integration of $\dot{x}_\perp^2$ with substitution of $\ddot{x}_\perp$ from Eq.(10) one finds

$$\Delta\sigma[-T, -\tau] + \Delta\sigma[\tau, T] = 2\sqrt{2}\left(\sqrt{z(-T)} - \sqrt{z(-\tau)} + \sqrt{z(T)} - \sqrt{z(\tau)}\right)$$

$$+ 2\epsilon T + \frac{1}{2}\left(\dot{\rho}\rho|_{-\tau} - \dot{\rho}\rho|_{-T} + \dot{\rho}\rho|_T - \dot{\rho}\rho|_\tau\right). \qquad (A11)$$

Using the explicit expressions (13),(14) for $\ddot{x}_\perp(t)$ we calculate the sum of (A9) and (A11)

$$\Delta\sigma[-T,T] = 2\sqrt{2}\left(\sqrt{z(T)}+\sqrt{z(-T)}\right)+2\epsilon T - \frac{1}{2}\sqrt{3}\omega\rho^2. \qquad (A12)$$

Finally summing (A12) with the action before the scattering (A3) ($t = -T$) one finds the action after the scattering ($t = T$),

$$\sigma(\rho,z(T)) = 2\sqrt{2z(T)}+(\epsilon+\epsilon_{nm})T - \frac{1}{2}\sqrt{3}\omega\rho^2 + \int_0^\rho p_\rho d\rho. \qquad (A13)$$

We take into account that due to Eq.(13) at $t = T$ the trajectory comes to the initial point: $\ddot{x}_\perp(T) = \ddot{x}_\perp(-T)$. For the initial state (A2) all the trajectories start from different $\rho(-T)$ but from identical $z(-T)$. However, they come to different z, since due to Eqs.(A4),(A7) $z(T)$ is a function of ρ. Therefore it is more correct to write $z_\rho(T)$. Thus Eq.(A13) gives the phase of the semiclassical wave function (A2) on the line $(\rho, z(\rho))$.

The evaluation of $B(\rho,z)$ in (A2) is very simple. The momentum before the scattering is practically the same as that after the scattering: $p \approx p_z \gg p_\rho$. The classical trajectory comes to the same point ρ. Therefore from the current conservation equation

$$div(|B|^2\vec{p}) = 0, \qquad (A14)$$

we can conclude that $|B(\rho,z_\rho(T))| \approx |B(\rho,z(-T))|$. Moreover, it is easy to verify that the trajectory (11),(12) touches the caustic $4k+2$ times. Therefore

$$B(\rho,z_\rho(T)) \approx -B(\rho,z(-T)). \qquad (A15)$$

According to Eqs.(A2),(A3),(A13), and (A15) the wave function on the line $(\rho, z_\rho(T))$ after the scattering is equal to

$$\psi(\rho,z_\rho(T)) = -\frac{1}{\sqrt{p_z}}exp\left[i\left(2\sqrt{2z_\rho(T)}+(\epsilon+\epsilon_{nm})T-\frac{1}{2}\sqrt{3}\omega\rho^2\right)\right]\cdot\psi_{nm}(\rho). \quad (A16)$$

$\psi_{nm}(\rho)$ is the wave function of the transverse motion. To find the $S^{(0)}$ matrix we should decompose (A16) into the states (8) at $z = z_\rho(T)$. This gives

$$<l,m|S^{(0)}|n,m> = -exp\left(i(\epsilon_{lm}+\epsilon_{nm})T\right)\cdot<\psi_{lm}|exp\left(-i\frac{\sqrt{3}}{2}\omega\rho^2\right)|\psi_{nm}>. \quad (A17)$$

This expression exactly coincides with the results (23),(24).

REFERENCES

[1]   G. R. Welch, M. M. Kash, Chun-ho Iu, L. Hsu, and D. Kleppner, Phys. Rev. Lett. **62**, 1975 (1989).

[2]   Chun-ho Iu, G. R. Welch, M. M. Kash, L. Hsu, and D. Kleppner, Phys. Rev. Lett. **63**, 1133 (1989).

[3]   D. Kleppner, Chun-ho Iu, and G. R. Welch, Comments At. Mol. Phys. **25**, 301 (1991).

[4]   H. Friedrich and D. Wintgen, Phys. Rep. **183**, 37 (1989).

[5]   D. Delande, A. Bommier, and J. C. Gay, Phys. Rev. Lett. **66**, 141 (1991).

[6]   Q. Wang and C.H.Greene, Phys. Rev. A **44**, 1874 (1991).

[7]   P. F. O'Mahony and F. Mota-Furtado, Phys. Rev. Lett. **67**, 2283 (1991).

[8]   Chun-ho Iu, G. R. Welch, M. M. Kash, D. Kleppner, D. Delaand J. C. Gay, Phys. Rev. Lett. **66**, 145 (1991).

[9]   M. Yu. Kuchiev and O. P. Sushkov, Phys. Lett. A **158**, 69 (1991).

[10] M. Yu. Kuchiev and O. P. Sushkov, Phys. Rev. A, **47** 3426 (BR) (1993).

[11] E. B. Bogomolny, Comments At. Mol. Phys. **25**, 67 (1991).

[12] L. D. Landau and E. M. Lifshitz, Quantum Mechanics. Nonrelativistic Theory (Pergamon Press, Oxford, 1958).

[13] M. Yu. Sumetsky, Zh. Eksp. Teor. Fiz. **83**, 1661 (1982); (Sov. Phys. JETP).

[14] A. Bohr and B. R. Mottelson, Nuclear Structure, Vol. 1 (Benjamin, New York, 1969).

[15] D. L. Shepelyansky, in 'Quantum Chaos – Quantum Measurement' eds. P. Cvitanovich, I. Percival, A. Wirzba, p. 81–7 (Kluwer Academic Publishers, Dordrecht, 1992).

The semiclassical helium atom

D. Wintgen, K. Richter, and G. Tanner

Fakultät für Physik der Universität, Hermann-Herder-Str. 3, 7800 Freiburg, Germany

(Received 12 November 1991; accepted for publication 13 January 1992)

Recent progress in the semiclassical description of two-electron atoms is reported herein. It is shown that the classical dynamics for the helium atom is of mixed phase space structure, i.e., regular and chaotic motion coexists. Semiclassically, both types of motion require separate treatment. Stability islands are quantized via a torus–quantization-type procedure, whereas a periodic-orbit cycle expansion approach accounts for the states associated with hyperbolic electron pair motion. The results are compared with highly accurate *ab initio* quantum calculations, most of which are reported here for the first time. The results are discussed with an emphasis on previous interpretations of doubly excited electron states

I. INTRODUCTION

The failure of the Copenhagen School to obtain a reasonable estimate of the ground–state energy of the helium atom (see e.g. the old review by van Vleck[1]) and of the H_2^+–molecule[2] was a cornerstone in the evolution of quantum mechanics. The pessimistic point of view concerning a semiclassical treatment of two–electron atoms is summarized in the book of Born[3]:

"... the systematic application of the principles of the quantum theory ... gives results in agreement with experiment only in those cases where the motion of a single electron is considered; it fails even in the treatment of the motion of the two electrons in the helium atom.

This is not surprising, for the principles used are not really consistent... A complete systematic transformation of the classical mechanics into a discontinuous mechanics is the goal towards which the quantum theory strives."

Nowadays we know the essential shortcomings of the old quantum theory:

(i) the role of conjugate points along classical trajectories and their importance for the approach to wave mechanics (which was not developed at those times) were not properly accounted for; and

(ii) the precise role of periodic trajectories when the classical dynamics are non–integrable or even chaotic was unknown.

The pessimistic point of view dominated the research for several decades and there were no successful attempts to attack the problem until Leopold and Percival[4] in 1980 gave a reasonable estimate of the ground–state energy of the helium atom, ignoring however item (ii). Nowadays, a proper semiclassical treatment of the helium atom is still an outstanding problem of the semiclassical theory. The helium atom therefore remains the essential touchstone of semiclassical mechanics, even though considerable progress in the development of the formal theory has been achieved within the last years, most of which is documented in this issue of CHAOS.

A semiclassical description of two–electron atoms is also highly desirable, because most parts of the spectral regions are still unexplored, both experimentally and quantum theoretically. From a conceptual point of view highly accurate quantum calculations are not too difficult to perform. However, the high dimensionality of the problem combined with the vast density of states can make the calculations cumbersome and elaborate. In addition, one has to deal with singular potentials, long–ranged interactions, and typically many open decay–channels, all of which prevents the success of brute–force methods. Furthermore, the problem of understanding the structure of the quantum solutions still remains after solving the Schrödinger equation. Again, simple interpretation of classical and semiclassical methods assists in illuminating the structure of the solutions. Classical calculations may also help to uncover local integrals of motions or adiabatic coordinates. Exploiting such properties may facilitate (approximate) quantum calculations considerably.

The necessary ingredient for any semiclassical analysis is a proper understanding of the underlying classical dynamics. Unfortunately, this information is highly nontrivial to obtain. The equations of motion are multidimensional, non–integrable, and singular, hence far away from an easy–to–do–job. In addition, the independent particle case $1/Z = 0$ (Z is the nuclear charge) is highly degenerate, which prohibits an application of the KAM theory to derive an independent particle limit. In other words, the phase space structure of the hydrogenic motion of two independent electrons depends on an (infinitesimal) perturbation and not only on the zero–order Hamiltonian itself. As a matter of fact, until recently it was even unknown whether the motion of two–electron atoms is ergodic or not.[5]

In this contribution we review on recent progress in the classical and semiclassical description of two–electron atoms. We show that the classical phase space is of mixed structure, i.e. regular and irregular motion of the electron pair co–exist. Roughly, the angular type of motion (i.e. bending motion of the electron pair relative to the nucleus) is mostly stable, whereas radial motion is mostly (but not always) unstable. The radial instability typically leads to ionization of one electron (we restrict the analysis to energies below the three–particle breakup threshold). A semiclassical treatment has to distinguish between fully stable (i.e. stable in all dimensions) and (partly) unstable motion. Fully stable motion allows for approximate torus quantization, and this applies to electron pair motion, where both electrons are located on the same side of the nucleus in a near–collinear configuration.[6] The classical motion for near–collinear configurations with both electrons on different sides of the nucleus turns out to be fully chaotic. In this case the semiclassical Gutzwiller theory combined with the cycle expansion method yields good results.[7]

We will not discuss in this contribution how to solve the Schrödinger equation, even though we partly "review" on quantum results which are at present not available in the literature. Our main concern lies in demonstrating the power of semiclassical methods for two–electron atoms and how the semiclassical results compare with highly accurate quantum results.

II. CLASSICAL MOTION IN HELIUM

There are only few rigorous results about the general classical three–body Coulomb problem. The reason for the lack of popularity of quantitative classical studies is obvious: the equations of motion are multi–dimensional, non–integrable, and singular. In addition, the independent particle case $1/Z = 0$ (Z is the nuclear charge) is highly degenerate, which prohibits a direct application of the KAM theorem to derive a proper independent particle limit. Quantitative analyses of the problem are being developed at present.[5–11]

An essential ingredient for the classical analysis of the three–body Coulomb problem is the regularization of the equations of motion.[12] For a nucleus with charge Z and infinite mass the Hamiltonian reads (atomic units used, $e = m_e = 1$):

$$H = \frac{\mathbf{p}_1^2 + \mathbf{p}_2^2}{2} - \frac{Z}{r_1} - \frac{Z}{r_2} + \frac{1}{r_{12}}. \tag{1}$$

The electron–nucleus distances are given by r_i, $i = 1,2$, and the distance between the electrons is r_{12}. Whenever an inter–particle distance vanishes (particle collision) the potential energy diverges. There is a striking difference in the topology of the various collisions. In analogy to the motion of the electron in the hydrogen atom, the motion can be regularized for *binary collisions*, where only one inter–particle distance vanishes. However, the *triple collision* $r_1 = r_2 = r_{12} = 0$ cannot be regularized, i.e. these solutions have branch points of infinite order.[13] A numerically convenient method to regularize the binary collisions can be found in Ref. 5.

The energy E and the total angular momentum $\mathbf{L}$ are constants of motion. Furthermore, the Hamiltonian (1) is invariant under reflection $(\mathbf{r}_1,\mathbf{r}_2) \to (-\mathbf{r}_1, -\mathbf{r}_2)$ and particle exchange $(\mathbf{r}_1,\mathbf{r}_2) \to (\mathbf{r}_2,\mathbf{r}_1)$. The Hamiltonian (1) is homogeneous in coordinates and momenta and the equations of motion can be scaled to energy independent form. The accumulated action along a classical path is then

$\widetilde{S}(E) = 2\pi zS$ with $z = (-E)^{-1/2}$ and $2\pi S$ the action at energy $E = -1$.

Here we shall focus on total angular momentum $\mathbf{L} = 0$, for which the motion of the electrons is confined to a plane fixed in configuration space. This removes three of the total of six degrees of freedom, and we take the three inter–particle distances r_i as dynamical variables. It is convenient to replace these by the *perimetric coordinates*[14]

$$x = r_1 + r_2 - r_{12}, \; y = r_1 - r_2 + r_{12}, \; z = -r_1 + r_2 + r_{12}, \tag{2}$$

with $x,y,z \geqslant 0$. The perimetric coordinates treat all inter–particle distances democratically. The discrete symmetries of the Hamiltonian (1) are readily identified as symmetry planes in the perimetric coordinate set. Collinear motion with both electrons on different sides of the nucleus is confined to the $x \equiv 0$ plane. Collinear motion with both electrons localized on the same side of the nucleus is given by either $y \equiv 0$ or $z \equiv 0$. Finally, motion on the so–called *Wannier ridge*[15] $r_1 \equiv r_2$ takes place in the $y \equiv z$ plane. The electron motion in the symmetry planes becomes essentially two–dimensional. The third degree of freedom is taken into account by linearizing the equations of motion around the symmetry plane.

Here we will focus on near–collinear configurations only, but as we will see this is already enough to uncover the difficulties of the full problem and to draw some definitive conclusions about it. Our main concern is to unravel the structure and the organization of the periodic orbits. They are the main ingredients of modern multi–dimensional semiclassical theories as discussed in the next sections. The radial motion along the Wannier ridge of symmetrical electron configurations $r_1 \equiv r_2$ is (except for the so–called Langmuir orbit[5]) extremely unstable and of minor importance for a semiclassical treatment. Therefore we will discuss this type of classical motion only briefly.

A. The $Z^{2+}e^-e^-$ configuration

Consider a collinear arrangement of a nucleus of charge Z and of two electrons, both being on the same side of the nucleus. The fundamental periodic motion of such a configuration is a coherent oscillation of both electrons with the same frequency but, as it turns out, with large differences in their individual radial amplitudes and velocities as shown in Fig. 1(a) for $Z = 2$ helium. The outer electron appears to stay nearly frozen at some fixed radial distance. For this reason we label the orbits as *frozen planet configurations*. The minimal nuclear charge to bind an electron in this type of collinear configuration is $Z > 1$; otherwise the outer electron potential is purely repulsive. On the other hand, $1/Z$ must be non–zero, i.e. the repulsive electron–electron interaction is of crucial importance for the formation of these states. Thus the configurations considered here cannot be described within an independent particle model and are of non–perturbative nature. The high degree of classical dynamical localization of the outer electron is mostly pronounced for helium and becomes weaker for larger integer values of Z.

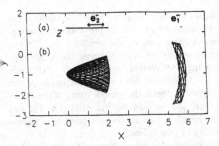

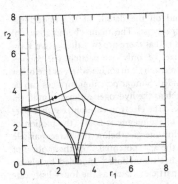

FIG. 1. (a) The straight line motion of the electron pair for the frozen planet PO, while a nonperiodic but regular trajectory in its neighborhood is shown in (b). Energy scaled units $(-E)r$ are used.

FIG. 3. Equipotential lines and boundary of the classical allowed region for the collinear electron configuration with both electrons on opposite sides of the nucleus ($Z = 2, E = -1$). The PO '$+ - - -$' is also shown.

Remarkably, the periodic orbit (PO) of Fig. 1(a) is linearly stable with respect to variations in the initial conditions. This is demonstrated in Fig. 1(b) which shows the resulting (regular) motion of the electrons when they are initially in a slightly off–collinear arrangement. The inner electron moves on perturbed Kepler ellipses around the nucleus, while the outer electron remains trapped at large radial distances following the slow angular oscillations of the inner electron.

For collinear configurations the motion is confined to the three–dimensional energy shell of a four–dimensional subspace of the full phase space. It is convenient to visualize the phase space structure by taking Poincaré surfaces of section. Such a section is shown for helium in Fig. 2. The phase space position $\{r_1, p_1\}$ of the outer electron is monitored each time the inner electron approaches the nucleus ($r_2 = 0$). The PO shown in Fig. 1(a) appears as the elliptic fixed point in the center of the extended torus structure. Near the fixed point the motion of the outer electron is nearly harmonic, but for large radial distances the tori are deformed according to the almost Keplerian motion of the outer electron. The non–closed manifolds surrounding the

tori represent (regular) trajectories for which the outer electron ionizes with $p_1 \to (2E_1 + 2/r_1)^{1/2}$ (E_1 is the asymptotic excess energy of the ionizing electron). Recalling the additional stability of the bending degree of freedom (i.e. motion off the collinear arrangement), the fundamental PO of Fig. 1 is embedded in a fully six–dimensional island of stability in phase space.

The near–integrability of the three–body Coulomb problem for asymmetric configurations as shown in Figs. 1 and 2 is a remarkable fact, which nevertheless was unknown until recently. The stability of the outer electron with respect to radial motion can be understood in a static model (e.g. by fixing the inner electron at its classical expectation value or its outer classical turning point) but the stability with respect to the bending degree of freedom is somewhat surprising and its origin is purely dynamical. It is also unexpected and surprising that these classical configurations are extremely stable against autoionization, which is allowed energetically. Intuitively, one would expect that the inner electron "kicks" the loosely bound outer electron out because the electron–electron interaction $1/r_{12}$ is maximized in such a collinear configuration. However, as we will see in the next section, instabilities of the system emerge mostly from the (non–regularizable) triple collisions, where all inter–particle distances vanish.

B. The $e^- Z^{2+} e^-$ configuration

Configurations where the electrons move on opposite sides of the nucleus are energetically favored because the electron–electron interaction is minimized. Quantum mechanically, these are the (resonant) states in which $- \langle \cos \Theta \rangle$ is close to unity. Here, Θ is the angle between r_1 and r_2. These states are dominantly excited in single–photon transitions from the ground state.[16]

Equipotential lines for this type of collinear electron arrangement are shown in Fig. 3 together with a typical periodic trajectory. The system ionizes if either $r_1 \to \infty$ or $r_2 \to \infty$. As a matter of fact, the topology of the equipotential lines and of the boundary of the classically allowed

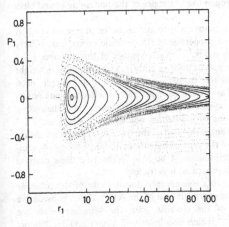

FIG. 2. Poincaré surface of section ($r_2 = 0$) for collinear configurations with both electrons on the same side of the nucleus.

region do not depend on the details of the underlying particle–particle interactions. The main characteristic of the potential surfaces is that there are two alternative ways of the system to ionize, i.e. only one particle (or equivalently one degree of freedom) can ionize whereas the other particle remains bounded. Similar potential surfaces can be found in problems such as the hydrogen atom in a uniform magnetic field[17] (where the electron can ionize either parallel or anti–parallel to the magnetic field), the motion of triatomic molecules for energies below the three–particle breakup threshold,[18] the x^2y^2 potential,[19] or the motion of ballistic electrons in heterojunctions[20] modeled by elastic pinball scattering of particles (such as the four–disk scattering system[17] or the hyperbola billiard[21]).

The classical motion of the collinear helium atom with the electrons on different sides of the nucleus turns out to be fully chaotic, even though we cannot rigorously prove this. A system is called "chaotic" if all PO are linearly unstable and their number proliferates exponentially with the action (or some other length characteristic). The exponential proliferation becomes obvious if the PO can be mapped onto a tree of symbols as, e.g., for the anisotropic Kepler problem (AKP)[2,22] or the diamagnetic Kepler problem (DKP).[17] Our numerical findings on the collinear motion of the helium atom suggest that the PO obey a binary coding.

To characterize the motion on the potential surface, Fig. 3, we introduce a symbolic description of the trajectories by recording the sequence $\{i_j\}, \ldots, i_{-1}, i_0, i_1, i_2, \ldots,$ of electron collisions with the nucleus, i.e., $r_i = 0$. Starting at an end point in Fig. 3 the PO is then coded by the periodically continued string of symbols $\ldots 12122121 \ldots$. There appears to be no restriction on allowed symbol sequences, but for a PO the length of the periodic symbol sequence must be even.

The restrictions on the possible POs can be overcome by de-symmetrizing the motion and considering the motion in the *fundamental domain*,[17,23] which is only half of the configuration space shown in Fig. 3 with an elastically reflecting wall at $r_1 \equiv r_2$. This classical procedure corresponds to the separation of discrete symmetries in quantum mechanics and to the symmetrization of the Green function in the semiclassical theory. Here, the discrete symmetry corresponds to the exchange of electron coordinates $r_1 \leftrightarrow r_2$ (Pauli principle). All information (either classically or quantum mechanically) is contained in the de-symmetrized motion of the fundamental domain,[24] to which we will restrict ourselves from now on.

When there is a discrete symmetry only an initial segment of the periodic orbit needs to be considered. After some fraction of its period, the orbit will pass through an image (under the discrete symmetry group) of the initial point. The further evolution may then be obtained from symmetry images of the initial segment. The action of the orbit and the period, being scalars, are simply additive under the symmetry transformation. The stability matrix, however, is sensitive to the type of symmetry transformation.[24] An example for such a symmetric PO is the trajectory shown in Fig. 3. A redefinition of the coding

scheme accounts for the symmetry of the PO: a collision is denoted by the symbol ' + ' if the previous collision was by the *same* electron, and by the symbol ' − ' if the collision before was by the *other* electron. Using this fundamental coding the symbol string for the PO reads ' + − − − '. Now we have to add an image of the fundamental code to obtain the symbol string of the PO in the full domain, just as with the PO itself. The PO has (topological) length 4, because its code consists of repetitions of a string of four symbols.

We now assume that the collinear PO *not* involving triple collisions can be mapped one–to–one onto the binary symbols $\{ +, - \}$. This conjecture is supported by numerical results summarized in Table I, where we list all PO up to symbol length 6. Some of these orbits are shown in Fig. 4.

Apart from the missing orbit ' + ' (which parallels the AKP and the DKP problem) all PO exist for the symbol sequences of Table I. The coding takes care automatically of the discrete symmetries of PO. The maximal number of conjugate points (within the collinear configuration) of a PO is given by its symbol length, and the Morse index α by twice the symbol length. The type of fixed point is determined whether the number of ' − ' in the sequence is odd or even. All stability exponents u listed in Table I are strictly positive, i.e. all the orbits are unstable with respect to the motion *within* the collinear arrangement (radial correlation). The linearized motion *off* the symmetry plane (angular correlation) is however stable and characterized by the winding number γ, i.e. the eigenvalues $\exp(\pm 2\pi i \gamma)$ of the stability matrix.[24]

The missing *fundamental orbit* (i.e., orbit of length 1) ' + ' can be formally assigned to a PO for which one electron is removed to infinity whereas the second electron is moving on a degenerate Kepler ellipse. The orbit does not give rise to resonant structures which are formed within a finite reaction zone around the nucleus. A 'naive' WKB–quantization of this trajectory, however, gives the correct energies of the two–particle breakup thresholds, i.e. the (hydrogen-like) energies of the remaining bound electron.

An alternative way to introduce the coding is to exploit the discrete properties of the periodic orbits, particularly the Morse index.[17,24] Each symbol of the alphabet $\{ +, - \}$ carries some discrete additive or multiplicative weight for the discrete properties of the PO. For example, each symbol is associated with *one* pair of self–conjugate points along the trajectory, which allows a definition of the coding via collisions with potential boundaries.[17] Also, each ' − ' symbol changes the sign of the trace of the orbit's stability matrix and the type of hyperbolicity, respectively. A symbolic description must be able to describe these discrete properties, otherwise it is useless.

The coding scheme introduced above of the collinear helium atom parallels that of the AKP. There, consecutive crossings of the electron with the symmetry axis are recorded,[2,22] which also leads to a binary coding. The orbits coming from a collision with the nucleus (*collision manifold*) generate the partitioning of the phase space into

TABLE I. Various properties of the collinear periodic orbits of the helium atom. u is the stability exponent, i.e. the Liapunov exponent times the action S of the orbit. The Morse index α for the motion in the symmetry plane and the winding number γ for the linearized motion off the symmetry plane are given in the next columns. The type of fixed point (FX) is denoted by H for hyperbolic orbits and by IH for hyperbolic orbits with reflection (taken from Ref. 7).

No	Code	S	u	γ	α	FX
1	+	...	...	...	2	H
2	−	1.829 00	0.6012	0.5393	2	IH
3	+ −	3.618 25	1.8622	1.0918	4	IH
4	+ + −	5.326 15	3.4287	1.6402	6	IH
5	+ − −	5.394 52	1.8603	1.6117	6	H
6	+ + + −	6.966 77	4.4378	2.1710	8	IH
7	+ + − −	7.041 34	2.3417	2.1327	8	H
8	+ − − −	7.258 49	3.1124	2.1705	8	IH
9	+ + + + −	8.566 19	5.1100	2.6919	10	IH
10	+ + + − −	8.643 07	2.7207	2.6478	10	H
11	+ + − + −	8.937 00	5.1563	2.7292	10	H
12	+ + − − −	8.946 19	4.5932	2.7173	10	IH
13	+ − + − −	9.026 90	4.1765	2.7140	10	H
14	+ − − − −	9.071 79	3.3424	2.6989	10	IH
15	+ + + + + −	10.138 74	5.6047	3.2073	12	IH
16	+ + + + − −	10.216 74	3.0324	3.1594	12	H
17	+ + + − + −	10.570 67	6.1393	3.2591	12	H
18	+ + + − − −	10.576 29	5.6766	3.2495	12	IH
19	+ + − + − −	10.706 99	5.3252	3.2520	12	IH
20	+ + − − + −	10.706 99	5.3252	3.2520	12	IH
21	+ + − − − −	10.743 04	4.3317	3.2332	12	H
22	+ − + − − −	10.878 55	5.0002	3.2626	12	H
23	+ − − − − −	10.910 15	4.2408	3.2467	12	IH

cells which are uniquely labeled by the binary code. Analogously, the collision manifold for the collinear helium atom is represented by the trajectories coming out of a triple collision, where all inter–particle distances vanish.

The collision manifold plays a peculiar role in the classical description of the collinear helium atom. If the singularity of the Coulomb potential is smoothed (for example by a non–vanishing total angular momentum L of the three–body complex) the orbits starting and ending in a triple collision become periodic trajectories. The appropriate coding is then ternary and equivalent to the DKP problem[17] and to problems having similar potential surfaces, as discussed at the beginning of this section. Eventually, if the smoothing is too strong, the symbolic tree of orbits is pruned and some orbits do no longer exist or become stable. The further fundamental orbit '0' of the ternary coding is the symmetric stretch motion of the electrons with $r_1 \equiv r_2$. The (in-phase) symmetric stretch trajectory is better known in atomic physics as *Wannier ridge* configuration, where it plays an important role in the Wannier theory of three–particle breakup (for $E > 0$).

In the Wannier configuration of non–vanishing angular momentum the electrons move on closed Kepler ellipses with $r_1 = -r_2$. An important property of the Wannier orbit (as well as the other orbits of the collision manifold) is that its stability exponent diverges as it approaches the triple collision.[5] This is shown in Fig. 5 where we plot (for various values of the nuclear charge Z) the stability index λ_α for the radial correlation against a logarithmic scale of the scaled total angular momentum. The diverging stability index for the Wannier orbit has important semiclassical consequences: the orbit does not give rise

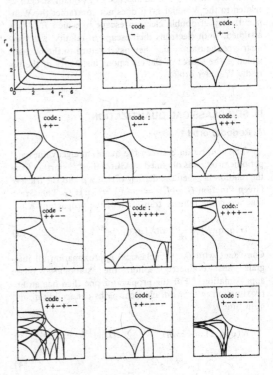

FIG. 4. Periodic orbits of the collinear helium atom ($\Theta = \pi$). The upper left figure shows some equipotential lines and the symmetric stretch motion (Wannier configuration) along the symmetry line $r_1 \equiv r_2$.

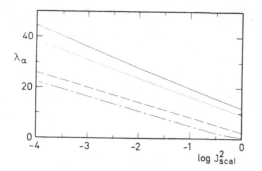

FIG. 5. The stability exponent λ_α for the Wannier PO as a function of the scaled total angular momentum J_{scal} for various Z-values, $Z = 0.4, 0.5, 1, 5, 100$ (from above). The scaled angular momentum is related to the real angular momentum L by $J_{scal} = L/L_{max}$, where $L_{max} = 2(Z - \frac{1}{4})/\sqrt{-E}$ is the maximal angular momentum at energy E.

to resonant structures in the density of states (see also the section about semiclassical quantization). This classical prediction is more remarkable considering that since decades the credo of electron pair motion along the Wannier ridge plays an important role in the interpretation of doubly excited states.[25–30] The absence of resonant structures related to the Wannier orbit does not contradict the Wannier theory of double electron escape in which $E \geqslant 0$ is assumed. Both electrons then escape to infinity in a symmetric configuration and they never return to the nucleus, which is the source of the enormous instability of the periodic Wannier orbit.

III. SEMICLASSICAL QUANTIZATION

A. Periodic orbit theory

The connection between the quantum eigenvalues and periodic orbits was obtained by Gutzwiller[2,31] starting from the relation between the density of states ρ and the trace of Green function G, $\rho(E) = -(1/\pi)\text{Im tr } G$. The Green function is the Fourier transform of the propagator K,

$$G(\mathbf{q}_2, \mathbf{q}_1; E) = \frac{1}{i\hbar} \int dt K(\mathbf{q}_2, \mathbf{q}_1; t) e^{iEt/\hbar}. \quad (3)$$

Consistent with the semiclassical approximation all integrals are evaluated using the stationary phase approximation.[32] For the propagator one then has an approximation in terms of classical paths connecting $\mathbf{q}_1$ and $\mathbf{q}_2$ in time t,

$$K(\mathbf{q}_2, \mathbf{q}_1; t) = (2\pi i\hbar)^{-N/2} \sum_{\text{paths}} |D_W|^{1/2} \times \exp(iW_{p,21}/\hbar$$
$$- i\pi\nu_p/2) \quad (4)$$

with $W_{p,21} = \int_1^2 L dt$ the Lagrangian action,

$$D_W = \det\left(-\frac{\partial^2 W_{p,21}}{\partial \mathbf{q}_2 \partial \mathbf{q}_1}\right) \quad (5)$$

the determinant of second derivatives, and ν_p the number of caustics, i.e. it counts the number of zeros of the determinant (5) along the path p.

A semiclassical expression for the Green function is the integral (3) with the (exact) propagator replaced by its semiclassical approximation (4). Stationary phase approximation of the Fourier integral yields

$$G(\mathbf{q}_2, \mathbf{q}_1; E) = G_0 + \frac{1}{i\hbar(2\pi i\hbar)^{(N-1)/2}} \sum_p |D_S|^{1/2}$$
$$\times \exp[iS_p(E)/\hbar - i\pi\nu_p'/2], \quad (6)$$

where now the sum extends over classical paths of nonvanishing length connecting $\mathbf{q}_1$ and $\mathbf{q}_2$ at a fixed energy E, irrespective of the time it takes; S_p is the classical action $\int_p \mathbf{p} d\mathbf{q}$ of Maupertuis,

$$D_S = \det\begin{vmatrix} \dfrac{\partial^2 S_p}{\partial \mathbf{q}' \partial \mathbf{q}} & \dfrac{\partial^2 S_p}{\partial E \partial \mathbf{q}} \\[2mm] \dfrac{\partial^2 S_p}{\partial \mathbf{q}' \partial E} & \dfrac{\partial^2 S_p}{\partial E^2} \end{vmatrix} \quad (7)$$

is the determinant of second derivatives, and the index ν_p' counts the number of caustics on the energy shell (which, dependent on the sign of $\partial^2 W/\partial t^2$, may differ from ν_p by unity). The paths of zero length contribute differently and are contained in G_0. They are unimportant for the following discussion.

Finally, to obtain the density of states we have to integrate the diagonal elements of the Green function over position space. The phase of G is stationary if the final and initial momenta coincide, which is the condition that the trajectory be periodic. In the neighborhood of every closed path a coordinate system with q_1 along the path and $q_2, \cdots q_N$ perpendicular to it may be introduced. Using the factorization of the determinant D_S and the fact that up to second order in the deviations from the trajectory the action only depends on the stability matrix of the classical path, one finds

$$\frac{1}{(2\pi i\hbar)^{(N-1)/2}} \int dq_2 \cdots dq_N |D_{S_p}|^{1/2}$$
$$\times \exp[iS_p(\mathbf{q})/\hbar - i\nu'_p\pi/2]$$
$$= \frac{1}{|\dot{q}_1|} \frac{\exp(iS_p/\hbar - i\mu_p\pi/2)}{|\det(M-1)|^{1/2}}, \quad (8)$$

where S_p is the action along the periodic orbit, M is the stability matrix around the orbit, and the phase shift μ_p is the Morse index of the PO.[24]

Since the trace of the stability matrix is independent of the position along the path, there remains the integral $\int dq_1/\dot{q}_1$, which by $dq/\dot{q} = dt$ is the period of the orbit. Allowing for multiple traversals of a particular PO, we finally find for the contribution of one *primitive* (i.e. nonrepeated) periodic orbit (PPO)

$$\frac{-i}{\hbar} T_p \sum_{r=1}^{\infty} \frac{e^{(iS_p/\hbar - i\mu_p\pi/2)r}}{|\det(M^r - 1)|^{1/2}}. \quad (9)$$

The trace over G_0 can be calculated as an asymptotic series in powers of $\hbar$.[33] The leading term is given by the size of the energy shell,

$$\text{tr } G_0 = \int \frac{d\mathbf{p}d\mathbf{q}}{h^N} \delta[E - H(\mathbf{p},\mathbf{q})]. \quad (10)$$

This function depends smoothly on energy.

B. Quantization of elliptic islands: the frozen planet configurations

As shown in the previous section, the frozen planet periodic orbit is linearly stable for helium. The two pairs of eigenvalues of the stability matrix M are then complex numbers on the unit circle, i.e. $\lambda_R = \exp(\pm 2\pi i \gamma_R)$ and $\lambda_\Theta = \exp(\pm 2\pi i \gamma_\Theta)$. For trajectories close to the periodic orbit the frequency ratios of the radial and angular motion transverse to the periodic orbit are given by the winding numbers $\gamma_R = 0.0677$ and $\gamma_\Theta = 0.4616$, respectively. Expanding the determinant into geometric series, the contribution (9) of the frozen planet periodic orbit to the density of states is

$$\rho_{fpo} \sim \sum_{r=1}^{\infty} \sum_{k,l=0}^{\infty} \exp 2\pi i r \left[\frac{S}{\hbar} - \frac{\mu'}{4} - \left(l + \frac{1}{2}\right)\gamma_R \right.$$
$$\left. - 2\left(k + \frac{1}{2}\right)\gamma_\Theta\right], \quad (11)$$

where the number of conjugate points along the trajectory are already contained in the winding number. The additional phase shift $\mu' = 2$ comes from the singularities in the Green function which are related to the vanishing total velocity at the turning points of the electrons and to binary collisions, for which the Jacobian of the transformation from (six–dimensional) Euclidean coordinates to an appropriate internal coordinate set [e.g. the perimetric coordinates (2)] vanishes. The non–Euclidean character of the internal coordinates is also responsible for the additional factor of 2 appearing for the contribution of the motion Θ perpendicular to the symmetry plane.[31]

The sum over the repetitions r in Eq. (11) is a geometric series which can be summed analytically. Thus Eq. (11) yields a triple– WKB formula with three quantum numbers n, k, l,

$$S(E) = 2\pi\hbar[n + \tfrac{1}{2} + (l + \tfrac{1}{2})\gamma_R + (2k + 1)\gamma_\Theta]. \quad (12)$$

Using the classical scaling property for the action and rearranging Eq. (12) results in a triple–Rydberg formula for the energies converging to the three–body breakup threshold,

$$E_{nkl} = -\frac{S^2}{[n + \tfrac{1}{2} + (l + \tfrac{1}{2})\gamma_R + (2k + 1)\gamma_\Theta]^2}, \quad (13)$$

with $S = 1.4915$ the scaled action of the periodic orbit. The semiclassical quantum numbers n, k and l reflect the approximate separability of the associated semiclassical wave functions in the local coordinates $\{q_i\}$ of the periodic orbit. Nodal excitations along the orbit are described by n, whereas k and l count the excitations perpendicular to the orbit.

Before applying Eq. (12) or (13) blindly, one should realize the inherent approximations and restrictions of these formulas. For integrable systems the equations actually represent an approximation of the torus– (or EBK–) quantization procedure, where the actions of the irreducible circuits on the tori are quantized separately.[34] Here these actions are approximated harmonically through the properties of the fixed point (periodic orbit) in the center of the elliptic island. The advantage of such an approach is that it is also applicable for non–integrable systems as long as the elliptic island surrounding the periodic orbit is large enough to support many eigenstates, i.e. its phase space volume is large compared to $(2\pi\hbar)^N$. The obvious disadvantage is that we cannot expect the approximations to be of good quality if the phase space volume of the island is small, or if the phase space structure (e.g. non–elliptic deformations) varies strongly over small phase space distances. However, due to the scaling properties for the present system such restrictions limit the applicability only to transversal excitations k, l, but not to n. Roughly, the maximal meaningful values for k and l increase proportional to $\sqrt{n}$ because the nodal structures of the wave functions parallel and perpendicular to the orbit scale with $1/\hbar$ and $1/\sqrt{\hbar}$, respectively[35] (due to the scaling property n takes over the role of $1/\hbar$).

In the derivation of the triple–WKB formula (12) we consistently expanded all expressions to leading order in $\hbar$. We then expect the absolute semiclassical error to be of the order $\hbar^2$ and the relative error to be of the order $\hbar/n$ (for $n \gg k,l$),

$$S(E_{nkl})/2\pi\hbar[n + c(k,l)] = 1 + \beta\frac{\hbar}{n}, \quad (14)$$

which shows the semiclassical limit $\hbar \to 0$ to be equivalent to $n \to \infty$. No general theory is available at present to estimate the error constant β. Nevertheless, the energy eigenvalues predicted with the simple semiclassical formula (13) should become exact in the semiclassical limit of high excitations with an error vanishing proportional to n^{-4}.

Note that the triple–Rydberg formula (13) yields real energies. In the lowest semiclassical approximation presented here the wave functions are square integrable and represent exactly bound states. These states can autoionize semiclassically by dynamical tunneling,[36] but the decay widths for such processes decrease exponentially with the nodal excitation along the orbit. The formula applies to both symmetrical and antisymmetrical states of electron exchange (i.e. to the spectroscopic $^{2S+1}L^\pi = {}^1S^e$ and $^3S^e$ series). Again, dynamical tunneling lifts this doublet degeneracy and the exchange energies vanish exponentially, but the precise determination of the splitting is beyond the scope of the lowest order semiclassical treatment.

Table II summarizes the positions and widths of frozen planet resonances ($n, k = 0, l = 0$) with n ranging from 2 to 10 together with the predictions of the simple semiclassical formula (13). Considering the rather large basis sets nec-

TABLE II. Energies E_{nkl} and total decay widths $\Gamma/2$ for planetary states with total angular momentum $L=0$ and nodal quantum numbers $k = l = 0$. They are given for both symmetry classes $^1S^e$ and $^3S^e$. The predictions of the semiclassical formula (13) are given as E_{scl}.

	$^1S^e$		$^3S^e$		
n	$-E$	$\Gamma/2$	$-E$	$\Gamma/2$	$-E_{scl}$
2	0.257 371 61	0.000 010 57	0.249 964 61	0.000 006 78	0.247 92
3	0.141 064 15	0.000 011 63	0.140 088 48	0.000 004 40	0.139 35
4	0.089 570 80	0.000 002 02	0.089 467 82	0.000 000 17	0.089 145
5	0.062 053 55	0.000 000 56	0.062 041 27	0.000 000 03	0.061 887
6	0.045 538 66	0.000 000 20	0.045 539 24	0.000 000 37	0.045 458
7	0.034 842 64	0.000 000 36	0.034 843 85	0.000 000 14	0.034 798
8	0.027 517 59	0.000 001 18	0.027 519 28	0.000 000 02	0.027 491
9	0.022 284 57	0.000 000 54	0.022 283 66	0.000 000 03	0.022 265
10	0.018 411 98	0.000 000 05	0.018 411 89	0.000 000 03	0.018 400

essary to obtain the accurate quantum results (up to ~ 5500 basis states used) the accuracy of the semiclassical results, which are obtained on a pocket calculator, are rather impressive. In Fig. 6 we plot the semiclassical error for the quantum defect–like quantity μ_n,

$$\mu_n = N_{\text{eff}} - n, \qquad (15)$$

where the effective quantum number N_{eff} is defined as the denominator in Eq. (13), i.e.

$$N_{\text{eff}} = S/\sqrt{-E}. \qquad (16)$$

From Fig. 6 we deduce the leading term of the semiclassical error in the quantum defect μ_n to be $\beta = -0.0366$. Hence, the error is of the order of what we expected, but the prefactor β is rather small.

As predicted by the semiclassical theory, the energy splittings $\Delta E_\pm$ between the parity doublets listed in Table II indeed decrease exponentially. The number of equal significant digits roughly increase by one for each additional node n. The widths of the resonances also decrease exponentially but they fluctuate rather largely around this general trend. The exponential stability of the quantum states is remarkable considering the vastly increasing number of open channels into which the states can decay; the $(n,k,l) = (10,0,0)$ state, e.g., is coupled to 55 continuum

channels. The extreme stability against (non–radiative) decay is a direct consequence of the semiclassical nature of these states.

A direct examination of the nodal structure of the associated wave functions is a more stringent test than comparing energy eigenvalues. Figure 7(a) depicts the condi-

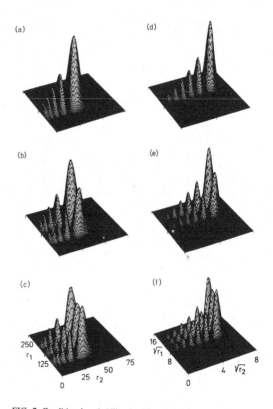

FIG. 7. Conditional probability densities of $(n,0,l)$ frozen–planet states with $n = 6$. The angle Θ between $\mathbf{r}_1$ and $\mathbf{r}_2$ is fixed to $\Theta = 0$. The axes have a linear (left part) and a quadratic scale (right part), respectively. The states shown belong to $l = 0$ (a,d), $l = 1$ (b,e), and $l = 2$ (c,f). Only the parts $r_1 > r_2$ are shown. The full wave function is symmetric in r_1 and r_2.

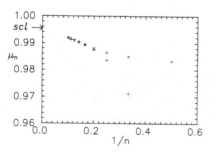

FIG. 6. Quantum defect μ_n as defined by Eq. (15). Both symmetry classes, $^1S^e$ (+) and $^3S^e$ ($\times$), are shown. The semiclassical limit is marked by an arrow.

TABLE III. Coefficients c_j of the cycle expansion, see Eq. (19) and text.

	$|c_0|$	$|c_1|$	$|c_2|$	$|c_3|$	$|c_4|$	$|c_5|$	$|c_6|$
$\prod (1 + a^j t_{PPO})$	1	0.740	0.394	0.866	1.055	1.515	1.937
$\prod (1 - a^j t_{PPO})$	1	0.740	0.394	0.283	0.204	0.130	0.081

tional probability distribution of the wave function for the (6,0,0)–state for the collinear arrangement $r_{12} = r_1 - r_2$. The off–collinear part of the probability density, not shown here, decreases exponentially indicating a zero–point motion in the bending degree of freedom. This zero–point motion is expressed by the assignment $k = 0$. The coordinate r_1 (r_2) denotes the radial distance of the outer (inner) electron. The outer electron probability is strongly localized in the region $r_1 \approx 125$, reflecting the classical localization of the 'frozen' electron. Note also the large differences in the radial extents r_j. The nodal excitations are all directed along the frozen planet PO, which is a nearly straight line along the frozen–planet radius indicated by an arrow in the figure. Recalling the typical quadratic spacing of nodal lines in Coulombic systems, we achieve nearly constant nodal distances by using quadratically scaled axes as done in part (d). The number of nodes along the orbit is 6 in agreement with the semiclassical predictions. The wave function only has a zero–point distribution perpendicular to the orbit (in the symmetry plane of collinear motion), which agrees with the semiclassical local coordinate classification $(n,k,l) = (6,0,0)$.

Wave functions with nodal excitations transverse to the orbit preserving the collinear character of the (quantum) motion are shown in parts (b), (e) and (c), (f) of Fig. 7. They correspond to the $l = 1$ and $l = 2$ nodal excitations of the $n = 6$ manifold of states. Their energies differ only slightly due to the small winding number γ_R. A striking property of the wave functions shown is their nearly rectangular nodal structure and a near–separability in individual–particle coordinates r_1, r_2, i.e., the total wave function is approximately a product of wave functions for r_1 and r_2. It is widely accepted that two particles (or equivalently two degrees of freedom) are uncorrelated if the wave functions are products of single particle coordinates. Nevertheless, from our semiclassical analysis it is obvious that the radial motion of the electrons *is* highly correlated: it is the electron–electron interaction which is responsible for the dynamical localization of the outer electron. Thus Fig. 7 demonstrates that even if the motion is highly correlated the wave function may (approximately) separate in (independent) single–particle coordinates. It is only the other way round which generally holds: if the electrons are independent, then the wave functions separate in single–particle coordinates.

C. Quantizing chaotic dynamics: cycle expansion for near–collinear configurations

The classical dynamics for collinear configurations with both electrons on different sides of the atoms turns out to be fully chaotic and the (approximate) torus quan-

tization described in the previous section cannot be applied. We now have to sum over the contributions (9) of *all* periodic orbits. This leads to the so–called Gutzwiller trace formula, which for the present system reads[7] (the product representation [2,37] is used)

$$\prod_n (E - E_n) \sim \prod_{PPO} \prod_{k=0}^{\infty} \prod_{m=0}^{\infty} (1 - t_{PPO}^{(k,m)}). \quad (17)$$

The weight $t_{PPO}^{(k,m)}$ of each PPO is given by

$$t_{PPO}^{(k,m)} = (\pm 1)^k a^j \exp[2\pi i z S - i a \pi /2 \\ - (k + \tfrac{1}{2})u - 4\pi i (m + \tfrac{1}{2})\gamma], \quad (18)$$

where all classical quantities are given in Table I. The plus sign applies to hyperbolic PPO and the minus sign to hyperbolic PPO with reflection. The bookkeeping indices $a = 1$, j are only introduced for convenience and will be discussed below.

The formal expression (17) relates the product over quantum eigenvalues with a product over periodic orbits. Unfortunately, the zeros of the right hand side cannot be naively identified with the zeros of the left hand side, because the eigenvalues E_n are located beyond the abscissa of absolute convergence of the rhs.[38] The problem of finding semiclassical approximations for the energies E_n from the diverging product over periodic orbits is a topic of several contributions to this issue of CHAOS and will not be discussed in detail here.

We use the *cycle expansion*[23,39] to evaluate the semiclassical expression over PO. The idea of the cycle expansion is to expand the infinite product (17) into a power series $\Sigma_j c_j a^j$ of the bookkeeping index a. For $k = m = 0$ this reads (j equals the symbol length of the PO)

$$\prod_{PPO} (1 - t_{PPO})$$

$$= 1 - t_+ - t_- - (t_{+-} - t_+ t_-)$$

$$- (t_{++-} - t_+ t_{+-}) - (t_{+--} - t_- t_{+-})$$

$$- \cdots . \quad (19)$$

Except for the fundamental orbits ' + ' and ' − ' each orbit contribution is accompanied by a compensating term pieced together from shorter orbits. Thus terminating the expansion at a given symbol length effectively means a re–summation of *all* orbits, with the approximation that the longer orbits are shadowed to increasing accuracy by the shorter ones. The absolute convergence of the re–grouped Dirichlet series (19) does of course not change. However, if each term t_{ab} together with its shadowing term $t_a t_b$ is viewed as a *single* entry d_{ab} then the series (19)

TABLE IV. Total binding energies E and effective quantum number $N_{\text{eff}} = 1/\sqrt{E}$ for $^1S^e$ states obtained by WKB quantization of the fundamental orbit ' − ', by the cycle expansion, and by full quantum solutions (taken from Ref. 7).

$(n_\lambda,n_\mu)_\nu$	$(Nl,N'l')$	N_{eff}			Energies	
		WKB	Cycle	QM	Cycle	·QM
$(0,0)_0$	$1s1s$	0.568	0.584	0.587	2.932 ·	2.904
$(0,2)_0$	$2s2s$	1.115	1.134	1.134	0.778	0.778
$(0,2)_1$	$2s3s$		1.308 ·	1.302	0.585	0.590
$(0,4)_0$	$3s3s$	1.662	1.684	1.682	0.353	0.354
$(0,4)_1$	$3s4s$		1.883	1.886	0.282	0.281
$(0,6)_0$	$4s4s$	2.208	2.243	2.231	0.199	0.201
$(0,6)_1$	$4s5s$		2.456	2.456	0.166	0.166
$(0,6)_2$	$4s6s$		2.574	2.575	0.151	0.151
$(0,8)_0$	$5s5s$	2.755	2.783	2.780	0.129	0.129
$(0,8)_1$	$5s6s$		3.025	3.020	0.109	0.110
$(0,8)_2$	$5s7s$		3.154	3.159	0.101	0.100
$(0,10)_0$	$6s6s$	3.302	3.343	3.329	0.0895	0.0902
$(0,10)_1$	$6s7s$		3.586	3.580	0.0778	0.0780
$(0,10)_2$	$6s8s$		3.733	3.733	0.0717	0.0718
$(0,12)_0$	$7s7s$	3.849	3.903	3.883	0.0657	0.0663
$(0,12)_1$	$7s8s$		4.140	4.138	0.0583	0.0584
$(0,12)_2$	$7s9s$		4.305	4.301	0.0540	0.0541
$(0,14)_0$	$8s8s$	4.395	4.429	4.411	0.0510	0.0514
$(0,14)_1$	$8s9s$		4.689	4.686	0.0455	0.0455
$(0,14)_2$	$8s10s$		4.865	4.865	0.0423	0.0423

converges absolutely. This is illustrated in Table III which shows the coefficients c_j of the cycle expanded product ($k = m = 0$) with the semiclassical weights replaced by their absolute values, i.e. $t_{\text{PPO}} = \exp(-\lambda/2)$. Obviously, the coefficients of the first row (determining the abscissa of absolute convergence of the unexpanded product) diverge exponentially with a ratio $c_{j+1}/c_j \approx 1.3$, whereas the coefficients of the second row converge exponentially with a ratio $c_{j+1}/c_j \approx 0.65$. A more careful analysis shows that the unexpanded product converges absolutely for $\Im\sqrt{-E} > 0.027$, i.e. only sufficiently far in the upper half of the complex energy plane, whereas the resonance poles are located close to the real energy axis. The cycle expansion, however, converges in the energy region, where the resonances are located (typically $\Im\sqrt{-E} > -0.01$).

The products over k and m in Eq. (17) originate from the expansion of the Gutzwiller amplitudes (9) into geometric series.[37] They have to be treated differently, because the stability characteristic is different for the two directions perpendicular to the orbit. Similar to the treatment for the (doubly) stable frozen planet orbit we identify m as a semiclassical quantum number for the stable bending degree of freedom. For the expansion of the remaining product we set the bookkeeping index j to $(2k + 1)$ times the symbol length of the PO. The present calculations are carried out including all orbit contributions up to $j = 6$. In Table IV we show our results for some doubly excited $^1S^e$ states with $m = 0$. For labeling the states we use the molecular-orbital (MO) classification $(n_\lambda,n_\mu)_\nu$ derived from an adiabatic treatment of the inter–electron vector **R**.[40] The MO quantum numbers accurately describe the nodal surfaces of the quantal wave functions for fixed inter–electron distances and moderate electron excitations $(N,N' < 6)$.[41,42] For convenience, we also give the "independent particle" labeling $(Nl,N'l')$, i.e. the configuration which comes closest in an independent particle description. Here, $N,N' \geqslant N$ roughly correspond to the principal quantum numbers of the electrons.

It is more natural to compare the effective quantum numbers $N_{\text{eff}} = E^{-1/2}$ than the binding energies E themselves, but both values are given in the table. We find that the cycle expansion results are mostly good to within 1% or better. This is better than might be expected; in fact, accurate quantum results for the very high lying doubly excited states ($N > 6$) are presently not available in the literature. For published data the results are comparable or even superior to elaborate (adiabatic) hyperspherical calculations.[43,44]

Generally, the energies obtained in the cycle expansion (as well as in the quantum mechanical calculations) are complex valued and we have only tabulated the real part of these energies. The widths of the resonances (i.e. the imaginary part of their energies) are still smaller than the semiclassical error in the real part and there is no likely reason why the imaginary part should be more accurate.

One may wonder why the semiclassical analysis works fine even down to the ground state, which has a smooth and—more or less—structureless wave function. The reason is that the PO do not only contain the information about the underlying dynamics, but they also "know" the size of the phase space. It is this property which often leads to rather good results when a fundamental (e.g. the shortest) orbit is naively quantized.

Although the results for the semiclassical energies are very satisfactory, it is even more valuable that the (semi–) classical analysis provides an insight of what the electrons are actually 'doing' in the highly correlated states. In Table IV we also list the results of the simplest cycle approach

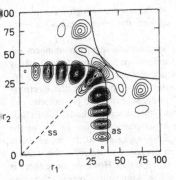

FIG. 8. Contour plot of the conditional probability distribution $|\Psi_{NN'}(r_1,r_2,r_{12}=r_1+r_2)|^2$ for the intra-shell wave function $N = N' = 6$ corresponding to the collinear arrangement $\Theta = \pi$ of the electrons. The axes have a quadratic scale to account for the wave propagation in Coulombic systems, where nodal distances increase quadratically. The fundamental orbit ' – ' (AS) as well as the symmetric stretch motion (SS) along the Wannier ridge are overlaid.

including only the fundamental PO ' – ', which is nothing but a WKB quantization of the orbit (but including the zero–point motion for the perpendicular degrees of freedom). Quantization of this fundamental asymmetric stretch PO gives rather accurate results for the doubly excited *intra–shell resonances* $N = N'$. For the other states $N \neq N'$ the inclusion of all the orbits of Table I is essential and the simplified WKB approach cannot yield them. These results indicate that the intra–shell resonances are associated with the asymmetric stretch like motion of the fundamental PO ' – ' rather than the symmetric stretch motion along the Wannier ridge. This conclusion is in striking contrast to the common viewpoint expressed in the literature (see, e.g., Refs. 25–30), but in line with recent suggestions.[5,8,42]

An inspection of the quantum mechanical intra–shell wave functions confirms the semiclassical conclusion on the fundamental electron motion. In Fig. 8 we show, e.g., the probability distribution of the $N = N' = 6$ state which is clearly localized along the fundamental orbit ' – ' (AS) and *not* along the Wannier ridge $r_1 \equiv r_2$ (SS). The classical probability along the trajectory is largest (as well

as the quantal wave function) where it passes the Wannier saddle point, but the motion is directed perpendicular to the ridge.

The WKB treatment of the fundamental orbit also provides the dynamical origin of the double Rydberg formula[26]

$$E_N = -(Z-\sigma)^2/(N-\mu)^2 \qquad (20)$$

for the intra–shell resonances. We find $Z - \sigma = S_-$ and $\mu = 1 - \gamma_- - \alpha_-/4$ (i.e. $\sigma = 0.1710$ and $\mu = -0.0393$) which fits well with the semi-empirically derived values of $\sigma = 0.1795$ and $\mu = -0.0597$.[45]

The classical analysis also applies to the $^3S^e$ states, i.e. those which are anti–symmetric with respect to the exchange of particle (configuration space) coordinates. For the semiclassical Green function we now have to take the Dirichlet boundary condition along the symmetry line $r_1 \equiv r_2$ instead of the von Neumann boundary condition. Thus each time a trajectory crosses the symmetry line we have an additional phase loss of π. Again the coding takes care automatically of the additional total phase loss: each symbol ' – ' of the electron pair motion is associated with a crossing of the symmetry line.

Table V gives the results for the low–lying 'intra–shell' $^3S^e$ states. The overall agreement with the quantum results is again remarkable. From the (semi–) classical analysis one expects the states to be of similar nature as those of $^1S^e$ symmetry. This is verified in Fig. 9 for the state $(n_\lambda,n_\mu)_\nu = (0,11,0)_0$. Obviously, the wave function images the same type of electron pair motion as Fig. 8 for the $^1S^e$ symmetry. The only difference is that the wave function has an odd number of nodes along the PO and hence a node at $r_1 = r_2$. The wave functions shown are completely symmetric with respect to 'individual' electron excitations. This demonstrates that the independent particle label $(Nl,N'l') = 6s7s$ [as well as the popular $N(K,T)^A N'$ labeling scheme[27]] is not useful classifying the internal structure of the state shown in Fig. 9. The MO quantum numbers, however, do give a proper classification of the state. The number of nodes along the PO is given by n_μ, whereas nodal excitations perpendicular to the orbit are labeled by ν. The quantum number n_λ describes the bending degree of freedom and is identical to the semiclassical quantum number m (both of which are zero for the states discussed so

TABLE V. Same as Table IV, but for $^3S^e$ states.

		N_{eff}			Energies	
$(n_\lambda,n_\mu)_\nu$	$(Nl,N'l')$	WKB	Cycle	QM	Cycle	QM
$(0,1)_0$	$1s2s$	0.842	0.712	0.678	1.972	2.175
$(0,3)_0$	$2s3s$	1.388	1.296	1.288	0.596	0.603
$(0,5)_0$	$3s4s$	1.935	1.870	1.866	0.286	0.283
$(0,7)_0$	$4s5s$	2.482	2.438	2.430	0.168	0.169
$(0,9)_0$	$5s6s$	3.029	2.989	2.989	0.1119	0.1119
$(0,11)_0$	$6s7s$	3.575	3.545	3.544	0.0796	0.0796
$(0,13)_0$	$7s8s$	4.122	4.107	4.097	0.0593	0.0596
$(0,15)_0$	$8s9s$	4.669	4.641	4.649	0.0464	0.0462

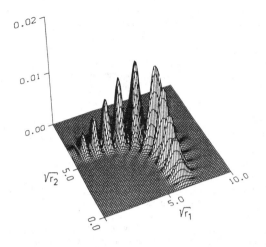

FIG. 9. Conditional probability distribution $|\Psi_{NN'}(r_1,r_2,r_{12}=r_1+r_2)|^2$ for the wave function $(n_\lambda,n_\mu)_v = (0,11)_0$.

far). In addition, for *fixed* inter–electron distance r_{12} the internal wave function approximately separates in MO coordinates.[42]

The choice $m = 0$ in our calculations means that the associated semiclassical wave functions are localized in the symmetry plane of collinear motion with only a zero–point motion perpendicular to it. This approach is justified by the stability of the classical motion perpendicular to the plane. In fact, the quantum wave functions show the same behavior. Putting $m \neq 0$ gives states with $n_\lambda = m$ in the MO description, but the linearization of the motion perpendicular to the plane becomes a rather crude approximation if $N \gg m$ does not hold. The results for intra–shell states of $^1S^e$ symmetry are summarized in Table VI. The semiclassical error is about 10% for the lowest lying state, for which there is no nodal excitation along the PO but one nodal excitation perpendicular to the symmetry plane. For higher lying states the semiclassical error again drops below 0.5%.

IV. ADIABATIC VERSUS CHAOTIC MOTION

Typically, classical chaotic motion is connected with the loss of (approximate) symmetries and associated quantum numbers.[46,47] One may then wonder, why all the states reported in this contribution can still be labeled with ap-

proximate quantum numbers, even though we exploited the intrinsic chaotic classical dynamics to calculate semiclassical eigenvalues.

Different sets of approximate quantum numbers describing three–body Coulomb systems were derived, partly starting from the independent particle description and applying degenerate perturbation theory for the electron–electron interaction.[48] More sophisticated perturbative calculations include group–theoretical methods.[49] Adiabatic expansions using the hyperspherical radius $\mathcal{R}$ [25,27] or the inter–electron radius $R = r_{12}$ (Refs. 40, 50) as adiabatic coordinates have been proposed to describe the internal structure of doubly excited states. However, since there is no obvious geometrical or kinematical reason, there are only few justifications why these coordinates can be treated adiabatically. Nevertheless, energies derived from both adiabatic approaches, hyperspherical and MO, yield quite accurate results for the states of maximal polarization along the inter–electron axis, i.e. those states for which $-\langle\cos\Theta\rangle$ is close to unity.[28,30,40,43,44] However, energies are not a sensitive test of the validity of the underlying assumptions. For example, highly accurate energy values for the intra–shell resonances can also be obtained within a diabatic approach,[51] even though the corresponding wave functions are totally inappropriate to describe the nodal structure of the states.

The possibility of an adiabatic description of the (regular) frozen planet configurations is immediately obvious from the consideration of the classical motion of the electrons. For the frozen planet PO the action of the accumulated action of the outer electron is about 2×10^4 smaller than the action of the inner electron. Since the frozen electron is strongly localized in configuration space, the outer electron radius r_1 suggests itself as an adiabatic coordinate. For fixed r_1 the remaining inner electron Hamiltonian is separable in (molecular) prolate spheroidal coordinates, which allows the labeling of the resulting adiabatic potential curves with a complete set of quantum numbers. For two–electron atoms the goodness of the adiabatic approximation for the frozen planet configurations has been analyzed in Ref. 52. If the outer electron is replaced by a heavy particle of the same charge (e.g. an anti–proton) then the adiabatic approach should be even more efficient. The anti–protonic analog of the frozen planet states were proposed recently as possible anti–matter traps.[53,54] The overlap of the anti–protonic wave function with the nucleus is extremely small, which prevents the three–body complex from decay via the strong interaction.

TABLE VI. Same as Table IV, but for $^1S^e$ states with $n_\lambda = 1$.

$(n_\lambda,n_\mu)_v$	$(Nl,N'l')$	Energies		
		WKB	Cycle	QM
$(1,0)_0$	$2p2p$	0.745	0.701	0.622
$(1,2)_0$	$3p3p$	0.344	0.337	0.317
$(1,4)_0$	$4p4p$	0.197	0.191	0.188
$(1,6)_0$	$5p5p$	0.1277	0.1260	0.1233
$(1,8)_0$	$6p6p$	0.0894	0.0873	0.0869

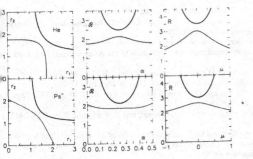

FIG. 10. The asymmetric stretch–type PO ' − ' in individual particle coordinates r_1, r_2, hyperspherical coordinates $\mathscr{R} = (r_1 + r_2)^{1/2}$, $\alpha = \operatorname{atan}(r_1/r_2)/\pi$, and MO–coordinates $R = r_{12}$, $\mu = (r_1 - r_2)/R$. The boundaries of the classically allowed regions are drawn as thick lines. The upper half of the figure applies to helium (He), the lower part to the positronium negative ion (Ps⁻). For a better comparison Z-scaled coordinates r_i/Z are used.

An adiabatic description of intra–shell states or near–collinear configurations with both electrons on opposite sides is not immediately obvious. Recently it has been shown however,[41,42] that the nodal structure of (moderately) doubly excited intra–shell states is accurately described by the MO–quantum numbers derived from an adiabatic treatment of the inter–electron distance. Again, for *fixed* $\mathbf{R} = \mathbf{r}_1 - \mathbf{r}_2$ the electronic center of mass (ECM) coordinate $\mathbf{r} = (\mathbf{r}_1 + \mathbf{r}_2)/2$ separates in molecular coordinates[50] leading to the full set of MO–quantum numbers used in Tables IV–VI. The adiabatic treatment does *not* imply that the actual wave functions separate (even approximately) in R and r; the nodal structure of the *full* wave function may show strong mixing of the ECM–coordinates with R (which actually is the case[42] for the intra–shell resonances discussed here). To illustrate this consider the wave function

$$\Psi(R,x,y) = \phi_R(R)\phi_x(Rx)\phi_y(Ry). \tag{21}$$

For each value of R the wave function separates exactly in x, y, leading to a rectangular nodal pattern in x and y. The full wave function, however, mixes R with x and y. The wave function (21) may be perfectly adiabatic, nevertheless it is non–separable in full space.

The classical analysis of collinear electron configurations and the subsequent semiclassical treatment with the cycle expansion uncovers the asymmetric stretch–type PO ' − ' as the fundamental electron motion for the intra–shell states of helium. Figure 10 shows the PO in the different adiabatic coordinate sets discussed above. For all the three coordinate sets the symmetry plane of collinear configuration results in a constant value of one coordinate, $\Theta \equiv \pi$ for the individual and hyperspherical coordinates, and $\lambda = (r_1 + r_2)/R \equiv 1$ for the MO–coordinates. Even though the individual radial distances r_1, r_2 vary largely, both the hyperradius $\mathscr{R}$ and the inter–electron distance R change only slightly along the trajectory. These results may ex-

plain to some extent why the intra–shell resonances can be treated adiabatically in $\mathscr{R}$ or R to good approximation.

The adiabatic behavior becomes even more pronounced for the molecular–like positronium negative ion (Ps⁻ $= e^- e^+ e^-$), as is shown in the lower half of Fig. 10. The fundamental PO ' − ' is stable for Ps⁻, even though the stability island surrounding the orbit is rather small. From the shape of the fundamental PO in MO–coordinates, it is not too surprising, that the results of adiabatic molecular quantum calculations are rather accurate in this case.[55]

The adiabatic behavior is somewhat hidden in the classical/semiclassical treatment described in the previous sections. As we have seen, the WKB–quantization of the fundamental PO ' − ' accounts for the intra–shell resonances $N = N'$, which is the first member of a Rydberg series $N' \geqslant N$ of resonances converging to the single–particle escape threshold leaving the He⁺–ion in a Stark–type polarized hydrogenic state with principal quantum number $n = N$. For most of the states considered here ($N \leqslant 8$), the energy differences between the exterior scaling N and the interior Rydberg scaling N' are rather large, which, translated via the correspondence principal, leads to largely different time scales in the system. But as can be seen from Table I the distribution of periods (actions) of the PO is rather smooth. How do the PO then reflect the different time scales? Amusingly, it is the nearly perfect hyperbolicity of the system which is responsible for the regularity (and thus adiabaticity) of the spectra. The inclusion of more and more orbits in the cycle expansion does not destroy the exterior scaling, which is already described by the fundamental PO ' − '. The longer orbits are nearly perfectly shadowed by the compensating terms of shorter orbits, i.e.

$$t_{ab} - t_a t_b = t_{ab}\{1 - \exp[2\pi i z(S_a + S_b - S_{ab})$$
$$- (u_a + u_b - u_{ab})/2]\} \approx 0. \tag{22}$$

The relevant quantities are then the small *differences* in the periods of the orbits and their shadowing parts. Inclusion of these contributions yield the other members of the Rydberg series tabulated in Table IV.

The energetic separation of the different Rydberg series for $N < 6$ combined with the quasi–separability of the bending degree of freedom is the reason why approximate quantum numbers exist. Due to the energetic separation, wave functions cannot mix with those originating from other series and they become rather simply structured similar to what one would expect for 'regular' wave functions. Near $N \geqslant 6$ the different Rydberg series begin to overlap energetically, and it is this region where we expect the breakdown of approximate quantum numbers and the appearance of irregular spectra and wave functions, reflecting the intrinsic chaotic dynamics of the electron motion (see also Ref. 9).

V. SUMMARY AND CONCLUSIONS

At present, semiclassical theories undergo a rapid and exciting evolution. The present issue of CHAOS documents part of this development. In this contribution we applied many of the new ideas to the problem of two–electron atoms.

The classical dynamics of the collinear helium atom with both electrons on different sides is fully chaotic. An application of the Gutzwiller formula for the full three–dimensional problem combined with the cycle expansion yields a number of resonances with high accuracy. The interpretative ability of the methods illuminates the structure of the quantal motion. The analysis shows that the near–collinear intra–shell resonances are associated with the (fundamental) asymmetric stretch like motion of the electron pair. Semiclassically, this observation is nearly trivial. The result is nevertheless remarkable, in that it has been widely believed for decades that these resonances are associated with the in–phase symmetric stretch motion of the electron pair along the Wannier ridge.

The classical dynamics of the collinear helium atom with both electrons on the same side of the nucleus is fully stable. Approximate torus quantization yields very accurate results for the positions of the associated quantum mechanical resonances. The semiclassical formalism also accounts semi–quantitatively for the decay widths and degeneracies of doublet states. The structure of the wave functions corresponds to what one would expect from considering the classical motion.

We are certainly at the beginning of refining the semiclassical methods for multi–dimensional methods like the PO–quantization approach. Important further developments will probably include the description of dynamical tunneling processes and a refined semiclassical consideration of discrete symmetries, which may allow the calculation of exponentially small decay widths or multiplet–degeneracies. An important (even though presumably formidable) step would be the inclusion of higher order terms of $\hbar$ in the semiclassical PO-theory. Other methods than the cycle expansion which drastically reduce the (exponentially growing) classical input of the semiclassic PO–quantization are also highly desirable. Even without such refinements, it is likely that the combined classical/semiclassical analysis of few–body systems such as the helium atom will uncover some more surprises in the near future.

ACKNOWLEDGMENTS

We gratefully acknowledge the collaboration with G. S. Ezra, from which parts of the presented results originate. We thank J. S. Briggs and J.-M. Rost for many discussions on three–body Coulomb problems, and P. Cvitanović and B. Eckhardt for motivating us to "cycle tours." The research was supported by the Deutsche Forschungsgemeinschaft (Wi877/2 and Wi877/5), and partly by the SFB 276 located in Freiburg and by NORDITA in Copenhagen.

[1] J. H. van Vleck, Philos. Mag. **44**, 842 (1922).

[2] M. C. Gutzwiller, *Chaos in Classical and Quantum Mechanics* (Springer, New York, 1990).

[3] M. Born, *Vorlesungen über Atommechanik* (Springer, Berlin, 1925). English translation: The Mechanics of the Atom (Ungar, New York, 1927).

[4] J. G. Leopold and I. C. Percival, J. Phys. B **13**, 1037 (1980).

[5] K. Richter and D. Wintgen, J. Phys. B **23**, L197 (1990).

[6] K. Richter and D. Wintgen, Phys. Rev. Lett. **65**, 1965 (1990); J. Phys. B **24**, (1991), L565.

[7] G. S. Ezra, K. Richter, G. Tanner, and D. Wintgen, J. Phys. B **24**, L413 (1991).

[8] J.-H. Kim and G. S. Ezra, in *Proceeding of the Adriatico Conference on Quantum Chaos*, edited by H. A. Cerdeira et al. (World Scientific, Singapore, 1991).

[9] R. Blümel and W. P. Reinhardt, in *Directions in Chaos*, edited by B. L. Hao et al. (World Scientific, Hong Kong, 1991), Vol. 4.

[10] B. Eckhardt, preprint Universität Marburg; Habilitationsschrift, Universität Marburg (1991).

[11] J. Müller, J. Burgdörfer, and D. W. Noid, preprint University of Tennessee.

[12] C. Marchal, *The Three–Body Problem* (Elsevier, Amsterdam, 1990).

[13] C. L. Siegel, Ann. Math. **42**, 127 (1941).

[14] H. M. James and A. S. Coolidge, Phys. Rev. **51**, 857 (1937).

[15] G. Wannier, Phys. Rev. **90**, 817 (1953).

[16] M. Domke, C. Xue, A. Puschmann, T. Mandel, E. Hudson, D. A. Shirley, G. Kaindl, C. H. Greene, H. R. Sadeghpour, and H. Petersen, Phys. Rev. Lett. **66**, 1306 (1991).

[17] B. Eckhardt and D. Wintgen, J. Phys. B **23**, 355 (1990).

[18] R. Schinke and V. Engel, J. Chem. Phys. **93**, 3252 (1990).

[19] P. Dahlqvist and G. Russberg, J. Phys. A **24**, 4763 (1991).

[20] D. Weiss, M. L. Roudes, A. Menschig, P. Grambow, K. von Klitzing, and G. Weimann, Phys. Rev. Lett. **66**, 2790 (1991).

[21] M. Sieber and F. Steiner, Phys. Lett. A**148**, 415 (1990); M. Sieber, Dissertation, Universität Hamburg (1991).

[22] G. Tanner and D. Wintgen, Chaos **2**, 53–59 (1992).

[23] P. Cvitanović and B. Eckhardt, Phys. Rev. Lett. **63**, 823 (1989).

[24] B. Eckhardt and D. Wintgen, J. Phys. A **24**, 4335 (1991).

[25] U. Fano, Phys. Rep. **46**, 97 (1983).

[26] A. R. P. Rau, J. Phys. B **16**, L699 (1983).

[27] S. Watanabe and C. D. Lin, Phys. Rev. A **34**, 823 (1986).

[28] H. R. Sadeghpour and C. H. Greene, Phys. Rev. Lett. **65**, 313 (1990).

[29] P. G. Harris, H. C. Bryant, A. H. Mohagheghi, R. A. Reeder, C. Y. Tang, J. B. Donahue, and C. R. Quick, Phys. Rev. A **42**, 6443 (1990).

[30] H. R. Sadeghpour, Phys. Rev. A **43**, 5821 (1991).

[31] M. C. Gutzwiller, J. Math. Phys. **8**, 1979 (1967); **10**, 1004 (1969); **11**, 1791 (1970); **12**, 343 (1971).

[32] M. V. Berry and K. T. Mount, Rep. Prog. Phys. **35**, 315 (1972).

[33] A. Voros and B. Grammaticos, Ann. Phys. (NY) **123**, 359 (1979).

[34] W. H. Miller, J. Chem. Phys. **63**, 996 (1975).

[35] E. B. Bogomolny, Physica D (Amsterdam) **31**, 169 (1988).

[36] M. J. Davies and E. J. Heller, J. Chem. Phys. **75**, 246 (1981).

[37] A. Voros, J. Phys. A **21**, 685 (1988).

[38] B. Eckhardt and E. Aurell, Europhys. Lett. **9**, 509 (1989).

[39] R. Artuso, E. Aurell, and P. Cvitanović, Nonlinearity **3**, 325 and 361 (1990).

[40] J. M. Rost and J. S. Briggs, J. Phys. B **24**, 4293 (1991).

[41] J. M. Rost, J. S. Briggs, and J. M. Feagin, Phys. Rev. Lett. **66**, 1642 (1991).

[42] J. M. Rost, R. Gersbacher, K. Richter, J. S. Briggs, and D. Wintgen, J. Phys. B **24**, 2455 (1991).

[43] N. Koyama, H. Fukuda, T. Motoyama, and M. Matsuzawa, J. Phys. B **19**, L331 (1986).

[44] H. Fukuda, N. Koyoma, and M. Matsuzawa, J. Phys. B **20**, 2959 (1987).

[45] Q. Molina, Phys. Rev. A **39**, 3298 (1989).

[46] B. Eckhardt, Phys. Rep. **163**, 205 (1988).

[47] H. Friedrich and D. Wintgen, Phys. Rep. **183**, 37 (1989).

[48] U. Fano and A. R. P. Rau, *Atomic Collisions and Spectra* (Academic, London, 1986).

[49] D. E. Herrick, Adv. Chem. Phys. **52**, 1 (1983).

[50] J. M. Feagin and J. S. Briggs, Phys. Rev. Lett. **57**, 984 (1986); Phys. Rev. A **37**, 4599 (1988).

[51] J. M. Rost and J. S. Briggs, J. Phys. B **22**, 3587 (1989).

Stretched helium: a model for quantum chaos in two-electron atoms

R. BLÜMEL

Department of Physics, University of Delaware,
Newark, Delaware 19716, USA

W. P. REINHARDT

Department of Chemistry, University of Washington,
Seattle, WA 98195, USA

Abstract

We show that the one-dimensional model of helium (stretched helium) is a chaotic scattering system. We provide evidence for the existence of a repetitive sequence of three dynamical regimes as a function of the excitation energy of the model atom. The three dynamical regimes are: Ericson, Wigner and Rydberg. They are characterized by $\Gamma \gg s$, (Ericson), $\Gamma \stackrel{<}{\sim} s$ (Wigner) and $\Gamma \ll s$ (Rydberg), respectively, where Γ is the width of the autoionizing resonances and s is their mean (local) spacing.

1 Introduction

The three-body system of celestial mechanics is undoubtedly one of the outstanding unsolved problems in theoretical physics[1]. For now more than three hundred years it has plagued physicists and mathematicians alike. The importance of the problem can hardly be overstated. Our own solar system can be considered as an assembly of three-body systems, the most important such sub-system, to us anyway, being the Earth–Moon–Sun system. With Newton's work as a starting point, many astronomers and eminent mathematicians tried to solve the three-body problem analytically. Despite enormous progress in the course of the last two centuries no real breakthrough occurred. It was Poincaré at the end of the last century[2] who discovered the reason for the insolubility of the three-body problem: chaos. This fact was like a jolt in the physics community. It means that in general (apart from well-known special solutions like Lagrange's equilateral triangle solution or Euler's collinear solution) the three-body problem cannot be solved analytically[3].

It is interesting to point out that a conceptually similar development occurred in the theory of parallels in the mathematical discipline of plane geometry. After millenia of attempts to prove Euclid's fifth axiom, the impossibility of such a proof was demonstrated by Lobachevsky, Bolyay and Gauss. In mathematics this seemingly "negative" result opened the door for fruitful resarch in non-Euclidean geometry. Therefore, we should not be discouraged by Poincaré's result. It marks the beginning of a fascinating subject: the investigation of deterministic chaos in three-body systems.

Chaos is not confined to the motion of celestial bodies. Slowly at first, but like an explosion during the last twenty years, chaos was identified in nearly every corner of science. Chaos is everywhere. Not even atomic physics escaped the grip of chaos[4]. Traditionally thought to be "protected" from chaos by the mitigating effects of wave mechanics[5-11], it was only recently appreciated that multiply-excited atoms behave like strongly correlated miniature solar systems which clearly show chaotic effects.

In this chapter the helium atom will serve as a paradigm of a chaotic three-body system in atomic physics. It will turn out that chaos theory – classical as well as quantum mechanical[12] – is the natural tool for a thorough discussion of many-electron atoms in the semiclassical regime. Not only does this theory help in the characterization of spectra and wavefunctions, it also makes specific predictions about the existence of new dynamical regimes[13]. Even thermodynamic concepts will find their way into the atomic theory of three-body systems as chaos theory predicts the existence of a "gas of resonances"[13,14].

In order to get acquainted with the dynamical systems view of atomic physics, it is not necessary to study the three-dimensional helium atom. It was shown that a one-dimensional version of the helium atom captures many essential features of the complete three-dimensional atom[13,15]. Treating a one-dimensional model enables us to focus on the dynamics of the system without being hampered by complications arising from angular momentum and fine structure. Although, admittedly, much physics is lost in reducing the helium atom to one dimension it is surprising how much physics is, in fact, preserved. In this respect the one-dimensional helium atom behaves like the one-dimensional model of hydrogen atoms in a strong microwave field[16-18]. The one-dimensional hydrogen atom captures much of the physics of the three-dimensional experiments[19,20]. Surprisingly, a comparision between hydrogen ionization experiments and the predictions of the one-dimensional model makes sense even on the quantitative level[21]. In fact, it was shown recently that the one-dimensional model works in the case of photoionization of helium in a strong laser field[22].

In this spirit this chapter is organized in the following way. In section 2 we present the one-dimensional model of the helium atom, the "stretched helium" model. Recently much research work was focussed on the investigation of this model[13,15,23-6]. Some aspects of the classical and quantum mechanics of stretched helium are reviewed briefly in section 3. An interesting view of stretched helium as a chaotic scattering system is presented in section 4. In section 5 we present evidence for the existence of three quantum regimes as a function of excitation energy. Section 6 concludes the chapter.

2 The model

One of the first experiments in atomic physics to show the importance of chaos on the atomic level was the microwave ionization experiment by Bayfield and Koch[19]. Interpreting this experiment on the basis of perturbation theory turned out to be near impossible since typically multiphoton processes of the order of 100 have to be considered. On the other hand Leopold and Percival were the first to show that a classical Monte Carlo approach[27] was successful in reproducing experimentally obtained ionization thresholds. Soon it was demonstrated[28,29] that on the classical level chaos occurs in this system, and ionization thresholds were successfully interpreted as chaos thresholds[16–18,20,21,29,30]. To our knowledge this was the first time that a dynamical systems point of view added to our understanding of a strictly quantum strictly atomic physics system. Moreover, it was demonstrated that in the microwave ionization system a full three-dimensional treatment is not necessary to obtain qualitative and even semiquantitative results. The one-dimensional "surface state electron" Hamiltonian[16,17,29]

$$h(x, p; Z) = \begin{cases} p^2/2 - Z/x & \text{for } x > 0 \\ \infty & \text{for } x \leq 0. \end{cases} \tag{2.1}$$

is sufficient to develop insight into the microwave ionization problem.

The simplest time independent atomic physics system capable of exhibiting chaos on the classical level is the helium atom. In analogy to the driven hydrogen atom briefly discussed above, it is enough to consider a one-dimensional version of the helium atom in order to introduce the dynamical systems point of view of the helium atom. A one-dimensional model of helium ($Z = 2$) can be constructed by "glueing" two one-dimensional hydrogen atoms together, joining them at the origin which is occupied by the nucleus of the helium atom:

$$H' = \frac{p_1^2}{2} + \frac{p_2^2}{2} - \frac{Z}{x_1} - \frac{Z}{x_2} + \frac{1}{x_1 + x_2}. \tag{2.2}$$

The situation is illustrated in Fig. 1. Electron 1 is confined to the left half space, electron 2 is confined to the right half space. We define $x_1, x_2 > 0$ so that electron 1 is, in fact, located at $-x_1$. This way the spatial dynamics of the two electrons takes place in the first quadrant of the x_1, x_2 plane. Singlet and triplet states of helium correspond to symmetric and antisymmetric wavefunctions, repectively. We will work in Z scaled coordinates $x \to x/Z$, $p \to pZ$. This transformation yields:

$$H = \frac{p_1^2}{2} + \frac{p_2^2}{2} - \frac{1}{x_1} - \frac{1}{x_2} + \frac{\epsilon}{x_1 + x_2} \tag{2.3}$$

with $H = H'/Z^2$ and $\epsilon = 1/Z$. For $\epsilon \neq 1/2$ this Hamiltonian describes the isoelectronic sequence of two-electron atomic ions. In order to construct eigenstates of (2.3) it is useful to consider the eigenstates of the single particle

Hamiltonian (2.1). It possesses a Rydberg series of bound states

$$\varphi_n(x) = < x \mid n > = n^{-3/2} \left(\frac{2x}{n}\right) L_{n-1}^{(1)}\left(\frac{2x}{n}\right) \exp(-x/n), \qquad (2.4)$$

and a continuum which consists of the states

$$\varphi_k(x) = < x \mid k > = \frac{2kx}{\sqrt{k(1 - \exp(-2\pi/k))}} \exp(-ikx) \,_1F_1(1 + \frac{i}{k}, 2; 2ikx). \quad (2.5)$$

They can be used to construct the eigenstates of (2.3). Forming product states from (2.4) and (2.5), symmetric and antisymmetric bound–bound, bound–continuum and continuum–continuum states can be constructed which diagonalize (2.3) for $\epsilon = 0$. Although the Hamiltonian (2.3) looks rather innocuous, its dynamics is far from trivial. Many researchers produced evidence[13,15,24–6)] that on the classical level the Hamiltonian (2.3) is completely chaotic for $\epsilon \neq 0$. It appears to be integrable only for $\epsilon = 0$. In the integrable case the classical equations of motion for x_1, p_1, x_2, p_2 can be integrated with the help of a simple canonical transformation

$$\left. \begin{array}{ll} \theta_i = 2\eta_i - \sin(2\eta_i), & \frac{d\theta_i}{d\eta_i} = 4\sin^2(\eta_i); \quad 0 \leq \eta < \pi \\ x_i = 2\,n_i^2 \sin^2(\eta_i); & p_i = \frac{1}{n_i}\cot(\eta_i), \quad i = 1, 2. \end{array} \right\} \qquad (2.6)$$

Since the problem seems to be completely chaotic for $\epsilon \neq 0$ no integrating canonical transformation can exist in this case. We have to resort to numerical methods to integrate the equations of motion. Expressing the dynamical variables x_1, p_1, x_2, p_2 in terms of the action-angle variables $n_1, \theta_1, n_2, \theta_2$ defines the "bound space projected" approach[21,31)] to the one-dimensional helium problem. This is so because the action and angle variables n and θ are only defined for bounded motion. The quantum analog of this classical projection method is to work in a Hilbert space which is spanned by products of the square normalizable states (2.4) only. Since this procedure defines an exact correspondence between the classical and the quantum models we will use this approach in the rest of the paper. It has to be emphasized that a basis constructed from the states (2.4) is not complete. The calculation of resonance widths, for instance, is beyond the scope of this approach. We found, however, that although the "bound space projected" basis (2.4) is incomplete, it yields very accurate results for the locations of autoionizing resonances as long as their widths are small. This condition is met for the energy regime investigated in section 5 where we consider the transition between the regular regime at low energies and the chaotic regime at moderately high excitation energies. In this regime the ratio of the decay widths and the separation of resonance energies is small[13)]. Thus, the bound space projected approach is justified. Moreover, the bound space projected calculations take only a fraction of the time required by calculations in a complete L^2 basis[13)]. Therefore, we felt that the bound space projected approach is adequate as a method for providing support for the ideas developed in this chapter.

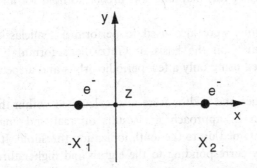

Fig. 1. Schematic sketch of the stretched helium geometry. The nucleus of charge Z is assumed to be at rest at $x = 0$ (infinite mass limit). Electron 1 is confined to the left half space, electron 2 to the right. Both electrons can interact via the repulsive electron–electron Coulomb force across the "barrier" at $x = 0$.

3 Brief review of stretched helium results

It is quite possible and we fully expect that the one-dimensional helium atom will acquire the same status as a paradigm in many-electron atoms as has already been attained by the one-dimensional hydrogen atom in microwave ionization[18] or the kicked rotor[5–8,11,32] in dynamical localization. Therefore, it is useful to explore the stretched helium model in every detail so that the results are handy for drawing qualitative analogies or pointing out discrepancies with the real three-dimensional helium atom. Although many results are known by now, much remains to be done, especially as far as analytical proofs are concerned for properties of this system which by now can only be asserted on numerical grounds. One such property is the absence of any regular islands in the classical phase space. Although by no means established analytically, there is currently no evidence for the existence of any regular islands[13,15,26]. Therefore it is conjectured that the one-dimensional stretched helium atom is fully chaotic[13,15,26]. How dangerous such a claim may be was demonstrated by Dahlqvist and Russberg[33] who were able to identify a small stable island for the $x^2 y^2$ problem previously conjectured to be fully chaotic by many authors. On the other hand the absence of stable islands was proved analytically for a system closely related to the present problem: the kicked one-dimensional hydrogen atom[34,35]. It is therefore not entirely out of the question to prove or disprove the chaoticity of stretched helium analytically.

Results on the periodic orbit structure of stretched helium seem to corroborate the chaoticity claim. All periodic orbits found so far are unstable and can be enumerated one-by-one by a binary code[13,15,24–6]. The remarkable feature of the periodic orbit structure is that no pruning is necessary. All imaginable finite binary sequences seem to correspond to a periodic orbit and vice versa. The one-to-one property of this mapping was proved to hold for all sequences up to length six[26].

The periodic orbits were also used to perform a semiclassical quantization of stretched helium[26] on the basis of Gutzwiller's formula[12,36,37]. Excellent results were reported using only a few periodic orbits and a special resummation technique[26].

Excitation energies and widths were also calculated within the framework of a complex rotation L^2 approach[13]. Results on real and imaginary parts of resonances were obtained up to the tenth ionization threshold. It was established that the resonances corresponding to the eighth and higher thresholds move in the complex plane as a function of ϵ (see (2.3)) not unlike the particles in a highly correlated gas or liquid[13]. We found indications that for increasing excitation energy the gas expands into the complex plane away from the real axis. Thus, for sufficiently high excitation energy a condition may be reached where the separation of the real parts of the resonance energies is comparable with their imaginary parts. This condition is expected to occur at energies around the 30th ionization threshold and corresponds to a situation of overlapping resonances. In this case scattering cross sections are expected to exhibit wild oscillations which cannot be associated any more one-to-one with individual resonances. This condition was first encountered in nuclear physics in heavy ion collisions at relatively high excitation energies. The phenomenon is nowadays referred to as "Ericson fluctuations" since it was predicted to occur by T. Ericson in the 1960's[38,39]. Recently it was established that a system which exhibits chaotic scattering on the classical level is likely to show Ericson fluctuations in the semiclassical regime[40–4].

In the next section we will show that chaotic scattering indeed occurs in the stretched helium model. In section 5 we will establish the existence of a third dynamical regime between the regular and the Ericson regimes. It is characterized by relatively small resonance widths but strongly correlated resonance energies which leads to level repulsion.

4 Stretched helium: a chaotic scattering system

In order to establish the connection between stretched helium and chaotic scattering, we first consider the potential energy in (2.3):

$$V(x_1, x_2) = -\frac{1}{x_1} - \frac{1}{x_2} + \frac{\epsilon}{x_1 + x_2}. \qquad (4.1)$$

Some equipotential lines for (4.1) are shown in Fig. 2 for $\epsilon = 1/2$. This potential is strongly reminiscent of "elbow" scattering systems investigated analytically, numerically and experimentally[45] in connection with chaotic scattering.

The potential shown in Fig. 2 clearly exhibits two asymptotic entry (exit) channels at $x_1 \to \infty$ and $x_2 \to \infty$, respectively. Scattering potentials of the type

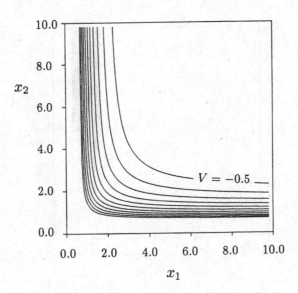

Fig. 2. Equipotential lines for the potential

$$V(x_1, x_2) = -\frac{1}{x_1} - \frac{1}{x_2} + \frac{\epsilon}{x_1 + x_2}$$

for $V = -1.5$ to $V = -0.5$ in steps of $\Delta V = 0.1$ for $\epsilon = 1/2$. The potential V resembles "elbow" scattering geometries studied in the literature.

shown in Fig. 2 also occur in the theoretical description of chemical reactions. Since complicated dynamics and chaotic scattering were reported to occur in these systems[45−8] it is natural to expect chaotic scattering to occur in the potential (4.1). In order to prove this point we launched 800 scattering trajectories at $n_1 = 1$, $\theta_1 = j \cdot (2\pi/801)$, $j = 1, 2, ..., 800$, $x_2 = 10$ and $p_2 = -0.5$. For all these trajectories we calculated the time T it takes to reach the "asymptotic" region which we defined to be $x_1 \geq 10$ or $x_2 \geq 10$. The result is shown in Fig. 3. The delay times T vary smoothly with the initial phase θ_1 for $0 < \theta_1 < 4.5$. At around $\theta_1 \approx 4.55$ we see a cusp-like behavior which is due to the occurrence of a triple collision, i.e., x_1 and x_2 both tend to zero simultaneously. For $4.7 < \theta_1 < 5.7$ we see a very complicated dependence of T on θ_1 which indicates the presence of chaotic scattering. Moreover, the delay time T seems to be unbounded at particular values of θ_1. In Fig. 3 we cut the delay time at $T = 200$ for graphical reasons.

It is interesting to speculate how the presence of classical chaotic scattering will manifest itself on the quantum mechanical level. In the case of a "molecule" scattered off the inhomogeneous electric field of charged wires it was demonstrated[40] that classical chaotic scattering yields a quantum scattering matrix which to a first approximation can be considered a representative of Dyson's circular en-

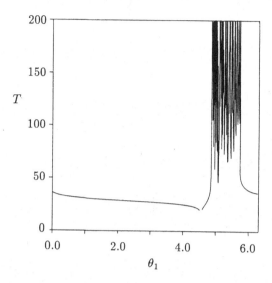

Fig. 3. Distribution of delay times for a chaotic scattering experiment. The first electron is initially in a bounded orbit with $n_1 = 1$. The second electron is lauched towards the helium nucleus with $x_2 = 10$ and $p_2 = -0.5$. Shown is the time the resulting scattering complex stays "bounded" (i.e., x_1 and $x_2 < 10$) as a function of the initial angle θ_1 of electron 1. Irregular behavior is seen for $4.7 < \theta_1 < 5.7$ indicating the presence of chaotic scattering.

semble of random unitary symmetric matrices[49–52]. Since in the stretched helium system time reversal symmetry is active, we expect to see the same behavior for the scattering matrix describing electron scattering off the one-dimensional He$^+$ ion[13].

According to the arguments presented at the end of the previous section, we also expect to find an energy regime which exhibits Ericson fluctuations[38,39]. Following first conjectures[53] it was shown that this regime is also important in mesoscopic systems[54]. In the case of the (one-dimensional) helium atom a complication arises in that close to thresholds the autoionizing resonances become sharper and sharper. Therefore, close to thresholds the imaginary parts of the scattering resonances are expected to be very small and the fluctuations, as a function of the energy, are expected to cross over into a resonance regime which we will call the "Rydberg regime". The Rydberg regime is characterized by clearly distinguishable isolated resonances. Therefore, as a function of the excitation energy we expect to observe an alternating sequence of Rydberg and Ericson regimes. The Ericson regime is expected to be developed between thresholds whereas the Rydberg regime will take over close to thresholds. In the cross-over region between the Ericson regime and the Rydberg regime, where the imaginary parts of the resonances slowly decrease to zero, we expect to see yet another regime which we will call the "Wigner regime". In this regime the spacings of the S-matrix poles will be larger than the imaginary parts of the resonances which allows us to speak of "energy levels"[55]. We expect that these levels exhibit level repulsion and a Wignerian nearest neighbor statistics. In analogy to dissipative systems[52,55,56] we expect the resonance poles in the complex plane to exhibit Ginibre statistics[52,55-57].

According to the discussion above, quantum mechanically we expect to see the following picture as a function of the excitation energy of the one-dimensional helium atom. Below the first ionization threshold we have a regular Rydberg series of bounded states. Even the states above the first threshold, but below the second will be very regular. The same is expected up to the fifth threshold: regular sequences of "states" (autoionizing resonances) only sporadically broken by "intruder states" from a sequence above[13]. For resonances above the fifth threshold we expect to see an increasingly well-developed Wigner regime between thresholds. But since the resonances are still relatively close to the real axis[13] we do not expect to see the Ericson regime yet. But we do expect to see a sequence of regular resonances close to thresholds. According to some very preliminary estimates[13] the Ericson regime is expected to manifest itself around the 30th ionization threshold from whereon the energy region between ionization thresholds is expected to display the full repetitive sequence of Ericson → Wigner → Rydberg regimes.

In the following section we will present preliminary evidence for the existence of the Wigner regime in stretched helium.

5 Dynamical regimes in stretched helium

In this section we will present preliminary evidence for the occurrence of a Wigner regime in doubly excited one-dimensional helium. It occurs at moderate excitation energy characterized by relatively small resonance widths. As discussed in section 2, the most elementary model for stretched helium is the "independent particle model", which totally neglects all correlations between the two electrons. Its energy spectrum is given by:

$$e_{nm} = -\frac{1}{2n^2} - \frac{1}{2m^2}. \tag{5.1}$$

In this formula n and m are the actions of the two independent electrons, respectively. Structurally, and on the crudest level, the true spectrum of one-dimensional helium is very similar to the spectrum (5.1). Therefore, the independent particle model will serve as a point of reference in the following discussion.

For $n \to \infty$ the energy (5.1) converges to the mth ionization threshold $e_m = -1/2m^2$. The energy levels of the one-dimensional helium atom including the electron–electron correlation energy will be denoted by E_{nm}. It is useful to stratify the helium spectrum by normalizing the helium levels to the mth ionization threshold:

$$Y_{nm} = \frac{1}{\sqrt{-2E_{nm}}}.$$ (5.2)

It is possible to calculate the levels of stretched helium with considerable accuracy in a complete L^2 basis[13,15,26]. For the following calculations, however, we decided to use a cheaper, only slightly less accurate, approach namely diagonalization of (2.3) in the (incomplete) basis (2.4). This approach was called the "bound space projected" approach in section 2. Comparing the results of this approach with an L^2 approach the relative accuracy of energy levels around and below the 10th threshold is typically better than 1%.

Figs. 4–6 show energy levels obtained for the class of symmetric wavefunctions. The bar graph in Fig. 4 shows the levels obtained in the bound space projected approach from thresholds $m = 1$ to $m = 5$. Regular sequences of states can be identified accumulating at the respective ionization thresholds. The bar graphs in Figs. 5 and 6 show the levels corresponding to the thresholds $m = 6, ... 10$ and $m = 11, ... 15$, respectively. It can be seen that the level sequences between thresholds become more and more irregular. Also, the space between thresholds tends to fill in more equally as the excitation energy increases. Theoretically, the density of states diverges at all ionization thresholds. While our calculations reproduce this behavior very well for low m thresholds (see Fig. 4), the density of states seems to thin out for higher m thresholds (see Figs. 5 and 6). This behavior is due to numerical restrictions. Only 406 basis states were used to diagonalize (2.3). Including more basis states increases the density of states at individual ionization thresholds.

On the basis of the bar graphs shown in Figs. 4–6, we performed a statistical analysis of nearest neighbor spacings. Fig. 4 shows the result of the spacing statistics for energy levels up to the 5th ionization threshold. Plotted is the probability distribution $P(s)$ of spacings s (normalized to the mean spacing). The resulting probability distribution mainly reflects the regular sequences of states which converge to the ionization thresholds. Therefore, the probability distribution is peaked at $s = 0$ and does not correspond to any of the established universality classes[49–52].

The situation looks dramatically different for the spacings distribution of levels from the 6th to the 10th threshold (see Fig. 5). It can be seen that the distribution starts to move towards a Wignerian statistics.

Fig. 6 represents the level distribution for thresholds from $m = 11$ to $m = 15$. Here, the distribution is already very close to Wignerian.

As discussed in the previous section, care has to be taken as to which levels to include in the statistical analysis. Levels close to the ionization thresholds are members of the Rydberg regime and should be excluded when focussing on the Wigner regime. We have presently not established any rigorous criteria for

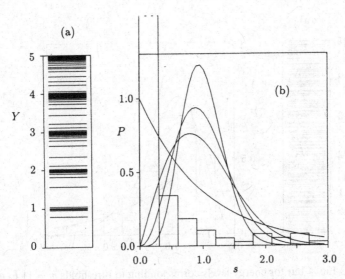

Fig. 4. Energy levels of one-dimensional helium and their statistical analysis: (a) stratified energy levels for thresholds $m = 1$ to $m = 5$; (b) Nearest neighbor spacing statistics of the levels shown in (a). The histogram corresponds to the numerical data. The smooth lines are theoretical spacing distributions which correspond to the Poissonian, Wignerian, Unitary and Symplectic ensemble, respectively.

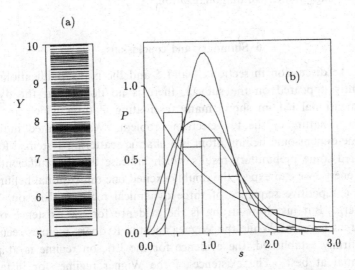

Fig. 5. Same as Fig. 4 but for energy levels corresponding to thresholds $m = 6$ to $m = 10$.

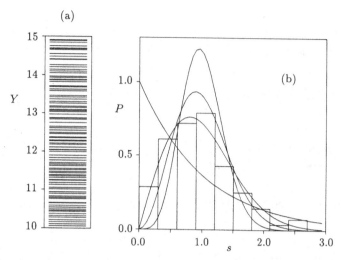

Fig. 6. Same as Fig. 4 but for energy levels corresponding to thresholds $m = 11$ to $m = 15$.

the boundaries between the different regimes. On the other hand, our numerical calculations do not reproduce more than a few states in the Rydberg regime (see bar graphs in Figs. 5 and 6) anyway. Therefore, in our present calculations, none of the states shown in the bar graphs in Figs. 4–6 was excluded and the level statistics shown in Figs. 4–6 was performed on the basis of all the states obtained from the 406 state numerical diagonalization.

6 Summary and conclusions

Following the discussion in sections 4 and 5 and the many publications which have recently appeared on the subject, there is no doubt that the dynamics of one-dimensional helium shows many fascinating features which are due to the nonlinear nature of the two-electron problem. We presented indications that the one-dimensional helium atom is a chaotic scattering system. Moreover, we presented some preliminary results which indicate that for sufficiently high excitation energy we can expect the doubly excited one-dimensional helium atom to exhibit a repetitive sequence of three dynamical regimes: Ericson, Wigner and Rydberg. But just how strong is the evidence for the existence of these three different regimes? While the Wigner and the Rydberg regimes seem to be relatively firmly established, the evidence for the Ericson regime is at present circumstantial at best. The existence of the Wigner regime, for instance, is strongly supported by the facts presented above. Also, there seems to be no doubt about the existence of the Rydberg regime. In reference 13, using an L^2 approach, regular sequences of states with monotonically decreasing widths were found which converge to ionization thresholds. These states can be identified with states of the Rydberg regime. The existence of the Ericson regime, however, was extrapolated on the basis of the behavior of the widths of states associated with ionization thresholds $m \leq 10$. But since the Ericson regime has so far been found in all chaotic scattering systems, we have no doubt that it will be present in the chaotic one-dimensional model of the helium atom.

We are convinced that the stretched helium model is an interesting dynamical system which shows highly complex behavior both on the classical and the quantum levels. If a one-dimensional version of the helium atom already shows such a rich dynamical behavior how much more structure is waiting to be explored in the "real" helium atom?

Acknowledgments

R. B. is grateful for financial support by the Deutsche Forschungsgemeinschaft and for a grant by the University of Delaware Research Foundation. W. P. R. gratefully acknowledges support of the United States National Science Foundation.

REFERENCES

[1]   P. Holmes, "Poincaré, Celestial Mechanics, Dynamical-Systems Theory and Chaos", Phys. Rep. **193**, 137–163 (1990).

[2]   H. Poincaré, "Les Methodes Nouvelles de la Mechanique Celeste", Gauthier Villars, Paris, (1892).

[3]   A. J. Lichtenberg and M. A. Lieberman, "Regular and Stochastic Motion", Applied Mathematical Sciences **38**, Springer, New York (1983).

[4]   H. Friedrich and D. Wintgen, "The Hydrogen Atom in a Uniform Magnetic Field – an Example of Chaos", Phys. Rep. **183**, 37–79 (1989).

[5]   G. Casati, B. V. Chirikov, F. M. Izraelev, and J. Ford, in "Stochastic Behavior in Classical and Quantum Hamiltonian Systems", Lecture Notes in Physics, **93**, p. 334, Springer, New York (1979).

[6]   B. V. Chirikov, F. M. Izraelev, and D. L. Shepelyansky, "Dynamical Stochasticity in Classical and Quantum Mechanics", Sov. Sci. Rev. Sect. **C2**, 209–267 (1981).

[7]   D. L. Shepelyansky, "Localization of Quasienergy Eigenfunctions in Action Space", Phys. Rev. Lett. **56**, 677–680 (1986).

[8]   R. Blümel and U. Smilansky, "Quantum Mechanical Suppression of Chaos", Physics World, Vol. 3, No. 2, 30–34 (1990).

[9]   R. Blümel and U. Smilansky, "Quantum Mechanical Suppression of Classical Stochasticity in the Dynamics of Periodically Perturbed Surface-State-Electrons", Phys. Rev. Lett. **52**, 137–140 (1984).

[10] R. Blümel and U. Smilansky, "Suppression of Classical Stochasticity by Quantum Mechanical Effects in the Dynamics of Periodically Perturbed Surface-State Electrons", Phys. Rev. **A30**, 1040–1051 (1984).

[11] S. Fishman, D. R. Grempel, and R. E. Prange, "Chaos, Quantum Recurrences, and Anderson Localization", Phys. Rev. Lett. **49**, 509–512 (1982).

[12] M. C. Gutzwiller, "Chaos in Classical and Quantum Mechanics", Springer Verlag, New York (1990).

[13] R. Blümel and W. P. Reinhardt, "Where is the Chaos in Two-Electron Atoms?" in *Directions in Chaos*, Vol. 4, edited by D. H. Feng and Y. M. Yuan, World Scientific, Singapore (1992).

[14] X. Yang and J. Burgdörfer, "Molecular Dynamics Approach to the Statistical Properties of Energy Levels", Phys. Rev. Lett. **66**, 982–985 (1991).

[15] J.-H. Kim and G. S. Ezra, "Periodic Orbits and the Classical-Quantum Correspondence for Doubly-Excited States of Two-Electron Atoms", in Proceedings of the Adriatico Workshop on Quantum Chaos, 1990.

[16] D. L. Shepelyansky, in "Chaotic Behavior in Quantum Hamiltonian Systems", edited by G. Casati, Plenum Press, New York (1985), p. 187.

[17] R. V. Jensen, "Stochastic Ionization of Surface-State Electrons: Classical Theory", Phys. Rev. **A30**, 386–397 (1984).

[18] G. Casati, B. V. Chirikov, D. L. Shepelyansky, and I. Guarneri, "Relevance of Classical Chaos in Quantum Mechanics: The Hydrogen Atom in a Monochromatic Field", Phys. Rep. **154**, 77 (1987).

[19] J. E. Bayfield and P. M. Koch, "Multiphoton Ionization of Highly Excited Hydrogen Atoms", Phys. Rev. Lett. **33**, 258–261 (1974).

[20] K. A. H. van Leeuwen, G. v. Oppen, S. Renwick, J. B. Bowlin, P. M. Koch, R. V. Jensen, O. Rath, D. Richards, and J. G. Leopold, "Microwave Ionization of Hydrogen Rydberg Atoms: Experiment versus Classical Dynamics", Phys. Rev. Lett. **55**, 2231–2234 (1985).

[21] R. Blümel and U. Smilansky, "Microwave Ionization of Highly Excited Hydrogen Atoms", Z. Phys. **D6**, 83–105 (1987).

[22] R. Grobe and J. H. Eberly, "Photoelectron Spectra for a Two-Electron System in a Strong Laser Field", Phys. Rev. Lett. **68**, 2905–2908 (1992).

[23] S. Watanabe, "Kummer-Function Representation of Ridge Travelling Waves", Phys. Rev. **A36**, 1566–1574 (1987).

[24] K. Richter and D. Wintgen, "Analysis of Classical Motion on the Wannier Ridge", J. Phys. **B23**, L197–201 (1990).

[25] B. Eckhardt, "Two-Electron Escape at Positive Energies: Wannier's Analysis Revisited", Univ. Marburg, preprint (1991).

[26] G. S. Ezra, K. Richter, G. Tanner, and D. Wintgen, "Semiclassical Cycle Expansion for the Helium Atom", J. Phys. **B24**, L413–L420 (1991).

[27] J. G. Leopold and I. C. Percival, "Ionization of Highly Excited Atoms by Electric Fields. III Microwave Ionization and Excitation", J. Phys. **B12**, 709–721 (1979).

[28] B. I. Meerson, E. A. Oks, and P. V. Sasarov, "Stochastic Instability of an Oscillator and the Ionization of Highly-Excited Atoms Under the Action of Electromagnetic Radiation", JETP Lett. **29**, 72–75 (1979).

[29] R. V. Jensen, "Stochastic Ionization of Surface-State Electrons", Phys. Rev. Lett. **49**, 1365–1368 (1982).

[30] R. Blümel and U. Smilansky, "Ionization of Excited Hydrogen Atoms by Microwave Fields: A Test Case for Quantum Chaos", Phys. Scr. **40**, 386–393 (1989).

[31] R. Blümel, C. Hillermeier, and U. Smilansky, "Classical and Quantum Dynamical Regimes in the Bound Space Projected Dynamics of Strongly Driven H Rydberg Atoms", Z. Phys. **D15**, 267–280 (1990).

[32] F. M. Izrailev, "Simple Models of Quantum Chaos: Spectrum and Eigenfunctions", Phys. Rep. 196, 299–392 (1990).

[33] P. Dahlqvist and G. Russberg, "Existence of Stable Orbits in the x^2y^2 Potential", Phys. Rev. Lett. **65**, 2837–2838 (1990).

[34] C. F. Hillermeier, R. Blümel, and U. Smilansky, "Ionization of H Rydberg Atoms: Fractals and Power-law Decay", Phys. Rev. **A45**, 3486–3502 (1992).

[35] R. Blümel, "Exotic Fractals and Atomic Decay", Proceedings of the International School of Physics "Enrico Fermi", Varenna, North Holland, Amsterdam (1993).

[36] M. C. Gutzwiller, "Periodic Orbits and Classical Quantization Conditions", J. Math. Phys. **12** , 343–358 (1971).

[37] M. C. Gutzwiller, "The Quantization of a Classically Ergodic System", Physica **D5**, 183–207 (1982).

[38] T. Ericson, "Fluctuations of Nuclear Cross Sections in the Continuum Region", Phys. Rev. Lett. **5**, 430–431 (1960).

[39] T. Ericson, "A Theory of Fluctuations in Nuclear Cross Sections", Ann. Phys. **23**, 390–414 (1963)

[40] R. Blümel and U. Smilansky, "Classical Irregular Scattering and its Quantum Mechanical Implications", Phys. Rev. Lett. **60**, 477–480 (1988).

[41] R. Blümel and U. Smilansky, "A Simple Model for Chaotic Scattering. II. Quantum Mechanical Approach", Physica **D36**, 111–136 (1989).

[42] R. Blümel and U. Smilansky, "Random-Matrix Description of Chaotic Scattering: Semiclassical Approach", Phys. Rev. Lett. **64**, 241–244 (1990).

[43] U. Smilansky, "The Classical and Quantum Theory of Scattering" in Proceedings of the 1989 Les Houches Summer School on "Chaos and Quantum Physics", eds M. J. Giannoni, A. Voros and J. Zinn-Justin, North Holland, Amsterdam (1992).

[44] R. Blümel, "Quantum Chaotic Scattering", in Directions in Chaos, Vol. 4, edited by D. H. Feng and J. M. Yuan, World Scientific, Singapore (1992).

[45] E. Doron, U. Smilansky, and A. Frenkel, "Experimental Demonstration of Chaotic Scattering of Microwaves", Phys. Rev. Lett. **65**, 3072–3075 (1990).

[46] D. W. Noid, S. K. Gray, and S. A. Rice, "Fractal Behavior in Classical Collisional Energy Transfer", J. Chem. Phys. **84**, 2649–2652 (1986).

[47] C. C. Rankin and W. H. Miller, "Classical S-Matrix for Linear Reactive Collision of $H + Cl_2$", J. Chem. Phys. **55**, 3150–3156 (1971).

[48] P. Gaspard, "Scattering and Resonances: Classical and Quantum Dynamics", Proceedings of the International School of Physics "Enrico Fermi", Varenna, (1992).

[49] F. J. Dyson, "Statistical Theory of the Energy Levels of Complex Systems", J. Math. Phys. **3**, 140–175 (1962).

[50] C. E. Porter, "Statistical Theory of Spectral Fluctuations", Academic, New York (1965).

[51] M. L. Mehta, "Random Matrices", Academic, New York (1991).

[52] F. Haake, "Quantum Signatures of Chaos", Springer, Berlin (1991).

[53] R. Blümel and U. Smilansky, "Quantenmechanik des irregulären Streuens", Phys. Bl. **45**, 379–381 (1989).

[54] R. A. Jalabert, H. U. Baranger, and A. D. Stone, "Conductance Fluctuations in the Ballistic Regime: A Probe of Quantum Chaos?", Phys. Rev. Lett. **65**, 2442–2445 (1990).

[55] R. Grobe, F. Haake, and H. J. Sommers, "Quantum Distinction of Regular and Chaotic Dissipative Motion", Phys. Rev. Lett. **61**, 1899–1902 (1988).

[56] R. Grobe and F. Haake, "Universality of Cubic-Level Repulsion for Dissipative Quantum Chaos", Phys. Rev. Lett. **62**, 2893–2895 (1989).

[57] J. Ginibre, "Statistical Ensembles of Complex, Quaternion and Real Matrices", J. Math. Phys. **6**, 440–449 (1965).

PART THREE

Semiclassical approximations

Semiclassical theory of spectral rigidity

By M. V. Berry, F.R.S.

H. H. Wills Physics Laboratory, Tyndall Avenue, Bristol BS8 1TL, U.K.

(Received 20 February 1985)

The spectral rigidity $\Delta(L)$ of a set of quantal energy levels is the mean square deviation of the spectral staircase from the straight line that best fits it over a range of L mean level spacings. In the semiclassical limit ($\hbar \to 0$), formulae are obtained giving $\Delta(L)$ as a sum over classical periodic orbits. When $L \ll L_{\max}$, where $L_{\max} \sim \hbar^{-(N-1)}$ for a system of N freedoms, $\Delta(L)$ is shown to display the following universal behaviour as a result of properties of very long classical orbits: if the system is classically integrable (all periodic orbits filling tori), $\Delta(L) = \frac{1}{15}L$ (as in an uncorrelated (Poisson) eigenvalue sequence); if the system is classically chaotic (all periodic orbits isolated and unstable) and has no symmetry, $\Delta(L) = \ln L/2\pi^2 + D$ if $1 \ll L \ll L_{\max}$ (as in the gaussian unitary ensemble of random-matrix theory); if the system is chaotic and has time-reversal symmetry, $\Delta(L) = \ln L/\pi^2 + E$ if $1 \ll L \ll L_{\max}$ (as in the gaussian ortho-gonal ensemble). When $L \gg L_{\max}$, $\Delta(L)$ saturates non-universally at a value, determined by short classical orbits, of order $\hbar^{-(N-1)}$ for integrable systems and $\ln(\hbar^{-1})$ for chaotic systems. These results are obtained by using the periodic-orbit expansion for the spectral density, together with classical sum rules for the intensities of long orbits and a semiclassical sum rule restricting the manner in which their contributions interfere. For two examples $\Delta(L)$ is studied in detail: the rectangular billiard (inte-grable), and the Riemann zeta function (assuming its zeros to be the eigenvalues of an unknown quantum system whose unknown classical limit is chaotic).

1. Introduction

Several statistical measures of the regularity of sequences of eigenvalues were introduced to describe the energy levels of many-particle systems such as nuclei (Porter 1965). Recently these spectral measures have been employed for bound systems with few freedoms, to explore the ways in which the distribution of quantal energy levels reflects integrability or chaos in the underlying classical trajectories. It was expected, and found, that classically integrable systems have levels that are locally uncorrelated and well described by a Poisson distribution; in contrast, classically chaotic systems have levels with strong local repulsion, well described by the eigenvalues of matrices drawn randomly from appropriate ensembles (for reviews see Berry 1983, 1984; Bohigas & Giannoni 1984).

My purpose in this paper is to explain the semiclassical origin and the limits of validity of these two types of spectral universality, by deriving theoretical expressions for one of the spectral statistics, namely the rigidity. This will be defined in (5) in terms of the spectral staircase

$$\mathcal{N}(E) \equiv \sum_n \Theta(E - E_n), \tag{1}$$

where $E_n = E_1, E_2 \ldots$ is the eigenvalue sequence and Θ denotes the unit step function. We also require the spectral density

$$d(E) \equiv \mathrm{d}\mathcal{N}(E)/\mathrm{d}E = \sum_n \delta(E - E_n). \tag{2}$$

For a system with N freedoms the local averages of these functions are

$$\langle \mathcal{N}(E) \rangle = \Omega(E)/h^N; \quad \langle d(E) \rangle = (d\Omega(E)/dE)/h^N, \tag{3}$$

where $\Omega(E)$ is the classical phase-space volume enclosed by the surface with energy E, given in terms of the Hamiltonian $H(q, p)$ by

$$\Omega(E) = \int \mathrm{d}^N q \int \mathrm{d}^N p \, \Theta(E - H(q, p)). \tag{4}$$

In the cases of interest here, $N \geqslant 2$.

The averages denoted by $\langle \ \rangle$ in (3) refer to all levels in an energy range that is classically small, i.e. small in comparison with E, but semiclassically large, i.e. large in comparison with the mean level spacing $\langle d \rangle^{-1} \sim \hbar^N$. This mean level spacing will play an important role in what follows, and will be called the inner energy scale.

The rigidity $\Delta(L)$ is now defined as the local average of the mean square deviation of the staircase from the best fitting straight line over an energy range corresponding to L mean level spacings, namely

$$\Delta(L) \equiv \left\langle \min_{(A, B)} \frac{\langle d(E) \rangle}{L} \int_{-L/2\langle d \rangle}^{L/2\langle d \rangle} \mathrm{d}\epsilon \, [\mathcal{N}(E + \epsilon) - A - B\epsilon]^2 \right\rangle. \tag{5}$$

This function was introduced by Dyson & Mehta (1963) (they called it Δ_3 to distinguish it from two less useful statistics). Minimizing over A and B leads to

$$\Delta(L) = \left\langle \left\{ \frac{\langle d \rangle}{L} \int_{-L/2\langle d \rangle}^{L/2\langle d \rangle} \mathrm{d}\epsilon \, \mathcal{N}^2(E + \epsilon) - \left[\frac{\langle d \rangle}{L} \int_{-L/2\langle d \rangle}^{L/2\langle d \rangle} \mathrm{d}\epsilon \, \mathcal{N}(E + \epsilon) \right]^2 \right. \right.$$
$$\left. \left. - 12 \left[\frac{\langle d \rangle^2}{L^2} \int_{-L/2\langle d \rangle}^{L/2\langle d \rangle} \mathrm{d}\epsilon \, \epsilon \, \mathcal{N}(E + \epsilon) \right]^2 \right\} \right\rangle. \tag{6}$$

When $L \ll 1$, the fact that $\mathcal{N}(E)$ is a staircase leads to the limit $\Delta \to \frac{1}{15}L$ whatever distribution the levels have (provided this is non-singular). Therefore the spectral rigidity gives no information about the very finest scales corresponding to the spacings between neighbouring levels. Its usefulness lies in the way it describes correlations over level sequences longer than the inner energy scale (which corresponds to $L = 1$).

We shall demonstrate the existence of two universality classes of rigidity, extending from $L \sim 1$, corresponding to the inner energy scale, to a value $L_{\max}$

corresponding to an outer energy scale h/T_{min}, where T_{min} is the period of the shortest classical closed orbit. Thus

$$L_{max} \equiv h\langle d\rangle/T_{min} \sim \hbar^{-(N-1)} \tag{7}$$

greatly exceeds unity, in spite of the fact that the outer energy scale is of order $\hbar$ and hence classically small. Now we can be more precise about the local averages denoted by $\langle \ \rangle$: these correspond to energy ranges much larger than the outer scale but still classically small, for example energy ranges of order $\hbar^{\frac{1}{2}}$.

The first universality class occurs for classically integrable systems. In these, as will be shown in §3, the Poisson form $\frac{1}{15}L$ extends from $L = 0$ to L_{max}. For $L > L_{max}$, $\Delta(L)$ reaches a saturation value (not universal), Δ_∞, which can be regarded as a measure of the totality of the spectral fluctuations on all scales. This extends earlier work (Berry & Tabor 1977a) in which Poisson statistics were shown to describe local fluctuations, and, as will be shown explicitly in §4 it explains recent numerical results of Casati *et al.* (1985) on the rectangular billiard.

The second universality class occurs for classically chaotic systems. In these, as will be shown in §6, $\Delta(L)$ increases only logarithmically in the range $1 \lesssim L < L_{max}$, which indicates long-range rigidity in the level distribution. The fact that for systems with time-reversal symmetry the coefficient of the logarithm is twice what it is when there is no such symmetry is given a simple semiclassical explanation in §8. For $L > L_{max}$, $\Delta(L)$ reaches a saturation value (not universal), Δ_∞, much smaller than in the integrable case. When applied to the 'level sequence' consisting of the imaginary parts of the zeros of the Riemann zeta function (§7), our results are consistent with what little is known and conjectured about the asymptotics of this sequence.

In deriving these results from (6), semiclassical methods are essential. The reason is that for local spectral statistics to attain well defined limiting values, classically small energy ranges must contain many levels, and this happens only as $\hbar \to 0$ (for scaling systems such as billiards, this is equivalent to $E \to \infty$). The semiclassical technique employed here (in §2) is the representation of the spectral density $d(E)$ as a sum over all the periodic orbits of the classical system, introduced and developed by Gutzwiller (1967, 1969, 1970, 1971, 1978) and Balian & Bloch (1972, 1974).

When applied to the calculation of spectral rigidity, the periodic-orbit technique requires two further ingredients. The first is a classical sum rule for the orbit intensities, recently discovered by Hannay & Ozorio de Almeida (1984), and the second is a new semiclassical sum rule derived in §5.

The arguments and conclusions of this paper complement those of Pechukas (1983). He obtained all the spectral statistics, but made use of a statistical assumption about the wavefunction. I make no such statistical assumption, but discuss only $\Delta(L)$.

2. Rigidity in terms of periodic orbits

Periodic-orbit theory gives the semiclassical spectral density as

$$d(E) = \langle d(E) \rangle + d_{\text{osc}}(E), \tag{8}$$

where $d_{\text{osc}}(E)$ is a sum over classical periodic orbits, each of which contributes an oscillatory function and whose combined effect is to produce, by constructive interference, a sequence of singularities (such as the δ functions in (2) or approximations to them) minus $\langle d \rangle$. We shall write d_{osc} in a form appropriate to systems that are completely integrable or completely chaotic. In an integrable system, closed orbits are not isolated but form $(N-1)-$ parameter families filling N-dimensional phase-space tori. A chaotic system we define as one that is ergodic and for which, in addition, all closed orbits are isolated and therefore unstable. The oscillatory contribution in (8) is

$$d_{\text{osc}}(E) = \frac{1}{\hbar^{\mu+1}} \sum_j A_j(E) \exp\{iS_j(E)/\hbar\}. \tag{9}$$

Detailed descriptions and discussions of this formula were given by Berry (1983, 1984); here we need the following facts. In the sum, j labels all distinct periodic orbits including all multiple traversals, positive and negative (but not zero). It will be of crucial importance that negative traversals correspond to retracings, where the orbit is followed backwards in time, and not to time-reversed orbits (the latter exist only for systems with time-reversal symmetry). The exponent μ is $\frac{1}{2}(N-1)$ for integrable systems and zero for chaotic ones. This difference is important and for integrable systems the periodic orbits on a torus combine coherently to produce much stronger spectral oscillations than an isolated orbit of a chaotic system, as well as giving the first hint that the level statistics will be different too. The (real) amplitudes A_j will be discussed later. In the exponent, the phase contains the action $S_j(E)$, defined for m traversals as

$$S_j(E) \equiv m\left\{\oint \boldsymbol{p} \cdot d\boldsymbol{q} + \alpha\hbar\right\}, \tag{10}$$

where the integral is over a single traversal and $\alpha\hbar$, which will play no part in what follows, gives focusing corrections such as Maslov indices. Because of the negative traversals, d_{osc} is real.

In (9) the energy dependence of the oscillations is determined by the orbit periods $T_j(E)$ because

$$T_j(E) = dS_j(E)/dE. \tag{11}$$

The longest oscillation, giving the largest scale of spectral fluctuations, comes from the shortest orbit and has 'wavelength' given by the outer scale h/T_{min}, already defined. Thus the constructive interference that gives δ functions, whose mean spacing is the inner scale $\langle d \rangle^{-1} \approx \hbar^N$, is determined by very long orbits, with periods $T \sim \hbar^{-(N-1)}$.

To incorporate the oscillations (9) into the rigidity formula (6) we use the fact that the energy range $L/\langle d \rangle$ is classically small (although it may be semiclassically large) to write

$$S_j(E+\epsilon) \approx S_j(E) + \epsilon T_j(E) \tag{12}$$

and ignore the ϵ dependences of A_j and $\langle d \rangle$. Thus the spectral staircase is

$$\mathcal{N}(E) = \langle \mathcal{N}(E) \rangle + \mathcal{N}_{\text{osc}}(E),$$

where

$$\mathcal{N}_{\text{osc}}(E+\epsilon) = \frac{-i}{\hbar^\mu} \sum_j \frac{A_j}{T_j} \exp\{i(S_j + \epsilon T_j)/\hbar\}. \tag{13}$$

The integrals in (6) are now elementary and give

$$\varDelta(L) = \left\langle \frac{1}{\hbar^{2\mu}} \sum_i \sum_j \frac{A_i A_j}{T_i T_j} \exp\{i(S_i - S_j)/\hbar\} \right.$$
$$\left. \times [F(y_i - y_j) - F(y_i) F(y_j) - 3F'(y_i) F'(y_j)] \right\rangle, \tag{14}$$

where

$$y_j \equiv LT_j/2\hbar\langle d \rangle, \tag{15}$$

$$F(y) \equiv \sin y/y, \tag{16}$$

and primes denote differentiation.

To arrive at (14) we made use of the result

$$\langle \exp\{iS_j/\hbar\} \rangle \to 0 \quad \text{as} \quad \hbar \to 0, \tag{17}$$

which holds because the local averaging is over an energy range much greater than the outer scale. It is tempting to think that the same principle of destructive interference will eliminate the non-diagonal terms $i \neq j$ in (14). This is the case for integrable systems (as will be shown in §5), but for chaotic systems the proliferation of pairs of very long orbits with action differences $(S_i - S_j) < \hbar$ is important and we shall see that local averaging does not diagonalize the sum.

However, the restriction to pairs with $(S_i - S_j)/\hbar < 1$ does have the effect that in the functions F in (14) we can set $y_i = y_j$. The reason is that, for long orbits,

$$S_j \to T_j N\Omega/(\mathrm{d}\Omega/\mathrm{d}E) \tag{18}$$

(Hannay & Ozorio de Almeida 1984), and together with (3) this implies

$$|y_i - y_j| \to \frac{|S_i - S_j|}{\hbar} \frac{L}{2N\mathcal{N}}, \tag{19}$$

which vanishes because $L \ll \mathcal{N}$. Thus the rigidity becomes

$$\varDelta(L) = \frac{2}{\hbar^{2\mu}} \int_0^\infty \frac{\mathrm{d}T}{T^2} \phi(T) G(LT/2\langle d \rangle \hbar), \tag{20}$$

where

$$G(y) \equiv 1 - F^2(y) - 3(F'(y))^2 \tag{21}$$

and

$$\phi(T) \equiv \left\langle \sum_i \sum_j^+ A_i A_j \cos\{(S_i - S_j)/\hbar\} \delta\{T - \tfrac{1}{2}(T_i + T_j)\} \right\rangle \tag{22}$$

in which the $+$ on the summation denotes restriction to positive traversals (i.e. $T_j > 0$).

Figure 1 shows the important function $G(y)$. Because G is small if $y \leqslant 1$, it selects from the sum (22) only those pairs of orbits whose average period exceeds $2\langle d \rangle \hbar/L$; therefore G will be called the orbit selection function. Such selection is physically reasonable because $\varDelta(L)$ is defined by (5) in terms of deviations from a linear approximation to the staircase over an energy range $L/\langle d \rangle$: the periodic-orbit sum (13) shows that this linear approximation is determined by orbits with $T_j < \langle d \rangle \hbar/L$, and deviations from it by orbits with $T_j > \langle d \rangle \hbar/L$. The effects of this orbit selection depend strongly on the value of L in comparison with unity and $L_{\max}$ (equation 7). For the shortest closed orbit, (7) shows that when $L = L_{\max}$ the argument of G is $y = \pi$.

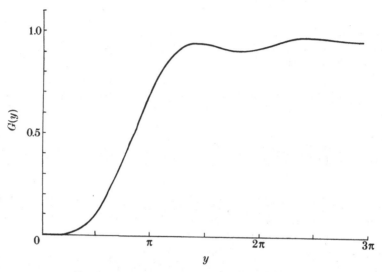

FIGURE 1. Orbit selection function defined by (21).

It will be important to know the large-T limiting form of the diagonal sum in (22), namely

$$\phi_{\mathrm{D}}(T) \equiv \langle \Sigma^+ A_j^2 \, \delta(T - T_j) \rangle. \tag{23}$$

This is the number density of orbits with periods near T, weighted with intensities A_j^2. As $T \to \infty$, the density of periodic orbits increases and the intensities decrease: as a power-law for integrable systems and exponentially for chaotic ones. The results of the competition between these two tendencies was calculated by Hannay & Ozorio de Almeida (1984) as a consequence of their extension of the important idea that very long periodic orbits are uniformly distributed in phase space (see, for example, Parry and Pollicot 1983; Parry 1984). They found that

$$\phi_{\mathrm{D}}(T) \to (\mathrm{d}\Omega/\mathrm{d}E)/(2\pi)^{N+1} \quad \text{(integrable)} \tag{24}$$

and

$$\phi_{\mathrm{D}}(T) \to T/4\pi^2 \quad \text{(chaotic)}. \tag{25}$$

(The integrable result (24) had previously been found by Berry & Tabor (1977a), in a different way.)

These formulae correspond to the amplitudes A_j combining incoherently as in (23), which is the correct procedure if no symmetry enforces strict degeneracy among the orbit actions S_j. When such degeneracy does exist, the appropriate amplitudes must be combined coherently, and we shall see in §8 that this has important repercussions for time-reversal symmetry. In writing (25) we have, for simplicity, ignored multiple traversals ($M > 1$ in 10), because these give contributions to ϕ_{D} that vanish as $T \to \infty$ (this will be illustrated in §7).

3. INTEGRABLE SYSTEMS

For integrable systems, global action-angle variables exist, with each set of actions $I = \{I_1...I_N\}$ denoting a phase-space torus. The Hamiltonian can be written $H(I)$ and the frequencies $\omega = \{\omega_1...\omega_N\}$ on the torus I are given by $\omega = \nabla_I H(I)$. The periodic orbits at energy E are knots on the torus, with winding numbers $M = \{M_1...M_N\}$ for the N irreducible cycles. These winding numbers constitute the label j in the periodic-orbit sums of the previous section. Each M defines a resonant torus I_M, which is one whose frequencies are commensurable, and hence a period T_M, by

$$\omega(I_M) = 2\pi M/T_M; \quad H(I_M) = E. \tag{26}$$

A convenient form for the orbit amplitudes A_M is that given by Berry & Tabor (1977b):

$$A_M^2 = \frac{(2\pi)^{N-1}}{T_M^N |\omega \cdot \partial I_M/\partial T_M \det\{\partial\omega_i/\partial I_j\}_M|}. \tag{27}$$

For the rigidity, (20)–(22), together with the diagonal average to be justified later, give the topological sum

$$\Delta(L) = \frac{2}{\hbar^{N-1}} \sum_M^+ \frac{A_M^2}{T_M^2} G(LT_M/2\langle d\rangle \hbar). \tag{28}$$

Thus $\Delta(L)$ is a sum of weighted scaled orbit selection functions (figure 1), one for each resonant torus.

When $L \ll L_{max}$, G selects only long orbits and the topological sum can be evaluated by using the continuum limit (24). This gives

$$\Delta(L) = \frac{\hbar}{(2\pi\hbar)^N \pi} \frac{d\Omega}{dE} \int_0^\infty \frac{dT}{T^2} G(LT/2\langle d\rangle \hbar)$$

$$= \frac{L}{2\pi} \int_0^\infty \frac{dy}{y^2} G(y). \tag{29}$$

The integral, from (21) and (16), equals $\frac{2}{15}\pi$, so that

$$\Delta(L) = \tfrac{1}{15}L \quad (L \ll L_{max}). \tag{30}$$

This supports the claim that the local spectra of classically integrable systems belong to the universality class of uncorrelated level sequences; of course 'local' means $L \ll L_{max}$.

When $L \gg L_{max}$, the orbit selection function in (28) is unity for all orbits M, and Δ attains a saturation value given by the convergent sum

$$\Delta_\infty = \frac{2}{\hbar^{N-1}} \sum_M^+ \frac{A_M^2}{T_M^2}. \tag{31}$$

Although this is semiclassically large, the r.m.s. fluctuations $\Delta_\infty^{\frac{1}{2}}$ in the staircase are still much smaller than the mean height $\langle \mathcal{N}\rangle$ of the staircase itself (equation (3)), by a factor $\hbar^{\frac{1}{4}(N+1)}$. Saturation of the rigidity has been observed in numerical calculations by Seligman *et al.* (1985) (for an integrable polynomial Hamiltonian) and Casati *et al.* (1985) (for an integrable billiard (see the next section)).

In the crossover region $L \sim L_{\max}$, (28) predicts a few weak oscillations as L increases to reveal the contribution of the shortest orbit with period $T_{\min}$. The slowest oscillations in $\Delta(L)$ have an L-wavelength of $L_{\max}$.

4. AN EXAMPLE: BILLARDS IN A RECTANGLE

For a particle of mass m moving freely within a rectangle with sides a, b and impenetrable walls, the quantal energy levels are

$$E_{l,n} = \hbar^2 \pi^2 (l^2 \alpha^{-\frac{1}{2}} + n^2 \alpha^{\frac{1}{2}})/2mab, \tag{32}$$

where

$$\alpha \equiv a^2/b^2. \tag{33}$$

We assume without loss of generality that $a \geqslant b$, i.e. $\alpha \geqslant 1$. Classically, the action Hamiltonian for this two-dimensional integrable system is

$$H(I_1, I_2) = \pi^2 (I_1^2/a^2 + I_2^2/b^2)/2m \tag{34}$$

with frequencies

$$\omega = \pi^2 (I_1/a^2, I_2/b^2)/m. \tag{35}$$

It follows from (26) that the period of the closed orbit with topology $\boldsymbol{M} = (M_1, M_2)$ is

$$T_{\boldsymbol{M}} = [2m(M_1^2 a^2 + M_2^2 b^2)/E]^{\frac{1}{2}}. \tag{36}$$

Three of these orbits are illustrated in figure 2. The resonant tori whose orbits have these periods have actions

$$\boldsymbol{I_M} = 2m (M_1 a^2, M_2 b^2)/\pi T_{\boldsymbol{M}}. \tag{37}$$

In terms of these quantities it is easy to calculate the torus amplitudes (27)

$$A_{\boldsymbol{M}}^2 = m^2 a^2 b^2/\pi^3 E T_{\boldsymbol{M}}. \tag{38}$$

It might be thought that time-reversal symmetry of the periodic orbits might cause them to contribute twice to the sum (28) for the rigidity. But this is not the case, because although each orbit on the torus $\boldsymbol{I_M}$ is distinct from its time-reverse if neither M_1 nor M_2 is zero, both these orbits are included on the same torus, which is a four-sheeted geometric object with inversion symmetry about $\boldsymbol{p} = 0$ in momentum space, and hence time-reversal symmetry. If either M_1 or M_2 is zero (for example the orbit $(0,1)$ in figure 2) the orbit is self-retracing and hence is its own time-reverse. The corresponding torus has only two sheets rather than four and so counts $\frac{1}{2}$ in amplitude and $\frac{1}{4}$ in intensity (for a detailed discussion of this phenomenon see Appendix C of Richens & Berry (1981)).

It is natural to express $\Delta(L)$ in terms of the scaled energy

$$\mathscr{E} \equiv E\langle d \rangle \ (= \langle \mathscr{N}(E) \rangle), \tag{39}$$

corresponding to a mean level spacing of unity, which by (28) gives

$$\Delta(L) = \frac{\mathscr{E}^{\frac{1}{2}}}{\pi^{\frac{5}{2}}} \sum_{M_1=0}^{\infty} \sum_{M_2=0}^{\infty} \frac{\delta_{\boldsymbol{M}} G(y_{\boldsymbol{M}})}{(M_1^2 \alpha^{\frac{1}{2}} + M_2^2 \alpha^{-\frac{1}{2}})^{\frac{3}{2}}}, \tag{40}$$

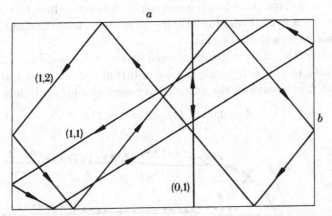

FIGURE 2. Rectangular billiard showing closed orbits with three different pairs of winding numbers (M_1, M_2). The orbit $(0, 1)$ is self-retracing; the other two are not.

where
$$\delta_M \equiv \begin{cases} 0 & \text{if } M_1 = M_2 = 0, \\ \tfrac{1}{4} & \text{if one of } M_1 \text{ and } M_2 \text{ is zero}, \\ 1 & \text{otherwise} \end{cases} \tag{41}$$

and
$$y_M = L\{\pi(M_1^2 \alpha^{\frac{1}{2}} + M_2^2 \alpha^{-\frac{1}{2}})/\mathscr{E}\}^{\frac{1}{2}}. \tag{42}$$

The shortest periodic orbit has winding numbers $(0, 1)$ and (7) leads to
$$L_{\max} = (\pi\mathscr{E})^{\frac{1}{2}} \alpha^{\frac{1}{4}}. \tag{43}$$

When $L \gg L_{\max}$, $G \approx 1$ for all M and so the saturation rigidity is
$$\varDelta_\infty = \frac{\mathscr{E}^{\frac{1}{2}}}{\pi^{\frac{5}{2}}} \sum_{M_1=0}^{\infty} \sum_{M_2=0}^{\infty} \frac{\delta_M}{(M_1^2 \alpha^{\frac{1}{2}} + M_2^2 \alpha^{-\frac{1}{2}})^{\frac{3}{2}}}. \tag{44}$$

This convergent series is easy to sum numerically. For α close to unity it varies slowly with α and so can be approximated by its value when $\alpha = 1$, which is (Zucker 1974)
$$\varDelta_\infty \overset{(\alpha=1)}{=} \mathscr{E}^{\frac{1}{2}} \pi^{-\frac{5}{2}} [\zeta(\tfrac{3}{2}) \beta(\tfrac{3}{2}) - \tfrac{1}{2}\zeta(3)] = 0.0947 \,\mathscr{E}^{\frac{1}{2}}, \tag{45}$$

where ζ and β denote the number-theoretic series
$$\zeta(s) \equiv \sum_{n=1}^{\infty} \frac{1}{n^s}; \quad \beta(s) \equiv \sum_{n=1}^{\infty} \frac{(-1)^{n+1}}{(2n-1)^s}. \tag{46}$$

These results will now be compared with quantal calculations of $\varDelta(L)$ by Casati *et al.* (1985). They used the levels $\alpha l^2 + n^2$ and so our $\mathscr{E}$ is their energy multiplied by $\pi/4\alpha^2$. Figure 3 shows the comparison for two energies, between $\varDelta(L)$ computed from (40) for $\alpha = 1$ and their quantal calculations for an ensemble of α-values between 0.9 and 1.2 (a range in which the different theoretical $\varDelta$-curves are almost indistinguishable). The agreement is good in the universal crossover and saturation ranges of L, although their oscillations are slightly stronger (this might be the result of their different averaging procedure).

Casati *et al.* also show (non-local) averages of $\Delta(L)$ over the whole energy range from zero to $\mathscr{E}$, for $\alpha = \frac{1}{3}\pi$. From (45), their curves ought to saturate at $\frac{2}{3} \times 0.0947\,\mathscr{E}^{\frac{1}{2}}$, and comparison shows that they do.

Finally, Casati *et al.* show a graph of $\Delta(L)$ in which L consists of the range from the ground state to the Lth level. In our notation this corresponds to choosing $\mathscr{E} = \frac{1}{2}L$ so that L is always in the saturation range and (45) predicts

$$\Delta = 0.0947\,L^{\frac{1}{2}}/\sqrt{2} = 0.067L^{\frac{1}{2}}. \tag{47}$$

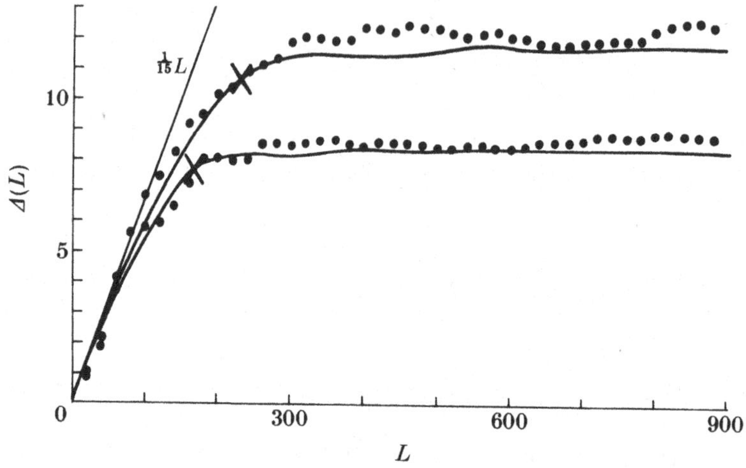

FIGURE 3. The full curves show $\Delta(L)$ computed from (40) for $\alpha = 1$ and $\mathscr{E} = 10\,500 \times \frac{1}{4}\pi$ (lower curve) and $\mathscr{E} = 20\,500 \times \frac{1}{4}\pi$ (upper curve) by using 1250 closed orbits. The circular points show data from Casati *et al.* (1985). The theoretical crossover values $L_{\max}$ are 161 and 225 and are indicated by crosses. The straight line shows the local universal Poisson rigidity $\frac{1}{15}L$.

Their curve is fitted by $0.063L^{\frac{1}{2}}$, which is a close agreement when one considers that the theoretical formulae have here been applied to values of L that are certainly not classically small.

5. A SEMICLASSICAL SUM RULE

The periodic-orbit sum (9) can be at best conditionally convergent because it represents the spectral density, with δ singularities at the energy levels. The delicate conspiracy of amplitudes A_j and phases $S_j/\hbar$ by which this is achieved is not fully understood, but there is some theoretical evidence that the semiclassical approximation (9) is in fact capable of reproducing singularities when infinitely long orbits are included. For integrable systems, we refer to Norcliffe & Percival (1968), Balian & Bloch (1972), and Berry & Tabor (1976). For chaotic systems, we refer to the Selberg identity (reviewed by Hejhal (1976) and McKean (1972)), which shows that for manifolds of constant negative curvature (9) is exact, and also to a study by Gutzwiller (1980) and an example to be given in §7 here. The purpose of the present section is to derive an identity that must be satisfied by the function $\phi(T)$ (equation (22)) that appears in the rigidity formula (20), to ensure that the periodic-orbit sum (9) has the correct density of singularities.

By analytically continuing the energy to $E \to E + i\eta$ and using the representation of the spectral density (2) as the imaginary part of the causal Green function, $d(E)$ can be expressed as the limit $\eta \to 0$ of the Lorentzians

$$d_\eta(E) = -\frac{1}{\pi} \operatorname{Im} \sum_n \frac{1}{E - E_n + i\eta}. \tag{48}$$

If
$$\eta \langle d \rangle \ll 1 \tag{49}$$

the Lorentzians do not overlap, and so

$$d_\eta^2(E) = \frac{\eta^2}{\pi^2} \sum_n \frac{1}{[(E - E_n)^2 + \eta^2]^2}. \tag{50}$$

It now follows from

$$\frac{2\eta^3}{\pi} \int_{-\infty}^{\infty} \frac{dx}{(x^2 + \eta^2)^2} = 1 \tag{51}$$

that
$$d(E) = \lim_{\eta \to 0} 2\pi\eta \, d_\eta^2(E). \tag{52}$$

By taking local averages and using the representation (8),

$$\langle d(E) \rangle = \lim_{\eta \to 0} 2\pi\eta \, \langle d_{\text{osc}, \eta}^2(E) \rangle. \tag{53}$$

The semiclassical formula for $d_{\text{osc}, \eta}(E)$, analogous to (9) for $d_{\text{osc}}(E)$, involves only positive traversals of periodic orbits, and gives

$$\langle d(E) \rangle = \lim_{\eta \to 0} \frac{4\pi\eta}{\hbar^{2\mu+2}} \langle \sum_i \sum_j{}^+ A_i A_j \cos\{(S_i - S_j)/\hbar\} \exp\{-\eta(T_i + T_j)/\hbar\} \rangle \tag{54}$$

$$= \lim_{\eta \to 0} \frac{4\pi\eta}{\hbar^{2\mu+2}} \int_0^\infty dT \, \phi(T) \exp\{-2\eta T/\hbar\}, \tag{55}$$

where (12) has been used with $\epsilon = i\eta$ and where ϕ is defined by (22).

Asymptotic inversion of the Laplace transform and use of (49) now gives

$$\phi(T) \to \langle d \rangle \hbar^{2\mu+1}/2\pi \quad \text{if} \quad T \gg \hbar \langle d \rangle. \tag{56}$$

This is the semiclassical sum rule. It guarantees that the amplitudes and phases of very long orbits generate the mean level density, and hence shows how late terms in the representation (8) and (9) determine the first term: an 'analytic bootstrap' reminiscent of that introduced in one dimension by Voros (1983).

If we define a new variable to measure time in relation to the inner energy scale, i.e.

$$\tau \equiv T/h\langle d \rangle, \tag{57}$$

and write
$$\phi(T) \equiv \langle d \rangle \hbar^{2\mu+1} K(\tau)/2\pi, \tag{58}$$

then the sum rule (56) gives

$$K(\tau) \to 1 \quad \text{when} \quad \tau \gg 1. \tag{59}$$

The physical interpretation of $K(\tau)$ is that this function is the spectral form factor, defined as the Fourier transform of the correlation function of the spectral density:

$$K(\tau) = \langle d \rangle^{-2} \int_{-\infty}^{\infty} \mathrm{d}L \, \langle d(E - L/2 \langle d \rangle) \, d(E + L/2 \langle d \rangle) \rangle \exp\{2\pi i L \tau\}. \tag{60}$$

(apart from a δ function at $\tau = 0$). In terms of K, the pair correlation function of the levels is

$$g(L) = 1 - \frac{1}{\pi L} \int_0^{\infty} \mathrm{d}\tau \, \sin\{2\pi L \tau\} \, K'(\tau). \tag{61}$$

For *integrable* systems, (56) can be written

$$\phi(T) \to (\mathrm{d}\Omega/\mathrm{d}E)/(2\pi)^{N+1} \quad \text{if} \quad T \gg \hbar \langle d \rangle, \tag{62}$$

which is identical to the asymptotic value (24) of the diagonal sum. Thus neglect of off-diagonal terms in ϕ is indeed justified for these systems, as asserted previously. Moreover, the form factor $K(\tau)$ is unity not only when $\tau \gg 1$ (as in (59)), but also down to the much smaller value

$$\tau_{\min} = T_{\min}/h \langle d \rangle \quad (\ll 1). \tag{63}$$

It then follows from (61) that the pair correlation is unity, which implies lack of level correlation, for sequences of length $L \ll L_{\max}$, consistent with the behaviour already found for $\Delta(L)$ in §3.

6. CHAOTIC SYSTEMS WITHOUT TIME-REVERSAL SYMMETRY

For chaotic systems the diagonal approximation to $\phi(T)$, which must be valid for any given $T \gg T_{\min}$ if $\hbar$ is small enough, is (25). Together with (59) and (63) this implies that the form factor $K(\tau)$ defined by (57) and (58) has the behaviour

$$K(\tau) \to \begin{cases} \tau & (\tau_{\min} \ll \tau \ll 1), \\ 1 & (\tau \gg 1). \end{cases} \tag{64}$$

The rigidity is given by (20) as

$$\Delta(L) = \frac{1}{2\pi^2} \int_0^{\infty} \frac{\mathrm{d}y}{y} \frac{K(y/\pi L)}{y/\pi L} G(y). \tag{65}$$

When $L \ll 1$, K can be replaced by unity (because of (64)) to give for $\Delta(L)$ the correct limiting form $\frac{1}{15} L$ (cf. (29) and (30) and the discussion below (6)). This result could not have been obtained without the semiclassical sum rule.

When $1 \ll L \ll L_{\max}$, it is possible to divide the integration range of (65) into two parts by choosing a value of Y that satisfies $1 \ll Y \ll L$, so that, from (64) and figure 1,

$$\frac{K(y/\pi L)}{y/\pi L} \approx 1 \quad \text{if} \quad y < Y \tag{66}$$

and

$$G(y) \approx 1 \quad \text{if} \quad y > Y.$$

Thus

$$\Delta(L) = \frac{1}{2\pi^2}\left[\int_0^Y \frac{dy}{y}\,G(y) + \int_Y^\infty \frac{dy}{y}\,\frac{K(y/\pi L)}{y/\pi L}\right]. \tag{67}$$

The definitions (16) and (21) lead to

$$\int_0^Y dy\, G(y)/y = \ln Y + \gamma + \ln 2 - \tfrac{9}{4}, \tag{68}$$

where γ is the Euler constant $0.577\ldots$. Integration by parts gives

$$\int_Y^\infty \frac{dy}{y}\,\frac{K(y/\pi L)}{y/\pi L} = -\ln(Y/\pi L) - \int_0^\infty d\tau \ln \tau\,\frac{d}{d\tau}\left(\frac{K(\tau)}{\tau}\right). \tag{69}$$

Thus the rigidity is

$$\Delta(L) = (\ln L)/2\pi^2 + D \quad (1 \ll L \ll L_{\max}), \tag{70}$$

where

$$D = \frac{1}{2\pi^2}\left[\ln 2\pi + \gamma - \tfrac{9}{4} - \int_0^\infty d\tau \ln \tau\,\frac{d}{d\tau}\left(\frac{K(\tau)}{\tau}\right)\right]. \tag{71}$$

Equation (70) is precisely the asymptotic rigidity of the gaussian unitary ensemble (g.u.e.) of random-matrix theory (Mehta 1967), i.e. $\Delta(L)$ averaged over the spectra of large Hermitian matrices whose elements are gaussian random variables with statistics invariant under unitary transformations. The logarithmic dependence and correct prefactor are a direct consequence of the diagonal sum rule (25) given by Hannay & Ozorio de Almeida (1984), which applies when all closed orbits are isolated and unstable, with no action degeneracies. Without the semiclassical sum rule, however, the additive constant D would be infinite (because (69) would then be illegitimate).

Without specifying $K(\tau)$ (or what is equivalent, $\phi(T)$) more closely than (64), the value of the constant D cannot be determined, and I know no direct semiclassical arguments based on the definition (22) by which this can be achieved. However, with the simplest interpolation, namely

$$K_0(\tau) = \begin{cases} \tau & (\tau \leqslant 1), \\ 1 & (\tau \geqslant 1), \end{cases} \tag{72}$$

the integral in (71) is -1 and

$$D = (\ln 2\pi + \gamma - \tfrac{5}{4})/2\pi^2 = 0.0590, \tag{73}$$

which is exactly the correct constant given by random-matrix theory for the g.u.e.! The reason is that (72) is the exact form factor of the g.u.e. (Mehta 1967). However, this must be regarded as a remarkable coincidence, because where there is time-reversal symmetry we shall see (§8) that the simplest interpolation fails to give the exact result. In any case, D is small and not very sensitive to $K(\tau)$. To illustrate this, the interpolations

$$K_1(\tau) = \tau/(1+\tau), \quad K_2(\tau) = \tau/(1+\tau^2)^{\frac{1}{2}}, \quad K_3(\tau) = 2\tau\,\pi^{-1}\arctan(\pi/2\tau) \tag{74}$$

give

$$D_1 = 0.0083, \quad D_2 = 0.0434, \quad D_3 = 0.0312. \tag{75}$$

The discontinuity in slope of the 'correct' form factor (71) is very surprising. It implies that as $\hbar \to 0$ the double sum (22) for $\phi(T)$ has an abrupt transition at $T = h\langle d\rangle$, between the diagonal (25) and that given by the semiclassical sum rule (56). The origin of this 'semiclassical phase phase-transition' is at present obscure. In the next section we shall present a curious example of it. (It should also be remarked that the g.u.e. repulsion between *neighbouring* levels, which causes $g(L)$ (equation (61)) to vanish as L^2 when $L\to 0$, cannot be obtained from the semiclassical arguments leading to (64), but is of course implicit in (72).)

The preceding results explain local universality of the rigidity when $L \ll L_{max}$, and the logarithmic behaviour (70) has recently been observed in computations for a chaotic system without time-reversal symmetry by Seligman et al. (1985).

When $L \gg L_{max}$, short orbits, and hence $\tau \sim \tau_{min}$ (equation (63)) give important contributions to the saturation value, which is given by (20) as

$$\varDelta_\infty = 2\int_0^\infty \mathrm{d}T\, \phi(T)/T^2. \qquad (76)$$

This can be expressed in terms of the short (non-universal) orbits and the (universal) density of the long ones by introducing an 'intermediate' period T_I such that

$$T_{min} \ll T_I \ll h\langle d\rangle. \qquad (77)$$

Then

$$\varDelta_\infty = 2\sum_{T_j < T_I} \frac{A_j^2}{T_j^2} + \frac{1}{2\pi^2}\int_{T_I/h\langle d\rangle}^\infty \frac{\mathrm{d}\tau}{\tau^2} K(\tau)$$

$$= 2\sum_{T_j < T_I} \frac{A_j^2}{T_j^2} + \frac{1}{2\pi^2}\ln\left\{\frac{h\langle d\rangle}{T_I}\right\} - \frac{1}{2\pi^2}\int_0^\infty \mathrm{d}\tau \ln\tau \frac{\mathrm{d}}{\mathrm{d}\tau}\left(\frac{K(\tau)}{\tau}\right). \qquad (78)$$

A useful approximation to $\varDelta_\infty$, valid up to an additive constant, can be obtained by replacing T_I by T_{min}, thereby extrapolating the continuous approximation (25) down to the shortest orbits. This gives (by also using (7), (72) and (3))

$$\varDelta_\infty \approx \frac{1}{2\pi^2}\ln\{\mathrm{e}\,L_{max}\} = \frac{(N-1)}{2\pi^2}\ln\left\{\frac{1}{\hbar}\left(\frac{e}{T_{min}}\frac{\mathrm{d}\Omega}{\mathrm{d}E}\right)^{1/(N-1)}\right\}, \qquad (79)$$

and shows that for chaotic systems the semiclassical spectral fluctuations increase only logarithmically with $\hbar^{-1}$ and so are much weaker than for integrable systems (cf. equation (31)).

7. EXAMPLE: RIEMANN'S ZETA FUNCTION

According to the Riemann hypothesis (Edwards 1974), the non-trivial zeros of the function $\zeta(s)$, defined in (46), all lie on the line $\mathrm{Re}\,s = \frac{1}{2}$. It is a natural conjecture, apparently first made by Hilbert and Polya, that the imaginary parts of these zeros are the eigenvalues of a linear operator (for a discussion, see Hejhal 1976). In this section I shall present evidence supporting the view that if this operator is regarded as the Hamiltonian of some (unknown) bound quantum-mechanical system, then in the classical limit the corresponding (unknown) dynamical system has trajectories that are chaotic and without time-reversal symmetry. By applying the semiclassical theory developed in this paper it will then be possible: to explain the local g.u.e. statistics that the Riemann zeros display (as was originally conjectured by Montgomery (1973) and as observed in computations reported by Bohigas & Giannoni (1984) and attributed by them to Odlyzko); to obtain an interesting formula involving prime numbers; and to predict the mean square fluctuations of the Riemann staircase.

Pavlov & Fadeev (1975) (see also Lax & Phillips (1976) and Gutzwiller (1983)) have discovered a scattering system (a leaky surface of constant negative curvature) whose phase shifts are given by the zeta function with Re $s = 1$, and whose classical limit is chaotic. It is not clear what relation, if any, exists between their work and that described here.

The starting point is the formula for $\zeta(s)$ as a product over primes p:

$$\zeta(s) = \prod_p (1 - p^{-s})^{-1}. \tag{80}$$

Defining
$$s \equiv \tfrac{1}{2} + iE, \quad \zeta(\tfrac{1}{2} - iE) \equiv D(E), \tag{81}$$

we have
$$\ln D(E) = -\sum_p \ln(1 - \exp\{iE \ln p\}/p^{\frac{1}{2}}). \tag{82}$$

Assuming the Riemann hypothesis, and noting that $\ln D$ vanishes as $\operatorname{Im} E \to +\infty$, we see that this function jumps by $-i\pi$ as each Riemann zero $E = E_n$ is traversed from above, so that the oscillatory part of the spectral staircase is

$$\mathcal{N}_{\mathrm{osc}}(E) = -\pi^{-1} \lim_{(\eta \to 0)} \operatorname{Im} \ln D(E + i\eta)$$

$$= -\frac{1}{\pi} \operatorname{Im} \sum_p \sum_{m=1}^{\infty} \frac{\exp\{im \ln p\}}{mp^{\frac{1}{2}m}}. \tag{83}$$

The oscillatory part of the spectral density is therefore
$$d_{\mathrm{osc}}(E) = -\frac{1}{\pi} \sum_p \frac{\ln p[\cos(E \ln p) - p^{-\frac{1}{2}}]}{1 - 2\cos(E \ln p)/p^{\frac{1}{2}} - p^{-1}} \tag{84}$$

or
$$d_{\mathrm{osc}}(E) = -\frac{1}{2\pi} \sum_p \sum_{m=-\infty}^{\infty}{}' \ln p \exp\{-|m|\ln p/2\} \exp\{iEm \ln p\}, \tag{85}$$

where the prime on the second summation means that $m = 0$ is omitted.

In the form (85) the analogy with the periodic-orbit formula (9) is clear: each primitive orbit corresponds to a prime p and is traversed m times; the action is

$$S_{m,p} = Em \ln p. \tag{86}$$

If E is regarded as the energy, the period is

$$T_{m,p} = \mathrm{d}S/\mathrm{d}E = m \ln p \tag{87}$$

and the amplitude is

$$A_{m,p} = -(\ln p \exp\{-|m| \ln p/2\})/2\pi. \tag{88}$$

Equation (86) shows that in this unknown dynamical system $\hbar^{-1}$ scales with E, so that we can set $\hbar = 1$ and regard $E \to \infty$ as the semiclassical limit. The amplitudes decay exponentially with period, as they must for a chaotic system (in contrast to an integrable one). Moreover the diagonal part $\phi_{\mathrm{D}}(T)$ of the sum (22) is, by the prime number theorem,

$$\phi_{\mathrm{D}}(T) = \sum_{m=1}^{\infty} \sum_{p} A_{m,\,p}^2 \,\delta(T - T_{m,\,p})$$

$$= \frac{1}{4\pi^2} \sum_{m=1}^{\infty} \sum_{p} \frac{\ln^2 p}{p^m} \,\delta(T - m \ln p)$$

$$\xrightarrow{(T \to \infty)} \frac{T}{4\pi^2} \sum_{m=1}^{[T/\ln 2]} \exp\{-T(1 - m^{-1})\}/m^2 \approx \frac{T}{4\pi^2}, \tag{89}$$

which agrees exactly with the diagonal sum rule (25) for periodic orbits in a chaotic non-degenerate system.

The local average of the Riemann staircase is (Montgomery 1976)

$$\langle \mathcal{N}(E) \rangle = \frac{E}{2\pi} \left(\ln\left\{ \frac{E}{2\pi} \right\} - 1 \right) - \frac{7}{8}, \tag{90}$$

corresponding to the average density

$$\langle d(E) \rangle = \frac{1}{2\pi} \ln\left\{ \frac{E}{2\pi} \right\}. \tag{91}$$

The inner energy scale is thus $2\pi/\ln\{E/2\pi\}$, which vanishes in the semiclassical limit $E \to \infty$, in contrast with the outer energy scale $2\pi/T_{\min} = 2\pi/\ln 2$.

Figure 4 a, b illustrates how well the periodic-orbit sum, with (equation (84)) and without (equation (85)) with $|m| = 1$) repetitions, is capable of reproducing δ functions at the Riemann zeros and $-\langle d(E) \rangle$ between them. (I am not claiming that (84) or (85) is an efficient way to calculate the positions of the zeros, because it is not: to roughly locate a zero near E requires about $(E/2\pi)/\ln\{E/2\pi\}$ primes, compared to $(E/2\pi)^{\frac{1}{2}}$ terms of the Riemann–Siegel formula (Edwards 1974), which would locate the same zero much more accurately.)

If on the basis of the identifications (86)–(88) and the diagonal sum (89) the existence of a chaotic non-degenerate dynamical system underlying the Riemann zeros is accepted, the arguments of preceding sections can be applied, and show that the zeros have the spectral rigidity of the g.u.e. In particular, for the spectral form factor (58), the semiclassical analysis gives the limits in (64). These limits agree with what was proved (for $\tau < 1$) and conjectured (for $\tau > 1$) by Montgomery (1973), namely that the Riemann zeros have a $K(\tau)$ that tends to the g.u.e. form (72) as $E \to \infty$.

When expressed in terms of closed orbits (primes and powers of primes) by using (58) and (22) this leads to a formula involving prime numbers that can be tested numerically. The test is most easily made by using not $K(\tau)$ but its integral, which from (72) is expected to be

$$S(\tau) \equiv 2 \int_0^\tau d\tau' \, K_0(\tau') = \begin{cases} \tau^2 & (\tau \leqslant 1), \\ 2\tau - 1 & (\tau \geqslant 1). \end{cases} \tag{92}$$

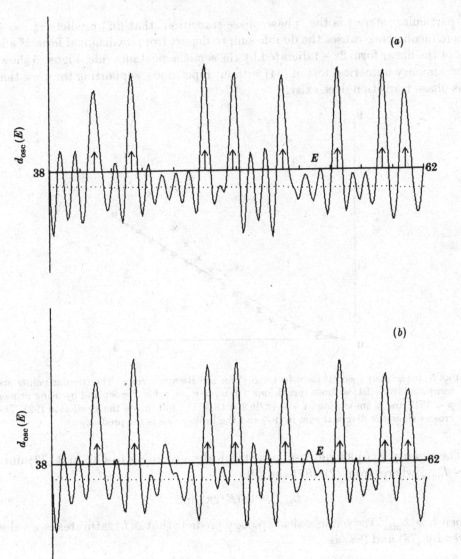

FIGURE 4. Spectral density of the Riemann zeros calculated with 150 primes (primitive closed orbits) from (*a*) equation (84) (i.e. including all repetitions, (*b*) equation (85) with $|m| = 1$ (i.e. without repetitions). In both figures the arrows denote the exact positions of the zeros and the dotted curve below the E-axis is $-\langle d(E) \rangle$.

The local spectral average $\langle\ \rangle$ in (22) is conveniently implemented by gaussian smoothing over an energy range η that is classically small but semiclassically large, i.e.

$$2\pi |\ln 2 \ll \eta \ll E. \tag{93}$$

The closed-orbit formula can now be written down as

$$S(\tau) = \lim_{(E \to \infty)} \frac{2}{(\ln\{E/2\pi\})^2} \sum_{p_1, m_1, p_2, m_2}^{p_1^{m_1} p_2^{m_2} < (E/2\pi)^{2\tau}} \frac{\ln p_1 \ln p_2}{p_1^{\frac{1}{2}m_1} p_2^{\frac{1}{2}m_2}} \cos\left(E \ln\{p_1^{m_1}/p_2^{m_2}\}\right)$$
$$\times \exp\left(-\eta^2 \ln^2\{p_1^{m_1}/p_2^{m_2}\}\right). \tag{94}$$

Of particular interest is the 'phase phase-transition' that (92) predicts at $\tau = 1$, where incoherence causes the double sum to depart from its diagonal form τ^2 and adopt the linear form $2\tau - 1$ dictated by the semiclassical sum rule. Figure 5 shows a preliminary numerical test of (94) without repetitions, supporting the view that this phase transition does exist.

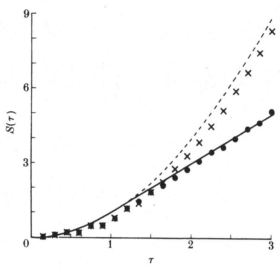

FIGURE 5. Integrated spectral form factor $S(\tau)$ for the Riemann zeros. The circular points are computed from (94) without repetitions, i.e. $m_1 = m_2 = 1$, for $E = 90$, and by using primes $p < 3517$ and a smoothing of $\eta = 4\pi/\ln 2 = 18.3$; the full line is the prediction (92). The crosses show the diagonal sum in (94) and the broken line is the prediction τ^2.

The rigidity $\Delta(L)$ should rise logarithmically according to (70) and (73) until $L \sim L_{max}$ where, from (7), (87) and (91)

$$L_{max} = \ln(E/2\pi)/\ln 2. \tag{95}$$

When $L \gg L_{max}$, the semiclassical analogy predicts that $\Delta(L)$ saturates at a value given by (78) and (88) as

$$\Delta_\infty = \left(\sum_{p < p_I} p^{-1} + \ln\ln\{E/2\pi\} - \ln\ln p_I + 1 \right)/2\pi^2$$

$$= (\ln\ln\{E/2\pi\} + 1.2615)/2\pi^2. \tag{96}$$

This shows that the r.m.s. staircase fluctuations grow as $(\ln\ln E)^{\frac{1}{2}}$ (it has been conjectured (Montgomery 1976, 1977) that the largest fluctuations grow as $(\ln E/\ln\ln E)^{\frac{1}{2}}$).

The saturation fluctuations predicted by (96) grow very slowly: for $E = 10^3$, $\Delta_\infty = 0.146$ (and $L_{max} = 7.3$); for $E = 10^6$, $\Delta_\infty = 0.190$ (and $L_{max} = 17.3$). To reach $\Delta_\infty = 0.5$ requires $E = 10^{2379}$, and to reach $\Delta_\infty = 1$ – an r.m.s. staircase deviation from $\langle \mathcal{N}(E) \rangle$ of only one level – requires $E = 10^x$ where $x \approx 5 \times 10^7$. This slow growth, indicating very slow approach to the semiclassical limit, would seem to rule out any direct test of (96).

8. Chaotic systems with time-reversal symmetry

When there is time-reversal symmetry, every orbit that is not self-retracing must be combined coherently with its time-reverse in the sum (22), because both orbits have the same action and period. The counterpart of the chaotic classical diagonal sum rule (25) must incorporate this coherence. Thus instead of $A_j^2 + A_j^2 = 2A_j^2$ we have $(A_j + A_j)^2 = 4A_j^2$, so the right-hand side of (25) must be multiplied by two. This means that the form factor $K(\tau)$ defined by (58) has the limiting form 2τ, rather than τ, when $\tau_{\min} \ll \tau \ll 1$. The semiclassical sum rule is, however, unaffected by the symmetry, so that instead of (6) the conditions on $K(\tau)$ are

$$K(\tau) \to \begin{cases} 2\tau & (\tau_{\min} \ll \tau \ll 1), \\ 1 & (\tau \gg 1). \end{cases} \tag{97}$$

The rigidity $\Delta(L)$ is given by (65) with this new form factor, and the argument parallels that in §6.

When $L \ll 1$, the limit $\frac{1}{15}L$ is regained, and is as before a consequence of the semiclassical sum rule.

When $1 \ll L \ll L_{\max}$, the analogues of (67), (70) and (71) are

$$\Delta(L) = \frac{1}{\pi^2}\left[\int_0^Y \frac{dy}{y} G(y) + \frac{1}{2}\int_Y^\infty \frac{dy}{y} \frac{K(y/\pi L)}{y/\pi L}\right]$$

$$= \frac{1}{\pi^2}\ln L + E \tag{98}$$

with
$$E = \frac{1}{\pi^2}\left[\ln 2\pi + \gamma - \frac{9}{4} - \frac{1}{2}\int_0^\infty d\tau \ln \tau \frac{d}{d\tau}\left(\frac{K(\tau)}{\tau}\right)\right]. \tag{99}$$

Equation (98) is precisely the asymptotic rigidity of the gaussian orthogonal ensemble (g.o.e.) of random-matrix theory (Dyson & Mehta 1963), that is $\Delta(L)$ averaged over the spectra of large real symmetric matrices whose elements are gaussian random variables with statistics invariant under orthogonal transformations. The logarithmic dependence and correct prefactor again follow from the Hannay & Ozorio de Almeida (1984) sum rule when this is modified to include the orbital degeneracy.

Without specifying $K(\tau)$ more closely than (97), the additive constant E cannot be determined and, as in §6, there is no obvious semiclassical way of doing this. Moreover, the simplest interpolation, analogous to (72), namely

$$K_1(\tau) = \begin{cases} 2\tau & (\tau \leqslant \frac{1}{2}), \\ 1 & (\tau \geqslant \frac{1}{2}), \end{cases} \tag{100}$$

gives
$$E = 0.067, \tag{101}$$

which is not the correct g.o.e. constant. To obtain this, it is necessary to use the correct form factor, which is (Mehta 1967)

$$K_0(\tau) = \begin{cases} 2\tau - \tau \ln\{1 + 2\tau\} & (\tau \leqslant 1), \\ 2 - \tau \ln\left\{\dfrac{1+2\tau}{2\tau-1}\right\} & (\tau \geqslant 1), \end{cases} \tag{102}$$

which gives $E = (\ln 2\pi + \gamma - \frac{5}{4} - \frac{1}{8}\pi^2)/\pi^2 = -0.00695.$ (103)

For comparison, the smooth interpolation

$$K_2(\tau) = 2\tau/(1+2\tau)$$ (104)

gives $E = -0.054.$ (105)

The 'correct' form factor (102) is also discontinuous, but in contrast to (72) the discontinuity is only in the third derivative. Therefore, the sum (22) for $\phi(t)$ also has a 'semiclassical phase phase-transition' when there is time-reversal symmetry, with a higher order than the phase transition when there is no such symmetry.

The result (98) explains the local universality observed in many numerical experiments on chaotic systems with time-reversal symmetry (see the review by Bohigas & Giannoni 1984), which holds for $L \ll L_{max}$. When $L \gg L_{max}$, the rigidity saturates non-universally, at a value approximately given by the analogue of (79), namely

$$\Delta_\infty = \pi^{-2} \ln\{eL_{max}\} - \frac{1}{8}$$

$$= \frac{(N-1)}{\pi^2} \ln\left\{\frac{1}{\hbar}\left(\frac{e}{T_{min}}\frac{d\Omega}{dE}\right)^{1/(N-1)}\right\} - \frac{1}{8}.$$ (106)

9. Discussion

The study reported here suggests a number of questions and directions for further investigation.

We have considered only second-order statistics ($\Delta(L)$ and $K(\tau)$). It is natural to enquire about higher statistics such as many-level distributions and the distribution of spacings between neighbouring levels. Direct generalization of the methods used here would involve multiple sums analogous to (22), and diagonal and partially diagonal analogues of the sum rules (24), (25) and (56). At present it is not clear how this could be accomplished.

Even the rigidity has been studied only in the integrable and chaotic extremes. In view of the recent interest in spectral statistics of systems that show a transition to chaos as a parameter is varied (Robnik 1984, Seligman *et al.* 1985; Meyer *et al.* 1984; Berry & Robnik 1984), it is desirable to extend the theory to cover these more general cases. In the range $L \ll L_{max}$, it follows as a rough approximation from the periodic orbit theory that $\Delta(L)$ is the sum of two contributions, one from the isolated unstable orbits in the chaotic region of phase space and one from the closed orbits in the region where there are tori (because of the finite resolution imposed by $\hbar$, orbits in this region can be considered as filling resonant 'tori' even though in reality they are isolated with near-marginal stability). The conjecture that $\Delta(L)$ will be approximately additive has also been made by Seligman *et al.* (1985). As L increases, the contribution of the orbits in the integrable component should increase in relative importance, and ought to completely dominate fluctuations in the saturation régime $L \gg L_{max}$.

Another interesting case is that of non-integrable systems with non-isolated periodic orbits. One important example is the stadium billiard of Bunimovich

(1974) (see also Berry 1981 *a*), a chaotic system in which all periodic orbits are isolated with the exception of those bouncing perpendicularly between the straight sides, which form a one-parameter family. This single orbit (and its repetitions) will not spoil the logarithmic universality of $\Delta(L)$ for $L \ll L_{\max}$ because the orbit selection function $G(y)$ (figure 1) will eliminate it. But its contribution will increase rapidly with L and it will dominate the saturation régime $L \gg L_{\max}$; arguments from §4 then give, for a stadium whose parallel sides have length a and are separated by b,

$$\Delta_\infty = (2\pi E)^{\frac{1}{2}} b^2 \zeta(3)/8\hbar\pi^3 a \tag{107}$$

(the factor $\zeta(3)$ incorporates repetitions). Analogous behaviour is predicted for the Sinai billiard (Berry 1981 *b*), for which there are non-isolated orbits whose finite number depends on the radius of the central disc. The emergence and dominance of particular non-isolated orbits as L increases through $L_{\max}$ is a good example of transition from universal to non-universal spectral behaviour.

Another example, whose theoretical treatment is less clear, is billiards in irrational-angled polygons. These systems have been conjectured to be ergodic (Hobson 1976), but they have zero Kolmogorov entropy (Sinai 1976) and so are not chaotic. This behaviour stems from their closed orbits, which are almost all non-isolated and moreover marginally stable. The arguments of this paper strongly suggest that the spectral fluctuations should be much stronger than for the chaotic systems considered in §§6, 8, i.e. $\Delta(L)$ should rise faster than logarithmically, in spite of a numerical study of the level spacings distribution of triangles (Berry & Wilkinson 1984) suggesting g.o.e. behaviour.

It is at first sight surprising that for the chaotic systems considered in §§6, 8 the Kolmogorov entropy S did not appear in the rigidity formulae. S is a non-universal quantity and so might be expected to contribute to $\Delta(L)$ when $L \sim L_{\max}$. However, S^{-1}, being the time for the separation between two nearby orbits to grow by a factor e, is of the same order of magnitude as $T_{\min}$ (for billiards the ratio is a geometrical factor), which does appear in the formulae (cf. (79) and (106)). Of course there remains the possibility that S might contribute directly to a higher-order statistic.

Another problem is the derivation of the semiclassical sum rule (56) directly from the definition (22). The derivation in §5 was based on (53), which is the condition for the spectral density to contain the correct density of singularities. A direct derivation from (22) would require detailed knowledge, at present lacking, of correlations between the actions of long orbits, which conspire in the off-diagonal terms of the sum to cancel the growth in the diagonal terms and cause $\phi(T)$ to saturate at the value (56). Such knowledge might also explain the 'phase phase-transitions' in the form factors (72) and (102).

The semiclassical sum rule is only the first in an infinite hierarchy of similar relations, obtained as the result of generalizing (53) by expressing a δ function as the limit of its analytic continuation raised to a power, namely

$$\langle d(E) \rangle = \lim_{\eta \to 0} \frac{(4\pi\eta)^{l-1} \Gamma^2(l)}{\Gamma(2l-1)} \langle d^l_{\mathrm{osc},\,\eta}(E) \rangle. \tag{108}$$

These higher-order sum rules might help in understanding higher-order spectral statistics. Their existence is related to a more fundamental and very remarkable 'bootstrap' property of the periodic-orbit sum (9): as a consequence of the spectrum being determined by the singularities of d_{osc}, which in turn are generated by very long classical periodic orbits, any finite number of terms may be deleted from (9) without destroying the spectral information it contains.

I thank Dr J. Hannay and Dr M. Robnik for helpful discussions.

Note added in proof (4 *June* 1985). Dr A. Voros has pointed out to me that (80), on which the closed-orbit sum (85) for the density of Riemann zeros is based, does not converge if Res < 1. Therefore the ability of (85) to discriminate individual zeros, as illustrated in figure 4, might deteriorate as E increases, and could fail altogether when $E > 2\pi \exp(4\pi) \sim 2 \times 10^6$.

REFERENCES

Balian, R. & Bloch, C. 1972 *Ann. Phys.* **69**, 76–160.
Balian, R. & Bloch, C. 1974 *Ann. Phys.* **85**, 514–545.
Berry, M. V. 1981*a* *Eur. J. Phys.* **2**, 91–102.
Berry, M. V. 1981*b* *Ann. Phys.* **131**, 163–216.
Berry, M. V. 1983 Semiclassical mechanics of regular and irregular motion. In *Chaotic behavior of deterministic systems* (Les Houches Lectures, vol. XXXVI, ed. G. Iooss, R. H. G. Helleman & R. Stora, pp. 171–271. Amsterdam: North-Holland.
Berry, M. V. 1984 Structures in semiclassical spectra: a question of scale. In *The wave–particle dualism* (ed. S. Diner, D. Fargue, G. Lochak & F. Selleri), pp. 231–252. Dordrecht: D. Reidel.
Berry, M. V. & Tabor, M. 1976 *Proc. R. Soc. Lond.* A **349**, 101–123.
Berry, M. V. & Tabor, M. 1977*a* *Proc. R. Soc. Lond.* A **356**, 375–394.
Berry, M. G. & Tabor, M. 1977*b* *J. Phys.* A **10**, 371–379.
Berry, M. V. & Robnik, M. 1984 *J. Phys.* A **17**, 2413–2421.
Berry, M. V. & Wilkinson, M. 1984 *Proc. R. Soc. Lond.* A **392**, 15–43.
Bohigas, O. & Giannoni, M. J. 1984 Chaotic motion and random-matrix theories, In *Mathematical and computational methods in nuclear physics* (ed. J. S. Dehesa, J. M. G. Gomez & A. Polls). *Lecture Notes in Physics* vol. 209, pp. 1–99. New York: Springer-Verlag.
Bunimovich, L. A. 1974 *Funct. Anal. Appl.* **8**, 254–255.
Casati, G., Chirikov, B. V. & Guarneri, I. 1985 *Phys. Rev. Lett.* **54**, 1350–1353.
Dyson, F. J. & Mehta, M. L. 1963 *J. Math. Phys.* **4**, 701–712.
Edwards, H. M. 1974 *Riemann's Zeta Function.* New York and London: Academic Press.
Gutzwiller, M. C. 1967 *J. Math. Phys.* **8**, 1979–2000.
Gutzwiller, M. C. 1969 *J. Math. Phys.* **10**, 1004–1020.
Gutzwiller, M. C. 1970 *J. Math. Phys.* **11**, 1791–1806.
Gutzwiller, M. C. 1971 *J. Math. Phys.* **12**, 343–358.
Gutzwiller, M. C. 1978 In *Path integrals and their applications in quantum, statistical and solid-state physics* (ed. G. J. Papadopoulos & J. T. Devreese), pp. 163–200. New York: Plenum.
Gutzwiller, M. C. 1980 *Phys. Rev. Lett.* **45**, 150–153.
Gutzwiller, M. C. 1983 *Physica* 7D, 341–355.
Hannay, J. H. & Ozorio de Almeida, A. M. 1984 *J. Phys.* A **17**, 3429–3440.
Hejhal, D. A. 1976 *Duke math. J.* **43**, 441–482.
Hobson, A. 1976 *J. Math. Phys.* **16**, 2210–2214.
Lax, P. D. & Phillips, R. S. 1976 *Scattering theory for automorphic functions.* Princeton University Press.
McKean, H. P. 1972 *Communs pure. appl. Math.* **25**, 225–246.

Mehta, M. L. 1967 *Random matrices and the statistical theory of energy levels*. New York and London: Academic Press.

Meyer, H.-D., Haller, E., Köppel, H. & Cederbaum, L. S. *J. Phys.* A **17**, L831–836.

Montgomery, H. L. 1973 *Proc. Symp. pure Math.* **24**, 181–193.

Montgomery, H. L. 1976 *Proc. Symp. pure Math.* **38**, 307–310.

Montgomery, H. L. 1977 *Communs Math. Helv.* **52**, 511–523.

Norcliffe, A. & Percival, I. C. 1968 *J. Phys.* B **1**, 774–83.

Parry, W. 1984 *Ergod. Th. Dynam. Syst.* **4**, 117–134.

Parry, W. & Pollicott, M. 1983 *Ann. of Math* **118**, 573–591.

Pavlov, B. S. & Fadeev, L. D. 1975 *Soviet Math.* **3**, 522–548.

Pechukas, P. 1983 *Phys. Rev. Lett.* **51**, 943–946.

Porter, C. E. 1965 *Statistical theories of spectra: fluctuations*. New York: Academic Press.

Richens, P. J. & Berry, M. V. 1981 *Physica* 1D, 495–512.

Robnik, M. 1984 *J. Phys.* A **17**, 1049–1074.

Seligman, T. H. & Verbaarschot, J. J. M. 1985 *Phys. Lett.* A **108**, 183–187.

Seligman, T. H., Verbaarschot, J. J. M. & Zirnbauer, M. R. 1985 *J. Phys.* A (In the press.)

Sinai, Ya. G. 1976 *Introduction to ergodic theory*. Princeton University Press.

Voros, A. 1983 *Ann. Inst. H. Poincaré* **39**, 211–338.

Zucker, I. J. 1974 *J. Phys.* A **7**, 1568–1575.

Semiclassical structure of trace formulas

R. G. LITTLEJOHN

Department of Physics, University of California,
Berkeley, California 94720 USA

Abstract

Trace formulas provide the only general relations known connecting quantum mechanics with classical mechanics in the case that the classical motion is chaotic. In particular, they connect quantal objects such as the density of states with classical periodic orbits. In this chapter, several trace formulas, including those of Gutzwiller and Balian and Bloch, Tabor, and Berry, are examined from a geometrical standpoint. New forms of the amplitude determinant in asymptotic theory are developed as tools for this examination. The meaning of caustics in these formulas is revealed in terms of intersections of Lagrangian manifolds in phase space. The periodic orbits themselves appear as caustics of an unstable kind, lying on the intersection of two Lagrangian manifolds in the appropriate phase space. New insight is obtained into the Weyl correspondence and the Wigner function, especially their caustic structures.

1 Introduction

This chapter concerns the trace formulas of Gutzwiller[1] and Balian and Bloch[2], which express the density of states of a bound quantal system as a sum over the periodic orbits of the corresponding classical system, and closely related trace formulas, such as that of Tabor[3] for the density of quasistates in a time-periodic system, and that of Berry[4] for the scars of Wigner functions in phase space. The purpose of this chapter is to explore the semiclassical structures of such formulas, i.e., the geometrical objects in the classical phase space associated with them and the interplay between these objects and the corresponding wave fields. The basic framework for such a study is the theory of Maslov[5], which associates the asymptotic properties of wave fields with Lagrangian manifolds in phase space. The application of this theory to trace formulas is unusual and interesting, and involves several novel elements.

The trace formulas considered here are significant because they are among the very few theoretical results of any generality which connect quantum mechanics with classical mechanics in the case that the classical motion is chaotic. These formulas work equally well for classically integrable systems, and in that case they have been shown by Berry and Tabor[6] to be equivalent to the usual Bohr-Sommerfeld methods (appropriately called "torus quantization" by Berry[7]). However, the usual methods fail for chaotic systems, mainly because of the lack of well behaved, invariant Lagrangian manifolds in phase space. For this reason, it is easy to get the impression that standard wave asymptotics or WKB theory does not apply at all to chaotic systems. We will show, however, that there is a rich geometrical structure associated with trace formulas, even in chaotic systems. For example, it turns out that Gutzwiller's periodic orbits define a certain, unstable kind of caustic, represented by the intersection of two Lagrangian manifolds in the appropriate phase space. We will also show that the density of states can itself be regarded as a wave function, in which the periodic orbit terms correspond to the branches of a WKB formula as in standard asymptotic theory.

We will be especially interested in Gutzwiller's[1] method of deriving his trace formula, since it connects most immediately with the geometry of phase space. The methods of Tabor[3] and Berry[4] for deriving their trace formulas are essentially identical to Gutzwiller's, and may be described as a determined application of the stationary phase approximation. Reviews of the trace formula, such as those of Berry[7] and Ozorio de Almeida[8], have basically followed these methods. Balian and Bloch[2] also used similar methods, but built them around an elegant formalism based on Laplace transforms in the quantity $1/\hbar$. From the standpoint of phase space geometry, however, the work of Balian and Bloch does not seem as useful a place to start as that of Gutzwiller. In addition, Balian and Bloch make some assumptions about classical mechanics which are certainly not correct, such as the idea that one can avoid caustics and multivalued solutions to the Hamilton–Jacobi equation simply by invoking complex energies. We will not be concerned at all in this chapter with the Selberg trace formula[9], although it continues to attract a great deal of interest, because it is an exact result obtained by special methods for a special system, and we are interested here in asymptotic results of general applicability.

One of the original goals of this work was to simplify Gutzwiller's derivation of his trace formula, especially the difficult manipulations of amplitude determinants. This led to a general examination of amplitude determinants and WKB wave functions, which are discussed in Sec. 2. We are careful to distinguish a particular solution of the Hamilton–Jacobi equation from a complete solution, the former being represented geometrically by a single, isolated Lagrangian manifold in phase space, and the latter by a foliation of phase space into Lagrangian manifolds. In the case of a particular solution, the amplitude determinant represents a density on the isolated Lagrangian manifold, according to the standard picture of Maslov[5]; but in the case of a complete solution, the provocative work of Miller[10] reveals connections between the solution of the Hamilton–Jacobi equation and that of the amplitude transport equation, based on concepts of unitarity. This is the idea we pursue in Sec. 2 in order to develop a representation of amplitude determinants in terms of complete sets of classically commuting observables. An outgrowth of this study is what appears to be a new expression for amplitude determinants, in terms of Poisson brackets connecting two complete sets of commuting observables. This expression is given below in Eq. (16), and part of its value to us is that it casts amplitude determinants into a form whose invariant meaning in phase space is manifest.

One section has been omitted from this version of this chapter for the sake of brevity. It contained a geometric interpretation of Gutzwiller's derivation of his trace formula. Its content was somewhat outside the main development of the chapter, and can be skipped without great loss of continuity.

The main theme of Sec. 3 is the use of the Hilbert-Schmidt scalar product of operators to interpret traces in terms of the asymptotic theory normally used for wave functions. This approach leads naturally to a doubled phase space with a doubled version of the Poisson bracket. The doubled phase space is a well known device, explained for example by Abraham and Marsden[11], for representing canonical transformations, and in this role it has an interesting interplay with the asymptotics of matrix elements of operators, especially unitary operators. The accumulation of evidence presented in Sec. 3 is, I believe, convincing that the doubled phase space is the necessary and correct medium for understanding the geometrical aspects of trace formulas, as well as many features of the Wigner function and Weyl correspondence.

Finally, in Sec. 4, some conclusions are drawn and suggestions made for future work.

The work reported on in this chapter began as an attempt to simplify Gutzwiller's derivation of his trace formula, and to reveal whatever deeper structures might underlie it. In the process of carrying out this program, however, there emerged much of the geometrical structure belonging to the mathematical theory of Fourier integral operators. For example, the idea of using the doubled phase space with the doubled symplectic form of Eq. (37) is one of the founding ideas of this theory, and is due to Hörmander[12]. Later, an apparently independent derivation of Gutzwiller's trace formula was made by Duistermaat and Guillemin[13], who interpreted the periodic orbits in terms of the intersections of a given Lagrangian manifold with the diagonal, representing the identity. Also, the role of intersecting Lagrangian manifolds in geometric quantization is discussed by Blattner and Kostant[14]. These and other issues are developed and discussed at greater length by Weinstein[15] and Guillemin and Sternberg[16].

Unfortunately, these mathematical developments have taken place in almost complete isolation from the more applied community, which is interested in trace formulas as a tool for understanding wave chaos. Therefore this chapter should not be taken as the development of new mathematics (although it is hoped that experts in the relevant fields will find some useful ideas), but rather as an application of existing mathematics to some problems in wave chaos.

2 Structure of semiclassical wave functions and matrix elements

In this section we investigate the phase space structure associated with semi-classical wave functions and matrix elements. We are especially interested in the general solution of the Hamilton–Jacobi equation, expressed in terms of complete sets of commuting observables, and in the restrictions which a given solution of the Hamilton–Jacobi equation imposes on the solutions of the amplitude transport equation. This line of inquiry leads naturally to Miller's formula for semiclassical wave functions, Eq. (8) below. Although many wave functions of interest in semiclassical mechanics fall under Miller's formula, not all of them do, and we provide the appropriate generalization, given in Eq. (12).

Next we turn to a reformulation of Miller's formula, in which the amplitude determinant is expressed purely in terms of Poisson brackets, thereby revealing its invariance under canonical transformations in a manifest manner. The new formulation of Miller's formula is displayed below in Eq. (16). The Poisson bracket version of the amplitude provides interesting perspectives on caustics, which we discuss. The Miller formula must be modified in the presence of caustics, with most of the changes occurring in the amplitude determinant. There are as many modifications as there are caustic types; these include the standard catastrophes[17], but also include caustics which are not catastrophes because they are not stable with respect to small perturbations. The latter class of caustics is especially of interest to us, because the periodic orbits in the Gutzwiller trace formula are precisely caustics of this kind. We work out the appropriate modifications to the Miller formula in the presence of such caustics, again expressing the amplitude in terms of Poisson brackets, and display the result in Eq. (27).

2.1 Semiclassical wave functions

Let us take the ordinary, time-independent Schrödinger equation $\hat{H}\psi = E\psi$ as a specific example to work with, although later we may want to replace $\hat{H}$ by some other operator. We apply standard, multidimensional WKB theory to this equation, and write $\psi(x) = \Omega(x)\exp[iS(x)/\hbar]$, giving us the Hamilton–Jacobi equation for the action $S(x)$,

$$H\left(x, \frac{\partial S}{\partial x}\right) = E, \tag{1}$$

and the amplitude transport equation for the amplitude $\Omega(x)$,

$$\frac{\partial}{\partial x}\left[\Omega(x)^2 \frac{\partial H(x, p(x))}{\partial p}\right] = 0, \tag{2}$$

where $p(x) = \partial S(x)/\partial x$. In these equations we write simply x for the configuration space coordinates $(x_1,\ldots,x_f)$, where f is the number of degrees of freedom, making no attempt to distinguish notationally between the one-dimensional and multidimensional cases. The required contractions and scalar products will ususally be obvious; where they are not, we will explicitly insert indices. We will use a similar notation for the momentum p and velocity v and other variables which are naturally interpreted as f-vectors. We will also sum over repeated indices, except as noted.

A particular solution $S(x)$ of the Hamilton–Jacobi equation is always the generating function of an invariant Lagrangian manifold in phase space. Lagrangian manifolds are f-dimensional surfaces in the $2f$-dimensional phase space on which the symplectic form vanishes; their theory is discussed by Arnold[18], and their use in WKB theory is explained by Maslov and Fedoriuk[5], Percival[19], and Delos[20]. Here we will simply note a few basic facts about them and about their generating functions. First, a Lagrangian manifold is invariant, i.e., mapped into itself by the flow, if and only if it is a subset of some energy shell $H(x, p) = E$. A convenient way of constructing an invariant Lagrangian manifold is to choose an arbitrary, $(f - 1)$-dimensional Lagrangian manifold L_0 in a surface of section, regarded as a phase space of $(f - 1)$ degrees of freedom in its own right, and to let L_0 move with the flow into the energy shell, sweeping out an f-dimensional manifold in the energy shell. It turns out that this manifold is Lagrangian in the full $2f$-dimensional phase space. The generating function $S(x)$ of a Lagrangian manifold is the function such that if (x, p) is on the Lagrangian manifold, then $p = \partial S(x)/\partial x$; it is generally multivalued, and its branches will be denoted by the index r, as in $S_r(x)$.

A so-called "complete solution" of the Hamilton–Jacobi equation[21] is an f-parameter family of solutions, $S = S(x, a)$, with $a = (a_1, \ldots, a_f)$. We will regard $S(x, a)$ as the generating function of a canonical transformation, in which the new momenta (they could equally well be new coordinates) are $A_1(x, p), \ldots, A_f(x, p)$. Here we use the capital letter A for the new momenta, regarded as functions of (x, p), and the lower case letter a for the values of these functions. We will denote the generalized coordinates conjugate to A by $\alpha = (\alpha_1, \ldots, \alpha_f)$, so that $\alpha_i = \partial S(x, a)/\partial a_i$. Then for each value of a, the Lagrangian manifold $p = p(x, a) = \partial S(x, a)/\partial x$ is the simultaneous contour surface, $A_i(x, p) = a_i$, $i = 1, \ldots, f$, of the new momenta; and the f-parameter family of Lagrangian manifolds arising by varying the as is a foliation of phase space into Lagrangian manifolds. The αs serve as coordinates on the Lagrangian manifolds. Since the energy E is constant on each of these Lagrangian manifolds, it is a function of a, i.e., $E = E(a)$. Sometimes it is convenient to choose one of the new momenta to be equal to the Hamiltonian itself, so that E is equal to one of the a's; this is not necessary, however.

We also note that any particular function $S(x)$ obtained from $S(x, a)$ by fixing a, i.e., the generating function of a particular Lagrangian manifold, is a simultaneous solution of the f Hamilton–Jacobi equations,

$$A_i\left(x,\frac{\partial S}{\partial x}\right)=a_i,\qquad i=1,\ldots,f.\tag{3}$$

This is almost obvious; if (x,p) is a point on the Lagrangian manifold labelled by a, then $A(x,p)=a$. Thus, when we think we are solving the single Hamilton–Jacobi equation (1), we are actually solving f simultaneous Hamilton–Jacobi equations, one for each of a set of f classically commuting observables $A_1,\ldots,A_f$. We will refer to such a set of classical observables as a *complete set*. Conversely, if we ask for the general solution of the simultaneous Hamilton–Jacobi equations (3), we find that it is $S(x,a)$, unique to within an additive constant. The constant can be regarded as a convention for the choice of initial point for the contour integral of $p\,dx$ giving $S(x,a)$.

Even when we have only an isolated, particular solution $S(x)$ of the Hamilton–Jacobi equation (1), it can always be imbedded in an f-parameter family of solutions. We will actually do this below when we consider the Gutzwiller trace formula for the density of states. Therefore we can say that every solution of the Hamilton–Jacobi equation is represented by some complete set of commuting classical observables; and, by extension, the same applies to every wave function of semiclassical interest. In particular, it applies to the semiclassical propagator and Green's function; it also applies, as we shall show, to Gutzwiller's formula for the density of states and to Berry's formula for scars of Wigner functions in phase space. The identification of the complete set of commuting classical observables associated with a semiclassical wave function is an important component of its semiclassical interpretation; it is, however, a component which has hitherto been missing in trace formulas.

Let us now suppose that some definite, complete solution $S(x,a)$ of the Hamilton-Jacobi equation has been found, and let us ask for the most general solution of the amplitude transport equation consistent with this $S(x,a)$. The amplitude transport equation is a continuity equation involving the density $\rho(x)=\Omega(x)^2$, regarded as a density in configuration space; one can equally well work with a density $\sigma(\alpha)$ on the Lagrangian manifold, given by

$$\sigma(\alpha)=\rho(x)\left|\det\frac{\partial x}{\partial\alpha}\right|.\tag{4}$$

The general solution of the amplitude transport equation is conveniently represented in terms of an initial density on some $(f-1)$-dimensional initial value surface, from which the density is transported along orbits. The initial value surface can be regarded as being in configuration space or in the Lagrangian manifold, depending on whether one wishes to work with $\rho(x)$ or $\sigma(\alpha)$. By choosing an initial density and transporting it along orbits for each Lagrangian manifold, we obtain an f-parameter family of solutions $\Omega(x,a)$ of the amplitude transport equation.

Therefore the general solution $\Omega(x,a)$ of the amplitude transport equation would seem to involve the selection of an arbitrary initial density, one for each Lagrangian manifold. Indeed, if the Lagrangian manifolds are topologically trivial, i.e., homeomorphic to $\mathbb{R}^f$, as in a scattering problem, then this conclusion is correct. But more generally, Lagrangian manifolds often have the topology of $\mathbb{R}^{f_1} \times (S^1)^{f_2}$, with $f_1 + f_2 = f$, i.e., f_1 lines crossed with f_2 circles[11]. In the extreme case $f = f_2$, the Lagrangian manifold is an f-torus. Therefore it may happen that an orbit leaving the $(f-1)$-dimensional initial surface on a Lagrangian manifold will return again to this initial surface, usually at a different point from where it left. Continuity then demands that the transported value of the density be equal to the initial density at a point on the initial surface where an orbit returns. In this way, restrictions may be placed on the choice of initial density when the Lagrangian manifolds are topologically nontrivial. In the extreme case that the Lagrangian manifold is an f-torus supporting ergodic orbits, the initial density (and therefore the density everywhere else) is determined to within a multiplicative constant; it is simply $\sigma(\alpha) = \text{const.}$

In this chapter we will mostly be interested in the opposite extreme of $f_1 = f$, i.e., topologically trivial Lagrangian manifolds, so our amplitude transport equation will have many possible solutions, even for a given $S(x,a)$. For the same reason, however, we will not need to worry about quantization, since our topologically trivial Lagrangian manifolds will automatically support wave functions which are single-valued on the manifold. In other words, for the applications we will consider, the quantities a will be allowed to take on continuous values.

Even when many solutions of the amplitude transport equation exist, there is one that especially stands out, namely

$$\Omega(x,a) = \left| \det \frac{\partial^2 S(x,a)}{\partial x \partial a} \right|^{1/2} = \left| \det \frac{\partial \alpha}{\partial x} \right|^{1/2}. \tag{5}$$

Expressed in terms of the density on the Lagrangian manifold, this is the solution $\sigma(\alpha) = 1$. A unique feature of this solution, not shared by other solutions of Eq. (2), is that it is actually a simultaneous solution of f amplitude transport equations,

$$\frac{\partial}{\partial x} \left[\Omega(x,a)^2 \frac{\partial A_i(x,p(x,a))}{\partial p} \right] = 0, \qquad i = 1, \dots, f, \tag{6}$$

where $p(x,a) = \partial S(x,a)/\partial x$, i.e., one for each of the commuting observables A_i. $\Omega(x,a)$ satisfies these equations because each of the As, regarded as a Hamiltonian, generates a flow which is a simple displacement in the corresponding α, and a density $\sigma(\alpha)$ which is constant is obviously invariant under these flows. Conversely, if we were to seek a simultaneous solution of Eq. (6), the solution is given by Eq. (5), and is unique to within a multiplicative constant.

In many applications the complete set of commuting classical constants of motion $A_1(x,p), \dots, A_f(x,p)$ which emerge from a complete solution of the Hamilton–

Jacobi equation are the classical counterparts of a set of commuting quantal constants of motion, $\hat{A}_1,\ldots,\hat{A}_f$. Since these quantal observables commute with $\hat{H}$ and with each other, they possess simultaneous eigenstates which are also eigenstates of the Hamiltonian. We shall denote one of these eigenstates by $|a\rangle$, with $a = (a_1,\ldots,a_f)$ now interpreted as a vector of eigenvalues. As a result, $\psi(x) = \langle x|a\rangle$ is an eigenfunction of the Hamiltonian.

Therefore we might consider an indirect approach to finding the eigenfunctions of the Hamiltonian, in which we first seek the simultaneous eigenfunctions of the $\hat{A}$s. If we do this by semiclassical means, we are led to the f simultaneous Hamilton–Jacobi equations (3) and the f simultaneous amplitude transport equations (6). As we have seen, the solutions of these equations give a wave function $\psi(x)$ which is unique to within an overall multiplicative constant; this constant, whose magnitude comes from the multiplicative constant for $\Omega(x,a)$ and whose phase comes from the additive constant for $S(x,a)$, can be different for different Lagrangian manifolds, i.e., it is a function of a. Its magnitude can be determined by demanding orthonormality, which we apply in the continuum sense, i.e.,

$$\langle a|a'\rangle = \int dx\, \langle a|x\rangle\langle x|a'\rangle = \delta(a - a'). \tag{7}$$

This is equivalent to demanding that $\langle x|a\rangle$ be the component of a unitary transformation matrix, taking us from the x-representation to the a-representation. The remaining arbitrary phase factor can be regarded as a phase convention for the states $|a\rangle$.

By evaluating the integral of Eq. (7) by the stationary phase approximation, we obtain the final expression for the simultaneous eigenfunction of the $\hat{A}$s,

$$\langle x|a\rangle = \frac{1}{(2\pi i\hbar)^{f/2}} \sum_r \left|\det \frac{\partial^2 S}{\partial x \partial a}\right|^{1/2}$$
$$\times\ \exp\left[\frac{i}{\hbar} S(x,a) - i\mu\frac{\pi}{2}\right]. \tag{8}$$

Here we have introduced the usual Maslov index μ to specify the proper phase shifts between branches r; both S and μ depend on r. We have also introduced for convenience an overall phase factor of $\exp(-if\pi/4)$; there remains an additional arbitrary phase which is a function of a. Apart from this, the answer is unique. Of course, this same result will be obtained whenever we seek a simultaneous eigenfunction of a set of f commuting $\hat{A}$s, whether or not they also commute with $\hat{H}$.

In this chapter we will not be much concerned with Maslov indices, so we will simply write μ for them wherever they occur, with no implication that the different μs are equal. The Maslov index in the Gutzwiller trace formula is considered as a separate issue in papers by Robbins[22] and by Creagh, Robbins and Littlejohn[23].

The realization that the requirement of unitarity leads to an essentially unique determination of the amplitude was evidently first made by Miller[10], so we will refer to Eq. (8) as Miller's form of the semiclassical matrix element $\langle x|a \rangle$. Actually, Miller's results were more general than Eq. (8), for he gave formulas for the unitary matrix elements $\langle b|a \rangle$, connecting any two representations. Although the modifications required for Miller's general result are easy (one simply replaces x wherever it occurs in Eq. (8) by b), nevertheless the implications are far reaching, for Miller's results demonstrate a kind of covariance of semiclassical theory under canonical transformations, exactly mirrored by the covariance of quantum mechanics under unitary transformations. In spirit, Miller's results are fundamentally geometrical. In this chapter, there will be special emphasis placed on the property of unitarity in examining the structure of trace formulas, and it will be seen to play a role in unusual contexts.

We will later see examples of complete sets of classically commuting observables, $A_1(x, p), \ldots, A_f(x, p)$, which are either multivalued or discontinuous in phase space. These are the functions one obtains in classical mechanics by attempting to create constants of motion by demanding that the functions be constant along orbits; except in special circumstances, the contour surfaces of these constants do not have simple imbeddings in phase space, but rather have complicated self-intersections. Such classical observables do not have any clear quantal counterparts, so it is really not meaningful to talk about simultaneous eigenfunctions $\langle x|a \rangle$. Nevertheless, it is possible to construct formal semiclassical expressions of the form of Eq. (8), based on the Lagrangian manifolds associated with such classical observables, and we will find it convenient to refer to these expressions with the same kind of bra-ket notation as the $\langle x|a \rangle$ occurring in Eq. (8). This notation is especially convenient for discussing Green's functions.

Although the wavefunction $\psi(x) = \langle x|a \rangle$ of Eq. (8) can sometimes represent an eigenfunction of $\hat{H}$, not every eigenfunction of $\hat{H}$ has the form of Eq. (8), even in the formal sense of the preceding paragraph. This is in spite of the fact that solving the original Hamilton–Jacobi equation (1) always leads to some complete set of As. The reason is that the trivial solution of the original amplitude transport equation (2), given by Eq. (5), is not generally the only solution. To find the general solution of Eq. (2), let us write $\rho(x) = |\det(\partial \alpha/\partial x)|$ for the square of the trivial amplitude shown in Eq. (5). This $\rho(x)$ satisfies the amplitude transport equation in the form

$$\frac{\partial}{\partial x}[\rho(x)v(x)] = 0, \tag{9}$$

where the velocity field $v(x)$ is given by

$$v(x) = \frac{\partial H(x, p(x))}{\partial p}, \tag{10}$$

with $p(x) = \partial S(x)/\partial x$. (Here we work with a single Lagrangian manifold, and suppress the dependence on a.) We now let $\rho'(x)$ be any other solution of Eq. (9), and write $\rho'(x) = g(x)\rho(x)$. It follows that

$$v(x)\frac{\partial g(x)}{\partial x} = 0, \tag{11}$$

i.e., that g is constant along orbits in configuration space. Choosing some such function $g(x)$, we can convert it into a function $G(x, p)$, defined for (x, p) points on the Lagrangian manifold in question by setting $G(x, p(x)) = g(x)$; the resulting $G(x, p)$ is then constant along orbits on the Lagrangian manifold in phase space. Finally, by carrying out this procedure for the whole family of Lagrangian manifolds, we obtain a function $G(x, p)$ which is defined over a whole finite region of phase space, and which is a constant of motion. We conclude that the most general semiclassical eigenfunction of $\hat{H}$ has the form

$$\psi(x) = \sum_r F(x, p_r(x, a)) T_r, \tag{12}$$

where T_r is the r-th term in Eq. (8), where $F(x, p)$ is a constant of motion (the square root of G), and where $p_r = \partial S_r(x, a)/\partial x$.

Notice that the constant of motion F in Eq. (12) need not commute with the As. Indeed, if it does commute with the As, then its value $F(x, p)$ on the Lagrangian manifold $A = a$ is simply a function of the as, and F can be absorbed into the normalization constant for $\langle x|a\rangle$. This would return us to the Miller form of the matrix element, Eq. (8). Therefore F in Eq. (12) provides a nontrivial modification to the Miller matrix element only when it does not commute with the As. One consequence of this is that in one degree of freedom, the semiclassical energy eigenfunctions are uniquely determined by the family of Lagrangian manifolds alone (without specifying the form of the Hamiltonian), because all constants of motion commute with H. But in higher degrees of freedom, there may be more than one way to associate wave functions with families of Lagrangian manifolds. This can happen even in bound state problems, for which the Lagrangian manifolds are tori, if the motion is not ergodic on the tori in a finite volume of phase space. Such nonergodic motion occurs in so-called degenerate classical systems, such as the two-dimensional isotropic harmonic oscillator or the hydrogen atom, in which phase space can be foliated into invariant tori in more than one way.

2.2 Poisson bracket form for the amplitude

Let us return to Eq. (8), and seek an expression for the unitary matrix element $\langle a|b\rangle$, where $\hat{B} = (\hat{B}_1, \dots, \hat{B}_f)$ is a collection of f new operators which commute with one another (but not necessarily with the $\hat{A}$s), with eigenvalues $b = (b_1, \dots, b_f)$ and classical counterparts $B_1(x, p), \dots, B_f(x, p)$. We do this by writing down the semiclassical formula for $\langle x|b\rangle$, analogous to Eq. (8), and by applying the stationary phase approximation to the integral $\int \langle a|x\rangle \langle x|b\rangle$. It is now necessary to distinguish the two generating functions, call them $S_A(x, a)$ and $S_B(x, b)$, producing (with definite values of a and b) two distinct Lagrangian manifolds $p = p_A(x, a)$ and $p = p_B(x, b)$ in phase space. We will call these the A-manifold and the B-manifold. We will also denote the generalized coordinates conjugate to the Bs by $\beta = (\beta_1, \ldots, \beta_f)$, which we use as coordinates on the B-manifold.

The computation of the integral is straightforward, except for the amplitude determinant, the reciprocal of which we write as

$$\det\left(\frac{\partial^2 S_A}{\partial x \partial a}\right)^{-1} \det\left(\frac{\partial^2 S_B}{\partial x \partial b}\right)^{-1} \det\left(\frac{\partial^2 S_A}{\partial x \partial x} - \frac{\partial^2 S_B}{\partial x \partial x}\right). \tag{13}$$

This, however, can be cast into the form

$$\det_{k\ell}\left[\left(\frac{\partial p_i}{\partial x_j}\right)_b \left(\frac{\partial a_k}{\partial p_j}\right)_x \left(\frac{\partial b_\ell}{\partial p_i}\right)_x - \left(\frac{\partial p_i}{\partial x_j}\right)_a \left(\frac{\partial a_k}{\partial p_i}\right)_x \left(\frac{\partial b_\ell}{\partial p_j}\right)_x\right], \tag{14}$$

where the quantities being held fixed in the derivatives are explicitly indicated. This in turn can be reduced to

$$\det_{k\ell}\left[\left(\frac{\partial a_k}{\partial x_j}\right)_p \left(\frac{\partial b_\ell}{\partial p_j}\right)_x - \left(\frac{\partial a_k}{\partial p_j}\right)_x \left(\frac{\partial b_\ell}{\partial x_j}\right)_p\right] = \det_{k\ell}\{A_k, B_\ell\}, \tag{15}$$

where the curly brackets are the usual Poisson bracket. We will simply write this final determinant as $\det\{A, B\}$. Altogether, we have

$$\langle a|b\rangle = \frac{1}{(2\pi i\hbar)^{f/2}} \times \sum_r \frac{\exp\left\{\frac{i}{\hbar}\left[S_B(x, b) - S_A(x, a)\right] - i\mu\frac{\pi}{2}\right\}}{\left|\det\{A, B\}\right|^{1/2}}. \tag{16}$$

The novel element in this formula is the expression of the amplitude determinant in terms of Poisson brackets.

The branches of the sum in Eq. (16) are the intersections of the A-manifold with the B-manifold, which for now we assume to take place at isolated points; this is the generic situation, since two f-dimensional manifolds in $2f$-dimensional space usually intersect in 0-dimensional points. Therefore each intersection has an (x, p) value, giving us the x coordinate at which to evaluate the actions $S_A(x, a)$ and $S_B(x, b)$, and the (x, p) coordinates at which to evaluate the Poisson brackets. This situation is illustrated in Fig. 1.

Since the left side of Eq. (16) obviously does not depend on any special features of the quantal x-representation, it is satisfying that the amplitude determinant on the right appears in a form which is manifestly a phase space invariant, i.e., independent of any special properties of the classical (x, p) coordinates. In this respect, this version of the amplitude determinant is an improvement over that shown in Eq. (8). It also has the practical consequence of allowing us to compute the amplitude in any canonical coordinates, which sometimes simplifies calculations.

As for the phase on the right side of Eq. (16), we cannot expect it to be

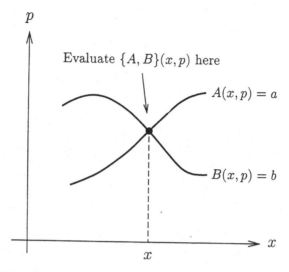

Fig. 1. The semiclassical matrix element $\langle a|b \rangle$ is expressed in terms of the intersections of the Lagrangian manifolds $A(x, p) = a$ and $B(x, p) = b$. The Poisson brackets of the amplitude determinant are evaluated at the intersection points, and the actions in the exponent are evaluated at the x coordinates of the intersection points.

a phase space invariant, because it depends on the phase conventions for the states $|a\rangle$ and $|b\rangle$. If our Lagrangian manifolds intersect in more than one point, however, as illustrated in Fig. 2, then the *relative* phase $(S_{B2} - S_{A2}) - (S_{B1} - S_{A1})$ can be expected to be a phase space invariant, because it is not affected by any overall phase factor. Indeed, it is easy to see that the relative phase is simply the symplectic area enclosed between the two intersections. We may note that in many degrees of freedom, where the Lagrangian manifolds are more than one-dimensional, the symplectic area is independent of the path chosen to form the loop.

The Maslov index in Eq. (16) can also be given an invariant interpretation in terms of the geometry of the Lagrangian manifolds; this is a little more subtle, and we will not go into it here.

Equation (16) shows that caustics occur when $\det\{A, B\} = 0$ at a point of intersection of the A-manifold and B-manifold. The geometrical meaning of this is that the two manifolds are tangent at the point of contact, as illustrated in Fig. 3. In f degrees of freedom, the caustic can be of any order from 1 to f, the order being determined by the corank of the matrix $\{A_k, B_\ell\}$, i.e., the number of linearly independent null eigenvectors it possesses. These null eigenvectors are related in a simple way to the directions in phase space which are simultaneously tangent to the A-manifold and the B-manifold at their point of intersection. To

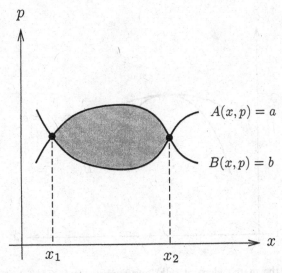

Fig. 2. The relative phase between branches, $(S_B(x_2, b) - S_A(x_2, a)) - (S_B(x_1, b) - S_A(x_1, a))$, is a phase space invariant; it is the symplectic area enclosed by the two intersections of the two Lagrangian manifolds.

see this, let c_ℓ be a null eigenvector of the matrix of Poisson brackets, so that

$$\{A_k, B_\ell\}c_\ell = 0. \tag{17}$$

Now associate this eigenvector with a phase space vector X^μ, defined by

$$X^\mu = c_\ell \, \Gamma^{\mu\nu} \frac{\partial B_\ell}{\partial \xi^\nu}, \tag{18}$$

where Greek indices run from 1 to $2f$ and Latin indices run from 1 to f, where $\xi = (x, p)$ is the $2f$-vector of phase space coordinates, and where Γ is the usual cosymplectic form. For given ℓ, the phase space vector $\Gamma^{\mu\nu}(\partial B_\ell/\partial \xi^\nu)$ is the flow vector in phase space arising from treating B_ℓ as a Hamiltonian; it represents simply a displacement in the β_ℓ coordinate on the B-manifold, and so is tangent to the B-manifold. Therefore X^μ, which is a linear combination of these vectors, is also tangent to the B-manifold. However, X^μ is also tangent to the A-manifold, since by Eq. (17) we have

$$X^\mu \frac{\partial A_k}{\partial \xi^\mu} = c_\ell\{A_k, B_\ell\} = 0. \tag{19}$$

Thus, X^μ represents a direction in phase space in which the A-manifold and B-manifold are tangent to one another.

If $\det\{A, B\}$ should vanish at an intersection point, the most typical case would be that the intersection point would be isolated and $\{A_k, B_\ell\}$ would have a single null eigenvector there. Then the right side of Eq. (16) for the matrix element

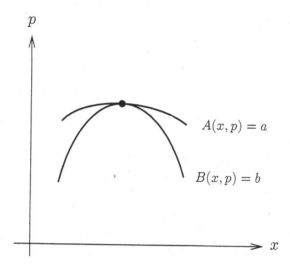

Fig. 3. Caustics of the matrix element $\langle a|b\rangle$ occur when the A-manifold and B-manifold are tangent at their point of intersection.

$\langle a|b\rangle$ would be replaced by an expression involving an Airy function. It would be straightforward to analyze this case, and to build the required expression out of elements whose invariant meaning in phase space is manifest.

For the applications we shall consider, however, the A- and B-manifolds intersect, not at an isolated point, but over a whole region of dimensionality $f_2 > 0$, with $f = f_1 + f_2$. We will call this f_2-dimensional intersection I, as illustrated in Fig. 4, so that $\{A_k, B_\ell\}$ has rank f_1 everywhere on I. For example, we will show in Sec. 3 that in the Gutzwiller trace formula for nonintegrable systems, I is identified with a periodic orbit, in which case $f_2 = 1$.

Figure 4 is misleading in one respect, namely that it suggests that the intersection I changes continuously under small changes in the A- and B-manifolds. Actually, such an intersection with dimensionality $f_2 > 0$ is unstable, and breaks up into isolated points under most perturbations of the manifolds. Indeed, the very existence of such nontrivial intersections seems to be related to symmetry; for example, in the standard Gutzwiller trace formula, it is related to conservation of energy.

In the cases of interest to us, the A- and B-manifolds intersect in an f_2-dimensional intersection I because f_2 of the As are identical with f_2 of the Bs (or because coordinate transformations of the As among themselves and of the Bs among themselves can bring this situation about). This is fortunate, because in this case the computation of the semiclassical matrix element $\langle a|b\rangle$ is easier than in the general case.

To carry out this computation, we write $A = (A_1, A_2)$, $B = (B_1, B_2)$, where the

$$A(x,p) = a \qquad\qquad B(x,p) = b$$

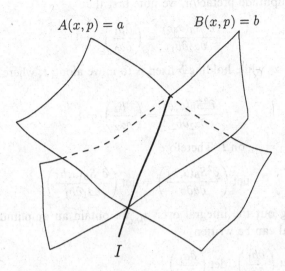

I

Fig. 4. A case of interest is where the A-manifold and B-manifold intersect in a region I of dimensionality $f_2 > 0$. In the Gutzwiller trace formula, I is the periodic orbit. This diagram is misleading, in the sense that it suggests that I is stable under small perturbations of the manifolds.

first and second members contain f_1 and f_2 functions, respectively, and where $B_2 = A_2$. With no loss of generality we also set $\alpha_2 = \beta_2$. In this way, α_2 or β_2 can be taken as coordinates on I, whereas both α_1 and β_1 are constant on I.

This coordinate system simplifies the calculation of the matrix element $\langle a|b\rangle$, which we carry out in the α-representation, performing the stationary phase approximation on $\int d\alpha \, \langle a|\alpha\rangle\langle\alpha|b\rangle$. The actions in the α-representation are the integrals of $a d\alpha$ along the respective manifolds; for the A-manifold we have $S_A(\alpha, a) = \alpha a$, so $\det(\partial^2 S_A/\partial\alpha\partial a) = 1$. Therefore the integral we must evaluate is

$$\int d\alpha \, \left|\det \frac{\partial^2 S_B(\alpha,b)}{\partial\alpha\partial b}\right|^{1/2} \exp\left\{\frac{i}{\hbar}\left[S_B(\alpha,b) - \alpha a\right]\right\}. \tag{20}$$

Let us write simply $S(\alpha)$ for the total action $S_B(\alpha, b) - S_A(\alpha, a)$ in the exponent; it is constant on I, because as we move along I, the increment in S_B cancels that in S_A. Therefore in expanding $S(\alpha)$ to second order about a point on I we have only the α_1 derivatives to take. For convenience we let I be specified by $\alpha_1 = 0$, so the expansion of $S(\alpha)$ is

$$S(\alpha) = S(I) + \frac{1}{2}\alpha_1 \frac{\partial^2 S}{\partial\alpha_1\partial\alpha_1}\alpha_1, \tag{21}$$

where the linear terms cancel because I is the stationary phase set.

As for the amplitude prefactor, we note first that

$$\frac{\partial^2 S_B(\alpha, b)}{\partial \alpha_2 \partial b_1} = \left(\frac{\partial \beta_1}{\partial \alpha_2}\right)_b = 0, \tag{22}$$

because to vary α_2 while holding b fixed is to move along I, where β_1 is constant. We also have

$$\frac{\partial^2 S_B(\alpha, b)}{\partial \alpha_2 \partial b_2} = \left(\frac{\partial \beta_2}{\partial \alpha_2}\right)_b = 1, \tag{23}$$

since we have $\alpha_2 = \beta_2$ on I. Therefore

$$\det\left(\frac{\partial^2 S_B(\alpha, b)}{\partial \alpha \partial b}\right) = \det\left(\frac{\partial^2 S_B(\alpha, b)}{\partial \alpha_1 \partial b_1}\right). \tag{24}$$

Now carrying out the integral over α_1, we obtain an amplitude determinant whose reciprocal can be written,

$$\det\left(\frac{\partial b_1}{\partial a_1}\right)_{\alpha_1} \det\left(\frac{\partial a_1}{\partial \alpha_1}\right)_{b_1}$$

$$= \det\left[-\left(\frac{\partial b_1}{\partial \alpha_1}\right)_{a_1}\right]$$

$$= \det\left[\left(\frac{\partial A_1}{\partial \alpha_1}\right)_{a_1}\left(\frac{\partial B_1}{\partial a_1}\right)_{\alpha_1} - \left(\frac{\partial A_1}{\partial a_1}\right)_{\alpha_1}\left(\frac{\partial B_1}{\partial \alpha_1}\right)_{a_1}\right] = \det\{A_1, B_1\}_1, \tag{25}$$

where in all the derivatives shown, $\alpha_2 = \beta_2$ and $a_2 = b_2$ are held fixed, and where the 1 subscript on the Poisson bracket indicates that the bracket is computed with respect to the f_1 canonical pairs (α_1, a_1) only. The right side is a determinant of an $f_1 \times f_1$ matrix of Poisson brackets. The 1 subscript on the Poisson bracket can be dropped, however, because the terms coming from the (α_2, a_2) canonical pairs vanish, due to the vanishing of the derivatives $\partial A_1/\partial \alpha_2$ and $\partial A_1/\partial a_2$. Therefore the Poisson brackets $\{A_{1k}, B_{1\ell}\}$ are standard Poisson brackets on the full phase space, and can be computed in any canonical coordinates.

These Poisson brackets are also independent of α_2, and can therefore be taken out of the α_2 integral. We show this by taking the α_2 derivative, which can be written as a Poisson bracket and transformed by using the Jacobi identity:

$$\frac{\partial}{\partial \alpha_{2i}}\{A_{1k}, B_{1\ell}\} = \{\{A_{1k}, B_{1\ell}\}, A_{2i}\} = -\{\{B_{1\ell}, A_{2i}\}, A_{1k}\} - \{\{A_{2i}, A_{1k}\}, B_{1\ell}\} = 0. \tag{26}$$

The final equality follows because $\{A_{2i}, A_{1k}\} = 0$, since all the As commute with one another, and because $\{B_{1\ell}, A_{2i}\} = \{B_{1\ell}, B_{2i}\} = 0$, since all the Bs also commute.

The final result is

$$\langle a|b \rangle = \frac{1}{(2\pi i\hbar)^{(f+f_2)/2}} \times \sum_r \frac{\exp\left\{\frac{i}{\hbar}\left[S_B(I) - S_A(I)\right] - i\mu\frac{\pi}{2}\right\}}{|\det\{A_1, B_1\}|^{1/2}} \int_I d\alpha_2. \tag{27}$$

The integral shown is just the α_2-volume of I. In Sec. 3 we will show that the Gutzwiller formula for the density of states of a classically nonintegrable system is a matrix element of this kind.

3 Semiclassical structure of trace formulas

The algebraic manipulations involved in the stationary phase approximation follow a few well defined patterns, which are essentially the same from one application to the next. The geometrical structures associated with these operations, however, may be quite different. This is immediately apparent when we contemplate Gutzwiller's stationary phase calculation of the trace of the Green's function, and try to understand it in terms of the theory discussed in Sec. 2, in which stationary phase points are associated with intersections of Lagrangian manifolds. The immediate problem is that it is not evident how two Lagrangian manifolds are associated with a single Green's function, or how periodic orbits are to be identified with the intersection of anything with anything else.

To clarify the role of the stationary phase approximation in the process of taking traces, we will work with the vector space of operators, in much the same manner as we have previously worked with the vector space of Schrödinger wave functions. We may regard this "operator space" as a kind of doubled Hilbert space; more precisely, for a suitable class of operators it is the tensor product of the Hilbert space of Schrödinger wave functions with its dual. For example, if a basis $|n\rangle$ of Schrödinger wave functions is chosen, then we have an associated basis $|n\rangle\langle m|$ of operators.

For a given operator A, we will talk about the associated "doubled wave function" or "wave function of the operator A," defined by

$$\psi_A(x, x') = \langle x|A|x'\rangle. \tag{28}$$

This wave function will be regarded as living on a doubled configuration space, with coordinates (x, x'). By change of basis, it can also be regarded as living on other doubled spaces, such as (x, p'), (p, x'), or (p, p'). In terms of these doubled wave functions, the Hilbert-Schmidt scalar product of operators can be written

$$\text{Tr}(A^\dagger B) = \int dx\, dx'\, \psi_A(x, x')^* \, \psi_B(x, x'). \tag{29}$$

In this manner, traces of operators appear as scalar products of wave functions, and can be incorporated into the formalism of Sec. 2.

We will also be interested in the semiclassical expressions for the doubled wave functions of various operators. When these expressions exist, they are of the form

$$\psi(x, x') = \sum_r \Omega(x, x') \exp\left[\frac{i}{\hbar} S(x, x') - i\mu\frac{\pi}{2}\right], \tag{30}$$

for some amplitude $\Omega(x, x')$ and action $S(x, x')$. Such a semiclassical expression is interpreted geometrically in terms of a Lagrangian manifold in a doubled phase space of $4f$ dimensions, whose coordinates are (x, p, x', p'). For example, the stationary phase points of the integral in Eq. (29) can be viewed as intersections of two Lagrangian manifolds in the doubled phase space. As we will show below, Gutzwiller's periodic orbits are intersections of this kind.

We begin this section by examining the doubled wave functions of unitary operators and their semiclassical representation in the doubled phase space. Unitary operators not only possess the simplest doubled wave functions, but also provide us with everything we need to study a variety of trace formulas. It turns out that Lagrangian manifolds in the doubled phase space play a dual role; not only do they support doubled wave functions, but they are also graphs of canonical transformations. The interplay between these roles runs throughout the asymptotics of doubled wave functions, and provides a richness that has no analog in the case of ordinary (single) wave functions and phase spaces. We finally use the geometrical structures in the doubled phase space to reduce a variety of results, including a trace formula due to Tabor[3], an intermediate result of Berry's[4] on the scars of Wigner functions in phase space, and the Gutzwiller-Balian-Bloch trace formula, to special cases of the asymptotic scalar products presented in Eqs. (16) and (27).

3.1 Unitary operators and the doubled phase space

We will primarily be concerned with unitary operators and their doubled wave functions. When these wave functions have a semiclassical limit, it is given by standard WKB theory, most conveniently expressed in terms of Miller's formula for the scalar product $\langle a|b \rangle$. A matrix element $\langle x|U|x' \rangle$ of a unitary operator can always be written as a scalar product of the form $\langle a|b \rangle$ by interpreting the operator U in a passive sense (i.e., as representing a change of basis) rather than an active sense (i.e., as representing a unitary mapping from one basis to another). This is precisely what we did for the propagator in deriving Eq. (31). Therefore unitary operators provide us with examples of doubled wave functions, in which we can read off the amplitude $\Omega(x, x')$ and action $S(x, x')$. We will now consider three specific examples.

The first is the propagator $K(x, t; x', t')$, in which the times t, t' are considered parameters and are required to satisfy $t > t'$. The propagator is the doubled wave function of the unitary time evolution operator $U(t, t')$, so we may write $\psi_U(x, x')$ for it. Furthermore, we have a ready-made semiclassical expression for this wave function, namely, the Van Vleck formula. From this we can easily read off the amplitude $\Omega(x, x')$ and action $S(x, x')$.

A second example is simply the identity I, which has the doubled wave function

$$\psi_I(x, p') = \frac{\exp(ixp'/\hbar)}{(2\pi\hbar)^{f/2}}. \tag{31}$$

By writing this in the (x, p')-representation, we see that the semiclassical expression is exact, and we can again read off the amplitude and action. Had we used the (x, x')-representation, we would have had a caustic, i.e., the wave function would be a delta function concentrated on the line $x = x'$ in the doubled configuration space.

A final unitary operator of interest is $W(\bar{x}, \bar{p})$, where $(\bar{x}, \bar{p})$ are parameters. It is defined by

$$\psi_W(x, p') = \langle x|W(\bar{x}, \bar{p})|p'\rangle = \frac{1}{(2\pi\hbar)^{f/2}} \exp\left\{\frac{i}{\hbar}\left[2\bar{p}(x - \bar{x}) + p'(2\bar{x} - x)\right]\right\}. \quad (32)$$

The doubled wave function shown is essentially the kernel of the Weyl transform, for if $\hat{A}$ is an operator and $A(\bar{x}, \bar{p})$ the corresponding Weyl symbol[36], then

$$A(\bar{x}, \bar{p}) = 2^f \text{Tr}\left[W(\bar{x}, \bar{p})^\dagger \hat{A}\right]. \quad (33)$$

Again, in order to avoid caustics, we choose the (x, p') representation in writing Eq. (32), and again, the semiclassical approximation is exact.

All three of these unitary operators, $U(t, t')$, I, and $W(\bar{x}, \bar{p})$, correspond in a simple way to classical canonical transformations. As shown by Miller[10], this correspondence is realized through the actions $S(x, x')$ or $S(x, p')$, treated either as a F_1- or F_2-type generating function[37], respectively, for a canonical transformation connecting (x, p) with (x', p'). One slight subtlety is that (x', p'), which appear as initial variables in the propagator, must be treated as "new" variables, whereas the final (x, p) are "old" variables.

For example, the action of the Van Vleck formula is Hamilton's principal function, the F_1-type generating function taking us from final (x, p) to initial (x', p'); and the action of the doubled wave function for the identity operator, Eq. (31), is $F_2(x, p') = xp'$, the generator of the classical identity canonical transformation. As for the unitary operator $W(\bar{x}, \bar{p})$, the generating function is

$$S(x, p') = F_2(x, p') = 2\bar{p}(x - \bar{x}) + p'(2\bar{x} - x), \quad (34)$$

which generates the "averaging" canonical transformation, given implicitly by

$$\bar{x} = \frac{x + x'}{2}, \qquad \bar{p} = \frac{p + p'}{2}. \quad (35)$$

Let us now consider the doubled phase space as a medium for interpreting the semiclassical expressions for our doubled wave functions. There is a certain subtlety concerning this space which requires some elaboration.

To begin, let us take a semiclassical representation for a doubled wave function, as in Eq. (30), and let us plot the wavefronts or contours of $S(x, x')$ in the doubled configuration space. This is illustrated schematically in Fig. 5. The gradient of $S(x, x')$ in the doubled configuration space is a vector perpendicular to the wave fronts; let us provisionally write $p = \partial S/\partial x$, $p' = \partial S/\partial x'$ for its components. These are functions of (x, x'), and give us a momentum field on the configuration

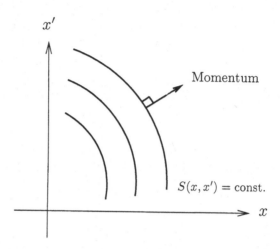

Fig. 5. The momentum can be defined as the gradient of the action, in this case in the doubled configuration space. If, however, p and p' are defined respectively as the old and new momenta under the canonical transformation $(x, p) \rightarrow (x', p')$, then the components of this momentum vector are $(p, -p')$.

space. Then in accordance with concepts from WKB theory which encompass all kinds of wave equations, not just those of quantum mechanics, we construct a phase space with coordinates (x, p, x', p'), within which our specific momentum field for our specific wave function is represented by a Lagrangian manifold. In this phase space, the variables conjugate to (x, x') are (p, p'), respectively, as defined above; it is in terms of these variables that Hamilton's equations for the rays take on their standard form and that Poisson brackets are computed in the standard way.

On the other hand, we are also identifying the action $S(x, x')$ with the F_1-type generating function of a canonical transformation, for which the generating function relations are $p = \partial S / \partial x$, $p' = -\partial S / \partial x'$. The point of this is that the equation for p' has a minus sign relative to the definition of p' in the preceding paragraph. The reason for this discrepancy is that there are now two conflicting interpretations for the symbol p': it is either the momentum conjugate to x' in the doubled phase space, or else it is the new momentum in the canonical transformation. We can choose only one of these interpretations, so by convention we will take the latter: our p' will be the new momentum in the canonical transformation. (We abandon the denfinition of p' in the preceding paragraph.)

This means that in the doubled phase space, the momentum conjugate to x' is $-p'$, and that the vector perpendicular to the wavefronts in the doubled configuration space has components $(p, -p')$. Therefore the action differential on the doubled phase space is

$$dS = p\,dx - p'\,dx',$$ (36)

and the symplectic form is

$$\omega_D = dp \wedge dx - dp' \wedge dx'.$$ (37)

A Lagrangian manifold in the $4f$-dimensional doubled phase space is a $2f$-dimensional manifold on which this form vanishes. Finally, if $f(x,p,x',p')$ and $g(x,p,x',p')$ are any two functions on the doubled phase space, then their Poisson bracket is given by

$$\{f,g\}_D = \left(\frac{\partial f}{\partial x}\frac{\partial g}{\partial p} - \frac{\partial f}{\partial p}\frac{\partial g}{\partial x}\right) - \left(\frac{\partial f}{\partial x'}\frac{\partial g}{\partial p'} - \frac{\partial f}{\partial p'}\frac{\partial g}{\partial x'}\right) = \{f,g\} - \{f,g\}',$$ (38)

where the D subscript indicates the doubled Poisson bracket, where the primed Poisson bracket is taken only with respect to the variables (x',p'), and where the unprimed Poisson bracket is taken only with respect to (x,p).

We now effectively have three phase spaces, an unprimed, a primed, and a doubled, which we may denote by Φ, Φ', and Φ_D respectively. We may also denote the creation of Φ_D out of the other two, in accordance with Eq. (38), by $\Phi_D = \Phi \otimes \Phi'^*$. The reason for this notation is analogy: if we denote the Hilbert space of Schrödinger wave functions by X, then the Hilbert space of operators is $X \otimes X^*$. We may also note that the minus sign in the nonstandard Poisson bracket of Eq. (38) comes originally from a complex conjugation of a phase.

The doubled phase space with the symplectic structure of Eq. (37) is a well known device for representing the geometrical structure of canonical transformations and their generating functions. It is discussed, for example, by Abraham and Marsden[11]. It is possible to give a geometrical representation of a canonical transformation without going to the doubled phase space; for example, a foliation of the $2f$-dimensional single phase space into an f-parameter family of Lagrangian manifolds does this. But such a representation does not treat the old and new variables on an equal footing, nor does it readily reveal the relationship among the various kinds of generating functions for a given canonical transformation. To satisfy these goals, the doubled phase space is necessary.

The doubled phase space is used to represent canonical transformations in the following manner. Suppose we are given a canonical transformation, say,

$$\begin{aligned} x &= X(x',p'), \\ p &= P(x',p'). \end{aligned}$$ (39)

Then the set of points (x,p,x',p') in the doubled phase space which satisfy these equations, i.e., the graph of the canonical transformation, is a $2f$-dimensional surface in this space. The same surface can equally well be specified by the various generating function relations; for example, if the canonical transformation

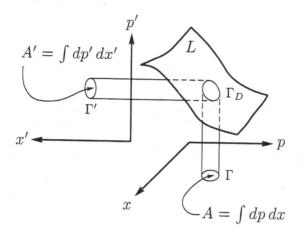

Fig. 6. The graph of a canonical transformation $(x, p) \rightarrow (x', p')$ in the doubled phase space is a Lagrangian manifold L with respect to the symplectic form of Eq. (37). Conversely, every Lagrangian manifold in this space with a nonsingular projection onto the (x, p) and (x', p') planes is the graph of a canonical transformation.

possesses an F_1-type generating function, then the equations

$$p = \frac{\partial F_1(x, x')}{\partial x}, \qquad p' = -\frac{\partial F_1(x, x')}{\partial x'} \qquad (40)$$

are equivalent to Eqs. (39), being simply an algebraic rearrangement of them.

It turns out that this $2f$-dimensional surface is a Lagrangian manifold in the doubled phase space with respect to the symplectic form of Eq. (37). This is most easily seen with the aid of a diagram such as Fig. 6. We let L be the graph of the given canonical transformation, and we choose a contractible closed curve in L, bounding a 2-dimensional region Γ_D. This region is projected onto the (x, p) and (x', p') phase spaces, producing respectively regions Γ and Γ'. Since L is the graph of the canonical transformation, the region Γ' is the image of the region Γ under the canonical transformation. But since this transformation is canonical, the respective symplectic areas A, A' of Γ, Γ', measured in the unprimed and primed phase spaces, must be equal. This in turn implies

$$\int_{\Gamma_D} \omega_D = A - A' = 0. \qquad (41)$$

Since this is true for arbitrary contractible Γ_D on L, we must have $\omega_D = 0$ on L, and therefore L is Lagrangian with respect to ω_D.

Conversely, suppose we are given a Lagrangian manifold L in the doubled phase space. A region of L which has a nonsingular projection onto the (x, p) and (x', p') phase spaces provides a mapping between these phase spaces; and this mapping preserves the symplectic area of two-dimensional area elements.

Therefore the mapping is canonical, at least within the given regions. (One can show that a region of a Lagrangian manifold in the doubled space has a nonsingular projection onto (x, p) if and only if its projection onto (x', p') is also nonsingular.)

A given Lagrangian manifold in the doubled phase space may well possess points where the projections onto the (x, p) and (x', p') planes are singular. In the typical case, these singularities occur on a subset of L of lower dimensionality than $2f$, and represent the places where the branches of a multivalued canonical transformation coalesce. We may note, however, that the three canonical transformations discussed above, the time evolution, the identity, and the averaging transformation, are all single-valued, so the corresponding Lagrangian manifolds have nonsingular projections onto (x, p) and (x', p'). It is also possible to construct exceptional Lagrangian manifolds in the doubled phase space which have projections onto (x, p) and (x', p') which are singular everywhere, and which therefore do not represent canonical transformations.

The representation of canonical transformations by Lagrangian manifolds in the doubled phase space is useful for insight it offers into the various types of generating functions. For example, if the Lagrangian manifold corresponding to a given canonical transformation has a nonsingular projection onto the (x, x') plane, then it has an F_1-type generating function. In this case, the generating function of the canonical transformation coincides with the generating function of an isolated Lagrangian manifold, as was discussed below Eq. (2), if the latter concept is transplanted to the doubled phase space. Notice that the projection onto (x, x') may be multivalued and have singularities, even if the projection onto (x, p) and (x', p') does not. This circumstance occurs, for example, in the canonical transformation for the time evolution.

More generally, we can obtain a generating function of any type by projecting L onto some $2f$-dimensional plane, spanned by some subset of old and new xs and ps. The only requirement is that the projection must be nonsingular, and that the plane onto which we project must itself be Lagrangian with respect to the doubled symplectic form. For example, (x, p') is allowed, but (x, p) is not.

One of the most important applications of Lagrangian manifolds in physical problems is in the representation of multidimensional wave fields in the short wavelength limit. The preceding discussion of the doubled phase space has concerned another application, namely, an elegant and symmetrical means of treating canonical transformations and all of their generating functions. It is remarkable, therefore, that these two applications come together in the asymptotics of the matrix elements of operators, i.e., in our doubled wave functions. We see now that every matrix element with a semiclassical limit corresponds to a Lagrangian manifold in the doubled phase space, which in turn, if it has nonsingular projections onto the (x, p) and (x', p') subspaces, corresponds to a canonical transformation.

Although in this chapter we are primarily interested in doubled wave functions of unitary operators, other operators may also be considered, and they are also represented semiclassically by Lagrangian manifolds in the doubled phase space. It is interesting that some of these give rise to semiclassical wave functions which do not correspond to a canonical transformation, because their Lagrangian manifolds have projections onto (x, p) or (x', p') space which are singular everywhere. For example, let $\psi(x) = \langle x|\psi \rangle$ be an ordinary (single) wave function with a semiclassical representation in the usual form, and consider the projection operator $|\psi\rangle\langle\psi|$. The doubled wave function for this projection operator gives rise to the Lagrangian manifold,

$$p = f(x), \qquad p' = f(x') \tag{42}$$

in the doubled phase space, where $f(x) = \partial S(x)/\partial x$, $S(x)$ being the (single) action of $\psi(x)$. This Lagrangian manifold has a singular projection onto (x, p) space, because only certain (x, p) values are consistent with Eq. (42), and because once such an (x, p) value is given, the values of (x', p') are undetermined. The Lagrangian manifold of Eq. (42) does have an F_1-type generating function, however; it is $F_1(x, x') = S(x) - S(x')$, and it is an example of a generating function which does not generate a canonical transformation. Obviously such generating functions must be taken into account in a general theory of semiclassical matrix elements.

3.2 The Hilbert–Schmidt scalar product of unitary operators

We will now apply the results of Sec. 2, especially Eq. (16) for the matrix element $\langle a|b \rangle$, to the computation of $\mathrm{Tr}(V^\dagger U)$, in which U and V are unitary operators. In order to have a specific example in mind, we may identify V with the identity operator I, and U with $U(t, t')$, the unitary time-evolution operator for a time-dependent system, so as to compute $\mathrm{Tr}U(t, t')$. We will assume that the system has no constants of motion, either the Hamiltonian or anything else. The reason for this assumption is that it leads to the simplest trace formula providing a connection with classical periodic orbits, and therefore is a good place to start.

Actually, in computing $\mathrm{Tr}(V^\dagger U)$, it is sufficient to assume $V = I$, since we can always rewrite the operator product $V^\dagger U$, itself unitary, simply as U. This step simplifies the calculation somewhat, and, when we are done, we are always free to factorize U once again into the product of two unitary operators. Let us therefore proceed with this simplification.

In the semiclassical computation of $\mathrm{Tr}(I^\dagger U)$, we know that the asymptotic forms for the matrix elements of I and U are associated with two Lagrangian manifolds in the doubled phase space, which we denote by L_I and L_U; and that these in turn are associated with two canonical transformations, the I-transformation which is the identity, and the U-transformation, which we denote by $x = X(x', p')$, $p = P(x', p')$. We will denote the F_1-type generating function of the U-transformation by $S(x, x')$, which is also the action for the semiclassical expression for $\psi_U(x, x') = \langle x|U|x' \rangle$. The amplitude of this expression is given by

Miller's formula, Eq. (8), as

$$\Omega(x, x') = \left| \det \frac{\partial^2 S(x, x')}{\partial x \partial x'} \right|^{1/2}. \tag{43}$$

An immediate fact which emerges from a consideration of Eq. (16) is that the stationary phase points in the computation of $\text{Tr}(I^\dagger U)$ will be the intersections of L_I and L_U in the doubled phase space, i.e., the points (x, p, x', p') such that

$$x = x' = X(x', p'), \qquad p = p' = P(x', p'). \tag{44}$$

These points on the doubled space represent periodic orbits on the single space, in the sense that $(x, p) = (x', p')$ is a fixed point of the U-canonical transformation. Another immediate fact is that the action at one of these stationary phase points is simply $S(x, x')$, because the (x, x')-action of the identity transformation is zero. (Properly, one should work in a representation in which neither operator has a caustic; it does not matter which, just as in Eq. (16) the right hand side is obviously independent of the choice of the x-representation. However, the results are as quoted here.)

To proceed further with the application of Eq. (16), however, we must somehow represent the doubled wave function $\psi_U(x, x')$ as the simultaneous eigenfunction of some set of commuting operators, say, $\mathscr{B}$, with eigenvalues b and classical counterparts B; and similarly the doubled wave function for the identity, using operators $\mathscr{A}$, eigenvalues a, and classical counterparts A. Because we are now working in the doubled space, each one of these sets contains $2f$ members, and the operators $\mathscr{A}$, $\mathscr{B}$ are not ordinary operators, but rather "doubled operators," i.e., linear mappings of ordinary operators into other ordinary operators. Similarly, the sets of classical functions A, B are functions of (x, p, x', p'), and must Poisson commute within each set according to the doubled Poisson bracket.

We must do this because on the doubled phase space, the doubled wave functions are represented by two particular Lagrangian manifolds in isolation, which are not members of foliations. Therefore in order to apply Eq. (16), or rather its transcription to the doubled phase space, it is necessary to imbed these individual Lagrangian manifolds in foliations. The analog of this process in the case of the single phase space would be to go from a particular solution of the Hamilton–Jacobi equation, say $S(x)$, to a complete solution, say, $\tilde{S}(x, a)$, for which $S(x) = \tilde{S}(x, a)$ for some fixed value of a ($a = 0$ is convenient). On the doubled phase space, we may regard the particular Lagrangian manifolds L_I and L_U as the "physical" Lagrangian manifolds, and the others as being used simply for the sake of applying Eq. (16). Since (virtually) all Lagrangian manifolds in the doubled space correspond to canonical transformations, each of the two foliations will produce two families of canonical transformations, parameterized by a or b.

The proposed imbedding is not unique and seems artificial, since a straight-forward application of the stationary phase approximation, along the lines of Gutzwiller's original derivation, does not require it, but rather works (effectively) with the two physical Lagrangian manifolds and the densities on them. We will proceed anyway with the suggested approach for the following reasons. First, the manifest phase space invariance of Eq. (16) is compelling, and reason to pursue this formula to see how general it is. Second, when we imbed a particular La-grangian manifold in a foliation and use the foliation in the manner suggested, it turns out that only the members of the foliation in an infinitesimal neighborhood of the original Lagrangian manifold have an effect on the result (as one would ex-pect). That is, these neighboring Lagrangian manifolds and the manner in which they are specified are equivalent to positing a density on the original Lagrangian manifold; therefore, working with such a foliation in the neighborhood of the original Lagrangian manifold is an alternative to dealing with densities and am-plitude determinants. Finding a means of clarifying manipulations on amplitude determinants is a major goal of this chapter. Third, the foliation we will suggest below leads to an interesting and novel means of calculating Gutzwiller's periodic orbit amplitudes. Fourth, the Wigner-Weyl formalism provides us for free with a foliation of the desired kind, and the results of pursuing this line of investi-gation yield new insights into the Wigner-Weyl formalism (and into Berry's[4] trace formula for the scars of Wigner functions in phase space). And finally, the kind of thing we propose here is not unheard of in physics; for example, Dirac brackets are usually computed using coordinates defined on a space larger than the physical one.

Let us work with the manifold L_U, and find the desired imbedding; that for L_I will then follow easily. We require $2f$ functions $B_1(x, p, x', p'), \ldots, B_{2f}(x, p, x', p')$ which take on constant values on L_U, and which commute with one another under the doubled bracket. In addition, they should give the correct amplitude determinant $\Omega(x, x')$, shown in Eq. (43), in accordance with the doubled version of Miller's formula, Eq. (8). We satisfy the first two goals by applying some guesswork, and find

$$
\begin{aligned}
B_i(x, p, x', p') &= x_i - X_i(x', p'), \\
B_{i+f}(x, p, x', p') &= p_i - P_i(x', p'),
\end{aligned}
\tag{45}
$$

for $i = 1, \ldots, f$. These Bs take on constant values on L_U, namely $b = 0$; and their doubled Poisson brackets among themselves vanish, as verified by direct substitution into Eq. (38).

Consider now members of the foliation for which $b = (b_x, b_p) \neq 0$. The Lagrangian manifold given by $B = b$ represents a canonical transformation, given in terms of the one for $b = 0$ by

$$
\begin{aligned}
x &= X(x', p') + b_x, \\
p &= P(x', p') + b_p,
\end{aligned}
\tag{46}
$$

which can be regarded as the composition of the U-canonical transformation with a rigid displacement in phase space. This canonical transformation has the F_1-type generating function,

$$\tilde{S}(x, x'; b) = S(x - b_x, x') + b_p x, \tag{47}$$

as follows from the fact that the F_1-type generating function of the composition of two canonical transformations is just the sum of the F_1-type generating functions of the constituents. The tilde distinguishes the generating function parameterized by b from the the original one (we have $\tilde{S} = S$ when $b = 0$). Therefore if we apply the doubled version of Miller's formula, Eq. (8), to the computation of the simultaneous eigenfunction of the $2f$ doubled operators $\mathscr{B}$ corresponding to the classical Bs of Eq. (45), we find

$$\langle\!\langle x, x' | b \rangle\!\rangle = \text{const.} \sum_r \tilde{\Omega}(x, x'; b) \times \exp\left[\frac{i}{\hbar} \tilde{S}(x, x'; b) - i\mu\frac{\pi}{2}\right], \tag{48}$$

where we use double angle brackets for a double scalar product, and where

$$\tilde{\Omega}(x, x'; b)^2 = \det \begin{pmatrix} \dfrac{\partial^2 \tilde{S}}{\partial x \partial b_x} & \dfrac{\partial^2 \tilde{S}}{\partial x \partial b_p} \\[2mm] \dfrac{\partial^2 \tilde{S}}{\partial x' \partial b_x} & \dfrac{\partial^2 \tilde{S}}{\partial x' \partial b_p} \end{pmatrix} = \det \begin{pmatrix} \dfrac{\partial^2 S}{\partial x^2} & \mathbf{I} \\[2mm] -\dfrac{\partial^2 S}{\partial x \partial x'} & 0 \end{pmatrix} = \det \dfrac{\partial^2 S}{\partial x \partial x'}. \tag{49}$$

Here the tilde on $\tilde{\Omega}$ has the same meaning as on $\tilde{S}$; and when $b = 0$, we see that $\tilde{\Omega}$ agrees with the amplitude Ω of Eq. (43). Therefore we now have

$$\psi_U(x, x') = \langle x | U | x' \rangle = \text{const. } \langle\!\langle x, x' | b = 0 \rangle\!\rangle; \tag{50}$$

apart from normalization, which we deal with momentarily, we have completed the classical and semiclassical aspects of imbedding our wave function, Lagrangian manifold, and canonical transformation in the required families.

Let us now consider the doubled operators $\mathscr{B}$ corresponding to the classical Bs of Eq. (45). First we introduce certain doubled operators associated with ordinary operators. For example, if F is an ordinary operator, we can associate it with a doubled operator $\mathscr{L}_F$ by left multiplication, i.e,

$$\mathscr{L}_F G = FG, \tag{51}$$

where G is any ordinary operator. Similarly, we can associate F with another doubled operator $\mathscr{R}_F$ by right multiplication, i.e.,

$$\mathscr{R}_F G = GF. \tag{52}$$

Then we can represent our desired $\mathscr{B}$s in terms of left and right multiplication by

$$\begin{aligned} \mathscr{B}_i &= \mathscr{L}_{\hat{x}_i} - \mathscr{R}_{U^\dagger \hat{x}_i U}, \\ \mathscr{B}_{i+f} &= \mathscr{L}_{\hat{p}_i} - \mathscr{R}_{U^\dagger \hat{p}_i U}, \end{aligned} \tag{53}$$

for $i = 1, \ldots, f$. These $\mathscr{B}$s commute with one another, as is easily verified, and their simultaneous eigenoperator $U(b)$ is given by

$$\langle x|U(b)|x'\rangle = \exp(ixb_p/\hbar)\langle x - b_x|U|x'\rangle, \tag{54}$$

where $U = U(0)$. That is, we have

$$\mathscr{B}U(b) = bU(b). \tag{55}$$

We note that $U(b)$ is unitary for all values of b. Equation (54) immediately allows us to write down the semiclassical expression for $\langle x|U(b)|x'\rangle$ in terms of that for $\langle x|U|x'\rangle$, and it is precisely Eq. (48). Therefore we have now imbedded our original operator U in a family $U(b)$, whose semiclassical and classical representatives are those given above, and identified the members of the family as simultaneous eigenoperators of a complete set of commuting doubled operators.

There remains only the normalization. It is convenient not to use the precise transcription of Eq.(7) to the doubled space, but rather to demand that

$$\langle\!\langle b|b'\rangle\!\rangle = (2\pi\hbar)^f \delta(b - b'). \tag{56}$$

This has the advantage that the normalized, doubled wave function $\langle\!\langle x, x'|b\rangle\!\rangle$ is exactly the unitary matrix element $\langle x|U(b)|x'\rangle$, so that

$$\mathrm{Tr}\big[U(b)^\dagger U(b')\big] = (2\pi\hbar)^f \delta(b - b'). \tag{57}$$

Thus, the constants in Eqs. (48) and (50) are unity.

Repeating the calculation leading from Eq. (45) to (55) with U replaced by I, B by A, etc., we derive parallel results for the identity canonical transformation. The most important of these is

$$
\begin{aligned}
A_i(x, p, x', p') &= x_i - x'_i, \\
A_{i+f}(x, p, x', p') &= p_i - p'_i,
\end{aligned}
\tag{58}
$$

for $i = 1, \ldots, f$.

There now arises an interesting point. As discussed earlier, the use of a complete set of commuting observables such as the As of Eq. (58) or the Bs of Eq. (45), is an alternative to working with a density on a Lagrangian manifold. Nevertheless, it is of interest actually to compute the density on a Lagrangian manifold in the doubled phase space corresponding to a unitary operator. Any number of coordinate systems are available for expressing the density, such as (x, x'), (x, p'), etc., and it is easy to see that the density function with respect to one of these coordinate systems is just the corresponding amplitude or Van Vleck determinant. These density functions can be thought of as arising from projecting a density intrinsic to the Lagrangian manifold down onto some Lagrangian plane in the doubled phase space. There is, however, no reason why the density must be projected onto a Lagrangian plane; for example, it could be projected onto the (x, p)-plane. In the single phase space, there would be no point in projecting onto a plane which was not Lagrangian, because no corresponding representation would exist for the wave functions. But in the doubled phase space, it is interesting to do so, because the (x, p)-plane has its own intrinsic, invariant measure, namely, the Liouville measure $dp\,dx$. Not surprisingly, we find that the density on a unitary Lagrangian manifold in the doubled phase space, when projected onto the (x, p)-plane, is constant, i.e., it agrees with the Liouville measure there. Thus, one can regard the doubled wave functions of unitary operators as having the simplest semiclassical structure of any doubled wave functions, in the sense that everything is specified by the Lagrangian manifold alone, since the density is the natural measure provided by the geometry.

It is now easy to compute the semiclassical expression for $\mathrm{Tr}(I^\dagger U)$. Only the amplitude determinant requires any calculation; using Eqs. (45) and (58), we find

$$\det\{A, B\}_D = \det \begin{pmatrix} -\dfrac{\partial X}{\partial p'} & -\mathbf{I} + \dfrac{\partial X}{\partial x'} \\ \mathbf{I} - \dfrac{\partial P}{\partial p'} & \dfrac{\partial P}{\partial x'} \end{pmatrix} = \det(\mathbf{M} - \mathbf{I}), \qquad (59)$$

where $\mathbf{M}$ is the symplectic matrix $\partial(X, P)/\partial(x', p')$, and where we have taken the transpose after the first equality. This determinant is to be evaluated at the stationary phase points, i.e., the periodic orbits, for which $\mathbf{M}$ becomes the monodromy matrix. Altogether, the result is

$$\mathrm{Tr}\,U = \sum \frac{\exp\left[\frac{i}{\hbar}S(x, x) - i\mu\frac{\pi}{2}\right]}{|\det(\mathbf{M} - \mathbf{I})|^{1/2}}. \qquad (60)$$

The sum is taken over fixed points (x, p) of the U-canonical transformation.

A trace formula of this type was first derived by Tabor[3], who quantized the standard map by imbedding it in a continuous time system, and then used the stationary phase approximation to take the trace of $U(0, T)$, where T is the period of the mapping. Our result is slightly more general than Tabor's, in that we work in any number of degrees of freedom, and we make fewer assumptions about the unitary operator U. For example, even if U is a time-evolution operator, the system need not be time-periodic, or, if it is, the elapsed time $t - t'$ need not be a period. These constitute minor modifications to Tabor's results, the main point of our presentation being the geometrical structure we reveal.

Equation (60) cannot be immediately generalized to the case of time-independent systems, because, for such systems, the monodromy matrix $\mathbf{M}$ has an eigenvector of eigenvalue $+1$, namely the flow vector along the periodic orbit, $(\dot{x}, \dot{p})$. Therefore the denominator of Eq. (60) vanishes in this case, indicating a caustic. The caustic is of the type in which the two Lagrangian manifolds intersect over a region of dimensionality 1, as discussed in Sec. 2.2 and illustrated in Fig. 4. This is the case originally considered by Gutzwiller, and it will be discussed more fully below. For now we simply note that the intersection I of the two Lagrangian manifolds is just the periodic orbit, and that the appropriate scalar product formula is not Eq. (16), but rather Eq. (27).

Similarly, we must exclude the possibility of constants of motion in Eq. (50), because the symmetry groups associated with such constants map orbits into other orbits. In particular, they map periodic orbits into other periodic orbits, thereby creating a continuous family of nonisolated stationary phase points. Evidently, the Hamiltonian, when conserved, has the same effect on the structure of our trace formula as any other constant of motion.

Of course, we may have caustics of Eq. (50) even when $\partial H / \partial t \neq 0$ and no constants of motion exist, such as in the case of a periodic orbit of parabolic stability (two eigenvalues of **M** equal to $+1$). Typically such stationary phase points are still isolated, and the caustic is of the Airy function (fold catastrophe) type. The formalism based on the Lagrangian manifolds in the doubled phase space makes it clear that this caustic is no different from any other Airy function-type caustic in any other system; facts like this have not been evident in earlier approaches to trace formulas. (See, however, the analysis by Ozorio de Almeida and Hannay[38], which reveals a surprisingly complex caustic structure in stable periodic orbits whose higher order iterations are resonant.)

Finally, it is useful to replace U by $V^\dagger U$, as suggested earlier, and write out the result. It is

$$\text{Tr}(V^\dagger U) = \sum \frac{\exp\left\{ \frac{i}{\hbar} \left[S_U(x,x') - S_V(x,x') \right] - i\mu\frac{\pi}{2} \right\}}{|\det(\mathbf{M}_U - \mathbf{M}_V)|^{1/2}}. \tag{61}$$

The sum is over points (x', p') which, when mapped by the U-canonical transformation to (x, p), and then again by the inverse of the V-canonical transformation, return to (x', p').

3.3 The Weyl correspondence

Let us now consider the Weyl transform of a unitary operator U, which we denote by $W_U(\bar{x}, \bar{p})$. This problem has been considered by Berry[4] in the case that U is the time-evolution operator, as a first step in the derivation of his formula for the scars of Wigner functions in phase space. Our strategy will be to write the desired Weyl transform as the trace of the product of two unitary operators, as in Eq. (33), to which we can immediately apply Eq. (61), identifying V with the operator $W(\bar{x}, \bar{p})$ of Eq. (32).

In doing this an interesting point arises, namely that the parameters $(\bar{x}, \bar{p})$ of the kernel of the Weyl transform, i.e., the location in the (single) phase space at which the function W_U is to be evaluated, can be interpreted as a complete set of $2f$ commuting functions of (x, p, x', p'), specifying a foliation of the doubled phase space into Lagrangian manifolds. This follows by using Eq. (35) to define $(\bar{x}, \bar{p})$ as functions on the doubled phase space, and then by computing the doubled Poisson bracket. Therefore it is not necessary to imbed the unitary operator

$W(\bar{x},\bar{p})$ in a family or its corresponding Lagrangian manifold in a foliation, since that is already done for us. On the other hand, the foliation suppled to us for free by the Weyl formalism is really no different from the one we guessed in Eq. (45), for if we use the averaging canonical transformation to define functions A according to

$$
\begin{aligned}
A_i(x, p, x', p') &= x_i + x'_i = 2\bar{x}_i, \\
A_{i+f}(x, p, x', p') &= p_i + p'_i = 2\bar{p}_i,
\end{aligned}
\tag{62}
$$

for $i = 1, \ldots, f$, we see that we have precisely the form we guessed in Eq. (45).

The desired Weyl transform is now immediate. The result is

$$
W_U(\bar{x},\bar{p}) = 2^f \sum \frac{\exp\left\{\frac{i}{\hbar}\left[S(x, x') - \bar{p}(x - x')\right] - i\mu\frac{\pi}{2}\right\}}{|\det(\mathbf{M} + \mathbf{I})|^{1/2}},
\tag{63}
$$

where $S(x, x')$ is the action of the U-canonical transformation, and where $\mathbf{M} = \mathbf{M}_U$. This is the formula originally derived by Berry[4]. The sum is taken over points (x', p') which, when mapped by the U-canonical transformation, yield points (x, p) satisfying Eq. (35), Berry's "midpoint rule." The plus sign in the determinant in the denominator, in contrast to the minus sign in Eq. (49), is a simple consequence of replacing the identity canonical transformation by the averaging transformation in the computation of $\det\{A, B\}_D$, giving $\mathbf{M}_V = -\mathbf{I}$.

In Berry's calculation, the derivation of Eq. (63) is only the first (and easier) step. The second step consists of taking the Fourier transform in time of Eq. (63), to obtain the Weyl transform of the Green's function. Although the second step only involves a one-dimensional integral, it is rather intricate in its execution, leading ultimately to Airy function caustics in the neighborhood of periodic orbits.

In order to limit the scope of this chapter, we will not pursue this line of investigation here, but rather promise it in future work. We will, however, offer the following general comments on the relation of the Wigner-Weyl formalism to the doubled phase space and its doubled Poisson bracket, and then make some final comments on Berry's calculation.

The considerations raised here reveal several features of the Wigner function and Weyl correspondence which otherwise are obscure. For example, consider the obvious fact that the Wigner function can be regarded as a kind of "wave function" on phase space. A peculiar feature of this interpretation, however, is that $(\bar{x},\bar{p})$ behave like configuration space coordinates, not only because they are the variables upon which a wave function depends, but also because they play the role of commuting q's when the asymptotics of the Wigner function are considered. The present analysis shows that $(\bar{x},\bar{p})$ actually are commuting variables, i.e., under the doubled Poisson bracket, and can indeed be taken as a complete set of commuting variables on the doubled phase space. That is, they constitute one half of a canonical coordinate system on the doubled phase space;

the simplest choice for the other half is

$$\tilde{x} = x - x', \qquad \tilde{p} = p - p'. \tag{64}$$

These give the commutation relations

$$\{\tilde{x}_i, \tilde{x}_j\}_D = \{\bar{p}_i, \bar{p}_j\}_D = 0,$$
$$\{\tilde{x}_i, \bar{p}_j\}_D = \{\tilde{x}_i, \bar{p}_j\}_D = \delta_{ij}, \tag{65}$$

showing that $(\bar{x}, \bar{p}; \tilde{x}, \tilde{p})$ are canonical variables on the doubled phase space.

For another example, much is known about the caustic structure of Wigner functions, especially as a consequence of the work of Berry[39]; what the present analysis shows is that these caustics are singularities of the projection of one Lagrangian manifold in the doubled phase space onto another. In other words, the asymptotic structure of the Wigner function can be handled as a special case of Maslov's theory, a fact not completely evident when the stationary phase approximation is applied to the formulas of the Wigner-Weyl formalism. Indeed, Berry[39] regarded the similarities between his calculations and Maslov's theory as more superficial than real.

To elaborate on this point, we may contrast the perspective of this chapter on caustics, as properties of the intersections of two Lagrangian manifolds in phase space, with the usual perspective, in terms of the singularities of the projection of one Lagrangian manifold onto another. The two views are equivalent, of course; for example, in a simple case we consider $\psi(x) = \langle x|\psi \rangle$, an ordinary (single) wave function, and regard the caustics as being the singularities of the projection of the Lagrangian manifold L_ψ onto configuration space. The projection takes place along lines of constant x, themselves Lagrangian manifolds whose intersections with L_ψ gives the picture expressed in Fig. 1 and Eq. (16). Configuration space can be identified with the subset of phase space $p = 0$, also a Lagrangian plane; but more generally, any surface transverse to the projection would work as well. That is, configuration space may be thought of as the quotient space under the equivalence engendered by the projection.

Similarly, in the doubled phase space, the caustics of the Weyl transform of an operator can be interpreted in terms of the intersections of some Lagrangian manifold (L_U in Eq. (63)) with the Lagrangian manifolds corresponding to the averaging transformation; or in terms of the projection of the given Lagrangian manifold onto the plane $\tilde{x} = \tilde{p} = 0$. The projection takes place at constant $(\bar{x}, \bar{p})$, which also serve as coordinates on the plane $\tilde{x} = \tilde{p} = 0$. In the usual interpretation of the Weyl transform, this latter plane is identified with the (single) phase space, although it could equally well be identified with the Lagrangian plane specifying the identity canonical transformation, since $\tilde{x} = \tilde{p} = 0$ implies $x = x'$, $p = p'$. One final comment is that the so-called covariant Weyl symbol[36], obtained by using a Heisenberg operator $T(\tilde{x}, \tilde{p})$ in Eq. (33) instead of the operator $W(\bar{x}, \bar{p})$, corresponds to projecting the given Lagrangian manifold onto the complementary plane $\bar{x} = \bar{p} = 0$, which is also sometimes identified with the single phase space (or a second copy of it).

These considerations make it evident that the Airy functions and interference fringes found by Berry[4] for the scars of Wigner functions in phase space are a direct consequence of projecting a certain Lagrangian manifold in the doubled phase space onto the Lagrangian plane representing the identity transformation. The proper way to do this is to work in the extended, doubled phase space, with coordinates $(x, p, t, w; x', p', t', w')$, in which the Lagrangian manifold being projected is that representing the time evolution. This is so because in the extended, doubled space Berry's difficult time integral is subsumed under the same geometrical picture as all the other integrals. We hope to develop this picture in more detail in the future.

3.4 The Gutzwiller trace formula

We obtain the trace formula of Gutzwiller by reconsidering the trace of the unitary time-evolution operator, this time in the case that the Hamiltonian is conserved, but no other conserved quantities exist. It often happens in systems with these properties that all the periodic orbits are isolated, and we will assume that this is the case. We will set $t' = 0$ and write $\mathrm{Tr}U(t) = \mathrm{Tr}[I^{\dagger}U(t)]$ for the desired trace.

The manifolds L_U and L_I intersect in periodic orbits of period t, i.e., each point (x, p, x', p') on the intersection satisfies $(x, p) = (x', p')$, where (x, p) is on a periodic orbit of period t in the single phase space. Since the periodic orbits are isolated, the intersections are one-dimensional. This situation is illustrated in Fig. 7, although, as mentioned in Sec. 2, the picture is misleading in that it suggests that the intersection is stable under small perturbations (it is not). Indeed, it seems clear that the very existence of this unstable configuration is due to a symmetry of the system, in this case, symmetry under time-displacements. Other symmetries would have similar effects.

The complete set of commuting observables A and B of Eqs. (45) and (58) specify the manifolds L_I and L_U as before, but are not directly suitable for computing the amplitude determinant. Instead, we must follow the logic leading to Eq. (27), because L_I and L_U intersect in a one-dimensional curve. This requires us to find a new set of As and Bs.

To begin, we notice that the quantity $H(x, p) - H(x', p')$ generates orbits in the doubled phase space (with the doubled Poisson bracket being used to create Hamilton's equations) in which both (x, p) and (x', p') follow ordinary orbits in their respective phase spaces. Thus, if an initial point (x, p, x', p') of the doubled phase space is such that $(x, p), (x', p')$, regarded as two points in the single phase space, lie on the same orbit with a given elapsed time between them, then they stay on this same orbit forever with the same time difference. This means that if an initial point (x, p, x', p') of the doubled phase space lies on L_I, it stays on L_I;

Periodic Orbit

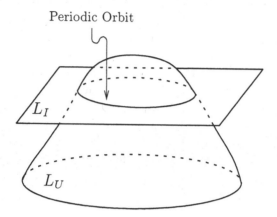

L_I

L_U

Fig. 7. In the Gutzwiller trace formula, the Lagrangian manifold in the doubled phase space representing the time evolution and that representing the identity canonical transformation intersect in the periodic orbits. These are the caustics of the system. The one-dimensional character of the intersection is not stable with respect to small perturbations in phase space.

and similarly if it lies on L_U. Therefore $H(x, p) - H(x', p')$ generates displacements along the intersection of L_U with L_I, and may be identified with the functions $A_2 = B_2$ used in Eq. (27). Furthermore, the variable α_2 of Eq. (27) can be identified with the elapsed time along the periodic orbit, so the integral is just the time required to go around the intersection of L_I with L_U. This is not necessarily the period t of the orbit, because the orbit may not be primitive; instead, we have

$$\int_{L_U \cap L_I} d\alpha_2 = T = \frac{t}{n}, \tag{66}$$

where T is the primitive period and n is the number of iterations of the primitive period in time t.

Next, we must find $2f - 1$ further As and $2f - 1$ further Bs which commute with one another and which are constant on the manifolds L_I and L_U respectively. The pattern established in Eqs. (45) and (58) suggests that we let the As be the difference between the old and new variables in a canonical coordinate system on the single phase space in which H is one of the coordinates. We write (η, τ, E) for these canonical coordinates, where $\eta = (y, p_y)$ are $2f - 2$ canonical variables in a surface of section, where τ is the elapsed time relative to the surface of section, measured along an orbit, and where E is interchangeable with H. We then take

$$A_\eta(\eta, \tau, E, \eta', \tau', E') = \eta - \eta',$$
$$A_\tau(\eta, \tau, E, \eta', \tau', E') = \tau - \tau',$$

$$A_E(\eta, \tau, E, \eta', \tau', E') \;=\; E - E', \tag{67}$$

where the first two equations specify the $2f - 1$ components of A_1, and the last is A_2. These As commute under the doubled Poisson bracket, and vanish on L_I.

Before proceeding to the Bs, some comments on the coordinate system (η, τ, E) are in order. These coordinates were used by Berry[4] to compute the scars of Wigner functions in phase space. Berry did not point out, however, that these coordinates are not canonical unless the surface of section variables η in one energy surface are chosen in a particular way relative to the ηs in nearby energy surfaces. In particular, the surface of section variables commonly employed in mechanical problems are generally not canonical on the full phase space. The problem is the Poisson bracket $\{\eta, \tau\}$, which does not vanish unless the ηs are chosen to be constant along the τ-trajectories. The fact that it is possible to choose η's which are simultaneously constant along both the τ-trajectories and the H-trajectories, i.e., so that $\{\eta, \tau\} = \{\eta, H\} = 0$, is due to the commutativity of the τ-flow and the H-flow. This in turn is a consequence of the relation $\{\tau, H\} = 1$. These facts are a straightforward consequence of Darboux's theorem[18, 40], in which H is initially chosen as one of the canonical coordinates.

In our analysis we will assume that the constructive program of Darboux's theorem has been followed, so that (η, τ, E) are canonical. We will also let the surface $\tau = 0$, which intersects the energy surfaces $H = E$ in the surfaces of section, serve as a branch cut for $\eta(x, p)$ and $\tau(x, p)$, so that these functions are single-valued on phase space. Note that $\tau(x, p)$ is time-independent, even though it signifies elapsed time.

It is also useful to consider branches of the functions η and τ other than the principal one. In particular, let $\eta = F(\eta', E')$ specify (say) the m-th return map in the surface of section, and let $T_{ret}(\eta', E')$ be the corresponding return time. Then we introduce the alternative branches of η and τ, given by

$$\hat{\eta}(\eta, E) \;=\; F(\eta, E),$$
$$\hat{\tau}(\eta, \tau, E) \;=\; \tau - T_{ret}(\eta, E), \tag{68}$$

which are the branches one obtains, not by following an orbit from a given point of phase space back to the most previous intersection with the surface of section, but rather back to the most previous and then forward in time m further intersections. The variables $(\hat{\eta}, \hat{\tau}, E)$ are also canonical variables on phase space, because of the preservation of Poisson brackets under Hamiltonian flows.

We may now write down the Bs, in obvious analogy to Eq. (45). We have

$$B_\eta(\eta, \tau, E, \eta', \tau', E') \;=\; \eta - \hat{\eta}(\eta', E'),$$
$$B_\tau(\eta, \tau, E, \eta', \tau', E') \;=\; \tau - \hat{\tau}(\eta', \tau', E') - t,$$
$$B_E(\eta, \tau, E, \eta', \tau', E') \;=\; E - E'. \tag{69}$$

The first $2f - 1$ of these constitute the B_1s and the last is B_2. These Bs commute and have the property that every point on L_U satisfies $B = 0$ for some m, and conversely; if (x, p, x', p') is such a point, m is the number of times the orbit connecting (x', p') with (x, p) crosses the surface of section. Therefore the intersections of L_I with L_U are given by $A = B = 0$ for some m, and points on one of these intersections are on a periodic orbit which crosses the surface of section m times. This m value must be divisible by n, the number of iterations of the primitive periodic orbit.

The amplitude determinant, which involves a $(2f - 1) \times (2f - 1)$ matrix of doubled Poisson brackets, may now be computed. Partitioning this matrix according to $2f - 1 = (2f - 2) + 1$, we have

$$\{A_1, B_1\}_D = \begin{pmatrix} \Gamma(\mathbf{I} - \dfrac{\partial \tilde{F}}{\partial \eta}) & \Gamma \dfrac{\partial T_{ret}}{\partial \eta} \\[2ex] -\dfrac{\partial \tilde{F}}{\partial E} & \dfrac{\partial T_{ret}}{\partial E} \end{pmatrix}, \tag{70}$$

where the tilde represents the transpose and where Γ is the $(2f - 2) \times (2f - 2)$ constant matrix representing the cosymplectic form. This matrix is to be evaluated on the periodic orbit, where $\eta = \eta'$ and $E = E'$.

This matrix may be simplified. On the periodic orbit we have $\eta = F(\eta, E)$ and $t = nT = T_{ret}(\eta, E)$, so we can eliminate η and solve for the primitive period T as a function of E. Taking differentials and solving for dT/dE, we have

$$n\frac{dT}{dE} = \frac{\partial T_{ret}}{\partial E} + \frac{\partial \tilde{F}}{\partial E}\left(\mathbf{I} - \frac{\partial \tilde{F}}{\partial \eta}\right)^{-1}\frac{\partial T_{ret}}{\partial \eta}. \tag{71}$$

This suggests that we multiply the first row of the matrix of Eq. (69) by the row vector $(\partial \tilde{F}/\partial E)(\mathbf{I} - \partial \tilde{F}/\partial \eta)^{-1}\Gamma^{-1}$ on the left, and add to the second row. We also write $\partial F/\partial \eta = \mathbf{M}^n$, where $\mathbf{M}$ is the linearized symplectic return map in the neighborhood of the primitive periodic orbit. The result is

$$\det\{A_1, B_1\}_D = n\frac{dT}{dE}\det(\mathbf{M}^n - \mathbf{I}). \tag{72}$$

It is now easy to write out the desired trace. The result is

$$\mathrm{Tr}\,U(t) = \sum_n \exp(-iE_n t/\hbar) = \frac{1}{\sqrt{2\pi i\hbar}}\sum \frac{T\exp\left[\dfrac{i}{\hbar}R(t) - i\mu\dfrac{\pi}{2}\right]}{\left|n\dfrac{dT}{dE}\right|^{1/2}|\det(\mathbf{M}^n - \mathbf{I})|^{1/2}}, \tag{73}$$

where E_n are the energy eigenvalues. The second sum is taken over all periodic orbits of period t, for which $R(t) = nR(T)$ is Hamilton's principal function, evaluated around the orbit. This formula generalizes Tabor's result to the case that H is conserved. We note that in one degree of freedom, in which the surface of section is vacuous, the determinant factor in the denominator is simply replaced by unity.

Equation (73) is easily converted into a formula for the density of states, by performing a Fourier transform in time of the Van Vleck formula by the stationary phase approximation, and then taking the imaginary part. In this way we obtain the Gutzwiller trace formula,

$$\rho(E) = \bar{\rho}(E) + \frac{1}{\pi\hbar} \sum \frac{T \cos\left[n\dfrac{S(E)}{\hbar} - \mu\dfrac{\pi}{2}\right]}{\left|\det(\mathbf{M}^n - \mathbf{I})\right|^{1/2}}, \tag{74}$$

where $\bar{\rho}(E)$ is the average density of states, where the sum is taken over all periodic orbits of energy E, and where $S(E)$ is the reduced action taken around the corresponding primitive periodic orbit. The fact that the Maslov index μ is proportional to the number of iterations of the primitive orbit is not obvious from this derivation (or from Gutzwiller's, either), but is proven by Robbins[22] and Creagh, Robbins, and Littlejohn[23].

It would be more in accordance with the geometrical philosophy of this chapter to examine the time integration in terms of Lagrangian manifolds in the extended, doubled phase space. We have not done this for several reasons, partly to avoid introducing another generalized phase space. Nevertheless, there can be no question that this is the space in which fully to understand the geometrical structure of the Gutzwiller trace formula. For example, the Maslov indices of the trace formula are ordinary Maslov indices on a Lagrangian manifold in this space, precisely in accordance with Maslov's general theory[5]; and the caustics which can occur in the trace formula are also best understood in this space.

A trivial yet interesting point about the transition from Eq. (73) to Eq. (74) by time integration is that the trace of the propagator (a complexified partition function) and the density of states are two representations of the same wave function, whose semiclassical expressions are supported by Lagragian manifolds in the time-energy phase plane. The Lagrangian manifolds in question are the time-energy curves for the periodic orbits, and the periodic orbit sum can be interpreted as a sum over the branches of a WKB wave function. Interesting examples of these Lagrangian manifolds have been presented by Baranger and Davies[41]. These Lagrangian curves can also be regarded as slices through other Lagrangian manifolds in a phase space of higher dimensionality.

4 Conclusions

We will now conclude by commenting on the results presented and raising questions for further investigation.

First let us consider the results of Sec. 2, in which the scalar product $\langle a|b \rangle$ is expressed in terms of intersections of Lagrangian manifolds in phase space. We have worked out the simplest case, in which the Lagrangian manifolds intersect transversally at isolated points. The next simplest case would be the one in which the Lagrangian manifolds intersect in partial or complete tangency, but still at isolated points. This case leads to the standard theory of caustics and catastrophes[17], but it would be interesting to see the geometrical elements involved expressed in terms of the A- and B-foliations. As discussed in Sec. 3, this case would be useful for providing new insights into the caustic structure of Wigner functions, a matter of some interest recently, since Berry[4] has shown that the scars of periodic orbits in phase space are precisely such caustics.

Another case, which we did develop in Sec. 2, is the one in which the A- and B-manifolds intersect in surfaces of dimensionality greater than zero. Here we assumed that there existed some coordinate transformations, replacing both the As and Bs by functions of themselves, such that some number of the new As would coincide with the same number of new Bs. This assumption leads to a simplification of the computation of the scalar product $\langle a|b \rangle$, in that the integrand becomes independent of α_2, as discussed in the derivation of Eq. (27). Although this assumption makes the case we considered a rather special one, it is the simplest case of higher dimensional intersections of Lagrangian manifolds, and it is also the one we need for the Gutzwiller trace formula. It also explains, in a sense, why Gutzwiller is able to do the time integral around the periodic orbit, i.e., why the integrand is independent of time.

But it is not the most general case of higher dimensional intersections of Lagrangian manifolds, as counterexamples will show. Therefore the question is raised, what is the deeper meaning of the existence of a coordinate transformation causing some As and Bs to coincide? The answer seems to be connected with symmetries, in the sense that any observable which can be expressed as a function of only the As, or alternatively of only the Bs, must commute with both of them; and the set of all such functions forms an Abelian group. No doubt the proper way to express the scalar product $\langle a|b \rangle$ of Eq. (27) would be in terms of some original collection of As and Bs, combined with the generators of the symmetry group. This would yield a much more elegant calculation of $\mathrm{Tr}U(t)$, in the case of a time-independent system, than the one we presented in Sec. 3, in that it would not be necessary to work with surface of section coordinates. Instead, we could work with the original xs and ps, and the generator of the symmetry group, $H(x, p) - H(x', p')$.

Perhaps one of our most interesting results, as simple as it is, is the realization of the close connection between the Green's function and the propagator in the surface of section. This shows, in a sense, why the surface of section monodromy matrix occurs in Gutzwiller's formula, while that for the full phase space occurs in Tabor's. In fact, apart from the period T of the orbit in Gutzwiller's formula and the reduction of the dynamics by one degree of freedom, the two formulas are identical. The factor T can be explained as arising from the fact that the surface of section evolution is really governed by H, the real Hamiltonian, and not by the Hamiltonian F for the surface of section evolution. This fact also explains why Gutzwiller's trace formula cannot be represented as a sum over the quasiphases of the nominal quantized surface of section mapping (a tempting idea not pursued because it fails.)

A natural extension of the work of Sec. 3 would be to include the effects of arbitrary symmetry groups. The extreme cases of complete integrability and complete chaos are well known, but intermediate cases are less so. Nonabelian symmetries, such as rotation, would be interesting to incorporate into the geometrical framework presented here; this would build on the work of Strutinskii and Magner[42], but operating from a rather different standpoint. Discrete symmetries in the trace formula have already been dealt with by Robbins[43]. There are also certain improvements which can be made in the elimination of degrees of freedom from the amplitude determinant of the trace formula, when symmetries exist. Finally, one should work in an extended phase space, in order to geometrize the time integrations and place them on an equal footing with all the others. These and other issues will be considered in the future.

Acknowledgements

I am especially grateful to the Institute for Theoretical Physics at the University of California, Santa Barbara, and to its director, Bob Schrieffer, for hosting me while much of this work was done; and to Eric Heller and Martin Gutzwiller for organizing the Workshop on Small Quantum Systems at the Institute. I am also grateful for useful discussions with many people, including Nandor Balazs, Stephen Creagh, Shmuel Fishman, Eric Heller, Martin Gutzwiller, Marcos Saraceno, André Voros, and Michael Wilkinson, and to Alan Weinstein for providing me with references on the theory of Fourier integral operators. I would also like to thank Bill Fine and the Theoretical Physics Institute of the University of Minnesota for further hospitality and support during the completion of this work.

This work was supported by the National Science Foundation under grant number NSF-PHY82-17853, supplemented by funds from NASA, by the Office of Energy Research, Office of Basic Energy Sciences, of the U. S. Department of Energy under contract number DE-AC03-76SF00098, by the National Science Foundation under grant number NSF-PYI-84-51276, and by the Theoretical Physics Institute of the University of Minnesota.

REFERENCES

[1] M. C. Gutzwiller, J. Math. Phys. **8**, 1979(1967); **10**, 1004(1969); **11**, 1791(1970); **12**, 343(1971).

[2] R. Balian and C. Bloch, Ann. Phys. **60**, 401(1970); **63**, 592(1971); **64**, 271(1971); **69**, 76(1972); **85**, 514(1974).

[3] M. Tabor, Physica **6D**, 195(1983).

[4] M. V. Berry, Proc. R. Soc. Lond. **A423**, 219(1989).

[5] V. P. Maslov and M. V. Fedoriuk, *Semi-Classical Approximation in Quantum Mechanics* (Reidel, Boston, 1981).

[6] M. V. Berry and M. Tabor, Proc. R. Soc. Lond. **A349**, 101(1976); J. Phys. A10, 371(1977).

[7]   M. V. Berry, in *Chaotic Behaviour in Deterministic Systems*, edited by G. Iooss, R. G. Helleman, and R. Stora (North Holland, Amsterdam, 1983) p. 171; and in the proceedings of the Les Houches School on Chaos and Quantum Physics, session no. 52, August 1989, to be published by North Holland.

[8]   Alfredo M. Ozorio de Almeida, *Hamiltonian Systems: Chaos and Quantization* (Cambridge University Press, 1988).

[9]   N. L. Balazs and A. Voros, Phys. Rep. **143**, 109(1986).

[10] W. H. Miller, Adv. Chem. Phys. **25**, 69(1974).

[11] R. Abraham and J. Marsden, *Foundations of Mechanics* (Benjamin/Cummings, Reading, Mass., 1978) 2nd ed.

[12] Lars Hörmander, Acta Math. **127**, 79(1971).

[13] J. J. Duistermaat and V. W. Guillemin, Invent. Math. **29**, 39(1975).

[14] Robert J. Blattner and Bertram Kostant, in *Géométrie Symplectique et Physique Mathématique*, proceedings of the Colloques Internationaux du Centre National de la Recherche Scientifique, Aix-en-Provence, no. 237 (Éditions du CNRS, Paris, 1975).

[15] Alan Weinstein, *Lectures on Symplectic Manifolds*, Conference Board of the Mathematical Sciences no. 29 (American Mathematical Society, Providence, R.I., 1976).

[16] V. Guillemin and S. Sternberg, *Geometric Asymptotics* (Amer. Math. Soc., Providence, R.I., 1977).

[17] M. V. Berry and C. Upstill, in *Progress in Optics*, edited by E. Wolf (North Holland, Amsterdam, 1980) v. XVIII, p. 250.

[18] V. I. Arnold, *Mathematical Methods of Classical Mechanics* (Springer, New York, 1978).

[19] I. C. Percival, Adv. Chem. Phys. **36**, 1(1977).

[20] J. B. Delos, Adv. Chem. Phys. **65**, 161(1986).

[21] L. D. Landau and E. M. Lifshitz, *Mechanics* (Pergamon, New York, 1960).

[22] Jonathan M. Robbins, Nonlinearity **4**, 343(1991).

[23] Stephen C. Creagh, Jonathan M. Robbins, and Robert G. Littlejohn, Phys. Rev. A **42**, 1907(1990).

[24] M. V. Berry and N. L. Balazs, J. Phys. **A12**, 625(1979); M. V. Berry, N. L. Balazs, M. Tabor, and A. Voros, Ann. Phys. **122**, 26(1979).

[25] J. H. Van Vleck, Proc. Natn. Acad. Sci. **14**, 178(1928).

[26] L. S. Schulman, *Techniques and Applications of Path Integration* (Wiley, New York, 1981).

[27] M. V. Berry and K. E. Mount, Rep. Prog. Phys. **35**, 315(1972).

[28] J. B. Keller, in *Calculus of Variations and its Applications*, edited by L. M. Graves (Mc Graw-Hill, New York, 1958).

[29] E. J. Heller, Phys. Rev. Lett. **53**, 1515 (1984).

[30] Marcos Saraceno, Ann. Phys. (N.Y.) **199**, 37(1990).

[31] A. Voros, J. Phys. **A21**, 685(1988).

[32] M. C. Gutzwiller, Physica **5D**, 183(1982).

[33] V. I. Man'ko, in *Lie Methods in Optics*, edited by J. S. Mondragón and K. B. Wolf (Springer, New York, 1986) p. 193.

[34] W. H. Miller, J. Phys. Chem. **83**, 960(1979).

[35] J. M. Robbins, S. Creagh, and R. G. Littlejohn, Phys. Rev. **A39**, 2838(1989).

[36] R. G. Littlejohn, Phys. Rep. **13**, 193(1986).

[37] Herbert Goldstein, *Classical Mechanics* (Addison-Wesley, Reading, Mass., 1980) 2nd ed.

[38] A. M. Ozorio de Almeida and J. H. Hannay, J. Phys. **A20**, 5873(1987).

[39] M. V. Berry, Phil. Trans. R. Soc. Lond. **287**, 237(1977).

[40] R. G. Littlejohn, J. Math. Phys. **20**, 2445(1979).

[41] M. Baranger and K. T. R. Davies, Ann. Phys. **177**, 330(1987).

[42] V. M. Strutinskii and A. G. Magner, Sov. J. Part. Nucl. **7**, 138(1976).

[43] J. M. Robbins, Phys. Rev. **A40**, 2128(1989).

ħ-Expansion for quantum trace formulas

P. GASPARD

Centre for Nonlinear Phenomena and Complex Systems,
Université Libre de Bruxelles,
Campus Plaine, Code Postal 231, Blvd du Triomphe,
B-1050 Brussels, Belgium.

Abstract

We present and review several results on the semiclassical quantization of classically chaotic systems. Using Feynman path integrals and the stationary phase method, we develop a new semiclassical theory for quantum trace formulas in classically hyperbolic systems. In this way, we obtain corrections to the Gutzwiller-Selberg trace formulas like asymptotic series in powers of the Planck constant. The coefficients of these series are expressed in terms of Feynman diagrams. We illustrate our method with the calculation of resonances for the wave scattering on hard disks.

1 Introduction

In recent years, important advances have been made in our understanding of the dynamical behaviours which are intermediate between the classical and the quantal regimes [1–7]. The general context of these advances is the current research in mesoscopic physics and chemistry, which is bridging the gap between the microscopic and the macroscopic worlds. Systems which have been particularly studied in this respect include highly excited or reacting atoms and molecules as well as nanometric semiconductor devices, among many other systems [8–19].

In this chapter, we consider quantum mechanics in the semiclassical regime where the actions of the dynamical processes in question are larger than the Planck constant. This eikonal or short-wavelength assumption allows us to solve the Schrödinger equation using information on the classical orbits of a system and more especially on their periodic orbits. In this way, we can understand in detail how classical mechanics emerges from quantum mechanics. Already, at the beginning of quantum mechanics, it has been shown how the classical equations are derived from Schrödinger's equation in the limit of short wavelengths. Feynman's formulation of quantum mechanics was decisive in deriving classical Hamilton variational principle from wave mechanics [20]. These early works were mainly concerned with the first variation of the classical action with respect to perturbations of the orbit. However, this first variation contains too little information to provide an effective method for solving the wave equation.

In the 1970s, Gutzwiller [2] developing previous works by Van Vleck [21], Morette [22], and others, showed that information on the stability of the classical orbits was essential to set up a semiclassical method of calculation of energy levels. Therefore, the second variation of the classical action was required to have access to the stability of the orbits. Orbit stability has been a specialized subject for a long time. However, when dynamical chaos became widely studied in the 1980s, it was commonly recognized that most periodic orbits of typical Hamiltonian systems are unstable of saddle type and that the number of unstable periodic orbits grows exponentially with their period in these chaotic systems [23]. Chaos was therefore characterized by quantities like the Lyapunov exponents and a statistical formalism was developed with new concepts like the Kolmogorov-Sinai entropy per unit time, the Ruelle topological pressure, and zeta functions [23,24].

If Gutzwiller's trace formula incorporates information from the second variation of the classical action we may wonder whether higher order variations would not be necessary to obtain a complete method of semiclassical quantization. In special geodesic flows on surfaces of constant negative curvature, the Gutzwiller trace formula is equivalent to the Selberg trace formula which establishes an exact relationship between the eigenvalues of the wave equation and the classical periodic orbits [2]. However, most hyperbolic systems cannot be reduced to one of these special geodesic flows on a surface of constant curvature.

Very recently, a general scheme to calculate systematically the corrections to Gutzwiller's trace formula has been developed in order to take this feature into account [25]. In this way, a semiclassical expansion in powers of the Planck constant has been obtained. The purpose of this chapter is to describe this new theory which we shall then illustrate with quantum billiards. The theory of the $\hbar$-expansion is expounded in Sec. 2. The treatment of quantum billiards is then presented in Sec. 3. The conclusions and perspectives are discussed in Sec. 4.

2 $\hbar$-Expansion of trace formulas

2.1 Generalities on quantum time evolution

Microscopic nonrelativistic systems are described by Schrödinger's equation

$$i\,\hbar\,\partial_t\,\psi = \sum_{j=1}^{f} -\frac{\hbar^2}{2m_j}\,\frac{\partial^2\psi}{\partial q_j^2} + V(q_1,\ldots q_f)\,\psi\,, \tag{1}$$

where ψ is the wavefunction and where we assume a time-independent Hamiltonian operator of a system with f degrees of freedom. Rescaling the positions, the masses may be set equal to unity ($m_j = 1$).

The potential may either grow indefinitely at large distances or reach a constant value at large separations between the particles. In the first case, the system is bounded and the spectrum of the Hamiltonian is discrete. In the second case, the system is of the scattering type and the Hamiltonian has a continuous energy spectrum besides its possible discrete spectrum. This latter case is very common since nuclei, atoms, or molecules usually ionize, radiate, or dissociate when they are excited at high enough energies.

The time evolution of a bounded or scattering system is controlled by the energy spectrum. For bounded systems with a discrete energy spectrum as well as for scattering systems, it is therefore necessary to solve the energy eigenvalue problem.

In scattering systems, the continuous spectrum is far from structureless. Its structures are revealed by studying the analytic properties of the S-matrix for complex energies. Accordingly, the S-matrix presents poles, or resonances, at complex energies $E_r = \varepsilon_r - i\Gamma_r/2$ with $\Gamma_r > 0$. Each resonance represents a metastable state of the system with a lifetime $\tau_r = \hbar/\Gamma_r$. The properties of the resonances are important in order to understand the reactivity of atomic or molecular systems. In the following, we shall thus be concerned with methods of calculating the resonances as well as the bound state energies.

2.2 Feynman path integrals

In bounded systems, the eigenvalues E_b can be obtained from the level density

$$n(E) = \sum_b \delta(E - E_b) = \text{tr } \delta(E - \hat{H}) = -\frac{1}{\pi} \text{ Im tr } \frac{1}{E + i0 - \hat{H}}, \quad (2)$$

which is given by the trace of a Dirac distribution involving the Hamiltonian operator (1). From known properties of the distributions with real E, the level density can be rewritten in terms of the resolvent of the Hamiltonian as shown in the last equality of (2) [26].

Similarly, the complex energies E_r of the scattering resonances of open systems can be obtained as the poles of a trace function like

$$g(E) = \text{tr}\left(\frac{1}{E - \hat{H}} - \frac{1}{E - \hat{H}_{as}}\right), \quad (3)$$

where E is here complex and $\hat{H}_{as}$ is some reference Hamiltonian like the one describing the asymptotic motion. The subtraction is necessary for the trace function to be defined.

Our purpose will be to calculate the trace of the resolvent by semiclassical methods. Introducing the evolution operator $\hat{U}(T) = \exp(-i\hat{H}T/\hbar)$ of the time-independent Hamiltonian (1), we have that [26]

$$\frac{1}{E - \hat{H}} = \frac{1}{i\hbar} \int_0^\infty dT \, \exp\left(\frac{i}{\hbar}ET\right) \hat{U}(T), \quad (4)$$

when Im $E > 0$. We conclude that we first have to calculate tr $\hat{U}(T)$.

A most convenient starting point is to express this trace as a Feynman path integral [20,27] using a discretization into N small time intervals $\Delta t = T/N \ll 1$ like

$$\operatorname{tr} \hat{U}(T) = \left(\frac{N}{2\pi i\hbar T}\right)^{Nf/2} \int d\mathbf{q}_0 \ldots d\mathbf{q}_{N-1} \exp\left[\frac{i}{\hbar} W(\mathbf{q}_0, \ldots, \mathbf{q}_{N-1})\right], \quad (5)$$

where the action function is given by

$$W(\mathbf{q}_0, \ldots, \mathbf{q}_{N-1}) = \frac{N}{2T} \sum_{n=0}^{N-1} (\mathbf{q}_{n+1} - \mathbf{q}_n)^2 - \frac{T}{N} \sum_{n=0}^{N-1} V(\mathbf{q}_n) + \mathcal{O}(N^{-2}). \quad (6)$$

The multiple integral is evaluated by the stationary phase method keeping all the corrections. The integral is expanded around the points where the phase is stationary, i.e. around the paths where the action — $W = \oint L(\mathbf{q}, \dot{\mathbf{q}})dt$ — is extremal. The equations resulting from this first variation of the action are Newton's equations discretized according to

$$\frac{\mathbf{q}_{n+1} - 2\mathbf{q}_n + \mathbf{q}_{n-1}}{\Delta t^2} = -\frac{\partial V}{\partial \mathbf{q}_n} + \mathcal{O}(\Delta t). \quad (7)$$

Because of the trace, the stationary solutions are closed paths with $\mathbf{q}(0) = \mathbf{q}(T)$. They are of two types:

(i) There are the fixed points for which $\mathbf{q}_n = \mathbf{q}_s$ at all times $n\Delta t$ and such that $\partial_\mathbf{q} V(\mathbf{q}_s) = 0$. The fixed points occur at the critical energies $E_s = V(\mathbf{q}_s)$. In typical systems, they are isolated at the bottom of the potential wells or at their saddles, but they may also be nonisolated in systems like billiards at $E_s = 0$.

(ii) There are the periodic orbits $\{p\}$. For the period T, a given fundamental period T_p may be repeated r times if $T = rT_p$. In anharmonic potentials, the fundamental periods vary with the energy: $T_p = T_p(E)$. When the time T is fixed we shall therefore find a discrete set of periodic orbits at the different energies $\{E_{pr}\}$ such that $T_p(E_{pr}) = T/r$ (see Fig. 1).

The explicit evaluation of the Feynman path integral at the stationary solutions involves the matrix of the second derivatives of the action W. Differences arise from whether the classical solution is a fixed point or a periodic orbit, and from whether the solution is isolated or not. This last question has been emphasized by Berry and Tabor [28]. In hyperbolic systems where the periodic orbits are all unstable of saddle type like the disk scatterers, the periodic orbits are isolated. However, the periodic orbits are nonisolated in integrable systems where they are of centre type and appear in continuous families [28]. Such families of marginally stable periodic orbits also arise in billiards as discussed in Sec. 3. In the following, we shall develop our systematic method for the case of an isolated and unstable periodic orbit.

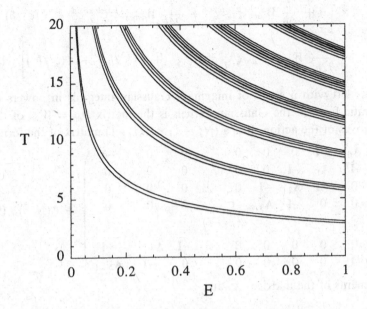

Fig. 1. Period versus energy for the periodic orbits of a point particle trapped between three disks of unit radius ($a = 1$) fixed at the vertices of an equilateral triangle of sides $R = 6$. The lowest fundamental periods and their repetitions are displayed for the motion in the symmetry reduced domain. The periods depend on the energy according to $L(m/2E)^{1/2}$ where L is the length of the orbit (here, $m = 1$).

2.3 Periodic orbit contribution

Periodic orbits are one-dimensional solutions of Newton's equations. Accordingly, one of the discretized positions takes an arbitrary value, i.e. q_{01}, while all the other positions are then determined by Newton's equations and the fixed value of T. The action function (6) is expanded in Taylor series around the classical solution. The $(Nf - 1)$-component vector $\{\xi^a\} = \{\mathbf{q}_n - \mathbf{q}_{\text{cl}}(n\Delta t)\}_{n=0}^{N-1}$ denotes the separation of the path with respect to the classical solution. An index like a will therefore stand for a double index $a = (i, m)$ where the first index i refers to the ith space component of $\mathbf{q}$ ($i = 1, \ldots, f$) while the second index m refers to the time $t = m\Delta t$ ($m = 0, \ldots, N - 1$). $W_{,a\ldots z}$ denotes the partial derivative $\partial_{\xi^a} \ldots \partial_{\xi^z} W$ evaluated at the classical solution $\xi^a = 0$ and W_{cl} is the corresponding classical action. The convention of summation over repeated indices is adopted. We have [25]

$$\text{tr } \hat{U}(T)\Big|_{\text{p.o.}} = \sum_{p,r} \left(\frac{N}{2\pi i\hbar T}\right)^{Nf/2} \exp\left(\frac{i}{\hbar}W_{\text{cl}}\right) \int dq_{01} \, d^{Nf-1}\xi \, \exp\left(\frac{i}{2\hbar}W_{,ab}\,\xi^a\xi^b\right)$$

$$\times \left[1 + \frac{i}{6\hbar} W_{,abc} \, \xi^a \xi^b \xi^c + \frac{i}{24\hbar} W_{,abcd} \, \xi^a \xi^b \xi^c \xi^d + \mathcal{O}(\xi^5/\hbar) \right.$$

$$\left. - \frac{1}{72\hbar^2} W_{,abc} \, W_{,def} \xi^a \xi^b \xi^c \xi^d \xi^e \xi^f + \mathcal{O}(\xi^7/\hbar^2) + \mathcal{O}(\xi^9/\hbar^3) \right]. \quad (8)$$

We are thus left with a series of imaginary Gaussian integrals in powers of ξ. The quadratic form of the Gaussian kernels is the matrix $D_{ab} = W_{,ab}$ of the second derivatives of the action. It is a $(Nf - 1) \times (Nf - 1)$ matrix of the form

$$\mathbf{D} = \frac{N}{T} \begin{pmatrix}
\tilde{\mathbf{A}}_0 & -\tilde{\mathbf{I}} & 0 & 0 & 0 & \cdots & 0 & 0 & -\tilde{\mathbf{I}} \\
-\tilde{\mathbf{I}}^T & \mathbf{A}_1 & -\mathbf{I} & 0 & 0 & \cdots & 0 & 0 & 0 \\
0 & -\mathbf{I} & \mathbf{A}_2 & -\mathbf{I} & 0 & \cdots & 0 & 0 & 0 \\
0 & 0 & -\mathbf{I} & \mathbf{A}_3 & -\mathbf{I} & \cdots & 0 & 0 & 0 \\
\vdots & \vdots & \vdots & \vdots & \vdots & \ddots & \vdots & \vdots & \vdots \\
0 & 0 & 0 & 0 & 0 & \cdots & -\mathbf{I} & \mathbf{A}_{N-2} & -\mathbf{I} \\
-\tilde{\mathbf{I}}^T & 0 & 0 & 0 & 0 & \cdots & 0 & -\mathbf{I} & \mathbf{A}_{N-1}
\end{pmatrix} + \mathcal{O}(N^{-2}), \quad (9)$$

where the elements of the matrices $\mathbf{A}_n$ are

$$A_{ij} = 2\,\delta_{ij} - \left(\frac{T}{N}\right)^2 \frac{\partial^2 V}{\partial q_i \partial q_j}(t), \quad (10)$$

with $t = n\Delta t$ and $i, j = 1, \ldots, f$. The matrix $\tilde{\mathbf{A}}_0$ is the same matrix as (10) but for $i, j = 2, \ldots, f$ and at $t = 0$. $\mathbf{I}$ denotes the $f \times f$ identity matrix while $\tilde{\mathbf{I}}$ is the $f \times (f - 1)$ matrix formed by removing the first line of a $f \times f$ identity matrix. T denotes the transpose. The corrections in N^{-2} are negligible since we shall take the limit $\Delta t \to 0$ at the end of the calculation and we shall keep only the expressions forming standard integrals over $T = N\Delta t$.

The first term in (8) contains the imaginary Gaussian integral

$$\int d^{Nf-1}\xi \, \exp\left(\frac{i}{2\hbar} D_{ab} \, \xi^a \xi^b \right) = \left[\frac{(2\pi i\hbar)^{Nf-1}}{\det \mathbf{D}} \right]^{1/2}, \quad (11)$$

which involves the following determinant [29]

$$\det \mathbf{D} = -\left(\frac{N}{T}\right)^{Nf} \dot{q}_{01}^2 \, \partial_E T \, \det[\mathbf{M}(T) - \mathbf{I}] \, [1 + \mathcal{O}(N^{-1})], \quad (12)$$

where $\mathbf{M}(T)$ is the linearized symplectic return map in the neighborhood of the prime periodic orbit p and over the full period T. Introducing the stability eigenvalues $\{\Lambda^{(k)}(T)\}$ of the periodic orbit p over the time T we have

$$\det[\mathbf{M}(T) - \mathbf{I}] = \prod_{k=1}^{f-1}\left[\Lambda^{(k)}(T) + \Lambda^{(k)}(T)^{-1} - 2 \right]. \quad (13)$$

These stability eigenvalues concern the perturbations which are transverse to the flow.

In the next terms of (8), higher derivatives appear which are given by

$$\frac{\partial^m W}{\partial q_{i_1}(t)\dots\partial q_{i_m}(t)} = -\frac{T}{N}\frac{\partial^m V}{\partial q_{i_1}\dots\partial q_{i_m}}(t) + \mathcal{O}(N^{-2}), \tag{14}$$

and zero otherwise. The higher integrals are performed like Gaussian averages of the powers of ξ. They involve the inverse matrix $\mathbf{D}^{-1}$ which exists for unstable isolated periodic orbits. Odd powers are vanishing. For even powers, we have [30]

$$\int d^{Nf-1}\xi\,\exp\left(\frac{i}{2\hbar}D_{ab}\,\xi^a\xi^b\right)\xi^{c_1}\xi^{c_2}\dots\xi^{c_{2L-1}}\xi^{c_{2L}}$$

$$= \left[\frac{(2\pi i\hbar)^{Nf-1}}{\det\mathbf{D}}\right]^{1/2}(i\hbar)^L\left[(\mathbf{D}^{-1})^{c_1c_2}\dots(\mathbf{D}^{-1})^{c_{2L-1}c_{2L}} +\dots\right], \tag{15}$$

where the latter dots denote all the other terms obtained by grouping the indices two by two.

Remembering the convention on the indices $a = (i, m)$ and $b = (j, n)$ and recovering the continuous time limit, the inverse matrix can be expressed as $(\mathbf{D}^{-1})^{ab} = G_{ij}(m\Delta t, n\Delta t)$ with $i, j = 1,\dots f$ in terms of the classical Green function associated with the periodic orbit around which the Gaussian integrals are evaluated. This Green function is a solution of

$$\frac{d^2}{dt^2}G_{ij}(t, t') + \frac{\partial^2 V}{\partial q_i\partial q_k}[\mathbf{q}(t)]\,G_{kj}(t, t') = -\delta_{ij}\,\delta(t - t'). \tag{16}$$

The fact that a line and a column are missing from $\mathbf{D}$ implies that the Green function must satisfy the following particular boundary conditions

$$G_{1j}(t_0, t') = G_{1j}(t_0 + T, t') = 0,$$
$$G_{ij}(t_0, t') = G_{ij}(t_0 + T, t'), \quad\text{for}\quad i = 2,\dots,f;\quad j = 1,\dots,f, \tag{17}$$

where t_0 is the initial time corresponding to the arbitrary coordinate q_{01}.

Gathering the results, the correction factor to the leading term takes the form

$$1 + \frac{i\hbar}{8}\frac{1}{T}\int_0^T dt_0\int_{t_0}^{t_0+T} dt\,\frac{\partial^4 V(t)}{\partial q_i\partial q_j\partial q_k\partial q_\ell}\,G_{ij}(t, t)\,G_{k\ell}(t, t)$$

$$+ \frac{i\hbar}{24}\frac{1}{T}\int_0^T dt_0\int_{t_0}^{t_0+T} dt\,dt'\,\frac{\partial^3 V(t)}{\partial q_i\partial q_j\partial q_k}\frac{\partial^3 V(t')}{\partial q_\ell\partial q_m\partial q_n}\left[3\,G_{ij}(t, t)\,G_{k\ell}(t, t')\,G_{mn}(t', t')\right.$$

$$\left. + 2\,G_{i\ell}(t, t')\,G_{jm}(t, t')\,G_{kn}(t, t')\right] + \mathcal{O}(\hbar^2). \tag{18}$$

Diagrams are associated with these integrals according to the following rules [30–32]. A vertex with m legs is associated with each $-\partial^m V(t)/\partial q_{i_1}\dots\partial q_{i_m}$. A line is associated with each Green function $G_{ij}(t, t')$ which joins two free legs either of the same vertex or between two different vertices. Integrals are then performed over the times of the different vertices. Finally, a time average $T^{-1}\int_0^T dt_0$ is carried out over the initial time t_0.

These diagrams have the following properties. In the Taylor expansion (8),

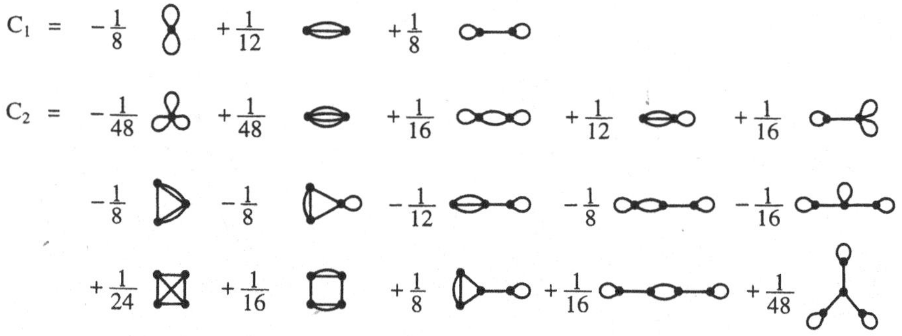

Fig. 2. Feynman diagrams and their numerical weights which enter into the first two coefficients of the semiclassical expansion of the logarithm of the amplitude associated with each periodic orbit in the trace formula (19).

we find terms with V vertices and ℓ legs. The number of legs ℓ is always even otherwise the term is vanishing. If there are $\ell = 2L$ legs there are $(2L-1)!!$ different ways of joining them two by two with L lines. Therefore many diagrams are generated but many of them are equivalent in the sense that they have the same topology and, as a consequence, the same numerical value. The difference $L - V$ gives the power of the Planck constant to which the diagram contributes. In the series, we find connected and disconnected diagrams. A very important property is that the series in the Planck constant involving all the diagrams can be transformed into the exponential of a series involving only the connected diagrams. Fig. 2 shows the connected diagrams and their numerical weights appearing in the coefficients of this new series in powers of $i\hbar$. The sign of the numerical weight of a diagram with V vertices is $(-)^V$.

Finally, the periodic orbits contribute to the trace the terms [25]

$$\text{tr}\,\hat{U}(T)\bigg|_{\substack{\text{p.o.} \\ T=rT_p}} = \sum_{p,r} \frac{T_p \exp\left[\frac{i}{\hbar}rW_p - i\frac{\pi}{2}r\mu_p + i\frac{\pi}{4}\,\text{sgn}\,\partial_E T_p + \sum_{n=1}^{\infty}(i\hbar)^n C_n(rT_p)\right]}{|2\pi\hbar\, r\,(\partial_E T_p)\,\det(\mathbf{M}_p^r - \mathbf{I})|^{1/2}},$$

(19)

where $W(T) = rW_p$. μ_p is the Morse index of the periodic orbit p defined as the number of changes of sign of the Jacobian [2]

$$J(t) = \frac{\partial(q_1,\ldots,q_f)}{\partial(t,p_{02},\ldots,p_{0f})},$$

(20)

along the unstable or the stable manifold of the periodic orbit of initial momentum $\mathbf{p}_0 = \dot{\mathbf{q}}_0$. C_n are the corrections given by the diagrams of Fig. 2. For geodesic flows on surfaces of constant negative curvature, all these corrections are vanishing $C_n = 0$ so that a Selberg trace formula is then recovered [2].

Defining $C_0(T)$ as the logarithm of the amplitude with the phase included, Eq. (19) can be written as

$$\text{tr } \hat{U}(T)\Big|_{\substack{\text{p.o.} \\ T=rT_p}} = \sum_{\substack{p,r \\ T=rT_p}} \exp\left[\frac{i}{\hbar}W(T) + \sum_{n=0}^{\infty} (i\hbar)^n C_n(T)\right]. \qquad (21)$$

We now turn to the trace of the resolvent of the Hamiltonian [cf. (4)]. The integral over the time T is again performed here by the stationary phase method. The critical points are: (1) $T = 0$ which leads to the Fermi-Thomas-Weyl-Wigner average level density; (2) the periods $T = rT_p$. The fixed points contribute extra terms. We shall here restrict ourselves to the contribution from the periodic orbits.

Each argument of the exponentials in (21) is expanded in a Taylor series around the stationary solutions $T = rT_p(E)$ determined from the condition $\partial_T W + E = 0$. We introduce the reduced action $S(E) = ET + W(T)$, from which $T_p = \partial_E S_p$ results [26]. Evaluating the Gaussian integrals as in (8) we obtain [25]

$$\text{tr } \frac{1}{E - \hat{H}}\Big|_{\substack{\text{p.o.} \\ T=rT_p}} = \sum_{\substack{p,r \\ T=rT_p}} \frac{T_p \exp\left(\frac{i}{\hbar}rS_p - i\frac{\pi}{2}r\mu_p\right)}{i\hbar|\det(\mathbf{M}_p^r - \mathbf{I})|^{1/2}}$$

$$\times \exp\left\{i\hbar\left[C_1(rT_p) - \frac{(\partial_E C_0)^2 + (\partial_E^2 C_0)}{2r(\partial_E^2 S_p)}\right.\right.$$

$$\left.\left. + \frac{(\partial_E^3 S_p)^2}{6r(\partial_E^2 S_p)^3} - \frac{\partial_E^4 S_p}{8r(\partial_E^2 S_p)^2}\right] + \mathcal{O}(\hbar^2)\right\}, \qquad (22)$$

where $S_p = S(E)/r$ and C_1 is the first correction of Fig. 2 which is evaluated at $T = rT_p(E)$ together with C_0. Eq. (22) gives the first correction in $\hbar$ to the Gutzwiller trace formula. The next corrections can be obtained systematically.

2.4 Fixed point contribution

The contribution from the fixed points is handled in a similar way. Fixed points are zero-dimensional solutions so that there is no need to remove one coordinate as we did for the periodic orbits. The vector $\{\xi^a\}$ is therefore of dimension Nf since the integral over q_{01} may also be expanded around the stationary value which is q_{s1}. Accordingly, the matrix D_{ab} of the second derivative of the action is now $Nf \times Nf$. Its determinant is

$$\det \mathbf{D} = \left(\frac{N}{T}\right)^{Nf} \prod_{k=1}^{f} \left(2\sinh\frac{\lambda_k T}{2}\right) [1 + \mathcal{O}(N^{-1})], \qquad (23)$$

where $\{\lambda_k\}$ are the Lyapunov exponents of the fixed points which are the eigenvalues of $\partial_{q_i}\partial_{q_j}V(\mathbf{q}_s)$.

The classical Green functions are solutions of Eq. (16) with $\mathbf{q}(t) = \mathbf{q}_s$ but here with the boundary condition $G_{ij}(t + T, t') = G_{ij}(t, t')$ for $i, j = 1, \ldots, f$. The correction terms are then of the same form as (18) and the diagrams are the same as in Fig. 2.

The contribution of the fixed points to the trace of the resolvent is then obtained after the integral over T is performed in (4). It turns out that this contribution disappears from the trace formula if the stationary phase method is used to integrate (4).

2.5 Average level density

As mentioned above, $T = 0$ is a critical point for the integral over T in (4). This contribution to the trace of the resolvent is well known to be due to paths of zero length as considered by Fermi, Thomas, Weyl, Wigner, and others [33,34]. In itself, this term of the trace has a divergent real part and only its imaginary part is finite in bounded systems. Like for the previous terms there are corrections in powers of the Planck constant. For smooth potentials, these are given by [34]

$$
\begin{aligned}
n_{\mathrm{av}}(E) \;=\; \int \frac{d^f q\, d^f p}{(2\pi\hbar)^f} \Bigg\{ & \delta(E - H_{\mathrm{cl}}) \\
+ \; \hbar^2 \Bigg[& \left(-\frac{1}{8m} \sum_k \frac{\partial^2 V}{\partial q_k^2} \right) \delta''(E - H_{\mathrm{cl}}) + \left(\frac{1}{24m} \sum_k \left(\frac{\partial V}{\partial q_k} \right)^2 \right. \\
+ \; \frac{1}{24m^2} & \left. \sum_{k,\ell} p_k\, p_\ell \frac{\partial^2 V}{\partial q_k \partial q_\ell} \right) \delta'''(E - H_{\mathrm{cl}}) \Bigg] + \mathcal{O}(\hbar^4) \Bigg\},
\end{aligned}
\tag{24}
$$

where δ'' and δ''' denote derivatives of the Dirac distribution to be taken after the integrals over $(\mathbf{q}, \mathbf{p})$. These terms are obtained using the Weyl–Wigner representation of operators as functions over the $(\mathbf{q}, \mathbf{p})$-space. In systems with Dirichlet, Neumann, or more complicated boundary conditions, corrections also appear which are in odd powers of $\hbar$ with respect to the leading term.

2.6 Semiclassical oscillations in the spectral structures

In summary, the semiclassical evaluation of the level density gives

$$
n(E) \;=\; n_{\mathrm{av}}(E) + n_{\mathrm{p.o.}}(E),
\tag{25}
$$

where n_{av} is the average level density (24). The contribution $n_{\mathrm{p.o.}}$ from the periodic orbits is of the form [2]

$$
n_{\mathrm{p.o.}}(E) \;=\; \sum_{p,r} A_{p,r}(E) \cos\left[\frac{1}{\hbar} r S_p(E) + \phi_{p,r}(E) \right],
\tag{26}
$$

where the amplitudes $A_{p,r}$ and the phases $\phi_{p,r}$ contain all the contributions in powers of the Planck constant $\hbar^n$ ($n \geq 0$). Let us also remark that the expression we calculated only gives the effects of the isolated periodic orbits. Nonisolated periodic orbits contribute terms which have the same dependency on $S_p(E)$ as in (26) but other amplitudes and phases. To understand (25), it is important to note that n_{av} contains terms in $\hbar^{-f}, \hbar^{-f+1}, \ldots$; and $n_{p.o.}$ contains terms in $\hbar^{-1}, \ldots$. In the case of integrable systems, the nonisolated periodic orbits would contribute terms in $\hbar^{-(f+1)/2}, \ldots$.

The mean level density is the most common among the two contributions to the exact level density since it is the leading one which is of crucial importance in equilibrium statistical mechanics. The mean level density is monotonous in energy. On the other hand, the contributions from the periodic orbits oscillate with energy. These spectral oscillations have been observed experimentally in several atomic and molecular systems such as hydrogen-like atoms in a magnetic field near their ionization threshold or in continuous bands of electronically excited and dissociating molecules [10–14]. These oscillations are also present in large but finite systems like nuclei [35] or sodium clusters [36] where they are evidence for energy shells or supershells of the constituent fermions. The systematic analysis of these semiclassical structures in energy spectra is becoming a standard method for obtaining information on the periods of the classical orbits emerging from quantum dynamics [10–14].

In the next section, we shall see that the semiclassical trace formulas can also be used to make predictions on the energy levels and on the resonances.

2.7 Semiclassical calculation of the eigenenergies

In order to calculate the energy levels or the resonances, it is necessary to perform the summation over the periodic orbits and, especially, over their repetitions. It was shown that this summation over r leads to the introduction of the Selberg or of the Ruelle zeta functions at the condition that the corrections in $(i\hbar)^n$ are neglected which is an approximation for general hyperbolic systems. From (22) we get [7,37–41]

$$\operatorname{tr} \left. \frac{1}{E - \hat{H}} \right|_{\text{p.o.}} = \frac{\partial}{\partial E} \ln Z(E) + \mathcal{O}(\hbar^0), \tag{27}$$

the logarithmic derivative of Z being in $\hbar^{-1}$. The Selberg zeta function is defined by

$$Z(E) = \prod_{m_1,\ldots,m_{f-1}=0}^{\infty} \prod_{p} \left[1 - \frac{\exp\left(\frac{i}{\hbar} S_p - i\frac{\pi}{2}\mu_p\right)}{\prod_{k=1}^{f-1} |\Lambda_p^{(k)}|^{1/2} \Lambda_p^{(k) \, m_k}} \right], \tag{28}$$

where $\Lambda_p^{(k)} = \Lambda_p^{(k)}(rT_p)^{1/r}$ are the stability eigenvalues over the fundamental periods T_p (cf. 13). Since the energy levels or the scattering resonances are poles of the trace of the resolvant in (27) these poles are zeros of the Selberg zeta function (28) because of the logarithmic derivative. In this way, approximate values can be obtained for the eigenenergies.

This semiclassical method has been applied to He and two-electron ions like H^- where the classical dynamics of a collinear configuration is known to be chaotic [15–17]. The energies of a series of doubly excited autoionizing states have been obtained within 1–2% of their quantum mechanical values [15–17]. In the next section, we shall present the application of the zeta function technique to the scattering of a point particle on hard disks.

In bounded systems, however, the energies are real rather than complex and another method was proposed in which the eigenvalues are found as the real zeros of the so-called Riemann–Siegel lookalike formula or functional equation that incorporates information on the average staircase function together with an adequately truncated Selberg zeta function. We refer the reader to Refs. [42–45] for more information on this technique.

Methods using the Selberg zeta function (28) neglect the higher corrections in the Planck constant. The inclusion of these corrections will, in general, modify the form of the function $Z(E)$ according to the dependency of the corrections C_n on the repetition number r. In hyperbolic systems, the classical Green functions $G_{ij}(t, t')$ exhibit the clustering property, i.e. they decrease to zero when $|t - t'| \to \infty$ together with $T \to \infty$ The decrease is often exponential and controlled by the Lyapunov exponents of the periodic orbit. According to this property, the coefficients $C_n(r T_p)$ will, in general, be proportional to the repetition number r plus some possible subdominant behaviours in $r^\gamma \exp(-\lambda_k r T_p)$. An example of resummation in the two-disk scatterer [25] will be presented in the next section where we shall illustrate the previous theoretical methods in the case of billiards.

3 Quantum billiards

3.1 Wave equation

Billiards occupy a special position among the dynamical systems because of the relative simplicity of their dynamics where hard particles fly in free motion between elastic collisions. The Schrödinger equation is therefore identical with the free particle equation

$$(\Delta_f + k^2)\, \psi = 0\,, \tag{29}$$

where $k = \hbar^{-1}(2mE)^{1/2}$ is the wavenumber. The specificity of the system is expressed by the boundary condition such as the Dirichlet boundary condition $\psi = 0$ taken on the impenetrable border of the billiard.

The border of the billiard is formed by the walls of the cavity for billiards like the Bunimovich stadium [46], the Lorentz [47] or the Sinai billiards [48]. For the hard sphere or hard disk gases, the border is composed not only by the walls of the container but also of the hypersurface of the multiple collisions between the particles [49]. Moreover, the bosonic or fermionic character of the particles implies that the wavefunction is either symmetric or antisymmetric for the exchange of the particles, spin components included.

Billiards may be bounded or open. In the later case, the billiard is a scattering system and its volume is infinite. The quantum dynamics can then be studied in terms of the scattering resonances which are the complex poles of the S-matrix or of the Green function. There are also semibounded billiards like the hyperbola billiard which have an infinite area but still a discrete energy spectrum. A billiard may be integrable like the circular and elliptic billiards or hyperbolic if all its periodic orbits are unstable.

Microwave cavities as well as electronic nanometric semiconductor devices have been built in order to investigate how the transition from regular to chaotic ray motions influences the wave properties of the billiards [18,19].

3.2 Multiple scattering expansion

The resolution of the Helmholtz equation (29) requires the Green function satisfying

$$(\Delta_f + k^2)\,\mathscr{G} = \frac{2m}{\hbar^2}\,\delta^f(\mathbf{q}-\mathbf{q}')\,, \tag{30}$$

together with the Dirichlet boundary condition that $\mathscr{G}(\mathbf{q},\mathbf{q}';k)=0$ for $\mathbf{q}$ belonging to the border of the billiard. Introducing the free Green function

$$\mathscr{G}_0(\mathbf{q},\mathbf{q}';k) = \frac{m}{2i\hbar^2}\left(\frac{k}{2\pi|\mathbf{q}-\mathbf{q}'|}\right)^{(f/2)-1} H^{(1)}_{(f/2)-1}(k|\mathbf{q}-\mathbf{q}'|)\,, \tag{31}$$

where $H^{(1)}_\nu(z)$ is the first Hankel function, the full Green function $\mathscr{G}$ can be obtained by a multiple scattering expansion like [35,50]

$$\hat{\mathscr{G}} = \sum_{m=0}^{\infty} \hat{Q}^m\,\hat{\mathscr{G}}_0 = (1-\hat{Q})^{-1}\,\hat{\mathscr{G}}_0\,. \tag{32}$$

$\hat{Q}$ is the operator [50]

$$(\hat{Q}f)(\mathbf{q}) = 2\oint dS'\,\frac{\partial}{\partial n'}\mathscr{G}_0(\mathbf{q},\mathbf{q}')\,f(\mathbf{q}')\,, \tag{33}$$

acting on functions f defined on the border of the billiard. dS' is the infinitesimal element of surface of the border. $\partial/\partial n$ is the gradient with respect to the normal pointing toward the interior of the billiard.

3.3 Semiclassical evaluation of the level density

In bounded billiards, the level density is then given by the trace of the Green function according to (2). The integrals of the multiple scattering expansion (32) can be evaluated by the stationary phase method. The free Green functions and their gradients are expanded in asymptotic series in k^{-1}. The stationary solutions turn out to be the paths for which the total length is extremal. The method can be developed systematically for classically chaotic billiards to evaluate semiclassically the level density which takes the form [48,49]

$$n(E) \; = \; n_{\mathrm{av}}(E) \; + \; n_{\mathrm{p.o.}}^{(\alpha)}(E) \; + \; n_{\mathrm{p.o.}}^{(\beta)}(E) \, . \tag{34}$$

The first term is the standard mean level density

$$n_{\mathrm{av}}(E) \; = \; \frac{m\mathscr{V}}{2\pi\hbar^2\Gamma(f/2)} \left(\frac{mE}{2\pi\hbar^2}\right)^{(f/2)-1} + \; \mathcal{O}\!\left[\hbar^{-(f-1)}\,E^{(f-3)/2}\right] , \tag{35}$$

where $\mathscr{V}$ is the f-volume of the billiard. The next term involves the $(f-1)$-surface of the border while the following terms in $\hbar^{f-k}$ with $k = 1, 2, \ldots$ contain the $(f-k)$-surfaces integrals of the lower-dimensional curvatures or discontinuities of the border.

Nonisolated periodic orbits. As mentioned above, the nonisolated periodic orbits have an important contribution to the level density as [28,48,49]

$$n_{\mathrm{p.o.}}^{(\alpha)}(E) \; = \; \frac{m}{2\pi^2\hbar^2} \left(\frac{mE}{2\pi^2\hbar^2}\right)^{(f-3)/4} \sum_{\mathbf{M}} \frac{\mathscr{V}_{\mathbf{M}}}{\mathscr{L}_{\mathbf{M}}^{(f-1)/2}} \, \cos\!\left[\hbar^{-1}(2mE)^{\frac{1}{2}}\mathscr{L}_{\mathbf{M}} - \frac{\pi}{4}\,\beta_{\mathbf{M}}\right]$$
$$+ \; \mathcal{O}\!\left[\hbar^{-(f-1)/2}\,E^{(f-5)/4}\right] . \tag{36}$$

$\mathbf{M}$ is a set of f integers labelling the families of nonisolated periodic orbits. $\mathscr{L}_{\mathbf{M}}$ is the length of the nonisolated periodic orbits of type $\mathbf{M}$. $\mathscr{V}_{\mathbf{M}}$ is the volume of the family $\mathbf{M}$ of nonisolated periodic orbits. $\mathscr{V}_{\mathbf{M}}$ decreases when $\mathscr{L}_{\mathbf{M}}$ increases. $\beta_{\mathbf{M}}$ is a phase which depends on the boundary conditions.

Isolated periodic orbits. The next contribution to the level density is due to the unstable periodic orbits. They are countable and their number grows exponentially with their period when the topological entropy is positive. Up to a correction of relative order $\hbar$, the contributions from the isolated periodic orbits can be written in terms of the Selberg zeta function (28)

$$n_{\mathrm{p.o.}}^{(\beta)}(E) \; = \; -\frac{1}{\pi}\,\operatorname{Im}\frac{\partial}{\partial E}\,\ln Z(E) \; + \; \mathcal{O}\!\left(\hbar^0\,E^{-1}\right) , \tag{37}$$

where the leading term is of $\mathcal{O}(\hbar^{-1}E^{-1/2})$. However, the next corrections — which are overlooked in this expression — cannot be neglected at low energy as described below.

Comparing the different contributions to the level density, we see that

$$\frac{n_{\mathrm{p.o.}}^{(\alpha)}(E)}{n_{\mathrm{av}}(E)} \; = \; \frac{n_{\mathrm{p.o.}}^{(\beta)}(E)}{n_{\mathrm{p.o.}}^{(\alpha)}(E)} \; = \; \mathcal{O}\!\left[\hbar^{(f-1)/2}\,E^{-(f-1)/4}\right] . \tag{38}$$

The contribution from the nonisolated periodic orbits is therefore midway between the average density and the oscillations from the isolated orbits [48,49].

A consequence of (38) is that the amplitude of each oscillating term decreases when the number of degrees of freedom increases, as expected.

3.4 Semiclassical calculation of the scattering resonances

Because of (32), the scattering resonances — which are the poles of the Green function — can be calculated as the zeros of the following Fredholm determinant [41,51]

$$0 = \det(1 - \hat{Q}) = \exp\left(-\sum_{m=1}^{\infty} \frac{1}{m} \operatorname{tr} \hat{Q}^m\right)$$

$$= \exp\left[-\sum_{m=1}^{\infty} \frac{2^m}{m} \oint dS_1 \dots dS_m \frac{\partial \mathscr{G}_0}{\partial n_1}(\mathbf{q}_m, \mathbf{q}_1) \dots \frac{\partial \mathscr{G}_0}{\partial n_m}(\mathbf{q}_{m-1}, \mathbf{q}_m)\right], \quad (39)$$

which is a series of multiple integrals carried out over points of the border. Unlike in smooth Hamiltonian systems, the functional determinant can here be expressed in terms of standard integrals rather than Feynman path integrals. In the semiclassical limit of short wavelengths, the integrals can be evaluated by the stationary phase method. Keeping only the leading terms from the imaginary Gaussian integrals shows that $\det(1 - \hat{Q}) \sim Z(E)$ [51].

We have calculated the scattering resonances of the three-disk scatterer at this k^0-approximation which is already very good for strong scatterers [38–41]. We assume that the disks are of unit radius and centered at the vertices of an equilateral triangle of side $R = 6$. Fig. 3 shows the semiclassical scattering resonances in the complex plane of the wavenumber. Because of the threefold symmetry of the scatterer we find resonances belonging to the three representations (A_1, A_2, E) of the symmetry group C_{3v} [38–41]. Superimposed on the semiclassical resonances we have plotted the exact quantum resonances as crosses in a window ranging from $\operatorname{Re} k = 8$ to $\operatorname{Re} k = 18$ and with $\operatorname{Im} k > -0.29$ where the agreement is excellent. At higher but not too high wavenumbers, the agreement is known also to be good but it deteriorates at low wavenumbers where we may expect a failure of the semiclassical approximation. In this region, it is thus important to go beyond the Gutzwiller approximation [25].

3.5 Asymptotics in the wavelength

Using a technique which is very similar to the one developed in Sec. 2 it is possible to calculate the corrections in inverse powers of the wavenumber, k^{-n} playing here the role of $ħ^n$ [25].

There are corrections of different types. Let us suppose that we are concerned with the classical periodic orbit of length L which makes m collisions on the border. We have to collect all the terms of the exponential in (39) which have the phase $\exp(ikL)$. Such terms appear in the integral m where we find the leading

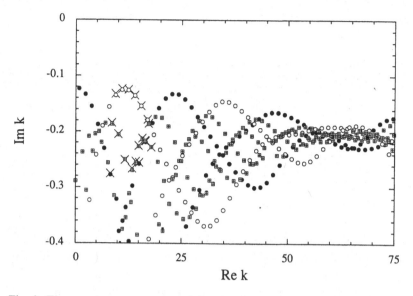

Fig. 3. The quantum resonances of the three-disk scatterer ($R = 6$ and $a = 1$) in the complex plane of the wavenumber and compared with the semiclassical approximation (37) calculated with the curvature expansion [39] in the fundamental domain up to period 6. The crosses are the exact poles [38]. The semiclassical A_1 poles are denoted by filled circles, the A_2 ones by open circles, and the E ones by crossed squares (from Ref. [41]).

contribution with corrections in $k^{-1}, k^{-2}, \ldots$. These corrections are given by diagrams like those in Fig. 2 plus extra diagrams coming from the fact that the integrand already has a nontrivial dependency on the wavenumber. Moreover, we also find corrections in $k^{-1}, k^{-2}, \ldots$ in the integral $m + 1$, corrections in $k^{-2}, k^{-3}, \ldots$ in the integral $m + 2$, and so on. These important corrections are due to paths where one, two, or more segments are vanishing.

This numerical calculation is carried out in Ref. [25] up to the k^{-2}-corrections for the two-disk scatterer when the distance between the centres of the disks is $R = 6$ and for disks of unit radius. Summing over the tail of the series in the repetition number $r \to \infty$ we obtain that the Fredholm determinant (39) vanishes according to [25]

$$0 = 1 - \frac{1}{\Lambda^{1/2}} \exp\left[ikL + ik^{-1}C_1 - k^{-2}C_2 + \mathcal{O}(k^{-3})\right], \qquad (40)$$

where $L = 8$ is the total length of the orbit, $\Lambda = 97.989795$ is its stability eigenvalue, and $C_1 = 0.625000$, $C_2 \simeq -0.75$ are the two first correction coefficients. Table 1 shows the exact quantum mechanical values of the resonances obtained in Ref. [52] and compared with the values of the standard k^0-approximation, of the k^{-1}-approximation, and of the k^{-2}-approximation [25]. Because the corrections are given by inverse powers of k Eq. (40) misses the lowest resonance which is

Table 1. Complex wavenumbers $k = \mathrm{Re}\,k + i\,\mathrm{Im}\,k$ of the lowest A_1 resonances of the two-disk scatterer with interdisk distance $R = 6$ and unit radius.

Resonances	k^0-approx.	k^{-1}-approx.	k^{-2}-approx.	Exact (Ref. [52])
1	0.7854	0.6936	—	0.7259753
	$-0.2866i$	$-0.3303i$	—	$-0.2611556i$
3	2.3562	2.3231	2.3190	2.3274167
	$-0.2866i$	$-0.2907i$	$-0.2738i$	$-0.2788150i$
5	3.9270	3.9071	3.9062	3.9087714
	$-0.2866i$	$-0.2880i$	$-0.2819i$	$-0.2832238i$
7	5.4978	5.4836	5.4833	5.4842448
	$-0.2866i$	$-0.2873i$	$-0.2842i$	$-0.2845793i$
9	7.0686	7.0575	7.0574	7.0578923
	$-0.2866i$	$-0.2870i$	$-0.2851i$	$-0.2853333i$

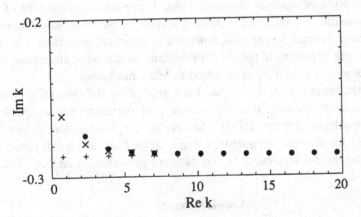

Fig. 4. The A_1 quantum resonances of the two-disk scatterer ($R = 6$ and $a = 1$) in the complex plane of the wavenumber. The crosses are the exact poles [52]. The pluses are the poles at the Gutzwiller k^0-approximation while the filled circles are the poles at the k^{-2}-approximation (40) [25].

outside the domain of validity of (40). Nevertheless, we observe the successive improvements in accuracy which occur on $\mathrm{Re}\,k$ at the k^{-1}-approximation and then on $\mathrm{Im}\,k$ at the k^{-2}-approximation. We are now able to reproduce the structure of the resonances at low wavenumbers which was overlooked at the standard Gutzwiller approximation (see Fig. 4).

4 Conclusion

In this chapter, we have presented and reviewed several results obtained using semiclassical methods in quantum mechanics.

We have described a new method of calculating $\hbar^n$ corrections to quantum trace formulas beyond the Gutzwiller approximation. Although these corrections are vanishing in special geodesic flows on surfaces with a constant negative curvature, they become necessary in general hyperbolic systems for the semiclassical expansion to be complete [25]. The series in the Planck constant are asymptotic series whose domain of validity must be discussed but we think that appropriate truncations of the series can improve the current numerical results on semiclassical quantization, in particular, of doubly excited states of He and H^- [15–17]. A semiclassical theory is also available for the calculation of weak electric or magnetic field effects on the eigenstates [17].

We have here discussed the application of the $\hbar$-expansion to the traces of the propagator and of the Green function. However, the $\hbar$-expansion can also be applied to the propagator and the Green function themselves [27,30]. Tunnelling effects as well as hopping between potential energy surfaces can, in principle, be treated by inclusion of complex classical orbits. Many other applications of the method are possible, for instance, to matrix Hamiltonians, to broken symmetry problems in field theories, to partition functions in statistical mechanics, but also, to Fokker–Planck equations in the weak noise limit, as well as to electromagnetic wave propagation, to acoustics, or to vibrating plate mechanics.

Recently, the $\hbar$-expansion theory has been applied to the classically chaotic three-disk scatterer, showing that the accuracy of periodic-orbit quantisation can be systematically improved [53]. Moreover, the $\hbar$-expansion theory has been extended to the case of quantum maps, as well as to equilibrium-point quantisation where it is equivalent to the standard perturbation method [54].

Acknowledgements

We thank Professors G. Nicolis, I. Prigogine, and S. A. Rice for support and encouragement in this research. The author is "Chercheur Qualifié" at the National Fund for Scientific Research (Belgium). Part of this research was supported by EEC Contract SC1*CT91-0711.

REFERENCES

[1]   G. Casati and J. Ford, Eds., *Stochastic Behavior in Classical and Quantum Hamiltonian Systems*, Lect. Notes Phys. **93** (Springer, Berlin, 1979).

[2]   M. C. Gutzwiller, J. Math. Phys. **12**, 343 (1971); J. Phys. Chem. **92**, 3154 (1988); *Chaos in Classical and Quantum Mechanics* (Springer, New York, 1990).

[3]   F. Haake, *Quantum Signatures of Chaos* (Springer, Berlin, 1991).

[4]   L. E. Reichl, *The Transition to Chaos* (Springer, New York, 1992).

[5] A. M. Ozorio de Almeida, *Hamiltonian systems: Chaos and Quantization* (Cambridge University Press, Cambridge, 1988).

[6] M.-J. Giannoni, A. Voros, and J. Zinn-Justin, Eds., *Chaos and Quantum Physics* (North-Holland, Amsterdam, 1991).

[7] P. Cvitanović, Ed., *Periodic Orbit Theory*, Chaos, Vol. 2, Nos. 1–4 (1992).

[8] W. H. Miller, Adv. Chem. Phys. **25** (1974) 69.

[9] S. A. Rice and P. Gaspard, Israel J. Chem. **30** (1990) 23; S. K. Gray, S. A. Rice, and D. W. Noid, J. Chem. Phys. **84** (1986) 3745; S. K. Gray, S. A. Rice, and M. J. Davis, J. Phys. Chem. **90** (1986) 3470.

[10] R. Schinke and V. Engel, J. Chem. Phys. **93** (1990) 3252.

[11] B. R. Johnson and J. L. Kinsey, Phys. Rev. Lett. **62** (1989) 1607.

[12] H. Friedrich and D. Wintgen, Phys. Rep. **183** (1989) 37.

[13] J. Main, G. Wiebusch, A. Holle, and K. H. Welge, Phys. Rev. Lett. **57** (1986) 2789.

[14] D. Wintgen, Phys. Rev. Lett. **58** (1987) 1589.

[15] G. S. Ezra, K. Richter, G. Tanner, and D. Wintgen, J. Phys. B: At. Mol. Opt. Phys. **24** (1991) L413.

[16] D. Wintgen, K. Richter, and G. Tanner, Chaos **2** (1992) 19.

[17] P. Gaspard and S. A. Rice, Phys. Rev. A **48** (1993) 54.

[18] R. Blümel and U. Smilansky, Phys. Rev. Lett. **60** (1988); 477; *ibid.* **64** (1990) 241; E. Doron, U. Smilansky, and A. Frenkel, Phys. Rev. Lett. **65** (1990) 3072; Physica D **50** (1992) 367.

[19] R. A. Jalabert, H. U. Baranger, and A. D. Stone, Phys. Rev. Lett. **65** (1990) 2442; C. M. Marcus, A. J. Rimberg, R. M. Westervelt, P. F. Hopkins, and A. C. Gossard, Phys. Rev. Lett. **69** (1992) 506.

[20] R. P. Feynman and A. R. Hibbs, *Quantum Mechanics and Path Integrals* (McGraw-Hill, New York, 1965).

[21] J. H. Van Vleck, Proc. Natl. Acad. Sci. USA **14** (1928) 178.

[22] C. Morette, Phys. Rev. **81** (1951) 848.

[23] A. J. Lieberman and A. J. Lichtenberg, *Regular and Stochastic Motion* (Springer, New York, 1983).

[24] D. Ruelle, Commun. Math. Phys. **125** (1989) 239.

[25] P. Gaspard and D. Alonso, Phys. Rev. A **47** (1993) R3468.

[26] M. V. Berry and K. E. Mount, Rep. Prog. Phys. **35** (1972) 315.

[27] L. S. Schulman, *Techniques and Applications of Path Integration* (Wiley, New York, 1981).

[28] M. V. Berry and M. Tabor, J. Phys. A: Math. Gen. **10** (1977) 371.

[29] R. G. Littlejohn, J. Math. Phys. **31** (1990) 2952.

[30] B. DeWitt, *Supermanifolds*, 2nd ed. (Cambridge University Press, Cambridge, 1992).

[31] C. DeWitt-Morette, Ann. Phys. **97** (1976) 367 (1976).

[32] J. Iliopoulos, C. Itzykson, and A. Martin, Rev. Mod. Phys. **47** (1975) 165.

[33] E. P. Wigner, Phys. Rev. **40** (1932) 749.

[34] B. Grammaticos and A. Voros, Ann. Phys. **123** (1979) 359.

[35] R. Balian and C. Bloch, Ann. Phys. **69** (1972) 76.

[36] J. Pederson, S. Bjørnholm, J. Borggreen, K. Hansen, T. P. Martin, and H. D. Rasmussen, Nature **353** (1991) 733.

[37] A. Voros, J. Phys. A: Math. Gen. **21** (1988) 685.

[38] P. Gaspard and S. A. Rice, J. Chem. Phys. **90** (1989) 2225, 2242, 2255; **91** (1989) E3279.

[39] P. Cvitanović and B. Eckhardt, Phys. Rev. Lett. **63** (1989) 823.

[40] S. A. Rice, P. Gaspard, and K. Nakamura, Adv. Class. Traj. Methods **1** (1992) 215.

[41] P. Gaspard, in: *Quantum Chaos*, G. Casati, I. Guarneri, and U. Smilansky, Eds., Proceedings of the International School of Physics "Enrico Fermi", Varenna, 1991 (North-Holland, Amsterdam, 1993).

[42] M. V. Berry and J. P. Keating, J. Phys. A **23**, 4839 (1990).

[43] E. B. Bogomolny, Comments At. Mol. Phys. **25** (1990) 67.

[44] G. Tanner, P. Scherer, E. B. Bogomolny, B. Eckhardt, and D. Wintgen, Phys. Rev. Lett. **67** (1991) 2410.

[45] R. Aurich, C. Matthies, M. Sieber, and F. Steiner, Phys. Rev. Lett. **68**, 1629 (1992).

[46] L. A. Bunimovich, Commun. Math. Phys. **65** (1979) 295.

[47] P. Cvitanović, P. Gaspard, and T. Schreiber, Chaos **2** (1992) 85.

[48] M. V. Berry, Ann. Phys. **131**, 163 (1981).

[49] P. Gaspard, in: *Quantum Chaos – Quantum Measurement*, P. Cvitanović, I. Percival, and A. Wirzba, Eds. (Kluwer, Dordrecht, 1992).

[50] R. Balian and C. Bloch, Ann. Phys. **60** (1970) 401.

[51] M. V. Berry and M. Wilkinson, Proc. R. Soc. London Ser. A **392** (1984) 15; T. Harayama and A. Shudo, Phys. Lett A **165** (1992) 417.

[52] A. Wirzba, Chaos **2** (1992) 77.

[53] D. Alonso and P. Gaspard, Chaos **3** (1993) 601.

[54] P. Gaspard, Proceedings of the 1993 Yukawa Int. Seminar "Quantum and Chaos: How Incompatible?" Suppl. Progr. Theor. Phys. (1994).

Pinball scattering

B. ECKHARDT, G. RUSSBERG
Fachbereich Physik der Philipps-Universität,
Renthof 6, D-3550 Marburg

P. CVITANOVIĆ, P. E. ROSENQVIST
Niels Bohr Institutet, Blegdamsvej 17, DK-2100 København Ø

P. SCHERER
Institut für Vegetative Physiologie der Universität Köln,
Robert Koch Str. 39, D-5000 Köln 41

Abstract

Classical and semiclassical periodic orbit expansions are applied to the dynamics of a point particle scattering elastically off several disks in a plane. Fredholm determinants, zeta functions, and convergence of their cycle expansions are tested and applied to evaluation of classical escape rates and quantum resonances. The results demonstrate the applicability of the Ruelle and Gutzwiller type periodic orbit expressions for chaotic systems.

1 Introduction

At the heart of semiclassical descriptions of chaotic systems is the Gutzwiller trace formula which relates the eigenvalue spectrum of the Schrödinger operator to the periodic orbits of the underlying classical system [1]. This relationship between the classical and the quantum properties can be viewed as a generalization of the Selberg trace formula which relates the spectrum of the Laplace-Beltrami operator to geodesic motion on surfaces of constant negative curvature [2]. Whereas the Selberg trace is exact, the Gutzwiller trace, derived within a stationary phase approximation, is only approximate, valid in a suitable semiclassical limit.

In one-dimensional systems the trace formula recovers the standard WKB quantization rules, which yield easy and sometimes quite accurate estimates for the quantum eigenvalues [3]. For systems with more than one degree of freedom a classical system can exhibit chaos. The simple WKB quantization fails and evaluation of the trace formulas can become rather difficult; in fact, it is often easier to do the full quantum calculation and to obtain the periods of classical periodic orbits from the quantum data by a Fourier transform [4]. Perhaps the main difficulty inherent in the periodic orbit quantization is the fact that for chaotic systems the number of periodic orbits grows exponentially with time, and formulas such as the Gutzwiller trace formula diverge in the regime of physical interest [5]. One aim of the present contribution will be to summarize techniques that have been developed recently to overcome such divergences [6, 7, 8].

From the vantage point of the dynamical systems theory, the trace formulas (both the exact Selberg and the semiclassical Gutzwiller trace formula) fit into a general framework of replacing phase space averages by sums over periodic orbits [9]. For classical hyperbolic systems this is possible since the invariant density can be represented by all periodic orbits, with weights related to their instability. The semiclassical periodic orbit sums differ from the classical ones only in phase factors and stability weights; such differences may be traced back to the fact that in quantum mechanics the amplitudes rather than the probabilities are added. However, it must be emphasized that for generic nonhyperbolic systems (which we shall not discuss here), with mixed phase space and marginally stable orbits, such summations are at present hard to control, and it is not clear that the periodic orbit sums should necessarily be the computational method of choice.

For hyperbolic systems, the cycle or curvature expansions of Fredholm determinants and zeta functions have proven extremely useful in evaluating eigenvalues quickly and accurately. For classical dynamics, demonstrations of the success of this method abound [7, 10]. We shall here focus on an ideal quantum system, scattering off three (or more) disks in the plane [11]. This system was investigated in a series of papers by Gaspard and Rice [12, 13], and, independently, by us. We shall present here our (mostly previously unpublished) results. We will demonstrate that for this nontrivial system cycle expansions offer an accurate test of Gutzwiller semiclassical quantization. We shall skirt some of the more technical issues; the reader can pursue them by perusing the references.

The outline of the chapter is as follows. In the next section we introduce the model and discuss the relevant aspects of its classical dynamics. The quantum results are given in section 3, and the cycle expansions are discussed in section 4. Methods for computation of periodic orbits are sketched in appendices A and B.

2 Classical pinball

The model that we shall discuss here is simple, yet physical and instructive. One can use it to illustrate and teach, in clearly physically motivated steps, almost everything one needs to know about deterministic chaos: from Smale horseshoes, Cantor sets, Lyapunov exponents, symbolic dynamics, discrete symmetries, bifurcations, pruning and diffusion, all the way to transfer operators, thermodynamic formalism, and classical and quantum zeta functions. We shall concentrate here on semiclassical calculations and tests of Gutzwiller type periodic orbit formulas.

Our classical pinball model consists of a point particle and three identical circular disks in the plane (fig. 1(a)). The point particle is scattered elastically off the disks and moves freely between collisions. The dynamics with one or two disks is simple (there is either no or one trapped trajectory), but with three or more disks there are infinitely many trapped trajectories, forming a repeller. This repeller can

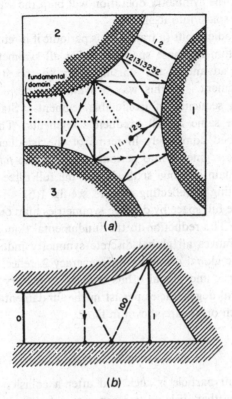

Fig. 1. The scattering geometry for the disk radius : separation ratio $R : d = 1 : 2.5$; (a) the three disks, with $\overline{12}$, $\overline{123}$ and $\overline{121232313}$ cycles indicated; (b) the fundamental domain, *i.e.*, a wedge consisting of a section of a disk, two segments of symmetry axes acting as straight mirror walls, and an escape gap. The above cycles restricted to the fundamental domain are now the two fixed points $\overline{0}$ and $\overline{1}$, and the $\overline{100}$ cycle.

be in principle observed by measurements such as irregularly fluctuating outgoing angles *vs.* impact parameter (the irregular or chaotic scattering [14]), but such measurements are difficult and very sensitive to small perturbations. Much more robust are the global averages of quantities such as the mean trapping time, which we shall discuss in what follows.

2.1 Symmetries of the model

As the three disks are equidistantly spaced, the system has C_{3v} symmetry. Applying an element (identity, rotation by $\pm 2\pi/3$, or reflection) of this symmetry group to any trajectory yields another dynamically acceptable trajectory. Symmetry operations map *nonsymmetric* orbits into different orbits of the same shape, and for a *symmetric* orbit, the symmetry operation will map the set of points making up the orbit in phase space into itself.

For symmetric periodic orbits (a trajectory is periodic if it returns to the starting position and momentum in phase space) some or all symmetry operations act like a shift in time, advancing the starting point to the starting point of a symmetry related segment. In this way a symmetric periodic trajectory can be subdivided into a sequence of irreducible segments. Stability, action and traversal time are the same for all irreducible segments. The global periodic orbits can be described completely in terms of the irreducible segments, by folding the irreducible segments into periodic orbits in the *fundamental domain*. The fundamental domain is a one sixth slice of the full three-disk system, with the symmetry axes acting as reflecting mirrors, see fig. 1(b).

Orbits related in the full space by discrete symmetries map onto a single fundamental domain orbit. The reduction to the fundamental domain desymmetrizes the dynamics and removes all global discrete symmetry induced degeneracies: rotationally symmetric global orbits have degeneracy 2, reflectionally symmetric ones have degeneracy 3, and global orbits with no symmetry are 6-fold degenerate. The time-reversal degeneracies persist in the fundamental domain as well. Some examples of such orbits are shown in fig. 2.

2.2 Symbolic coding

The motion of a point particle is such that after a collision with one disk it either continues to another disk or it escapes to infinity. Labelling the disks 1, 2 and 3, this suggests associating with every trajectory a sequence of labels, indicating the disks with which the particle collides. The collision sequence will be finite for a scattering orbit, coming in from infinity and escaping after a finite number of collisions, and it will repeat periodically for a (trapped) periodic orbit. Arguments used in the usual horseshoe construction show that among the infinitely long sequences and the infinitely long unstable trapped orbits there is a one-to-one relationship; there exists an orbit to every (allowed) infinite sequence and every (allowed) infinite sequence labels a unique orbit.

There is one obvious restriction to the possible sequences, namely that two consecutive symbols must not be identical, since the particle cannot collide twice in succession with the same disk. In addition, there are relabelling symmetries, relating for instance the periodic orbits $\overline{12}$, $\overline{23}$, and $\overline{13}$, which are mapped into the same fundamental domain orbit. (A bar over a sequence indicates periodic repetitions; it will often be omitted when it is clear from the context that we are dealing with periodic orbits.) By replacing the absolute disk labels by relative symbols, indicating only the orientation of the motion (clockwise or anticlockwise), both the symbol repetitions and the symmetry degeneracy are removed. We shall use the symbol 1 to indicate that the orientation after collision is kept, and the symbol 0 to indicate that it is reversed. Depending

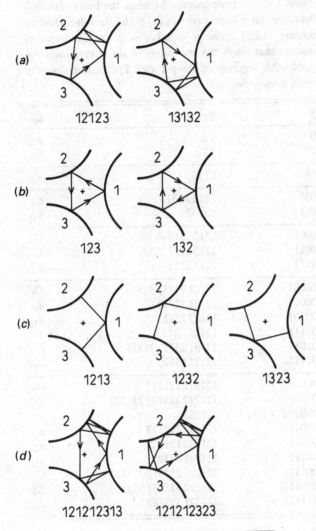

Fig. 2. Some examples of three-disk cycles: (a) $\overline{12123}$ and $\overline{13132}$ are mapped onto each other by σ_{23}, the flip across 1 axis; this cycle has degeneracy 6 under C_{3v} symmetries. Similarly (b) $\overline{123}$ and $\overline{132}$ and (c) $\overline{1213}$, $\overline{1232}$ and $\overline{1323}$ are degenerate under C_{3v}. (d) The orbits $\overline{121212313}$ and $\overline{121212323}$ are related by time reversal but not by any C_{3v} symmetry.

Table 1. C_{3v} correspondence between the binary labelled
fundamental domain prime cycles $\tilde{p}$ and the full three-disk
ternary $\{1,2,3\}$ cycles p, together with the C_{3v} transformation that maps the end point of the $\tilde{p}$ cycle into the
irreducible segment of the p cycle. The degeneracy of p
cycle is $m_p = 6n_{\tilde{p}}/n_p$.

$\tilde{p}$	p	$g_{\tilde{p}}$
0	1 2	σ_{12}
1	1 2 3	C_3
01	12 13	σ_{23}
001	121 232 313	C_3
011	121 323	σ_{13}
0001	1212 1313	σ_{23}
0011	1212 3131 2323	C_3^2
0111	1213 2123	σ_{12}
00001	12121 23232 31313	C_3
00011	12121 32323	σ_{13}
00101	12123 21213	σ_{12}
00111	12123	e
01011	12131 23212 31323	C_3
01111	12132 13123	σ_{23}
000001	121212 131313	σ_{23}
000011	121212 313131 232323	C_3^2
000101	121213	e
000111	121213 212123	σ_{12}
001011	121232 131323	σ_{23}
001101	121231 323213	σ_{13}
001111	121231 232312 313123	C_3
010111	121312 313231 232123	C_3^2
011111	121321 323123	σ_{13}

on the symmetry of the global orbit, periodically continued binary string labels
correspond either to the full periodic orbit or to a repeating irreducible segment
(examples are shown in fig. 1). If the disks are sufficiently far apart there are
no further restrictions on symbols, and all periodic binary sequences are realized
as allowed periodic orbits. Table 1 lists some of the shortest binary symbol
strings, together with the corresponding full three-disk symbol sequences and
orbit symmetries.

2.3 Periodic orbits

There is only one length scale in the system, the ratio of the center-to-center
separation to the disk radius $d : R$. The energy is a quadratic function of

momenta, $H = p^2/2m$, so motion at different energies E and E_0 is related by the scaling $p_E \to p_0\sqrt{E/E_0}$ for momenta, $t_E \to t_0\sqrt{E_0/E}$ for times, and

$$S(E) = L\sqrt{2mE} = S(E_0)\sqrt{E/E_0} \tag{1}$$

for the actions, where L is the geometrical length of the orbit. The eigenvalues of the jacobian transverse to a periodic orbit (see below) are invariant under the above energy rescaling. These observations will be useful below in the semiclassical context where the energy in (1) will combine with $\hbar$ to the relevant quantum variable, the wave number $k = \sqrt{2mE}/\hbar$.

The motion between collisions is completely characterized by an angle s (or arclength in case of a general billiard) marking the point of collision along a disk and the impact parameter $b = b'/R = \sin\phi$ measured in units of the radius, with ϕ the incidence angle (the angle between the outgoing particle and the outgoing normal to the billiard edge). Because of symmetry, we can always select the disk 1 as the disk of current collision and disk 3 as the origin of the particle. Ingoing coordinates then are (φ_{in}, b_{in}) and outgoing coordinates are $(\tilde{\varphi}_{out}, \tilde{b}_{out})$, where the $\tilde{\ }$ indicates that these coordinates refer to the next collision disk. When working in the fundamental domain they still need to be mapped back onto disk 2. Accordingly, we have two types of collisions:

0: the particle returns to the disk it came from
1: the particle continues to the next disk.

The corresponding maps are (the angle s is measured clockwise relative to the line connecting the centers of disks 1 and 3)

$$T_0 : \begin{cases} \varphi_{out} = -\varphi_{in} + 2\arcsin b_{in} \\ b_{out} = -b_{in} + \dfrac{d}{R}\sin\varphi_{out} \end{cases} \tag{2}$$

for reflection and

$$T_1 : \begin{cases} \varphi_{out} = \varphi_{in} - 2\arcsin b_{in} + \varphi_{shift} \\ b_{out} = b_{in} - \dfrac{d}{R}\sin\varphi_{out} \end{cases} \tag{3}$$

with $\varphi_{shift} = 2\pi/3$ for the case of continuation. Each map has a fixed point, corresponding to the orbits $\overline{0}$ and $\overline{1}$. Longer periodic orbits are fixed points of sequences of maps, *e.g.*

$$T_0 T_0 T_1 T_0 T_1 x_{10100} = x_{10100}, \tag{4}$$

(note that in our convention the maps are applied in reverse order compared to the symbolic sequence).

The jacobian of the single collision map is given by

$$J_i = \frac{\partial T_i(\varphi_{out}, b_{out})}{\partial(\varphi_{in}, b_{in})} \tag{5}$$

and the cycle jacobian J_p is given by the product of jacobians for the bounces around the cycle p. As the dynamics is phase-space volume preserving, $\det J = 1$, and the eigenvalues depend only on $tr(J)$:

$$\Lambda_\pm = \frac{1}{2}\left(tr(J) \pm \sqrt{tr(J)^2 - 4}\right). \tag{6}$$

The sign of the eigenvalue depends on the number of collisions along the cycle (see appendix B). For the '0' symbol there are two bounces in the fundamental domain: one with the disk and one with the reflecting wall. Since the wall can be regarded as a disk of infinite radius, the trace changes sign twice and thus the eigenvalues are positive. Symbol '1' corresponds to one bounce with the disk but two wall bounces and hence the eigenvalues of the '1'-cycle are negative. For an arbitrary fundamental domain cycle, the eigenvalue sign is given by $(-1)^{n_1}$, where n_1 is the number of '1's in the binary string corresponding to the cycle.

The exact lengths and eigenvalues of $\overline{0}$, $\overline{1}$ and $\overline{10}$ cycles follow from elementary geometrical considerations (we set the disk radius $R = 1$ throughout). For the fundamental domain $\overline{0}$ (the 2-cycle of the complete 3-disk space) and $\overline{1}$ (the 3-cycle of the complete 3-disk space) fixed points we obtain

$$
\begin{array}{lllll}
\overline{0}: & \varphi_0 = 0 & b_0 = 0 & L_0 = d - 2 & \Lambda_0 = (d-1) + \sqrt{d(d-2)} \\
\overline{1}: & \varphi_1 = 0 & b_1 = \frac{1}{2} & L_1 = d - \sqrt{3} & \Lambda_1 = -(\frac{2}{\sqrt{3}}d - 1) - \sqrt{\frac{2}{\sqrt{3}}d(\frac{2}{\sqrt{3}}d - 2)}
\end{array}
, \tag{7}
$$

and for the $\overline{10}$-cycle we obtain

$$
\begin{aligned}
\overline{10}: \quad L_{10} &= \sqrt{1 + (2d - \sqrt{3})^2} - 2, \\
tr(J_{10}) &= \frac{L_{10}(L_{10} + 1)(L_{10} + 2)}{\sqrt{3}d/2 - 1} + 2L_{10} + 2.
\end{aligned}
\tag{8}
$$

Λ_{10} follows from (6). Longer cycles require numerical evaluation by methods such as the multipoint shooting or orbit length minimization, described in the appendix A. Formulas for evaluation of the cycle jacobians are given in the appendix B. A typical set of the periodic orbit data, for $d : R = 6$ and length ≤ 6, is listed in table 2.

2.4 Classical escape rate

The interesting part of the classical scattering dynamics is generated by the chaotic repeller, formed by the trapped (periodic and aperiodic) orbits. The closure of this set in the phase space is of zero Lebesgue measure, and almost every trajectory entering the vicinity of the repeller eventually escapes. However, some will be trapped for very long times; the quantity that characterizes this phenomenon is the mean trapping time or its inverse, the *escape rate*. The escape rate can be extracted from the trace of the classical Liouville operator [15], which we interpret as follows. The probability of returning to the starting point in phase

Table 2. All periodic orbits up to six bounces for the three-disk fundamental domain at $d : R = 6$. The columns list the topological length of the cycle, its expanding eigenvalue Λ_p, the length of the orbit (for billiards this is the same as the action), and the binary code for the cycle. Note that the two period 6 orbits $\overline{001011}$ and $\overline{001101}$ are degenerate due to the time reversal symmetry, but are not related by any discrete spatial symmetry.

period	Λ_p	Action	code
1	9.898979485566	4.000000000000	0
1	$-1.177145519638 \times 10^1$	4.267949192431	1
2	$-1.240948019921 \times 10^2$	8.316529485168	01
3	$-1.240542557041 \times 10^3$	12.321746616182	001
3	$1.449545074956 \times 10^3$	12.580807741032	011
4	$-1.229570686196 \times 10^4$	16.322276474382	0001
4	$1.445997591902 \times 10^4$	16.585242906081	0011
4	$-1.707901900894 \times 10^4$	16.849071859224	0111
5	$-1.217338387051 \times 10^5$	20.322330025739	00001
5	$1.432820951544 \times 10^5$	20.585689671758	00011
5	$1.539257907420 \times 10^5$	20.638238386018	00101
5	$-1.704107155425 \times 10^5$	20.853571517227	00111
5	$-1.799019479426 \times 10^5$	20.897369388186	01011
5	$2.010247347433 \times 10^5$	21.116994322373	01111
6	$-1.205062923819 \times 10^6$	24.322335435738	000001
6	$1.418521622814 \times 10^6$	24.585734788507	000011
6	$1.525597448217 \times 10^6$	24.638760250323	000101
6	$-1.688624934257 \times 10^6$	24.854025100071	000111
6	$-1.796354939785 \times 10^6$	24.902167001066	001011
6	$-1.796354939785 \times 10^6$	24.902167001066	001101
6	$2.005733106218 \times 10^6$	25.121488488111	001111
6	$2.119615015369 \times 10^6$	25.165628236279	010111
6	$-2.366378254801 \times 10^6$	25.384945785676	011111

space after a time t is $\delta(\mathbf{x} - \mathbf{x}_t)$, so integrating over all phase space points yields the total recurrence probability

$$\sum_{p.o.} \frac{T_p}{|\det(1 - \mathbf{J}_{p.o.})|} \delta(t - T_{p.o.}),$$

where *p.o.* indicates that the sum extends over all periodic orbits, including multiple traversals; their period time (single traversal) is T_p, and the jacobian (or the monodromy matrix) of the mapping transverse to the trajectory is $\mathbf{J}_{p.o.}$. The above expression is correct if the system is hyperbolic, *i.e.*, if all periodic points are isolated and their stability eigenvalues are strictly bounded away from unity. In this case the sum decays exponentially with time, $\propto e^{-\gamma t}$, and the leading pole of its Laplace transform

$$\Omega(s) = \sum_{p.o.} \frac{T_p}{|\det(1 - \mathbf{J}_{p.o.})|} e^{sT_{p.o.}} \tag{9}$$

yields the decay rate $s = \gamma$. Now distinguish a primitive periodic orbit p from its r-th traversal, $p.o. = rp$, and write

$$\Omega(s) = \sum_{p} \sum_{r=1}^{\infty} \frac{T_p}{|\det(1 - \mathbf{J}_p^r)|} e^{rsT_p} . \tag{10}$$

For a Hamiltonian two-degrees-of-freedom system, $\mathbf{J}_p$ is a $[2 \times 2]$ matrix with unit determinant. If the cycle is unstable, the eigenvalues Λ_p and $1/\Lambda_p$ are real, and we denote the expanding eigenvalue by Λ_p. The denominator can then be expanded in a geometric series

$$1/|\det(\mathbf{J}_p - 1)| = |\Lambda_p|^{-1}(1 - 1/\Lambda_p)^{-2} = |\Lambda_p|^{-1} \sum_{j=0}^{\infty} (j+1)\Lambda_p^{-j} .$$

Performing the r summation and interchanging sums and logarithms one ends up with $\Omega(s) = (\partial/\partial s) \ln F(s)$, where $F(s)$ is the *classical Fredholm determinant*

$$F(s) = \prod_{p} \prod_{j=0}^{\infty} \left(1 - |\Lambda_p|^{-1} \Lambda_p^{-j} e^{sT_p}\right)^{j+1} . \tag{11}$$

As $\Omega(s)$ is a logarithmic derivative, its poles are given by the zeros and poles of $F(s)$. Denoting the classical weight of the cycle p by

$$t_p = z^{n_p} e^{sT_p}/|\Lambda_p| \tag{12}$$

and defining *dynamical zeta functions* [9]

$$1/\zeta_j = \exp\left(-\sum_{p} \sum_{r=1}^{\infty} \frac{1}{r}(t_p/\Lambda_p^j)^r\right) = \prod_{p} \left(1 - t_p/\Lambda_p^j\right) , \tag{13}$$

the Fredholm determinant can be written as an infinite product over $1/\zeta_j$:

$$F(s) = \prod_{p} \prod_{j=0}^{\infty} (1 - t_p/\Lambda_p^j)^{j+1} = \prod_{j=0}^{\infty} 1/\zeta_j^{j+1} . \tag{14}$$

We have introduced a bookkeeping variable z raised to the power of the topological length (number of disk collisions in a cycle) in order to be able systematically to expand the infinite products in terms of increasing topological cycle length.

Postponing a discussion of both the cycle expansions and their convergence until we have introduced the corresponding quantum objects, we present the results for the classical escape rate, computed by using $1/\zeta_0$ from eq. (13), in fig. 3. For sufficiently separated disks, already period 2 cycles yield results in good agreement with the Monte Carlo simulation estimates, demonstrating that

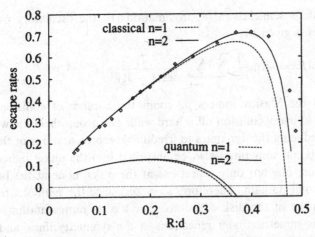

Fig. 3. Classical escape rates and bounds on quantum escape rates, *i.e.* the zeros of the zeta functions (11) and (17) with absolute weights (19), as functions of the disk-disk separation. The top two curves show the classical escape rate. Zeros computed from fixed points alone are shown as dotted lines, zeros from fixed points and period two orbits as full lines, and the Monte Carlo estimates of classical escape rates [12, 18] are marked by diamonds. The classical escape rate should approach 0 as $R : d \to 0.5$; in this limit cycle expansions are expected to converge poorly due to pruning and intermittency effects. The lower two curves show s_c, the abscissa of absolute convergence of the Gutzwiller trace formula, which serves as a crude lower bound on the imaginary part of the semiclassical quantum resonances. Though this bound becomes negative near $R : d \approx 0.33$, the semiclassical resonances do remain below the real energy axis, see fig. 4.

the zeta function formalism offers a powerful method for evaluation of escape rates.

3 Quantum pinball

For the three-disk system, the explicit expressions of the outgoing waves in terms of the ingoing waves using the quantum scattering matrix S have been given by Gaspard and Rice [13]. Resonances are related to complex zeros of $\operatorname{tr} S^{\dagger}(dS/dE)$, the generalization of the concept of the density of states to scattering systems [16]. We shall first evaluate these complex zeros in the semiclassical approximation, and then compare them with the exact quantum-mechanical resonances.

3.1 The semiclassical density of states

In the Gutzwiller's semiclassical approximation [1], the oscillatory part of the density of states is given by

$$\rho(E) = -\mathrm{Im}\frac{1}{i\pi\hbar}\sum_{p}\sum_{r=1}^{\infty}\frac{T_p}{|\det(1-J_p^r)|^{1/2}}e^{ir(S_p/\hbar-\mu_p\pi/2)}. \qquad (15)$$

For billiards the Maslov indices μ_p count the number of collisions, with a phase loss of π at every collision off a hard wall. The group-theoretic weights associated with reducing the dynamics to irreducible representations of the discrete symmetry group can sometimes also be absorbed into the phase indices. In the A_1 representation, one has only reflections at the disks, none at the boundaries of the fundamental domain. Therefore, $\mu_p = 2n_p$, since the symbol string length counts the number of the disk collisions. For the A_2 representation, the wave function is antisymmetric under reflections at the symmetry lines, and one can associate one additional reflection with each occurrence of the symbol 0, and two reflections with each occurrence of the symbol 1. The net effect is an additional overall minus sign, if the number of 0s in the symbol string of the orbit is odd. However, in general (for example, for the two-dimensional E representation of the C_{3v} discrete group) a full group-theoretic decomposition in terms of group characters is required [19]. Furthermore, special attention should be paid to the orbits that run along the borders of the fundamental domain. The three-disk system studied here does not have boundary orbits, but the four-disk system, for example, does have such orbits.

As mentioned above one can replace $S_p(E)/\hbar = L_p k$, where L_p is the geometric length of the orbit, and k is the wave number. Expressing the density as a function of k, and manipulating the denominator as in sect. 2.4, one finds

$$\rho(k) = -\mathrm{Im}\frac{1}{\pi}\frac{\partial}{\partial k}\log Z(k) \qquad (16)$$

where $Z(k)$ is the *quantum Selberg zeta* function [17]

$$Z(k) = \prod_{j=0}^{\infty}\prod_{p}(1 - e^{iL_p k - i\mu_p\pi/2}|\Lambda_p|^{-1/2}\Lambda_p^{-j}). \qquad (17)$$

The quantum Selberg zeta function can also be expressed as a product over dynamical zeta functions (13),

$$Z(k) = \prod_{j=0}^{\infty}1/\zeta_j, \qquad (18)$$

this time with cycles weighted by semiclassical weights

$$t_p = z^{n_p}e^{iS_p/\hbar - i\mu_p\pi/2}/|\Lambda_p|^{1/2}. \qquad (19)$$

We have evaluated several hundreds of semiclassical and exact quantum resonances in the three symmetry subspaces A_1, A_2 and E; some of them are listed

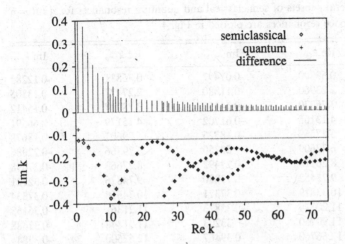

Fig. 4. Semiclassical scattering resonances (diamonds) compared with the exact quantum scattering resonances (crosses) in the A_1 subspace. The lines in the upper half of the diagram indicate the (geometrical) difference between the semiclassical and quantum resonances for all resonances with Im $k > -0.4$, magnified by a factor 10. The dotted line at $k_c = -0.121556$ indicates the semiclassical abscissa of convergence; all semiclassical resonances lie below this line, but the first two quantum resonances lie above it. For the semiclassical calculation, all orbits up to symbolic length 5 have been used in the cycle expansion for ζ_0^{-1}.

in table 3 and plotted in fig. 4. The accuracy and numerical convergence of the semiclassical estimates came as a surprise [8].

3.2 "Quantum escape rate"

The region of absolute convergence of the Gutzwiller trace formula in the $k = \kappa - is$ complex plane is determined by the convergence of the sum of absolute values of the terms in the series (15). The sum converges absolutely for $s < s_c$, where s_c is called the abscissa of absolute convergence. Since s_c is the value of s for which the sum

$$\tilde{\Omega}(s) = \sum_{p.o.} \frac{T_p}{|\det(1 - \mathbf{J}_{p.o.})|^{1/2}} e^{sT_{p.o.}}, \tag{20}$$

diverges, one can, by analogy with the determination of the classical escape rate by the divergence of sum (9), interpret the abscissa of absolute convergence s_c as a "quantum escape rate". In fact, s_c is only a lower bound on the escape rate; the correct rate of a decay of a given initial wave function is given by a superposition of complex resonances evaluated below. Within the same short cycles truncations as in the classical case, one finds the s_c curves shown in the lower half of fig. 3.

Table 3. Several subsets of semiclassical and quantum resonances for $d : R = 6$. The A_1 subspace resonances are plotted in Fig. 4.

	Re k_{QM}	Im k_{QM}	Re k_{zeta}	Im k_{zeta}
	0.69800	-0.07497	0.75831	-0.12282
	2.23960	-0.11880	2.27427	-0.13305
	3.76270	-0.14756	3.78787	-0.15412
	4.13165	-0.61702	4.15179	-0.66591
	5.27569	-0.18325	5.29607	-0.18678
	6.77609	-0.22750	6.79366	-0.22986
	8.26114	-0.27492	8.27663	-0.27698
	9.73452	-0.13880	9.74826	-0.32121
	10.33819	-0.37371	10.34656	-0.37834
	11.20210	-0.35823	11.21361	-0.36168
A_1	11.90760	-0.33223	11.91448	-0.33488
	12.66760	-0.39467	12.67500	-0.39841
	13.47692	-0.29412	13.48266	-0.29623
	14.13370	-0.42883	14.13680	-0.42956
	15.04170	-0.25552	15.04705	-0.25762
	16.59706	-0.21700	16.60244	-0.21889
	18.14115	-0.18280	18.14647	-0.18426
	19.67567	-0.15653	19.68084	-0.15761
	21.20308	-0.13958	21.20807	-0.14032
	22.72484	-0.13187	22.72966	-0.13235
	24.24120	-0.13284	24.24588	-0.13311
	25.75156	-0.14130	25.75605	-0.14153
	7.93363	-0.15129	7.94561	-0.15526
	9.45604	-0.13196	9.46661	-0.13458
	10.97616	-0.12325	10.98563	-0.12504
	12.49347	-0.12432	12.50215	-0.12550
	14.00693	-0.13468	14.01501	-0.13534
	15.51469	-0.15329	15.52229	-0.15368
A_2	17.01453	-0.17810	17.02178	-0.17836
	18.50574	-0.20544	18.51273	-0.20573
	19.99011	-0.23236	19.99681	-0.23275
	20.54565	-0.31325	20.54989	-0.31495
	21.46993	-0.25775	21.47596	-0.25817
	22.10098	-0.28557	22.95261	-0.28199
	22.94703	-0.28164	23.66203	-0.26365
	18.85038	-0.17271	18.56037	-0.17350
	19.44833	-0.25471	19.45385	-0.25454
E	20.37203	-0.17461	20.37745	-0.17502
	20.94214	-0.28849	20.94718	-0.28767
	21.88904	-0.19045	21.89438	-0.19059

Since the trace formulas are convergent in the domain of absolute convergence, one cannot have zeros of the zeta function or resonances of the S-matrix in that domain. In particular, for sufficiently large $d : R$, all resonances will stay a finite distance from the real axis. In contrast to the classical case, for $d : R$ less than about 2.8 the bound s_c actually becomes negative. As we shall see below, the semiclassical resonances do lie below the real axis, as they should; but this serves as a reminder that the resonances (and the energy eigenvalues for bound systems) are being evaluated in a region where cycle expansions are only conditionally convergent, and one has to be very careful in ordering terms in such expansions.

The same absolute convergence arguments can be applied to the dynamical zeta functions. Their logarithmic derivatives correspond to sums over orbits of form

$$-\frac{\partial}{\partial k} \ln \zeta_j = i \sum_p L_p \frac{e^{iL_pk - i\mu_p\pi/2}}{|\Lambda_p|^{1/2}\Lambda_p^j} \, . \tag{21}$$

Due to the extra powers of Λ_p's present in $1/\zeta_j$ cycle weights, the corresponding abscissas of absolute convergence form an ordered sequence $s_j < s_{j-1} < \cdots < s_0$. The resonances closest to the real axis - which will be noticable as the sharpest resonances - should be due to zeros in $1/\zeta_0$ since the leading zeros of other $1/\zeta_j$ have larger imaginary parts. We often find it convenient to use this fact and restrict our numerical work to the leading zeros of $1/\zeta_0$ rather than the full Selberg zeta functions.

3.3 Quantum resonances

We evaluate the exact quantum mechanical resonances by the method described by Gaspard and Rice [13], which we have improved by implementing a symmetry reduced code. A comparison of resonances obtained from cycle expansions truncated to the 14 cycles of periods ≤ 5 with the exact quantum values is given in table 3 and in fig. 4, together with the difference between the semiclassical and quantum resonances $\delta = \sqrt{|k_{QM} - k_{sc}|}$. Numerically this difference seems to decrease with increasing Re k, i.e., with approach to the semiclassical limit. As the semiclassical approximation ignores terms of higher order in $\hbar$, one expects on general grounds this difference to approach a nonzero constant. Our data are insufficient to estimate the asymptotic behaviour and to bound it away from zero.

Also shown in fig. 4 is the abscissa of absolute convergence s_c. No semiclassical resonance lies above it. Two of the lowest exact quantum resonances, for which the semiclassical approximation error is largest, do lie above s_c, but that is acceptable as the bound is semiclassical.

4 Cycle expansions

The periodic orbit formulas for classical and semiclassical escape rates and resonances introduced above are in practice evaluated by expanding the appropriate
zeta functions and determinants as *cycle expansions*:

$$F(z) = \sum_{j=0}^{N_{max}} C_j z^j , \qquad (22)$$

and investigating their zeros and radii of convergence as functions of truncations
to finite numbers of shortest cycles. The bookkeeping variable z that keeps
track of the topological cycle length n_p is set to $z = 1$ in actual calculations.
The evaluation of cycle expansions is facilitated by understanding the analytic
properties of $F(z)$, by judicious use of the symmetries of classical dynamics, and
by topology guided rearrangements of terms in the expansions. As we shall see
below, the main virtue of cycle expansions is their fast convergence.

4.1 Symmetry factorizations

Discrete symmetries of the classical dynamics play a role with which we are familiar from quantum mechanics; as they commute with the evolution operators, they
can be used to decompose them and factorize the associated determinants [19]:

$$1/\zeta^{3-disk} = \prod_\alpha 1/\zeta_\alpha^{d_\alpha} .$$

The product is over the d_α-dimensional irreducible representations α of the
symmetry group, in this case C_{3v}, with two one-dimensional representations A_1,
A_2 and a pair of two-dimensional representations E. The factorization relates
each fundamental domain orbit to the corresponding degenerate set of full space
orbits as follows:

symmetry	full space		A_1	A_2	E
rotation	:	$(1-t_{rot}^3)^2$ =	$(1-t_{rot})$	$(1-t_{rot})$	$(1+t_{rot}+t_{rot}^2)^2$
reflection	:	$(1-t_{ref}^2)^3$ =	$(1-t_{ref})$	$(1+t_{ref})$	$(1-t_{ref}^2)^2$
none	:	$(1-t_{non})^6$ =	$(1-t_{non})$	$(1-t_{non})$	$(1-t_{non})^4$

Fundamental domain cycles up to length 5 are listed in table 1, together
with the symmetry factors that map them into the corresponding global orbit
irreducible segments; these determine which of the above factorizations apply to
a given cycle. Substituting the shortest cycles into the zeta functions, we obtain
for the completely symmetric A_1 subspace:

$$
\begin{aligned}
1/\zeta_{A_1}(z) = \ & (1-zt_0)(1-zt_1)(1-z^2t_{01})(1-z^3t_{001})(1-z^3t_{011}) \\
& (1-z^4t_{0001})(1-z^4t_{0011})(1-z^4t_{0111})(1-z^5t_{00001})(1-z^5t_{00011}) \\
& (1-z^5t_{00101})(1-z^5t_{00111})(1-z^5t_{01011})(1-z^5t_{01111})\ldots . \qquad (23)
\end{aligned}
$$

In the example at hand, with complete symbolic dynamics and no pruning rules, the cycle expanded zeta function is obtained by expanding the infinite product as a power series in z:

$$
\begin{aligned}
1/\zeta_{A_1}(z) = \ & 1 - zt_0 - zt_1 - z^2[(t_{01} - t_1 t_0)] \\
& -z^3[(t_{001} - t_{01} t_0) + (t_{011} - t_{01} t_1)] \\
& -z^4[(t_{0001} - t_0 t_{001}) + (t_{0111} - t_{011} t_1) \\
& +(t_{0011} - t_{001} t_1 - t_0 t_{011} + t_0 t_{01} t_1)] - \cdots .
\end{aligned}
\tag{24}
$$

For the A_2 subspace cycles with an odd number of 0s pick up an additional minus sign:

$$
\begin{aligned}
1/\zeta_{A_2}(z) = \ & 1 + zt_0 - zt_1 + z^2[(t_{01} - t_1 t_0)] \\
& -z^3[(t_{001} - t_{01} t_0) - (t_{011} - t_{01} t_1)] \\
& +z^4[(t_{0001} - t_0 t_{001}) + (t_{0111} - t_{011} t_1) \\
& -(t_{0011} - t_{001} t_1 - t_0 t_{011} + t_0 t_{01} t_1)] - \cdots .
\end{aligned}
\tag{25}
$$

The E subspace cycle expansion takes a somewhat less obvious form [19]:

$$
\begin{aligned}
1/\zeta_E = \ & (1 + zt_1 + z^2 t_1^2)(1 - z^2 t_0^2)(1 + z^3 t_{100} + z^6 t_{100}^2)(1 - z^4 t_{10}^2) \\
& (1 + z^4 t_{1001} + z^8 t_{1001}^2)(1 + z^5 t_{10000} + z^{10} t_{10000}^2) \\
& (1 + z^5 t_{10101} + z^{10} t_{10101}^2)(1 - z^{10} t_{10011})^2 \cdots \\
= \ & 1 + zt_1 + z^2(t_1^2 - t_0^2) + z^3(t_{001} - t_1 t_0^2) \\
& +z^4 \left[t_{0011} + (t_{001} - t_1 t_0^2)t_1 - t_{01}^2 \right] \\
& +z^5 \left[t_{00001} + t_{01011} - 2t_{00111} + (t_{0011} - t_{01}^2)t_1 + (t_1^2 - t_0^2)t_{100} \right] + \cdots
\end{aligned}
\tag{26}
$$

All our numerical results are obtained by determining the zeros of finite cycle length truncations of the above cycle expansions, or the corresponding ones for the Fredholm determinats and quantum Selberg zeta functions. The crucial question that we now turn to is - how good are such truncations?

5 Convergence of cycle expansions

While various periodic orbit formulas, such as the Gutzwiller trace formula and the quantum Selberg zeta function, are formally equivalent, in practice determinants have much better convergence properties than the traces. This can be understood on two levels: a geometrical one, as shadowing of long cycles by shorter ones, or more abstractly, as a consequence of analytic properties of the Fredholm determinants. Particularly strong results exist for nice, "Axiom A" hyperbolic systems, for which the dynamical zeta functions are holomorphic [9], and the Fredholm determinants are entire functions [20].

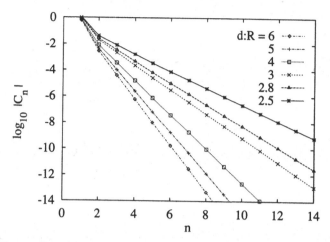

Fig. 5. The coefficients of the cycle expansion for $1/\zeta_0$ for six $d : R$ ratios (from top to bottom: $d : R = 6, 5, 4, 3, 2.8, 2.5$). Clearly visible is the exponential decay; the slope yields the location of the leading pole of $1/\zeta_0$.

5.1 Cycle shadowing and convergence

The important feature to note in the cycle expansion (24) is that the contributions t_0 and t_1 from the two fixed points stand isolated, while all others come in groups. This is seen particularly clearly if the weights are set to $t_p = 1$; in this case the zeta function reduces to the topological polynomial, *i.e.*, the generating polynomial for counting the numbers of topologically distinct cycles. For a complete binary coding without pruning rules, the topological polynomial is just $1 - 2z$, so all longer orbits appear in groups with signs and the numbers of terms such that they cancel. The success of the cycle expansion relies on the extent to which these cancellations survive when t_p are set equal to the true periodic orbit weights.

The effect of such cancellations is illustrated by fig. 5 which demonstrates that the coefficients in the cycle expansion of the dynamical zeta function $1/\zeta_0$ with quantum weights (19) fall off exponentially. The coefficients are evaluated for wave number $k = 0$ and for a range of $d : R$ ratios. Note that also for the cases $d : R = 2.5$ and 2.8, which lie outside the domain of absolute convergence (see fig. 3), there is no qualitative change in the behaviour of the coefficients.

The rate of fall-off of the cycle expansion coefficients can be estimated by observing that for subshifts of finite type [6, 7] the contributions from longer orbits can always be grouped in shadowing combinations such as $t_{a^j b} - t_a t_{a^{j-1} b}$, involving a long cycle $\overline{a^j b}$ and a pseudo-cycle built by shadowing $\overline{a^j b}$ by $\overline{a}$ followed by $\overline{a^{j-1} b}$. These orbits are periodic approximations to an orbit homoclinic to $\overline{a}$.

Table 4. Demonstration of curvature compensations in $t_{ab} - t_a t_b$ for the three disk fundamental domain cycles at $d : R = 6$, table 1.

n	$t_{ab} - t_a t_b$	$L_{ab} - (L_a + L_b)$	$\log\left[\frac{\Lambda_a \Lambda_b}{\Lambda_{ab}}\right]$	ab-a*b
2	$-5.23465150784 \times 10^{-4}$	$4.85802927371 \times 10^{-2}$	$-6.29365864467 \times 10^{-2}$	01-0*1
3	$-7.96028600139 \times 10^{-6}$	$5.21713101432 \times 10^{-3}$	$-9.82663364947 \times 10^{-3}$	001-0*01
4	$-1.03326529874 \times 10^{-7}$	$5.29858199419 \times 10^{-4}$	$-1.26966635483 \times 10^{-3}$	0001-0*001
5	$-1.27481522016 \times 10^{-9}$	$5.35513574697 \times 10^{-5}$	$-1.55176109954 \times 10^{-4}$	00001-0*0001
6	$-1.52544704823 \times 10^{-11}$	$5.40999882625 \times 10^{-6}$	$-1.83824278428 \times 10^{-5}$	000001-0*00001
2	$-5.23465150784 \times 10^{-4}$	$4.85802927371 \times 10^{-2}$	$-6.29365864467 \times 10^{-2}$	01-0*1
3	$5.30414752996 \times 10^{-6}$	$-3.67093656690 \times 10^{-3}$	$7.71831060288 \times 10^{-3}$	011-01*1
4	$-5.40934261680 \times 10^{-8}$	$3.14925761316 \times 10^{-4}$	$-9.23436155345 \times 10^{-4}$	0111-011*1
5	$4.99129508833 \times 10^{-10}$	$-2.67292822795 \times 10^{-5}$	$1.00342411247 \times 10^{-4}$	01111-0111*1
6	$-4.39246000586 \times 10^{-12}$	$2.27087116266 \times 10^{-6}$	$-1.03941678234 \times 10^{-5}$	011111-01111*1

Substituting $t_{a^j b}$ from (13) one finds

$$\frac{t_{a^j b} - t_a t_{a^{j-1} b}}{t_{a^j b}} = 1 - e^{i(S_a + S_{a^{j-1} b} - S_{a^j b})/\hbar} e^{-i\pi(\mu_a + \mu_{a^{j-1} b} - \mu_{a^j b})/2} \left| \frac{\Lambda_a \Lambda_{a^{j-1} b}}{\Lambda_{a^j b}} \right|^{-1/2} .$$

The phase factors cancel since the number and the kind of symbols involved in both terms coincide (though for the E representation of C_{3v} this is admittedly not obvious). Furthermore, since with increasing j segments of $\overline{a^j b}$ come closer to $\overline{a}$, the differences in action and the ratio of the eigenvalues converge exponentially with the eigenvalue of the orbit $\overline{a}$,

$$S_a + S_{a^{j-1} b} - S_{a^j b} \approx const \times \Lambda_a^{-j}, \quad |\Lambda_a \Lambda_{a^{j-1} b} / \Lambda_{a^j b}| \approx \exp(-const \times \Lambda_a^{-j})$$

Expanding the exponentials one thus finds that this term in the cycle expansion is of the order of

$$t_{a^j b} - t_a t_{a^{j-1} b} \approx const \times t_{a^j b} \Lambda_a^{-j} . \tag{27}$$

Even though the number of terms in a cycle expansion grows exponentially [7], the shadowing cancellations improve the convergence by an exponential factor compared to trace formulas, and extend the radius of convergence of the periodic orbit sums. Table 4 shows some examples of such compensations between long cycles and their pseudo-cycle shadows.

The shadowing cancellations require that a long cycle and the associated pseudo-cycles (products of shorter cycles following the same symbol sequence) have nearly the same weight *and* a relative minus sign. The first requirement is guaranteed by the hyperbolicity and the smoothness of the flow. The second requirement implies that, contrary to the "semi-classical" intuition, the cycle expansions are expected to converge for *low energies*. For sufficiently high wave numbers k the differences in actions $S_p - S_{shadow} = k(L_p - L_{shadow})$ can be of order

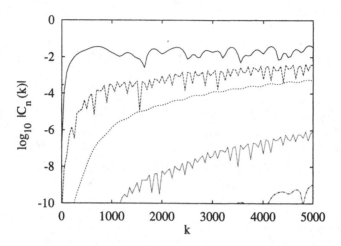

Fig. 6. Energy dependence of the cycle expansion coefficients for $d : R = 6$. Shown are (from top to bottom) the absolute values of the coefficients C_4, C_6, C_8, C_{10} and C_{12} (barely visible in the lower right corner) as a function of the wave number k.

π or higher, in which case $t_p - t_{shadow} \approx 2t_p$, rather than $t_p - t_{shadow} \approx t_p/|\Lambda_p|$ expected at $k = 0$. This is illustrated in fig. 6, which shows the energy dependence of the coefficients ($\text{Im} k = 0$) in the cycle expansion of the quantum Selberg zeta function (17). It is encouraging to note that the shadowing cancellations persist for very large intervals in k, so for the three-disk system studied here the lowest few terms in the cycle expansion suffice for evaluation of thousands of the lowest semiclassical resonaces.

5.2 Convergence of Fredholm determinants

While the above shadowing analysis of cycle expansions implies exponential convergence, cycle expansions can actually converge even faster than exponentially. If the dynamical evolution can be cast in terms of a transfer operator multiplicative along the flow, if the corresponding mapping (for example, the return map for a Poincaré section of the flow) is analytic, and if the topology of the repeller is given by a finite Markov partition, then the Fredholm determinant (14) with classical weight (12) is *entire*. This has been recently proven by H.H. Rugh [20] (earlier mathematical literature dealt only with the expanding directions, not the full hyperbolic flow). In this case the cycle expansion coefficients (22) fall off asymptotically as $C_n \approx \Lambda^{-n^{3/2}}$. This estimate is in agreement with our numerical results for the three-disk repeller, see fig. 9. However, as it is not known how quickly the asymptotics should set in, such numerical results can be misleading, and preasymptotic oscillations can already be observed in simple one-dimensional

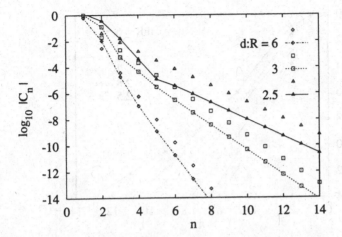

Fig. 7. The coefficients of the cycle expansion for the quantum Selberg zeta function $Z_{qm}(z)$ (connected with lines) and for $1/\zeta_0$ (not connected) for ratios $d : R = 6$ (diamonds), 3 (squares) and 2.5 (triangles). Note that the asymptotic slopes for the quantum Selberg zeta functions and the dynamical zeta functions are the same; the double pole present in the dynamical zeta function persists as a single pole in the Selberg product.

repellers. In the present case, such preasymptotic oscillations are noticable in data for larger disk-disk spacings.

5.3 Poles of dynamical zeta functions

The exponential decay of the coefficients for $1/\zeta_0$ indicates the presence of a pole. Numerical investigations [21] of both the classical and the quantum dynamical $1/\zeta_j$ functions for two-dimensional Hamiltonian flows indicate that a $1/\zeta_j$ function has a *double* pole coinciding with the leading zero of $1/\zeta_{j+1}$. Consequently $1/\zeta_0$, $1/\zeta_0\zeta_1$ and the quantum Selberg zeta function all have the same leading pole, and coefficients in their cycle expansions fall off exponentially with the same slope, fig. 7. Multiplying the quantum Selberg zeta function by $(1 - z/z_1)$, where z_1 is the leading zero of $1/\zeta_1(z)$, one obtains faster, but still exponential decay in the coefficients, indicating further poles down in the complex plane (see fig. 8).

The double pole is not as surprising as it might seem at first glance; indeed, the theorem that establishes that the classical Fredholm determinant (14) is entire implies that the poles in $1/\zeta_j$ must have right multiplicities in order that they be cancelled in the $F = \prod 1/\zeta_j$ product [22]. More explicitly, $1/\zeta_j$ can be expressed in terms of weighted Fredholm determinants

$$F_j = \exp\left(-\sum_p \sum_{r=1}^{\infty} \frac{1}{r} \frac{(t_p/\Lambda_p^j)^r}{(1 - 1/\Lambda_p^r)^2}\right) \tag{28}$$

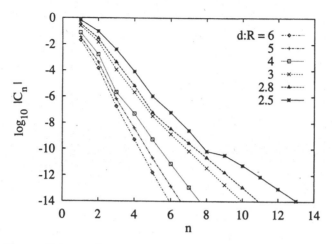

Fig. 8. The coefficients of the cycle expansion for the quantum Selberg zeta function $Z_{qm}(z)$ after multiplication with $(1 - z/z_1)$ where z_1 is the zero of $\zeta_1^{-1}(z)$. The coefficients still decay only exponentially but with much steeper slope, indicating the presence of yet another pole.

by inserting the identity

$$1 = \frac{1}{(1 - 1/\Lambda)^2} - \frac{2}{\Lambda} \frac{1}{(1 - 1/\Lambda)^2} + \frac{1}{\Lambda^2} \frac{1}{(1 - 1/\Lambda)^2}$$

into the exponential representation (13) of $1/\zeta_j$. This yields

$$1/\zeta_j = \frac{F_j F_{j+2}}{F_{j+1}^2} \, , \tag{29}$$

and we conclude that for two-dimensional Hamiltonian flows the dynamical zeta function $1/\zeta_j$ has a *double* leading pole coinciding with the leading zero of the F_{j+1} Fredholm determinant.

The effect of such convergence properties of the coefficients on the calculation of classical and quantum escape rates is demonstrated in table 5 and fig. 9.

6 Conclusion

As we have shown in the above, the three-disk system is an ideal model for tests of periodic orbit expansions in hyperbolic systems, not only in the classical but also in the semiclassical context. With a little bit of geometry one can obtain very good estimates for the classical and quantum escape rates as a function of the separation : radius ratio (fig. 3), demonstrating the accuracy of approximating the repeller by a few scales. With numerically obtained periodic orbits up to period 14 one can test the analyticity properties of quantum Selberg zeta functions and

Table 5. Classical and quantum escape rates computed from the Fredholm determinant F_{cl} (11), the quantum Selberg zeta function Z_{qm} (17), and the dynamical zeta function $1/\zeta_0$, as function of the maximal cycle length. Due to the presence of the same pole in both quantum zeta functions, the convergence of the quantum Selberg zeta function is not significantly better than the convergence of the dynamical zeta function. See also fig. 9.

	n	F_{cl}	Z_{qm}	$1/\zeta_0$
	1	0.39	0.11	0.119
	2	0.4105	0.12153	0.12152
	3	0.410338	0.1215574	0.121556
$d : R = 6$	4	0.4103384074	0.121557625	0.12155760
	5	0.4103384077696	0.1215576283	0.121557627
	6	0.410338407769346482	0.1215576284	0.1215576284
	7	0.4103384077693464892	0.1215576284	0.1215576284
	8	0.410338407769346489338468	.	.
	9	0.410338407769346489338461613074	.	.
	10	0.410338407769346489338461613078192	.	.
	1	0.41	-0.076	0.019
	2	0.72	0.041	0.038
	3	0.675	0.04052	0.0403
	4	0.67797	0.040575	0.04054
$d : R = 3$	5	0.677921	0.0405789	0.040575
	6	0.6779227	0.04057935	0.040578
	7	0.6779226894	0.040579405	0.0405793
	8	0.6779226896002	0.0405794102	0.04057940
	9	0.677922689599532	0.0405794108	0.0405794099
	10	0.67792268959953606	0.0405794108	0.0405794107

Fredholm determinants; to our surprise we found that quantum Selberg zeta functions have poles. Their presence spoils the faster than exponential convergence typical of the classical Fredholm determinants; whether their analyticity can be improved is still being investigated. In the case of sufficiently separated three disks, this may not seem to be terribly important (one already has good exponential convergence), but the more poorly converging cycle expansions for spectra of bounded systems also seem to have poles.

Quite generally, despite considerable progress (see, for example, the periodic orbit theory theme issue of *CHAOS* [23]), the semiclassical quantization of bounded systems is still not a routine calculation: all known bounded systems have either pruning, marginally stable periodic orbits and/or accumulating sequences of orbits, and the quantum Selberg zeta functions have poles that degrade the convergence of cycle expansions. We remain optimistic, and believe that in the near future many of these problems will be overcome.

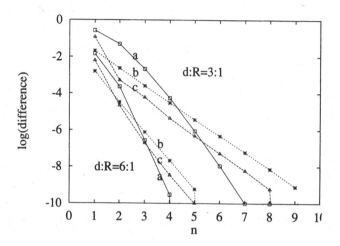

Fig. 9. Convergence of zeros of zeta functions towards the asymptotic values as more and more orbits are included. Shown are the convergence of the classical Fredholm determinant escape rate estimates, the quantum Selberg zeta function lowest resonance estimates, and the quantum $1/\zeta_0$ lowest resonance estimates for ratios $d : R = 6$ and $d : R = 3$. The zero obtained for $n = 14$ has been taken as the asymptotic value. Note the faster than exponential convergence for the classical Fredholm determinant, and the exponential approach for the quantum zeta functions.

Acknowledgements

The present collaboration was initiated at the Workshop on Chaos in Dynamical Systems, the Max-Planck Institute für Mathematik, Bonn, April 1988. P.C. is grateful to the Carlsberg Fundation for support, and to I. Procaccia for the hospitality at the Weizmann Institute, where part of this work was done. B.E. was supported in part by the National Science Foundation under grant PHY82-17853, supplemented by funds from the National Aeronautics and Space Administration. G.R. acknowledges support through the Alexander von Humboldt Foundation. It is a pleasure to thank E.B. Bogomolny, F. Christiansen, P. Gaspard, P. Grassberger, S. Fishman, R. Mainieri, K. Müller, I. Percival, H.H. Rugh, U. Smilansky, and D. Wintgen for stimulating discussions and fruitful collaborations.

APPENDIX

A: Numerical determination of periodic orbits

Fixed-point searches based on direct solution of the fixed-point condition (4) as an initial value problem are numerically rather unstable. Methods such as the multipoint shooting or orbit length minimization, described here, are considerably more robust and faster. Other methods for determination of n-disk periodic orbits are given in refs. [24, 25]. A preliminary step to either calculation is preparation of a list of all distinct allowed prime periodic symbol sequences; an example of such list is given in table 1.

Multipoint shooting method

In the multipoint shooting approach one treats the N cycle points (for example, the five cyclic permutations of $\overline{00101}$) as independent degrees of freedom, and solves the system of equations

$$
\begin{aligned}
T_1 x_{10100} &= x_{01001} \\
T_0 x_{01001} &= x_{10010} \\
T_1 x_{10010} &= x_{00101} \qquad\qquad \text{(A1)} \\
T_0 x_{00101} &= x_{01010} \\
T_0 x_{01010} &= x_{10100}
\end{aligned}
$$

using the $2N$-dimensional Newton method.

In most of our computations, we fix the ratio $d : R = 6$ and use the fixed points of the maps T_i as starting guesses; the Newton method works quickly and reliably for all the period lengths tested (all orbits of length $N \leq 15$). Once a periodic orbit has been found, orbits for different $d : R$ ratios may be obtained by (adiabatically) varying the ratio, and using the old orbit points as starting guesses in the Newton method. This works well for $d : R$ larger than about 2.4. For smaller values, some orbits change rather quickly and require very small step sizes. In addition, for ratios below $d : R = 2.04821419\ldots$ one has to worry about pruning, *i.e.*, the possibility that the minimal length trajectories are blocked by intervening disks. (This problem has no easy solution and is not treated here.)

Orbit length minimization method

The simplest method for determining billard cycles is given by the principle of least action, or, equivalently, by minimizing the length of an approximate orbit that visits a given sequence of disks. In contrast to the multipoint shooting method which requires variation of $2N$ phase-space points, minimization of the cycle length requires variation of only N bounce positions s_i, $i = 1, 2, \ldots, N$.

Let the points (x_i, y_i) denote the centers of the N nonintersecting disks with radii R_i. The length (or equivalently, the period or the action) of the approximate cycle is given by

$$
L = \sum_{i=1}^{N} l_i = \sum_{i=1}^{N} [(\Delta x_i)^2 + (\Delta y_i)^2]^{1/2}
$$

where $\Delta x_i \equiv x_{i+1} - x_i + R_{i+1} \cos(s_{i+1}) - R_i \cos(s_i)$, $x_{N+1} \equiv x_1$, and similarly for

Δy_i. The cycle length varies with variation of s_i as

$$(\vec{\nabla}L)_i = \frac{\partial l_{i-1}}{\partial s_i} + \frac{\partial l_i}{\partial s_i} = R_i \left(\sin(s_i) \left(\frac{\Delta x_i}{l_i} - \frac{\Delta x_{i-1}}{l_{i-1}} \right) - \cos(s_i) \left(\frac{\Delta y_i}{l_i} - \frac{\Delta y_{i-1}}{l_{i-1}} \right) \right).$$
(A2)

The minimization is achieved by recursive implemention of the following algorithm:

    (i) Select an initial set of bounce positions $\vec{s}_0$.

    (ii) Evaluate $\vec{\nabla}L|_{\vec{s}=\vec{s}_0}$.

    (iii) Minimize L along the tangent space spanned by the above gradient, *i.e.*, minimize the function $L(\vec{s}_0 + \vec{\nabla}L|_{\vec{s}=\vec{s}_0} \cdot \vec{t})$ with respect to the variation $\vec{t}$.

    (iv) Use the bounce points $\vec{s}_1$ so determined as the starting point for the next iteration of the algorithm, until the desired accuracy is attained.

    (v) If the dynamics is pruned, check that the final minimal length orbit does not penetrate any of the disks.

The orbit minimization algorithm works very well in practice.

B: Cycle stability for billiards

Consider a two-dimensional billiard with phase space coordinates (q_1, q_2, p_1, p_2). Let t_k be the instant of the k-th collision of the billiard with the billiard boundary, and $t_k^\pm = t_k \pm \epsilon$, ϵ positive and infinitesimal. Setting the mass and the velocity equal to 1, we impose energy conservation by parametrizing the momentum direction by angle θ, $(q_1, q_2, \sin\theta, \cos\theta)$. Now parametrize the two-dimensional neighbourhood of a trajectory segment between $(k-1)$-th and k-th collisions by $\delta x = (\delta\theta, \delta z)$, where

$$\delta z_k = \delta q_1 \cos\theta_k - \delta q_2 \sin\theta_k,$$

is the coordinate variation transverse to the k-th segment of the flow. Using $dq_i/dt = p_i$, we obtain the equations of motion for the linearized neighbourhood

$$\frac{d}{dt}\delta\theta = 0, \quad \frac{d}{dt}\delta z = \delta\theta.$$

Let $\delta\theta_k = \delta\theta(t_k^+)$ and $\delta z_k = \delta z(t_k^+)$ be the local coordinates immediately after the k-th collision, and $\delta\theta_k^- = \delta\theta(t_k^-)$, $\delta z_k^- = \delta z(t_k^-)$ immediately before. Integrating the free flight from t_{k-1}^+ to t_k^- we obtain

$$\begin{aligned}
\delta\theta_k^- &= \delta\theta_{k-1} \\
\delta z_k^- &= \delta z_{k-1} + \tau_k \delta\theta_{k-1}, \qquad \tau_k = t_k - t_{k-1},
\end{aligned}$$
(A3)

and the transverse jacobian is given by

$$\mathbf{J}_T(x_k) = \begin{bmatrix} 1 & 0 \\ \tau_k & 1 \end{bmatrix}.$$

At incidence angle ϕ_k (the angle between the outgoing particle and the outgoing normal to the billiard edge), the incoming transverse variation δz_k^- projects onto an arc on the billiard boundary of length $\delta z_k^- / \cos \phi_k$. The corresponding incidence angle variation $\delta \phi_k = \delta z_k^- / R_k \cos \phi_k$, $R_k = $ the local radius of curvature, increases the angular spread to

$$
\begin{aligned}
\delta \theta_k &= -\delta \theta_k^- - \frac{2}{R_k \cos \phi_k} \delta z_k^- \\
\delta z_k &= -\delta z_k^- ,
\end{aligned}
\tag{A4}
$$

so the jacobian associated with the reflection is

$$
\mathbf{J}_R(x_k) = - \begin{bmatrix} 1 & r_k \\ 0 & 1 \end{bmatrix}, \qquad r_k = \frac{2}{R_k \cos \phi_k}.
$$

The jacobian of a cycle p of length n_p is given by

$$
\mathbf{J}_p = (-1)^{n_p} \prod_{k=1}^{n_p} \begin{bmatrix} 1 & r_k \\ 0 & 1 \end{bmatrix} \begin{bmatrix} 1 & 0 \\ \tau_k & 1 \end{bmatrix}.
\tag{A5}
$$

As $\det \mathbf{J} = 1$, the sign of the leading eigenvalue depends only on the trace of the determinant: $\Lambda = \frac{1}{2}(trJ \pm \sqrt{tr^2 J - 4})$, and by (A5) the trace after n compositions of the determinants has the sign $(-1)^n$, *i.e.*, the eigenvalues flip sign at each collision. This yields a convenient way of finding the correct sign of the stabilities in the fundamental domain, since a straight wall can be considered as the limit of a disk whose radius tends to infinity.

An alternative approach to the eigenvalue evaluation is based on observation that the [2x2] volume preserving matrix multiplication can be achieved by iteration of linear fractional maps;

$$
\mathbf{M}_i = \begin{bmatrix} c_i & a_i \\ d_i & b_i \end{bmatrix}, \quad \det \mathbf{M}_i = 1, \quad \rightarrow \quad T_i(z) = \frac{a_i + b_i z}{c_i + d_i z},
$$

$$
\mathbf{M}_{ij} = \mathbf{M}_j \mathbf{M}_i \quad \rightarrow \quad T_i T_j(z) = \frac{a_i c_j + b_i a_j + (a_i d_j + b_i b_j)z}{c_i c_j + d_i a_j + (c_i d_j + d_i b_j)z}
$$

If we represent the translations and the reflections by

$$
\begin{aligned}
\mathbf{J}_R(x_k) &\quad \rightarrow \quad R_k(z) = r_k + z, \\
\mathbf{J}_T(x_k) &\quad \rightarrow \quad T_k(z) = \frac{1}{\tau_k + \frac{1}{z}},
\end{aligned}
$$

the k-th segment of the trajectory is represented by

$$
T_k R_k(z) = \frac{1}{\tau_k + \dfrac{1}{r_k + z}}.
\tag{A6}
$$

For the cycle p the iteration yields

$$
z = T_1 R_1 T_2 R_2 \cdots T_{n_p} R_{n_p}(z) = \frac{a_p + b_p z}{c_p + d_p z}.
\tag{A7}
$$

For dispersing billiards $r_m > 0$, so all coefficients are positive. The expanding eigenvalue satisfies $|\Lambda_p| + 1/|\Lambda_p| = b_p + c_p$, so the cycle eigenvalue is a root of the quadratic equation

$$|\Lambda_p| = \frac{b_p + c_p + \sqrt{(b_p + c_p)^2 - 4}}{2} = b_p - d_p z_p. \tag{A8}$$

alternatively given as the root z_p of the linear fractional representation fixed point condition (A7). The sign of Λ_p is $(-1)^{n_p}$.

Eqs. (A3) and (A4) can be rewritten as

$$\frac{\delta z_k^-}{\delta \theta_k^-} = \frac{\delta z_{k-1}}{\delta \theta_{k-1}} + \tau_k$$

$$\frac{\delta \theta_k}{\delta z_k} = \frac{\delta \theta_k^-}{\delta z_k^-} + \frac{2}{R_k \cos \phi_k}, \tag{A9}$$

leading to the continued fraction recursion (A6)

$$\kappa_{k-1} = \cfrac{1}{\tau_k + \cfrac{1}{r_k + \kappa_k}}, \qquad \kappa_k = -\frac{\delta \theta_k}{\delta z_k}. \tag{A10}$$

for the Sinai-Bunimovič curvatures [26, 27]. In other words, the Sinai-Bunimovič continued fraction method of evaluating curvatures (used, for example, in ref. [25]) is identical to multiplying [2x2] jacobian matrices, the method by which we evaluate the stabilities.

Interpretation: imagine a set of projectiles leaving a point (q_1, q_2) in all directions, parametrized by angle θ; they generate a "horocycle" in the configuration space, a set of all points reached by time t. A $\delta \theta$ wedge of angles stretches into a horocycle arc $\delta z = t \delta \theta$, and $\delta \theta / \delta z = 1/t$ is the local curvature of the horocycle [26, 27].

REFERENCES

[1] M.C. Gutzwiller, *Chaos in Classical and Quantum Mechanics* (Springer, New York 1990)

[2] A. Selberg, *J. Ind. Math. Soc.* **20**, 47 (1956)

[3] M.V. Berry and K.E. Mount, *Rep. Prog. Phys.* **35**, 315 (1972)

[4] D. Wintgen, *Phys. Rev. Lett.* **58**, 1589 (1987)

[5] B. Eckhardt and E. Aurell, *Europhys. Lett.* **9**, 509 *(1989)*

[6] P. Cvitanović, *Phys. Rev. Lett.* **61**, 2729 (1988)

[7] R. Artuso, E. Aurell and P. Cvitanović, *Nonlinearity* **3**, 325 (1990)

[8] P. Cvitanović and B. Eckhardt, *Phys. Rev. Lett.* **63**, 823 (1989)

[9] D. Ruelle, *Statistical Mechanics, Thermodynamic Formalism* (Addison-Wesley, Reading 1978)

[10] F. Christiansen, G. Paladin and H.H. Rugh, *Phys. Rev. Lett.* **65**, 2087 (1990)

[11] B. Eckhardt, *J. Phys.* A **20**, 5971 (1987)

[12] P. Gaspard and S. Rice, *J. Chem. Phys.* **90**, 2225 (1989)

[13] P. Gaspard and S. Rice, *J. Chem. Phys.* **90**, 2242 (1989); **90**, 2255 (1989)

[14] B. Eckhardt, *Physica* **D 33**, 89 (1988)

[15] P. Cvitanović and B. Eckhardt, *J. Phys.* **A 24**, L237 (1991)

[16] R. Balian and C. Bloch, *Ann. Phys. (NY)* **85**, 514 (1974)

[17] A. Voros, *J. Phys.* **A 21**, 685 (1988)

[18] P.E. Rosenqvist (unpublished)

[19] P. Cvitanović and B. Eckhardt, *Nonlinearity* **6**, 277 (1993)

[20] H.H. Rugh, *Nonlinearity*,**5**, 1237 (1992)

[21] B. Eckhardt and G. Russberg, *Phys. Rev.* **E 47**, 1578 (1993)

[22] P. Cvitanović and P.E. Rosenqvist, *A new determinant for quantum chaos*, Proc. SISSA workshop "From Classical to Quantum Chaos", Trieste (July 1992)

[23] P. Cvitanović, ed., *Periodic Orbit Theory - theme issue*, *CHAOS* **2**, 1-158 (1992)

[24] T. Harayama and A. Shudo, *J. Phys.* **A 25**, 4595 (1992)

[25] P. Gaspard and D. Alonso Ramirez, *Phys. Rev.* **A 45**, 8383 (1992)

[26] Ya.G. Sinai *Usp. Mat. Nauk* **25**, 141 (1970)

[27] L. Bunimovič, *Comm. Math. Phys.* **65**, 295 (1979)

Logarithm breaking time in quantum chaos

G. P. BERMAN

*Center for Nonlinear Studies, MS-B258 Los Alamos National Laboratory,
Los Alamos, New Mexico 87545, U.S.A.*

*Kirensky Institute of Physics; Research and Educational Center for Nonlinear Processes,
Krasnoyarsk Polytechnical Institute; 660036 Krasnoyarsk, Russia*

G. M. ZASLAVSKY

*Courant Institute of Mathematical Sciences and Physical Department, New York University,
251 Mercer Street, New York, NY 10012, U.S.A.*

Abstract

The breaking time $\tau_\hbar$ of the semiclassical approximation for the quantum chaos situation is discussed for two kinds of models. The first model is a periodically kicked oscillator, for which the logarithmic estimation $\tau_\hbar = \ell n(const./\hbar)$ has been obtained. The second model is an ensemble of N atoms interacting with light in the resonant cavity. For this model, estimation gives: $\tau_\hbar = \ell n(const.N)$. Both examples develop our earlier logarithmic estimation for the breaking time [1], and extend the possibility of observing the phenomena of quantum chaos, particularly for macroscopic systems.

1 Introduction

Recently, the criteria for chaos to occur have been understood mainly for simple (1.5 or 2 degrees of freedom) dynamical Hamiltonian systems which are considered classically [2-4]. The main conditions are the local instability of trajectories and the limiting region of phase space (at least in one direction). Quantum chaos deals mainly with investigations of quantum dynamical systems which are chaotic in the classical limit. In particular, the behavior of such systems in the quasiclassical region is not completely understood. The fundamental problem that we are concerned with here is the time-scale $\tau_\hbar$ of the classical-quantum correspondence when the condition of strong chaos in the classical limit is fulfilled. An understanding of this problem would allow us to define one of the main parameters of quantum chaos - the characteristic time-scale of the quasiclassical approach. For a wide class of time-dependent problems, expressions for $\tau_\hbar$ have been obtained in Refs [1],[5]–[10]. The meaning of this parameter is as follows: Consider, for example, a quantum nonlinear oscillator in the quasiclassical region with the wave function localized initially in action in the domain $\delta I \sim \hbar$ (I is the characteristic action of the system). If strong chaos occurs in the classical limit, then for a rather short time $\tau_\hbar$ the wave packet spreads over the phase volume: $\Delta I \sim \hbar exp(\lambda \tau_\hbar)$ where λ is the characteristic Lyapunov exponent. Therefore, for the time-scale $\tau_\hbar$, one has: $\tau_\hbar \sim \lambda^{-1}\ell n(\Delta I/\hbar) = (\ell n\kappa)/\lambda$, where κ is of the order of the number of quanta of characteristic wave packet width. So, the time-scale $\tau_\hbar$ arises because of the simultaneous interaction of the effects of quantum dynamics and the classical local instability in phase space. Of course, such a simple consideration is not rigorous, and thorough investigation is necessary for different systems and initial conditions. In fact, the existence of the time-scale

$$\tau_\hbar = C_1 \ell n(C_2/\hbar) \tag{1.1}$$

where $C_{1,2}$ are constants has been observed and discussed in detail for some typical models of quantum chaos [1,5–13]. The time-scale $\tau_\hbar$ may be one of the universal and fundamental characteristics of quantum chaos accessible to experimental observation.

This chapter deals with the investigation of quantum chaos for two types of quantum systems in which the unperturbed Hamiltonian describes linear oscillations. Namely, we consider: (1) a kicked quantum linear oscillator; and (2) an ensemble of two-level atoms interacting with a self-consistent electromagnetic field in a resonant cavity. It is shown that under the conditions of strong chaos in the classical limit, the time-scale $\tau_\hbar$ has the form (1.1) for both systems, and specifically for the second problem $\kappa \sim N$ where N is the number of atoms. Then $\tau_\hbar \sim C_1 \ell n(C_2 N)$, and as $N \gg 1$, this system represents significant interest for the experimental observation of the time-scale $\tau_\hbar$.

2 Kicked quantum linear oscillator

Consider the Hamiltonian of the system in the form [14]:

$$H = a^+ a + 1/2 + \kappa \, cos\left[(k/\sqrt{2})(a^+ + a)\right] \sum_{n=-\infty}^{\infty} \delta(t - nT) \tag{2.1}$$

where a^+, a are the creation and annihilation boson operators ($[a, a^+] = 1$); κ, k, T are dimensionless parameters and t is dimensionless time. It is easy to find that the evolution operator $\hat{T}$ has the form:

$$\left. \begin{array}{l} \hat{T} = exp\left[-iT(a^+ a + 1/2)\right] \cdot exp\left\{-i\kappa \, cos\left[(k/\sqrt{2})(a^+ + a)\right]\right\} \\ \psi_n = \hat{T}\psi_{n-1} = \hat{T}^n \psi_0; \ (\psi_n \equiv \psi(t_n - 0)); \ t_n = nT \end{array} \right\} \tag{2.2}$$

where ψ_n is the solution of Schrödinger equation.

2.1 Classical limit

The classical variant of the model (2.1) was introduced and studied in Refs [15] and [16] (see also review [17]). In the classical limit, the Heizenberg operator a_n

must be replaced by the c-number α_n

$$\alpha_{n+1} = exp(-iT)\left\{\alpha_n + i\kappa(k/\sqrt{2})sin\left[(k/\sqrt{2})(\alpha_n^* + \alpha_n)\right]\right\} \qquad (2.3)$$

Analysis of (2.3) has shown that for arbitrary values of the parameter

$$K = K_H|sin\,T| = \kappa k^2|sin\,T| \qquad (2.4)$$

there exists a stochastic particle dynamics in a certain region of phase space. This region forms a so-called stochastic web of q-fold symmetry if the resonant condition $T = 2\pi r/q$ occurs and $K << 1$ (r and q are intergers). When $K >> 1$ strong chaos occurs. Special q-fold symmetry of the chaotic pattern in the phase space reveals the corresponding properties of quantum dynamics. Particularly, it is shown that there is no localization of the wave function at least for the $q = 3, 4, 6[14]$.

2.2 Quantum dynamics of observable values

To compare a classical and a quantum dynamics, we consider here as an example, the behavior of a quantum characteristic function

$$D_n(ik/\sqrt{2}) \equiv\, < \psi_n|exp(ik\eta)|\psi_n > \qquad (2.5)$$

where $\eta = (1/\sqrt{2})(a^+ + a)$ is a dimensionless oscillator coordinate. Using the evolution operator $\hat{T}$ (2.2) for this function, the following recurrent relation can be derived [14]:

$$D_n(\gamma_0) =\, < \psi_n|exp(\gamma_0 a^+ - \gamma_0^* a)|\psi_n > =\, \sum_{m_1,...,m_n=-\infty}^{\infty} J_{m_1}(2\kappa\, sin\, \xi_1) \cdot$$

$$J_{m_2}(2\kappa\, sin\, \xi_2) \cdots J_{m_n}(2\kappa\, sin\, \xi_n) \cdot D_0(\gamma_n) \qquad (2.6)$$

where $J_m(x)$ is the Bessel function, $D_0(\gamma_n) =\, < \psi_0|exp(\gamma_n a^+ - \gamma_n^* a)|\psi_0 >$ is the initial condition, $|\psi_0 >$ is an arbitrary initial wave function, and the following definitions are used:

$$\left.\begin{array}{l} \gamma_0 = ik/\sqrt{2}, \quad \gamma_\ell = \gamma_{\ell-1}exp(iT) + im_\ell k/\sqrt{2} \\ \xi_\ell = -(k/2\sqrt{2})(\gamma_\ell + \gamma_\ell^*); \quad (\ell = 1, ..., n). \end{array}\right\} \qquad (2.7)$$

In a classical limit, it is necessary to replace $sin\, \xi_\ell$ in (2.6) by ξ_ℓ, then we have from (2.6), (2.7):

$$D_n^{(cl)}(\gamma_0) = \sum_{m_1,...,m_n=-\infty}^{\infty} J_{m_1}\{K_H\, sin\, T\}J_{m_2}\{K_H[sin(2T) + m_1 sin\, T]\} \cdots$$

$$J_{m_n}\{K_H[sin(nT) + m_1 sin(n-1)T + \cdots + m_{n-1} sin\, T]\}D_0^{(cl)}(\gamma_n) \qquad (2.8)$$

where $K_H = \kappa k^2$ (see (2.4)). The initial condition $D_0^{(cl)}(\gamma_n)$ in the classical limit is determined by the following expression:

$$D_0^{(c\ell)}(\gamma_n) = \int\limits_{-\infty}^{\infty} d^2\alpha P(\alpha, \alpha^*) exp(\gamma_n \alpha^* - \gamma_n^* \alpha)$$

where $P(\alpha, \alpha^*)$ is an initial distribution function. Now we shall compare quantum dynamics (2.6) with classical dynamics (2.8), assuming that for $\hbar \to 0$ the quantum initial condition transforms to the classical one.

2.3 *Quantum boundary of stochasticity*

It is seen from (2.6) that the dynamics is completely quantum at the times $t \geq T(n \geq 1)$ under the condition $|\xi_1| \geq 1$, which gives according to (2.7)

$$k^2 |sin\ T|/2 = K/2\kappa \geq 1$$

In the region in which parameters $K \gg 1$ and $\kappa \gg 1$ where the inverse inequality $K/2\kappa \ll 1$ is satisfied, the dynamics of the quantum system approximately coincides with the classical one for some finite time $t > T$ (or even $t \gg T$). So, we shall call the equality: $2\kappa = K$ the quantum boundary of stochasticity (QBS), which was introduced in Ref. [18] for the quantum kicked nonlinear oscillator.

2.4 *The time of quasiclassical description for strong chaos*

Consider the case of strong chaos in the classical limit when $K \gg 1$. To find the time-scale $\tau_\hbar$ in which quantum dynamics approximately coincides with classical chaotic dynamics, we compare expressions (2.6) and (2.8). The applicability of the classical expression (2.8) for describing quantum dynamics at finite times $n > 1$ means that in (2.6) at least the condition $|\xi_\ell| \ll 1 (\ell = 1, ..., n)$ should be satisfied and all functions $sin\ \xi_\ell$ in (2.6) may be approximated by ξ_ℓ. Then, taking into account that at $|m_\ell| \gg 2\kappa|\xi_\ell|$ the Bessel function $J_{m_\ell}(2\kappa|\xi_\ell|)$ is exponentially small, we have from (2.6) an estimate of the characteristic number $m_\ell(P = 1, ..., n)$ of terms in (2.6)

$$\left.\begin{aligned}
|m_1| &\sim 2\kappa|\xi_1| = K_H|sin\ T| = K \\
|m_2| &\sim 2\kappa|\xi_2| = K_H|sin\ 2T + m_1 sin\ T| \\
|m_n| &\sim 2\kappa|\xi_n| = K_H|sin(nT) + m_1 sin(n-1)T + \cdots + m_{n-1} sin\ T|
\end{aligned}\right\} \quad (2.9)$$

At strong chaos ($K \gg 1$) we have from (2.9)

$$|m_1| \sim K \gg 1, \quad |m_2| \sim K^2, \quad \cdots, \quad |m_n| \sim K^n = exp(n\ell nK).$$

Then we have the estimate for the value

$$|\xi_n| \sim |m_n|/2\kappa \sim exp(n\ell nK)/2\kappa.$$

So, the condition $|\xi_n| < 1$ gives

$$t < \tau_\hbar = \ell n(2\kappa)/\ell n K; \quad (2\kappa >> K >> 1) \tag{2.10}$$

Thus, in the region of parameters K and κ (2.10), the dynamics of the quantum system at times $t < \tau_\hbar$ approximately coincides with the chaotic dynamics of classical system, and the time-scale $\tau_\hbar$ has the form (1.1).

2.5 Weak quantum chaos

When $K << 1$ and dynamical chaos is weak in the classical limit, the estimate for $\tau_\hbar$ (we designate it τ_ω) can be presented in the form: $t < \tau_\omega = 2\kappa/K_H^2 \sim 1/\hbar$. So, in the case $K << 1$, the difference between quantum and classical dynamics manifests itself at times considerably exceeding $\tau_\hbar$ (2.10): $\tau_\omega >> \tau_\hbar$.

3 Quantum chaos of atoms in a resonant cavity

It is clear that determination of the time-scale $\tau_\hbar$ is possible only in systems in a quasiclassical region ($\kappa >> 1$). One such system is an ensemble of atoms in a resonator, interacting with self-consistent electromagnetic field. We then present the Hamiltonian in the form:

$$H = \hbar\omega a^+ a + \hbar\omega_0 \sum_{j=1}^{N} S_j^z +$$
$$(16\pi d_0^2 \hbar\omega/V)^{1/2} \sum_{j=1}^{N} (aS_j^+ + a^+ S_j^- + \mu a^+ S_j^+ + \mu a S_j^-) \tag{3.1}$$

where a^+, a are the creation and annihilation operators of photons in a resonator ($[a, a^+] = 1$); $\hat{S}_j = (S_j^z, S_j^+, S_j^-)$ are the operators of dimensionless spin of the value $1/2$ describing the dynamics of j-th two-level atom: ω is the frequency of a chosen mode in the resonator; ω_0, d_0 are the frequency and dipole moment of atomic transition; V is the value of the resonator volume. The parameter μ can take two values $\mu = 0, 1$. The Dicke model [19] corresponds to $\mu = 0$, *i.e.*, to neglecting the nonresonant interaction. In Ref. [20], the complete problem (3.1) was considered in the semiclassical approximation, and it was shown that chaos is due to the presence of terms with $\mu \neq 0$. Subsequent investigations of the model proposed in Ref. [20] confirmed the existence of chaos in it and a significant change of the quantum properties of the problem on passage through the region of strong chaos [21-27]. The measure of dynamical chaos in the semiclassical approximation is characterized by the dimensionless constant of interaction of atoms with the field: $\Lambda = (16\pi\rho_0 d_0^2/\hbar\omega)^{1/2}$, where $\rho = N/V$ is the density of atoms. When $\Lambda << 1$, the stochasticity arises in the exponentially small layer in the vicinity of separatrix; at $\Lambda > 1$ chaos takes a global character. In what follows, we shall be mainly interested in the case in which the energy of the atoms in order of magnitude coincides with the energy of the field in the resonator: $\hbar\omega|\alpha|^2 \sim \hbar\omega_0 N/2$, where $|\alpha|^2 \sim n$ is a characteristic number of photons in the field mode. As $\omega = \omega_0$, then $n \sim N >> 1$, and the system is populated in the quasiclassical region. In the next section, we shall derive a set of closed c-number equations for Hamiltonian (3.1) which describe the dynamics of observable values. As a first step, we introduce the new operators of "collective" motion

$$A = a/\sqrt{N}; \quad A^+ = a^+/\sqrt{N}; \quad \hat{S} = (1/N)(S^z, S^+, S^-) \qquad (3.2)$$

and present Hamiltonian (3.1) in the form

$$H = \hbar\omega N[A^+A + S^z + \Lambda(AS^+ + A^+S^- + \mu A^+S^+ + \mu AS^-)] \qquad (3.3)$$

where we put $\omega_0 = \omega$.

3.1 The equations of motion for observable variables in coherent states

We now introduce boson ($|Z>$) and spin ($|\xi>$) coherent states for the field and atoms at $t = 0$. We have [28-30]:

$$|Z >= exp(NZA^+ - NZ^*A)|0 >; \qquad (3.4)$$

$$|\xi >= (1 + |\xi|^2)^{-J} exp(\xi \hat{J}^+)|J, -J >; \quad (0 \le |Z|, |\xi| < \infty) \qquad (3.5)$$

where $A|Z >= Z|Z >$ and when constructing spin state $|\xi > (3.5)$ the following operator of total spin was used:

$$J = \sum_{j=1}^{N} S_j; \left(J^+ = J^x + iJ^y; [J^2, H]\right.$$
$$= 0; J^2|J, M >= J(J + 1)|J, M >; M = -J, ..., J\right).$$

Consider the time evolution of the system described by Hamiltonian (3.3) and at the initial time $t = 0$ in the coherent state $|Z, \xi >\equiv |Z > |\xi >$ [11]. Let

$$f = f(A^+, A, S^z, S^+, S^-)$$

be an arbitrary operator. Then we have by averaging the Heisenberg equation for f in the coherent state $|Z, \xi >$

$$\left. \begin{array}{l} d <f> /dt =< \xi, Z|\dot{f}|Z, \xi >= (i/\hbar) < \xi, Z|Hf - fH|Z, \xi >; \\ <f> =< f > (Z^*, Z; \xi^*, \xi, t) \equiv < \xi, Z|f(A^+, A, S^z, S^+, S^-)|Z, \xi > \end{array} \right\} \qquad (3.6)$$

Using the techniques developed in Refs [11], [28]-[33], we derive a closed equation for the c-number function

$$\left. \begin{array}{l} <f> (Z^*, Z; \xi^*, \xi, \tau); \quad (\tau = \omega t; \mu = 1) \\ \partial <f> \partial \tau = \hat{K} <f>; \quad \hat{K} = \hat{K}_{c\ell} + (1/N)\hat{K}_q \end{array} \right\} \qquad (3.7)$$

where

$$\left. \begin{array}{l} \hat{K}_{c\ell} = i\left\{ Z^* \frac{\partial}{\partial Z^*} - Z\frac{\partial}{\partial Z} + \Lambda\frac{J}{N}\frac{2}{(1+|\xi|^2)} \left(\xi^*\frac{\partial}{\partial Z^*} - \xi\frac{\partial}{\partial Z}\right) + \right. \\ +\Lambda\left(Z^*\frac{\partial}{\partial \xi^*} - Z\frac{\partial}{\partial \xi}\right) + \xi^*\frac{\partial}{\partial \xi^*} - \xi\frac{\partial}{\partial \xi} + \Lambda\left(Z^*\xi^2\frac{\partial}{\partial \xi} - Z\xi^{*2}\frac{\partial}{\partial \xi^*}\right) \\ +\Lambda\left[Z\xi^2\frac{\partial}{\partial \xi} - Z^*\xi^{*2}\frac{\partial}{\partial \xi^*} + Z\frac{\partial}{\partial \xi^*} - Z^*\frac{\partial}{\partial \xi} + \frac{J}{N}\frac{2}{(1+|\xi|^2)}\left(\xi\frac{\partial}{\partial Z^*} - \xi^*\frac{\partial}{\partial Z}\right)\right]\right\}; \\ \hat{K}_q = -i\Lambda\left(\xi^{*2}\frac{\partial^2}{\partial Z^*\partial \xi^*} - \xi^2\frac{\partial^2}{\partial Z\partial \xi}\right) + i\Lambda\left(\frac{\partial^2}{\partial Z^*\partial \xi^*} - \frac{\partial^2}{\partial Z\partial \xi}\right) \end{array} \right\} \qquad (3.8)$$

Using A^+, A, S^+, S^Z as operator f we have equations (3.7) for $< A^+ > < A >$, $< S^+ >, < S^Z >$ with the following initial conditions:

$$< A_0^+ >= Z^*; < A_0 >= Z; < S_0^+ >= \frac{J}{N} \frac{2\xi^*}{(1+|\xi|^2)}; < S_0^Z >= -\frac{J}{N} \frac{(1-|\xi|^2)}{(1+|\xi|^2)}$$

We now discuss some properties of operator $\hat{K}$ in (3.7). The operator $\hat{K}_{cl}$ in (3.7) describes the semiclassical limit

$$N \to \infty, J \to \infty; J/N = const.$$

and corresponds to the substitution of all operators in Heizenberg equations by c-numbers. All quantum corrections are described by the operator $\hat{K}_q$ in (3.7). As the operator $\hat{K}_{cl}$ includes only the first derivatives, the semiclassical limit can be considered using the method of characteristics.

3.2 Perturbation method for quantum corrections

To derive the quantum corrections to the classical solution, we shall use the method of perturbation by a small parameter $1/N$. We introduce a correlation function for two arbitrary operators F and G: $P(F,G)/N =< FG > - < F >< G >$. Using (3.7), we have the equation for P:

$$\partial P/\partial \tau = \hat{K}_{cl}P + i\Lambda \left\{ \xi^2 \frac{\partial^+}{\partial(Z,\xi)} - \xi^{*2}\frac{\partial^+}{\partial(Z^*,\xi^*)} + \frac{\partial^+}{\partial(Z^*,\xi^*)} - \frac{\partial^+}{\partial(Z,\xi)} \right\} (< F >, < G >) + \frac{1}{N}\hat{K}_q P \qquad (3.9)$$

where

$$\frac{\partial^+(a,b)}{\partial(Z,\xi)} \equiv \frac{\partial a}{\partial z}\frac{\partial b}{\partial \xi} + \frac{\partial a}{\partial \xi}\frac{\partial b}{\partial Z}$$

and $P(\tau = 0) = 0$. Now we present the dynamical c-number equations for observable values in a form which includes the correlation functions P of the type (3.9). Introduce for this new operators

$$A^+ = A_x + iA_y; \quad S^{\pm} = S^x \pm iS^y.$$

Then, we have the set of c-number equations:

$$\left.\begin{array}{l} d < A_{x,y} > /d\tau = \mp < A_{y,x} > + \Lambda(\mu \mp 1) < S^{y,x} >; \\ d < S^{x,y} > /d\tau = \mp < S^{y,x} > \pm 2(1 \mp \mu)\Lambda < A_{y,x} >< S^z > \pm \\ \qquad \pm \frac{2}{N}(1 \mp \mu)\Lambda P(A_{y,x}, S^z); \\ d < S^z > /d\tau = 2\Lambda\{(\mu - 1) < A_y >< S^x > + (\mu + 1) < A_x >< S^y > + \\ \qquad + \frac{1}{N}(\mu - 1)P(A_y, S^x) + \frac{1}{N}(\mu + 1)P(A_x, S^y)\} \end{array}\right\} \qquad (3.10)$$

The system (3.9), (3.10) represents the complete set of exact c-number equations for observable values and correlation functions P. In the general case ($\mu = 1$), the system has only two integrals of motion H and

$$\hat{J}^2 = N\{(S^x)^2 + (S^y)^2 + (S^z)^2\}$$

In the resonant case ($\mu = 0$), the additional integral exists

$$W = A^+A + S^z$$

and this allows us to integrate the system. To solve the system of equations (3.9), (3.10) numerically the following perturbation method was used. When calculating the correlation functions P in (3.9), we neglected the last term $\sim 1/N$ and substituted the functions $<F>$ and $<G>$ with their semiclassical solutions (which were obtained from (3.10) when neglecting correlation functions P). Then the quantum solutions of (3.10) were found numerically taking into consideration correlation functions P.

3.3 The results of numerical calculations

Most attention in the numerical experiment was paid to the derivation of the time-scale $\tau_\hbar$ when quantum solution differs significantly from the semiclassical one. This time $\tau_\hbar$ was determined according to the following criterion [11]. Let us consider the vector

$$\mathbf{D}(\tau) = \left(<A_x>, <A_y>, <S^x>, <S^y>, <S^z>\right).$$

In semiclassical limit $N \to \infty, J \to \infty, J/N = const.$ the length of this vector is $D_{c\ell}(\tau) \equiv |\mathbf{D}_{c\ell}(\tau)|$. We now introduce the function

$$\delta(\tau) = |\mathbf{D}(\tau) - \mathbf{D}_{c\ell}(\tau)|/|\mathbf{D}_{c\ell}(\tau)|$$

which determines the value of quantum corrections to the semiclassical solution if $\delta(0)$ is small (below $\delta(0) = 0$). The time-scale $\tau_\hbar$ was derived from the equation $\delta(\tau_\hbar) = const.$ where in the numerical experiment $const. = 0.02$ and N reached: $N \sim 10^9$. Fig. 1 shows the function $<S^z(\tau)>$ in a resonant approximation (integrable case, $\mu = 0$) and in the semiclassical limit (classical field). In this case, regular nonlinear oscillations occur. The power-law dependence $\tau_\hbar(N)$ for this case is shown in Fig. 2.

Now we come to the nonintegrable case ($\mu = 1$) with strong chaos in a semiclassical limit. The chaotic time dependence $<S^z(\tau)>$ is shown in Fig. 3 for the semiclassical limit. The main results of this part of our work are presented in Figs. 4 and 5. Fig. 4 shows the logarithmical-like dependence $\tau_\hbar(N) \sim C_1\ell n(C_2N)$ for this case ($\mu = 1$). In Fig. 5 we have presented the time dependence of the quantum correlation function

$$P(\tau) = |P(A_y, S^x)| + |P(A_x, S^y)| + |P(A_y, S^z)| + |P(A_x, S^z)|$$

As can be seen, in the case of strongly developed chaos in the semiclassical limit (in the semiclassical approximation $P(\tau) = 0$ for arbitrary time τ), the quantum correlator $P(\tau)$ grows in time according to a law, which is close to exponential.

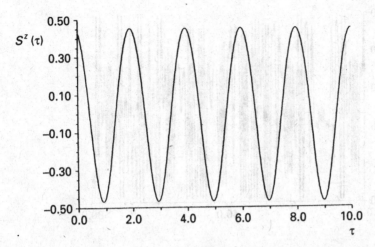

Fig. 1. Time dependence $< S^z(\tau) >$ in the integrable ($\mu = 0$) and semiclassical limit (field is considered classically); the initial conditions $< A_x(0) >=< A_y(0) >= 1$; $< S^x(0) >= 0.25$; $< S^y(0) >= 0.05$; $< S^z(0) >= 0.43$; $\Lambda = 1$.

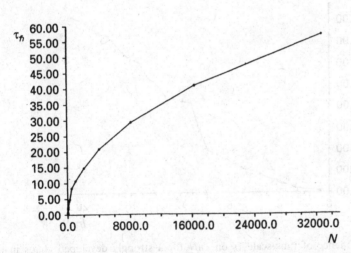

Fig. 2. The dependence of the time-scale $\tau_\hbar$ on the number of atoms N (quasiclassical parameter of the field) in the integrable case ($\mu = 0$). (Parameters and initial conditions are as in Fig. 1.)

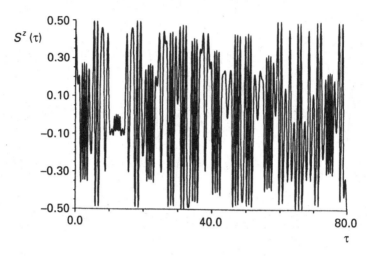

Fig. 3. The time dependence $< S^z(\tau) >$ for semiclassical chaotic motion ($\mu = 1$; $\Lambda = 1$). (Parameters and initial conditions are as in Fig. 1.)

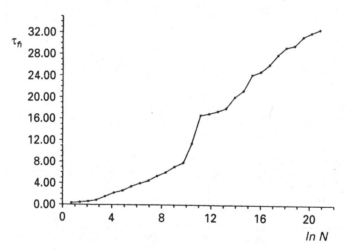

Fig. 4. The dependence of time-scale $\tau_\hbar$ on $\ell n N$ for a strongly developed chaos in the semiclassical limit ($\mu = 1$; $\Lambda = 1$). (Parameters and initial conditions are as in Fig. 1.)

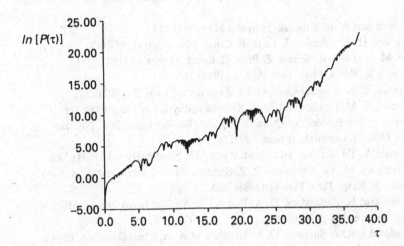

Fig. 5. The time dependence of the quantum correlator $P(\tau)$ for strongly developed chaos in the semiclassical limit ($\mu = 1$; $\Lambda = 1$). (Parameters and initial conditions are as in Fig. 1.)

4 Conclusion

We have considered the peculiarities of quantum dynamics in the important class of dynamical systems in which perturbation and nonlinearity are described by the same terms in the Hamiltonian of interaction. Namely, we have considered the quantum kicked linear oscillator and quantum Dicke model with nonresonant terms. It has been shown that when strongly developed chaos takes place in the semiclassical limit (the radiation field is considered classically), the quantum corrections already play a significant role on the characteristic time-scale $\tau_\hbar \sim C_1 \ell n(C_2 \kappa)$. For the kicked linear oscillator, the quasiclassical parameter $\kappa \sim 1/\hbar$. The considered quantum model of N atoms in a resonator interacting with a self-consistent field taking into account nonresonant terms represents a significant interest as the parameter κ is big enough and is proportional to the number of atoms $N(\kappa \sim N) : \tau_\hbar \sim \ell n N$. This circumstance can be rather important for the verification of the time-scale $\tau_\hbar$ in real experiments.

REFERENCES

[1]   G. P. Berman and G. M. Zaslavsky, Physica 91A (1978) 450.
[2]   B. V. Chirikov, Phys. Reports 52 (1979) 263.
[3]   A. Lichtenberg and M. Lieberman, Regular and Stochastic Motion (Springer, Berlin, 1983).
[4]   G. M. Zaslavsky, Chaos in Dynamic Systems (Harwood Academic Publ., NY, 1985).
[5]   G. M. Zaslavsky, Phys. Reports, 80 (1981) 157.

[6] B. V. Chirikov, F. M. Izrailev and D. L. Shepelyansky, Soviet Scient. Rev. 2C (1981) 208.

[7] G. P. Berman and A. R. Kolovsky, Physica 8D (1983) 117.

[8] H. Frahm and H. J. Mikeska, Z. Phys. B. Cond. Matter, 60 (1985) 117.

[9] F. Haake, M. Kus and R. Scharf, Z. Phys. B. Cond. Matter 66 (1987) 381.

[10] M. Toda and K. Ikeda, Phys. Lett. A, 124 (1987) 165.

[11] G. P. Berman, E. N. Bulgakov and G. M. Zaslavsky, Chaos, 2 (1992) 257.

[12] G. P. Berman, F. M. Izrailev and A. R. Kolovsky, Physica A 152 (1988) 273.

[13] B. V. Chirikov, The Problem of Quantum Chaos (Budker Institute of Nuclear Physics, 630090 Novosibirsk, Russia, 1992).

[14] G. P. Berman, V. Yu. Rubaev and G. M. Zaslavsky, Nonlinearity, 4 (1991) 543.

[15] G. M. Zaslavsky, M. Yu. Zakharov, R. Z. Sagdeev, D. A. Usikov and A. A. Chernikov, Zh. Eksp. Teor. Fiz., 91 (1986) 500.

[16] A. A. Chernikov, R. Z. Sagdeev, D. A. Usikov, M. Yu. Zakharov and G. M. Zaslavsky, Nature, 326 (1987) 559.

[17] G. M. Zaslavsky, R. Z. Sagdeev, D. A. Usikov and A. A. Chernikov, Sov. Physics Uspekhi, 30 (1987) 436.

[18] G. P. Berman and G. M. Zaslavsky, Physica 111A (1982) 17.

[19] R. H. Dicke, Phys. Rev., 93 (1954) 99.

[20] P. I. Belobrov, G. M. Zaslavsky and G. Kh. Tartakovsky, Sov. Phys. JETP 44 (1976) 945.

[21] R. Graham and M. Hohnerbach, Z. Phys. B., 57 (1984) 233.

[22] R. Graham and M. Hohnerbach, Phys. Lett., 101A (1984) 61.

[23] R. F. Fox and J. Eidson, Phys. Rev., 34A (1986) 482.

[24] R. F. Fox and J. Eidson, Phys. Rev., 36A (1987) 4321.

[25] R. F. Fox, in: The Ubiquity of Chaos, Ed. S. Krasner (American Association for the Advancement of Science, 1990) 105.

[26] N. Klenner, J. Weis and M. Doucha, J. Phys., 19C (1986) 4673.

[27] M. B. Cibils, Y. Cuche, W. F. Wreszinski, J-P. Amiet and H. Beck, J. Phys. A. Gen., 23 (1990) 545.

[28] R. J. Glauber, Phys. Rev., 131 (1963) 2766.

[29] J. M. Radcliffe, J. Phys. 4A (1971) 313.

[30] R. J. Glauber and F. Haake, Phys. Rev., 13A (1976) 357.

[31] A. M. Perelomov, Sov. Phys. Usp., 20 (1977) 703.

[32] Yu. A. Sinitsyn and V. M. Tsukernik, Phys. Lett., 90A (1982) 339.

[33] O. B. Zaslavsky, Ukr. Fiz. Zh., 29 (1984) 419.

Semiclassical Propagation: How Long Can It Last?

M. A. Sepúlveda,[1] S. Tomsovic,[2] and E. J. Heller[1],[2]

[1]Department of Chemistry, University of Washington, Seattle, Washington 98195
[2]Department of Physics, University of Washington, Seattle, Washington 98195
(Received 6 April 1992)

The Van Vleck–Gutzwiller propagator is a fundamental quantity in semiclassical theory whose validity was recently demonstrated to extend beyond the time previously thought feasible, i.e., well past the time after which classical chaos has mixed the phase space on a scale smaller than Planck's constant. Little justification was given for this seeming contradiction of the usual vision of semiclassical errors. Though perhaps nonintuitive, we find that standard arguments, properly applied to chaotic dynamics, do explain the long-time accuracy.

Semiclassical approximations to the Schrödinger equation remain important in a large variety of contexts. They play the dual role of computational tools (when exact calculations are too difficult or unnecessary) and sources of insight and intuition, even if numerical solutions are available. However, classical chaos often spoils the utility of semiclassical methods. Gutzwiller [1] gave a formal connection between periodic orbits (embedded in chaos) and eigenvalues (the trace formula). Although the trace formula is not a practical tool and even divergent, it has been the guiding light in the search for more servicable approaches. A large effort to "quantize chaos," over many years, has begun to come to fruition. Recent progress has been dramatic, in both the time domain [2–4] and the energy domain [5–8]. Historically, however, the great bulk of the effort in semiclassical methods has taken place in the energy representation.

In 1928, Van Vleck [9] gave the time-dependent coordinate space propagator which was later modified by Gutzwiller to extend beyond caustics [1]. The Van Vleck–Gutzwiller expression is arguably the progenitor of all other semiclassical formulas: WKB wave functions, the Gutzwiller trace formula, and other energy domain quantities are obtained from it by a stationary phase Fourier transform from time into energy. It is therefore manifestly important to understand the accuracy of the semiclassical propagator and the time scale and mechanism for breakdown. In two rather different chaotic systems, the baker's map [4] and the stadium billiard [3], the semiclassical time propagation of localized initial states was shown to yield accurate dynamics for a much longer time than had been anticipated, based on simple (too simple it turns out) arguments about the classical time for mixing on the scale of Planck's constant. The mixing time for the exponentially unstable dynamics, such that most cells of size h in phase space are accessed by a typical initial state, goes as $\ln(1/h)$. The reasons for the good results well beyond the "logarithmic time" were suggested in the simple, locally linear structures which develop in the evolving phase space Lagrangian manifolds (this was actually exploited in the numerical method used to evaluate the semiclassical expressions). In the stadium case, some 30 000 branches [30 000 terms in the sum over

contributions in Eq. (1) below] of the manifold passed through the region of the initial state. The distance between separate branches of the manifolds was vastly smaller than the distance across a circular phase space cell, yet the semiclassical correlation function was accurate.

The often quoted "truth" that structures on a scale finer than Planck's constant cannot have quantal significance was thus found to be only half true: Such fine structures cannot be seen individually in the quantum amplitudes but collectively they *can* yield accurate and usable semiclassical amplitudes.

How is it that the fine structures contain elements of the correct quantum amplitudes? Here we show that proper consideration of the evolving classical phase space structure, together with standard rules for the domain of accuracy of semiclassical methods, explains the observed long time accuracy. It is unnecessary to average the semiclassical and quantal results to obtain useful comparisons, as was done in Refs. [10,11].

The Van Vleck–Gutzwiller formula is [1,9,12]

$$G(q,q_0;t) \approx G_{sc}(q,q_0;t)$$

$$= \left[\frac{1}{2\pi i \hbar}\right]^{d/2} \sum_j \left|\mathrm{Det}\left(\frac{\partial^2 S_j(q,q_0;t)}{\partial q \, \partial q_0}\right)\right|^{1/2}$$

$$\times \exp[iS_j(q,q_0;t)/\hbar - \tfrac{1}{2} i\pi v_j] . \quad (1)$$

In this expression, the sum over j is for all trajectories connecting q_0 to q in time t, d is the number of degrees of freedom; the determinantal prefactor is the square root of the classical probability for the $q_0 \rightarrow q$ process, and the phase is the classical action $S_j(q,q_0;t)$. An index v_j based on the caustic structure of the evolving manifold $q(t)$ is due to Gutzwiller [1]; see also Maslov and Fedoriuk [12] and a very recent and concise discussion by Littlejohn [13]. The action $S_j(q,q_0;t)$ is the time integral of the Lagrangian

$$S_j(q,q_0;t) = \int_0^t dt' \{p(t') \cdot \dot{q}(t') - H(p(t'),q(t'))\} \quad (2)$$

along the jth classical path (H is the classical Hamiltonian).

Accuracy of semiclassical propagation revolves around the stationary phase approximation. The key to understanding the errors is to regard the initial Lagrangian manifold as primal [12], representing the semiclassical Green's function as an integral over all initial momenta [2,14] corresponding to the initial state $|q_0\rangle$ at $t=0$ for fixed q_0. For the Van Vleck–Gutzwiller expression, one can derive an alternative formulation

$$G_{sc}(\mathbf{q},\mathbf{q}_0;t) = \left(-\frac{i}{4\pi^2\hbar^2}\right)^{d/2}\int d\mathbf{p}_0 \left|\mathrm{Det}\left(\frac{\partial\mathbf{p}_t}{\partial\mathbf{p}_0}\right)\right|^{1/2}$$

$$\times\exp[i\mathbf{p}_t\cdot\mathbf{q}/\hbar + i\tilde{S}(\mathbf{p}_0,\mathbf{q}_0;t)/\hbar - \tfrac{1}{2}i\pi\nu(\mathbf{p}_0)],$$

(3)

where $\tilde{S}(\mathbf{p}_0,\mathbf{q}_0;t) = S(\mathbf{q}_t,\mathbf{q}_0;t) - \mathbf{p}_t\mathbf{q}_t$ and $S(\mathbf{q}_t,\mathbf{q}_0;t)$ is the usual coordinate space action and $\mathbf{q}_t$ is considered to be a function of $\mathbf{p}_0,\mathbf{q}_0$. Note that the sum over separate contributions to the amplitude from $|q_0\rangle$ to $|q\rangle$ is absent, since all the stationary phase contributions are included in the integral over initial values of momentum. After some time t, the current values of the position and momentum are given parametrically (in p_0 for fixed q_0) as

$$\mathbf{q}_t = \mathbf{q}_t(\mathbf{q}_0,\mathbf{p}_0),\quad \mathbf{p}_t = \mathbf{p}_t(\mathbf{q}_0,\mathbf{p}_0).$$

(4)

For convenience we specialize to a two-dimensional phase space. For a chaotic system, the path traced out by the parametric equations grows exceedingly complicated as time increases. The Lagrangian manifold τ_t ($=\{q_t,p_t\}$) begins to track the homoclinic and heteroclinic oscillations of pieces of unstable manifolds, following further along their winding arms as time increases. Nonetheless, we conduct our analysis in terms of standard results for the accuracy of stationary phase integrals. The integral, Eq. (3), has stationary phase points in p_0 whenever $q_t(q_0,p_0)=q$. Near a fold in the Lagrangian manifold, two such stationary phase points coalesce, and in the usual manner the stationary phase integral becomes inaccurate if the stationary phase points are close enough together that an area less than h is enclosed in the sector between the line q and τ_t.

Figure 1 shows the shape of a segment of the $q_0=0$ manifold near $p_0=0$ after several iterations of the standard map [15], except that we have unfolded the map by not applying the 2π modulo condition in angle. This has the advantage of simplifying the structure locally in phase space; the overlap with a localized state is now obtained by replicating that state periodically, as is shown by the shaded disk. We consider the Green's function first, i.e., $\langle q|q_t\rangle$. The figure has regions blackened out that violate the area h rule for a particular value of h. Some of the blackened areas are standard textbook caustics; others are thin "tendrils" [10]. However, no matter how thin a tendril has become nor how many folds upon folds have been generated by the dynamics, the sole criterion for accuracy of a given pair of coalescing station-

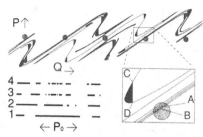

FIG. 1. Upper: Part of the $q_0=0$ manifold unfolded in the angle variable after several iterations. Left: Fragment of the initial q-manifold after removing "bad" segments for times 1, 2, 3, and 4. Right: Detail of the upper diagram showing the coherent state and the semiclassically inaccurate region in dark.

ary phase points in the integration variable p_0 is their distance as measured by the phase accumulation of the exponent between the points in question; this accumulation should be 2π or greater, which translates into the area h in the phase plane. Thus, even though the points marked C,D in Fig. 1 are not separated by 2π, and are generating inaccuracies if $|q\rangle$ should happen to cut through those regions, the contribution from A and B encloses area greater than h and is accurate even though the distance between A and B is minuscule on the scale of $h^{1/2}$. Note that subsequent evolution will fold the Lagrangian manifold further but cannot reduce the area enclosed between A and B (which have been chosen to lie on a stable manifold and will move exponentially closer together). The Poincaré–Cartan theorem of dynamics also guarantees that the area preservation and phase difference between A and B hold even if the phase plane is a surface of section. We emphasize that developing folds, once formed, remain and collapse upon themselves with further evolution while preserving their h area. This has the effect that caustic behavior becomes increasingly nonlocal.

As a global measure of the accuracy of the semiclassical propagator, we adopt the following criterion: Ranges of p_0 which contribute to such caustic zones are eliminated, and the fraction F_p of initial p_0 remaining good constitutes a figure of merit. This is a reasonable measure, because all "good" regions translate into pieces of the evolving Lagrangian manifold which yield accurate amplitudes.

At first it would seem that the good regions of initial p_0 would disappear exponentially fast, since the folds and thus the caustics will proliferate exponentially. However, since the length of the manifold is increasing exponentially in tandem with the folds (typically with the same exponent), folds which develop later each correspond to an exponentially smaller piece of the initial range of p_0. Figure 1 illustrates how the ranges of good initial p_0 become more and more like a Cantor set.

In coordinate space, the inaccurate regions of overlap are almost everywhere if one considers (1) the black regions showing violation of the area h rule in the coordinate space amplitude, and (2) the modulo 2π condition, which folds Fig. 1 over on itself. Also, it should be noted that diffraction errors on the "dark side" of the black regions extend the region of inaccuracy. Still, the Van Vleck–Gutzwiller Green's function can accurately propagate states: In spite of its failure in coordinate space, Fig. 2 shows that a coherent state, represented by the disk, is faring quite well. Propagation of an extended state in amplitude space by explicitly integrating over the badly behaved semiclassical Green's function thus works. The major part of the correlation function $\langle a|q_t\rangle$, where $|a\rangle$ is the coherent state is coming from the replicas on either side of the center, and these are entirely in "safe" zones. The zones of poor amplitude for a "circular $\Delta p = \Delta q$" coherent state are reasonably estimated by a superposition of the p and the q error zones. If a coherent state comes too close or enters such a zone, it will yield somewhat inaccurate amplitudes, although the magnitude of the error cannot climb to very large values as it does in coordinate or momentum space because of singularities in the Van Vleck–Gutzwiller determinant.

From the phase space, Lagrangian manifold analysis it is quite clear that the accuracy of semiclassical amplitudes is very nonuniform; it would be possible to find inaccurate regions almost immediately, while other zones are well behaved far past the logarithmic time. Very often, smooth state correlation functions are physically the desired quantity; other times, as when determining energy eigenvalues, for example (by Fourier transform of a correlation function [3]), one has a choice of states. This makes the present results of far-reaching consequence for applications.

We turn to the dependence of the fraction F_p on h. At a fixed time, a finite number of folds will have developed. As $h \to 0$, $F_p \to 1$. For a given fold, suppose we transform canonically to the coordinate system p',q' so that the fold is aligned along the q' axis. This gives, to second order in p_0, $q' = \gamma + ap_0 + \beta p_0^2$, $p' = a'p_0$. The line corresponding to $|q_0\rangle$ is now rotated by the transformation,

and cuts the fold at an angle. If the area enclosed is h, the range of p_0 corresponding to "bad" parts of the fold is easily shown to go as $h^{1/3}$, at least for sufficiently small h, where the folds are isolated and the quadratic expansion in p_0 holds. This dependence has been checked for the standard map; see Fig. 3.

The time dependence of F_p is more problematical and system dependent, but we can make some headway by using the Smale-like horseshoe construction as a model of the tangles developing in the Lagrangian manifold. Consider a p-like (horizontal) Lagrangian manifold which folds once to a U shape, lying on its side, making one caustic in coordinate space. After compression by a factor of 2 and stretching by the same factor, it is folded again, making a total of three caustics. The nth cycle yields $2^{n+1} - 1$ total caustics, but the stretching by the factor 2^n means that each successive caustic spans a smaller range of initial manifold, by a factor $\frac{1}{2}^n$. A constant fraction of the initial manifold therefore lands in caustics at each step, whether or not it has previously been part of another caustic. Thus the fraction removed is not confined just to the remaining good regions of the initial manifold, but is applied apparently at random to the whole of the initial manifold. The differential equation describing this removal procedure is

$$-\frac{dF_p}{dt} = a\lambda h^{1/3}F_p, \quad \to F_p(t) = \exp(-a\lambda h^{1/3}t), \quad (5)$$

where λ is the Lyapunov exponent and a a system-dependent proportionality constant. Thus the good initial manifold disappears relatively slowly and the half-life for good semiclassical propagation scales as $h^{-1/3}$.

The area rule leads to the remarkable conclusion that stronger chaos may actually *help* the semiclassical accuracy by producing larger folds with more area enclosed. This might lead to the worry that the generic "soft" chaos systems would be problematical, but some recent calculations show good agreement though much more study is required.

The present Letter reconciles the standard theories of semiclassical mechanics with the intriguing findings of unexpected long time accuracy for chaotic systems. The

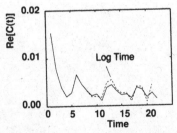

FIG. 2. Comparison of an exact quantum and semiclassical calculation of $C(t) = \langle a|p_0(t)\rangle$ for the standard map ($K = 2.14$). The semiclassical is the dashed line and the quantum is the solid line. The initial state is the p state at 0; the coherent state was localized at $x_0 = \pi$, $p_0 = 0$, and $h = 0.00159$.

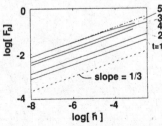

FIG. 3. The fraction F_p of the "good" manifold remaining in $|p_0\rangle$ is shown as a function of h for different iterations of the mapping ($t = 1$ to 5).

understanding of the source of errors, in our case the "lack of errors," opens a new door in the field of applications of semiclassical techniques for classically chaotic systems. Perhaps of more importance is to realize that with so much research emphasis on semiclassical techniques in the energy domain, its fundamental precursor, the time Green's function, still deserves far more exploration.

We acknowledge important discussions with P. W. O'Connor. This research was supported by the National Science Foundation under Grant No. CHE-9014555. One of us (M.A.S.) acknowledges support from the Ministerio de Educación y Ciencia of Spain.

---

[1] M. C. Gutzwiller, J. Math. Phys. **12**, 343 (1971), and references therein; *Chaos in Classical and Quantum Mechanics* (Springer-Verlag, New York, 1990).

[2] E. J. Heller, J. Chem. Phys. **94**, 2723 (1991).

[3] S. Tomsovic and E. J. Heller, Phys. Rev. Lett. **67**, 664 (1991).

[4] P. W. O'Connor, S. Tomsovic, and E. J. Heller, Physica (Amsterdam) **55D**, 340 (1992); P. W. O'Connor, doctoral dissertation, University of Washington, 1991 (unpublished).

[5] P. Cvitanovic and B. Eckhardt, Phys. Rev. Lett. **63**, 823 (1989).

[6] G. Tanner, P. Scherer, E. B. Bogomolny, B. Eckhardt,

and D. Wintgen, Phys. Rev. Lett. **67**, 2410 (1991); R. Aurich, C. Mathies, M. Sieber, and F. Steiner, Phys. Rev. Lett. **68**, 1629 (1992).

[7] M. V. Berry and J. P. Keating, J. Phys. A **23**, 4839 (1990).

[8] E. B. Bogomolny, Comments At. Mol. Phys. **25**, 67 (1990).

[9] J. H. Van Vleck, Proc. Natl. Acad. Sci. U.S.A. **14**, 178 (1928).

[10] M. V. Berry, N. L. Balazs, M. Tabor, and A. Voros, Ann. Phys. (N.Y.) **122**, 26 (1979).

[11] Z. V. Lewis, Ph.D. thesis, University of Bristol, 1982 (unpublished); see also M. V. Berry, Ann. N.Y. Acad. Sci. **357**, 183 (1983).

[12] V. P. Maslov and M. V. Fedoriuk, *Semiclassical Approximations in Quantum Mechanics* (Reidel, Dordrecht, 1981), English translation.

[13] Robert G. Littlejohn (to be published).

[14] W. H. Miller, J. Chem. Phys. **53**, 1949 (1970); **53**, 3578 (1970); W. H. Miller, Adv. Chem. Phys. **25**, 69 (1970).

[15] The standard map is a 2D mapping of the cylinder onto itself defined by the following equations: $p_{n+1} = p_n + K \sin(x_n)$, $x_{n+1} = x_n + p_{n+1}$ (mod2π). The parameter K determines the chaoticity of the mapping. We selected $K = 2.14$ in the calculations shown in this Letter. For information on the quantized version see Chirikov's contribution in *Chaos and Quantum Physics*, Proceedings of the Les Houches Summer School, Session LII, edited by M.-J. Giannonni, A. Voros, and J. Zinn-Justin (Elsevier, New York, 1991), and references therein.

The Quantized Baker's Transformation

N. L. Balazs

Department of Physics, State University of New York at Stony Brook,
Stony Brook, New York 11794

AND

A. Voros*

Service de Physique Théorique, Institut de Recherche Fondamentale
CEA-CEN Saclay, 91191 Gif-sur-Yvette Cedex, France

Received March 9, 1988; revised September 26, 1988

A quantum analogue of the Baker's transformation is constructed using a specially developed quantization procedure. We obtain a unitary operator acting on an N-dimensional Hilbert space, with N finite (and even), that has similar properties to the classical baker's map, and reduces to it in the classical limit, which corresponds here to $N \to \infty$. The operator can be described as a very simple, fully explicit $N \times N$ matrix. Generalized Baker's maps are also quantized and studied. Numerical investigations confirm that this model has non-trivial features which ought to represent quantal manifestations of classical chaoticity. The quasi-energy spectrum is given by irrational eigenangles, leading to no recurrences. Most eigenfunctions look irregular, but some exhibit puzzling regular features, such as peaks at coordinate values belonging to periodic orbits of the classical Baker's map. We compare the quantal and classical time-evolutions, as applied to initially coherent quasi-classical states: the evolving states stay in close agreement for short times but seem to lose all relationship to each other beyond a critical time of the order of $\log_2 N \sim -\log \hbar$. © 1989 Academic Press, Inc.

Introduction

It is commonly believed that the essential features of chaotic behaviour in the classical Hamiltonian systems are basically understood [1].

The situation is rather different if one turns to the quantal transcription of classical dynamical systems. Classical chaos appears in bound systems, which implies that the quantal Hamiltonian operator has a discrete spectrum. Then, however, all the standard diagnostic tools used in classical dynamics seem to imply the absence of chaotic behaviour. Indeed, the time evolution of any state is quasi-periodic; hence the Liapunov exponents and the different entropies all vanish, there is no mixing and no decay of time correlations, etc. For this reason, it is correct to state that the literal definitions of classical chaos fail to specify different classes of quantal systems [2].

* Member of CNRS, France.

One must then address three other questions:

(a)   where and how in the quantal description is the information buried that the system before quantization was chaotic?

(b)   how does chaoticity reappear in the classical limit?

(c)   how far can the classical solutions be used in the quantal description of the system?

For us the problem of so-called quantum chaoticity resides entirely in the analysis and answer to these (or equivalent) questions.

One must also observe the practical importance of these problems. In quantum theory there is no classification scheme of the solutions comparable to the classical theory, ordered along increasing complexity. This fact, coupled with the lack of understanding of the semi-classical approximations, puts the study of many Schrödinger equations beyond the reach of any analytical scheme. For this reason we believe in the importance of studying simple models which at first sight seem to bear no relation to problems in physics; what they are related to is the structure of the mathematical problem-setting.

The understanding of classical chaoticity was greatly aided by the construction and investigation of suitably simple models from which most accidental complications, unrelated to chaoticity itself, have been removed. Of these models the discrete area-preserving maps occupy a special position. On the one hand, they are the direct descendants of dynamical systems evolving continuously in time, via the Poincaré section method; on the other hand, they lead directly to the standard descriptions of general chaotic behaviour in terms of entropy, Markov partitions, etc. In addition, they effectively reduce the number of degrees of freedom and allow one to develop the full theory without explicitly integrating the equations of motion.

A similar understanding of quantal chaoticity is not available. The subject is particularly hampered by the lack of easily comprehensible models, especially ones which might qualify as being generic. By analogy with the classical situation, we expect that the study of *quantum maps* will prove particularly useful. These are the quantal analogs of classical area-preserving maps: *a quantum map is a suitable unitary transformation describing the evolution of the quantum states during one discrete time step.* Sometimes we can conceive of this evolution as a succession of stroboscopic pictures of a continuous time evolution; however, this is not always possible or desirable (either in the classical or in the quantal case).

The construction and study of these quantum maps is still in its infancy, and only a handful of models do exist. In this work we construct and study a new example, the quantized version of the celebrated baker's transformation [3]. In fact the mere existence of this model is interesting by itself, since the standard quantization techniques (Schrödinger, or Feynman, etc.) give no clue of its necessary existence. Moreover, the resulting quantum map is very simple; at the same time it cannot be treated analytically.

Section I specifies the classical model in a form which prepares for quantization. In Section II we construct the quantal equivalent; the compact phase space is replaced by a *finite-dimensional* vector space of states, and the classical baker's map is replaced by a unitary matrix B acting on the states. In Section III we analyse B, verify that it has all the required properties, and exhibit a variety of generalized B operators. In Sections IV and V we study and summarize the properties of the eigenvalues and eigenvectors of B, as obtained numerically for dimension N ranging from 2 to 128. We also construct analogs to the coherent states displayed in this vector space, and study the time evolution of some.

We find that: the eigenvalues of B are not rational roots of unity, and that neither their distribution nor that of their nearest-neighbour separations exhibit any simple structure; the eigenvectors in the q (or p) representation are of a varied nature; most have no apparent regular structure, but some carry the imprint of short classical periodic orbits, and a few have, rather unexpectedly, a very regular appearance. The time evolution of a coherent wave packet imitates closely the classical evolution in phase space up to a time T of the order of magnitude of $\log N$, while afterwards the classical and quantal time evolutions seem to lose all relationship to each other.

There are other deeper analytical features associated with the nature of the semi-classical approximations. These are under active investigation and will be dealt with in another paper.

I. Classical Model

One of the great achievements of the study of classical dynamical systems was the introduction of suitable models (or caricatures) which represent the essential features of classes of motions without being hampered by the idiosyncrasies of particular examples [1].

The baker's transformation is such a successful model. It replaces the canonical equations of motion with a simple area-preserving transformation of a rectangle in the p–q plane, the phase plane. The model acts as a prototype for most properties of classical chaotic systems. One easily computes the associated Liapunov exponent and Kolmogorov–Sinai entropy. It is the simplest example which describes a deterministic motion as Bernoulli shifts; hence it is also the prototype of symbolic dynamics. In fact one may say that all the properties of the motion present are linked to *some feature of chaoticity*, in the sense that if we were to remove all features linked to chaotic behaviour *nothing* would remain, instead of an integrable system.

The model is as follows [1].

In Fig. 1 the P, Q rectangle in the p–q plane is filled uniformly with the phase fluid. Since p and q represent canonically conjugate variables the motion of the phase fluid must be *area preserving*. (In fact this is the *only* relevant property of the

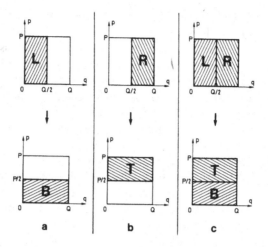

FIG. 1.   The classical baker's transformation. (a) the L → B transformation given by Eq. (1a); (b) the R → T transformation given by Eq. (1b); (c) the full transformation mapping the entire P, Q rectangle to itself.

canonical equations of motion which will be exploited.) The motion is now given by a simple transformation sending the point (q, p) into (q', p')

$$\begin{pmatrix} q' \\ p' \end{pmatrix} = \begin{pmatrix} 2q \\ p/2 \end{pmatrix}, \qquad 0 \leqslant q < Q/2, \qquad (1a)$$

$$\begin{pmatrix} q' \\ p' \end{pmatrix} = \begin{pmatrix} 2q - Q \\ (p + P)/2 \end{pmatrix}, \qquad Q/2 \leqslant q < Q. \qquad (1b)$$

The sequence of pictures on Fig. 1 gives the visual representation of the transformation; (1a) transforms points, or states in the left vertical half-strip L into the bottom horizontal half-strip B. Similarly. (1b) transforms the right half-strip R into the top horizontal half-strip T. Thus the motion of the phase point proceeds in discrete time steps (of one unit) according to (1a) and (1b). The transformation is piecewise linear and inhomogeneous; it stretches the horizontal, q direction by a factor of 2, and it compresses the vertical, p direction by the same factor; thus it is a *hyperbolic* map. The piecewise nature of the transformation makes it globally discontinuous at the boundaries of the strips. The first property of the transformation is needed to make the motion exponentially unstable; the second property is required to confine the phase points to the original (P, Q) rectangle, given that the compression and dilation directions were parallel to the sides. We observe that the presence of the discontinuity is what makes the transformation non-trivial, and mixing; at the same time it affects only a set of points of measure zero. (This is similar to the motion found in billiards, which is also discontinuous in the phase space). Because of these discontinuities of the motion in the phase space it is best to think of the time evolution as really discrete, and not as a sequence of stroboscopic pictures of a continuous development. The transformation is clearly invertible; thus each phase point will retrace its (discrete) path applying the reciprocal of the transformation. However, this motion cannot be made to correspond to an actual forward motion by some change of the p's alone. Thus the motion does not possess the time-reversal property (although it is invertible). (By time reversal property we mean that an actual motion generated by the map turns into another *actual* motion with the change $t \rightarrow -t$, $p \rightarrow -p$.) The map, however, *is* symmetric under the combination of time reversal and the discrete canonical transformation $(q \rightarrow p$, $p \rightarrow -q)$, an *anti-canonical* symmetry leading to an anti-unitary one in the quantal version.

In view of the great success of this model, a quantal version is highly desirable, in order to see if and how the classical chaotic features manifest themselves when the motion becomes quantized.

II. The Quantal Baker's Transformation

The quantization procedure consists of two steps [3]. We first specify, following Weyl [4], the kinematics, i.e., the states and the operators which are associated with the description of the system. Then we turn our attention to dynamics, i.e., the time evolution of these objects.

Kinematics

The classical phase is compact; thus the number of phase-cells $N = PQ/2\pi\hbar$ is finite and gives the dimensionality of the vector-space needed to describe all possible quantum states of the system. This implies that only those classical situations can be quantized where PQ is an integer multiple of $2\pi\hbar$. (This is a simple example of the more general notion of quantizable phase-space [5].) We then define the canonically conjugate p and q operators as *translation* operators acting on each other's eigenstates. To do this conveniently it is useful to extend the labelling of the p and q eigenstates *periodically* in order to treat all these eigenstates on the same footing. (Thus we may extend the classical picture by periodically repeating the P, Q rectangle.)

Then, the (normalized) position eigenvectors are $|n\rangle$, with eigenvalues $q_n = 2\pi\hbar n/P$, $n = 0, 1, ..., N-1$, where $|n + N\rangle = |n\rangle$; the (normalized) momentum eigenvectors are $|m\rangle$, with eigenvalues $p_m = 2\pi\hbar m/Q$, $m = 0, 1, ..., N-1$, where $|m + N\rangle = |m\rangle$. The transformation functions $\langle n|m\rangle$ are thus defined as

$$\langle n|m\rangle = \frac{1}{\sqrt{N}} e^{2\pi inm/N} = \sqrt{\frac{2\pi\hbar}{PQ}} e^{iq_n p_m/\hbar}. \qquad (2a)$$

Hence the passing from the q to the p representation corresponds now to the *discrete* Fourier transformation F_N on N sites, with the matrix representation

$$(F_N)_{mn} = \frac{1}{\sqrt{N}} e^{-2\pi imn/N}. \tag{2b}$$

A general state ψ in the q representation is given by

$$\langle n| \rangle = \psi_n, \tag{2c}$$

with the normalization

$$\sum |\psi_n|^2 = 1, \tag{2d}$$

while ψ in the p representation is given by

$$\langle m| \rangle = \tilde{\psi}_m = \sum_{n=0}^{N-1} \frac{1}{\sqrt{N}} e^{-2\pi imn/N} \psi_n = (F_N\psi)_m. \tag{2e}$$

We defined here immediately the *periodic* position and momentum operators by giving their eigenvalues and eigenfunctions. Thus, if we would call the usual position and momentum operators $\hat{q}$, $\hat{p}$, we defined immediately $e^{2\pi i\hat{q}/Q}$, $e^{2\pi i\hat{p}/P}$. (We stress that the eigenvalues p_m, q_n cannot be obtained by a simultaneous diagonalization of operators; hence the formal (p_m, q_n) lattice has no direct physical meaning and must not be thought of as some sort of discretized classical phase space.)

If we let P and Q tend to infinity, keeping $\hbar$ fixed, the whole formalism goes over to that of usual quantum theory on the infinite line. If we keep Q and $\hbar$ fixed and let P tend to infinity we obtain the usual quantum theory on a circle of circumference Q. However, the limit of greatest interest to us leaves P, Q fixed while $\hbar$ goes to zero (or N to infinity). This leads to the classical limit on the fixed, compact PQ rectangle with the opposite edges identified, corresponding thus to a torus. In our case this periodic extension of the original rectangle *is a natural part of the quantization scheme*.

(If P and Q are fixed we may put them equal to unity by suitable scaling and this will indeed be done in the numerical work and in the figures. However, in most of the general discussion we will retain the P and Q labels for dimensional clarity).

Dynamics, Preliminaries

In general, the classical dynamics defined by a conservative map is a one-to-one, area-preserving transformation of the phase space into itself. Its quantized equivalent shall be a unitary transformation of the Hilbert space of states (and not, for example, of the density operator itself), which has the symmetry properties of the classical system, and reproduces in the classical limit the original area-preserving transformation.

Nothing is known *in general* about the existence and constructibility of such unitary transformations. In fact only a few of these quantum maps have been devised; moreover, most relied on conceiving the maps as snapshots of a continuous time evolution [6–8]. An additional complication arises if the classical phase space is *compact*; then the space of quantum states is finite dimensional, and all standard quantization methods are inapplicable. (The compactness of phase space is *imperative* here to *guarantee* the chaotic behaviour of the classical conservative motion.)

Among conservative maps on compact phase spaces only very few have been quantized [9–12], and among these only one family which cannot be conceived as snapshots of a continuous time evolution, the linear transformations of the torus, i.e., the generalized Arnold cat maps [11, 12]. Here, too, the quantization of the map corresponds to the construction of a unitary operator acting on the vector space already specified. However, this particular quantization relies on the classical map being *linear without discontinuities*, and consequently it cannot be generalized. These features also imply special number theoretical relations in the quantal description, which regularize the motion, obliterating many of the chaotic properties associated with the classical evolution.

Thus, the construction of a quantal baker's transformation constitutes a new and autonomous problem.

We will accomplish this quantization by adapting stepwise the construction of the classical baker's map to actually produce a *unitary* transformation on the space of quantum states. This way we automatically incorporate the crucial properties of the classical map, and essentially guarantee the correct classical limit. While this procedure is natural if the classical map is linear, it is by no means obvious that it can work at all if the classical map is only *piecewise* linear, since the unitarity of the quantal evolution is a global property. (Indeed, for certain other piecewise continuous classical maps an altogether different scheme was required to produce a consistent quantal description [8, 10].)

Of course, no quantization scheme is unique, and it may be interesting to explore alternative possibilities, for example to seek a classical physical process which generates the classical map, and quantize that process.

Dynamics, Explicit Construction

We now construct such a unitary transformation B, which is in a sense the simplest quantized baker's transformation. In the next section we shall generalize this simple example in many different ways.

The classical baker's map begins by decomposing the phase space into two disjoint parts L (Left) and R (Right), the dividing line being at $q = Q/2$. Then each part is separately mapped into the two disjoint parts B (Bottom) and T (Top), the dividing line being now at $p = P/2$ (where $L \to B$, $R \to T$). Each map separately compresses p and dilates q.

In the quantal approach we imitate stepwise this procedure. Hence, first we decompose the vector space into two orthogonal subspaces, $\mathscr{L}$, $\mathscr{R}$ with vectors Ψ^L, resp. Ψ^R. We require that in the *position* representation the components $\Psi_n^L \equiv 0$ for $n \geqslant N/2$, and $\Psi_n^R \equiv 0$ for $n < N/2$. To maintain complete symmetry between $\mathscr{L}$ and $\mathscr{R}$ we henceforth restrict N to be *even*.

One iteration must transfer the vectors Ψ^L and Ψ^R into two *other* disjoint subspaces $\mathscr{B}$ and $\mathscr{T}$ with vectors Φ^B and Φ^T having this time in the *momentum* representation the components $\tilde{\Phi}_m^B = 0$ for $m \geqslant N/2$, $\tilde{\Phi}_m^T = 0$ for $m < N/2$. The transformations connecting these subspaces are A^{BL} and A^{TR} with $\Phi^B = A^{BL}\Psi^L$, $\Phi^T = A^{TR}\Psi^R$. A^{TR}, A^{BL} must embody the notions of stretching in q and compressing in p. We now construct A^{BL}.

The stretching part is easy to accomplish on the *even* components, Φ_{2n}^B in the q representation, by requiring that

$$\Phi_{2n}^B = \frac{1}{\sqrt{2}} \Psi_n^L, \qquad n \leqslant N/2 - 1 \tag{3a}$$

$$\text{(with } \Psi_n^L \equiv 0 \quad \text{for} \qquad n \geqslant N/2). \tag{3b}$$

The remaining components of Φ^B we specify through the p representation in which the compression becomes a dilation if the transformation is applied *backwards*. Thus we require that

$$\tilde{\Psi}_{2m}^L = \frac{1}{\sqrt{2}} \tilde{\Phi}_m^B, \qquad m \leqslant N/2 - 1 \tag{4a}$$

$$\text{(and } \Phi_m^B \equiv 0 \text{ for } m \geqslant N/2), \tag{4b}$$

in full symmetry with Eqs. (3) read backwards. Equations (3a), (4a), and (4b) are $3N/2$ conditions on the N vector components ((3b) is a condition on Ψ and not on Φ). Thus, a priori, even the *existence* of such a transformation is by no means guaranteed. Nevertheless, the conditions turn out to be *compatible* (see Appendix A) and define *uniquely* the transformation matrix A^{BL}

$$A_{n'n}^{BL} = \begin{cases} 2^{-1/2}\delta_{n'/2,\,n} & (0 \leqslant n < N/2, \ n' \text{ even}) \\ 2^{1/2}N^{-1}\left[1 + i\cot\left(2\pi N^{-1}\left(\frac{n'}{2} - n\right)\right)\right] & (0 \leqslant n < N/2, \ n' \text{ odd}) \\ 0 & (N/2 \leqslant n < N). \end{cases} \tag{5}$$

(Although the transformation initially applied $\mathscr{L}$ to $\mathscr{B}$, and hence in the q representation would be described by an $N \times (N/2)$ matrix, we immediately extended it formally to an $N \times N$ matrix, filling in the missing slots by 0.)

The matrix A^{BL} has all the required properties; (a) it is a *unitary* transformation from $\mathscr{L}$ onto $\mathscr{B}$ (see Appendix A) (b) it shifts the values of the wavefunction in each representation to the *dilated* or *compressed* values of the arguments as prescribed by (3), (4); (c) it gives thus the correct *classical* $L \to B$ map as $N \to \infty$, PQ fixed; (d) by construction it has the *symmetry* property that if $q \to p$, $p \to q$ the map goes to its reciprocal, just as the classical map does.

In an identical fashion one may construct the transformation A^{TR}; $\Phi^T = A^{TR}\Psi^R$ mapping unitarily $\mathscr{R}$ onto $\mathscr{T}$, giving

$$A_{n'n}^{TR} = \begin{cases} 2^{-1/2}\delta_{n'/2,\,n-N/2} & (N/2 \leqslant n < N,\, n' \text{ even}) \\ 0 & (0 \leqslant n < N/2) \\ -2^{1/2}N^{-1}\left[1 + i\cot\left(2\pi N^{-1}\left(\dfrac{n'}{2} - n\right)\right)\right] & (N/2 \leqslant n < N,\, n' \text{ odd}). \end{cases} \quad (6)$$

(Similarly to A^{BL}, this matrix was immediately extended to be an $N \times N$ matrix.)

Since any Ψ can be expressed as $\Psi = \Psi^L + \Psi^R$, the quantized baker's transformation is given by

$$\Phi = B\Psi = A^{BL}\Psi^L + A^{TR}\Psi^R, \qquad \text{with} \quad B_{n'n} = A_{n'n}^{BL} + A_{n'n}^{TR}$$

or

$$B_{n'n} = \frac{1}{\sqrt{2}} \times$$

0	1	2	$\cdots$	$(N/2-1)$	$N/2$	$(N/2+1)$	$\cdots$	$(N-1)$	n / n'
0	0	0	$\cdots$	0	1	0	0	$\cdots$ 0	0
$C_{1/2}$	$C_{-1/2}$	$C_{-3/2}$	$\cdots$	$C_{(3-N)/2}$	$-C_{1/2}$	$-C_{-1/2}$	$-C_{-3/2}$	$\cdots -C_{(3-N)/2}$	1
0	1	0	$\cdots$	0	0	1	0	$\cdots$ 0	2
$C_{3/2}$	$C_{1/2}$	$C_{-1/2}$	$\cdots$	$C_{(5-N)/2}$	$-C_{3/2}$	$-C_{1/2}$	$-C_{-1/2}$	$\cdots -C_{(5-N)/2}$	3
0	0	1	$\cdots$	0	0	0	1	$\cdots$ 0	4
$\vdots$	$\vdots$	$\vdots$		$\vdots$	$\vdots$	$\vdots$		$\vdots$	$\vdots$
0	0	0	$\cdots$	1	0	0	0	$\cdots$ 1	$N-2$
$C_{(N-1)/2}$	$C_{(N-3)/2}$	$C_{(N-5)/2}$	$\cdots$	$C_{1/2}$	$-C_{(N-1)/2}$	$-C_{(N-3)/2}$	$-C_{(N-5)/2}$	$\cdots -C_{1/2}$	$N-1$

$$(7a)$$

where

$$C_{n'/2-n} = \frac{2}{N}\left[1 + i\cot 2\pi N^{-1}(n'/2 - n)\right] \qquad \text{(for } n' \text{ odd,)} \quad (7b)$$

satisfies the periodicity condition $C_{\nu + N/2} = C_{\nu}$. Because of this periodicity, the successive odd half rows of the matrix are linked to each other by a shift (while the even half rows are obviously so linked). In addition the left and right halves of each odd row differ only by sign.

The left and right transformations can also be joined with an arbitrary phase factor (see also next section). In the present paper we have chosen the relative phase as given in (7). Another simple choice replaces A^{TR} with $-A^{TR}$. We have verified by numerical computations that those features of the transformation which are studied in this paper do not essentially change by this replacement, although the *analytical* properties of the operator B may be strongly affected.

The quantization principle which we applied to the baker's map can also be understood in terms of the signal encoding technique used in stereo broadcasting, i.e., frequency multiplexing (see Appendix B).

In the subsequent sections we shall analyze the properties of the transformation B.

III. ANALYSIS OF THE B OPERATOR

A convenient block matrix expression can be provided for the partial transformations A^{BL}, A^{TR} and hence for the full baker's operator B.

The partial transformation A^{BL} which maps Ψ^L into Φ^B is entirely specified by Eqs. (4a), (4b) which read as

$$\tilde{\Phi}^B_m = \frac{\sqrt{2}}{\sqrt{N}} \sum_{m=0}^{N/2-1} e^{-2\pi i \cdot 2mn/N} \; \Psi^L_n \qquad (m < N/2),$$

$$\tilde{\Phi}^B_m = 0 \qquad\qquad\qquad (m \geqslant N/2).$$

In matrix form, this is equivalent to

$$\tilde{\Phi}^B = \begin{pmatrix} F_{N/2} & 0 \\ 0 & 0 \end{pmatrix} \psi^L, \tag{8a}$$

where $F_{N/2}$ is the $N/2 \times N/2$ matrix of the discrete unitary Fourier transformation *on N/2 sites* (we recall that N is even). Likewise, the mapping from Ψ^R to Φ^T is equivalent to the matrix transformation

$$\tilde{\Phi}^T = \begin{pmatrix} 0 & 0 \\ 0 & F_{N/2} \end{pmatrix} \psi^R, \tag{8b}$$

and the full baker's map can be represented as

$$\tilde{\Phi} = \begin{pmatrix} F_{N/2} & 0 \\ 0 & F_{N/2} \end{pmatrix} \psi. \tag{8c}$$

Equations (8a)–(8c) are *mixed* representations of the corresponding operators since they involve the source vector Ψ in the position representation and the image vector Φ in the momentum representation. Because $\tilde{\Phi} = F_N \Phi$ from Eq. (2c), we immediately obtain the B matrix in the position representation as

$$B = F_N^{-1} \begin{pmatrix} F_{N/2} & 0 \\ 0 & F_{N/2} \end{pmatrix}. \tag{9}$$

The attractive features of this form are:

(a) B appears as the product of two *basic, unitary* matrices; the wealth of the model resides in the *non-commutation* of the two;

(b) this form suggests a *vast variety of generalizations*. For example, arbitrary phase factors can multiply the two blocks $F_{N/2}$. The blocks can also be made more numerous, unequal in size, and shifted off the diagonal; these more complicated models have interesting classical limits which are generalized baker's transformations (see below). (Even further generalizations can be conceived by increasing the number of factors and the complexity of each in the defining formula (9).)

Properties

We now list some *general* properties of B, consequences of Eq. (9) (the more *peculiar* features which should reflect the quantal manifestations of chaoticity are buried deeper and will be analyzed during the discussion of the numerical results):

(a) unitarity;
(b) the anti-unitary symmetry

$$F_N B F_N^{-1} = (B^{-1})^*. \tag{10}$$

(This follows from (9) with $F_{N/2}^{-1} = F_{N/2}^*$.)

The classical motion did not possess time-reversal invariance. In the quantal description this results in a B different from its time-reversed form, which is $(B^{-1})^*$. Equation (10) is the quantal analog of the classical anti-canonical symmetry $p \to q$, $q \to p$, $t \to -t$, stating that B expressed in the p representation, $F_N B F_N^{-1}$, coincides with the time-reversed matrix in the q representation, $(B^{-1})^*$. (In general, for a Hamiltonian H, anti-unitary symmetry is defined as $UHU^{-1} = H^*$ for some unitary operator U [13]. For a finite time evolution operator like B this results in

$$UBU^{-1} = (B^{-1})^*,$$

which gives Eq. (10) when $U = F_N$.)

These properties of B, though unrelated to chaoticity, have important consequences for the eigenvalues and eigenvectors. Since B is unitary, all eigenvalues are of the form $e^{i\theta}$ with the eigenangles θ being real, and the associated eigenvectors form a complete orthonormal basis.

The lack of time-reversal invariance suggests that the eigenfunctions are complex, which is fully confirmed by the numerical results. It follows from Eq. (10) that the eigenvectors can be so specified that in the p representation they should coincide with their complex conjugate in the q representation, i.e., $\tilde{\Psi} = \Psi*$.

By contrast, the symmetry of the classical map about the center of the PQ rectangle ($q \rightarrow Q - q, p \rightarrow P - p$) is *not* preserved under quantization, being broken by the discretization of the p, q values. While the preservation of the *other* classical properties is an essential criterion for the correctness of the quantal baker's map, the loss of this particular symmetry is not critical, and moreover irrelevant to the main thrust of this paper. In fact, we can develop very *asymmetric* classical baker's transformations and quantize them, preserving the other, essential features of the model.

Classical Limit

The discussion of the kinematics has already shown that the classical limit $\hbar \rightarrow 0$ must be taken as $N \rightarrow \infty$ with P, Q fixed, implying that p and q will take continuous but bounded ranges of values in the limit. This is *different* from the usual situation, where $\hbar$ tends to 0 with the ranges of p and q being *unbounded*.

There are two different aspects of the classical limit to be discussed in general: (a) the limit of the equations of motion, (b) the limit of their solutions.

The first one is the easier question. If a quantization procedure was used to establish a correspondence from classical to quantum dynamics, the classical limit of the equations can be conceived in full generality by using this correspondence in reverse.

However, the basic and unresolved questions which link the chaotic aspects of the classical motion to features of the quantal description reside entirely in class (b).

In the present subsection we address ourselves to point (a), the recovery of the classical baker's transformation from the unitary operator B, as $N \rightarrow \infty$. The second group of questions can at present be tackled only numerically, and this will be done in Section V.

If B is written in the block form (9), the classical limit is intuitively immediate. The discrete Fourier summations tend to continuous Fourier integrals on their respective ranges, which stay fixed in the limiting process. The resulting integral kernels tend (in the well-known oscillatory manner) to the Dirac delta function kernels describing the classical transformations (1a), (1b).

This result is insensitive to the possible phase factors attached to the blocks $F_{N/2}$, implying the existence of (at least) a one-parameter family of quantal baker's operators sharing the same classical limit. (Most numerical results will replace these phase factors with unity.)

If we use several blocks F_{N_1}, F_{N_2}, $\cdots$ with $N_1 + N_2 + \cdots = N$, we can perform a similar limiting process leading to generalized classical baker's transformations, where $N \rightarrow \infty$ with all ratios N_k/N having fixed limits. (In this way, moreover, congruence conditions on N are no longer required.) The resulting correspondence is best explained by a diagram (see Fig. 2).

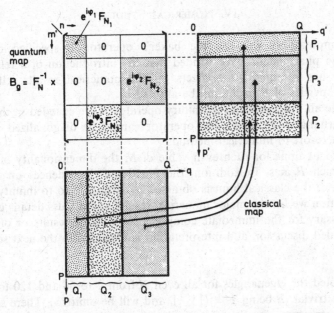

FIG. 2. An example of a generalized quantal baker's transformation with three blocks, and its classical limit. The quantal baker's operator B_g is displayed through the matrix $F_N B_g$, i.e., with matrix elements in the mixed representation $\langle m'|B_g|n\rangle$; optional phase factors $e^{i\varphi_k}$ are shown as well. The classical transformation, obtained in the limit $\sum N_k = N \to \infty$, connects the two phase rectangles partitioned so that $Q_k/Q = P_k/P = \lim N_k/N$ (these being the respective compression factors). (The phase rectangles are drawn *here* with their p axes pointing *downwards*, in order to agree with the conventional ordering of the row indices m' in the matrix and at the same time exhibit the semi-classical relationship $p' \sim Pm'/N$. The shaded portions in the matrix show this and the conjugate relationship $q \sim Qn/N$.)

An alternative approach to the classical limit uses Wigner's function and its equation of motion. This method was originally conceived for the standard p, q operators of quantum mechanics, which have the whole real axis as spectra. Its adaptation to the torus as phase space has been accomplished [11] but at the expense of complications. The Wigner function is now defined on a *discrete lattice* which is 4-fold redundant, and its classical limit is *not* understood. In addition, this lattice is *not mapped into itself under the classical map*, save for the *single* special case of linear automorphisms (i.e., "cat maps"). Thus, this Wigner lattice cannot be conceived as a discretized classical phase space; even more, no lattice invariant under the classical map will in general exist. This shows that the Wigner formalism, as it stands, will not lead to a straightforward understanding of the quantization, or of the classical limit, for *any other map* on a compact phase space. Any other association of a similar kind, like Kirkwood's function [14], has difficulties of its own, including the non-invariance of the lattice supporting the function.

IV. Numerical Studies

(Most computations will use the baker's operator B specified by Eq. (9) containing no phase factors. We verified that the introduction of a relative phase factor (-1) has no qualitative effects on the numerical results, although the *analytical* properties of B might still be affected.)

In principle all properties of the unitary operator B are encoded in the eigenvectors and eigenangles. However, the operator cannot be diagonalized analytically and we must resort to numerical methods.

This has to be done for each even value of N, the dimensionality of the vector space on which B acts. In addition, the functional dependences on N are also important since the classical limit is obtained by letting N go to infinity.

In this section we list the major numerical results and discuss details only insofar as it is necessary for the immediate comprehension of the results or of their uses. A more detailed discussion and interpretation is relegated to the next section.

Eigenangles

We computed the eigenangles for all even N from 4 to 64 and 120 to 128. (The $N = 2$ case is trivial, B being $2^{-1/2}(\begin{smallmatrix} 1 & 1 \\ 1 & -1 \end{smallmatrix})$, and will be omitted). There seems to be no special structure in the eigenangle distributions such as degeneracies, symmetries, recursions, repetitions. Contrary to the cat maps, we do not find any arithmetical relations or dependences on number theoretical properties of N [11, 12, 15]. Eigenangles which are rational fractions of 2π seem to be exceptional, the only ones being 0, π for $N = 6$, and $-\pi/2$ for $N = 8$. All other eigenangles are irrational fractions of 2π; hence the time evolution of a state is *quasi*periodic, in contrast to the quantized cat map, which is periodic in time [11]. We also diagonalized some of the generalized baker's operators with similar results.

The distributions of the eigenangles and of their nearest-neighbour spacings are shown in Figs. 3, 4, and will be discussed in the next section.

Eigenfunctions

In contrast to the eigenvalues one expects that the eigenfunctions may contain more useful information about the underlying chaotic properties of the classical model. In the present case they are also easy to compute and to display on account of (a) the discrete nature of the coordinates, and (b) the finite number of eigenvectors.

As observed before, the eigenfunctions must be complex and each must also be the complex conjugate of its own Fourier transform. Thus they are essentially the same in the q and p representation. Using the antiunitary symmetry (Equation (101)), we have also constructed a representation in which all eigenvectors are *real*, but made no use of it here.

The eigenfunctions seem to belong to different classes according to their symmetries (exact or approximate, or none) and according to their regular or irregular appearance (see Figs. 5–7 and next section).

Time Evolution

Since chaotic features are revealed classically by the long time evolution of the system, we have studied numerically the time evolution of an initially coherent state, both from the quantal and classical points of view. In the present problem, we define a coherent state centered on $q = 0$, $p = 0$, as the ground state of the operator $[2 - \cos 2\pi(\hat{q}/Q) - \cos 2\pi(\hat{p}/P)]$. (This is a special case of Harper's operator. We notice that it is periodic in p and q, and reduces to the harmonic oscillator Hamiltonian when $\hat{p}$ and $\hat{q}$ are "small"). This coherent state is its own Fourier transform, is positive, and minimizes the periodicized uncertainty product (with the dispersion $(\varDelta q)^2$ being defined as the expectation value of $(1 - \cos 2\pi\hat{q}/Q)$, and similarly for p); hence it is the best quasi-classical pure state available, centered at $\langle q \rangle = \langle p \rangle = 0$ and periodic.

The results about the time evolution of coherent states, and their relation to chaoticity and to the classical limit, will be discussed in the next section (see Fig. 8).

V. ANALYSIS OF THE RESULTS

Here we present the numerical evidence pertaining to the eigenangles, eigenfunctions, and time-dependent *solutions*. In particular we want to discuss those features which are liable to have a semi-classical interpretation.

Eigenangle Distributions

In the study of the eigenvalues of Hamiltonians associated with classically chaotic systems, it is customary to study the statistical properties of their distributions, because no explicit expressions are available for the individual eigenvalues. In particular, it has been conjectured that the fluctuations of the eigenvalues around their mean distribution can be used as a quantal indication whether the system is classically chaotic or not [16].

For quantum maps one has only the *unitary evolution operator* for one time step, from which one can or cannot construct a time independent *Hamiltonian*. If one can, then *its* eigenvalues are fundamental, and the eigen*angle* spectrum is obtained by wrapping the eigenvalues around the unit circle; then the nearest-neighbour *eigenangle* spacing is likely to have a Poissonian distribution. However, the existence of a classical Hamiltonian in one degree of freedom makes the *classical map* integrable; thus, for a classical *chaotic map* (of one degree of freedom) no such Hamiltonian can exist. Accordingly, in the quantized version the fundamental spectral entities must be the eigen*angles* themselves. In other words, classical chaoticity negates any guiding principle according to which the eigenangles can be unwrapped from the unit circle (adding suitable multiples of 2π) to form the energy spectrum of a putative Hamiltonian. For this reason we must study the *distribution of the eigenangles themselves*. Fewer numerical results are available about *their* distributions, but they support the validity of the same classification scheme using random matrix ensembles [17].

(a)

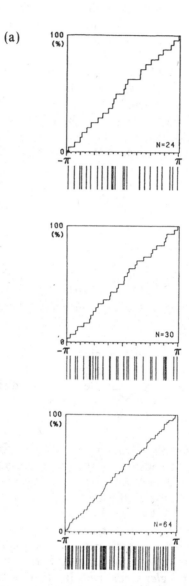

FIG 3. The spectrum of the quantal baker's transformation for a sample of values of *N*. (a) The *eigenangles* and their cumulative density; (b) the cumulative density of the nearest-neighbour *spacings*, with comparison curves: Poisson distribution (solid line), GOE distribution (dotted line), GUE distribution (dashed line) [16].

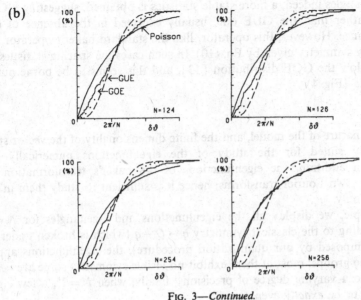

FIG. 3—*Continued.*

In Fig. 3a we show some examples: the *eigenangles* themselves (which exhibit no degeneracies); then their *spectral staircases*, which approximate closely the straight line associated with a uniform distribution, exhibiting thus no preference for particular regions. (We surmise that the actual uniform distribution constitutes here the classical limit, as the Weyl distribution does for the eigenvalues of a billiard, and the Thomas–Fermi distribution for standard Hamiltonians. Alternatively, this uniformity might be a purely statistical effect without any deeper dynamical meaning, simply indicating the lack of any preferred direction on the unit circle.)

In Fig. 3b we show some cumulative distributions of nearest-neighbour *spacings*. There we find that the *spacing statistics vary as a function of the dimensionality*, shifting all the way between Poisson-type and random-matrix type (GOE, GUE) distributions [16], through intermediary distributions.

However, when the classical model has a spatial symmetry like here, it is commonly understood that the statistical studies must be applied to each symmetry class *separately*. In our problem, this symmetry is broken under quantization, which generates complications: we must either select the exactly symmetric or antisymmetric wavefunctions (which leaves us too few objects for a suitable statistical treatment), or have further rules to divide into classes the eigenfunctions with approximate symmetry.

Alternatively, we may destroy this spatial symmetry in the classical model altogether, as for instance in the following generalized quantal baker's map

$$B' = F_N^{-1} \begin{pmatrix} F_{2N/3} & 0 \\ 0 & F_{N/3} \end{pmatrix}, \qquad N = 3, 6, 9.... \tag{11}$$

For its eigenangles indeed, a more stable statistics is obtained, suggesting a GOE distribution, rather than the GUE one, usually expected in the absence of time reversal invariance. However, this operator, like the standard baker's operator, has the anti-unitary symmetry given by Eq. (10). In such cases the spacing statistics are expected to follow the GOE distribution [13], and this seems to be borne out by the present data (Fig. 4).

Eigenfunctions

The discrete nature of the model, and the finite dimensionality of the vector space are particularly suited for the study of the eigenfunctions numerically and exhaustively. In addition, the eigenfunctions of the baker's transformation are essentially their own Fourier transforms; hence it is sufficient to study them in one representation only.

As an example, we display all the eigenfunctions and eigenangles for $N = 30$ (Fig. 5). According to the classical symmetry $q \to Q - q$ (which is broken under the discretization imposed by our quantization procedure), the eigenfunctions appear to fall into three groups: most of them exhibit no visible symmetry; some are *nearly* even or odd (to a varying degree of precision); finally, when $N = 2^k$, a few eigenfunctions seem to be *exactly* even or odd.

Although we have no objective criteria for regularity, we observe that most eigenfunctions have an overall erratic appearance, for instance the eigenfunction

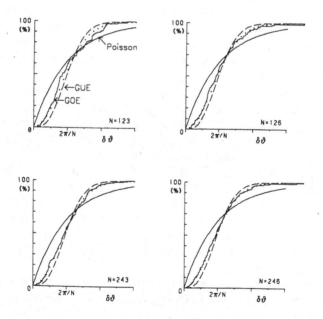

Fig. 4. The cumulative density of the nearest-neighbour spacings for an asymmetric baker's operator B' (Eq. (11)).

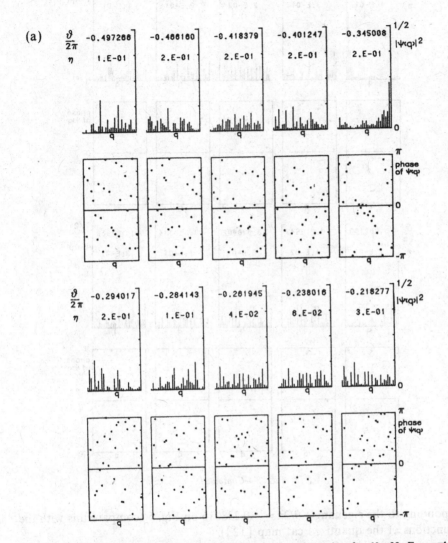

FIG. 5. All 30 eigenfunctions of the quantal baker's transformation for $N = 30$. For each eigenfunction $\Psi(q)$, we show: its normalized *probability density* $|\Psi(q)|^2$; its *phase*, adjusted so that $\Psi = \Psi^*$; and two numbers, $\theta|2\pi$, the associated *eigenangle* in units of 2π, and η, a figure of merit for the degree of *asymmetry* of the eigenfunction around the midpoint $q = Q/2$. (We use $\eta = |\Psi(Q/2)| = |\Psi_{N/2}|$ because we observed empirically a systematic correlation between this special value and the overall lack of odd *or even* symmetry of the eigenfunction around the midpoint).

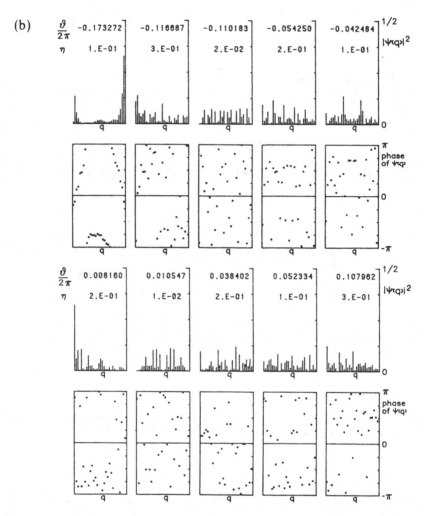

FIG. 5—*Continued.*

corresponding to the eigenangle $\theta/2\pi = -0.238016$ on Fig. 5 (compare this with the eigenfunctions of the quantized cat map [12]).

However, a surprising feature (which seems uncorrelated with the division into symmetry classes) is the appearance of a *variety* of *regular patterns*, which persist as a function of N and can be observed on many eigenfunctions, suggesting the presence of regular families. For instance, some eigenfunctions exhibit *peaks* at characteristic locations, regularly or symmetrically arranged; some others show a regular rise and fall on the larger scale, with fine modulations superimposed (some of these modulations are also quite regular). We now suggest some explanation for the presence of the peaks and modulation patterns.

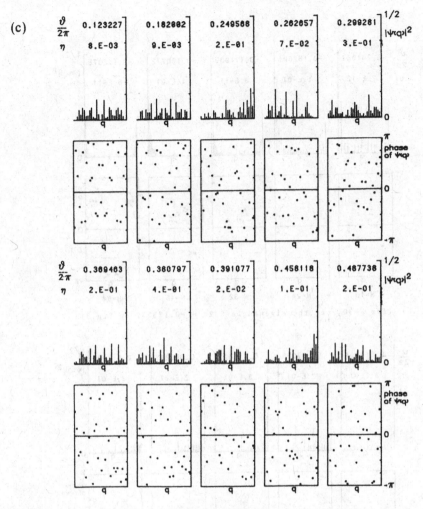

FIG. 5—*Continued.*

The sharpest peaks appear to be *associated with the* (discrete) *periodic orbits of the classical map*. A classical orbit is a sequence of discrete points because the map proceeds in discrete time steps. The peaks are actually visible above particular short periodic orbits encountered in q space. These could be the scars of periodic orbits on eigenfunctions as conjectured by Heller [18].

The shortest periodic orbit of the classical map is its (unique) fixed point at the origin ($q = 0$, $p = 0$). Indeed, several eigenfunctions have strong peaks at $q = 0$ (and hence at $p = 0$ as well, being essentially their own Fourier transforms) (see Fig. 6a). The next periodic orbit of period 2 ($q_0 = \frac{1}{3}$, $p_0 = \frac{2}{3}$; $q_1 = \frac{2}{3}$, $p_1 = \frac{1}{3}$) imprints the eigenfunctions shown on Fig. 6b: the peaks associated with the two periodic orbits of

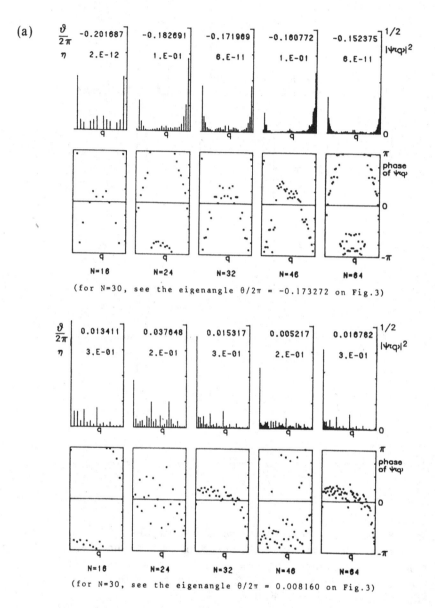

FIG. 6. Some families of eigenfunctions peaked at periodic points of the classical map. (a) at $q = 0$ (or $q = Q$); (b) at $q = Q/3$ and $2Q/3$. The similarity in shapes can be further enhanced in (a) by changing the normalization to increase the scale of $|\Psi|^2$ by N.

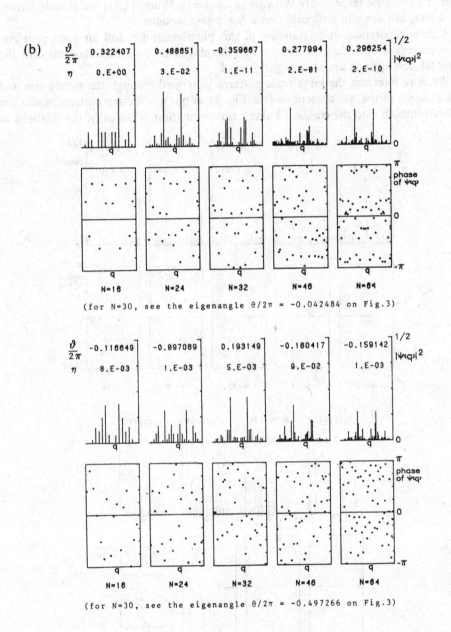

(b)

(for N=30, see the eigenangle θ/2π = -0.042484 on Fig.3)

(for N=30, see the eigenangle θ/2π = -0.497266 on Fig.3)

FIG. 6—Continued.

period 3 (passing respectively through $(q_0, p_0) = (\frac{1}{7}, \frac{4}{7})$ and $(\frac{3}{7}, \frac{6}{7})$) are already harder to detect, but are still noticeable on a few eigenfunctions.

A clear theoretical understanding of this phenomenon is still an open problem. The present model may be particularly helpful since both the classical and the quantal descriptions are so straightforward.

We may interpret the large-scale patterns generated through the steady rise and fall of some of the wavefunctions (see Fig. 7) as quasi-invariant patterns under the dilatation built into the classical baker's transformation. Classically, the dilations in

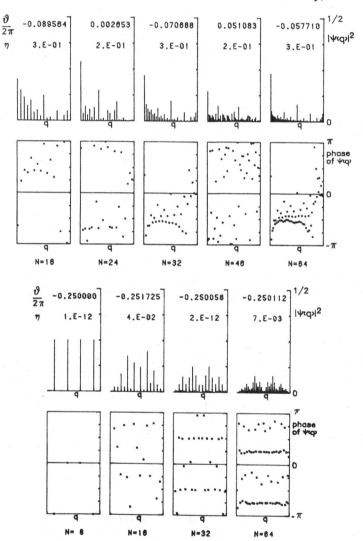

FIG. 7. Some families of eigenfunctions showing regular large scale patterns (the very remarkable last family is only observed at dimensions $N = 2^k$).

the compact phase space induce the *mixing*, which obliterates any characteristic pattern in the stationary distributions, making them uniform. It is thus remarkable that in the *quantal* version, even for large N, such patterns should persist in a stationary state. The selection of these particular eigenfunctions by analytical means remains an intriguing problem.

Time Evolution

It is a truism that many of the *classical* chaotic features are defined and detected through the long-time evolution of a dynamical system. At the same time the quantal time evolution is quasi-periodic. Indeed it has been argued that for a quantum system with a discrete spectrum all standard, classical definitions of chaoticity (Liapunov exponents, entropy, mixing, decay of time correlations, arithmetic complexity, etc.) exclude chaotic behaviour. Then, however, the question arises how to reconcile these two seemingly contradictory behaviours with the existence of a classical limit.

In order to do this we now study the time evolution of a quantum state which corresponds initially to a small, compact blob of phase fluid; this state is an adaptation of the coherent state to a discrete, periodic vector space. We use this idea since an alternative method, the use of Wigner's function, is plagued with mathematical complications [11]. The use of Kirkwood's function [14], which is complex, gives rise to other difficulties. (See the discussion of the classical limit in Section III.)

The sequence of pictures in Fig. 8 shows simultaneously the time evolution of a quantum state with $N = 64$ and that of a classical state. At time zero the quantum state is a coherent state centered at $q = Q/4$, $p = P/4$ ($P = Q = 1$); the classical state in turn is a phase-blob with the same center and width as the quantal probability distribution, Fig. 8 (0). The quantal probability distributions in p and q are shown as vertical bars over the corresponding axes; the horizontal unit square located between these axes shows the classical phase-blob in perspective. Both the classical and quantal probability distributions develop now according to their own laws. We find that through two time steps (Fig. 8(1), (2)) they develop hand in hand showing, among other things, that the classical limit is yet correctly embedded in the description. However, *as the evolution progresses discrepancies arise.*

For example, already at the sixth iterate (Fig. 8(6)) the classical q distribution is quite uniform while the quantal q distribution exhibits a ragged appearance. At the same time in the p distribution both the quantal and classical distributions prefer the same values of the momenta. This disparity between the behaviour of the q and p distributions (which arises because the transformation treats them very asymmetrically) is temporary. As the iterations progress the classical distribution in p and q becomes more and more uniform (corresponding to classical mixing); at the same time the quantal distributions continue to exhibit a fine structure which becomes similar for the p and q distributions and persists under repeated iterations, albeit with constantly changing fine details (Fig. 8(8), (9)).

This behaviour appears to be independent of the location of the initial coherent

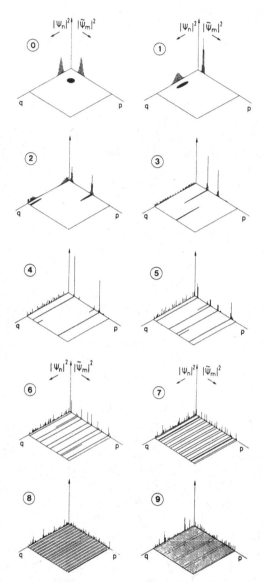

Fig. 8. The *quantal* time evolution, for $N = 64$, of the q and p probability densities of a discrete coherent state, with the corresponding *classical* time evolution in the P, Q square (seen in perspective). The initial state (0) and its evolution through nine time steps ((1)–(9)) are shown. (The dotted structure of the classical distribution at late times is only a computer artifact.)

state in phase space (including the case where this state lies on the singular dividing line, being thus immediately divided by the first iteration!).

These persistant erratic fluctuations arise as interference patterns. The doubling and cutting process which generates the classical transformation influences the wavefunctions at the even and odd positions differently. These then interfere and produce a fine pattern. The scale length of this pattern is doubled (in the q representation) at each iteration, while a new fine pattern is also generated by interference. The continuation of this process ultimately *produces all possible scale lengths*. The p distribution, being associated with a Fourier transform, must then contain the reciprocal of all these scale lengths, and will thus be similarly erratic.

The same set of pictures also describes the *backward* time evolution if the labels q and p are interchanged, because (a) the particular initial state chosen was its own Fourier transform, and (b) the quantized baker's map has the anti-unitary symmetry given by Eq. (10).

It is commonly believed that the linearity of the Schrödinger equation and the nonlinearity of the classical dynamical equations result ultimately in completely different asymptotic time behaviours for each. If this is so, the question arises how one reconciles this difference with the existence of the classical limit.

Some time ago we conjectured that *new characteristic times* appear in the quantal evolution of dynamical system [7, 19]. Of these the one associated with the hyperbolic fixed points of the classical motion is the most important, and is of the form $T = (\text{Liapunov exponent})^{-1} \log \hbar^{-1}$. For times less than T the classical and quantal time evolutions go hand in hand in the sense that the semi-classical propagator and the true propagator are qualitatively the same, while for times greater than T the disparities increase and persist. (For our purposes it is sufficient to consider the semiclassical propagator as the one which evolves coherent states according to the classical motion). Consequently, if the classical system exhibits chaotic behaviour for times less than T, the quantum evolution will reflect these features up to times $t \sim T$, though they will eventually disappear as time progresses.

This region of disparity $t \gtrsim T$ will, however, be pushed out to infinity in the classical limit. Thus we can reconcile the seeming disappearance of chaotic features in the quantal description, their presence in the classical description, and the existence of the classical limit. Then it also follows that if we desire a semi-classical description up to a given time, it is essential to treat Planck's constant as a quantity which will become *exponentially small* with the time limit.

We must stress that the exponential smallness of Planck's constant, or the logarithmic dependence of the critical time on $\hbar$ are the features which correspond to the extreme sensitivity of the classical motion arising through the hyperbolic fixed points. Applying these considerations to the present problem, we find $\omega = \log 2$, $\hbar \sim 1/N$ giving the critical time or iteration length

$$T \sim \log_2 N. \tag{12}$$

The existence of such a critical time and its dependence on N are indeed numerically verifiable in the present model.

Conclusion

We have succeeded in constructing a quantal transcription of the baker's transformation. Although our quantization scheme proceeded from mathematical simplicity and was not forced upon us through physical analogies, the result embodies features which are also desirable from the physical point of view. Therefore we believe that it provides a sensible and in some sense optimal choice among the possible quantal versions of the classical model. Like its classical analog, it enjoys a great formal simplicity; indeed, its matrix is *quite elementary* from both algebraic and numerical viewpoints, being the product of two discrete *Fourier matrices*. At the same time, the mere fact that the two factors do not commute makes the description sufficiently rich to encode features related to classical chaoticity. No systematic properties are manifest in the eigenangles and their distribution; most eigenfunctions (though not all) are without a regular structure; a new characteristic time scale appears in the motion, whose logarithmic dependence on Planck's constant reflects the exponential instability of the classical motion.

The analytical properties connected with the semi-classical features, which are under investigation, are more elusive. Nevertheless we believe that from the models presently available, this one combines the simplicity and the richness of characteristic features in a particularly suitable fashion.

Appendix A

We construct the transformation matrix $A_{n'n}^{BL}$ (see Eq. (5)).

The $N = 2N'$ sites in the q representation are labeled by $n = 0, 1, ..., N' - 1$; $N', N' + 1, ..., N - 1$, the first N' labels referring to the left, and the second N' labels referring to the right. A general normalized left state is defined by the conditions,

$$\Psi_n^L \text{ arbitrary for } n = 0, ... N' - 1; \tag{A.1}$$

$$\Psi_n^L = 0 \text{ for } n = N', ..., N - 1. \tag{A.2}$$

$$\sum_{n=0}^{N-1} |\Psi_n^L|^2 = 1. \tag{A.3}$$

The Fourier transform $\widetilde{\Psi_m^L}$ is given by

$$\widetilde{\Psi_m^L} = \frac{1}{\sqrt{N}} \sum_{n=0}^{N-1} \Psi_n^L e^{-2\pi i m n/N}. \tag{A.4}$$

On account of the $N^{-1/2}$ factor, the vector $\widetilde{\Psi_m^L}$ is also normalized to unity.

We now construct the transformed vector Φ_n^B, obtained from Ψ_n^L and obtain thereby the matrix $A_{n'n}^{BL}$.

This vector $\Phi_{n'}^B$ is conveniently specified through its Fourier components $\widetilde{\Phi}_m^B$,

$$\Phi_{n'}^B = \frac{1}{\sqrt{N}} \sum_{m=0}^{N-1} \widetilde{\Phi}_m^B\, e^{2\pi i m n'/N}, \tag{A.5}$$

where the mapping stipulates (Eqs. (4a), (4b)) that

$$\widetilde{\Phi}_m^B = \sqrt{2}\, \widetilde{\Psi}_{2m}^L, \qquad m = 0, \ldots, N'-1 \tag{A.6}$$

$$\widetilde{\Phi}_m^B = 0, \qquad m = N', \ldots, N-1 \tag{A.7}$$

(the $\sqrt{2}$ factor is required to insure unitarity, as verified later).

Substituting $\widetilde{\Psi}_{2m}^L$ from Eq. (A.4) into Eq. (A.6), we immediately obtain

$$\widetilde{\Phi}_m^B = \frac{\sqrt{2}}{\sqrt{N}} \sum_{n=0}^{N'-1} \Psi_n^L\, e^{-2\pi i n \cdot 2m/N}. \tag{A.8}$$

If we replace N with $2N'$, we observe that (a) the $\sqrt{2}$ factors cancel, and that (b) the resulting summation, up to $(N'-1)$, defines the discrete Fourier transform on N' sites, which is *unitary* due to the presence of the $N'^{-1/2}$ factor. This, then, settles the correctness of the $\sqrt{2}$ factor in Eq. (A.6).

Further inserting $\widetilde{\Phi}_m^B$ from Eq. (A.8) into Eq. (A.5), we obtain the final expression

$$\Phi_{n'}^B = \frac{\sqrt{2}}{N} \sum_{m=0}^{N'-1} \sum_{n=0}^{N'-1} e^{2\pi i m(n'-2n)/N}\, \Psi_n^L \tag{A.9}$$

$$= \sum_{n=0}^{N'-1} A_{n'n}^{BL}\, \Psi_n^L. \tag{A.10}$$

We evaluate now $A_{n'n}^{BL}$ by performing the summation over m in Eq. (A.9). If n' is even,

$$\sum_{m=0}^{N'-1} e^{2\pi i m(n'-2n)/N} = N' \delta_{n'/2,\, n}. \tag{A.11}$$

If n' is odd,

$$\sum_{m=0}^{N'-1} e^{2\pi i m(n'-2n)/N} = \frac{e^{i\pi(n'-2n)}-1}{e^{2\pi i(n'-2n)/N}-1} = \frac{-2}{e^{2\pi i(n'-2n)/N}-1} = 1 + i \cot \frac{\pi}{N}(n'-2n). \tag{A.12}$$

Consequently,

$$A^{\text{BL}}_{n'n} = \frac{1}{\sqrt{2}} \delta_{n'/2,\,n} \qquad\qquad \textit{if } n' \textit{ is even,} \qquad\qquad \text{(A.13)}$$

$$= \frac{\sqrt{2}}{N} \left(1 + i \cot \frac{2\pi}{N} (n'/2 - n) \right) \quad \textit{if } n' \textit{ is odd.} \qquad \text{(A.14)}$$

We see now that the remaining equation (3a) is automatically obtained by virtue of Eq. (A.13), exhibiting the required compatibility.

Appendix B

The action of the quantized baker's map can also be usefully compared to the signal encoding technique known as *frequency multiplexing*. Here we interpret the variable q as the *time* (hence p as the *frequency*), and the two components Ψ^L and Ψ^R of the wavefunction Ψ as two separate *signals*; we want to transmit them on a single channel within the given time interval $[0, Q]$ and in the frequency band $[0, P]$. One possibility uses the combined signal given by the wave vector Ψ itself: Ψ^L and Ψ^R are then transmitted *sequentially*, each signal occupying its own half of the entire time interval. Sometimes, however (as in stereo broadcasting with two sound channels), it is preferable to transmit both signals *simultaneously*. Then frequency multiplexing can be used for this purpose. First, both signals are *time-stretched* to extend over the same total time range $[0, Q]$. We cannot *yet* send the two signals simultaneously, since only their superposition (or sum) would be thus transmitted, all information about their difference being lost. However, each stretched signal now fills only the *bottom half* of the total frequency band $[0, P]$; hence we may shift the stretched right signal to the *top half-band* (multiplying it by the high frequency wave $\exp(iPq/2)$, called the *subcarrier*). If *this modulated* right signal is combined with the (stretched) left signal, then the two components can be separated again at the receiving end, now thanks to their disjoint *frequency* ranges (this is the standard operating principle of a stereo broadcast decoder); thus the information is preserved.

The quantized baker's transformation is the adaptation of this idea to signals *periodic* both in time and in frequency. The even rows of the B matrix stretch both components Ψ^L and Ψ^R, and superpose them at the even sites, leaving the odd sites empty. These are then filled by an interpolation process, where it is required that the stretched Ψ^L should only have "low" frequency components (in the "bottom" range), and the stretched Ψ^R only "high" ones (in the "top" range); the frequency shift is given by a fast modulation of the *sign*, which is reversed at every *odd* site only. The unitarity of the resulting matrix B reflects the conservation of information by this transformation of the signal.

It is this new feature of a doubly periodic signal which leads to a chaotic map in the classical limit, by making the phase space compact. In turn, the implied *discrete* nature of the signal (both in time and frequency) is exploited in an essential way to succeed in the actual construction of the quantized map.

ACKNOWLEDGMENTS

One of us (N.L.B.) expresses his gratitude to the National Science Foundation (U.S.A.), the Commissariat à l'Énergie Atomique, and the Ministère des Relations Extérieures (France) for partial financial support. He is also grateful to the Service de Physique Théorique de Saclay for their extensive and kind hospitality. The other author (A.V.) is grateful to the State University of New York at Stony Brook for their kind hospitality and support, and to A. Grossmann (Marseille) for fruitful discussions. We also thank Mrs. C. Verneyre (Saclay) for her help in computer programming, and finally the Referee for his incisive and constructive suggestions.

REFERENCES

1. E.g., V. I. ARNOLD AND A, AVEZ, "Ergodic Problems of Classical Mechanics," Benjamin, New York, 1968; A. J. LICHTENBERG AND M. A. LIEBERMAN, "Regular and Stochastic Motion," Springer, New York, 1983.
2. J. FORD, in "Directions in Chaos" (Hao Bai-lin, Ed.), Vol. II, p. 128, World Scientific Publishing, Singapore, 1988.
3. N. L. BALAZS AND A. VOROS, *Europhys. Lett.* **4** (1987), 1089. [We must apologize for (unessential) errors in two figures: in Fig. 1 the left–right inverted spectrum was used for $N = 34$, and in Fig. 2 the $N = 128$ cumulative density of spacings was shown in the diagram marked $N = 64$.]
4. H. WEYL, "The Theory of Groups and Quantum Mechanics," pp. 272–288, Methuen, London, 1931; J. SCHWINGER, *Proc. Natl. Acad. Sci. USA* **46** (1960), 570; reprinted in J. Schwinger, "Quantum Kinematics and Dynamics," pp. 63–72, Benjamin, New York, 1970.
5. J.-M. SOURIAU, "Structure des Systèmes Dynamiques," pp. 319–325, Dunod, Paris, 1970.
6. G. CASATI, B. V. CHIRIKOV, F. M. IZRAILEV, AND J. FORD, "Stochastic Behavior in Classical and Quantum Hamiltonian Systems" (G. Casati and J. Ford, Eds), p. 334, Springer, New York, 1979.
7. M. V. BERRY, N. L. BALAZS, M. TABOR, AND A. VOROS, *Ann. Phys. (N.Y.)* **122** (1979), 26.
8. R. GRAHAM, *Z. Phys. B* **59** (1985), 75.
9. K. NAKAMURA, Y. OKAZAKI, AND A. R. BISHOP, *Phys. Rev. Lett.* **57** (1986), 5; H. FRAHM AND H. J. MIKESKA, *Z. Phys. B* **65** (1986), 249; M. KUS, R. SCHARF, AND F. HAAKE, *Z. Phys. B* **66** (1987), 129.
10. N. L. BALAZS AND A. VOROS, *Phys. Rep.* **143** (1986), 109, Sect. VIII.5.
11. J. H. HANNAY AND M. V. BERRY, *Physica D* **1** (1980), 267.
12. B. ECKHARDT, *J. Phys. A* **19** (1986), 1823.
13. M. ROBNIK AND M. V. BERRY, *J. Phys. A* **19** (1986), 669.
14. N. L. BALAZS AND B. K. JENNINGS, *Phys. Rep.* **104** (1984), 347.
15. I. C. PERCIVAL AND F. VIVALDI, *Physica D* **25** (1987), 105.
16. O. BOHIGAS AND M.-J. GIANNONI, "Mathematical and Computational Methods in Nuclear Physics" (J. S. Dehesa, *et al.*, Eds.) p. 1, Springer, New York, 1984.
17. J. V. JOSE, *Phys. Rev. Lett.* **56** (1986), 290; F. M. IZRAILEV, *Phys. Rev. Lett.* **56** (1986), 541, and preprint 88–45 (Novosibirsk, 1988).
18. E. HELLER, *Phys. Rev. Lett.* **53** (1984), 1515; E. B. BOGOMOLNYI, *JETP Lett.* **44** (1986), 561.
19. M. V. BERRY AND N. L. BALAZS, *J. Phys. A* **12** (1979), 625.

Classical Structures in the Quantized Baker Transformation

M. SARACENO

Institute for Theoretical Physics, University of California,
*Santa Barbara, California 93106 and *Departamento de Física,*
Comisión Nacional de Energía Atomica, 1429 Buenos Aires, Argentina

Received August 29, 1989

We study the role of periodic trajectories and other classical structures on single eigenfunctions of the quantized version of the baker's transformation. Due to the simplicity of both the classical and the quantum description a very detailed comparison is possible, which is made in phase space by means of a special positive definite representation adapted to the discreteness of the map. A slight but essential modification of the original version described by Balasz and Voros (*Ann. Phys.* **190** (1989), 1) restores the classical phase space symmetry. In particular, we are able to observe how the whole hyperbolic neighborhood of the fixed points appears in the eigenfunctions. New scarring mechanisms related to the homoclinic and heteroclinic trajectories are observed and discussed. © 1990 Academic Press, Inc.

I. INTRODUCTION

In this paper we provide a detailed analysis of the classical structures that we are able to recognize in single eigenfunctions of a strongly chaotic system, the quantized version of the baker transformation [1], thus contributing to an answer to the question "what are single eigenfunctions of classically chaotic systems made of?" If we were to ask this question for an integrable system the answer would be immediate: the wave functions peak on the invariant tori quantized by discrete values of the actions. Thus each wave function shows the imprint of one torus which becomes more and more sharply defined as the classical limit is approached. To make this statement precise one has, of course, to represent the wave function in phase space, either as a Wigner or as a Husimi distribution. In systems which show soft chaos, i.e., a mixture of integrable and chaotic motion, the statement is still basically true for the regular portions of phase space. However, when the phase space is strongly chaotic no simple answer to the above question exists.

If an uncertainty ΔE is allowed in the energy by considering a smoothed contribution from several neighboring eigenfunctions or equivalently only finite times $\Delta T = \hbar/\Delta E$ are considered in the establishment of a stationary quantum distribution, then several results can be derived about the structure of eigenfunctions. In

---

* Permanent address.

particular Heller [2], Bogomolny [3], and Berry [4] have provided the basic explanation and the shapes that can be expected for the scars of short periodic trajectories that seem to dominate the landcape of many eigenfunctions. These scars modify the natural hypothesis of Berry [5] and Voros [6] that predicted a uniform distribution over the available phase space. If, however, the smoothing ΔE is reduced so as to uncover the structure of single eigenstates the situation is much less clear. On the mathematical side, serious convergence problems for the periodic orbit sum [7] makes very difficult and sometimes meaningless any attempt at summation. (However, recently an efficient summation technique using the curvature expansion has been proposed [8].) Recently, Ozorio de Almeida has investigated the role of homoclinic motion [9] and indeed his work has provided a strong motivation for our analysis here. A recent review of some of these issues has been published [10].

In view of the fact that no fully satisfactory theory is available it is therefore of interest to try to identify those classical structures that have a strong influence on chaotic eigenfunctions. The objective is of course to provide some insight as to how to judiciously combine these structures so as to produce semiclassical single eigenstates.

The quantum baker map recently proposed by Balazs and Voros [1], is ideally suited for this objective. From the classical side, its periodic orbit and manifold structure are almost trivial and therefore can be easily recognized in the eigenfunctions. On the quantum side, its study consists in the diagonalization of a finite unitary matrix so that all its eigenfunctions can be studied with standard techniques.

The second advantage of the baker map is that chaos is there in its essence, (i.e., hyperbolicity and mixing) but at the same time at its minimum complication. There are no remnants of integrability, as in the kicked rotor maps [11], nor number theoretical degeneracies, as in the cat maps [12]. It is thus the model whose detailed study can most probably prove to be generic, i.e., to provide clues as to the structure of eigenvalues and eigenfunctions in "hard" [13] chaotic systems.

The paper is organized as follows. In Section II we review the symbolic dynamics and the classical structures of the baker transformation in a way suitable to our purposes. In Section III we introduce two new elements in the quantum map. Although we follow closely the elegant construction of Ref. [1], we modify the map to restore its symmetries and by means of a special representation we study it in phase space. Both additions are essential for a detailed comparison with the classical structures. In Section IV we study the wave functions and the time evolution of the map. Besides the scars of simple periodic trajectories we are able to identify new scarring mechanisms related to homoclinic and heteroclinic families of trajectories.

II. Classical Structure of the Baker's Transformation

In this section we review the well known classical elements of the baker's map [14] in the way that we will need them in the comparison with the quantum results. The map is defined as

$$q' = 2q - [2q]$$
$$p' = \frac{p + [2q]}{2},$$

(2.1)

where the bracket indicates the integer part and where q, p are thought of as coordinates in a *compact* phase space which for convenience we take to be the unit square. The action of (2.1) is that of the baker squeezing the dough (in p) and stretching it (in q). The additional action of "cutting and putting on top" is what brings in the mixing character of the transformation. Thus both elements of generic chaotic motion (hyperbolicity and mixing) are present in the simplest possible context. The fact that the map is area preserving is the only reminder that we are dealing with canonical equations of motion.

II.1 *Symbolic Dynamics*

What makes the motion of (2.1) easy to analyze is the fact that it can be represented as a Bernoulli shift [15] in a sequence of binaries. In fact, it is easy to see that, if we represent q and p as binary decimals

$$q = \sum_{i=1}^{\infty} a_i \left(\tfrac{1}{2}\right)^i$$
$$p = \sum_{i=1}^{\infty} b_i \left(\tfrac{1}{2}\right)^i,$$

(2.2)

where a_i, b_i are the binary digits 0 and 1, then the transformation (2.1) is equivalent to the following symbolic procedure [16]:

(i)   Represent a point p, q in phase space as a doubly infinite binary sequence

$$(\cdots b_i \cdots b_2 b_1 \bullet a_1 a_2 \cdots a_i \cdots).$$

(2.3)

The dot represents the present.

(ii)   The action of the map shifts the dot to the right by one digit as can be seen from (2.1).

The notation will be that a doubly infinite sequence without the dot is the symbolic description of the whole trajectory from $T = -\infty$ to $T = +\infty$. The different points on the trajectory are obtained by placing the dot somewhere along it and reading off the p and q coordinates according to (2.2).

Having mapped the action of (2.1) into such a simple symbolic motion it is now straightforward to analyze completely the dynamical elements of the transformation. Let us enumerate them for reference.

(a)   Fixed points.

The only fixed points correspond to the sequences formed of all 0's or all 1's. According to (2.2) they are the points $(0, 0)$ and $(1. 1)$.

(b)   Periodic points.

The periodic sequences are obviously obtained by infinitely repeating a *finite* binary pattern and can therefore be put in one-to-one correspondence with the integers. Explicitly we can obtain the coordinates q, p of the periodic trajectories as follows:

(i)   Take a binary pattern $\gamma = (a_1, ..., a_L)$ consisting of L binaries a_i. Associated to γ there is a unique integer $n_\gamma = 2^{L-1}a_1 + 2^{L-2}a_2 + \cdots a_L$. The periodic trajectory is then described by $(\cdots \gamma\gamma\gamma \cdots)$.

(ii)   The coordinates of the initial periodic point are obtained in terms of n_γ by a simple calculation using (2.2),

$$q_1^\gamma = \frac{n_\gamma}{2^L - 1},$$
$$0 < n_\gamma < 2^L, \qquad (2.4)$$
$$p_1^\gamma = \frac{R(n_\gamma)}{2^L - 1},$$

where $R(n_\gamma)$ is obtained from n_γ by reversing the order of its bits.

(iii)   The other points on the trajectory are obtained by *cyclical* shifts of the bits of n_γ and $R(n_\gamma)$. Clearly all periodic points have rational coordinates.

Thus the calculation of any periodic trajectory is trivial and can be accomplished by a few lines of code. In Fig. 1 we show all trajectories calculated in this way up to period $L = 6$.

(c)   Stable and unstable manifolds.

The squeezing and stretching of the transformation affects all points of the map in the same way and therefore through any periodic point the stable and unstable manifolds meet at right angles and are parallel to the coordinate axes. The Lyapunov multipliers for each periodic point are 2 and $\frac{1}{2}$ and therefore the stability exponent of a trajectory of length L is

$$\lambda = L \ln 2. \qquad (2.5)$$

The symbolic description of orbits in the stable manifold of a periodic trajectory γ is

$$W_\gamma^- = (\text{anything } \gamma\gamma\gamma \cdots) \qquad (2.6)$$

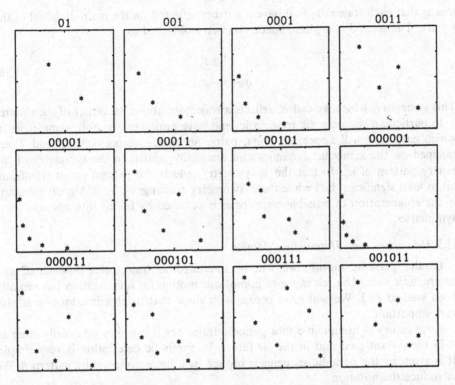

FIG. 1. Periodic trajectories of the baker's transformation. All trajectories up to period $L = 6$ are given together with their symbolic binary description. Trajectories that are not symmetric with respect to any (or both) of the two diagonals will generate a new trajectory upon reflection.

and the unstable one is

$$W_\gamma^+ = (\cdots \gamma\gamma\gamma \text{ anything}). \tag{2.7}$$

Clearly W_γ^- will approach γ in the future while W_γ^+ will do so in the past.

(d) Chaotic trajectories.

These will be represented by sequences with no asymptotic periodicity and will correspond to the vast majority of initial conditions.

II.2 *Symmetries*

The transformation (2.1) has two basic symmetries [1]. The first is an anticanonical time reversal symmetry related to the interchange of p and q and the reversal $t \rightarrow -t$. We call this the T-symmetry and observe that geometrically it means that each trajectory will have a partner reflected on the main diagonal of the square. The second is a phase space symmetry obtained as

$$p \to 1 - p$$
$$q \to 1 - q. \tag{2.8}$$

This symmetry, which we call R, reflects a trajectory about the center of the square.

In particular, the periodic trajectories can have none, one, or both symmetries in which case they will appear in fours, pairs, or singles, respectively. R and T are mapped on the symbolic dynamics and are easily related to the symmetry of the binary pattern of n_γ. In fact the R-symmetry reflects the pattern (most significant bit to least significant bit) while the T-symmetry exchanges 1 by 0. A great economy in the enumeration of possible trajectories is achieved by taking into account these symmetries.

II.3 *Homoclinic and Heteroclinic Motion*

Besides periodic motion we will be interested in trajectories homoclinic and heteroclinic to it. The relevance of homoclinic motion for quantization has recently been studied [9]. We will have occasion to show that heteroclinic motion is also very important.

A trajectory is homoclinic to a periodic trajectory if it comes arbitrarily close to it in the distant past and in the far future. Its symbolic description is very simple. It is given by the periodic sequence γ broken by a finite nonperiodic pattern h. We introduce the notation

$$h_\gamma = (\cdots \gamma\gamma\gamma h \gamma\gamma \cdots). \tag{2.9}$$

It is clear that (2.9) is the symbolic description of a trajectory that both in the past and in the future is asymptotic to γ. We call the binary pattern h the *homoclinic excursion* from the trajectory γ.

Heteroclinic motion occurs when the past and the future of the trajectory are asymptotic to different periodic trajectories. In analogy to (2.9) we denote them by

$$h_{\gamma\gamma'} = (\cdots \gamma\gamma\gamma h \gamma'\gamma' \cdots) . \tag{2.10}$$

and in this case we call h a *heteroclinic connection*.

It is important to note that in both cases the finite binary h completely characterizes the way the trajectory drifts away from periodic motion. Simple excursions will be determined by short binaries. In the case of heteroclinic connections $h = 0$ is of course allowed.

From (2.6) and (2.7) it is obvious that the coordinates of the homo and heteroclinic trajectories occur at the intersections of the stable and unstable manifolds and that therefore they trace particular paths on the respective tangles. Thus for example for the fixed point $(0, 0)$ the unstable manifold is formed by horizontal lines with rational abscissas $p = n/2^k$ (because they have terminating binary expansions) and the stable one by vertical lines with $q = n'/2^k$. The intersections form a grid of homoclinic points and a particular h in (2.9) determines a path on this grid.

III. The Symmetric Quantum Baker Map

In this section we follow closely the rationale and rely heavily on the very clear and elegant construction of the quantum baker transformation given in Ref. [1]. However, for our purposes we introduce two new elements. First, we modify in a slight but essential way the unitary matrix given in Ref. [1] so as to restore the R symmetry of the classical map. Second, we introduce a representation by discrete coherent states adapted to the torus lattice that allows us to look at the eigenfunctions of the map directly in phase space and therefore compare them with the structures discussed in Section II. Both innovations are essential for a detailed comparison of classical and quantum results.

III.1 *Restoration of the R-Symmetry*

In Ref. [1] the baker map B was given in a convenient form as a product of two non-commuting unitary matrices

$$B = F_N^{-1} \begin{pmatrix} F_{N/2} & 0 \\ 0 & F_{N/2} \end{pmatrix}, \tag{3.1}$$

where

$$(F_N)_{kn} = \langle k | n \rangle = \frac{1}{\sqrt{N}} e^{-i(2\pi/N) kn} \qquad k, n = 0, ..., N-1 \tag{3.2}$$

is the discrete Fourier transform matrix on N sites. The notation $\langle k | n \rangle$ in (3.2) is meant to indicate that F_N is the transformation matrix between the discrete coordinate states $|n\rangle$ and the discrete momentum states $|k\rangle$.

The matrix B has the following general properties, as discussed in Ref. [1]:

    (a)   unitarity,

    (b)   antiunitary symmetry

$$F_N B F_N^+ = (B^{-1})^*. \tag{3.3}$$

This has the effect that the eigenfunctions in the $|n\rangle$ and $|k\rangle$ representation are complex conjugate of each other. This is the quantum counterpart of the classical T-symmetry. The matrix (3.1), however does not commute with the quantum R operator which is given in the $|n\rangle$ representation by

$$\langle n | R_N | n' \rangle = \delta(n + n' + 1) \qquad (\text{mod } N) \tag{3.4}$$

and is the matrix with ones on the secondary diagonal. The breaking of the R-symmetry can be traced to the way the kinematical phase space was set up in Ref. [1]. In fact the discrete Fourier transform (3.2) with integer k and n treats differently the $(0, 0)$ site and the (N, N) site and one has the feeling that sites at half integer points would fare much better. This is in fact the case and a symmetric map can be achieved in the kinematical construction of the phase space by imposing *antiperiodic* boundary conditions. Thus the position eigenstates are $|n\rangle$, $n = 0, ..., N-1$ but they are repeated according to

$$|n+N\rangle = -|n\rangle. \tag{3.5}$$

It is useful to define a unitary operator V which shifts the basis states anticyclically

$$\langle n|V = \langle n+1|$$
$$V^N = -1. \tag{3.6}$$

The operator V, being unitary, can be diagonalized in a basis that we call $|k\rangle$ $(k = 0, ..., N-1)$ and is of course the momentum basis. Its diagonal form is

$$\langle k|V|k'\rangle = \delta_{kk'}\, e^{i(2\pi/N)(k+1/2)}. \tag{3.7}$$

Similarly the states $|k\rangle$ can be shifted by another operator U

$$U|k\rangle = |k+1\rangle$$
$$U^N = -1. \tag{3.8}$$

Again U can be diagonalized and the basis $|n\rangle$ where this happens is where we started from. Thus we have two shift operators U and V whose diagonal forms are identical (but in different basis). The overlap between the $|k\rangle$ and $|n\rangle$ basis is given by

$$(G_N)_{kn} = \frac{1}{\sqrt{N}} e^{-i(2\pi/N)(k+1/2)(n+1/2)} \qquad k, n = 0, ..., N-1. \tag{3.9}$$

The U and V operators are analogous to the translation operators $\exp(ip\hat{q}/\hbar)$ and $\exp(iq\hat{p}/\hbar)$ in the continuous case. Their commutation relation is

$$UV = VU\, e^{-i2\pi/N}. \tag{3.10}$$

It should be noted that no infinitesimal version of this relationship can be obtained in the discrete case. The matrix G_N is unitary, antiperiodic, and has the property

$$G_N^2 = -R_N \tag{3.11}$$

as can be seen by a direct computation. It is this property, which is not shared by the usual discrete Fourier transform (3.2), that allows the restoration of the R-symmetry. It is easy to show that if we replace F_N by G_N in (3.1) we obtain a map with all the desired properties

$$\tilde{B} = G_N^{-1} \begin{pmatrix} G_{N/2} & 0 \\ 0 & G_{N/2} \end{pmatrix}. \tag{3.12}$$

(i) *R*-symmetry

$$R_N \tilde{B} R_N = \tilde{B}. \tag{3.13}$$

This property is established using the relations

$$R_N G_N^{-1} R_N = G_N^{-1} \tag{3.14}$$

$$R_{N/2} G_{N/2} R_{N/2} = G_{N/2} \tag{3.15}$$

which follow from (3.11).

(ii) *T*-symmetry

$$G_N \tilde{B} G_N^+ = (\tilde{B}^{-1})^*. \tag{3.16}$$

R_N and G_N are then the quantum operators for the *R*- and *T*-symmetries. Their action on the translation operators are

$$R_N U R_N = U^{-1}, \qquad R_N V R_N = V^{-1} \tag{3.17}$$

$$G_N U G_N^+ = V. \tag{3.18}$$

The classical limit is unchanged as the addition of $\frac{1}{2}$ in the exponent of the Fourier matrix becomes a $1/N$ correction when the limit $N \to \infty$ with $n/N = q$ fixed is taken.

There are several advantages that arise from the new baker operator. The main one is of course that the quantization maintains all the symmetries of the classical problem. The wave functions can be labeled by the parity (under *R*) and the numerical work can be done with matrices of half the size. If needed, the matrix $\tilde{B}$ can be explicitly computed in the $|n\rangle$ representation and results in a very simple form. (See Appendix.) On the other hand one of the attractive features of the original baker map is retained: in all the calculations that we have performed the eigenangles always resulted in irrational multiples of 2π and with no degeneracies. Also no special number theoretical dependence on *N* of the eigenangles was observed. We have concentrated our study of the wave functions and therefore we have not analyzed the eigenangle statistics of the new baker map in detail. However, we do not expect any surprise there. In Ref. [1] it was observed that the fact that the discretization broke the *R* symmetry meant that two weakly coupled symmetries were considered together in the distribution and as a result a change in the spacings distribution was observed as the dimensionality changed. To avoid this problem the spatial symmetry was completely destroyed by desymmetrizing the map and then good agreement with GOE statistics was obtained. This should not even be necessary here as the two parities are not coupled and their distributions can be analyzed separately.

III.2 *Discrete Phase Space Representation*

In this section we describe briefly the main line of thought leading to the phase space representation of wave functions on a discrete torus lattice. The construction turns out to be very interesting in itself and we intend to pursue its study elsewhere. Here we only use it as a way to display the wave functions in phase space by means of positive definite distributions very similar to the Husimi ones in the continuous case.

Using the operators U and V defined in (3.6) and (3.8) we can shift a wave packet to different points on the lattice and construct the family

$$|p, q\rangle = T(p, q)|\psi_0\rangle, \tag{3.19}$$

where p, q are integers and where

$$T(p, q) = e^{(i\pi/N) pq} U^p V^{-q} \tag{3.20}$$

is the unitary translation in phase space. The state $|\psi_0\rangle$ will be defined below.

These wave packets can then be used to define *positive definite* distributions

$$W_\psi(p, q) = \frac{1}{N} |\langle p, q| |\psi\rangle|^2. \tag{3.21}$$

The reader will have noted the close parallel with the usual construction of Husimi distributions in the continuous case. It is worthwhile to point out the differences:

(a) The discrete Fourier transform replaces the continuous one as the transformation bracket between momentum and position states.

(b) There is the possibility of imposing either periodic or antiperiodic boundary conditions. The latter, which is used here, leads to half integer values for position and momentum variables.

(c) The construction avoids completely the use of hermitian infinitesimal operators $\hat{P}$, $\hat{Q}$ which cannot be properly defined in the discrete case.

Before going on to the explicit construction of $|\psi_0\rangle$ we discuss some of the desirable properties that the distribution should have to be a useful phase space representation

(a) Periodicity

$$W(p, q) = W(p + N, q) = W(p, q + N) = W(p + N, q + N). \tag{3.22}$$

(b) *R*-symmetry

$$W_{R\psi}(p, q) = W_\psi(Rp, Rq) = W_\psi(N-p, N-q). \tag{3.23}$$

(c) *T*-symmetry

$$W_{T\psi}(p, q) = W_\psi(Tp, Tq) = W_\psi(q, p). \tag{3.24}$$

If these conditions are satisfied then the effect of the quantum symmetry operation on the wave function will be reflected in the corresponding classical symmetry on the distribution. In particular the distributions for the eigenfunctions of $\tilde{B}$ will have to be symmetric with respect to the two diagonals.

The conditions for the validity of (3.22), (3.23), and (3.24) translate into conditions on the state $|\psi_0\rangle$ as follows

(a) $\quad \langle n|\psi_0\rangle = -\langle n+N|\psi_0\rangle$

(b) $\quad R|\psi_0\rangle = -|\psi_0\rangle$ $\tag{3.25}$

(c) $\quad G_N|\psi_0\rangle = |\psi_0\rangle^*.$

Moreover we must require that $|\psi_0\rangle$ be as localized as possible in both the $|n\rangle$ and $|k\rangle$ representation. Thus we need to construct a localized antiperiodic wavepacket with a definite *R*-symmetry which is an eigenfunction of the discrete Fourier transform G_N. (We are also assuming that the wave packet is real.)

We have devised two methods to construct this state. The first defines it as

$$[U - U^+ + i(V - V^+)]|\psi_0\rangle = 0 \tag{3.26}$$

in analogy to the continuous condition $(\hat{q} + i\hat{p})|0\rangle = 0$ which defines the Gaussian wave packet of minimum uncertainty. The second, which was also used in Ref. [1], defines it as the ground state of a special Harper operator

$$\left(2 - \frac{U + U^+}{2} - \frac{V + V^+}{2}\right)|\psi_0\rangle = E_0|\psi_0\rangle. \tag{3.27}$$

This is analogous to $\frac{1}{2}(p^2 + q^2)|0\rangle = E_0|0\rangle$ for the continuous case.

We have not been able to find an analytical relationship between both states. In numerical work they are very similar when N is large and we have used both. Equation (3.26) is much easier to use as it is a simple recurrence relation. However, it can be very unstable and we found it difficult to maintain a good precision when solving it. In the work reported here we have used (3.27) which requires a full diagonalization but is numerically more reliable. The state $|\psi_0\rangle$ minimizes the product of periodicized uncertainties Δp, Δq, (defined as $\Delta_p^2 = \langle\psi_0((U - U^+)/2i)^2|\psi_0\rangle$ for (3.26) and as $\Delta_p^2 = \langle\psi_0(1 - (U + U^+)/2)|\psi_0\rangle$ for (3.27) and analogously for Δq). In both cases for large N we obtain

$$\Delta q = \Delta p = 1/\sqrt{N}$$

so that the wave packet $|p, q\rangle$ can be thought of as a "quantum point" of area $\sim 1/N$ centered at the discrete coordinates p, q.

In conclusion, for this section we point out that the methods presented here would also be useful in the phase space study of other discrete maps like the cat maps and the standard map. They provide a useful alternative to Wigner distributions in discrete phase space [12].

IV. OBSERVATION OF CLASSICAL STRUCTURES IN THE EIGENFUNCTIONS

In this section we provide a graphical survey of the classical structures that we have been able to observe in individual eigenstates. This survey can be made very detailed because both the classical and quantum mechanics of the model are so extremely simple. However, we also believe that it is generic to hyperbolic motion in bound systems and therefore the structures that we find should also be observable in other computations with the proviso that the convolutedness of the classical motion will render them very hard to recognize. A semiclassical theory of the baker transformation has not been formulated and therefore no numerical quantitative comparison has been attempted. The richness of the structures that we observe certainly warrants its development.

IV.1 *Time Evolution*

The coherent state representation allows for an excellent way to portray the quantum evolution of wave packets and the divergence between the quantum and classical motion. In Fig. 2 we show the coherent state representation of powers of the map $\tilde{B}$

$$R^T(p, q) = |\langle pq| \tilde{B}^T |pq \rangle|^2, \tag{4.1}$$

where T is the (integer) time. $R^T(p, q)$ is the probability that an initial wave packet centered at p, q will return to this point after a time T. This quantity, which is the time autocorrelation of a wave packet centered at p, q, is usually displayed for a fixed p, q as a function of time. Here we prefer to display the phase space dependence at different times. It is obvious that as long as the classical and the quantum evolution agree this quantity should be strongly peaked at the periodic trajectories of period T. This is clearly shown in Fig. 2 up to period $T = 6$. Each of the peaks clearly visible up to that time correspond to some of the periodic points of Fig. 1, as long as they are reflected with the R and T symmetries. At $T = 7$, however, a very sharp transition occurs and any connection to the classical trajectories is gone. There is a very simple argument that explains this behavior, if $N = 2^M$, the number of periodic points (i.e., points on some periodic trajectory of period $\leqslant M$) is exactly equal to the number of quantum states. Therefore if we think of wave packets moving on these trajectories as "quantum points" of area $\hbar = 1/N$, only trajectories up to length M can be accomodated without interference.

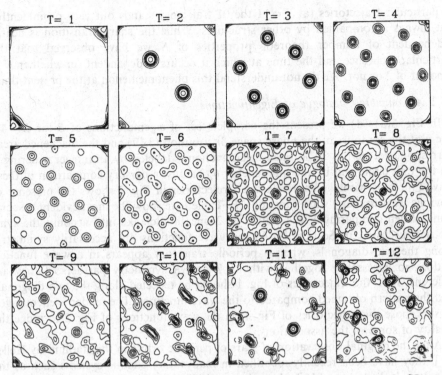

FIG. 2. Evolution of quantum packets. The contours of the quantity $R^T(p,q) = |\langle p,q| \hat{B}^T |p,q\rangle|^2$ are plotted in phase space. Note the sharp disappearance of the periodic peaks at $T = 7$.

An equivalent statement is to say that to distinguish the points on trajectories of length $L > M$ we must violate the uncertainty principle. This argument gives a different perspective to the conjecture [17] that the time at which classical and quantum time dependent phenomena diverge for chaotic systems will be of the order

$$T^* = \log_2 1/\hbar = M.$$

This is strikingly borne out by the results of Fig. 2. In fact, if we take into account the fact that the R-symmetry separates the quantum system in two multiplets of size $N/2$ then

$$T^* = \log_2 2^M/2 = M - 1.$$

This for $N = 128$ gives $T^* = 6$ which is exactly the time at which divergence occurs in Fig. 2. What is striking and quite unexpected, is the sharpness of the transition: while at $T = 6$ *all* the peaks are on periodic points, at $T = 7$ *none* are. Note also the sudden appearance of two large peaks (at $q = \frac{1}{2}$, $p = 0$ and at $q = 0$, $p = \frac{1}{2}$) completely unrelated to periodic motion. For longer times there seems to be recurrences of particular trajectories (at $T = 11$ the 01 trajectory stands out very prominently) but always accompanied by other structures. While the sharp transition is nearly independent of number theoretic properties of N we have observed that the particular recurrence and the time at which it occurs is dependent on whether N is a power of 2 or not. We do not understand this phenomemenon at the present time.

IV.2 *A Complete Catalogue of Eigenfunctions*

In Fig. 3(a)–(d) we present the complete set of eigenfunctions of $\tilde{B}$ for $N = 48$. We plot the contours of the phase space distribution (3.19) and the blackened areas correspond to the top 30%. Each eigenfunction is labeled by an integer ordering it on the unit circle and by the quasienergy $\theta_i/2\pi$. Just a cursory comparison of these wave functions with the short classical trajectories in Fig. 1 shows the presence of obvious scars in many wave functions. In Table I we show the most evident associations. When comparing Fig. 1 with Fig. 2 one should take into account the symmetries discussed in Section II. As the wave functions are necessarily symmetric along the two diagonals, when a periodic trajectory appears in a wave function it does so necessarily together with its T and R symmetry partners, if they are different. Then the trajectories of Fig. 1 should be reflected along the two diagonals and the pattern obtained compared to the wave functions. The reader is encouraged to superpose the trajectories of Fig. 1 on the wave functions of Fig. 2 to judge the quality of some of the associations.

After the original observation of scars of periodic orbits in the stadium by Heller [2], they have been recently identified in many other chaotic systems such as the quantum kicked rotor [18], the quartic oscillator [19], and the hydrogen atom both in intense magnetic [20] and microwave fields [21].

The presence of some states in the baker map scarred by periodic orbits was already noted in Ref. [1]. Our improved representation shows that it is a much more pervasive phenomenon and that many wave functions show strong imprints of periodic orbits. What is, however, more striking is the fact that there are states which, although not obviously related to periodic orbits, have a very simple structure. For examples, states $12-$, $16-$ and $19-$ in Fig. 3(c, d) are dominated by a single peak at $p = q = \frac{1}{2}$. Similarly, states $10+$, $15+$ and $22+$ in Fig. 3(a, b) show as a dominant structure two peaks at $p = 0$, $q = \frac{1}{2}$ and $p = \frac{1}{2}$, $q = 0$ (when the periodicity in p and q is taken into account). Even the more complicated states like $21+$, $19+$, $2-$, $3-$, $21-$ are not at all "random" and seem to have a well organized structure which can perhaps be associated to longer trajectories. The collection of wave functions in Fig. 3 points to the fact that one of the conclusions of Ref. [1], namely that "...most eigenfunctions (though not all) are without a regular structure..." should be reversed: in fact, most wave functions *have* a regular

---

FIG. 3. (a, b). Eigenfunctions of the baker operator for $N = 48$ and positive R-parity. The blackened area represents the distribution at 70% of its maximum and contours at 50 and 30% have been drawn. The wave functions are labeled by an integer ordering them on the unit circle and by the quasi-energies $\theta/2\pi$. (c, d) Same as (1, b) but negative R-parity.

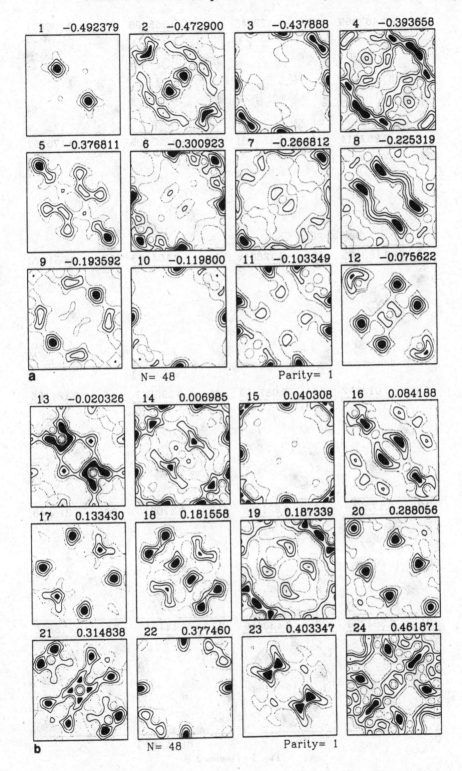

a   N= 48   Parity= 1

b   N= 48   Parity= 1

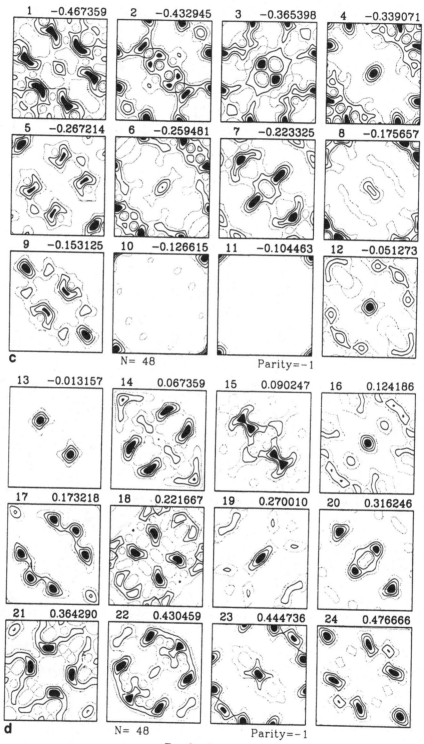

FIG. 3—*Continued*

TABLE I

Wave function	Periodic trajectory	Wave function	Periodic trajectory
1+	01	1−	001
2+	0011	5−	000111
5+	000111	7−	0011
9+	001	9−	0011
12+	000111	10−	0(1)
16+	0011	11−	0(1)
17+	001	13−	01
20+	000111	14−	000111
23+	01	15−	01
24+	001	20−	0011
		22−	001
		24−	000111

Note. Periodic trajectories that scar the eigenfunctions of Fig. 3(a–d). The wave functions are labeled by their order on the unit circle and by the R parity. The trajectories are labeled by their symbolic code (see Fig. 1). The trajectories that are not self-symmetric under R or T should be reflected to obtain the full pattern.

structure, which can be brought out by the phase space representation and is mostly (but not always) related to periodic orbits. The explanation of this structure in terms of classical elements is of course one of the outstanding tasks ahead in quantum chaos.

IV.3 *Hyperbolicity*

In this section we take up the study of the peculiar features of the wave functions not obviously related to periodic trajectories. For this purpose we study selected wave functions for a larger value of N ($=128$) so as to make the system more classical and increase the resolution of the plots. In Fig. 4(a, b) we display the wave functions related to the fixed point. The selection is somewhat arbitrary but we have included these eigenfunctions showing a large concentration at or near the origin. For clarity we have shifted the coordinates (using the periodicity of W) to display the fixed point at the center of the plot. In this way the fixed points at $(0, 0)$ and $(1, 1)$ are identified and the R and T symmetries are reflections on the two diagonals through the center of the square. The stable and unstable manifolds are also drawn. Taken as a family, these wave functions are very striking. The presence of the unstable fixed point seems to organize the wave functions in its neighborhood in a very similar way as an elliptic point would do in the stable case. While the wave functions of the latter would be concentrated on closed ellipses around the fixed point here we obtain branches of *hyperbolae* which are the flow lines close to the unstable point. Thus, not only is there a wave function concentrated at the fixed point (the very strong scar of the trajectory of period 1 at $n = 28$ in Fig. 4(b)) but there are many more that trace its hyperbolic neighborhood. This same phenomenon occurs for other periodic trajectories, although it is not so clear at the value of N that we have used.

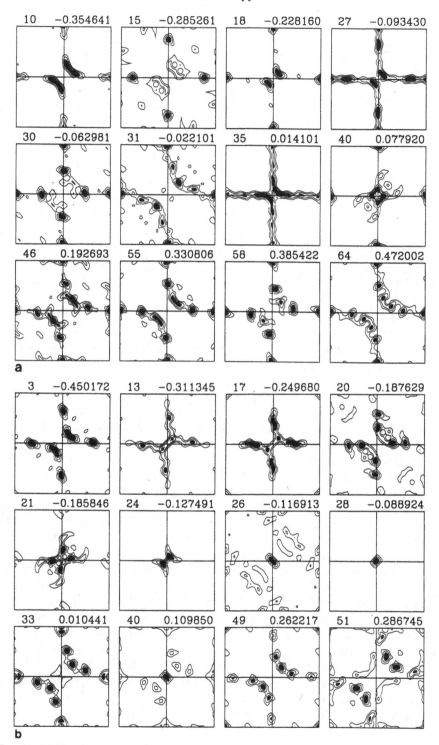

F<small>IG</small>. 4.   (a) Eigenfunctions quantized in the hyperbolic neighborhood of the fixed point for $N = 128$, $R = +1$. For emphasis the fixed point has been shifted to the center of the plot and the stable and unstable manifolds drawn. (b) Same as (a) but for $R = -1$.

Some of the hyperbolic states seem to "hug" and propagate along the stable and unstable manifolds. This has also been observed in other systems [18, 19]. However, note in our context that we cannot distinguish the manifolds from the very sharp hyperbolae close to them. It would also be possible to interpret the hyperbolae as very long periodic trajectories of a certain type (see, for example, the cases 00001 and 000001 in Fig. 1), but in that case there are many contributing trajectories in the same region and therefore they could only be collective scars of many trajectories accumulating near the fixed point [9]. We study this possiblity in detail in the next section.

IV.4 *Scars of Homoclinic and Heteroclinic Families*

In Section II.3 we defined the homoclinic trajectories. These are non-periodic trajectories, which, however, spend most of their time very close to a periodic one. They can be approximated very closely by families of long periodic trajectories as follows. Take the sequence of binary patterns

$$\gamma_h^{(k)} = (h \underbrace{\gamma \cdots \gamma}_{k \text{ times}}).$$

Each binary $\gamma_h^{(k)}$ when repeated indefinitely creates a periodic trajectory whose points accumulate close to γ and which makes a definite excursion also close to the homoclinic points. Thus the whole family will show an accumulation at the periodic points of γ and at selected homoclinic points (determined by the binary h). We have observed that many features of the wave functions, even down to minute details, reflect the patterns that these families create. In Figs. 5 and 6 we show examples of homoclinic motion near the period 2 trajectories. On the left we have plotted contours of the distribution linearly spaced from 100% down to 10% of the maximum. If we were to look only at the top 50% both pictures would be almost identical, showing a strong scar of the period 2 trajectory and some tendency to spread along the stable and unstable manifolds. The two distributions however differ in their

10   −0.327327

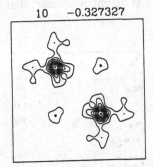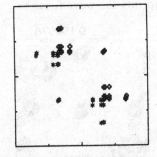

FIG 5. Scar of the homoclinic family ($\cdots \underline{01}\ 001\ \underline{01}\ \underline{01} \cdots$). The left part shows the contours of the distribution from 100 to 10% of the maximum. The right shows the periodic points of the trajectories (00101), (0010101), (001010101). The different symbols correspond to trajectories reflected by the R and T symmetries.

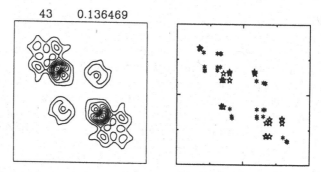

43   0.136469

FIG. 6. Scar of the homoclinic family (...01 01 0011 01 01 ···). See Fig. 5.

small components and it is there that homoclinic trajectories can be identified. On the right of Figs. 5 and 6 we show our interpretation in terms of families of homoclinic trajectories. In Fig. 5 we show the family of trajectories associated with the excursion 001. We have plotted the trajectories 00101, 0010101, and 001010101 corresponding to the beginning of the family $h_\gamma = 001_{01} = (001\underline{01}\ \underline{01}\ \cdots)$. We have also plotted (with different symbols) the R- and T-symmetry partners when they are different. It is clear that the pattern created is very similar to that of the wave function on the left. The same procedure in Fig. 6 associates the wave function to the family $h_\gamma = 0011_{01} = (0011\underline{01}\ \underline{01}\ \cdots)$. Again the reader is encouraged to superpose the patterns to judge the kind of detail that these families seem to reproduce.

In Figs. 7 and 8 we associate two eigenfunctions with heteroclinic families. Figure 7 is particularly impressive. The heteroclinic trajectory $h_{\gamma\gamma'} = 10_{0,\,1}$ is approximated by a sequence of long periodic trajectories of the type $(\underbrace{0\cdots 0}_{k}\,10\,\underbrace{1\cdots 1}_{k})$. They accumulate to the points $(0, 0)$ and $(1, 1)$ and to characteristic points on the axis and at $(\frac{1}{2}, \frac{1}{2})$. The wave function reproduces remarkably well all the details of the family, including the peculiar slope at which they approach the accumulation points. Note that although the wave function is rather simply structured, it cannot be associated to simple periodic trajectories. It definitely appears that it is the whole family that imprints the wave function.

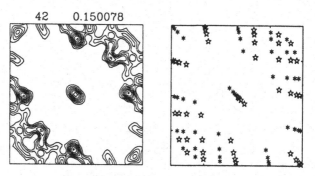

42   0.150078

FIG. 7. Scar of the heteroclinic family ($\cdots\underline{0}\,\underline{0}\,\underline{0}\,10\,\underline{1}\,\underline{1}\,\underline{1}\cdots$). See Fig. 5.

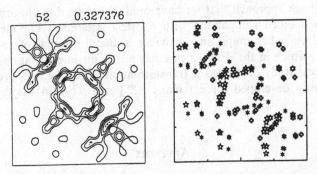

52   0.327376

FIG. 8. Scar of the heteroclinic family ($\cdots 0011\ \underline{01}\ \underline{01} \cdots$). See Fig. 5.

Figure 8 is another example of the same phenomena. The family $11_{0.01}$ has again the same detailed pattern as the wave function even down to values as small as 10% of its maximum. We have observed many more instances of these associations, both homo and heteroclinic. When either the asymptotic orbits or the excursion h are long the patterns become complicated and hard to recognize with the present value of N. The fundamental theoretical question as to why a particular family scars a given eigenfuncion, just as in the case of scars of periodic orbits, remains open and represents a major challenge.

Conclusion

We have presented for the first time a complete study of the wave functions of a system which classically shows "hard" [13] chaos. The most notable feature that emerges is their great regularity. Due to the extreme simplicity of both the classical and the quantum descriptions we are able to recognize many of the classical patterns that dominate the eigenfunctions. We find that short periodic trajectories scar many wave functions but they also have a strong influence on their whole hyperbolic neighbourhoods. Moreover it appears that *families* of periodic trajectories that can be very simply constructed from the classical symbolic dynamics create patterns that reproduce very accurrately the shape of eigenfunctions. Both these observations indicate that the classical structures that scar the wave function of chaotic systems are collective contributions from many periodic orbits. These collective contributions are related to homoclinic and heteroclinic motion. Their apparent simplicity warrants the effort towards quantitative calculation of their effects along the lines recently undertaken by Ozorio de Almeida [9].

We believe that the features observed in the wave functions of the baker transformation are *generic* of purely hyperbolic and mixing behavior. In most realistic cases the homoclinic tangle will be so convoluted that it will be difficult to uncover these features there. However, our results can be taken as a guide in the interpretation of the complicated and seemingly erratic patterns that have been observed [18, 19, 20].

The phase space approach that we have used, based on discrete torus lattices, is essential for a detailed comparison with classical structures. The method can be easily adapted to other maps and provides a useful alternative to the Wigner representation, which in the discrete case has a difficult interpretation [12].

A semiclassical study of the baker transformation is now needed, probably along the lines recently developed for cat maps [22] and work in the direction is in progress.

APPENDIX

Here we calculate explicitly the matrix elements of the $\tilde{B}$ operator. Using (3.9) and (3.12) we obtain

$$\langle n'|\tilde{B}|n\rangle = \frac{\sqrt{2}}{N}\sum_{k=0}^{N/2-1} e^{i(2\pi/N)(n'-2n-1/2)(k+1/2)} \qquad \text{for} \quad 0 \leqslant n \leqslant \frac{N}{2}-1 \qquad \text{(A.1)}$$

and

$$\langle n'|\tilde{B}|n\rangle = \frac{\sqrt{2}}{N}\sum_{k=0}^{N/2-1} e^{i(2\pi/N)(n'+1/2)(k+N/2+1/2)} e^{-i(2\pi/N)(k+1/2)(n+1/2)}$$

$$\text{for} \quad \frac{N}{2} \leqslant n \leqslant N-1. \qquad \text{(A.2)}$$

The evaluation of the geometric sums in (A.1) and (A.2) is straightforward and yields the explicit form for $\tilde{B}$

$$\langle n'|\tilde{B}|n\rangle = \frac{e^{i\phi_{n'n}}}{N\sin \pi/N(n'-2n-1/2)}, \qquad \text{(A.3)}$$

where the phase $e^{i\phi_{n'n}}$ is given by

$$e^{i\phi_{n'n}} = \frac{e^{i\pi n'}-i}{\sqrt{2}}, \qquad 0 \leqslant n \leqslant \frac{N}{2}-1$$

$$e^{i\phi_{n'n}} = \frac{e^{i\pi n'}+i}{\sqrt{2}}, \qquad \frac{N}{2} \leqslant n \leqslant N-1. \qquad \text{(A.4)}$$

Note that the denominator in (A.3) never vanishes.

The fact that $\tilde{B}$ commutes with R allows us to label the eigenfunctions with the eigenvalues ± 1 of R and reduce the diagonalization to matrices of dimension $N/2$. We provide here the details of this reduction.

The solutions of

$$\tilde{B}|\psi_i\rangle = \lambda_i|\psi_i\rangle = e^{i\theta_i}|\psi_i\rangle \qquad \text{(A.5)}$$

must have the form

$$|\psi_{\pm}\rangle = \frac{1}{\sqrt{2}}\begin{pmatrix} |u\rangle \\ \pm R_{N/2}|u\rangle \end{pmatrix} \tag{A.6}$$

which have the correct parity $R_N|\psi_{\pm}\rangle = \pm|\psi_{\pm}\rangle$ for arbitrary $|u\rangle$.

In (A.6) $|u\rangle$ is a vector of dimension $N/2$ and is chosen to have the same normalization as $|\psi\rangle$, i.e.,

$$\sum_{n=0}^{N-1}\langle\psi_{\pm}|n\rangle\langle n|\psi_{\pm}\rangle = \sum_{n=0}^{N/2-1}\langle u|n\rangle\langle n|u\rangle. \tag{A.7}$$

The equation for $|u\rangle$ is derived from inserting (A.6) into (A.5) and separating $\tilde{B}$ into $N/2 \times N/2$ blocks

$$\begin{pmatrix} \tilde{B}_{11} & \tilde{B}_{12} \\ \tilde{B}_{21} & \tilde{B}_{22} \end{pmatrix}\begin{pmatrix} |u\rangle \\ \pm R_{N/2}|u\rangle \end{pmatrix} = \lambda\begin{pmatrix} |u\rangle \\ \pm R_{N/2}|u\rangle \end{pmatrix}. \tag{A.8}$$

Thus $|u\rangle$ satisfies the eigenvalue equation

$$[\tilde{B}_{11} \pm \tilde{B}_{12}R_{N/2}]|u_{\pm}\rangle = \lambda_{\pm}|u_{\pm}\rangle. \tag{A.9}$$

The equation involving $\tilde{B}_{21}$ and $\tilde{B}_{22}$ can be shown to be redundant with (A.9) using the symmetry of $\tilde{B}$ (Eq. (3.13)). $\tilde{B}_{11}$ and $\tilde{B}_{12}$ are the appropriate blocks of the general expression (A.3).

The unitarity of $(\tilde{B}_{11} \pm \tilde{B}_{12}R_{N/2})$ follows directly from the unitarity of $\tilde{B}$. The solutions of (A.9) therefore yield two sets of $N/2$ eigenvalues on the unit circle with their corresponding set of orthonormal eigenfunctions.

Acknowledgments

Discussions with Alfredo Ozorio de Almeida were instrumental in shaping many of the ideas expressed in this paper. Martin Gutzwiller and Eric Heller provided input at many critical points. Finally, André Voros and N. Balazs, the quantum bakers, shared their insight and the excitement. The hospitality of the Institute for Theoretical Physics of S.B. is gratefully acknowledged.

This research was supported in part by the National Science Foundation under Grant No. PHY82-17853, supplemented by funds from the National Aeronautics and Space Administration, at the University of California at Santa Barbara.

References

1. N. L. Balazs and A. Voros, *Europhys. Lett.* **4** (1987), 1089; *Ann. Phys.* **190** (1989), 1.
2. E. J. Heller, *Phys. Rev. Lett.* **53** (1984), 515; *in* "Quantum Chaos and Statistical Nuclear Physics" (T. H. Seligman and H. Nishioka, Eds.), Lecture Notes in Physics, Vol. 273, pp. 162, Springer-Verlag, New York, 1986.

3. E. B. Bogomolny, *Phys. D* **31** (1988), 169.
4. M. V. Berry, *Proc. Roy. Soc. London A* **423** (1989), 219.
5. M. V. Berry, *J. Phys. A* **10** (1977), 2083.
6. A. Voros, *in* "Stochastic Behavior in Classical and Quantum Systems" (G. Casati and J. Ford, Eds.), Lecture Notes in Physics, Vol. 93, Springer-Verlag, Berlin, 1979.
7. A. Voros, *J. Phys. A* **21** (1988), 685.
8. P. Cvitanović and B. Eckhardt, Preprint ITP-89-241 Institute for Theoretical Physics, Santa Barbara, March, 1989.
9. A. M. Ozorio de Almeida, *Nonlinearity* **2** (1989), 519.
10. B. Eckhardt, *Phys. Rep.* **163** (1988), 205.
11. G. Casati, B. V. Chirikov, F. M. Israilev, and J. Ford, *in* "Stochastic Behavior in Quantum and Classical Hamiltonian Systems" (G. Casati and J. Ford, Eds.), Lecture Notes in Physics, Vol. 93, Springer-Verlag, Berlin, 1979.
12. J. H. Hannay and M. V. Berry, *Phys. D* **1** (1980), 267.
13. M. C. Gutzwiller, *J. Chem. Phys.* **92** (1988), 3154.
14. V. I. Arnol'd and A. Avez, "Ergodic Problems of Classical Mechanics," Benjamin, New York. 1968.
15. A. J. Lichtemberg and M. A. Lieberman, "Regular and Stochastic Motion," Springer-Verlag, New York, 1983.
16. R. L. Devaney, "An Introduction to Chaotic Dynamical Systems," Benjamin, Menlo Park, 1986.
17. M. V. Berry, N. L. Balazs, M. Tabor, and A. Voros, *Ann. Phys.* **122** (1979), 26.
18. G. Radons and R. E. Prange, *Phys. Rev. Lett.* **61** (1988), 1691.
19. R. L. Waterland, *et al.*, *Phys. Rev. Lett.* **61** (1988), 2733; B. Eckhardt, *et al.*, *Phys. Rev. A* **39** (1989), 3776.
20. D. Wintgen and A. Hönig, submitted for publication.
21. R. V. Jensen, M. M. Sanders, M. Saraceno, and B. Sundaram, submitted for publication.
22. J. Keating, Semiclassical properties of the cat maps, PhD thesis, University of Bristol, 1989.

Quantum nodal points as fingerprints of classical chaos

P. LEBŒUF[*]
Division de Physique Théorique[†],
Institut de Physique Nucléaire, F-91406 Orsay Cedex, France

A. VOROS
Service de Physique Théorique[‡],
Centre d'Etudes de Saclay, F-91191 Gif-sur-Yvette Cedex, France

This chapter deals with the *semiclassical analysis of the individual eigenfunctions* in a quantum system, especially when the classical dynamics is *chaotic* and the quantum *bound states* are considered. The situation is still barely understood, but analytic methods relevant to the problem have been steadily developing [1-4]. On the one hand, *quantum maps* have emerged as ideal dynamical models for basic' studies, with their ability to exhibit classical chaos within a single degree of freedom [5]. On the other hand, *phase space techniques* have become recognized as extremely powerful for describing quantum states; however, because these techniques concentrate routinely on the *density operators* (namely, the eigenprojectors in Wigner or Husimi representations [6]), they are still currently a long way from grasping the semiclassical shapes of the wavefunctions as such.

We argue that well-adapted representations of eigenfunctions are essential for semiclassical analysis and that they should incorporate all previous observations. First, the dynamical problem should be considered in the reduced form of a quantum map; then, its eigenstates should be analyzed in phase space; there, however, they should not be displayed as density operators but directly parametrized as wavefunctions.

This chapter essentially reviews an explicit realization of that program in one degree of freedom, in which the crucial ingredient is a *phase-space parametrization of 1-d wavefunctions* [7]. Every 1-d wavefunction is first expressed in a holomorphic (Bargmann) representation, then *factorized over the zeros of its Husimi function*, to end up being represented by a pattern of essentially $N \sim h^{-1}$ of those zeros in a 1–1 correspondence; at that point the semiclassical regime appears as a thermodynamic ($N \to \infty$) limit. Central to the whole approach are (hitherto unnoticed) *multiplicative properties* of the Husimi functions, only established at present upon certain phase spaces, *e.g.*, the torus $\mathbb{T}^2$ and the sphere S^2 (to mention only compact examples, the most relevant here). Over these, however, our

[*] Member of CNRS.
[†] Unité de Recherche des Universités Paris XI et Paris VI associée au CNRS
[‡] Laboratoire de la Direction des Sciences de la Matière du Commissariat à l'Energie Atomique.

framework encompasses *all 1-d wavefunctions, and in particular the eigenfunctions of classically chaotic quantum maps.*

Our main result about eigenfunctions concerns a characterization of their *semiclassical behavior* when they are portrayed as patterns of zeros. The nodal patterns for eigenfunctions, in their high-density regime, *condense on lines for classically integrable systems,* whereas *for classically chaotic quantum maps the zeros appear to diffuse fairly uniformly* over (almost) all of phase space [7]. We conjecture that this diffusion is genuinely realized and constitutes a quantal imprint (upon the eigenfunctions) of the chaotic classical dynamics.

1 A review of phase space methods

As our semiclassical discussion of quantum bound states will clearly proceed from the *phase-space approach,* we begin with a brief background review of the latter topic, in any number d of degrees of freedom.

Conventional wavefunction representations have well-known shortcomings in the semiclassical regime. The asymptotic behavior of eigenfunctions is simply *not extractible* from the bare Schrödinger equation $\hat{H}\psi = E\psi$ which is not a regular perturbation of classical dynamics, and its general description remains unknown. Even in special known cases, e.g.,

$$\psi \sim A(q)\,e^{iS(q)/\hbar} \quad (\hbar \to 0) \quad \text{(WKB)}, \tag{1}$$

a phase-space picture provides a clearer description, nonsingular and global, by which this semiclassical behavior is linked to the structure of the *classical invariant subsets in the energy shell* (of phase space). Specifically, WKB behavior requires these invariant sets to be *Lagrangian* submanifolds (embedded d-dimensional surfaces, of local equations $\{p = \nabla S(q)\}$) [8]. For classical bound systems, however, such Lagrangian surfaces, or *invariant tori,* are the hall-mark of completely integrable systems (including all conservative 1-d systems) and of small perturbations thereof (where some tori are preserved, by the KAM theorem). Instead, for the chaotic systems on which we focus, typical orbits are ergodic in the full energy shell, some are periodic and unstable, but neither type of orbit can generate embedded Lagrangian manifolds. Thus, WKB behavior is ruled out for the quantized eigenfunctions. At the same time there is no alternative semiclassical parametrization in sight ("the semiclassical eigenfunction puzzle").

Phase-space methods partly obviate this difficulty by shifting the semiclassical analysis onto the *density operator* $\rho = |\,\psi \, ><\, \psi\,|$, in representations where the quantum dynamics (the Heisenberg equation) explicitly appears as a deformation of classical dynamics (the Liouville equation); this allows a formal treatment order by order in $\hbar$. Among these "phase-space representations" the most popular ones, being real and symmetric, are given by the Wigner and Husimi functions;

respectively [6],

$$W_\psi(q,p;\hbar) = (2\pi\hbar)^{-d} \int \psi(q-r/2)\psi^*(q+r/2)\,e^{ipr/\hbar}\,d^dr, \tag{2}$$

$$H_\psi(q,p;\hbar) = (\pi\hbar)^{-d} \int W_\psi(q',p';\hbar)\,e^{-\left[(q-q')^2+(p-p')^2\right]/\hbar}d^dp'\,d^dq' \tag{3}$$

$$= \frac{|<z|\psi>|^2}{<z|z>} \tag{4}$$

where the following standard formulae,

$$|z> = e^{\bar{z}a^\dagger}|0>, \qquad a^\dagger = 2^{-1/2}\hbar^{-1}(\hat{q}-i\hat{p}), \qquad z = 2^{-1/2}(q-ip), \tag{5}$$

define the antianalytic plane coherent state $|z>$ labeled by z, admitting the position wavefunction

$$<q'|z> = (\pi\hbar)^{-d/4}e^{\left[-(\bar{z}^2+q'^2)/2+\sqrt{2}\,\bar{z}q'\right]/\hbar}. \tag{6}$$

We now review the main properties of the Wigner and Husimi representations for pure states.

	WIGNER MAP	HUSIMI MAP
Nonlinearity	quadratic with respect to ψ	
Range	nonlinear subset in much bigger	
	function space (doubling of variables)	
Invertibility	YES, but restricted to range of map	
Classical limit	YES, explicit convergence in weak topology	
Positivity	NO, deep oscillations	YES, more tame oscillations
	sweeping both signs	
Uniqueness	YES	NO

The main virtue built into both representations is that if the quantum state has a classical limit at all, it reveals itself through *explicit* convergence when $\hbar \to 0$ of $W_\psi(q,p;\hbar)$ (or H_ψ) towards a *classical density* (i.e., a phase-space probability distribution obeying the classical dynamics). Such convergence can only be realized *weakly*, i.e., upon smearing of W_ψ or H_ψ against smooth classical observables or in the sense of expectation values. Pointwise, W_ψ or H_ψ rapidly oscillate in (parts of) phase space; this is an unphysical artefact of "casting quantum wave mechanics into a classical mold" . Since the Husimi map (3) is the Wigner map convoluted with a Gaussian of width $\propto \sqrt{\hbar}$ in all directions of phase space, it shows more damped oscillations, confined to positive values by Eq.(4), whereas W_ψ oscillates into the negative range as well (see examples in [9]).

The *inversion* of the Wigner map $\psi \longrightarrow W_\psi$, to regenerate the wavefunction itself, becomes very difficult in the semiclassical limit. The Wigner representation, which doubles the number of variables, is realized in a highly redundant function space, in which the range of the Wigner map forms a very "thin" nonlinear "manifold". Both the equations of this manifold and the inversion formula $W_\psi \longrightarrow \psi$, which are well defined for fixed $\hbar$, acquire essential singularities at $\hbar = 0$ which make them totally unstable in the semiclassical regime of weak convergence for W_ψ. (Although the Husimi function can converge more strongly, the additional Gaussian smoothing in Eq.(3) can only make the inversion procedure $H_\psi \longrightarrow \psi$ more delicate.) Special cases, however, allow such an explicit algorithmic reconstruction of ψ from W_ψ order by order in $\hbar$: the WKB eigenfunctions can be recovered in semiclosed form to all orders in $\hbar$ precisely like that, starting from an expansion of W_ψ over a Lagrangian manifold [10,11].

Under a classically ergodic dynamics, by contrast, the Wigner functions of eigenstates usually appear to tend (distributionwise) to the *microcanonical density* [12,13]; for the Husimi functions the same holds in a stronger sense (measurewise) and this is, moreover, proved [14]. Thus, the eigenstates are assigned a definite semiclassical behavior; however, no specific asymptotic wavefunction expression is known to correspond to that type of behavior. Hence, in chaotic situations (which are ergodic), the semiclassical eigenfunctions are not determined any better by such arguments.

2 Reduced models

In an exploration of the quantum features arising from classically chaotic dynamics, the notion and the outcome of dynamical reduction are essential.

A classical system in d degrees of freedom can exhibit chaotic behavior only if it has $r < d$ conserved first integrals. Its dynamical reduction is then a process in which each constant of the motion is used up to integrate away one degree of freedom, resulting in a discrete-time system (a map) over the remaining $d - r$ degrees of freedom, a system which is now "irreducibly" nonintegrable. For a chaotic conservative Hamiltonian, in the simplest case of $d = 2$ and $r = 1$ (only energy is conserved), the reduced map can be a bounce map for a billiard, or a return map over a surface of section for any system; in any case it has one degree of freedom ($d = 1$, $r = 0$): it is an area-preserving map acting on a surface (the reduced phase space), which is *compact* for a bound motion. Compactness is also the simplest condition relative to the phase space itself (or *kinematical* condition) which ensures chaotic behavior provided the map is locally unstable (a *dynamical* condition). In fact, most theoretical studies of bound motion start directly at the reduced level, addressing maps on compact surfaces with or without precursors in 2-d Hamiltonians.

A theory of the quantum–classical correspondence should reasonably aim at displaying quantum structures in parallel to classical ones as far as possible. Then, just for the lowest-dimensional chaotic Hamiltonians ($d = 2$, $r = 1$), an ideally complete quantum description should incorporate: (*a*) specific *quantum kinematics* for compact phase spaces, *i.e.*, canonical constructions for Hilbert spaces of corresponding quantum states, and (b) a framework of *quantum dynamics* incorporating a quantum reduction procedure, to recast the diagonalization of a 2-*d* Schrödinger operator in terms of a 1-*d* return operator acting on a Hilbert space as found in (a). Such a comprehensive quantum theory is not yet available to provide a perfect analogy with the classical theory; only partial results exist, either formal semiclassical elaborations [15,16], or treatments of special examples. We choose to follow the latter course, and will specialize all our constructions to selected low-dimensional models, with a special mention when the classical chaotic behavior is not just observed but also *proven* (such models, though often of a formal character, are more reliable).

- *plane 2-d billiards:* these are fundamental on two grounds; (i) almost all the plane Hamiltonian systems with proven chaotic behavior are billiards (Sinai's billiard, stadium...); (ii) there exists a complete canonical quantization from billiards to wave equations incorporating a dynamical reduction scheme.

For a classical, plane and bounded billiard, let $(q, \cos \theta)$ parametrize each bounce as q (mod 1) = arc-length of the bounce point as a fraction of the boundary perimeter, θ = angle of the incident ray with the boundary. The billiard dynamics is then reducible to the *bounce map*

$$q, \cos \theta) \longrightarrow (q', \cos \theta') \quad (= \text{ parameters of the next bounce}), \qquad (7)$$

an area-preserving map over the *cylinder* phase space $C^2 = S^1 \times \mathbb{R}$ having the bounded accessible region $(q, \cos \theta) \in S^1 \times [-1, 1]$.

Likewise, a wave-mechanical billiard or Helmholtz equation with boundary conditions (*e.g.*, Dirichlet),

$$(\Delta + E)\Psi(x, y) = 0 \quad \text{inside}, \qquad \Psi = 0 \quad \text{on boundary}, \qquad (8)$$

reduces to an *integral equation* (involving the free Green's function G_0), for an auxiliary function $\psi(q)$ *on the boundary*,

$$\psi(q) = -2 \oint dq' \frac{\partial}{\partial n} G_0(E; q, q')\psi(q'), \qquad (9)$$

supplemented with an *explicitly invertible* linear mapping $\Psi(x, y) \longrightarrow \psi(q)$ [17,18]. Thus is achieved a *reversible compression* of the solution data, from two degrees of freedom into a single degree of freedom, that of the classical bounce map itself: the reduced solutions $\psi(q)$ are then the irreducible unknowns and become the prime targets for semiclassical analysis. However, actual developments of that scheme (initiated in [18,16]) lie well beyond the scope of this chapter, which we will restrict to examples of the type discussed next.

- *quantum maps:* we have seen that no exact reduced equation (like Eq.(9) for the billiard) is known for the bound states of a 2-*d* Schrödinger equation. An alternative approach for producing quantum-mechanical examples without facing explicit reduction is then to operate entirely at the reduced level and study *quantum maps*, *i.e.*, discrete-time quantum evolution operators obtained by *direct quantization of chaotic maps* [5] on *compact surfaces* (keeping here $d = 1$, the minimum possible value).

The *torus* $\mathbb{T}^2$ is for that purpose the most useful compact phase space, as it carries the most fundamental chaotic maps: the *cat* and *baker*'s maps which are *proven* chaotic (Bernoullian) [19], and also the ubiquitous *standard map* which is chaotic at large coupling [20]. The sphere S^2 also carries chaotic models such as the strongly kicked spin [21].

Last but not least, all these models are quantizable. The basic quantum signature of a compact phase space is for the corresponding Hilbert space $\mathscr{H}$ to have a *finite* dimension N: because each 1-d quantum state heuristically occupies a volume $\approx 2\pi\hbar$ in phase space, one expects

$$2\pi N\hbar \approx \text{area of phase space.} \tag{10}$$

It is natural to keep this area fixed in the semiclassical regime which is thus equivalent to $N \to \infty$.

Quantum maps are then simply unitary $N \times N$ matrices with a particular type of N-dependence in their entries, and semiclassical analysis deals with the behavior of their eigenvalues and eigenvectors as $N \to \infty$. The quantized standard map or kicked rotor [20], cat map [22], kicked spin [21], baker's map [23], and kicked Harper operator [24] each have their own physical or mathematical merits.

3 Finite-dimensional quantum structures

A preliminary but essential step, motivated by the preceding section, is to describe quantum and semiclassical formalisms corresponding to nonplanar phase spaces, especially compact surfaces. This step precedes the consideration of any specific dynamics and can be called *quantum kinematics*. Again, lacking a general theory of canonical quantization, we must limit ourselves to special examples. Fortunately, the situation is well understood for the crucially important phase spaces of the previous section: the torus $\mathbb{T}^2$ and the sphere S^2 [25]. Although it is noncompact, we also treat the cylinder C^2, which in addition to its own importance (for billiards, not treated here) also provides a helpful intermediate space between the plane and the torus.

Canonical quantizations

To avoid confusion in the following, double brackets are used to distinguish between the states of the usual phase space $\mathbb{R}^2$ and those defined on $\mathbb{T}^2$, S^2 or C^2.

- cylinder

The configuration space is $S^1 \approx \mathbb{R} \pmod 1$, hence a Schrödinger representation is provided by wavefunctions periodic up to a fixed but undetermined phase,

$$\psi(q + 1) = e^{i\theta_q}\,\psi(q). \tag{11}$$

This defines not one but a family of Hilbert spaces $\mathcal{H}^{\theta_q}$, consisting of the wavefunctions which obey (11) and have a finite norm under the scalar product

$$\langle\langle\phi\,|\,\psi\rangle\rangle = \int_0^1 \bar{\phi}(q)\psi(q)\mathrm{d}q = \sum_{m=-\infty}^{+\infty} \bar{\phi}_m\tilde{\psi}_m, \tag{12}$$

where $\tilde{\psi}_m$ denotes the discrete Fourier components of $|\psi\rangle$,

$$\tilde{\psi}(p) = \sum_{m=-\infty}^{+\infty} \tilde{\psi}_m\delta(p - p_m), \qquad \tilde{\psi}_m = \int_0^1 \psi(q)\,\mathrm{e}^{-ipq/\hbar}\,\mathrm{d}q, \tag{13}$$

carried by the lattice $\{p_m = (2\pi m + \theta_q)\hbar,\ m \in \mathbb{Z}\}$. (Representation labels will often be omitted, as in p_m for $p_m^{(\theta_q)}$.)

This structure applies, for instance, to the reduced billiard equation (9), with $\theta_q = 0$.

- torus

Applying the previous argument in position and in momentum space (both being S^1), we can require both wavefunctions ψ and $\tilde{\psi}$ to be periodic up to phases, replacing the condition (11) by

$$\psi(q + 1) = \mathrm{e}^{i\theta_q}\,\psi(q), \qquad \tilde{\psi}(p - 1) = \mathrm{e}^{i\theta_p}\,\tilde{\psi}(p), \tag{14}$$

which endows ψ with discrete components (as well as $\tilde{\psi}$). The resulting Hilbert space $\mathcal{H}^\zeta$, belonging to a family conveniently labeled by the complex variable $\zeta = \theta_p - i\theta_q$, consists of all wavefunctions of the form [26]

$$\psi(q) = \sum_{\nu=-\infty}^{+\infty} \mathrm{e}^{i\nu\theta_q} \sum_{n=0}^{N-1} \psi_n\delta\,(q - q_n - \nu), \qquad q_n = \frac{n + (2\pi)^{-1}\theta_p}{N}, \tag{15a}$$

amounting to

$$\tilde{\psi}(p) = \sum_{\mu=-\infty}^{+\infty} \mathrm{e}^{i\mu\theta_p} \sum_{m=0}^{N-1} \tilde{\psi}_m\delta\,(p - p_m + \mu), \qquad p_m = \frac{m + (2\pi)^{-1}\theta_q}{N}, \tag{15b}$$

through the (unitary) finite Fourier transformation

$$\tilde{\psi}_m = \sum_{n=0}^{N-1} (F_N)_{mn}\psi_n, \qquad (F_N)_{mn} = N^{-1/2}\,\mathrm{e}^{-2\pi i N p_m q_n}. \tag{16}$$

The scalar product on $\mathcal{H}^\zeta$ is given by

$$\langle\langle\phi|\psi\rangle\rangle = \sum_{n=0}^{N-1} \bar{\phi}_n\psi_n = \sum_{m=0}^{N-1} \bar{\phi}_m\tilde{\psi}_m. \tag{17}$$

The finite dimensionality N now follows from the imposed invariance of the Hilbert space under both translations (14), and it *exactly* satisfies [26]

$$N = (2\pi\hbar)^{-1}, \tag{18}$$

whereby h^{-1} becomes quantized to *integer values*. The relevant Hilbert spaces for the quantum cat and baker's maps are precisely of that type. However, a more general quantization is also conceivable, leading to *rational* values for h^{-1}, as used in Section 4 for the standard map.

- sphere

This is the phase space for a classical spin of fixed length s. The associated quantum Hilbert space is the irreducible spin-S representation space of finite dimension $N = (2S + 1)$ corresponding to $\vec{S}^2 = S(S+1)\hbar^2$ ($S = 0, 1/2, 1, ...$). With $s \sim \hbar S$ fixed, the classical limit can be described as $S \to \infty$. (We then scale s to unity for simplicity.)

A usual basis for that Hilbert space consists of the eigenstates $|m\rangle\rangle$ of S_z, $S_z |m\rangle\rangle = m\hbar |m\rangle\rangle$, $m = -S, \cdots, S$. Then an arbitrary state of Hilbert space can be expressed as

$$|\psi\rangle\rangle = \sum_{m=-S}^{S} \psi_m |m\rangle\rangle, \tag{19}$$

with the scalar product

$$\langle\langle\phi|\psi\rangle\rangle = \sum_{m=-S}^{S} \bar{\phi}_m \psi_m.$$

This Hilbert space applies, for example, to the kicked-spin model (relevant in solid state physics [21]) and to the schematic Lipkin shell model of nuclear physics [27].

Phase-space representations

Their basic motivation is to provide efficient tools for semiclassical analysis over the just defined N-dimensional Hilbert spaces as $N \to \infty$.

The construction of the Wigner representation, which imposes itself by symmetry on a linear phase space, becomes awkward on compact phase spaces. Its existence and uniqueness are unclear; one known construction on the torus exhibits uncanny nonlocal features inherited from the finite dimensionality of the Hilbert space [22]. Although these problems are intrinsically quite interesting, they make the Husimi representation more convenient for us: this only needs coherent states, which have been studied *ad nauseam*! [28]. Still, compact phase spaces confront coherent states with identical problems of existence and uniqueness: the many definitions of the usual plane coherent states become globally incompatible, while separately they give inequivalent generalizations. We have finally decided to preserve the *analyticity properties* of coherent states above all. For this we require, at the classical level, a phase space with a *compatible complex structure*, and at the level of quantum kinematics, an *antianalytic* family (with respect to that complex structure) of coherent states denoted $|z\rangle$. This formally suffices to define a *holomorphic* representation of wavefunctions by means of their

Bargmann transforms, $\psi(z) =< z \mid \psi >$, which are everywhere analytic functions of z; the Husimi functions then follow as $H_\psi(q,p) = \mid < z \mid \psi > \mid^2 / < z \mid z >$. We can then consider $\psi(z)$ as a sort of phase-space representation for the wave vector $\mid \psi >$.

The real issue is now to define the Bargmann transformation rigorously and to give a clean characterization of its *range* as a Hilbert space of analytic functions. This description reflects the classical geometry of phase space, and has to be specified case by case. For the phase space $\mathbb{R}^2$, this range is the classic Bargmann space [29] consisting of all the entire functions — necessarily of order ≤ 2 — such that $< \psi \mid \psi >= (2\pi\hbar)^{-1} \int_{\mathbb{R}^2} \mid \psi(z) \mid^2 e^{-\mid z \mid^2/\hbar} \, dq \, dp$ remains *finite*. On the cylinder and torus, antianalytic coherent states $\mid z \rangle\rangle$ can be obtained by brute-force periodicizations of the antianalytic coherent states $\mid z >$ on the plane (Eqs.(5)–(6)). On the other hand, for the sphere we use the well known Bloch or spin coherent states [28].

- *cylinder*

Periodicization of the kernel (6) using (11) leads to a Hilbert space of entire functions

$$\psi(z) = \int_0^1 dq' \psi(q')\langle\langle z \mid q'\rangle\rangle \tag{20a}$$

$$\langle\langle z \mid q'\rangle\rangle = \sum_{v=-\infty}^{+\infty} e^{iv\theta_q} < z \mid q' + v >=< z \mid q' > \theta_3 \left(\frac{i}{2\hbar} (q' - i\hbar\theta_q - \sqrt{2}z) \Big| \frac{i}{2\pi\hbar} \right), \tag{20b}$$

where θ_3 is the Jacobi theta function ([30], chap. 21)

$$\theta_3(v \mid \tau) = \sum_{v=-\infty}^{+\infty} e^{i\pi\tau v^2 + 2ivv} . \tag{21}$$

Eq.(11) is transformed under the mapping (20) into the quasiperiodic condition

$$\psi(z + 1/\sqrt{2}) = e^{i\theta_q} e^{\left(\frac{1}{2}+\sqrt{2}z\right)/2\hbar} \psi(z), \tag{22}$$

and the scalar product takes the form

$$\langle\langle \phi \mid \psi \rangle\rangle = 2 \int_{C^2} d^2z \, \bar{\phi}(z) \, \psi(z) \, e^{-\mid z \mid^2/\hbar},$$

with the inversion formula

$$\psi(q') = 2 \int_{C^2} d^2z \langle\langle q' \mid z \rangle\rangle \psi(z) \, e^{-\mid z \mid^2/\hbar} .$$

- *torus*

In this case, the further periodicization of (20) in the p direction and the use of $2\pi N\hbar = 1$ result in the finite sum

$$\psi(z) = \sum_{n=0}^{N-1} \psi_n \langle\langle z \mid q_n\rangle\rangle, \tag{23a}$$

$$\langle\langle z \mid q_n \rangle\rangle = (2N)^{1/4} \, e^{-2\pi N[(z^2+q_n^2)/2-\sqrt{2}zq_n]} \, \theta_3 \left(i\pi N \left[q_n - i\frac{\theta_q}{2\pi N} - \sqrt{2}z \right] \middle| iN \right),$$

(23b)

whereby wavefunctions are precisely represented by the *entire* functions on the plane satisfying the quasiperiodic conditions (cf. Eq.(14))

$$\psi(z + 1/\sqrt{2}) = e^{i\theta_q} \, e^{\pi N(\frac{1}{2}+\sqrt{2}z)} \, \psi(z) \,,$$

(24a)

$$\psi(z + i/\sqrt{2}) = e^{i\theta_p} \, e^{\pi N(\frac{1}{2}-i\sqrt{2}z)} \, \psi(z) \,;$$

(24b)

as analytic objects over the torus, these are not functions but sections of a complex line bundle of degree N. The Husimi functions, however, if defined as

$$H_\psi(q,p) = \frac{|\psi(z)|^2}{\langle z \mid z \rangle} = e^{-2\pi N|z|^2} |\psi(z)|^2 \,,$$

(25)

are strictly doubly-periodic, *i.e.*, defined as genuine functions over the torus.

The scalar product on the torus reads as

$$\langle\langle \phi \mid \psi \rangle\rangle = 2 \int_{\mathbb{T}^2} d^2z \, \bar{\phi}(z) \, \psi(z) \, e^{-2\pi N|z|^2} \,;$$

with the inversion to the usual representation (15a) given by

$$\psi_n = 2 \int_{\mathbb{T}^2} d^2z \langle\langle q_n \mid z \rangle\rangle \psi(z) \, e^{-2\pi N|z|^2} \,.$$

Next, we point out that the analytic structure of this representation coupled with the compactness of the phase space make a parametrization of wavefunctions in terms of their *phase-space zeros* most attractive. Denoting by Γ the boundary of the fundamental domain of analyticity, $[0,1/\sqrt{2}] \times [0,1/\sqrt{2}]$, it follows from the quasiperiodic conditions (24) that *every* function of the Bargmann space has N zeros in the fundamental domain,

$$(2\pi i)^{-1} \oint_\Gamma \frac{\psi'(z)}{\psi(z)} \, dz = N,$$

subjected to the constraint

$$\sum_{k=1}^{N} z_k = (2\pi i)^{-1} \oint_\Gamma z \frac{\psi'(z)}{\psi(z)} \, dz = \frac{1}{\sqrt{2}} \left[\frac{N}{2}(1-i) + \frac{\zeta}{2\pi} \right].$$

(26)

The unique state ("singlet") of the space $\mathscr{H}^\zeta$ for $N = 1$ is given from Eqs.(23) by

$$\psi_1(z) = 2^{1/4} \, e^{-\pi\left[z^2+\left(\frac{\theta_p}{2\pi}\right)^2-2\sqrt{2}z\frac{\theta_p}{2\pi}\right]} \, \theta_3 \left(i\pi \left[\frac{\zeta}{2\pi} - \sqrt{2}z \right] \middle| i \right),$$

(27)

with its single zero constrained by Eq.(26) to lie at

$$z_0 = \frac{1}{\sqrt{2}} \left[\frac{1}{2}(1-i) + \frac{\zeta}{2\pi} \right].$$

(28)

Eq.(26) illustrates the fact that the location of the zeros is a function of the parameters (θ_q, θ_p): when those parameters are varied each zero moves on the phase space torus, while the motion of the "center of mass" is governed by that equation. In particular, according to (28) the unique zero of $\psi_1(z)$ can be positioned at an *arbitrary* point of phase space by changing the periodicity conditions. The *sensitivity* of the zeros to such changes and the induced *coverings* of phase space by the zeros were exploited in [24] to characterize by means of a topological integer the eigenstates of classically chaotic quantum maps on the torus.

The displacement of the zeros also has a group-theoretical interpretation. The unitary action of the Weyl–Heisenberg group on the plane,

$$(T_{z'}\psi)(z) = e^{-|z'|^2/2\hbar} \, e^{\bar{z}'z/\hbar} \, \psi(z - z'),\tag{29}$$

naturally becomes, when the plane is quotiented to a cylinder (resp. a torus), a unitary mapping $T_{z'} : \mathscr{H}^{\theta_q} \longrightarrow \mathscr{H}^{\theta_q + p'/2}$ (resp. $\mathscr{H}^{\zeta} \longrightarrow \mathscr{H}^{\zeta + 2\pi N\sqrt{2}z'}$). This must not be confused with a continuous shift of the coherent state label z! On the torus, for instance, in every fixed space $\mathscr{H}^{\zeta}$ there exists a coherent state $|z\rangle\rangle$ with any continuous value of z, although the q or p eigenvalues are discrete. In the extreme case $N = 1$, the specification of the unique normalized state in $\mathscr{H}^{\zeta}$, $\psi_1(z) = \langle\langle z|q_0\rangle\rangle$ as given by Eq.(27), can also be read, after conjugation, as the specification of the single component of every coherent state $|z\rangle\rangle$, $\langle\langle q_0|z\rangle\rangle = \overline{\psi_1(z)}$ (except, for $N = 1$, at the singular vanishing point $z = z_0$).

A particularly interesting feature of the zeros of wavefunctions is that they can be used to *regenerate* those functions, in the most general context where these are entire, by means of Weierstrass or Hadamard factorization formulae [31]; the latter reconstructs any function of *finite order* up to $e^{P(z)}$ where $P(z)$ is a polynomial, and $\deg(P) \le r \le \text{ord}(f)$:

$$f(z) = e^{P(z)} \prod_{k=1}^{\infty} (1 - \frac{z}{z_k}) \, e^{\frac{z}{z_k} + \cdots + \frac{z^r}{r z_k^r}}.$$

Here, however, we wish to emphasize the special simplicity and virtue of Hadamard factorization *upon Bargmann functions when the phase space is compact* (i.e., after dynamical reduction, for continuous-time systems). As we show by examples, this compactness makes the factorization *finite* and *rigid* (free from undetermined constants); actually, the elements of Bargmann space are then similar to polynomials of a fixed degree (strictly speaking, they are the holomorphic sections of a complex line bundle over the compact phase space).

In the case of the torus, the Hadamard product for $\psi(z)$ is to be performed first over all the translates of each zero separately, then recombined so as to satisfy both quasiperiod conditions. The result is a *finite* multiplication formula from elliptic function theory,

$$\psi(z) = Z_\psi \, e^{2\pi \left(\sum_{j=1}^{N} (\bar{z}_j - \bar{z}_0) \right) z} \prod_{k=1}^{N} \Psi_1 (z + z_0 - z_k) = Z'_\psi \prod_{k=1}^{N} \left(T_{z_k - z_0} \Psi_1 \right)(z), \quad (30)$$

where Ψ_1 is the singlet state (27) in the space ($\zeta = 0$). Now the overall complex factor Z_ψ, which sets the normalization, is the only unspecified constant. Analogous factorization schemes have also been used in connection with the motion of 2-d electrons in a uniform magnetic field (the quantum Hall effect) [32,33] and in an arbitrary periodic magnetic field [34].

The factorization (30) constitutes the central mathematical formula around which further developments will be based (see also Eq.(38) below). It allows the reconstruction of wavefunctions from their zeros, thus establishing a one-to-one mapping between a ray of the (projective) Hilbert space and a particular distribution of zeros on $\mathbb{T}^2$. Underlying it, there is a multiplicative mapping $\mathcal{H}_N^\zeta \times \mathcal{H}_{N'}^{\zeta'} \longrightarrow \mathcal{H}_{N+N'}^{\zeta+\zeta'}$ given by pointwise multiplication of functions in Bargmann space. By construction, the associated Husimi function (25) will have the same zeros as $\psi(z)$. Therefore, this real and positive-definite function carries, through its zeros, the full description of the wavefunction. It, too, has a multiplicative structure, inherited from Eq.(30):

$$H_\psi(q, p) = C_\psi \prod_{1}^{N} H_{\Psi_1}(q + q_0 - q_j, p + p_0 - p_j). \quad (31)$$

The use of $\psi(z)$ provides a continuous representation of the finite-dimensional Hilbert space $\mathcal{H}$; because of that, it is a highly redundant representation. Alternatively, the $(N-1)$ independent complex zeros provide a "minimal" parametrization of ψ, remaining within a phase-space formalism. Ordinary representations like (15a) are equivalently "minimal" since they parametrize the wavefunctions by $(N-1)$ independent complex numbers plus a normalization. In fact, making an analogy with the polynomials (cf. the case of the sphere below), the change of representation $\{\psi_n\} \to \{z_k\}$ is the analog of changing from the coefficients to the zeros. We will see in the following, however, that the latter representation has, at least in the semiclassical limit, much nicer properties than the standard configuration representation.

- sphere

An equivalent factorization scheme in terms of zeros is mathematically much simpler in the case of a spherical geometry. For a spin-S system, the antianalytic Bloch coherent state centered at the spherical coordinates (θ, φ) is defined as [28]

$$| z \rangle\rangle = e^{\bar{z} S_+ / \hbar} | -S \rangle\rangle, \qquad S_+ = S_x + iS_y, \qquad z = \cot(\theta/2)\, e^{i\varphi}, \quad (32)$$

where the complex variable z labeling the coherent states is related to the coordinates (θ, φ) on the unit sphere by the stereographic projection from the

North pole. An analytic representation of an arbitrary state of the Hilbert space (19) in terms of ordinary polynomials of degree $2S = N - 1$ is then provided by

$$\psi(z) = \langle\langle z\,|\,\psi\rangle\rangle = \sum_{m=-S}^{S} \sqrt{\binom{2S}{S+m}}\, \psi_m\, z^{S+m}. \tag{33}$$

The scalar product and inversion formula with respect to the usual representation (19) are, respectively,

$$\langle\langle \phi\,|\,\psi\rangle\rangle = \left(\frac{2S+1}{\pi}\right) \int_{S^2} d^2z\, (1+|z|^2)^{-(2S+2)}\, \bar{\phi}(z)\psi(z),$$

$$\psi_m = \left(\frac{2S+1}{\pi}\right) \sqrt{\binom{2S}{S+m}} \int_{S^2} d^2z\, (1+|z|^2)^{-(2S+2)}\, \bar{z}^{S+m}\psi(z).$$

Since the coordinate z is singular for $\theta = 0$, the stereographic projection from the South pole should be considered as well, corresponding to differently normalized coherent states,

$$|z'\rangle\rangle' = e^{\bar{z}'S_-/\hbar}\,|+S\rangle\rangle, \qquad S_- = S_x - iS_y, \qquad z' = \tan(\theta/2)\, e^{-i\varphi}, \tag{34}$$

$$\psi(z') = {}'\langle\langle z'\,|\,\psi\rangle\rangle = \sum_{m=-S}^{S} \sqrt{\binom{2S}{S-m}}\, \psi_m\, z'^{S-m}, \tag{35}$$

or

$$\psi(z') = z^{-2S}\psi(z). \tag{36}$$

In analogy with Eqs.(24) for the torus, the latter coordinate-transformation formula means that these polynomials are not ordinary functions, but sections of a complex line bundle over the Riemann sphere; whereas the Husimi functions are now indifferently given by

$$H_\psi(\theta,\varphi) = \frac{|\psi(z)|^2}{\langle\langle z\,|\,z\rangle\rangle} = (1+|z|^2)^{-2S}\,|\psi(z)|^2 \\ = (1+|z'|^2)^{-2S}\,|\psi(z')|^2\,; \tag{37}$$

hence they are genuine functions on the sphere.

In this case, Eq.(30) is replaced by the usual factorization formula for polynomials,

$$\psi(z) = Z_\psi \prod_{k=1}^{2S}(z - z_k), \tag{38}$$

where the normalization factor Z_ψ can be related to the zeros by means of standard formulae connecting the coefficients of a polynomial to its zeros,

$$|Z_\psi| = \left[1 + \sum_{m=1}^{2S}|g_m|^2\right]^{-1/2}, \qquad g_m = \sqrt{\binom{2S}{S+m}} \sum_{1\le k_1<k_2<\cdots<k_m\le 2S} z_{k_1}z_{k_2}\cdots z_{k_m}.$$

As with the torus, an arbitrary state of this Hilbert space has then been represented by a distribution of $N - 1 = 2S$ points in the phase space, the zeros of the positive-definite Husimi function (37). It is amazing to realize that the parametrization (38) was used as early as 1932 by Majorana [35] to treat the motion of a spin-S atom moving in the vicinity of a point where the magnetic field vanishes (the so-called Landau–Zener effect).

The net outcome of the preceding paragraphs is a representation of quantum states by a finite set of $(N - 1)$ independent zeros or phase-space nodal points, with $N \rightarrow \infty$ in the semiclassical limit. As mentioned in Section 1, Husimi functions are known to concentrate, in that limit, onto the classically invariant sets. We therefore expect the morphological differences existing between the high-density regions of H_ψ for eigenfunctions of integrable and chaotic systems to be reflected in the way its zeros *distribute* in phase space. But the zeros also provide a faithful parametrization of the quantum states, and therefore they qualify better as fundamental objects for semiclassical analysis than the high-density features. This emphasizes a particularly interesting aspect of the parametrization by the zeros: it combines our phase space knowledge (and intuition) — via the Husimi distribution and the correspondence principle — with an exact (and "minimal") quantum description.

In the next section we substantiate these ideas by showing the distribution of zeros of eigenstates of some integrable and chaotic systems on the torus $\mathbb{T}^2$. Examples on the sphere S^2 have been investigated in [7], with essentially identical conclusions.

4 Numerical results and discussion

In the absence of a method to obtain eigenfunctions directly in their nodal-point parametrization, we have numerically computed eigenfunctions, and then their Bargmann transforms, for various 1-d systems. The simplest way to display the phase-space zeros is through *density plots* of the Husimi functions. Each Husimi function will be plotted twice:

- on the left, a linear density scale shows with the correct weight the high-density regions of H_ψ (in black); this plot obliterates, however, the delicate relief near zero level (in white).

- on the right, a logarithmic density scale sacrifices level resolution at high densities but brings out, with a suitable adjustment of the contrast factor, each zero as a small white dot.

Fig.1 shows the Husimi plots for $\Psi_1(z)$, the elementary building block of Eq.(30), when $(\theta_q, \theta_p) = (0, 0)$. Its only zero lies at the center of the square, as predicted by Eq.(28), but can be relocated at an arbitrary point by varying the phases. (From now on we assume that $(\theta_q, \theta_p) = (0, 0)$.)

Fig.2 shows the Husimi function for the coherent state centered at the origin, $\psi(z) = \langle\langle z \,|\, 0 \rangle\rangle$, for $N = 16$.

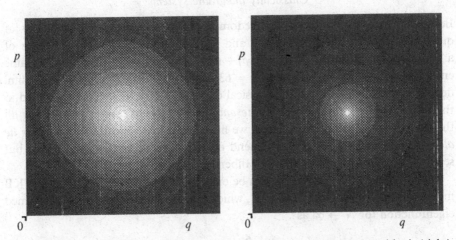

Fig. 1. Husimi function plots on the torus, in linear (left) and adjustable logarithmic (right) density scales — both using black for highest densities; all plots take $(\theta_q, \theta_p) = (0, 0)$. This double plot shows the singlet state ψ_1 ($N = 1$) (Eq.(27)); the Husimi function shows one zero and one maximum at exactly antipodal locations on the torus.

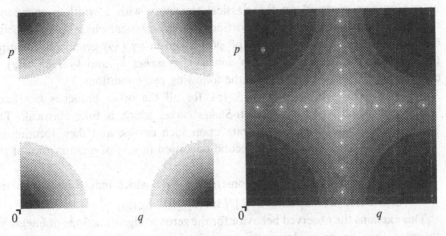

Fig. 2. As Fig.1, but showing the coherent state $|0\rangle\rangle$ for $N = 16$.

Classically integrable systems

Because we are in one degree of freedom, this corresponds to eigenfunctions of quantized *Hamiltonians*. In Fig.3(a) and Fig.3(b) we display two eigenstates of a Hamiltonian, $H(p,q) = 2 - \cos(2\pi p) - \sin(2\pi q) - \cos(6\pi q)$, at about the same energy $E \approx 1.54$, for $N = 21$ and $N = 63$ respectively. As N grows, the Husimi density concentrates around the classical curve $\mathscr{C} : \{H(p,q) = E\}$ (superposed in the figure), while its zeros *concentrate onto another set of curves*. In fact, for all the eigenstates of integrable systems we have studied, we found that *the zeros lie along some simple curves*, which depend on the classical system and energy, but seemingly not on the dimension of Hilbert space.

Both concentration properties can be explained in terms of the complex WKB method in the coordinate z [11,36,37], which expresses the Bargmann-transformed eigenfunction for $N \to \infty$ as

$$\psi(z) \sim \sum_{k=1}^{K} A_k(z) \, e^{2\pi N S_k(z)}, \tag{39}$$

where $\{S_k\}$ are branches of the classical complex action in the z variable. On the one hand, this result implies [36] that for integrable systems the high-density regions of the Husimi function can be approximated, in the semiclassical limit, by a Gaussian centered on the classical trajectory with a width proportional to $\sqrt{\hbar}$ and a height inversely proportional to the classical phase-space velocity. On the other hand, the vanishing of $\psi(z)$ as given by (39) generically requires the destructive interference of *two dominant branches* k_1 and k_2 [38], $\psi(z) \sim e^{2\pi N S_{k_1}(z)} - e^{2\pi N S_{k_2}(z)} = 0$, leading to the following two conditions:

(i) $\operatorname{Re} S_{k_1}(z) = \operatorname{Re} S_{k_2}(z) > \operatorname{Re} S_{k_n}(z)$, for all the other branches k_n. Each such equality defines an *anti-Stokes curve*, which is fully classical. The zeros of $\psi(z)$ then concentrate upon such curves, and their location on them is determined by the second condition (a sort of quantization of the anti-Stokes line);

(ii) $\operatorname{Im}\left[S_{k_1}(z) - S_{k_2}(z)\right] = (j + \text{const})/N \ (j \in \mathbb{Z})$, which makes zeros regularly distributed with spacing $d = \left[N \operatorname{Im}\left(S'_{k_1}(z) - S'_{k_2}(z)\right)\right]^{-1}$.

This explains the observed behavior for the zeros of eigenfunctions of integrable systems: they lie on a classically determined curve (anti-Stokes curve) which depends on the Hamiltonian and the energy (through $S_k(z)$), while N^{-1} (or $\hbar$) fixes the scale of the mean spacing between the zeros on that curve.

Conversely, a product like (30) or (38) over regularly spaced zeros can generate, in the large N limit, a rapidly oscillating function of the form (39) by a duality similar to the Poisson multiplication formula, or exponentiated summation formula, the prototype of which is $\sin(z/\hbar) = z/\hbar \prod_{k \neq 0}(1 - z/k\pi\hbar)$.

We can, moreover, make a general statement on the number of zeros enclosed by the classical trajectory $\mathscr{C}$. Since there is a Bohr–Sommerfeld quantization condition, $(2\pi i)^{-1} \oint_{\mathscr{C}} \psi'/\psi \, dz = \ell$ (=integer quantum number), there should be

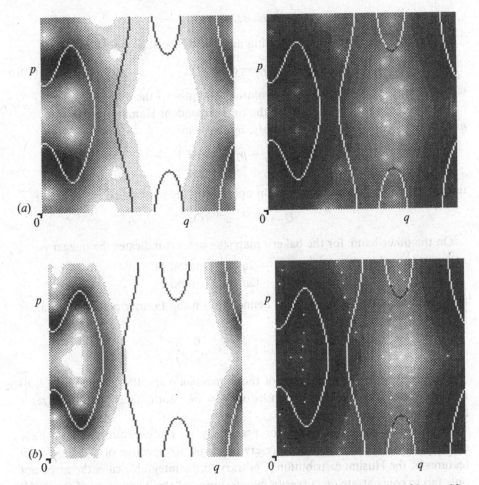

Fig. 3. As Fig.1, but showing two eigenfunctions at a fixed classical dynamics of integrable type: a Hamiltonian at a given energy (see text; the classical energy curve is superposed, and arrows at top left identify the $H < E$ region). The quantum parameter N is varied; (a) $N = 21$ (8th eigenfunction); (b) $N = 63$ (24th eigenfunction). Practical remark: in (or near) the integrable regime and for large N, the low densities are extremely depressed and require considerable log-scale amplification to reveal the zeros; this often results in noisy pictures and poor discrimination of the zeros; see also Figs.6(a)–(b),(e)–(f).

ℓ zeros in the region $H < E$ (at least if this region is topologically a disk) and $(N - \ell)$ above it. This result is a phase-space generalization of the counting theorem for the zeros of 1-d configuration-space wavefunctions.

Fully chaotic systems

The quantization of an area-preserving map yields the stationary equation

$$U \,|\, \psi \rangle\rangle = e^{i\omega} \,|\, \psi \rangle\rangle \,, \tag{40}$$

where U is the one-step (unitary) evolution operator of the map.

For the cat map [19], defined by the time-dependent Hamiltonian $H = p^2/2 - q^2/2 \times \sum_{j=-\infty}^{\infty} \delta(t - j)$ or, alternatively, by the map

$$q_{j+1} = q_j + p_j \quad \text{(mod 1)}$$
$$p_{j+1} = q_j + 2p_j \quad \text{(mod 1)},$$

the corresponding quantum evolution operator is the matrix [22]

$$U = e^{i\pi N \hat{q}^2} \, e^{-i\pi N \hat{p}^2} \,.$$

On the other hand, for the baker's map (the bracket indicates the integer part),

$$q_{j+1} = 2q_j - [2q_j] \quad \text{(mod 1)}$$
$$p_{j+1} = \left(p_j + [2q_j]\right)/2 \quad \text{(mod 1)},$$

the quantum evolution operator is defined [23] using Fourier matrices (notation of Eq.(16)):

$$\hat{U}_N = F_N^{-1} \begin{pmatrix} F_{N/2} & 0 \\ 0 & F_{N/2} \end{pmatrix} \,.$$

Fig.4 shows two eigenfunctions of those *quantum maps*, the cat map (Fig.4(a)) and the baker's map (Fig.4(b)), both for $N = 64$. Both states have their zeros highly spread out on the square.

That same observation holds for practically *all* the eigenfunctions we have computed for fully chaotic area-preserving maps, irrespective of the superficial features of the Husimi distribution. Contrary to the integrable case, the zeros not only fail to concentrate on curves (a manifestation of the inadequacy of the WKB form), but seem *to fill, like a gas, the whole available phase space*. (We have only found for the baker's map for $N = 2^k$, one isolated (arithmetical ?) exception of a state with a dominantly linear distribution of zeros.) In fact, Fig.4 and other plots [7] strongly suggest that the distributions of zeros representing eigenfunctions of quantized fully chaotic maps are, to first order in the semiclassical approximation, *uniform and isotropic* in the large (outside scars).

At this point, the crucial input of the total number of zeros, N, implies that the average spacing between nearest neighbors must be $O(N^{-1/2})$, against $O(N^{-1})$ in the integrable case. For the time being, these statements only indicate global trends, and may need refinements and corrections as better data are collected. Nevertheless they clearly point to a new semiclassical regime, drastically different from the integrable one.

Although we cannot prove it, this behavior of the zeros seems strongly tied with the ergodicity of H_ψ in the semiclassical limit, meaning here $\lim_{N\to\infty} H_\psi(q,p) = 1$ (the Liouville measure, cf. Section 2). This convergence can never take place

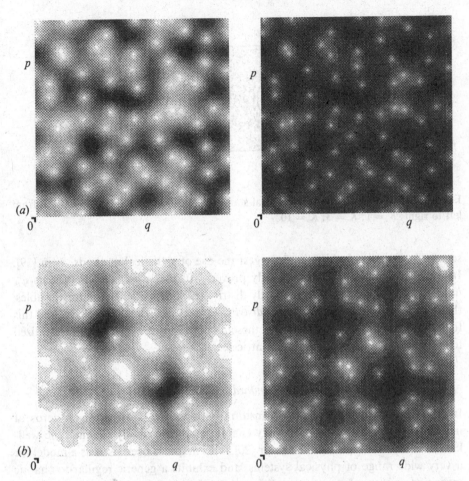

Fig. 4. As Fig.1, but showing eigenfunctions of classically chaotic quantum maps, with $N = 64$: (a) cat map; (b) baker's map — this eigenfunction is markedly scarred by the classical orbit of period 2: $((1/3, 2/3),(2/3, 1/3))$.

pointwise since H_ψ has N zeros, an increasing number. The next best possibility for H_ψ to uniformize in a weaker sense, in view of the factorized representation (31), is for the *zeros themselves* to become uniformly distributed in phase space.

The problem of the derivation, from a supposedly quasiuniform distribution for the zeros, of an explicit asymptotic form to be obeyed by $\psi(z)$ is important. The 1-*d* distributions of zeros in the integrable case were regular, and a "Poisson multiplication formula" could suggest a WKB result. By contrast, these 2-*d* patterns of zeros can equidistribute in many ways, regular (crystal-like) *or disordered*. In the former case, the product (30) can be expressed in automorphic functions but this occurs only exceptionally (for a few eigenfunctions of cat maps). We do not know how to analyze products like (30) or (38) if the factors are shifted in

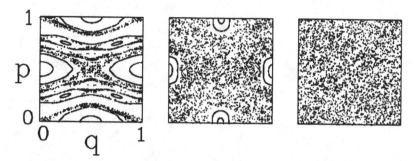

Fig. 5. Phase-space orbits of the classical standard map (41) for various couplings; from left to right: $K = 1$; $K = 3$; $K = 10$.

seemingly irregular fashion, and suggest the use of statistical methods, as in [39]. In this direction, it has been recently proved [40] that when the coefficients ψ_m in (33) (or (19)) are taken as normally-distributed independent random variables, then the zeros of (33) are *uniformly distributed* over the Riemann sphere, and this for an arbitrary N. We remark that these assumptions are also precisely realized for the eigenfunctions of random matrices.

The quantum standard map or kicked rotor

In order better to illustrate the transition between the distribution of zeros of integrable systems and those of fully chaotic maps, we finally examine the well-known standard map or kicked rotor [20]. This map has been used as a model for a very wide range of physical systems, and exhibits a generic regular-to-chaotic transition as a coupling constant K is increased. It is defined by the Hamiltonian

$$H = \frac{p^2}{2} + \frac{K}{4\pi^2} \cos(2\pi q) \sum_{j=-\infty}^{\infty} \delta(t - j).$$

Integrating Hamilton's equations between $(j-1/2)$ and $(j+1/2)$ and rescaling $p/2 \to p$, we get the map

$$q_{j+1} = q_j + p_j + p_{j+1} \quad \text{(mod 1)} \tag{41a}$$

$$p_{j+1} = p_j + \frac{K}{4\pi} \sin\left[2\pi\left(q_j + p_j\right)\right]. \tag{41b}$$

K controls the chaoticity of the map. For $K \to 0$, the dynamics is nearly integrable; for $K \approx 0.9716$, the last KAM torus winding around the cylinder in the q-direction disappears, and diffusion in the p direction starts; for $K \gg 1$, the volume occupied by stable trajectories is exponentially small [41], and the map can be treated as completely chaotic. In Fig.5 we show some trajectories of the map for several values of the coupling constant K.

The map (41) is invariant under the discrete translations

$$q' = q + 1 \tag{42a}$$
$$p' = p + 1 \tag{42b}$$

showing that the natural classical phase space is the 2-*d* torus. Many studies have concentrated, however, on the motion on the cylinder. In particular, it has been shown that for large values of the coupling constant ($K > 4.5$) the particle diffuses in momentum, $\langle p_j^2 \rangle \simeq D j$ (for j sufficiently large), with the diffusion constant $D = K^2/2$.

Associated with this map is the quantum evolution operator

$$U = e^{-i\hat{p}^2/4\hbar}\, e^{-iK \cos(2\pi\hat{q})/(4\pi^2\hbar)}\, e^{-i\hat{p}^2/4\hbar}. \tag{43}$$

The quantization can be pursued in two different ways (kinematics enters here). For irrational values of Planck's constant h, quantum mechanically the invariance of the dynamics under the discrete translation (42b) does not hold and the phase space is the cylinder; U is then represented by an infinite matrix. The main physical result emerging in connection with this model is the localization in momentum of the eigenstates of (43) [42], $\tilde{\psi}_m \sim e^{-|m-m_0|/l_\infty}$, a behavior that contrasts with the aforementioned classical diffusion (l_∞ = localization length).

A second possibility is to impose translational invariance in *both directions* of the fundamental domain $(q,p) \in [0,1] \times [0,1]$. This leads to the condition (18), which also sets the dimension of Hilbert space, and therefore of the matrix U. A more general quantization on the torus can be realized by imposing periodicity (up to phases) in the enlarged torus

$$q' = q + 1 \tag{44a}$$
$$p' = p + M, \qquad M \in \mathbb{N}, \tag{44b}$$

which can be implemented redefining Planck's constant as $(2\pi\hbar)^{-1} = N/M$. As $M \to \infty$, the enlarged phase space torus approaches the cylinder. The classical limit corresponds to $N \to \infty$, $M/N \to 0$. Fig.6 shows selected eigenfunctions of the standard map quantized with $N = 61$. Figs. 6(a)–(d) concern the basic case $M = 1$ for different values of K. For $K = 0$ (Fig.6(a)) (an eigenstate of $\hat{p}^2$, the integrable limit), the high-density regions of H_ψ are concentrated over two horizontal lines corresponding to $|p|$ = const; the zeros also align upon two horizontal lines. In Fig.6(b) we show an eigenstate for $K = 1$ having as classical support an invariant curve winding around the phase-space point $(q,p) = (0,1/2)$ (cf. Fig.5, left). Although the high-density regions are exponentially concentrated on a quantized curve, as for completely integrable systems, all the zeros are no longer aligned; instead, some tend to diffuse. This tendency is amplified for higher K (Fig.6(c), $K = 3$), where the zeros conserve a local alignment but already tend to fill the whole available phase space left out of the regions surrounding

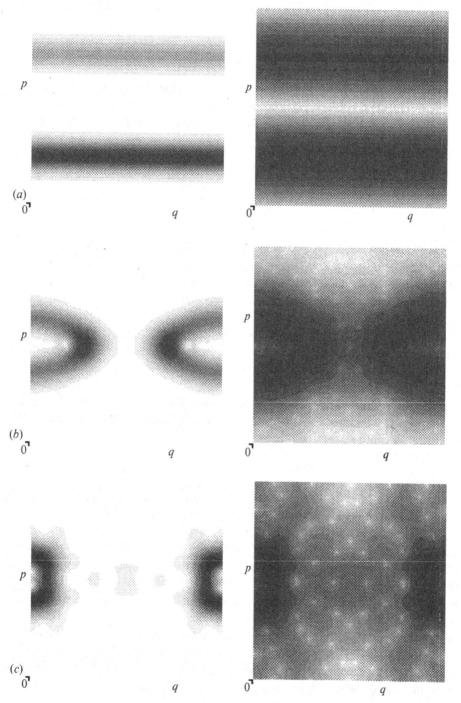

Fig. 6. Selected eigenfunctions of the quantized standard map, with $N = 61$; for (a)–(d) $M = 1$ is fixed (cf. Eq.(44b)) and the coupling constant is varied. (a) $K = 0$ (integrable case); (b) $K = 1$; (c) $K = 3$; (d) (next page) $K = 10$. The remark on Fig.3 applies to (a) and (b).

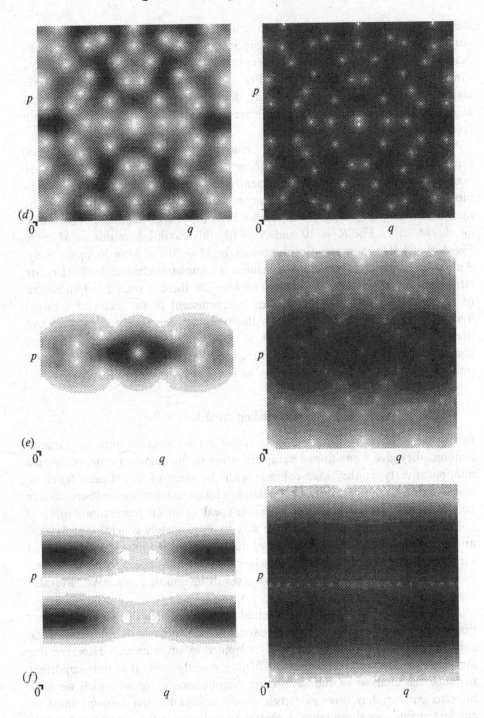

Fig. 6. (continued): for (d)–(f) the coupling is fixed, $K = 10$. (d) $M = 1$; (e) $M = 20$; (f) $M = 30$. The remark on Fig.3 applies to (e) and (f).

the quantized curve. This state, which is also supported by a classical invariant curve, differs from, *e.g.*, the one of Fig.3: the zeros, irrespective of the high-density behavior of H_ψ, *seem highly sensitive to the dominant underlying classical dynamics* (for $K = 3$, more than 80% of phase space is covered by chaotic trajectories, cf. Fig.5, center). Finally, for $K = 10$ (Fig.6(d)) even the local alignment has disappeared, and a diffusive behavior of the zeros typical of fully chaotic wavefunctions entirely prevails.

We have, moreover, analyzed the behavior of the wavefunctions and their zeros in the strongly chaotic regime ($K = 10$) for increasing M. It has already been pointed out [43] that for sufficiently large M the eigenstates localize in momentum. The transition to that regime can be estimated by requiring the localization length l_∞ to be much smaller than the size of the phase-space torus, i.e., $l_\infty / M \ll 1$. For $K = 10$ and $N = 61$, this condition results in $M \gg 6$. Figs.6(e) and 6(f) show two eigenfunctions for $M = 20$ and $M = 30$, respectively. Although the underlying classical dynamics is completely chaotic, both states are strongly localized in the p-direction. For $M = 20$, there is some local alignment of the zeros, while the overall pattern is reminiscent of the states of a mixed dynamics (cf Fig.6(b)). For $M = 30$, the zeros are completely aligned on two horizontal lines, like for $K = 0$ (Fig.6(a)). Thus, the effect of increasing M is not just a localization of the eigenstates in momentum, but also *a suppression of all traces of the classical chaotic dynamics*, as far as the zeros seem able to detect.

5 Concluding remarks

In the Bargmann representation, the zeros of a wavefunction enjoy very special features: they give a one-to-one parametrization of the quantum state, of discrete multiplicative type; they also coincide with the zeros of the Husimi function, which appear as quasiparticles in the classical phase space (they could specifically be nicknamed "zerons" for short – we also speak of *stellar* representation).

These parameters seem to pave the way to a completely general semiclassical analysis of eigenfunctions. In support of that statement, we have mainly observed the *diffusive behavior of the zeros for eigenfunctions in classically chaotic regimes* as opposed to their concentration on curves in integrable cases. (We have also observed intermediate regimes.)

All that remains very far from an actual semiclassical description of eigenfunctions in chaotic situations, where the easiest access to the zeros still seems to be … by the quantum wavefunction itself (computed by other means). However, this approach suggests new directions for further investigation. It is thus important to refine the analysis of the asymptotic distributions of zeros, which we have hitherto only crudely observed; their actual uniformity and isotropy must be checked much more thoroughly. A central related issue is the interaction between the zeros which expresses the original quantum dynamics, and how that reflects the classical dynamics. For example, if the phase space is the sphere and the quantum dynamics is specified by a Hamiltonian operator having the expression $\hat{H}(z, \partial_z) = \sum_j f_j(z) \partial_z^j$ in the variable z of Eq.(33), then the corresponding equation of motion for the zeros is [44]

$$\dot{z}_k = \frac{i}{\hbar} \sum_j j! f_j(z_k) \sum_{\substack{1 \le k_1 < k_2 < \cdots < k_{j-1} \le N \\ (\ne k)}} \frac{1}{(z_k - z_{k_1})(z_k - z_{k_2}) \cdots (z_k - z_{k_{j-1}})}, \quad k = 1, \cdots, N.$$

Other relations of the zeros to time-evolution problems have been stressed in [45].

The subsequent problem would be to construct effective asymptotic formulae for the eigenfunctions based on the observed distributions of zeros.

As for generalizations, the main difficulty lies with higher dimensions (*i.e.,* more than one degree of freedom *after reduction*), because the theory of analytic functions in several complex variables is much more involved.

REFERENCES

[1] B. Eckhardt, *Quantum Mechanics of Classically Non-Integrable Systems*, Phys. Rep. **163** (1988) 205-297.

[2] A. M. Ozorio de Almeida, *Hamiltonian Systems: Chaos and Quantization*, Cambridge University Press, Cambridge (1988).

[3] M.C. Gutzwiller, *Chaos in Classical and Quantum Physics*, Springer, New York (1990).

[4] M.-J. Giannoni, A. Voros and J. Zinn-Justin eds., *Chaos and Quantum Physics* (Session LII, Les Houches Summer School, August 1989), North Holland, Amsterdam (1991).

[5] M.V. Berry, N.L. Balazs, M. Tabor and A. Voros, *Quantum Maps*, Ann. Phys. (NY) **122** (1979) 26-63.

[6] M. Hillery, R.F. O'Connell, M.O. Scully and E.P. Wigner, *Distribution Functions in Physics: Fundamentals*, Phys. Rep. **106** (1984) 121-167; N.L. Balazs and B.K. Jennings, *Wigner's Function and Other Distribution Functions in Mock Phase Spaces*, Phys. Rep. **104** (1984) 347-391.

[7] P. Lebœuf and A. Voros, *Chaos-Revealing Multiplicative Representation of Quantum Eigenstates*, J. Phys. **A23** (1990) 1765-1774; A. Voros and P. Lebœuf, *Multiplicative Formulation of Quantum Mechanics...*, in: *Analyse Algébrique des Perturbations Singulières* 2 vols., (Proceedings, Marseille-Luminy Oct. 1991), Hermann, Paris (to appear 1994).

[8] V.P. Maslov, *Théorie des Perturbations et Méthodes Asymptotiques*, Dunod, Paris (1972).

[9] K. Takahashi, *Distribution Functions in Classical and Quantum mechanics*, Prog. Theor. Phys. suppl. **98** (1989) 109-156.

[10] A. Voros, *Asymptotic ℏ-expansions of Stationary Quantum States*, Ann. Inst. H. Poincaré **26A** (1977) 343-403.

[11] A. Voros, *Développements Semi-Classiques*, Thèse d'Etat, Université Paris-Sud, Orsay (1977, unpublished).

[12] A. Voros, *Semiclassical Approximations*, Ann. Inst. H. Poincaré **A24** (1976) 31-90.

[13] M.V. Berry, *Regular and Irregular Semiclassical Wave Functions*, J. Phys. **A10** (1977) 2083-2091.

[14] Y. Colin de Verdière, *Hyperbolic Geometry in Two Dimensions and Trace Formulas*, in: [4], pp. 305-330.

[15] P. Lebœuf and M. Saraceno, *Eigenfunctions of Non-integrable Systems in Generalised Phase Spaces*, J. Phys. **A23** (1990) 1745-1764.

[16] E.B. Bogomolny, *Semiclassical Quantization of Multidimensional Systems*, Nonlinearity 5 (1992) 805-866.

[17] R.E. Kleinman and G.F. Roach, *Boundary Integral Equations for the Three-Dimensional Helmholtz Equation*, SIAM Review **16** (1974) 214-236.

[18] M.V. Berry and M. Wilkinson, *Diabolical Points in the Spectra of Triangles*, Proc. Roy. Soc. **A392** (1984) 15-43.

[19] V.I. Arnold and A. Avez, *Problèmes Ergodiques de la Mécanique Classique*, Gauthier-Villars, Paris (1967) (English translation: Benjamin, New York (1968)).

[20] B.V. Chirikov, *Time-Dependent Quantum Systems*, in [4], pp. 443-545 and references cited therein.

[21] K. Nakamura, Y. Okazaki and A.R. Bishop, *Periodically Pulsed Spin Dynamics: Scaling Behavior of Semiclassical Wave Functions*, Phys. Rev. Lett. **57** (1986) 5-8; F. Haake, M. Kuś and R. Scharf, *Classical and Quantum Chaos for a Kicked Top*, Z. Phys. **B 65** (1987) 381-395.

[22] J.H. Hannay and M.V. Berry, *Quantization of Linear Maps on a Torus — Fresnel Diffraction by a Periodic Grating*, Physica **1D** (1980) 267-290.

[23] N.L. Balazs and A. Voros, *The Quantized Baker's Transformation*, Ann. Phys. (N.Y.) **190** (1989) 1-31.

[24] P. Lebœuf, J. Kurchan, M. Feingold and D. P. Arovas, *Phase-Space Localization: Topological Aspects of Quantum Chaos*, Phys. Rev. Lett. **65** (1990) 3076-3079.

[25] F.A. Berezin, *General Concept of Quantization*, Comm. Math. Phys. **40** (1975) 153-174.

[26] H. Weyl, *The Theory of Groups and Quantum Mechanics*, ch. IV-D, Dover, New York (1950).

[27] H.J. Lipkin, M. Meshkov and J. Glick, *Validity of Many-Body Approximation Methods for a Solvable Model*, Nucl. Phys. **62** (1965) 188-198.

[28] J.R. Klauder and B. Skagerstam, *Coherent States*, World Scientific, Singapore (1985); A. Perelomov, *Generalized Coherent States and their Applications*, Springer, New York (1986).

[29] V. Bargmann, *On a Hilbert Space of Analytic Functions and an Associated Integral Transform*, Comm. Pure Appl. Math. **14** (1961) 187-214 and **20** (1967) 1-101.

[30] E.T. Whittaker and G.N. Watson, *A Course of Modern Analysis*, Cambridge University Press, Cambridge (1965).

[31] A.S.B. Holland, *Introduction to the Theory of Entire Functions*, Academic Press, New York (1973).

[32] F.D.M. Haldane and E.H. Rezayi, *Periodic Laughlin–Jastrow Wave Functions for the Fractional Quantized Hall Effect*, Phys. Rev. **B31** (1985) 2529-2531.

[33] D.P. Arovas, R.N. Bhatt, F.D.M. Haldane, P.B. Littlewood, and R. Rammal, *Localization, Wave-Function Topology, and the Integer Quantized Hall Effect*, Phys.

Rev. Lett. **60** (1988) 619-622.

[34] B. A. Dubrovin and S.P. Novikov, *Ground States of a Two-Dimensional Electron in a Periodic Magnetic Field*, Sov. Phys. JETP **52** (1980) 511-516.

[35] E. Majorana, *Atomi Orientati in Campo Magnetico Variabile*, Nuovo Cimento **9** (1932) 43-50.

[36] J. Kurchan, P. Lebœuf and M. Saraceno, *Semiclassical Approximations in the Coherent-State Representation*, Phys. Rev. **A40** (1989) 6800-6813.

[37] A. Voros, *Wentzel–Kramers–Brillouin Method in the Bargmann Representation*, Phys. Rev. **A40** (1989) 6814-6825.

[38] J. Heading, *An Introduction to Phase-Integral Methods*, Methuen, London (1962).

[39] M.V. Berry, *Disruption of Wavefronts: Statistics of Dislocations in Incoherent Gaussian Random Waves*, J. Phys. **A11** (1978) 27-37.

[40] E. Bogomolny, O. Bohigas and P. Lebœuf, *Distribution of Roots of Random Polynomials*, Phys. Rev. Lett. **68** (1992) 2726-2729.

[41] B.V. Chirikov, *A Universal Instability of Many-Dimensional Oscillator Systems*, Phys. Rep. **52** (1979) 263-379.

[42] G. Casati, B.V. Chirikov, J. Ford and F.M. Izrailev, *Stochastic Behavior of a Quantum Pendulum under a Periodic Perturbation*, in: *Stochastic Behavior in Classical and Quantum Hamiltonian Systems*, (Proceedings, Como 1977), eds., G. Casati and J. Ford, Lecture Notes in Physics **93** (1979) 334-352, Springer, Berlin (1979).

[43] F. Izrailev, *Statistical Properties of Quantum Chaos*, in [4], pp. 771-792.

[44] P. Lebœuf, *Phase Space Approach to Quantum Dynamics*, J. Phys. **A24** (1991) 4575-4586.

[45] S. Adachi, *A Numerical Evaluation of the Semiclassical Coherent State Path Integral*, Ann. Phys. (NY) **195** (1989) 45-93.

Chaology of action billiards

A. M. OZORIO DE ALMEIDA

Instituto de Matemática Pura e Aplicada,
Estrada Dona Castorina, 110, 22460 Rio de Janeiro, RJ, Brazil

M. A. M. DE AGUIAR

Instituto de Física, "Gleb Wataghin",
Universidade Estudual de Campinas, 13081 Campinas, SP, Brazil

We review the construction of action billiards that correspond classically to the truncation of Hamiltonian matrices in quantum mechanics. For billiards in which the action variable moves freely between collisions with the boundary, the classical motion will be regular or chaotic depending only on the boundary shape, whereas a universal matrix describes all these quantum systems except for the truncations. The fluctuations of the energy spectrum of action billiards are found to follow the known universality classes depending on the character of the classical motion being regular or chaotic.

1 Introduction

The study of the semiclassical limit of quantum systems whose classical analogues exhibit chaotic trajectories has been termed "quantum chaology" by Michael Berry. The basis of this study has been the semiclassical propagator, that is a specific representation of the time-evolution operator, first derived by van Vlech (1928) directly from the Schrödinger equation and subsequently from Feynmann path integrals (see, e.g., Berry and Mount (1972) for a derivation and references). The semiclassical propagator is represented entirely in terms of the classical trajectories and their actions. These are canonically invariant, so it has been natural to assume that the semiclassical propagator describes systems whose evolution is not necessarily governed by the Schrödinger equation, or, in other words, whose Hamiltonian does not have the simple form

$$H = p^2/2 + V(q). \tag{1.1}$$

Though a generalization of path integrals and their semiclassical limit has been presented by Ozorio de Almeida (1992), in the absence of a proof of their convergence, there is great interest in the numerical investigation of the semiclassical limit of quantum systems with nonsimple Hamiltonians.

As most Hamiltonians with at least two degrees of freedom are nonintegrable, almost any system is a valid subject for quantum chaology. However, for practical purposes it is convenient to choose model systems with the following properties:

(i) the orbits on every energy surface are mainly chaotic and (ii) a large number of eigenvalues of the Hamiltonian matrix can be accurately obtained. When this is achieved, a large number of eigenvalues per energy interval can be calculated and placed in correspondence with the underlying classical dynamics. These two conditions are, however, very hard to reconcile: classical chaos is generally associated with nonlinearity and implies large nondiagonal matrix elements. As a consequence, a large number of basis states is needed to compute a much smaller number of "well converged" eigenvalues even for simple systems such as billiards.

In a series of papers (de Aguiar and Ozorio de Almeida, 1992; de Aguiar, 1992; Ozorio de Almeida and de Aguiar, 1992, hereafter refered to as I, II and III) we have presented a new class of quantum and classical systems that generalize smooth systems in analogy to the way that generalized functions (such as the unit step functions) generalize smooth functions. For these "action billiards" a finite block of the Hamiltonian matrix describes completely the subsystem within the billiard and, therefore the above property (ii) is always met. In other words, matrix truncation is made exact. It is the classical system that pays for this enormous simplification: truncating the Hamiltonian matrix is shown to generate a boundary in the action space, binding the motion inside the action billiard. As in common billiards, the conjugate (angle) variable suffers a discontinuous jump when the orbit hits the boundary. In I the connection rule for the angles at a boundary collision were derived and some examples were discussed. In the simple case of a harmonic oscillator, no orbit ever hits the boundary and the effect of truncation is just to discard part of the orbits of the original system. Therefore, in the case of nearly harmonic systems most orbits remain unaffected by the boundary and only a small portion of phase space has its dynamics changed. The relation between this part of phase space affected by the cut-off and the number of "converged" eigenvalues of the corresponding truncated matrix was discussed in II for a simple one-dimensional system.

The extreme case of the harmonic oscillator has a counterpart where every orbit hits the boundary and, therefore, the original untruncated system is completely different from the corresponding action billiard. These systems are the analogue of common billiards in configuration space and were discussed in detail in III. The quantization of these systems results in a universal matrix, with various truncations corresponding to each of the classical billiards.

In this chapter we provide a unified review of the theory of action billiards developed in our previous papers. The general outcome is a confirmation of standard semiclassical results for nonsimple chaotic systems. We also discuss in detail the semiclassical limit for the one-dimensional case and present new results for the stadium action billiard in terms of scars of periodic orbits. The chapter is organized as follows: in Section 2 we briefly review the theory of action billiards; in Section 3 we discuss in detail the one-dimensional systems. Examples of two-dimensional systems are given in Section 4 and in Section 5 we discuss our results.

2 Brief review of action billiards

In common billiards the motion is confined to a finite region in the configuration space. When a trajectory hits the billiard boundary the perpendicular momentum changes sign instantaneously while all other variables are kept constant. The jump at the boundary does not depend on the Hamiltonian inside the billiard.

Action billiards are completely analogous, with the advantage of providing finite matrices as their quantum analogues. Once a boundary has been defined in the action space, the angle variables suffer discontinuities when the trajectories hit it. The connection rule for the angles can be obtained by considering the limit of smooth cut-offs.

Although the classical definition of an action billiard is completely independent of its quantum mechanical properties, it is useful and instructive to keep in mind the original motivation for this study; namely, to get the classical analogue of truncated Hamiltonian matrices in the harmonic oscillator representation. Therefore, we shall discuss sharp boundaries as the limit of smooth (ever steeper) boundaries in connection with the corresponding quantum mechanical operators.

We start with one-dimensional systems. The coordinates are q, p and action and angle variables I, θ for a harmonic oscillator are defined as usual by

$$q = \sqrt{2I} \cos \theta$$

$$p = \sqrt{2I} \sin \theta$$

Let $\Phi_\lambda(I)$ be an analytical step function such that $\Phi_\lambda > 1$ outside a given region, $1 \geq \Phi_\lambda > 1 - \lambda$ inside and the width of the boundary in which $\lambda \geq \Phi_\lambda \geq 1 - \lambda$ shrinks as $\lambda \to 0$. Then, from the harmonic oscillator basis $|n>$, we can define the operator $\hat{\Phi}_\lambda$ from its matrix elements

$$< n'|\Phi_\lambda|n > \equiv \delta_{nn'}, \; \Phi_\lambda[(n+1/2)h]$$

and, in the limit $\lambda \to 0$ the classical step function $\Phi_0(I - \mathscr{L})$ generates the projection operator

$$\hat{\Phi}_o = \sum_{n=0}^{N} |n><n|$$

where $N = \mathscr{L}/h$, $\mathscr{L}$ being the classical boundary.

Therefore, the smoothed quantum action billiard is defined by

$$\hat{H}_\lambda = \hat{\Phi}_\lambda H \hat{\Phi}_\lambda \tag{2.1}$$

and its classical version by

$$H_\lambda = H(I, \theta)[\Phi_\lambda(I)]^2 \tag{2.2}$$

Both these Hamiltonians are defined on the extended phase space, so that the operators $\hat{p}$ and $\hat{q}$ can be taken as having continuous eigenvalues. It is only in the limit $\lambda \to 0$ that a sharp cut-off is obtained, but in a given calculation with fixed Planck's constant, we may consider that λ is small but finite.

Equations (2.1) and (2.2) can be generalized in the following way: let $A(I, \theta)$ be another Hamiltonian and $\xi_\lambda(I) = I - \Phi_\lambda(I)$ be the complementary step function. Then

$$\xi_0 = \sum_{n=N+1}^{\infty} |n><n|$$

and we define

$$\hat{H}_\lambda = \hat{\Phi}_\lambda \hat{H} \hat{\phi}_\lambda + \hat{\xi}_\lambda \hat{A} \hat{\xi}_\lambda \tag{2.3}$$

and

$$H_\lambda = H(I, \theta)[\Phi_\lambda(I)]^2 + A(I, \theta)[\xi_\lambda(I)]^2 \tag{2.4}$$

The matrix for $\hat{H}_0$ consists of two separate blocks, the eigenvalues of the first $\hat{\Phi}_0 \hat{H} \hat{\Phi}_0$ not being affected by those of the second. The topology of the classical orbits, however, may depend on $A(I, \theta)$ even in the limit $\lambda \to 0$. Note that for one degree of freedom energy conservation dictates the rule for connecting the angles at the boundary $I = \mathscr{L}$ as

$$H(\mathscr{L}, \theta_i) = H(\mathscr{L}, \theta_f) \tag{2.5}$$

where θ_i and θ_f are the angles before and after the collision respectively. To make things clearer, consider the example where

$$H(I, \theta) = I(1 + \cos 2\theta) \tag{2.6}$$

and the corresponding billiard is

$$H_0(I, \theta) = \begin{cases} H(I, \theta) & \text{for} \quad I \le \mathscr{L} \\ A & \text{for} \quad I > \mathscr{L} \end{cases} \tag{2.7}$$

where A is just a constant.

The orbits of H have two humps centred at $\theta = \pi/2$ and $3\pi/2$. The boundary at $I = \mathscr{L}$ cuts off part of both humps for trajectories having $E > \mathscr{L}(1-a)$. Then, an original orbit of H is cut into three pieces in the I versus θ plane and Equation (2.5) allows for two different ways of pasting them together: either $\theta_f = 2\pi/\theta_i$ or $\theta_f = \pi - \theta_i$. In the first case there will be two distinct rotation orbits of H_0; one with oscillations around $\theta = 0$ and another around $\theta = \pi$, while in the second case a single libration (θ going from 0 to 2π) is generated by the cut-off. For the smooth billiard

$$H_\lambda = H\Phi_\lambda^2(I - \mathscr{L}) + A\xi_\lambda^2(I - \mathscr{L}) \tag{2.8}$$

the two situations above correspond to different choices of the constant A: if $A < \mathscr{L}(1 + a)$ then in the limit $\lambda \to 0$ we get $\theta_f = 2\pi - \theta_i$ and if $A > \mathscr{L}(1 + a)$, then $\theta_f = \pi - \theta_i$, as discussed in detail in I. A schematic representation of both situations is presented in Fig. 2.1 for a smooth cut-off.

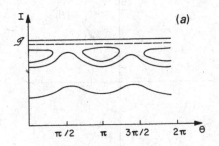
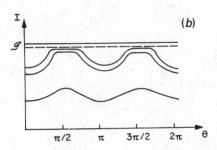

Fig. 2.1. Schematic representation of the phase space topology for the smooty billiard Equation (2.8). In part (*a*) $A < \mathcal{I}(1 + a)$ and in part (*b*) $A > \mathcal{I}(1 + a)$.

In two or more freedoms the condition (2.5) takes a slightly different form. For the simple case of rectangular billiards with boundaries at $I_1 = \mathcal{L}_1$ and $I_2 = \mathcal{L}$, then when an orbit hits $I_1 = \mathcal{L}_1$, for instance, the rule for the jumping angle θ_1 is

$$H(\mathcal{L}_1, I_{2i}, \theta_{1i}, \theta_{2i}) = H(\mathcal{L}_1, I_{2i}, \theta_{1f}, \theta_{2i})$$

In this case, the orbits of H_0 will be made of pieces of different orbits of H, all at the same energy shell.

For more complicated boundaries the rule is basically the same but the action variables I_1 and I_2 have to be changed by a linear canonical transformation into $I_\parallel$ and $I_\perp$, the components of the vector **I** that are parallel and perpendicular to the boundary point. Then, as in common billiards, $\theta_\parallel$ stays constant and $\theta_\perp$ suffers the jump according to

$$H(I_\parallel, I_\perp = \mathcal{L}, \theta_{\parallel i}, \theta_{\perp i}) = H(I_\parallel, I_\perp = \mathcal{L}, \theta_{\parallel i}, \theta_{\perp f})$$

In *I* numerical examples of both one- and two-dimensional systems have been exploited and we refer to that paper for further details.

The quantized action billiard with more than one freedom is obtained in exactly the same way as that for one freedom. All that is necessary is to interpret, q, p, I, θ and n at the beginning of this section as vectors.

3 Compact billiards in one-dimension

3.1 Classical mechanics

A regular billiard in one dimension corresponds to a particle confined by a square well. The potential $V(q)$ is zero between q_1 and q_2 and infinite outside. The classical motion preserves the energy $E = p^2/2$ and the momentum changes sign instantaneously when the particle hits the boundary. Quantum mechanically, the wave-functions go to zero outside the well and momentum is quantized. The phase space is, however, not limited and the number of eigenstates is infinite.

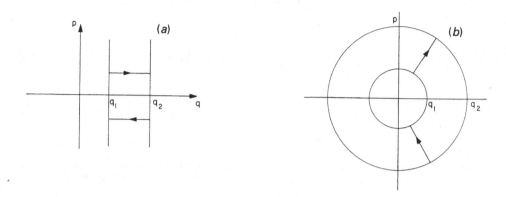

Fig. 3.1. (*a*) Ordinary one-dimensional billiard with walls at q_1 and q_2; (*b*) compactification of phase space generating the action billiard.

One way to compactify this region is to consider the vertical lines intersecting the horizontal axis to be approximations to circles centred on the origin as shown in Fig. 3.1. The greater the values of q_1 and q_2, the larger will be the region where this approximation holds. The advantage of this view is that now the anular region bounded by the two circles is finite. Switching to canonical polar coordinates in phase space (i.e. the action-angle variables of the harmonic oscillator):

$$p^2 + q^2 = 2I, \quad p/q = \tan\theta \qquad (3.1)$$

we obtain the simple equation $I = q_1^2/2 = \mathcal{L}_0$ and $I = q_2^2/2 = \mathcal{L}_2$ for the boundary circles.

All that remains is to choose the Hamiltonian for this compactified square well. For small E the free Hamiltonian is quite satisfactory, but the $E > q_1^2/2$ the character of the motion changes completely. It is therefore better to use the fact that in the action-angle variables the motion in the compactified well (Fig. 3.1) will look just like that of the ordinary well if we choose the Hamiltonian as an even function of θ. The simplest choice is

$$H(I, \theta) = -\cos\theta \qquad (3.2)$$

so that we obtain straight radial motion between the circles that is reversed at every collision, as shown in Fig. 3.1(*b*). For small energies, we have

$$H(I, \theta) - 1 + \theta^2/2 \qquad (3.3)$$

so that we recover the usual motion in a box by approximating the radial motion by that along horizontal lines.

In order to get a better feeling of the topology of the classical motion we define

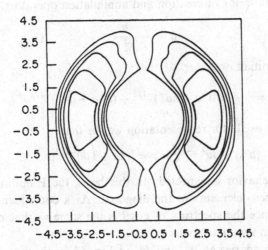

Fig. 3.2. Classical orbits for the smooth one-dimensional billiard. The orbits are always confined to the left or right half-ring. In this figure $\lambda = 0.1$ and the cut-offs are at $\mathscr{I}_1 = 2.0$ and $\mathscr{I}_2 = 8.0$

the smooth version of this billiard as

$$H_\lambda(I, \ \theta) = -\cos\theta \cdot \Phi_\lambda^2(I - \mathscr{L}_2) \cdot \Phi_\lambda^2(\mathscr{L}_1 - I) \tag{3.4}$$

where

$$\Phi_\lambda(I) = \frac{1 - \tanh(I/\lambda)}{2} \tag{3.5}$$

in a smooth step function, in accordance with Equation (2.2).

The trajectories $I = I(\theta)$ will be level curves of Equation (3.4). These are better visualized in the q, p plane, as shown in Fig. 3.2. It is clear from this figure that, for any finite λ, a typical orbit inside the billiard is confined either to the right or to the left branch of the ring. These branches are separated by a "vertical" orbit. At $\theta = \pi/2$ and $3\pi/2 \, (q = 0)$. As λ goes to zero the motion becomes exactly radial with the "circular" pieces traversed in zero time. These considerations will be important in the discussion of semiclassical quantization.

3.2 Quantum mechanics

The quantization of Hamiltonian (3.2) is not unique. In this chapter, as in III, we use Equations (3.1) to write

$$H = -q/\sqrt{q^2 + p^2}$$

and, quantizing q in terms of creation and annihilation operators a^+, a

$$\hat{q} = \frac{a + a^+}{\sqrt{2}}$$

we obtain the hermitian operator

$$\hat{H} = \frac{1}{\sqrt{2}}[a(a^+a + aa^+)^{-1/2} + (a^+a + aa^+)^{-1/2}a^+] \qquad (3.6)$$

In the harmonic oscillator representation we see that

$$(a^+a + aa^+)^{-1/2}|n> = [(2n + 1)\hbar]^{-1/2}|n>$$

and no singular behavior is expected. In this basis, the Hamiltonian matrix is tri-diagonal with zero elements in the diagonal. As a consequence the trace is zero. Moreover, since the spectrum of $\cos\theta$ is the same as that of $-\cos\theta$, the eigenvalues come in symmetry pairs.

The classical boundaries at $\mathscr{L}_1$ and $\mathscr{L}_2$ in Fig. 3.1 imply that the basis states entering in the quantum billiard go from $n = N_1$ to $n = N_2$; the total number of states is $N = N_2 - N_1 + 1$. If N is odd, one of the eigenvalues is zero.

The numerical diagonalization of the Hamiltonian matrix is straightforward and free of convergence problems. In the limit of $\hbar \to 0$, N goes to infinity and the mean density of states approaches the Weyl density

$$n_{Weyl}(E) = \int \delta[H(I, \theta) - E]dI\,d\theta$$

For the present case it is easy to check that

$$n_{Weyl}(E) = \frac{\mathscr{L}_2 - \mathscr{L}_1}{2\pi\hbar\sqrt{1 - E^2}} \qquad (3.7)$$

3.3 Semiclassical limit

The Bohr–Sommerfeld quantization condition applied to a square-well is

$$\oint p(q, E)dq = nh \qquad (3.8)$$

This has a geometric interpretation: the quantum energies correspond to orbits encircling a phase-space area which is an integral multiple of Planck's constant. When applied to the action billiard generated by Equation (3.2) this reads

$$\oint \theta dI = nh$$

The calculation, although trivial, since θ is constant between consecutive bounces, deserves some care. For orbits with low energy (actually with $E < 0$), the phase space area enclosed in one cycle is just $2(\mathscr{L}_2 - \mathscr{L}_1)\theta$, as can be seen from the smoothed picture, Fig. 3.2. However, for large energies (actually for $E > 0$) the orbits are entirely on the left branch of the ring defining the billiard (see Fig. 3.2

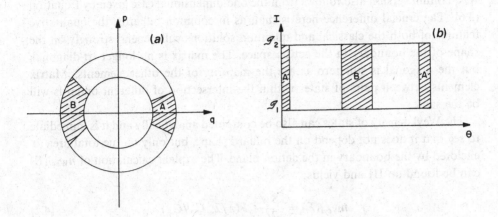

Fig. 3.3.

again), and the area after one cycle is $2(\mathscr{L}_\in - \mathscr{L}_\infty)(\pi - \theta)$. These two situations are illustrated in Fig. 3.3.

It follows immediately that the semiclassical spectrum is also symmetric but always has an even number of eigenenergies. This semiclassical restriction (not present in the quantum problem) can be understood if we let $\hbar$ vary and look at how the basis states originally outside the billiard come inside discontinuously in both the quantum and the semiclassical pictures. Quantum mechanically, the accretion of a new state happens every time the area $\mathscr{A} = 2\pi(\mathscr{L}_2 - \mathscr{L}_1)$ of the billiard is equal to Nh. At this point we add one more state and go from $N - 1$ to N energy levels. Semiclassically, however, we only get new states when half of this area is a multiple of h, or when $\mathscr{A}/h = 2M$. Then we go from $2(M - 1)$ to $2M$ states. What happened when $\mathscr{A}/h = 2M - 1$? At this point, half of the area available to get the new state is on the right branch of ring, while the other half is on the left branch. Therefore, neither side gets the missing state (which would correspond to $E = 0$).

Notice that the action billiard does not require its area to always be a multiple of h. This is possible because the phase space is itself not limited, but the Haliltonian operator is. However, this restriction that is mandatory for the quantization of compact phase spaces is needed for the correct semiclassical limit of the border state.

4 Two-dimensional systems

The natural extension of Hamiltonian (3.2) to two dimensions is

$$H(\theta_1, \theta_2) = -\cos\theta_1 - \cos\theta_2 \tag{4.1}$$

Its quantum version also follows from the one-dimensional case given by Equation (3.6). The crucial difference here is that, as in common billiards, the qualitative features of both the classical and quantum solutions will depend strongly on the shape of the boundary in the action space. The matrix is no longer tri-diagonal, but the diagonal is still zero as are the majority of the other elements. Matrix elements between pairs of states within the intersection of different billiards will be the same.

The Weyl density of states can also be computed analytically and it is immediate to see that it does not depend on the billiard shape, but only on the total area $\mathscr{A}$ enclosed by the boundary in the action plane. The explicit calculation of $n_{Weyl}(E)$ can be found on III and yields:

$$n_{Weyl}(E) = \frac{8\mathscr{A}}{h^2 C_+} F(\pi/2, \ C_-/C_+) \tag{4.2}$$

where

$$C_{\pm}^2 = (1 \mp |E|/2)^2$$

and

$$F(\pi/2, \ k) = \int_0^{\pi/2} \frac{d\theta}{\sqrt{1 - k^2 \sin^2 \theta}}$$

the complete elliptic integral of first kind.

To illustrate the basic features of the action billiards we have chosen two shapes with distinct classical dynamics: the rectangle and the Bunimovich stadium. In the first case the motion is known to be integrable while in the second it is ergodic.

The dimensions of these billiards will be kept constant throughout this section, and only $\hbar$ will vary. For a given $\hbar$, the harmonic oscillator basis states define a grid in the action space. The points on the grid enclosed by the billiard are the basis states to be used in the quantum billiard. Of course, the smaller h is, the larger will be the quantum matrix. Fig. 4.1 shows the rectangle and the desymmetrized stadium billiards to be considered here, while Figs. 4.2 and 4.3 show the respective nearest neighbor distribution (NND) for different values of $\hbar$. The spectra for the rectangle follow a Poisson distribution while the stadium has a GOE type NND of random matrix theory. In III we have also presented results for the circular billiard and we have checked that it also follows a Poisson distribution, as should be the case for integrable dynamics.

Once this general feature has been established we proceed to a finer study of the quantum properties of the Bunimovich stadium in connection with classical periodic orbits. According to Gutzwiller (1971), the density of states can be approximated by the Weyl term $n_{Weyl}(E)$, in the semiclassical limit $\hbar \to 0$, plus a series of oscillatory quantum corrections due to periodic orbits:

$$n(E) \simeq n_{Weyl}(E) + n_{osc}(E) \tag{4.3a}$$

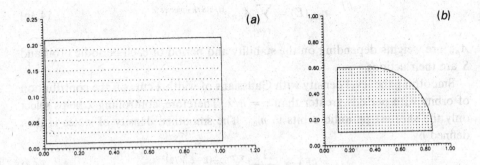

Fig. 4.1. Billiard enclosures for (a) the rectangle and ((b) the quarter-stadium. The grid points indicate the harmonic oscillator states entering in the quantum matrix.

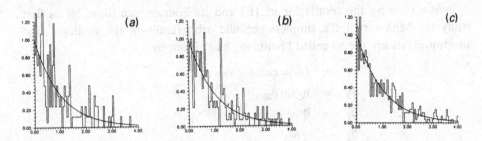

Fig. 4.2. Nearest neighbor distribution for the rectangle for different values of $\hbar$ and matrix dimension N: in (a) $\hbar = 1/25$, $N = 156$; in (b) $\hbar = 1/50$, $N = 561$; in (c) $\hbar = 1/70$, $N = 1065$.

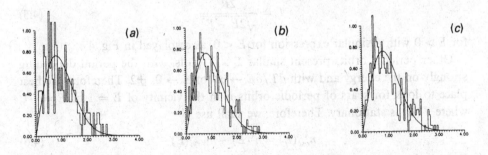

Fig. 4.3. Nearest neighbor distribution for the quarter-stadium for different values of $\hbar$ and matrix dimension N: in (a) $\hbar = 0.03$, $N = 372$; in (b) $\hbar = 0.023$, $N = 627$; and in (c) $\hbar = 0.02$, $N = 801$.

where

$$n_{osc}(E) = \sum_{p.o.} A_{p.o.} e^{(i/\hbar)S/\hbar + phases} \tag{4.3b}$$

$A_{p.o.}$ are weights depending on the stability and period of each periodic orbit and S are their actions.

Smoothing the level density with Gaussians of width λ cuts off the contribution of orbits with periods greater than $\tau = \hbar/\lambda$. Therefore, controlling λ helps select only the shortest periodic orbits in n_{osc}. The smoothed density of states is then defined by

$$n^\lambda(E) = \frac{1}{\lambda\sqrt{2\pi}} \sum_n e^{-(E-E_n)^2/2\lambda^2} \tag{4.4}$$

If Equations (4.3) indeed apply to the action billiards we expect that the Fourier transform of the oscillating part of n^λ will have peaks centred at the periods of the contributing orbits.

Before showing the results for $n^\lambda_{osc}(E)$ and its Fourier transform, let us first study the behavior of the simplest periodic orbit family in the stadium. The shortest orbits are the so called "bouncing ball" given by

$$
\begin{aligned}
I_1 &= I_{10} = const. \\
I_2 &= t \cdot \sin\theta_{20} \\
\theta_1 &= 0 \text{ or } \pi \\
\theta_2 &= \pm\theta_{20} \\
E &= -1 - \cos\theta_{20} \text{ or } E = 1 - \cos\theta_{20}
\end{aligned}
$$

The period of these orbits is given in terms of the energy E and the radius R of the semicircle forming the billiard by

$$T = \frac{2R}{\sqrt{2E - E^2}} \tag{4.5}$$

for $E > 0$ with a similar expression for $E < 0$, as displayed in Fig. 4.4.

Other periodic orbits present similar $E \times T$ plots, with the period depending strongly on the energy and with $\partial T/\partial E \to \infty$ for $E \to 0, \pm 2$. Therefore, the best place to look for scars of periodic orbits is in the vicinity of $E = 1$ or $E = -1$, where $T(E)$ is stationary. Therefore, we shall use

$$n^\lambda_{osc}(E) = n^\lambda(E) - \bar{n}_{Weyl}(E) \tag{4.6}$$

and

$$n^\lambda_{osc}(T) = \frac{1}{i\hbar} \int_{0.5}^{1.5} n^\lambda_{osc}(E) \, e^{iET/\hbar} dE \tag{4.7}$$

(the energy interval $\Delta E = 1$ around $E = 1$ was chosen rather arbitrarily just to avoid the points $E = 0$ and $E = 2$). Notice that the Weyl term was substituted by $\bar{n}_{Weyl}(E)$ in Equation (4.6). This is a numerically computed Weyl density that

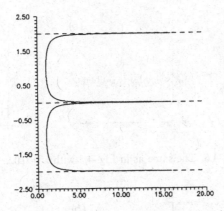

Fig. 4.4. Energy versus period plot for the "bouncing ball" orbit.

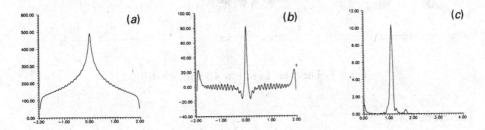

Fig. 4.5. (a) Smoothed density of states for $\hbar = 0.02$ and $\lambda = 0.03$; (b) $\bar{n}_{osc}$ for the same parameters; (c) Fourier spectrum of $\bar{n}_{osc}$.

smooths out the singularity of Equation (4.2) at $E = 0$ and also modifies slightly the end points near $E = \pm 2$. Since these regions do not enter Equation (4.7) we shall refer to III for a detailed discussion of $\bar{n}_{Weyl}$.

Figs. 4.5–4.7 show plots of $n^{\lambda}(E)$, $n^{\lambda}_{osc}(E)$ and the Fourier transform of $\bar{n}^{\lambda}_{osc}(E)$ for different values of λ (0.03, 0.02 and 0.01 respectively). Fig. 4.5(c) shows a single peak at $T = 1$, corresponding to the shortest orbit, the bouncing ball (notice that $R = 0.5$ in Equation (4.5)). As λ diminishes new orbits start to contribute to $\bar{n}^{\lambda}_{osc}$. Fig. 4.6(c) shows a second peak at $t = 1.75$, corresponding to the periodic orbit running horizontally on the top of the lower boundary of the billiard (this is the second shortest orbit). Finally Fig. 4.7(c) shows several other peaks: $T = 2$ represents two repetitions of the bouncing ball; $T = 3$ counts three repetitions of the bouncing ball; between $T = 2$ and $T = 3$ several other orbits (more complicated) can also be distinguished (see, for instance, Bogomolny (1988)).

The eigenfunctions $|\psi_n >$ of the Bunimovich stadium can also be studied. In this case the best representation for plotting $|\psi_n >$ is the action representation.

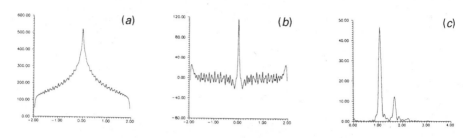

Fig. 4.6. The same as in Fig. 4.5 with $\lambda = 0.02$.

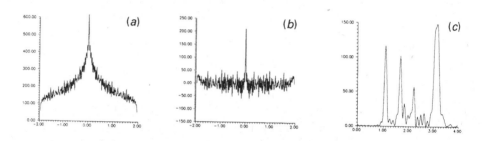

Fig. 4.7. The same as in Fig. 4.5 with $\lambda = 0.01$.

Therefore, given

$$|\psi_n > = \sum_{\mathbf{m}} C_{\mathbf{m}}^n \,|\mathbf{m} >,$$

where $|\mathbf{m} >$ is the harmonic oscillator basis states, we construct contour plots of $|C_{m_1,m_2}^n|^2$ in the I_1, I_2 plane. This is equivalent to plotting $|\psi(x, y)|^2$ in the x, y plane in the case of common billiards. Figs. 4.8–4.10 show examples of such plots, and again the influence of periodic orbits is neatly observed. In Figs. 4.8 and 4.9 the scar (see, for instance, Heller (1984)) of some simple orbits can be singled out, while Fig. 4.10 represents a chaotic eigenstate.

5 Conclusion

The quantization of billiards that we present here is radically different from the free motion between the billiard boundaries in position space to which one is accustomed. Indeed the Hamiltonian (4.1) cannot be separated into kinetic and potential terms. Nevertheless, the identical structure of periodic orbits to that of common billiards implies the same universality class for the short range spectral fluctuations, and the same long range oscillations connected with the shortest periodic orbits. The Hamiltonian, and hence its matrix elements, is universal for all these billiards. The spectral statistics depend only on the form of the

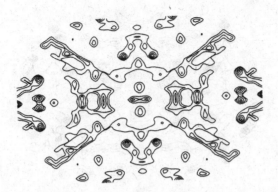

Fig. 4.8. Contour plot of $|\psi_n|^2$ in the action plane for the state 451 and $\hbar = 0.023$, showing the scar of the "bow-tie" periodic orbit.

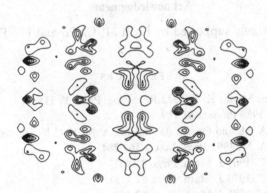

Fig. 4.9. The same as in Fig. 4.8 for the state 454, showing the scar of the "bouncing ball" periodic orbits.

double index truncation. For each fixed shape of the billiard we may approach the infinite matrix limit by increasing its dimensions or by taking $\hbar \to 0$. This may be an important test for any attempt to attribute statistical properties of the spectrum to specific features of the matrix.

The concept of action billiards has other uses besides this important verification of the generality of the connection between the periodic orbit structure and the statistical properties of the quantum energy spectrum. We have shown in I and II that for systems displaying soft chaos, so that the harmonic oscillator basis provides a convenient representation, most classical orbits will not collide with the billiard boundary. In such cases the action billiard provides a classical system that corresponds exactly to the quantum system with a finite Hilbert space obtained by truncating the Hamiltonian matrix.

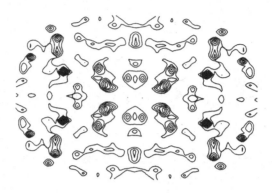

Fig. 4.10. The same as in Fig. 4.8 for the state 453, showing the a "chaotic" (largely spread) wave-function.

Acknowledgements

This work was partially supported by FAPESP, CNPq and FINEP.

REFERENCES

[1]   Berry M. V. and Mount K. E. 1972 Rep. Prog. Phys. **35** 315.
[2]   Bogomolny E. 1988 Physica **31** 169.
[3]   de Aguiar M. A. M. and Ozorio de Almeida A. M. 1992 Nonlinearity **5** 523.
[4]   de Aguiar M. A. M. 1992 Phys. Lett. **A164** 284.
[5]   Gutzwiller M. C. 1967 J. Math. Phys. **8** 1967.
[6]   Gutzwiller M. C. 1970 J. Math. Phys. **11** 1791.
[7]   Gutzwiller M. C. 1971 J. Math. Phys. **12** 343.
[8]   Heller H. J. 1984 Phys. Rev. Lett. **53** 515.
[9]   Ozorio de Almeida A. M. 1992 Proc. R. Soc. London **A439**, 139.
[10] Ozorio de Almeida A. M. and de Aguiar M. A. M. 1992 Chaos, Solitons and Fractals, **2** 377.
[11] van Vleck J. H. 1928 Proc. Natl. Acad. Sci. USA **14** 178.

PART FOUR

Level statistics and random matrix theory

Characterization of Chaotic Quantum Spectra and Universality of Level Fluctuation Laws

O. Bohigas, M. J. Giannoni, and C. Schmit

Division de Physique Théorique, Institut de Physique Nucléaire, F-91406 Orsay Cedex, France

(Received 2 August 1983)

It is found that the level fluctuations of the quantum Sinai's billiard are consistent with the predictions of the Gaussian orthogonal ensemble of random matrices. This reinforces the belief that level fluctuation laws are universal.

The statistical study of spectra of quantum systems is almost as old as quantum mechanics itself. One distinguishes two types of properties: *global* ones and *local* ones. An example of the former is provided by the density of levels as a function of excitation energy. In this Letter we shall discuss local properties, or more precisely, fluctuations (departures of the energy-level distribution from uniformity). We shall deal with time-independent systems and energies of stationary states.

There exists a well established theory to describe fluctuation properties of quantal spectra, namely the random matrix theory (RMT) initiated by Wigner, developed mainly by Dyson and Mehta,[1] and later extended by several authors.[2] Recently,[3] the predictions of RMT [specifically, the predictions of the Gaussian orthogonal ensemble (GOE)] have been compared in great detail with the whole body of available nuclear data coming mainly from *compound-nucleus resonances*. No discrepancy between theory and experiment has been detected. In particular the data have been shown to exhibit two of the salient phenomena predicted by the theory—the *level repulsion* (tendency of the levels to avoid clustering) and especially the *spectral rigidity* (very small fluctuation around its average of the number of levels found in an interval of given length), which is a property due to correlations between level spacings. We should also mention that recently a comparison between data of *atomic levels* and GOE has been performed.[4] Although the significance of the comparison is lower than in the nuclear case (due to the method used in Ref. 4 as well as to the relatively small number of available data), a good agreement between GOE predictions and experiment was found.

Once the ability of the theory (GOE) to predict the fluctuation properties exhibited by data is established, one could think that no major question in this field is still open. It is one purpose of this Letter to show that this is not the case. Indeed, as it will be discussed in what follows, the origin of the success of the theory as well as its domain of validity remain to be clarified.

In connection with the study of regular and irregular motion some very interesting results have been obtained in recent years. Strong arguments have been given which indicate that for integrable systems with more than one degree of freedom,[5] the nearest-neighbor spacing distribution $p(x)$ of the quantum energy levels should be Poisson-like $[p(x) = \exp(-x)]$ and the spacings should not be correlated.[6,7] In contrast, evidence of level repulsion has been put forward by studying numerically systems having two degrees of freedom and known to be chaotic[8] in the classical case (stadium,[9] Sinai's billiard[10]). Although this feature is clearly appealing, it is up to now only qualitative [one would need a spectrum with very many levels to get a precise evaluation of the behavior of $p(x)$ at the origin]. On the other hand, the information carried by $p(x \sim 0)$ is rather limited. In particular, it does not give any indication about the *correlations* between spacings which are responsible for the degree of regularity of the spectrum. The purpose of this Letter is to use some of the systematic tools developed in RMT to make a detailed comparison of the level fluctuations of the quantum Sinai's billiard (SB) with GOE predictions. The choice of a two-dimensional billiard is convenient for our aim for several reasons: (i) Billiards have the lowest possible number of degrees of freedom allowing for chaotic motion; (ii) for billiards, it is possi-

ble to make a precise separation between global and local properties [cf. the Weyl formula, Eq. (1)]; (iii) billiards have a discrete spectrum with an infinite number of eigenvalues and by computing a large number of them one can reach a high statistical significance of the results. Finally, SB is known to be strongly chaotic (K system) and there exists an efficient method to compute its eigenvalues.

We proceed as follows. We determine the eigenvalues $E_n = k_n^2/2m$ of the Schrödinger equation $(\Delta + k_n^2)\psi_n = 0$ for the "desymmetrized" SB [see upper right-hand corner of Fig. 1(a)] with Dirichlet boundary conditions by using the Korringa-Kohn-Rostoker method as described in Ref. 10. We compute several sets of eigenvalues $\{E_i(R)\}$ for different values of the parameter R (see caption of Fig. 1). By using the Weyl-type formula,[11] which gives the average number of levels up to energy E,

$$\bar{N}(E) = (\tfrac{1}{4}\pi)(SE - L\sqrt{E} + K), \qquad (1)$$

where S and L are, respectively, the surface and the perimeter of the billiard and K is a constant of the order of unity, we can map the spectrum $\{E_i(R)\}$ onto the spectrum $\{\epsilon_i(R)\}$ through $\epsilon_i(R) = \bar{N}(E_i(R))$. Each spectrum $\{\epsilon_i(R)\}$ has on the average a constant mean spacing $D(R)$ which will be taken as the energy unit. The cumulative density $n(\epsilon)$ of levels ϵ_i will therefore have a staircase shape fluctuating around a straight line of slope equal to unity. In order to investigate the fluctuations we study the nearest-neighbor spacing distribution $p(x)$ and the Dyson-Mehta statistic Δ_3. Δ_3 is defined, for a fixed interval $[x, x+L]$, as the least-squares deviation of the staircase function $n(\epsilon)$ from the best straight line fitting it:

$$\Delta_3(L,x)$$
$$= (1/L)\text{Min}_{A,B} \int_x^{x+L} [n(\epsilon) - A\epsilon - B]^2 \, d\epsilon. \qquad (2)$$

It provides a measure of the degree of rigidity of the spectrum: For a given L, the smaller Δ_3 is, the stronger is the rigidity, signifying the long-range correlations between levels. We proceed as described in Ref. 3: Given a stretch of levels on the ϵ axis, we compute $\Delta_3(L)$, for instance, for the intervals $[a, a+L]$, $[a+L/2, a+3L/2]$, $[a+L, a+2L]$, $[a+3L/2, a+5L/2]$,... until the stretch $[a,b]$ has been covered. If the spectrum fluctuations are translationally invariant on the ϵ axis, then the average value $\bar{\Delta}_3$ of Δ_3 will be independent of the chosen interval $[a,b]$ [equiva-

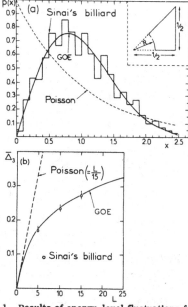

FIG. 1. Results of energy-level fluctuations for desymmetrized Sinai's billiards as specified in the upper right-hand corner of (a). 740 levels have been included in the analysis corresponding to the 51st to 268th level for $R = 0.1$, 21st to 241st level for $R = 0.2$, 16th to 194th level for $R = 0.3$, 11th to 132nd level for $R = 0.4$. (a) Results for the nearest-neighbor spacing distribution. (b) Results for the average value of the $\Delta_3(L)$ statistic of Dyson and Mehta for $L = 5$, 10, and 15. Curves corresponding to the Poisson case (stretch of uncorrelated levels) and to the random-matrix–theory predictions (GOE) are drawn for comparison. The error bars in (b) (one standard deviation) correspond to finite-sampling effects as predicted by GOE.

lently, the average of Δ_3 computed with Eq. (2) will not depend on the initial value x of the interval]. We have checked numerically that this is the case for the SB spectra, provided that the first lowest levels of each spectrum $\{\epsilon_i(R)\}$ are omitted (see caption of Fig. 1). This procedure seems appropriate if one observes the spectra corresponding to different values of R: Indeed, the smaller the value of R the larger the number of levels, starting from the ground state, that can be attained by perturbation theory from the spectrum $\{\epsilon_i(R=0)\}$ (triangular billiard). We emphasize that the fluctuation properties we are looking for will appear to be essentially different from the ones corresponding to the triangular billiard[12] and cannot be attained from it by perturbation theory. To increase the statistical significance of the results, four spectra $\{\epsilon_i(R)\}$ corresponding to different values of R will be analyzed as corresponding to a single stretch of 740 levels (see caption of Fig. 1). Care has been taken that one is dealing with "independent information": The different values of R should not be chosen too close to one another. Otherwise, two different spectra corresponding to R and $R+\delta R$ would be almost deducible one from the other and one would just be analyzing redundant information.

Let us now discuss the results. In Fig. 1(a) is shown the nearest-neighbor spacing distribution $p(x)$ which is compared to the GOE and Poisson predictions. As can be seen, the SB results follow very closely GOE not only for small spacings (level repulsion) but over the whole range of spacings. The variance of $p(x)$ for SB is 0.273 which is close to the GOE value 0.286 ± 0.015 (the error bar takes into account the finite sampling effects) and far from the Poisson value 1.0. We next consider quantities related with the spacing correlations. In Fig. 1(b) are shown the average values of $\Delta_3(L)$ for $L=5$, 10, and 15 for SB; they are close to the corresponding GOE values. We have also computed the correlation factor between two adjacent spacings. For SB we obtain -0.30, to be compared to -0.27 ± 0.04 (GOE) and 0.0 (Poisson). We can summarize the numerical results as follows: All fluctuation properties of SB investigated so far are fully consistent with GOE predictions.

Is this a surprising result? With a few inconclusive exceptions (see a discussion on small metallic particles, for instance in Ref. 2), the basic hypotheses leading to RMT have always been put forward by invoking the complexity of

the system. In other words, it has been taken as essential that one is dealing with a many-particle system (system with many degrees of freedom). Our results indicate that this is by no means a necessary condition. Indeed, the quantum chaotic system with two degrees of freedom studied here (a one-particle system in two dimensions) shows also GOE fluctuations. The present work should have further developments [for instance, when time-reversal invariance does not hold, the adequate model in RMT is the Gaussian unitary ensemble (GUE) and one should look for "simple" chaotic systems having GUE fluctuations]. It is an attempt to put in close contact two areas—random matrix physics and the study of chaotic motion—that have remained disconnected so far. It indicates that the methods developed in RMT to study fluctuations provide the adequate tools to characterize chaotic spectra and that, conversely, the generality of GOE fluctuations is to be found in properties of chaotic systems. In summary, the question at issue is to prove or disprove the following conjecture: Spectra of time-reversal–invariant systems whose classical analogs are K systems show the same fluctuation properties as predicted by GOE (alternative stronger conjectures that cannot be excluded would apply to less chaotic systems, provided that they are ergodic). If the conjecture happens to be true, it will then have been established the *universality of the laws of level fluctuations* in quantal spectra already found in nuclei and to a lesser extent in atoms. Then, they should also be found in other quantal systems, such as molecules, hadrons, etc.

We are grateful to C. Jacquemin and A. Pandey for valuable suggestions and discussions. The Division de Physique Théorique is a Laboratoire associé au Centre National de la Recherche Scientifique.

[1] *Statistical Theories of Spectra: Fluctuations*, edited by C. E. Porter (Academic, New York, 1965); M. L. Mehta, *Random Matrices and the Statistical Theory of Energy Levels* (Academic, New York, 1967).

[2] T. A. Brody, J. Flores, J. B. French, P. A. Mello, A. Pandey, and S. S. M. Wong, Rev. Mod. Phys. **53**, 385 (1981).

[3] R. U. Haq, A. Pandey, and O. Bohigas, Phys. Rev. Lett. **48**, 1086 (1982); O. Bohigas, R. U. Haq, and A. Pandey, in *Nuclear Data for Science and Technology*, edited by K. H. Böckhoff (Reidel, Dordrecht, Netherlands, 1983), pp. 809–813.

[4]H. S. Camarda and P. D. Georgopulos, Phys. Rev. Lett. **50**, 492 (1983).

[5]The harmonic-oscillator case is an exception and must be studied separately; see M. V. Berry and M. Tabor, Proc. Roy. Soc. London **356**, 375 (1977); O. Bohigas, M. J. Giannoni, and A. Pandey, to be published.

[6]Berry and Tabor, Ref. 5.

[7]M. V. Berry, in Lectures of the Les Houches Summer School, 1981 (North-Holland, Amsterdam, to be published).

[8]We use here the term chaotic in a qualitative way. In the classical case it designates a large class of systems whose trajectories are unstable with respect to the initial conditions. In fact, the specific systems we call chaotic in the Letter are K systems. Via the correspondence principle we cover the quantum case, for which the asymptotic behavior of the spectrum at high energies is of most relevance.

[9]S. W. McDonald and A. N. Kaufman, Phys. Rev. Lett. **42**, 1189 (1979); G. Casati, F. Valz-Gris, and I. Guarneri, Lett. Nuovo Cimento **28**, 279 (1980).

[10]M. V. Berry, Ann. Phys. (N.Y.) **131**, 163 (1981).

[11]H. P. Baltes and E. R. Hilf, *Spectra of Finite Systems* (Wissenschaftsverlag, Mannheim, 1976).

[12]For the triangular billiard ($R=0$) the system is integrable and the spectrum is asymptotically dominated by degeneracies (Ref. 10).

Quantum chaos, localization and band random matrices

F. M. IZRAILEV

Budker Institute of Nuclear Physics,
630090 Novosibirsk, Russia

Abstract

The known manifestation of quantum chaos is the so-called Wigner-Dyson distribution $P(s)$ for spacings between neighbouring levels in the spectrum. In the other limiting case of completely integrable systems, the distribution $P(s)$ turns out to be very close to the Poissonian one. In the present chapter, the influence of quantum effects on level statistics is studied for the case in which the corresponding classical systems are fully chaotic. The numerical study of the kicked rotator model with a finite number of states allows us to establish the link between the degree of localization and level repulsion. As a good model for this phenomenon of localization, the band random matrices are suggested. It is shown that such matrices can be used to describe statistical properties of localized quantum chaos.

1 Introduction

One of the important problems in the theory of quantum chaos is the description of statistical properties of systems using classical and quantum (semiclassical) parameters. Numerous studies (see the review [1] and references therein) have shown that the general situation is very complicated and no universal properties can be predicted when in the classical limit the motion is not fully chaotic. On the other hand, it has been proved numerically that for many classical models (for example, billiards, see [2]) with strong chaos, statistical properties both of energy (quasienergy) spectra and eigenfunctions are well described by the Random Matrix Theory (RMT) [3]. Until recently, it was believed that the only condition for the applicability of the RMT in describing the properties of quantum chaos, was the condition of strong classical chaos. However, it has been discovered [4] that the influence of quantum effects can be very strong and should be taken into account even in a deep semiclassical region.

In this chapter, the influence of quantum effects on statistical properties of both spectra and eigenfunctions is studied for the case in which classical chaos is very strong. The main result is that due to the influence of localization, the properties of quantum chaos can be quite far from the limiting ones described by the RMT. Nevertheless, this 'intermediate' or 'soft' quantum chaos was found to have universal scaling features. It is shown that a good model for this 'localized' quantum chaos is an ensemble of random matrices with a band structure, the latter is related to the degree of quantum localization.

2 Kicked rotator and dynamical localization

To study the influence of quantum effects on chaotic motion of dynamical systems, the model of the kicked rotator (KR) was proposed [4] which is given by the Hamiltonian:

$$\hat{H} = -\frac{\hbar^2}{2I}\frac{\partial^2}{\partial\theta^2} + \varepsilon_0 \cos\theta \cdot \delta_T(t) \tag{1}$$

Here $\delta_T(t) \equiv \sum_{m=-\infty}^{\infty} \delta(t - mT)$ is a periodic delta function with period T, the parameter ε_0 is the perturbation strength and I is the moment of inertia. In spite of the specific form of the external field in (1) this model appears to exhibit generic properties of quantum chaos (see, e.g., [5])

By integrating between succesive kicks the motion of this model can be described by the mapping for the ψ function in one period of the perturbation:

$$\psi(\theta, t + T) = \hat{U}\psi(\theta, t) \tag{2}$$

$$\hat{U} = \exp\left(i\frac{T\hbar}{4I}\frac{\partial^2}{\partial\theta^2}\right)\exp\left(-i\frac{\varepsilon_0}{\hbar}\cos\theta\right)\exp\left(i\frac{T\hbar}{4I}\frac{\partial^2}{\partial\theta^2}\right).$$

In the above expressions the ψ function is determined in the middle of free rotations, between two successive kicks. In such a representaion the evolution operator has the symmetric form which is more useful for the analysis. As is seen from (2), the dynamics of our model depends (apart from the initial state $\psi(\theta,0)$) on two parameters only,

$$\tau \equiv \frac{T\hbar}{I}; \quad k \equiv \frac{\varepsilon_0}{\hbar}. \tag{3}$$

It is convenient to write the evolution operator $\hat{U}$ in the unperturbed ($k = 0$) basis by representing the solution $\psi(\theta,t)$ in the form of an expansion in eigenfunctions of the angular momentum,

$$\psi(\theta, t) = \frac{1}{\sqrt{2\pi}} \sum_{n=-\infty}^{\infty} A_n(t)\exp(in\theta). \tag{4}$$

As a result, the mapping for the Fourier coefficients of ψ has the form:

$$A_n(t + T) = \sum_{m=-\infty}^{\infty} U_{nm}A_m(t), \tag{5}$$

$$U_{nm} = \exp\left(i\frac{\tau n^2}{4}\right)(-i)^{n-m}J_{n-m}(k)\exp\left(i\frac{\tau m^2}{4}\right).$$

It is seen that in the unperturbed basis the evolution operator has essentially the band structure due to the strong decay of Bessel function for $|n - m| \gtrsim k$.

To compare with the motion of the corresponding classical model, it is convenient to introduce another parameter:

$$K = \tau k = \frac{T\varepsilon_0}{I} \qquad (6)$$

which is of pure classical nature (there is no dependence on $\hbar$). As the second independent parameter we use the perturbation parameter $k \sim 1/\hbar$, which is a quantum one; it determines the size of the band ($\approx 2k$) for the unitary matrix U_{nm}. The classical limit corresponds to $k \to \infty$, $\tau \to 0$ with $K = const$.

The classical counterpart of KR is known as the *standard mapping* [6, 7] which can be obtained from the classical version of the Hamiltonian (1). It is well known that when the only parameter K in the classical model is large enough (practically, for $K \gtrsim 5$), the motion turns out to be strongly chaotic resulting in unbounded diffusion in momentum space: $< (P - P_o)^2 > = D_{cl}t$. Here, D_{cl} is the classical diffusion coefficient, P_0 is the initial momentum and time is measured by the number of kicks. As a result, the energy $E(t)$ in the classical model is increasing linearly in time, $E(t) = D_{cl}t/2 + E(0)$.

In contrast to classical diffusion, the motion of the quantum model (1) was found to have specific properties for any finite value of semiclassical parameter $k \gg 1$ [4], see, also, references in [5]. It was shown numerically that the correspondence of the quantum behavior to the classical diffusion holds only for some characteristic time $t \lesssim t^*$. After this time, for $t \gtrsim t^*$, the energy growth of the KR decreases with time, and for sufficiently large time the diffusion appears to be completely suppressed. This phenomenon (for generic values of τ, namely, for $\tau/4\pi \neq r/q$ with r, q integers, see [8]) was found to be a generic property and was termed *quantum suppression of classical chaos* [9, 10].

The explanation of this quantum suppression was given in [9, 10]: it is based on the properties of the quasienergy spectra and the structure of eigenfunctions (EF). The equation for quasienergies ε and EFs, $\varphi_n(\varepsilon)$, in the momentum representation has the form:

$$\exp(-i\varepsilon)\varphi_n(\varepsilon) = \sum_m U_{nm}(k, \tau)\varphi_m(\varepsilon), \qquad (7)$$

where $n, m \in (-\infty, \infty)$.

The main point of the approach of [9, 10] is that all the EFs for non-resonant values of τ turn out to be *exponentially localized* in the infinite momentum space of the system, $\varphi_n \sim \exp(-|n - n_o|/l_\infty)$ for $|n| \to \infty$ (see [11]). This means that for any finite k only a finite number of the exact states are excited by the generic initial wave packet $\psi(\theta, 0)$. Therefore, the diffusion in momentum space has to be bounded. This property of EFs is closely related to the type of quasienergy spectrum, namely, it has to be discrete in this case. Moreover, the statistics of the quasienergy spectrum turns out to be close to uncorrelated, which leads to the

Poissonian form for the distribution $P(s)$ of spacings between neighboring levels [12, 13]:

$$P(s) \sim \exp(-s). \tag{8}$$

One of the most important results of numerical simulations [14] is the remarkable relation between the average localization length l_∞ and the classical diffusion coefficient D_{cl}:

$$l_\infty \simeq \frac{D_{cl}(K)}{2\tau^2} \simeq \frac{k^2}{4}. \tag{9}$$

Here, the relation $D_{cl} \simeq K^2/2$ is used which is valid for large $K \gg 1$ (also, for $K \approx 5$, see [7]) and l_∞ is given by the number of unperturbed levels. Extensive numerical data have shown [9, 10] that the effective time of breakdown t^* for the energy diffusion is proportional to k^2, $t^* \sim k^2$.

Another very important result (see [11]) is that the localization of EFs in the KR may be associated with the well-known Anderson localization which takes place in some models of the solid state. However, this analogy is not rigorously true since the Anderson localization appears for models which have a disordered potential. In our case, there are no random parameters, and this is the principal difference. For this reason, the term *dynamical localization* is often used to emphasize the dynamical nature of localization.

3 Maximal quantum chaos and random matrix theory

Unlike the above model of a KR with an infinite number of states (the momentum space is unbounded), we pass now to the model with a *finite number of states*. This allows us to study the whole transition from weak to strong statistical properties of quantum chaos. This model can be derived from (5) by using the properties of the so-called *quantum resonance* [8] which occurs for rational values of the rescaled period τ, for $\tau/4\pi = r/q$ with integers r, q. In the classical limit, this model corresponds to the standard mapping on a torus of size $2\pi m_0/T$ in the momentum p. The quantization conditions for this model of *kicked rotator on a torus* (KRT) are $N = 2\pi m_0/T\hbar$ and $\tau = 4\pi r/N$ where N is the total number of states and m_0 is an integer. The classical limit is now $\tau \to 0, k \to \infty, N \to \infty$ with $K = const$ and $N\tau = const$ [10, 5].

As a result, the evolution unitary matrix U_{nm} is finite of size N and has the form [5]

$$U_{nm} = \exp\left(i\frac{\tau}{4}n^2\right) \frac{1}{N} \sum_{l=-N_1}^{N_1} \exp\left[-ik\cos\left(\frac{2\pi}{N}l\right)\right] \exp\left[-i\frac{2\pi}{N}l(n-m)\right] \exp\left(i\frac{\tau}{4}m^2\right).$$
$$\tag{10}$$

where $n, m = -N_1, ..., N_1$ and $N = 2N_1 + 1$. It is clear that in the limit of an infinite number of states, $N \to \infty$, the properties of the KR (1) are recovered.

Due to the additional parameter N in the KRT model (10) (compared to the

KR model (7)) one may expect to observe both very weak and very strong chaotic properties depending on the model parameters. As was shown in [15, 5], the related parameter which determines the transition from weak to strong quantum chaos is not the strength parameter k but the ratio of the localization length l_∞ to the total number of states N,

$$\Lambda = \frac{l_\infty}{N}. \tag{11}$$

This scaling conjecture is similar to that known in solid state physics [16] (see the discussion below) and numerically proved to be true in the dynamical model (10) only when the classical parameter K is strong enough and all statistical properties are expected to depend on quantum parameters only. It is clear that for $\Lambda \ll 1$ the properties of the model (10) are quite close to the original model (1). Unlike, in the opposite limit, $\Lambda \gg 1$, it is natural to expect that the statistical properties are very strong.

Following the idea [17, 2] that strong quantum chaos can be described by the random matrix theory (RMT) [3], the numerical study of statistical properties of spectrum and EFs has been performed in [18, 19] for the KRT model (10) for the case of very large values of $\Lambda \gg 1$ and $K \approx 5$. In this case, all EFs seem to be of very complicated structure (see the typical example in Fig. 1). To analyze the structure of EFs, the distribution of EF-components φ_n in the momentum representation was studied [19, 10] for all EFs of the matrix U_{nm} with very large $\Lambda \gg 1$. The RMT predicts a completely random structure of EFs with the distribution $W_N(\varphi_n)$ given in the form

$$W_N(\varphi_n) = \frac{\Gamma\left(\frac{N}{2}\right)}{\sqrt{\pi}\Gamma\left(\frac{N-1}{2}\right)} \left(1 - \varphi_n^2\right)^{\frac{N-3}{2}} \rightarrow \sqrt{\frac{N}{2\pi}} \exp\left(-\frac{N}{2}\varphi_n^2\right), \tag{12}$$

where the last expression (the Gaussian distribution for components φ_n) appears in the large N limit. Since the meaning of N in (12) is the number of independent components of the EFs, in our case it is equal to $N \rightarrow N_1 = (N-1)/2$, due to the parity conservation ($\varphi_n = \pm\varphi_{-n}$).

The numerical data in [19, 10] show good agreement with the prediction (12) of the random structure of EFs. In particular, the Gaussian character of the distribution of φ_n occurs with a high accuracy (using the χ^2-fit of numerical data). Moreover, when the size of the unitary matrix is not large enough ($N_1 = 25$ in (12)), numerical data [10] clearly give a correspondence to the first expression in (12), rejecting the Gaussian distribution due to a very poor ($< 10^{-6}$) confidence level for the fit (see details in [5]).

Concerning the properties of the quasienergy spectrum, the RMT predicts for the level spacing distribution $P(s)$ the form which is described, with good accuracy, by the Wigner-Dyson surmise:

$$P(s) = As^\beta \exp(-Bs^2), \tag{13}$$

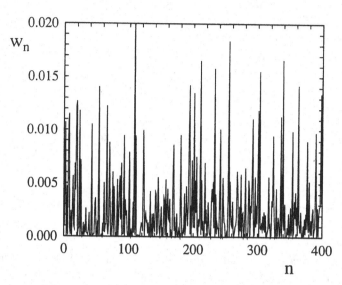

Fig. 1. A typical EF in the unperturbed basis for the model (10) is shown for $N = 801; N_1 = 400; k \approx 320; K \approx 5; \tau \approx 0.016$. On the vertical axis the probability $w_n = |\varphi_n|^2$ is plotted. Due to the symmetry $|\varphi_n|^2 = |\varphi_{-n}|^2$, only w_n with $n > 0$ are presented.

where A and B are normalizing constants, and β is a parameter which depends on the symmetry of the system and characterizes the repulsion between neighboring levels ($\beta = 1$, 2, or 4 for the Gaussian Orthogonal Ensemble (GOE), Gaussian Unitary Ensemble (GUE) and Gaussian Symplectic Ensemble (GSE), respectively, see [3]). In our case, the expected value of β is $\beta = 1$ because of the parity conservation and time reversal invariance in (7) and (10), see the discussion in [5]. However, other versions of the model have been proposed and studied in which the additional symmetries can be broken, resulting in the maximal repulsion $\beta = 2$ [18, 20, 21] and $\beta = 4$ [20]. In all these cases, a high correspondence of the numerical data with the analytical expression (13) was discovered.

Thus, we can see that under two conditions (strong classical chaos, $K \gg 1$, and delocalization of all states, $\Lambda \gg 1$) the statistical properties both of EFs and the quasienergy spectrum are the same as for random matrices. It is important to emphasize that our unitary matrix $U_{n,m}$ is not random by construction and depends on two parameters, τ and k, only. In this sense, the above two conditions can be regarded as the conditions for *maximal quantum chaos* which is described by the RMT.

Similar results are obtained for the so-called kicked top model [22], for which the distribution $P(s)$ of the quasienergy level spacings also manifests all the limiting types of distribution (13) with $\beta = 1, 2, 4$ depending on the symmetry of the model.

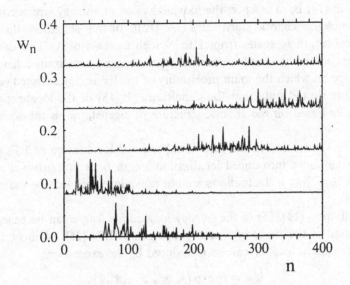

Fig. 2. Five of the $N_1 = 400$ EFs in the unperturbed basis, n, for the KRT model (10) are shown. On the vertical axis is the probability $w_n = |\varphi_n|^2$; on the horizontal axis is the number of unperturbed states, $n > 0$. The parameters are: $k \approx 32; K \approx 5.0; \tau \approx 0.16; \Lambda \approx 0.64$

4 Scaling properties of localized chaotic states

Now, we discuss the general situation for which the influence of localization effects should be taken into account, and in which therefore, the localization parameter Λ is not very large. A typical structure of eigenstates for this *intermediate* case is shown in Fig. 2 for the model (10) with $K \approx 5$.

The important peculiarity of such eigenstates is that on some scale they seem to have clear chaotic structure. Numerical data show that the 'effective size' of these *localized chaotic states* in the given (unperturbed) basis fluctuates very much depending on the specific eigenstate. Nevertheless, on average, the size of these EFs is smoothly decreasing with the decrease of Λ. To measure the scale on which the EFs are mainly located, in [15] a new definition of the average localization length was introduced (see also [23]):

$$l_N = N \exp(<\mathcal{H}> - \mathcal{H}_R),\qquad(14)$$

which is based on computing the *spectral entropy* of eigenstates:

$$\mathcal{H} = -\sum_{n=1}^{N} w_n \ln w_n; \qquad w_n = |\varphi_n|^2.\qquad(15)$$

The averaging in (14) is performed over all EFs of one matrix U_{nm} and over an ensemble of matrices with slightly different values of Λ. The normalizing coefficient $\mathcal{H}_R$ in (14) is, in essence, the maximal value of entropy which corresponds to the completely chaotic eigenstates (see (12)). In this definition, the averaged localization length l_N scales from 1 to N with increase of Λ. This definition is closely related to the simple approach that measures localization length as an effective size on which the main probability of the EF is concentrated (see details in [5]). One should note that the definition (14),(15) of the localization length is essentially based on the chaotic structure of eigenstates on the scale of their location.

It can be shown that in the case of exponential localization of EFs (when l_N, $l_\infty \ll N$) the above introduced localization length l_N is proportional to l_∞; the relation is $l_N \approx 2el_\infty$ if fluctuations are not taken into account (see the discussion in [15, 24]).

The definition (14),(15) of the *entropy localization length* can be generalized in the way used in the study of fractal properties of EFs [25]. Namely, the whole set of 'localization lengths' can be introduced by the expression:

$$l_{q,N} = N \exp\left(< \mathcal{H}_q > -\mathcal{H}_{q,\beta}\right). \tag{16}$$

Here, the quantity $\mathcal{H}_q$ is the so-called *Renyi information* defined by [26]

$$\mathcal{H}_q = \frac{1}{1-q} \ln\left(\sum_{n=1}^{N} |\varphi(n)|^{2q}\right), \tag{17}$$

where $q \geq 0$ is the continuous parameter. For the specific value $q = 1$ the expression (17) takes the form (15) and for $q = 2$ the definition (16),(17) is directly related to the participation ratio, the quantity commonly used in solid state physics. One should note that there is another way to make the averaging in the definition of the localization length, namely, $\tilde{l}_{q,N} = N < \exp(\mathcal{H}_q - \mathcal{H}_{q,\beta}) >$, which is also used in other studies. The proposed definition (16),(17) seems to have some advantage, for example, the ratio $l_{q,N}/N$ in the case of a fractal structure of EFs does not depend on q (for $q \ll N$).

The localization length $l_{q,N}$ introduced by (16),(17) may be associated with the generalized localization length $\xi_{q,N}$ in solid state models for finite samples of size N. Then, according to the approach developed in [16], some scaling for $l_{q,N}$ may be expected. Indeed, the localization in solid state physics in finite samples is related to the residual conductance of the samples themselves. The cornerstone of the scaling theory of localization is the assumption that conductance depends only on the ratio between the localization length ξ_∞ for the infinite sample and the size of the sample N. This scaling approach has been reformulated in [24] for the entropy localization length l_N in the following way. It was conjectured that for the dynamical model (10) the scaling

$$\beta_{loc} \equiv \frac{l_N}{N} = \tilde{f}\left(\frac{l_\infty}{N}\right) = f\left(\frac{k^2}{N}\right) \tag{18}$$

holds. Here, a new *localization parameter* β_{loc} is introduced which scales from

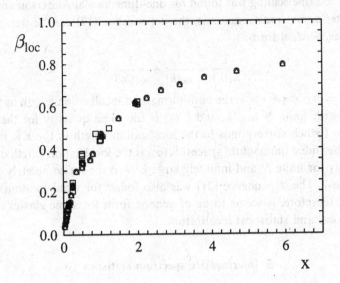

Fig. 3. The parameter β_{loc} is plotted versus the variable $x = k^2/N$ for $300 < N \leq 500$ (squares), $500 < N \leq 700$ (triangles) and $700 < N \leq 900$ (circles) for the model (10).

$\beta_{loc} \sim l_\infty/N \approx 0$ for very localized states ($l_\infty \ll N$), to $\beta_{loc} = 1$ for completely delocalized chaotic states. The last expression in (18) appears since $l_\infty \sim k^2$ for strong classical chaos ($K \gg 1$) and a large value of semiclassical parameter $k \sim 1/\hbar \gg 1$.

Numerical simulation with the model (10) has been performed to check this conjecture (see details in [24, 5]). For this, the quantity l_N was computed through (14),(15) for many different values of N and k in the ranges $200 \leq N \leq 860$ and $1 \ll k \leq 239$; the classical parameter was taken in all cases as $K \approx 5$. The data are shown in Fig. 3 which gives good evidence of scaling in the mean. In spite of the imperfection of the numerical study, for not very small values of k^2/N the numerical data allowed the analytical form of the scaling (18) to be established approximately. It was found in [24] that the scaling law may be described by the simple expression:

$$\beta_{loc} = \frac{\gamma x}{1 + \gamma x}; \qquad x \equiv \frac{k^2}{N} \approx 4\Lambda; \qquad \gamma \approx 0.8. \qquad (19)$$

It is convenient to introduce the new variables,

$$\ln(y) \equiv \ln\left(\frac{\beta_{loc}}{1 - \beta_{loc}}\right); \qquad \ln(x) \equiv \ln\left(\frac{k^2}{N}\right) \qquad (20)$$

in which the dependence (19) has the form of a straight line with a slope of one, $\ln(y) = \ln(x) + c_0$. Another representation of this dependence has been given in

[27] where the same scaling was found for one-dimensional Anderson and Lloyed models. After some transformations, the dependence (19) can be written in the *parameter independent* form:

$$\frac{1}{l_N(k)} = \frac{1}{l_\infty(k)} + \frac{1}{l_N(\infty)} \tag{21}$$

where $l_N(k) = \exp < (\mathscr{H}) >$ is the non-normalized localization length in the KRT model (10) with finite N and k, and $l_\infty(k)$ is the same quantity for the limiting case $N \to \infty$ (which corresponds to the localization length in the KR model (7) with the unbounded momentum space); $l_N(\infty)$ is the localization length defined in the same way for finite N and infinitely large $k \to \infty$ (or for completely extended chaotic states). The dependence (21) was also found for band random matrices (see below), therefore, it seems to be of generic form for some classes of model with dynamical and statistical localization.

5 Intermediate spectrum statistics

One of the important questions is the influence of localization on spectrum statistics. In what follows we discuss only the commonly used quantity, the level spacing distribution $P(s)$. It has been shown numerically [12, 13], in the case of strong localization this distribution is very close to the Poissonian one. In other extreme limit of maximal quantum chaos (see above) the form of $P(s)$ is given by the RMT, namely, it is well described by the Wigner-Dyson surmise (13). To study the spectrum statistics for localized quantum chaos (the so-called *intermediate statistics*) which appears in the KRT model (10) for not very large values of the parameter Λ, in [15] a new expression for $P(s)$ was suggested,

$$P(s) = As^\beta \exp\left[-\frac{\pi^2}{16}\beta s^2 - \frac{\pi}{2}\left(B - \frac{\beta}{2}\right)s\right], \tag{22}$$

where the parameters A and B are determined by the normalization conditions

$$\int_0^\infty P(s)\,ds = 1; \qquad \int_0^\infty sP(s)\,ds = 1. \tag{23}$$

The expression (22) has the *repulsion parameter* β which can be non-integer, unlike the Wigner-Dyson distribution (13). The peculiarity of the dependence (22) is that for $\beta = 1, 2, 4$ it appears to be quite close to the Wigner-Dyson law (13) (see [15, 5]). Also, the behavior of $P(s)$ for small $s \ll 1$ gives the right linear, quadratic and quartic repulsion for $\beta = 1, 2$ and 4, respectively. In addition, for $\beta = 0$ it is exactly Poissonian. Therefore, this expression may be used approximately to describe the whole transiton from uncorrelated (Poissonian) statistics to the limiting Wigner-Dyson statistics with the specific value of β related to the underlying symmetry of the system. The dependence (22) has a form that approximately takes into account the rigorous asymptotic expression for $P(s)$ for $s \to \infty$ [28], namely, the exponential decrease $\exp(-\beta\pi^2 s^2/16)$.

Detailed numerical investigation of the intermediate statistics (see review [5]) has shown that all the data for $P(s)$ can be fitted, with a high degree of accuracy, by expression (22) in a large range of values of $\Lambda \equiv l_\infty/N$ (for $\Lambda \gtrsim 0.05$). This means that for the case of strong classical chaos and in a deep semiclassical region the $(k \sim 1/\hbar \gg 1)$ the statistical properties of spectrum can indeed be described by the universal dependence (22) on the repulsion parameter β.

In analogy with the localization parameter β_{loc} (see (18)), it is natural to expect that the dependence of β on the model parameters k and N (actually, no dependence on K for $K \gtrsim 5$ was found) also has scaling of the form $\beta(k, N, K) \to \beta(\Lambda)$. This very important assumption was also confirmed by numerical data [5], namely, for different values of k and N in (10) but fixed $\Lambda = const$ the distribution $P(s)$ has the same form, with a high confidence level of the χ^2-fit.

Another intriguing result is the relation

$$\beta \approx \beta_{loc} \tag{24}$$

which seems to hold approximately, within the accuracy of the numerical study [15, 5]. Such a simple relation between the repulsion parameter β and the localization parameter β_{loc} is quite strange , especially when taking into account that there are other possibilities for defining the localization parameter by the relation $\beta_{q,loc} = l_{q,N}/N$ (see also Section 7).

Similar scaling properties for the dynamics of the model (10) have been found numerically and explained analytically (by using the semiclassical approximation) in [29], where the scaling dependence on the parameter Λ was discovered for other important dynamical and spectral quantities. This results in the conclusion that *all* statistical properties of the KRT model (10), both dynamical and spectral ones, appear to depend on one parameter, Λ, only.

6 Band random matrices as the model of localized quantum chaos

The unitary matrix U_{nm} for the KR models (7),(10) appears to have a band structure with the size of the band ($\approx 2k$) determined by the quantum parameter k. At the same time, the classical parameter K seems to be responsible for the degree of randomness of matrix elements inside this band. This has led in [18] to the conjecture that the statistical properties of dynamical models (7) and (10) for large $K \gg 1$ are also typical for the ensemble of Band Random Matrices (BRMs).

In what follows, we consider real symmetrical matrices which are defined by the parameter b determining the band size: $b = 1$ for diagonal, $b = 2$ for tridiagonal and $b = N$ for GOE matrices. Inside the band, for $|i - j| < b$, matrix elements A_{ij} are independent Gaussian numbers with the mean equal to zero and the variance corresponding to the probability density

$$\mathscr{P}(A_{ii}) = \sqrt{\omega/\pi} \exp(-\omega A_{ii}^2) \tag{25}$$

$$\mathscr{P}(A_{ij}) = \sqrt{2\omega/\pi} \exp(-2\omega A_{ij}^2) \qquad i < j$$

while outside the band all matrix elements equal zero.

The ensemble (25) is fully characterized by the three parameters ω, b and N; however, the first parameter only determines the size of the eigenvalues and is not relevant for describing statistical properties. As was shown numerically in [30] and proved analytically in [31], the eigenvalue density $\rho(E)$ of BRM in the limiting case of large $N \to \infty$ has the semicircular form

$$\rho(E) = \frac{2}{\pi r^2} \sqrt{r^2 - E^2}. \tag{26}$$

It has a second moment $\langle E^2 \rangle = r^2/4$ where $r^2 = 2F/(N\omega)$ and $F = \frac{1}{2}b(2N - b + 1)$ is the number of independent non-zero matrix elements.

This type of BRM ensemble may be regarded as a version of the more general ensemble of BRMs, which was introduced by Wigner in 1955 [32] (see next section). For a long time, these BRMs were completely forgotten, mainly, because of serious difficulties in the rigorous analysis.

The success of the existing theory of full random matrices [3] is based on the idea that for very complex systems, like heavy nuclei and many electron atoms, the interaction between particles is so strong that statistical description is possible. This idea, put forward by Wigner and Dyson, proved to be effective for many physical systems and shows once more in physics, that symmetries are the relevant features: the fluctuation properties seem to be relatively insensitive to the details of the interaction. It has recently become clear that classically chaotic dynamics is the underlying condition for the random matrix analogy to apply, even for systems with a few degrees of freedom (see e.g. [2]).

However, full random matrices can by used to describe only local statistical properties of spectra. To study global properties of Hamiltonians, one needs to take into account that in physical applications the interaction of unperturbed states always has a finite range which determines the band structure of the Hamiltonians. The BRM ensembles may therefore prove to be more effective than the standard random matrix ensembles and can be regarded as the generalization of common ensembles of random matrices.

The revival of interest in BRMs emerged in connection with possible applications to the quantum chaos problems [33, 34]. General semiclassical arguments have been given in [35] in support of the relevance of the BRM to generic Hamiltonian systems. Another source of interest in the BRM is related to solid state physics where band matrices are widely used to describe the dynamics of electrons in disordered solids.

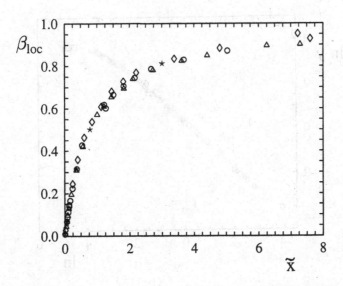

Fig. 4. The scaled localization length β_{loc} versus $\tilde{x} = b^2/N$ for $N = 200(\diamond)$, $N = 400(\triangle)$, $N = 600$ ($\circ$), $N = 800(\star)$.

In analogy with the KRM model (10), the parameter

$$\tilde{x} = \frac{b^2}{N} \qquad (27)$$

has been introduced in [36] to investigate the statistical properties of the BRM model (25).

First, we discuss the numerical data obtained in the study of eigenvectors (EV) [36]. The typical structure of EVs appears to be similar to that for the KRM model, see Figs. 1,2. The dependence of the localization parameter β_{loc} defined by (18), on the parameter $\tilde{x}$ is presented in Figs. 4,5. To compute the localization length l_N, the average over all EVs of a number of matrices with a given b and N has been taken [36]. The simulation with N ranging between 200 and 1000 shows remarkable scaling behavior for $\beta_{loc}(\tilde{x})$. Indeed, all the points in Fig. 4 fall on a smooth curve with a high degree of accuracy. To illustrate the scaling dependence, it is convenient to present the same data in the varaibles $\ln(y)$ and $\ln(\tilde{x})$ where y is defined by relation (20). These variables are more useful to analyze the dependence for both $\tilde{x} \ll 1$ and $\tilde{x} \gg 1$. From Fig. 5 the linear dependence is clearly seen over a large range $\tilde{x} \lesssim 10$. A fit of the numerical data in this range of $\tilde{x}$ gives the relation $\ln(y) = \alpha \ln(\tilde{x}) + c_1$ with $\alpha \approx 1$ and $c_1 \approx 0.35$. For $\beta(\tilde{x})$ this corresponds to the law (19) with $\gamma \approx 1.4$ (with the change $x \to \tilde{x}$).

For large values of $\tilde{x} \gtrsim 10$ the numerical data of Fig. 5 deviate from the straight line; however, they indicate that the scaling behaviour still holds. One

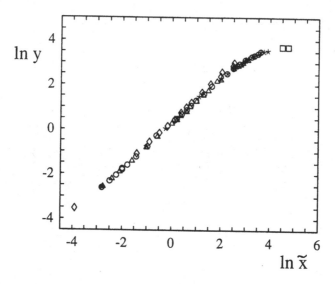

Fig. 5. A ln–ln plot of the data of Fig. 4 in the variables $\tilde{x}$ and y, see (20), together with additional data for large $\tilde{x}$ Here, the data for $N = 1000(\square)$ are also presented.

possible explanation of the different behaviour for large $\tilde{x}$ is that the condition $1 \ll b \ll N$ is not well fulfilled.

The analytical expression for the scaling found can be written in a form similar to (21):

$$\frac{1}{l_N(b)} = \frac{1}{l_\infty(b)} + \frac{1}{l_N(N)}, \tag{28}$$

with b instead of k, and where $l_N(N)$ has the meaning of $l_N(\infty)$ for the KRM model (localization length for extended and random states). Very recently [37] the above analytical scaling (28) has been proved rigorously (for the inverse participation ratio, which is directly related to the localization length $l_{q,loc}(E)$ with $q = 2$).

From expression (17) it is seen that the quantity $l_{q,loc}$ is defined through the moments of the distribution $\tilde{W}$ of components φ. Therefore, if the scaling holds for *all* values of q, it results in the scaling for the distribution $\tilde{W}(\varphi_n)$ itself. The scaling for $\tilde{W}$ has been numerically studied in [38] for different values of b and N. The data show that the form of $\tilde{W}$ remains the same if the parameter Λ is fixed. The problem of the determination of the analytical form of the dependence of the distribution $\tilde{W}$ on Λ is still open though some approximate expressions have been proposed in [38].

The physical meaning of the scaling parameter $\tilde{x}$ is of the same nature as for the KR models (7) and (10). Namely, the localization length for the infinite

BRM ($N \to \infty$) is proportional to b^2. This was also supported by estimates given in [39].

Rigorous mathematical investigation of BRM ensembles is very difficult since they are not rotational invariant under simple orthogonal rotations. Nevertheless, by using the powerful supersymmetry method rigorous results have recently been presented in [40] where it was proved that for a more general form of the BRM (with a smooth, but fast enough decrease of matrix elements away from diagonal) in the limit $N \to \infty, b \to \infty, b^2/N = const$, the localization length has the form $l_\infty \sim \rho^2 b^2$. Here $\rho(E)$ is the level density (26), therefore, there is a smooth dependence of ensemble average localization length on the energy, $l_\infty(E)$. In the first numerical experiments [30] the averaging over all energy intervals was performed, but recently [41, 42] the same scaling result was obtained for small energy windows, with the only difference that the coefficient γ was found to be dependent on the energy. Also, the same dependence (19) has been numerically found [43] for the generalized localization parameter $\beta_{q,loc}$ with $q = 1, 2, 3, 4, 5$.

In analogy with the KR model (10) it is natural to expect that the distribution $P(s)$ of spacings between neighboring eigenvalues of the BRM is essentially dependent on the parameter $\tilde{x}$ only, rather than on b and N independently. In one extreme case of diagonal matrices ($b = 1$) the spacings between eigenvalues are not correlated, resulting in the Poisson distribution for $P(s)$. In the opposite case of fully random matrices ($b = N$), the RMT predicts the Wigner-Dyson form (13).

To describe the intermediate statistics $P(s)$ for the BRM where $P(s)$ changes from the Poisson to the Wigner-Dyson distribution, in [30, 44] an improved version of the phenomenological dependence (22) has been used

$$P(s) = A s^\beta (1 + B\beta s)^{f(\beta)} \exp\left[-\frac{\pi^2}{16}\beta s^2 - \frac{\pi}{2}(1 - \frac{\beta}{2})s\right], \qquad (29)$$

where A and B are normalizing parameters and

$$f(\beta) = \frac{2^\beta (1 - \frac{\beta}{2})}{\beta} - 0.16874 \qquad (30)$$

is some characteristic function.

For $\beta = 0$ expression (29) reduces to the Poisson distribution and for $\beta = 1, 2, 4$ it approximates very closely the $P(s)$ distribution GOE, GUE and GSE, respectively. The distribution (29) is based on the Dyson Coulomb model [28] (see also [45]) and constructed in such a way that for large spacings, $s \gg 1$, and $\beta = 1, 2, 4$ it has the correct dependence for full random matrices. In addition to linear, $\sim s$, and quadratic, $\sim s^2$, terms in the exponent, a logarithmic term is introduced in (29), in accordance to the results of [46, 47]). Expression (29) is more complicated than (22), but it gives much better correspondence with the exact results of the RMT [48]. For example, for $\beta = 1$ the deviation from the exact dependence of $P(s)$ is less than 0.3% for small ($s \le 0.1$) and large

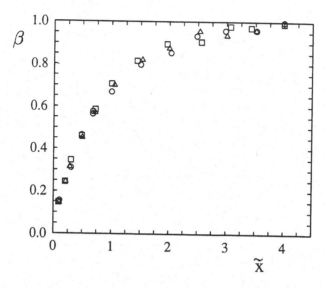

Fig. 6. The repulsion parameter β for intermediate statistics $P(s)$ versus $\tilde{x} = b^2/N$, for different values of N and b: $N = 400$ ($\square$), $N = 800$($\triangle$) and $N = 1600$($\circ$). Each value of β was obtained by fitting the numerical data for the level spacing distribution with expression (29). All values of β are within a 1% confidence level.

($s \geq 2$) spacings; it is less than 0.02% in the most important intermediate region $0.5 \leq s \leq 1.6$. This distribution is thus closer to the exact one than Wigner's distribution (13) itself. Agreement with the RMT is also very good for $\beta = 2$ and 4.

In numerical experiments [30] BRM matrices have been used with $N = 400, 800, 1600$ and different band sizes $b \gg 1$. The distribution $P(s)$ was obtained by averaging over the $P(s)$ for Q different random matrices with the same N and b ($Q = 50, 25, 12$ for $N = 400, 800, 1600$, respectively). Since the eigenvalue density is not uniform, the spacings have been normalized to the local density. To avoid the influence of large fluctuations caused by the finite size of the matrices, a number of eigenvalues at the edges of the semicircle distribution (20) have not been taken into account. As a result, for each N and b, the total number of spacings in the final distribution of $P(s)$ is approximately equal to $M = 16000$–17000.

The summarized data for different values of $\tilde{x}$ are given in Fig. 6. It can be seen that the scaling behavior for the repulsion parameter β occurs in a large range of the parameter $\tilde{x}$. This result indicates that fluctuations in the eigenvalue spectra of the BRM appear to have universal properties which can be described by a single parameter $\tilde{x}$.

7 Concluding remarks

When comparing the properties of the KRT model (10) with those of the BRM ensemble (25), a remarkable agreement is seen. This means that the BRM ensemble can be used as a good model of localized quantum chaos. It is impressive that for these two models (10) and (25), both localization (β_{loc}) and spectrum (β) parameters depend in the same way on one scaling parameter, $\Lambda = l_{\infty}/N$. This proves that for strong classical chaos ($K \gg 1$) random matrix models are valid if they take into account specific quantum features of considered systems.

Another possible application of the BRM model (25) is one-dimensional and quasi-one-dimensional models of disordered solids. Indeed, even for tridiagonal models with diagonal disorder (Anderson and Lloyed models which are quite different from (25)), exactly the same form of scaling law (21),(28) was recently found for the localization length l_N (see details in [27]). This indicates that scaling properties of the BRMs are general for quite a large class of systems, both dynamical and statistical.

It should be noted that BRMs of the type (25) are a special case of the more general ensemble of BRMs, which was introduced long ago by Wigner [32]. This ensemble is given by Hamiltonian matrices

$$H_{mn} = \frac{m}{\rho_0}\delta_{mn} + v_{mn}; \qquad v_{mn} = v_{nm}. \tag{31}$$

Here ρ_0 is the mean level density and the off-diagonal matrix elements are random and statistically independent with the zero mean $< v_{mn} >= 0$, and with the variance $< v_{mn}^2 >= v^2$ for $|m - n| < b$; the matrix elements outside the band are zeros, $v_{mn} = 0$. In [32] the simplest distribution for v_{mn} was chosen, namely, $v_{mn} = \pm v$ with random signs.

Such matrices represent conservative systems defined by the Hamiltonian $H = H_0 + V$ with an unperturbed part H_0 and perturbation V. The latter is assumed to be random in a finite range of the unperturbed states. This representation seems to be quite generic for Hamiltonian systems with an integrable part H_0 and a very complicated form of V (see [35]). Due to mathematical difficulties, no rigorous results for statistical properties have been obtained, only some limiting cases were analyzed by Wigner for the so-called weighted density of states [32]. New interest in the model (31) arose after studies [35] in which the general properties of eigenvectors and spectrum were analyzed.

It is seen that the BRM model (25) is a particular case of (31), it appears for finite N in the limit $\rho_0 \to \infty$ (also, a Gaussian distribution for v_{mn} is taken with $v^2 = 1/(4\omega)$). For *infinite matrices*, $N \to \infty$, the first important results have been obtained in [39, 49]. In particular, the dependence of the localization length l_{∞} on the parameters of the model was found [39], $l_{\infty} \sim b^2 f(\alpha b^{3/2})$ were $f(z)$ is some unknown function. Recently, this prediction has been proved rigorously in [40] by making use of an analogy with disordered models of the solid state under an external electric field.

It was also found [39] that the scaling parameter in the infinite model (31) has the form $\lambda = a_0 z f(z)$ with some constant a_0 and $z \equiv \alpha b^{3/2}$. This result allowed it to be shown numerically [49, 50] that the spectrum parameter β, found from the fit of level spacing distribution $P(s)$, depends only on λ and changes from 0 to 1 with an increase of λ.

To describe the structure of EVs, in the infinite model (31) the localization parameter β_{loc} can be also introduced [51] as:

$$\beta_{loc} = l_N / l_\perp. \tag{32}$$

Here $l_\perp$ is the maximal value of the localization length l_N, the latter being defined by common expressions (16),(17). This idea allowed the scaling behavior of β_{loc} as the function of λ to be found (see the details in [51]). The important result is that the form of the scaling $\beta_{loc}(\lambda)$ is completely different from (21) and (28), however, the puzzling relation $\beta_{loc} \approx \beta$ (see (24)) was again found numerically with a high degree of accuracy.

A much more complicated case is the model (31) with a *finite number N* of states. Preliminary numerical experiments [42] showed that the dependence of β_{loc} on the model parameters has a two-parameter scaling form, $\beta_{loc} = \beta_{loc}(\tilde{x}, \lambda)$. Here, one parameter, $\tilde{x} = b^2 / N$, is the same as for the model (25), and the other one, λ, is the same as for the infinite $N \to \infty$ model (31). Numerical data show the interplay between the two different regimes depending on the ratio $\tilde{x}/\lambda$; however, the analytical form for the function β_{loc} has still not been found.

Acknowledgements

The author is deeply indepted to his co-workers: G. Casati, B. Chirikov, M. Feingold, S. Fishman, J. Ford, A. Gioletta, I. Guarneri, L. Molinari, R. Sharf and D. Shepelyansky. The author also expresses his gratitude to his colleagues in the Physics Department of Milan University in Como for their hospitality during his visit when this paper was prepared.

REFERENCES

[1]   B. Eckhardt, Phys. Rep., 163 (1988) 205.

[2]   O. Bohigas and M.-J. Giannoni, Lect. Notes Phys. 209 (1984) 1.

[3]   Statistical Theories of Spectra: Fluctuations, ed. C.E. Porter (Academic Press, New York) 1965; M.L. Mehta, Random Matrices (Academic Press, New York) 1967; T.A. Brody, J. Flores, J.B. French, P.A. Mello, A. Pandey and S.S.M. Wong, Rev. Mod. Phys. 53 (1981) 385.

[4]   G. Casati, B.V. Chirikov, J. Ford and F.M. Izrailev, Lect. Notes Phys. 93 (1979) 334.

[5]   F.M. Izrailev, Phys. Rep. 196 (1990) 299.

[6]   B.V. Chirikov, Phys. Rep. 52 (1979) 263.

[7]   A.L. Lichtenberg and M.A. Liberman, Regular and Stochastic Motion (Springer, Berlin) 1983.

[8] F.M. Izrailev and D.L. Shepelyansky, Dokl. Akad. Nauk SSSR, 249 (1979) 1103 (Sov. Phys. Dokl., 24 (1979) 996); Teor. Mat. Fiz. 43 (1980) 417 (Theor. Math. Phys., 43 (1980) 553).

[9] B.V. Chirikov, F.M. Izrailev and D.L. Shepelyansky, Sov. Sci. Rev. C 2 (1981) 209.

[10] B.V. Chirikov, F.M. Izrailev, and D.L. Shepelyansky, Physica D 33 (1988) 77.

[11] S. Fishman, D.R. Grempel and R.E. Prange, Phys. Rev. Lett., 49 (1982) 508.

[12] F.M. Izrailev, Distribution of Quasienergy Level Spacings for Classically Chaotic Quantum Systems, preprint 84-63, Institute of Nuclear Physics, Novosibirsk, 1984.

[13] M. Feingold, S. Fishman, D.R. Grempel and R.E. Prange, Phys. Rev. B 31 (1985) 6852.

[14] D.L. Shepelyansky, Phys. Rev. Lett. 56 (1986) 677.

[15] F.M. Izrailev, Phys. Lett. A 134 (1988) 13; J. Phys. A 22 (1989) 865.

[16] P.A. Lee and T.V. Ramakrishnan, Rev. Mod. Phys., 57 (1985) 287; J.L. Pichard, J. Phys. C 19 (1986) 1519; J.L. Pichard and G. Andre, Europhys. Lett. 2 (1986) 477.

[17] G. Casati, F. Valz-Gris and I. Guarneri, Lett. Nuovo Cim. 28 (1980) 279.

[18] F.M. Izrailev, Phys. Rev. Lett. 56 (1986) 541.

[19] F.M. Izrailev, Phys. Lett. A 125 (1987) 250.

[20] R. Scharf, J. Phys., A 22 (1989) 4223.

[21] R. Blümel and U. Smilansky, Phys. Rev. Lett. 69 (1992) 217.

[22] M. Kus, R. Scharf and F. Haake, Z. Phys., B 66 (1987) 129; M. Kus, J. Mostowski and F. Haake, J. Phys. A 21 (1988) L1073; R. Scharf, B. Dietz, M. Kus, F. Haake and M. Berry, Europhys. Lett. 5 (1988) 383.

[23] F. Yonezawa, Journ. Non-Crystal. Solids, 35-36 (1980) 29; R. Blümel and U. Smilansky, Phys.Rev.Lett. 52 (1984) 137; R. Livi, M. Pettini, S. Ruffo, M. Sparpaglione and A. Vulpiani, Phys. Rev., A 31 (1985) 1039.

[24] G. Casati, I. Guarneri, F.M. Izrailev and R. Scharf, Phys. Rev. Lett. 64 (1990) 5.

[25] M. Schreiber, Physica A 167 (1990) 188.

[26] A. Renyi, Probability Theory (North Holland, Amsterdam) 1970.

[27] G. Casati, I. Guarneri, F.M. Izrailev, S. Fishman and L. Molinari, J. Phys., C4 (1992) 149.

[28] F.J. Dyson, J. Math. Phys. 3 (1962) 140; 157; 166.

[29] T. Dittrich and U. Smilansky, Nonlinearity 4/1 (1991) 59.

[30] G. Casati, F.M. Izrailev and L. Molinari, J. Phys. A 24 (1991) 4755.

[31] M. Kus, M. Lewenstein and F. Haake, Phys. Rev. A 44 (1991) 2800; G. Casati and V.L. Girko, Rand. Oper. Stoch. Eqs. 1 (1992) 1; M. Feingold, Europhys. Lett. 17 (1992) 97; S.A. Molchanov, L.A. Pastur and A.M. Khorunzhy, Teor.Mat.Fiz. 90 (1992) 163.

[32] E. Wigner, Ann. Math. 62 (1955) 548; 65 (1957) 203.

[33] B.V. Chirikov, Phys. Lett. A 108 (1985) 68.

[34] T.H. Seligman, J.J.M. Verbaarschot and M.R. Zirnbauer, Phys. Rev. Lett. 53 (1985) 215.

[35] M. Feingold, D.M. Leitner and O. Piro, Phys. Rev. A 39 (1989) 6507.

[36] G. Casati, F.M. Izrailev and L. Molinari, Phys. Rev. Lett. 64 (1990) 16.

[37] Y.V. Fyodorov and A.D. Mirlin, Phys. Rev. Lett., 69 (1992) 1093.

[38] K. Zyczkowski, M. Lewenstein, M. Kus and F.M. Izrailev, Phys. Rev. A 44 (1991) 73.

[39] M. Wilkinson, M. Feingold, and D.M. Leitner, J.Phys. A 24 (1991) 175.

[40] Ya.V. Fyodorov and A.D. Mirlin, Phys. Rev. Lett. 67 (1991) 2405.

[41] A. Gioletta, Thesis, Milan University, 1992.

[42] M. Feingold, A. Gioletta, F.M. Izrailev and L. Molinari, Phys. Rev. Lett. 70 (1993) 2936.

[43] S.N. Evangelou and E.N. Economou, Phys. Lett. A 151 (1990) 345.

[44] F.M. Izrailev, in "Quantum Chaos - Measurement", Eds. P. Cvitanovic, I. Percival and A. Wirzba, (Kluwer, Amsterdam) 1992, p. 89.

[45] F.M. Izrailev and R. Scharf, Phys. Lett. A 142 (1989) 89; R. Scharf and F.M. Izrailev, J. Phys. A 23 (1990) 963.

[46] J. Cloizeaux and M.L. Mehta, J. Math. Phys. 14 (1973) 1648.

[47] F. Dyson, Comm. Math. Phys. 47 (1976) 171.

[48] B. Dietz and F. Haake, Z.Phys. B 80 (1990) 153.

[49] M. Feingold, D. Leitner and M. Wilkinson, Phys. Rev. Lett. 66 (1991) 986.

[50] T. Cheon, Phys. Rev. Lett. 65 (1990) 529.

[51] G. Casati, B.V. Chirikov, I. Guarneri and F.M. Izrailev, Phys. Rev. E, 48 (1993) 1613.

Structural invariance in channel space: a step toward understanding chaotic scattering in quantum mechanics

T. H. SELIGMAN

Laboratorio de Cuernavaca, Instituto de Física,
University of Mexico (UNAM), Mexico City, Mexico

Abstract

We define a classical scattering process as a canonical map between the phase spaces of incoming and outgoing channels. Time-reversal invariance is introduced and the iterated scattering map as defined by Jung appears naturally. Using the theory of unitary representations of classical canonical transformations the transition to the quantum problem is achieved in the framework of a semi-classical approximation. Families of canonical scattering maps are defined by canonical transformations on channel space; the invariance properties of these families translate naturally into invariant ensembles of S-matrices that are unique according to Dyson. The concept of structural invariance of an iterated scattering map or Poincaré scattering map is introduced to take into account the generic features of the scattering system on hand. We shall argue that, if this invariance is present in our map, its unitary representation is a characteristic member of the ensemble and thus by way of ergodicity the statistical properties of the corresponding S-matrix are those of the circular ensemble known as COE.

1 Introduction

The statistical behaviour of scattering processes has been of interest for many years in nuclear physics [1], but recently it has acquired more general importance in the context of quantum manifestations of chaos [2]. In this context one question asked, refers to the statistical distribution of the eigenphases of the S-matrix. Others are of more practical interest, but are also more complicated, such as average fluctuating cross sections and their variances and correlations. The interplay between energy averages for specific models and ensemble averages for random matrix ensembles is usually the key point of such an analysis.

Blümel and Smilansky have shown [3] that, in the semi-classical limit, the two-point function of the eigenphase distribution coincides with the one of the circular ensemble of unitary and symmetric matrices [4] (referred to in what follows as COE), if the corresponding iterated scattering map [5] is chaotic.

We shall follow Ref. [6] and use the fact that the S-matrix is a unitary representation of the classical scattering map, which is a canonical transformation on channel space. This allows us to use the theory of unitary representations of classical canonical transformation [7,8] to approach the problem in a different way from previous authors. The concept of structural invariance of a classical map [9] will play a crucial role in obtaining non-trivial results.

In this chapter we shall first formulate the classical scattering in terms of classical canonical transformations. Both the scattering map and the iterated scattering map [5] will be discussed as canonical transformations on the phase space or symplectic manifold of scattering channels. Next we introduce the concept of structural invariance [9] on this phase space. It allows us to define invariant families of maps that should be generic if the structural invariance has been properly identified; generic here—as almost always—is a loose term that will acquire a little more rigour in the quantum part that we shall discuss next.

We use the theory of representations of classical canonical transformations in quantum mechanics [7,8], and show that the S-matrix in the semi-classical limit is nothing but the unitary representation of the above-mentioned scattering map. The structural invariance properties derived for a family of classical transformations can be translated to the quantum case. Then, based on the theory of random matrices [4,10], it becomes quite natural to associate a measure to an invariant family in order to obtain an associated ensemble of scattering matrices in quantum mechanics. Such invariant ensembles are well known to be of the COE or CUE type depending on the presence or absence of time-reversal invariance. Finally we have to contend with the problem of ergodicity, which, in principle, is quite well established for these ensembles [11], but the subsets of measure zero often contain relevant cases; these are eliminated by a careful choice of the structural invariance group that typically sends them to a smaller ensemble in which they are generic.

2 Structural invariance of the classical scattering map

For the sake of simplicity we shall first discuss the case of a structureless particle in a two dimensional configuration space as this is the simplest case where chaotic scattering can occur, but we wish to emphasize that the entire argument is trivially generalizable to more than two dimensions and composite particles.

The purpose is to describe a scattering process by a time independent and time-reversal invariant Hamiltonian that differs from the free particle Hamiltonian on a region of compact support only. Thus the energy is constant and the particles will be asymptotically free.

Consider a free particle in four dimensional phase space with coordinates (p_1, q_1, p_2, q_2). We can perform a canonical transformation

$$(p_1, q_1, p_2, q_2) \rightarrow (E, t, P_\phi, \phi) \tag{1}$$

to energy E, time t, angular momentum P_ϕ, and angle ϕ. In principle this transformation is two to one [12], but we shall be interested in particular subspaces.

As we chose a time independant Hamiltonian we consider the free particle for large times $t \to \infty$ and outgoing particles as the final channel space and the same for times $t \to -\infty$ and incoming particles as the initial channel space. These are typical surfaces of section that on their own form a phase space each. The choice of incoming and outgoing particles respectively chooses between the two Riemann like surfaces we have in phase space [12].

We shall call the outgoing and incoming spaces respectively M_o and M_i and they are characterized by coordinates

$$(E = \text{const.}, t = \pm\infty, P_\phi, \phi) \to (P_\phi, \phi). \tag{2}$$

These spaces are thus effectively two dimensional as E and t are fixed, and we shall consider them as such in what follows.

Scattering by a Hamiltonian of the above-mentioned type can be represented as a canonical transformation

$$C_S : M_i \to M_o, \tag{3}$$

which we will call a scattering map. This scattering map is defined for each energy, but we shall, as mentioned above, consider a fixed energy. By canonical transformation we here simply mean any map between phase spaces that leave the Poisson brackets, i.e. the symplectic structure invariant. Note that the scattering map is a one to one, map, up to subsets of measure zero, that are irrelevant to our argument.

As we are mainly interested in time-reversal invariant scattering processes we have to consider the time-reversal operation T. This operation is non-canonical in that it changes P_ϕ to $-P_\phi$ while simultaneously changing an outgoing into an incoming space or vice versa. Time-reversal invariance implies that $T C_S T C_S = 1$ is the identical transformation on M_i.

The iterated scattering map was introduced by Jung [5] in order to be able to use, for iterated maps, the surface of section implicit in the definition of channel space. In our language we shall define the iterative map $I_S = T C_F T C_S$, where C_F is the map induced by the free particle dynamics. This map can readily be iterated as it maps M_i onto itself, and its iteration is indeed identical to the iterated scattering map defined by Jung [5].

In general the phase space of channels is more complicated and of higher dimension, but the basic concepts "scattering map" and "iterated scattering map" remain unchanged; only the simple visualization of the iterated scattering map as a surface of section is lost due to the higher dimensionality.

Using the iterated scattering map we can next determine the structure of the phase space M_i under this map. In general this map may be chaotic, ordered or mixed in the sense that it has invariant tori and chaotic regions. Both may well be of non-zero measure. As we have specified compact support for the scattering potential as well as energy conserving dynamics, only a finite part of the phase space M_i will be non-trivially affected by the map. The size of this part will depend on energy, but this does not concern us as we work at fixed energy. On the remaining part of M_i the iterated map will act as an identity, as both C_S and C_F will correspond to free particle motion. We shall from here on omit this trivial part of phase space and take M_i (as well as M_o) to denote the compact part of phase space on which the maps act non-trivially.

We now consider the group G of all bijective canonical transformations of M_i onto itself. Surprisingly little is known [13] about this group despite of its fundamental role in classical mechanics. This has forced us to introduce some new concepts originally proposed in the context of the unitary time evolution operator [9] and which will prove equally useful for the problem on hand.

As mentioned above the iterated scattering map defines both chaotic and ordered areas on M_i and thus a certain structure on this phase space. We now define a subgroup of G that leaves this structure invariant. This subgroup is named the *structural pre-invariance group* of the map I_S. Clearly the iterated scattering map itself forms part of the so defined subgroup. Note that for a scattering process that is completely chaotic on M_i this subgroup becomes identical with the full group G of all bijective canonical transformations on M_i.

The structure induced by the map on phase space does not determine the structural properties of a scattering map completely; other properties such as the symmetries of the map induced by the symmetries of the Hamiltonian play an important role. We shall call the group of bijective canonical transformations that leave all structural properties of the map invariant the *structural invariance group* G_S of this map. Clearly this is a subgroup of the structural pre-invariance group defined in the previous paragraph. Unfortunately in contrast to the latter the invariance group is not well defined by what we have said up to now. The problem is to identify the relevant structures of the problem. The symmetries will be of foremost importance and we shall require that all transformations commute with the operations of the symmetry group of the problem. For the most common continuous symmetries, such as rotational invariance for example, this will not lead us beyond the pre-invariance group as conserved quantities and thus invariant tori are associated to such symmetries. Discrete symmetries on the other hand have no such consequences, but are still structurally very relevant. In classical mechanics they play a subordinate role, but we do known how important parity, for example, is in quantum mechanics. Other symmetries (e.g. time-reversal invariance or point symmetries) are equally relevant and some may not always be easy to identify.

In the case of time evolution over a period [9] the question of whether we are dealing with a primitive period or whether shorter periods are relevant, was of great importance. The equivalent question in the scattering problem refers to the occurrence of multiple scattering processes, i.e. to the question of whether a sequential description of the scattering process is possible. Note that this may happen both as true multiple scattering events or as some coupled multistep phenomenon. Such situations may well be relevant in practical cases in nuclear or molecular physics, such as, for example, nuclear precompound reactions.

Structural invariance of a scattering problem as introduced in this section is thus a concept that depends on our knowledge of the physical system and its essential properties. The most obvious ones refer to the phase space structure of the scattering map and the symmetries and approximate symmetries of the system. Yet dynamical properties are also relevant and it is not yet quite clear whether they can always be reduced to dynamical symmetries.

Note that here as in [9] for the time evolution only the structural pre-invariance is defined in terms of the canonical map itself. The full structural invariance requires knowledge of the dynamics of the system. As mentioned earlier very little is known about the group of bijective canonical transformations, and if some day the concept of structural invariance can be defined cleanly for maps it may be of great help to understand the structure of this group.

The elements C^α of G_S are canonical transformations acting on M_i. The corresponding transformation on M_o is $TC^\alpha T$, and we can define a family of scattering maps

$$C_{S^\alpha} = TC^{\alpha-1}TC_S C^\alpha, \tag{4}$$

with a corresponding family of iterative scattering maps I_{S^α}. These two families of scattering maps have by construction a remarkable invariance property. If we perform an arbitrary transformation from G_S on M_i and simultaneously the time-reversed transformation on M_o the family of transformations C_{S^α} remains invariant, and a similar argument holds for the iterative maps.

We have thus used the structural invariance group to construct a family of scattering systems with the same structural invariance properties. For what follows we would like to argue that all systems included in this family that have a smaller structural invariance group than the one we are interested in are irrelevant so that the family represents the case of interest in some sense. The obvious answer would be to define an ensemble and show that the undesirable cases form a subset of measure zero. Unfortunately among the many things unknown about bijective canonical transformations is an invariant measure even when they act on a compact manyfold as is the case here.

3 Structural invariance and ensembles of S-matrices

We now proceed to the quantum mechanical problem. To do this we follow the ideas of Dirac, Moshinsky and others [7,8] that allow us to associate unitary operators to canonical transformations. In general these unitary operators do not represent the quantum problem on hand "precisely" [13], as ordering problems will occur. In other words the representation will not be unique. On the other hand any unitary representation we may choose will, by construction, give the correct semi-classical limit [8]. We can therefore choose any representation in channel space and associate a canonical transformation to this unitary operator. We furthermore have to recall that the representation of conjugation by the time-reversal operator is simply the anti-unitary operation of canonical conjugation.

If the S-matrix S is the unitary representation of the classical scattering map C_S then $TC_STC_S = 1$ implies $S^*S = 1$ or, due to the unitarity of S, $S = S^t$. We are only interested in those parts of channel space, whose trajectories are affected by the non-free part of the Hamiltonian, which was chosen to have compact support. This part of channel space is, in turn, compact for fixed energy and for finite Planck constant will lead in the semi-classical limit to a finite dimensional S-matrix. If we denote the representation of C_{S^α} by S^α, we have a family of symmetric S-matrices that, at least in the semi-classical limit, represents the family of canonical scattering maps, which we constructed above. Thus we can choose

$$S^\alpha = U^{\alpha+}SU^\alpha. \tag{5}$$

Here S^α is the unitary representation of C^α, and the transpose arises from the inverse and time reversal.

For simplicity let us first assume that our scattering problem is completely chaotic and has no symmetries except time-reversal invariance and no other structural properties that should be taken into account. Then we have a structural invariance group $G_S = G$ that is equal to the group of all bijective canonical transformations on M_i. As mentioned above this group is still very little understood and no invariant measure is known. In consequence we cannot construct an ensemble from the family of scattering maps, which we defined in eq. (4).

On the other hand the corresponding S-matrices S^α can readily and naturally be endowed with a measure. This invariance of the family of scattering maps discussed above translates into an invariance of the family of S-matrices, such that if S^α is a member of the family so is $U^tS^\alpha U$ where U is any unitary transformation on channel space that in turn represents any bijective classical canonical transformation on M_i. For the measure $d\mu(S)$ on the space of unitary and symmetric matrices this invariance implies

$$d\mu(S) = d\mu(U^+SU) \tag{6}$$

which is exactly the relation that defines the invariant measure on this space whose uniqueness was shown by Dyson [14].

If we adopt this measure, our family of S-matrices forms an ensemble. This ensemble is indeed one of the classical circular ensembles, also known as classical domains [10], namely the invariant ensemble of unitary symmetric matrices (COE).

Thus the family of scattering maps developed from classical notions leads naturally to the COE, but we now have to establish under what circumstances we can expect the individual S-matrix on hand to be typical in the sense that we can use ergodicity. Ergodicity is here invoked not in the usual sense where time averages are replaced by ensemble averages. Rather we replace averages over eigenvalues by ensemble averages, and in this sense we proceed in a similar fashion as for the energy averages commonly considered in random matrix theory. Ergodic behaviour has been proven for the Gaussian orthogonal ensemble [11, 4], and this proof can be straightforwardly extended to the closely related COE. Thus a typical representative of the ensemble will have the same spectral fluctuations as the ensemble as the dimension of the matrix goes to infinity, as it does in the semi-classical limit.

Clearly there will be a subset of measure zero that does not fulfill the ergodic property. The first question is whether this subset is important. In fact this is the case, but we are able to understand the arising problem in terms of the structural invariance group introduced in the previous section. If, contrary to the assumption made above, our structural invariance group is smaller than the group of all canonical transformations on M_i the unitary representation of the elements of the structural invariance group will no longer cover the entire unitary group and thus we will not generate the COE, but other ensembles which we shall discuss later. Thus we expect to find with probability one either COE statistics for our individual S-matrix in the limit $n \to \infty$ or a failure in our assumption about the structural invariance group. What is missing at this stage is a clear recipe for finding the structural invariance group.

This leads us immediately to the next and well-known restriction. In quantum mechanics spaces that belong to irreducible representations of a symmetry group will induce independent invariant measures and thus lead to superpositions of COEs. This happens in close analogy to the bound case, and obliges us to consider minimal invariant subspaces of each chaotic region of channel space.

Even if we separate symmetry classes or eliminate symmetries to take this problem into account, this is not sufficient to ensure COE behaviour as other types of symmetries may also occur. Consider the map $I_{SS} = I_S I_S$ which is again a iterated scattering map. The corresponding S-matrix S^2 has eigenphases that are double the eigenphases of S according to the spectral theorem, and as they are confined to a circle, they will be stretched to cover this circle twice. If the eigenphases of S corresponded to a COE those of S^2 correspond to a superposition of two COEs. This point looks somewhat artificial for the scattering problem, but if we consider the monodromy operator of a periodic time dependent system as discussed in Ref. [9], this amounts to the requirement that T be a minimal period. We shall return to this point in the next section.

Even now we cannot be sure that we have identified all possible sources of non-generic behaviour, though we have eliminated the most obvious ones. We have thus to keep in mind that any property of the system that makes it possible to define separate Hilbert spaces corresponding to the same classically connected domain in phase space not connected by the quantum Hamiltonian will cause non-generic behaviour.

We have here focussed on a single chaotic domain; the generalisation to a map with several ordered and chaotic domains is quite straightforward. We follow Berry's suggestion that each chaotic domain will contribute a COE of a given density and the integrable parts will contribute a Poisson type spectrum [15]. As pointed out in Ref. [16] this approximation, will not hold over distances of the order of one level spacing or smaller because of tunneling; a short range level repulsion must always be expected. Also for the integrable parts we have to be careful of the oscillator anomaly. In this case it obviously includes the unscattered part of the incoming beam, as a plane wave is the continuum limit of the oscillator (a Poisson spectrum is known to yield a dense pure point spectrum in the same limit). Yet this problem is not too serious as it is basically limited to forward scattering plus diffraction and can be eliminated by standard methods from any experiment.

4 Some complementary considerations

We shall start this section by stating that it is not essential to the general understanding of the chapter, but that we feel there are a number of considerations that might be useful to the specialist reader. We shall address a few such points:

In section II we constructed a family of non-linear bijective canonical transformations, and stated that these belonged to a group so complicated that no invariant measure or even the existence of such is known. Then we proceeded in section 3 to consider their unitary representation and obtained a family uniquely endowed with an invariant measure. We know that in the latter family the ergodic theorem holds and that exceptional cases therefore are of measure zero in the corresponding ensemble. The reader more familiar with classical transformations may feel that this is a rash conclusion because the KAM theorem tells us that even finite perturbations conserve some KAM tori. This would seem to indicate that around each integrable "exceptional" map there are other non-generic ones such as to yield a measure different from zero in case such a measure exists. Actually this is not true, because the KAM theorem needs certain smoothness constraints on the perturbation, that are not implied when we talk about all bijective canonical transformations as pointed out in Ref. [9]. Note on the other hand that both the objection and the remark are not relevant in a practical sense, because problems are not selected at random, and we must explore their KAM structure in order to establish even the pre-invariance group. In this context it is very important to notice that "Cantori", i.e. fractal remainders of bearly broken tori, may play an important role [17].

We have restricted our considerations to the time-reversal invariant case which is, in practice, the relevant one for most scattering experiments in nuclear and molecular physics. Let us just recall that in the absence of time-reversal invariance and any other anti-unitary symmetry [18], we lose the correlation between incoming and outgoing channel space and the canonical unitary transformations on either become independent. The structural invariance and pre-invariance groups then become direct products of two groups of bijective canonical transformations, and the translation to the unitary representation leads directly to the Haar measure of the unitary group.

The most relevant remark refers to quantum effects that have no classical counterpart. For scattering theory no doubt the most important one is tunneling. Indeed we known that in nuclear and atomic experiments particularly when the compound part is well defined, i.e. when we expect a nice and pure COE, the transmission coefficient is different from one, i.e. quite far from its classical value. This casts some doubt on the practical usefulness of the entire argument. To develop this point let us look at the theory of ensembles of S-matrices as developed by Mello and myself; in this theory the ensemble is formed directly for S-matrices and not for some underlying quantities such as width and levels or Hamiltonians. Indeed we start with the COE for the S-matrix and find that we obtain an ensemble in very good agreement with experiment by choosing the minimum entropy measure with respect to the COE while fixing the transmission coefficients in the absence of direct reactions (and indeed all so-called optical parameters if we wish to allow for the latter). Thus we obtain the correct ensemble by introducing the tunneling parameters with minimum information deviation from the semi-classical result.

The localization effect well-known for the kicked rotator is another problem not easily accounted for in this context; it may well be of marginal relevance for the scattering problem, but it certainly causes a problem on grounds of principle. We may think of it as a purely quantum mechanical problem and then seek to tackle it using information theory in a fashion similar to the barrier transmission problem. If the effect has no classical background such a phenomenological approach will indeed be recommendable. Yet we look forward to the results of Smilansky and his group, in their quest for an explanation based on periodic orbit expansions. In the latter case we would expect this explanation to display some non-generic classical property that would reduce the structural invariance group.

5 Conclusions and outlook

The concept of structural invariance has been introduced to give some structure to the set of canonical maps such as the scattering or monodromy maps. From a mathematical point of view this may be a first step toward an understanding of the group of all (non-linear) bijective canonical transformations. But the relevant point in our context is the strong structure this induces in quantum mechanics via the theory of quantum representations of classical canonical transformations. Indeed we have been led directly to the classical domains of Cartan [10], that constitute the circular ensembles of random matrices. The big question that remains partially open is—as usual in this field—the identification of the non-generic cases.

Thus we have a clear indication that not only the two-point function but actually the entire circular ensemble is the correct statistical description of the scattering map. This implies that we have information on the fluctuations of the cross sections, which is the original purpose of statistical scattering theory. If we look for further implications of this result we should address the two basic dichotomies that prevail in the field.

In classical scattering theory it is well known that a chaotic classical scattering map does not necessarily imply a hyperbolic system. The results of Smilansky and coworkers on the two-point function of the eigenphase distribution indicate that we do not need hyperbolic properties to have the generic case that leads to a COE [3]; this is not clear from our argument though there is certainly no contradiction.

In quantum mechanics we have on one hand dynamical theories such as the one of Weidenmüller and his coworkers [19] that start from a GOE Hamiltonian which leads to an ensemble of energy dependent S-matrices that allows to calculate all measureable quantities. On the other hand we have the more phenomenological theory [20] that yields an ensemble of energy independant S-matrices and allows us to calculate cross section fluctuations and their correlations as long as all quantities are taken at a given energy.

Our argument yields the ensemble of S-matrices directly, and as mentioned above it seems to hold even for non-hyperbolic cases. The uniqueness theorem for the COE based on the unitary covariance that holds in the semi-classical limit (total absorption) tells us that the S-matrix ensemble at fixed energy must be the same in both cases. Thus we expect possible differences in the correlation function for different energies. It will therefore be of great interest to compare these correlations for hyperbolic and non-hyperbolic systems with chaotic scattering map.

In conclusions we may say that we have found the general connection between classical chaotic scattering and the circular ensembles of Cartan or in other words the connection between chaos and random matrix theory in the scattering case. The basic question that remains open relates to the fact that an invariant measure for the group of bijective canonical transformations is not known.

Acknowledgement

I am grateful F. Leyvraz, U. Smilansky and H. Weidenmüller and many useful discussions and to CONACYT for financial support.

REFERENCES

[1]   E.g. ed., T.H. Seligman and H. Nishioka, "Quantum Chaos and Statistical Nuclear Physics", Proceedings, Cuernavaca 1986, Springer, Heidelberg, 1986.

[2]   U. Smilansky, "The Classical and Quantum Theory of Chaotic Scattering", in Proceedings Les Houches 1989 (to be published Elsevier 1991).

[3]   R. Blümel and U. Smilansky, Phys. Rev. Lett. **60** (1988) 477.

[4]   T.A. Brody et al., Rev. Mod. Phys. **53** (1981) 385.

[5]   C. Jung, J. Phys. A **19** (1986) 1345

[6]   T.H. Seligman, Intl. Conf. Group Theory and its Applications Goslar 1991, World Scientific, Singapore, 1992.

[7]   P.A.M. Dirac, "The Principles of Quantum Mechanics", Oxford University Press, London/New York, 1947.

[8]   M. Moshinsky and C. Quesne, "Oscillator Systems" in Proc. XV Solvay Conf., Brussels 1971; P.A. Mello and M. Moshinsky, J. Math. Phys. **16** (1975) 2017; J. Deenen, M. Moshinsky and T.H. Seligman, Ann. Phys. (N.Y.) **127** (1980) 458.

[9]   F. Leyvraz and T.H. Seligman, Phys. Lett. A **168** (1992) 318.

[10] E. Cartan, Abh. Math. Sem. Univ. Hamburg **11** (1935) 116; L.K. Hua, Harmonic Analysis of Functions of Several Complex Variables in the Classical Domains" AMS, Providence 1963; Original: Science Press, Peking, 1958.

[11] A. Pandey, Ann. Phys. (N.Y.) **119** (1979) 170.

[12] M. Moshinsky and T.H. Seligman, J. Math. Phys. **22** (1981) 1338.

[13] F. Leyvraz and T.H. Seligman, J. Math. Phys. **30** (1989) 2512; *id* in Proceedings Cocoyoc 1991; AIP conference proceedings, 1992.

[14] F. Dyson, J. Math. Phys. **3** (1962) 1191

[15] M. Berry and M. Robnik, J. Phys. A **17** (1984) 2413.

[16] T.H. Seligman and J.J.V. Verbaarschot, J. Phys. A **18** (1985) 2227.

[17] L. Reichl in: ed. T.H. Seligman and H. Nishioka, "Quantum Chaos and Statistical Nuclear Physics", Proceedings, Cuernavaca 1986, Springer, Heidelberg, 1986, p. 251.

[18] M. Robnik in: ed. T.H. Seligman and H. Nishioka, "Quantum Chaos and Statistical Nuclear Physics", Proceedings Cuernavaca 1986, Springer, Heidelberg, 1986, p. 120.

[19] J.J.V. Verbaarshot, H. Weidenmüller and M. Zirnbauer, Phys. Rep. **161** (1985) 367.

[20] P. Mello, P. Pereyra and T.H. Seligman, Ann. Phys. (N.Y.) **161** (1985) 254.

Spectral properties of a Fermi Accelerating Disk

R. BADRINARAYANAN AND J. V. JOSÉ

Physics Department, Northeastern University,
Boston, MA 02115, U.S.A.

Abstract

We study the quantum mechanics of a two-dimensional version of the Fermi accelerator. The model consists of a free particle constrained to move inside a disk with radius varying periodically in time. A complete quantum mechanical solution of the problem is possible for a specific choice of the time-periodic oscillation radius. The quasi-energy spectral (QES) properties of the model are obtained from direct evaluation of finite-dimensional approximations to the time evolution operator. As the effective $\hbar$ is changed from large to small the statistics of the QES eigenvalues change from Poisson to circular orthogonal ensemble (COE). Different statistical tests are used to characterize this transition. The transition of the QES eigenfunctions is studied using the χ^2 test with v degrees of freedom. The Porter–Thomas distribution is shown to apply in the *COE* regime, while the Poisson regime does not fit the χ^2 with $v = 0$. We find that the Poisson regime is associated with exponentially localized QES eigenfunctions whereas they are extended in the COE regime.

1 Introduction

Although a complete understanding of the quantum manifestations of classical chaos (QMCC) is not yet in sight, significant progress has been made in obtaining partial answers to this question. This progress has primarily been achieved from studies of lower-dimensional models: two-dimensional for energy conserving models and one-dimensional when energy is not conserved. In particular, very few studies have been carried out in two-dimensional time-dependent problems, for even in the classical limit the theoretical analysis is nontrivial. It is the purpose of this chapter to consider a time-periodic two-dimensional problem which, because of the particular nature of the model, has a reduced degree of complexity and thus allows for a detailed analysis of its solutions.

The model considered here is a two-dimensional extension of the thoroughly studied Fermi accelerator [1]. The one-dimensional quantum problem was introduced by José and Cordery [2], who showed that, for a very specific form of the wall-oscillation function, the one-cycle evolution operator can be explicitly written and analyzed, without the complications of numerical time-ordering (which effectively precludes any thorough investigation). We present here a parallel quantum analysis of a two-dimensional oscillating disk. The model consists of a free particle bouncing elastically inside a disk with a radius that oscillates periodically in time. The periodicity in the Fermi accelerating disk (FAD) allows for the quantum mechanical use of Floquet's theory in terms of the one-period time evolution operator.

This model is of interest for several reasons: first, because the dynamics is not kicked, as in almost all other time-periodic models considered, *e.g.* the thoroughly investigated periodically kicked rigid rotator model (PKR) [3] and second, because the explicit form of the tight-binding-like model obtained for the dynamical equations decays algebraically instead of exponentially, as in the PKR. The FAD model then allows for the comparison of our results with those of the kicked models, for it is important to know if the corresponding results are generic. Thirdly, the possibility exists here that the spectrum of the evolution operator contains a continuous component [4] implying nonrecurrent behavior of the wavefunctions. Evidence suggestive of such a continuous spectrum has been reported [5]. This model is also shown to be equivalent to a particle subjected to a bounded, rotationally symmetric, inverted harmonic oscillator potential in the presence of time-periodic kicks. This last relationship allows for a connection to experiments in mesoscopic quantum dots [6].

One of the major advances in the understanding of the QMCC comes from the clear differences found between the eigenvalue spectral properties of models that classically exhibit chaos from those that do not [7]. The tools used to measure these differences are extensions of those developed in random matrix theory (RMT) [8]. An explicit connection between classical phase space solutions and their corresponding quantum counterparts provides further QMCC in terms of special solutions like scarred wave functions. In this chapter we concentrate on the analysis of the spectral properties of the time evolution operator and leave the phase-space correspondence to be reported elsewhere [9].

The outline of the chapter is as follows: In section 2 we define the *FAD*, both classically and quantum mechanically. In section 2.2 the quantum problem is shown to be almost completely integrable for a specific functional form of the oscillating radius. Section 3 presents the eigenvalue statistical properties in terms of different measures that include the nearest-neighbor probability distribution and its integrated form, the Δ_3 statistic and the two-point function Σ_2. In fitting the data from the Poisson to the COE regime we use the Brody function as well as the Izrailev tests to try and quantitatively parametrize the transition region. In section 3 we discuss the properties of the eigenfunctions in different regimes using the χ^2 distribution of v freedoms as a convenient parametrization of the results. Also in section 3 we discuss the localization properties of the quasi-energy eigenfunctions in terms of the properties of the time-averaged transition probabilities. Finally, in section 4 we present both our conclusions and also questions left for future studies.

2 The FAD model

The FAD consists of a free particle confined to move inside a two-dimensional disk whose radius oscillates periodically in time as $R(t) = R_0\theta(t)$, $\theta(t+T) = \theta(t)$, with period T. The walls are taken to represent infinite potential barriers for the motion of the particle, that is, the particle undergoes perfectly elastic collisions, while quantum-mechanically the particle's wavefunction vanishes identically at the boundary.

2.1 Classical FAD

The classical Hamiltonian is conveniently written in polar coordinates as

$$H = \frac{1}{2m}(p_r^2 + p_\phi^2),$$

where r is the radius, ϕ the azimuthal angle, and m the mass of the particle, to be taken as one. The resulting equations of motion are,

$$\frac{d^2r}{dt^2} = \frac{L^2}{m^2r^3}, \qquad 0 \le r(t) \le R_0\,\theta(t), \tag{1}$$

$$\frac{d\phi}{dt} = \frac{L}{mr^2} \qquad 0 \le \phi(t) \le 2\pi, \tag{2}$$

where L is the magnitude of the *conserved* orbital angular momentum. In principle, we can study the dynamics of the problem for any form of the wall oscillation function $\theta(t)$. However, as we shall show in the next subsection, the quantum problem dictates the appropriate $\theta(t)$ that allows an explicit evaluation of the evolution operator. The specific form of $\theta(t)$ is found to be

$$\theta(t) = \sqrt{1 + 2\delta\left|\mathrm{mod}(t, \frac{T}{2})\right|}, \tag{3}$$

where δ is the amplitude of the wall oscillation, and the period T will be taken to be unity from now on. Note that $\theta(t)$ is already discontinuous in its first derivative, which is crucial to having chaotic solutions.

As mentioned above, the disk representation is not the most appropriate one to discuss the classical–quantum correspondence of the solutions. For this reason we concentrate on the quantum solution in this chapter. However, to show that the model does indeed show a transition to chaos we have solved the classical equations of motion and obtained Poincaré's surfaces of section for different parameter values. Note that since ϕ is related to $r(t)$ by Eq. (2), if $r(t)$ shows irregular behavior, so will $\phi(t)$. The phase space plots in Fig. 1(a-b) show p_r vs r for two values of L and δ.

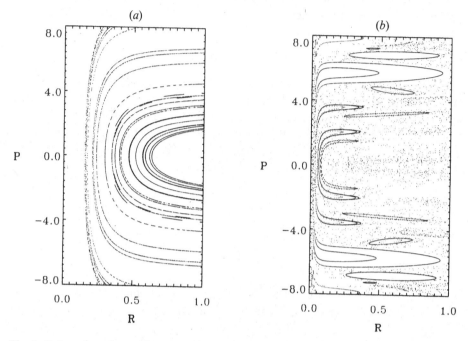

Fig. 1. Poincaré surfaces of section for the (r, p_r) plane. (a) Almost regular behavior for small wall amplitude ($\delta = 0.001$, $L = 1$). (b) Intermediate case where regular and chaotic motions coexist ($\delta = 1.0$, $L = 10$).

2.2 Quantum FAD

The quantum mechanical version of the FAD is defined in terms of the Schrödinger equation in polar coordinates

$$i\hbar \frac{\partial \Psi}{\partial t} = -\frac{\hbar^2}{2m} \left\{ \frac{\partial^2}{\partial r^2} + \frac{1}{r}\frac{\partial}{\partial r} + \frac{1}{r^2}\frac{\partial^2}{\partial \phi^2} \right\} \Psi(r, \phi, t), \tag{4}$$

with the boundary condition $\Psi\left(r(t) = R(t), \phi, t \right) = 0$, and normalization

$$\int_0^{R(t)} r\, dr \int_0^{2\pi} d\phi\, |\Psi(r, \phi, t)|^2 = 1. \tag{5}$$

These equations completely specify the quantum FAD model considered in this chapter. To find the general solution to the quantum FAD we expand the wave-function $\Psi(r, \phi, t)$ in terms of the natural complete set of basis functions ψ_{nl},

$$\Psi(r, \phi, t) = \sum_{l=-\infty}^{+\infty} \sum_{n=1}^{\infty} C_{nl}(t) \, \psi_{nl}(r, \phi, t), \tag{6}$$

where the ψ_{nl}s are the *instantaneous* orthonormalized eigenfunctions for a given $R(t)$, *i.e.* the solutions to

$$-\frac{\hbar^2}{2m} \nabla^2 \psi_{nl}(r, \phi, t) = E_{nl}(t) \, \psi_{nl}(r, \phi, t). \tag{7}$$

Here t is a parameter and ψ_{nl} satisfies the same boundary and normalization conditions as Ψ. These instantaneous solutions are given by

$$\psi_{nl}(r, \phi, t) = \frac{1}{\sqrt{\pi} \, R J_{l+1}(\beta_{nl})} J_l(\beta_{nl} \frac{r}{R}) \exp(i \, l \, \phi), \tag{8}$$

where J_l is the regular Bessel function of order l and β_{nl} is its nth zero. The corresponding instantaneous energies are

$$E_{nl}(t) = \frac{\hbar^2}{2m} \left(\frac{\beta_{nl}}{R_0 \, \theta(t)} \right)^2. \tag{9}$$

We now substitute the expansion (6) into the Schrödinger equation (4). After using orthogonality and known identities involving Bessel functions, one can evaluate the integral

$$\begin{aligned} K_l(m, n) &= \int_0^1 x^2 \, J_l(\beta_{ml} x) \frac{d}{dx} J_l(\beta_{nl} x) \, dx \\ &= \frac{\beta_{ml} \beta_{nl}}{\beta_{nl}^2 - \beta_{ml}^2} J_{l-1}(\beta_{ml}) J_{l-1}(\beta_{nl}) \qquad (m \neq n) \end{aligned}$$

so that the final result can be cast in the form

$$\begin{aligned} i \hbar \dot{C}_{nl} &= \sum_{m,k} \left[E_{nl} \delta_{k,l} \delta_{n,m} - i \hbar \frac{\dot{\theta}}{\theta} \frac{2 \beta_{mk} \beta_{nk}}{\beta_{mk}^2 - \beta_{nk}^2} \delta_{k,l}(1 - \delta_{m,n}) \right] C_{mk} \\ &\equiv \sum_{m,k} H_{nm;kl} \, C_{mk}. \end{aligned} \tag{10}$$

Equations (10) are exactly and completely equivalent to Eq. (3). The time development of the system is obtained from

$$C_{nl}(t) = \sum_{m,k} U_{nm;kl}(t, t') \, C_{mk}(t') \tag{11}$$

with the time evolution operator defined by

$$U_{nm;kl}(t, t') = \mathbf{T} \exp\left(-\frac{i}{\hbar} \int_{t'}^{t} H_{nm;kl}(s) \, ds \right), \tag{12}$$

where $\mathbf{T}$ symbolically represents the time-ordering operation. We can, however, get rid of the time-ordering operation if the effective Hamiltonian H factors into an overall time-dependent factor multiplying a time-independent part. For this to happen, we see from Eqs. (10) and (9) that we must have

$$\frac{\dot{\theta}}{\theta} \sim \frac{1}{\theta^2}. \tag{13}$$

An appropriate form for $\theta(t)$, satisfying the periodicity and time-reversal invariance conditions is the one given in (3). Using this form for $\theta(t)$ the matrix elements of the one-period time evolution operator are given by

$$U_{nl;mk}(t+T,t) = \exp\left(-\frac{i}{\hbar} H_{nl;mk} \int_t^{t+T} \frac{ds}{\theta^2(s)}\right). \tag{14}$$

Note that the time evolution matrix is nominally four-dimensional but because of the conservation of angular momentum it reduces to a two-dimensional matrix, for a given value of l.

The required zeros of the Bessel functions are calculated using the Newton–Raphson technique after bracketing the root using the fact that separation between successive zeros asymptotically approaches π [10]. This method is very efficient, since finding the roots even for the high zeros and large l values can be done to machine accuracy. The quasi-energy spectrum is determined once the zeros of the Bessel functions are found and the U-matrix is diagonalized numerically (with unitarity preserved to machine precision). Since U is unitary, its eigenvalues are unimodular, and we write

$$U(t+T,t)\,|\lambda_n\rangle = \exp(i\lambda_n T)\,|\lambda_n\rangle ; \qquad 0 \le \lambda_n \le 2\pi. \tag{15}$$

Complete knowledge of U implies *full* knowledge of the evolution of the system at all integer multiples of the period because $U(t+NT,t) = U^N(t+T,t)$. This product of Us is the quantum counterpart of the classical one-period maps. The λ_n, being invariant under unitary transformations, are representation-independent and the quasi-energy spectrum obtained this way is of high quality from a numerical point of view. For our calculations, we used a truncated basis of 400 states for calculating properties of the QES eigenfunctions, while we used a basis of 500 states to calculate the QES eigenvalues. We checked that the QES was stationary, *i.e.* the eigenvalues remain basically unchanged to within 1% in this case as we increase the size of the U-matrix.

As mentioned above, this representation of the FAD has its drawbacks when one is interested in the correspondence between classical and quantum problems. One of the reasons for the inadequacy of this representation is that to make this correspondence one uses a Husimi representation of the quasi-energy eigenfunctions. Now, for an angular momentum quantum number of $l=10$, for the typical parameters we chose, a Bessel function oscillates $\sim 10^3$ times for a disk of unit radius, and consequently the integrand oscillates too rapidly for reliable numerical evaluation. These difficulties are surmounted by performing a generalized nonlinear canonical transformation in both space and time coordinates. Detailed results of taking this approach are reported in Ref. 9.

3 Statistics of the quasi-energy eigenvalues of the FAD

As mentioned before, it has been found that one of the clear QMCC emerges when one compares the spectral properties of specific model systems as appropriate parameters are tuned to produce classically a transition from integrable to completely chaotic regimes. In this section we follow the general thinking developed in RMT to implement different tests to quantify the spectral properties of the FAD model. These properties are obtained from a direct diagonalization of the one-period time evolution matrix. The relevant tunable parameters in the problem are δ, related to the amplitude of the radial oscillation, $\delta/\hbar$ and L. For the results presented here we varied the value of L and $\delta/\hbar$, while we took the wall amplitude $\delta = 1$. This value corresponds to a nonperturbative value, where the kinetic energy, the centrifugal barrier and the energy imparted by the wall all have about equal magnitudes. Lastly, we rescale all lengths by δ and all energies by $\hbar$. The semiclassical limit of interest here is obtained when $l \uparrow \infty$, $\hbar \downarrow 0$, with

$$L = \hbar\sqrt{l^2 - \tfrac{1}{4}} = const.$$

Next we discuss the RMT tests and their application to the results obtained for the FAD.

3.1 Nearest-neighbor QES eigenvalue distributions.

A local measure often used in RMT is the distribution of nearest-neighbor energy level separations, $P(s)$, where $s = \lambda_{n+1} - \lambda_n$. In the extreme integrable and chaotic regimes it has been established that the $P(s)$ takes the Poisson or Wigner distribution forms,

$$\begin{aligned} P_P(s) &= \exp(-s) \\ P_W(s) &= \frac{\pi}{2} s \exp\left(-\frac{\pi}{4}s^2\right), \end{aligned} \tag{16}$$

respectively. A convenient and often successful parametrization of the $P(s)$ obtained in the transition between P_P to P_W is provided by the Brody interpolation formula [11]:

$$P_\alpha(s) = \gamma(\alpha + 1) s^\alpha \exp(-\gamma s^{\alpha+1}), \tag{17}$$

where $\gamma = \{\Gamma[(\alpha + 2)/(\alpha + 1)]\}^{\alpha+1}$ and $\Gamma(x)$ is the Gamma function. This distribution is normalized and, by construction, has mean spacing $\langle s \rangle = 1$. We recover the Poisson case taking $\alpha = 0$ and Wigner for $\alpha = 1$. A criticism of the Brody distribution is, however, that there is no first principles justification for its validity. The fact remains that it does fit the specific results found when considering explicit model systems.

Instead of first looking at the $P(s)$ we found that the fits are better if we start by calculating the integrated distribution

$$\Pi_\alpha(s) = \int_0^s ds' P_\alpha(s'), \tag{18}$$

where we used all the spacings calculated. The results for $\Pi_\alpha(s)$ are shown in Fig. 2. Next we use the Brody form to find the exponent α to fit the data to $\Pi_\alpha = 1 - \exp(-\gamma s^{\alpha+1})$. The α exponent is determined from a linear fit to $\ell n \ell n (1 - \Pi_\alpha)^{-1} = \ell n(\gamma) + (1 + \alpha)\ell n(s)$. The fits shown in Fig. 2 are obtained by using these values of α. These results clearly show the general trend. As the nonintegrability parameter $\delta/\hbar$ is increased, the $\Pi_\alpha(s)$ goes from being Poisson-like to Wigner-like. For small values of s (≤ 0.1), the Poisson limit is well fitted by the Brody function while for the COE the fit is not as good, as one would expect due to the level repulsion. This is significant since there has been some controversy as to the specific power law dependence of the $P(s)$ distribution for small values of s. The corresponding results for $P(s)$ are shown in Fig. 3, with the Brody fits using the α values obtained from $\Pi_\alpha(s)$. We found that the fits are better by analyzing the data this way rather than first fitting $P(s)$ since in the binning process we lose information.

An alternative phenomenological interpolation formula with some justification was proposed by Izrailev [12] who used the analogy between the partition function of a two-dimensional Coulomb gas of charged particles on a circle (the Dyson gas) and the joint probability distribution of the quasi-energy eigenvalues of the COE to construct an approximate form for $P(s)$ given by

$$P_\beta(s) = A \left(\frac{\pi}{2}s\right)^\beta \exp\left[-\frac{\beta}{4}\left(\frac{\pi}{2}s\right)^2 - \left(\frac{2B}{\pi} - \frac{\beta}{2}\right)\frac{\pi}{2}s\right], \tag{19}$$

where $A(\beta)$ and $B(\beta)$ are constants fixed from normalization and the condition $\langle s \rangle = 1$. This formula also reduces to the appropriate Poisson and Wigner limits for $\beta = 0$ and 1, respectively. As we discuss below, this interpolation formula is also good for the FAD but the Brody distribution provides a slightly better fit to our results. The interesting aspect of Izrailev's distribution is that in principle it provides a *quantitative* connection between the energy-level repulsion parameter β and the degree of localization of the QES eigenfunctions. The connection is established by considering the ensemble-averaged "entropy localization length" l_S for the u_n components of QES eigenfunctions for a chosen basis and defined as $\langle l_S \rangle = N \exp\left(\langle \mathcal{S}_N \rangle - \mathcal{S}_N^{COE}\right)$, where the information entropy

$$\mathcal{S}_N(u_1, .., u_N) = -\sum_{n=1}^N w_n \ln w_n, \qquad w_n = |u_n|^2. \tag{20}$$

Note that $\mathcal{S}_N$ is essentially the logarithm of the number of sites significantly populated by the corresponding eigenstate, and $\mathcal{S}_N^{COE}$ is the entropy corresponding to the random (COE) limit, introduced so that for completely chaotic states

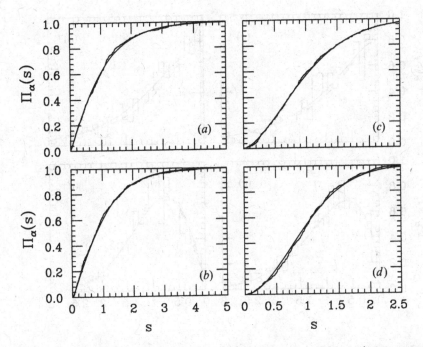

Fig. 2. (a)–(d) Transition from Poisson to Wigner for Π_α as the tuning parameter is changed. The solid lines in (a)–(d) correspond to a $\Pi_\alpha(s)$ fit with the α value obtained from the best straight line fit to $\ell n \ell n [1 - \Pi_\alpha(s)]^{-1}$. The Brody fit is essentially on top of the calculated data. The values of $(\hbar, L)$ were (a) (1.0, 10), (b) (0.5, 20), (c) (0.1, 100) and (d) (0.01, 1000) for $\delta = 1$. The repulsion parameters α for (a)–(d) are $0.10, 0.20, 0.60$ and 0.86, respectively.

the exact limiting value of $\langle l_S \rangle = N$. The conjecture is that the quantity $\langle l_S \rangle / N$, which varies between 0 for completely localized states to 1 for fully extended ones, is *identically* equal to the repulsion parameter β in (19).

To check the validity of this conjecture, we fitted the level-spacing distribution to $P_\beta(s)$ given in (3.4) and determined the value β_{hist} which minimized the χ^2. We also calculated the ensemble-averaged entropy localization length as defined above, and calculated the corresponding β_S, and compared the two. The results are presented in Fig. 4. We see that the agreement is good where the classical chaos is strong, but gets worse as the classical motion is more regular. Otherwise, the transition from COE to Poisson statistics is consistent with the general trends observed previously. We conclude therefore that while the Izrailev model is intuitively more appealing, the evidence to support it is not compelling enough (in our model at least) to prefer it over the Brody formula.

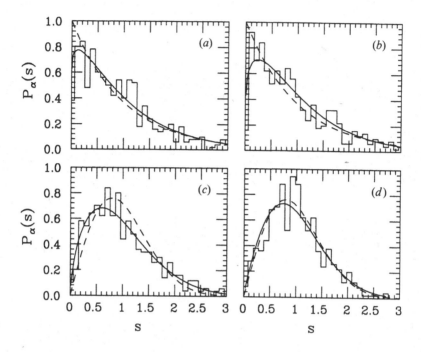

Fig. 3. (a)–(d) Nearest-neighbor spacing distribution $P(s)$ for the same parameters as in Fig. 2. The dashed lines in (a) and (b) corresponds to the Poisson limit, and to the COE case in (c) and (d). The solid lines in (a)–(d) correspond to a P_α fit with the α values obtained in the fits of fig. 2.

3.2 Δ_3 and Σ^2 correlations

We also calculated higher-order spectral correlations. The average number of levels in an interval of length L is $< n(L) >= (1/L)\sum_\alpha n(\alpha, L)$, where the $<>$ stands for spectral average, and $n(\alpha, L)$ is the number of levels in an interval of length L starting at α and ending at $\alpha + L$. More important are the various moments of the level distribution. The one considered here is the second moment of the average number of levels in a given stretch of length L of the spectrum, the $\Sigma^2(L)$ statistic [13]

$$< \Sigma^2(L) >= \left\langle (n(\alpha, L)- < n(\alpha, L) >)^2 \right\rangle. \tag{21}$$

Another often calculated statistic is the the Dyson–Mehta $\Delta_3(L)$ which measures the stiffness of the spectrum. This is defined by

$$\Delta_3(L,\alpha) = \frac{1}{L} min_{A,B} \int_\alpha^{\alpha+L} dx[\tilde{N}(x) - Ax - B]^2, \tag{22}$$

where $\tilde{N}(x)$ is the unfolded number density. In our case there is no need to unfold the spectrum. Δ_3 is just the least mean square deviation of $\tilde{N}(x)$ from

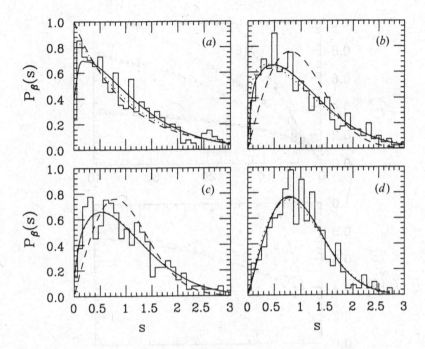

Fig. 4. (a)–(d) Fits of the data to the Izrailev formula $P_\beta(s)$ (solid line). The dots are the fit as given by β_S (see text for details). The best-fit and predicted values for (β_{hist}, β_S) are, (a) (0.20,0.05) (b) (0.22,0.12) (c) (0.48,0.50) and (d) (1.14,1.00).

the mean straight line behavior. This statistic is directly proportional to the $<\Sigma^2>$ by $\Delta_3(L) = (2/L^4) \int_0^L (L^3 - sL^2x + x^3)\Sigma^2(x)dx$, and thus can be calculated for the circular ensembles as well. The specific theoretical predictions for the averaged $<\Delta_3(L)> = (1/L)\sum_\alpha \Delta_3(L,\alpha)$, are $\Delta_3^{(COE)}(L) = (1/\pi^2)\ell nL - 0.007$, and $<\Delta_3^{(Poisson)}(L)> = (L/15)$. These results are correct in the asymptotic limit valid for $15 \leq L$.

In Fig. 5 we present our results for $<\Delta_3>$ and $<\Sigma^2>$. In these figures one clearly sees the transition from Poisson-like (dashes) to COE-like (solid line) behavior as $\delta/\hbar$ is varied. We note that the Δ_3 statistic does not saturate in the COE limit, even for the maximum interval L that we looked at, as would be expected from semiclassical arguments originally proposed by Berry. Furthermore, for the largest L considered the Poisson limit does not present the knee seen in other completely integrable systems. All in all the results shown in Fig. 5 are consistent with what we have come to expect for the transition region.

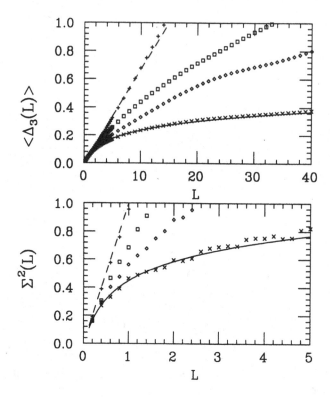

Fig. 5. (a) The $< \Delta_3(L) >$ and (b) the $\Sigma^2(L)$ statistics. The parameter values are the same as in Fig. 2 with (a) +, (b) □ , (c) ◇, and (d) for ×. The dashed line is the exact Poisson result while the continuous line is the corresponding COE RMT results. We observe a continuous transition from Poisson to COE. In the Poisson case the calculated values are slightly above the predicted behavior but it does not show the bending characteristic of picket fence spectra. In the COE regime no knee is found either.

3.3 Quasi-energy eigenfunctions

To study the behavior of the eigenfunctions we begin by discussing the statistics of the overlap of the eigenfunctions with the natural basis vectors. Several authors [14] have conjectured that as the classical motion changes from chaotic to regular, this distribution can be represented by a χ^2 distribution in v degrees of freedom, with v decreasing from 1 (the Porter–Thomas limit) to 0 (the Poisson limit):

$$P_v(y) = \frac{(v/2)^{v/2}}{\Gamma(v/2)} \ y^{v/2-1} \ \exp(-vy/2), \tag{23}$$

where $y \equiv |< \lambda|nl >|^2$, $|\lambda\rangle$ label the QES eigenfunctions and $|nl\rangle$ label a set of N orthogonal basis vectors. (The ys have been rescaled so that $\langle y \rangle = 1$.) We tested this hypothesis for the overlap strengths. The results, plotted on a

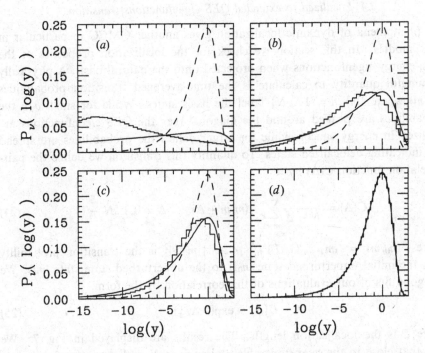

Fig. 6. (*a*)–(*d*) Distribution of amplitudes of QES eigenfunctions with the natural basis states for the same parameter values as in Fig. 2. Close to the COE limit (*d*) the amplitudes are nearly gaussian or Porter–Thomas randomly distributed. Away from this limit the distributions are not well fitted by the χ^2 distribution with a significant difference seen close to the Poisson limit. This discrepancy is explained in the text. The values of v from the fits are (*a*) 0.10, (*b*) 0.15, (*c*) 0.52, and (*d*) 1.

logarithmic scale and shown in Fig. 6, show the general trend of decreasing v as the corresponding classical system becomes more regular. However, we note that as we move from the COE to the Poisson limit the fits to the χ^2 get worse. Note especially the appearance of a sharp second peak well away from 0 as v decreases. This discrepancy is related to the fact that the results are basis-dependent when not in the COE limit. Equivalently, we do not expect to have a good fit to the χ^2 except perhaps if we take the special basis obtained from a semiclassical calculation. Even if we manage to get good agreement with the χ^2 for a properly chosen basis the result will not be generic and therefore the statistical analysis of amplitudes would lose its universal meaning. Universality does apply, however, in the COE limit.

3.4 Localized to extended QES eigenfunctions transition

The phenomena of dynamic localization gives another QMCC in particular in PKR models. In this section we check for the localization properties of the quasi-energy eigenfunctions when projected onto the natural basis. A physically meaningful quantity to calculate is the time-averaged transition probabilities $\overline{P}(n, m)$ where $\{|n\rangle, n = 1 \ldots N\}$ label the basis states. While for small $\delta/\hbar$ the probabilities are peaked around the diagonal (i.e. the QES eigenfunctions are localized in energy space), while for larger values the probabilities are spread out, indicating delocalized states. To quantify this transition we define the pair-correlation function $C(\Delta)$ by

$$C(\Delta) = \frac{1}{N - \Delta} \sum_{m=1}^{N-\Delta} \overline{P}(m, m + \Delta) \qquad \Delta = 0, 1 \ldots N - 1 \qquad (24)$$

where $\overline{P}(n_0, m) = \ell im_{T\to\infty}(1/T) \int_0^T |< n_0 \mid m >|^2$ is the transition probability from the initial unperturbed state $|n_0 l\rangle$ to the unperturbed state state $|ml\rangle$. We find good fits of our evaluations of this correlation by the form,

$$C(\Delta) \sim \exp(-\Delta/\xi), \qquad (25)$$

where ξ is the localization length. The results are displayed in Fig. 7. We note that close to the near-integrable regime $\xi \ll N$, indicating localization of the overlaps, while for larger $\delta/\hbar$, $\xi \geq N$. These results are consistent with the statistical analyses and with the expectation of a smooth transition from localized to extended solutions as we change from the Poisson to the COE regimes.

4 Discussion and outlook

We have carried out a detailed statistical analysis of the quasi-energy spectrum of a Fermi accelerating disk (FAD). We found clear evidence for the transition from Poissonian to COE-like behavior in the eigenvalue statistics when we changed the effective $\hbar$ from large to small. To make sure that these results relate to a transition from integrable to chaotic classical dynamics we need to compare the quantum to the classical answers explicitly. As mentioned above, this cannot be done efficiently in the present disk representation of the problem. This analysis is carried out in Ref. 9. The advantage of the present representation, over the one discussed there [9] is that the statistical analysis here can be carried out in detail. We have also connected the Poisson to COE transition to the corresponding localized to extended regime of the eigenfunctions. In the COE regime the Porter–Thomas distribution gives a good account of the statistical properties of the probability amplitudes.

We have also extended the techniques presented here to the corresponding spherical problem, both classically and quantum mechanically. So far we have found no surprises in the statistical analyses of the results. Of more interest then

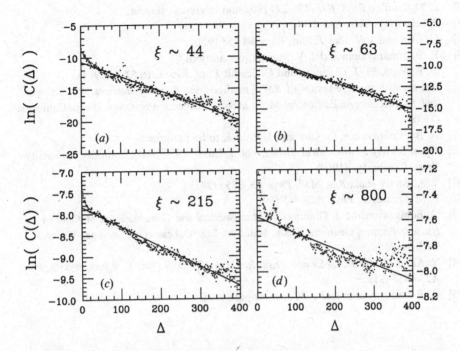

Fig. 7. (*a*)–(*d*) Logarithm of the pair-correlation function defined in (24) as a function of Δ for the same parameter values as in Fig. 2. The transition from localized to extended behavior, exemplified by the value of the localization length ξ as compared to the matrix sizes ($N = 400$ in this case).

is the study of the classical to quantum correspondence of the eigenfunctions, in their Husimi representation, which show scarred wavefunctions [15]. We leave the discussion of these results for a future publication.

Acknowledgements

We thank G. Farmelo for contributions in the initial stages of this research, and F. Leyvraz and T. Seligman for useful discussions on eigenvector statistics. This work has been supported by the Donors of the Petroleum Research Fund, administered by the American Chemical Society, under grant ACS-PRF#22036-AC6, and by the Office of Naval Research.

REFERENCES

[1]   For a review, see *Regular and Stochastic Motion* by A.J. Lichtenberg & M.A. Lieberman, Applied Mathematical Series, vol 38, Springer-Verlag, New York (1992).

[2]   J.V. José and R. Cordery *Phys. Rev. Lett.* **56** 290 (89).

[3] F. M. Izrailev *Phys. Rep.* **196** 299 (90), and references therein.

[4] P. Seba *Phys. Rev.* **A41** 2306 (90).

[5] G. Chu and J. V. José *J. Stat. Phys.* **68** 153 (92).

[6] R. Badrinarayanan and J. V. José (in preparation).

[7] O. Bohigas, M.-J. Giannoni and C. Schmit *Phys. Rev. Lett.* **52**, 1 (1984).

[8] For an authoritative review of *RMT* methods, see *Random Matrices: An Enlarged and Revised Second Edition*, by M. L. Mehta, Academic Press, San Diego, California (1991).

[9] R. Badrinarayanan, G. Chu and J. V. José, to be published.

[10] *Royal Society Mathematical Tables*, Vol. **7**, Ed. F.W.J. Olver. Cambridge University Press, Cambridge (1960).

[11] T. A. Brody *et.al.*, *Rev. Mod. Phys.* **53**, 385 (1981).

[12] F. M. Izrailev *J. Phys.* **A22** 865 (89).

[13] O. Bohigas and M. J. Giannoni in *Mathematical and Computational Methods in Nuclear Physics*, Granada, Spain, edited by J. S. Dehesa *et.al.*, Springer-Verlag, Berlin (1984).

[14] Y. Alhassid and R. D. Levine *Phys. Rev. Lett.* **57** 2879 (86); K. Zyczkowski *J. Phys.* **A23** 4427 (91).

[15] R. Badrinarayanan and J. V. Jose, unpublished.

Spectral properties of systems with dynamical localization

T. DITTRICH

Institut für Physik, Universität Augsburg,
Memminger Straße 6, D-86135 Augsburg, Germany

U. SMILANSKY

Dept. of Physics of Complex Systems, The Weizmann Institute of Science,
Rehovot 76100, Israel

Abstract

We study the correlations in the quasienergy spectra of systems with dynamical localization, using the quantum kicked rotor (QKR) as a paradigm. Two complementary approaches are taken: We first study the "local spectral density". For level separations below the mean distance, its two-point correlations are dominated by level attraction rather than level repulsion, a feature known from the electron spectra in disordered solids. We then turn to the unbiased spectra for the QKR on a finite-dimensional Hilbert space ("finite-sample approach"). They are characterized by a novel universal statistics which depends on the ratio γ of the localization length to the basis size. We derive semiclassical expressions for the two-point correlations which interpolate between COE behaviour for $\gamma \to \infty$ and Poissonian (lack of correlations) for $\gamma \to 0$. We show how the diffusive nature of the classical dynamics finds its expression in the quantal spectral correlations.

1 Introduction

One of the most impressive results in "quantum chaology" was that as soon as the corresponding classical dynamics becomes chaotic, the spectral fluctuations, in the limit $\hbar \to 0$, obey universal distribution laws, as predicted by random-matrix theory [1–3]. Recently, this quantum–classical correspondence has been extended to the domain of chaotic scattering, where the fluctuations in the S-matrix eigenphases were shown to follow the statistics of Dyson's circular orthogonal ensemble (COE) [4]. The periodic-orbit theory was the main tool for the semiclassical investigation of spectra and their relation to random-matrix theory.

There exists, however, a very important class of systems where the results of random-matrix theory for the canonical ensembles have no reason to hold, in spite of the underlying chaotic classical dynamics. The common feature for these systems is that classically they may explore an extended (infinite) phase space, where the chaotic dynamics induce phase-space diffusion. In the quantum domain, diffusion is suppressed by an interference effect akin to Anderson localization [5,6]. A basic property of the canonical ensembles is the invariance of their statistical properties under arbitrary changes of the representation. Localization, on the other hand, introduces a highly specific, representation-dependent structure of the eigenstates. In particular, it implies that eigenstates localized at sites separated sufficiently far become effectively uncorrelated [7]. As a consequence, the levels statistics is rendered Poissonian, despite the chaotic nature of the corresponding classical dynamics [7–9].

In the present chapter, we investigate the effects of classical diffusion and quantum localization on the corresponding spectral statistics by studying in detail, the quasienergy spectrum of the Quantum Kicked Rotor (QKR). It is a paradigm for systems whose chaotic motion induces diffusion which is supressed by Anderson localization. The question of how spectral statistics are influenced by localization, has a longer history in the quantum theory of disordered media than in quantum chaology. Therefore, it is natural to adopt concepts and methods of spectral analysis which have proven useful in the context of Anderson localization. The fact that randomness in the QKR is based on an underlying *deterministic* dynamics, can be exploited by using the powerful semiclassical techniques developed in the quantum chaology of time-independent and scattering systems.

Specifically, we investigate two-point correlation functions for two different types of quasienergy spectra, both of which are sensitive to spatial structures, and are inspired by analogous concepts in solid-state theory: the *local spectrum* and the *unbiased spectrum for a finite sample*. Our main result is that these correlation functions, in the case of the QKR, bear the marks of classical deterministic diffusion as well as of quantum localization. Dynamical localization is reflected in novel features of the spectral fluctuations, different from those derived for any one of Dyson's circular ensembles.

The local spectrum is obtained by assigning to each quasienergy value a weight, equal to the probability for the corresponding eigenstate to overlap with a local site, specified, e.g., as an eigenstate of the unperturbed Hamiltonian [5]. Correlations in the local spectrum are related by Fourier transform to the time evolution of the population of the state in which the system was prepared initially. This function, called the *staying probability*, contains essential information on the dynamics. In systems with localization, it decays diffusively during the initial stages of the evolution and approaches asymptotically the inverse of the number of accessed states. We utilize it to demonstrate that dynamical localization is reflected in the local spectrum by a tendency to display level *clustering* at small quasienergy separations. A similar result has recently been obtained in the context of Anderson localization [10–12]. There, level clustering could be explained by the occurrence, at avoided level crossings, of quasienergy eigenstates with two (or more) centers of localization, a mechanism initially proposed by Mott [13]. The apparent attraction is thus revealed to be solely due to a strong increase, with decreasing quasienergy separation, of the relative weight given to pairs of quasienergies in the local two-point correlation function. This is sufficient to overcompensate the repulsion of the bare levels. We suggest a corresponding interpretation for the case under study, thereby adding a new aspect to the analogy between quantum chaos and the quantum mechanics of random media.

The analogue of a finite sample is introduced, in the context of dynamical localization, by restricting the infinite-dimensional Hilbert space spanned by the unperturbed Hamiltonian to a finite-dimensional subspace. The time evolution in the truncated space of dimension L remains unitary and approaches the evolution of the unrestricted QKR for $L \to \infty$. The L quasienergies of the truncated system can be treated on equal footing by assigning the same weight to all of them.

In order to set the stage for the discussion of spectral properties, we have to introduce the notations and some basic, partly novel concepts to be used in the description of the classical and quantal dynamics of the kicked rotor. This will be done in Section 2. In particular, we propose a simple, one-parameter scaling theory for the time evolution of the staying probability. Section 3 is devoted to the study of the local spectrum. Its relation to the staying probability is used to translate the scaling *ansatz* developed in Section 2 into the functional form of the two-point correlations characterizing the local spectrum. In Section 4, we shall derive the spectral correlations for the unbiased spectra of finite samples. We summarize our results in Section 5.

The present chapter is a revised and shortened version of two papers which have been published some years ago [14,15]. We concentrate here on results concerning two-point correlation functions, and exclude sections which deal with space–quasienergy correlations and with the experimental aspects of the local spectrum.

2 Classical diffusion and quantum localization

The present study makes use of the kicked rotor [16] as a paradigm of the class of periodically driven systems with one freedom which display a chaotic classical dynamics. Since we intend to use classical quantities in semiclassical and quantal expressions in what follows, it is convenient to measure angular momenta and actions in units of $\hbar$ from the beginning. In such a notation, the Hamiltonian describing the kicked rotor is given by

$$H(l,\theta) = \frac{l^2}{2} + k\cos\theta \sum_n \delta(t - n\tau), \tag{1}$$

with a cylindrical classical phase space $-\infty < l < \infty$, $0 \le \theta < 2\pi$.

The kicked rotor possesses two basic symmetries which we state for further reference. It is invariant under time reversal (T: $t \to -t, l \to -l$), and reflection (P: $\theta \to 2\pi - \theta, l \to -l$). We shall denote by "conjugation" (C) the product of these two symmetries (C = PT: $t \to -t, \theta \to 2\pi - \theta$).

A version of the standard map [16], the classical stroboscopic map correspond-ing to the Hamiltonian (1), is obtained by integrating the equations of motion between $t_n = n\tau + \epsilon$ and $t_{n+1} = (n+1)\tau + \epsilon$, $\epsilon \to 0^+$,

$$l_{n+1} = l_n + k\sin(\theta_{n+1}),$$
$$\theta_{n+1} = (\theta_n + \tau l_n)\,\mathrm{mod}\,2\pi. \qquad (2)$$

In addition to the symmetries mentioned above, this map is periodic not only in θ but also in the angular momentum l, with period $2\pi/\tau$.

The dynamics generated by the classical standard map becomes globally chaotic if the classical stochasticity parameter $K = k\tau$ exceeds the threshold $K_c \approx 1$. A basic approximation of the time evolution in this regime, providing a simple picture of the long-time dynamics, is its description as a diffusion process. An ensemble of classical trajectories, prepared initially at $l = l_0$ and spread homogeneously in θ, will be distributed in angular momentum, at a time n, according to [17]

$$P_{cl}(l_0 \to l;n) = \frac{1}{\sqrt{2\pi D n}}e^{-(l-l_0)^2/(2Dn)}, \qquad (3)$$

where $D \approx k^2/2$ is the diffusion constant. This description, however, becomes valid only from the time n_d on, given by the time scale on which correlations in the classical dynamics decay. It can roughly be estimated as the inverse of the positive Lyapunov exponent associated with the classical map.

The quantum map, obtained from (1) by canonical quantization [17], is defined by the Floquet operator

$$U = \exp\left(-ik\cos\hat{\theta}\right)\exp\left(-i\tau\frac{\hat{l}^2}{2}\right). \qquad (4)$$

The angular-momentum and angle operators $\hat{l}$, $\hat{\theta}$, respectively, obey the commu-tation relation $[\hat{l},\hat{\theta}] = -i$. Eigenstates and eigenvalues of the angular-momentum operator are given by $\hat{l}|l\rangle = l|l\rangle$, $l = 0,\pm1,\pm2,\ldots$; they are collectively referred to as the unperturbed basis throughout this chapter. The Floquet eigenvectors and the corresponding quasienergies will be denoted by $|\alpha\rangle$ and ω_α, respectively. Both families of eigenstates span the infinite-dimensional Hilbert space forming the quantal analogue of the cylindrical classical phase space.

The localization of the Floquet eigenvectors is intimately connected with the measure of the corresponding quasienergy spectrum. If the spectrum is pointlike, the Floquet states are localized. Otherwise they are delocalized, i.e., they are not normalizable. There exists a rigorous criterion to distinguish between these two situations: The function

$$P_s(n) = \left\langle \left|\sum_\alpha e^{i\omega_\alpha n}|\langle\alpha|l\rangle|^2\right|^2\right\rangle \qquad (5)$$

gives the mean probability to return to an initial state specified by the angular momentum l. The angle brackets indicate averaging over a sufficiently large number of initial states $|l\rangle$. The symbol $\mathcal{L}$ implies integration and summation over the continuous and the pure point components of the spectrum. The quantity

$$\xi^{-1} = \lim_{N \to \infty} \frac{1}{N} \sum_{n=0}^{N-1} P_s(n) \tag{6}$$

provides the criterion of localization. If ξ^{-1} vanishes, the spectrum is continuous, whereas if ξ^{-1} is finite, the spectrum contains a pointlike component [18], and hence implies normalizability of at least a subset of the Floquet eigenstates.

This theorem allows for a simple physical interpretation. The function $P_s(n)$ gives the average population left at a time n, in an initial state prepared as an angular-momentum eigenstate $|l\rangle$. This probability can vanish asymptotically for long times only if the eigenstates of the Floquet operator cover the entire space, i.e., if they are delocalized. When they are localized, $P_s(n)$ must remain finite on average, and provides an estimate of the number of quasienergy states overlapping appreciably with the initial state. Indeed, by substituting eq.(5) into eq.(6), we get

$$\xi^{-1} = \left\langle \sum_\alpha |\langle \alpha | l \rangle|^4 \right\rangle. \tag{7}$$

This is the definition of the mean *inverse participation ratio*. For systems with exponential localization, the participation ratio ξ is proportional (with a proportionality factor of the order of 1) to the localization length.

There is ample numerical evidence (but no rigorous proof) in support of the observation that the QKR does localize for sufficiently irrational $\tau/2\pi$ [19–21]. We shall restrict the discussion from now on to this, generic case, where the Floquet eigenstates are normalizable and the staying probability approaches a non-zero constant.

The function $P_s(n)$ is of special interest. It allows us to formulate a scaling theory for the saturation of the classical diffusion due to quantum localization in the QKR. Following the basic analysis of this crossover proposed in ref. [22], we distinguish an initial phase of the time evolution where the quantal dynamics essentially mimics the classical, and an asymptotic phase dominated by the effects of localization.

The diffusion approximation for the classical dynamics (see eq.(3)) predicts an algebraic decay of the staying probability,

$$P_s^{cl}(n) \approx \frac{1}{\sqrt{2\pi nD}}, \qquad n > n_d \gtrsim 1. \tag{8}$$

Eq.(8) gives the correct time dependence also for the initial decay of the corresponding function in the quantal case. However, it does not take into account a basic quantum-coherence effect, known as "weak localization" in the context of disordered media: The quantal staying probability is enhanced by a factor

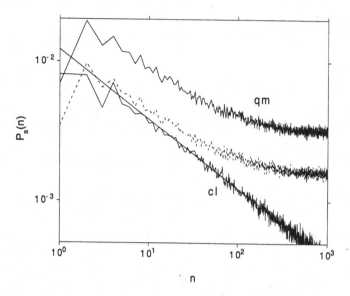

Fig. 1. Evolution, over the first 10^3 time steps, of the staying probability for the quantum-mechanical (qm) and classical (cl) versions of the kicked rotor. Parameter values are $K = 20$, and $\tau/2\pi = 0.05/((\sqrt{5}-1)/2)$ in the quantum-mechanical case, so that $k = K/\tau = 39.3$. The dashed graph represents the quantum-mechanical function, reduced by a factor of 2. The straight line indicates the asymptotic decay $\propto 1/\sqrt{n}$ of the staying probability in the classical case.

of 2, compared with the classical case, due to constructive interference between contributions from pairs of trajectories which are related by the "conjugation" symmetry C. Sometimes, one finds cases where the pair degenerates into a single trajectory which is self conjugate. Such cases become progressively rarer as n increases and the effective enhancement factor approaches 2. Hence, the quantal staying probability in the initial time regime, is given by

$$P_s(n) \approx \sqrt{\frac{2}{\pi n D}}, \text{ for small } n, \text{ but } n > n_d. \tag{9}$$

For large n, on the other hand, the discreteness assumed for the quasienergy spectrum implies that, in the mean,

$$P_s(n) \to \xi^{-1}, \text{ for } n \to \infty \tag{10}$$

(see eq.(6)).

In Fig. 1, we compare the time evolution of the staying probability $P_s(n)$ for the classical and the quantum-mechanical versions of the kicked rotor. In the classical case, after a non-generic initial phase extending over several time steps, the staying probability decays according to the algebraic law (8), indicated by the

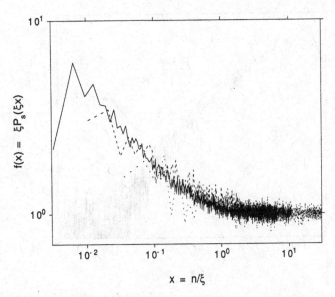

Fig. 2. Scaled staying probability $\xi P_s(\xi_x)$, as a function of the scaled time $x = n/\xi$, for the QKR at three different values of the effective Planck's constant. Parameter values are $K = 20$, and $\tau/2\pi = 0.05/g$ (solid line), $0.1/g$ (dashed line), $0.2/g$ (dotted line), where $g = (\sqrt{5} - 1)/2$. The correponding values of the quantal parameter $k = K/\tau$ are $k = 39.3$ (solid line), 19.7 (dashed line), 9.8 (dotted line).

straight line. The corresponding quantum-mechanical function approaches the classical one as an asymptote for short times, if the weak localization enhancement of the staying probability is taken into account (the dashed graph represents the quantum-mechanical data, reduced by a factor of 2).

We introduce a scaling *ansatz* by assuming that the function $P_s(n)$ for all $n > n_d$ depends on the two parameters k and τ of the Floquet operator (4) only via a *single* scaling parameter, ξ. The two asymptotic forms (9) and (10) are reconciled if $\xi \sim D$ and, for $n > n_d$,

$$P_s(n) = \frac{1}{\xi} f(n/\xi), \tag{11}$$

where $f(x)$ is a scaling function which satisfies

$$f(x) \approx \begin{cases} \sqrt{2c/\pi x} & x \leq 1, \\ 1 & x > 1, \end{cases} \tag{12}$$

and c denotes a constant to be determined from numerical data.

We have checked the validity of the scaling *ansatz* (12) numerically. Representative data are depicted in Fig. 2. Additional information about the scaling function is obtained by plotting $f(x) - 1$ (Fig. 3): This function emphasizes the

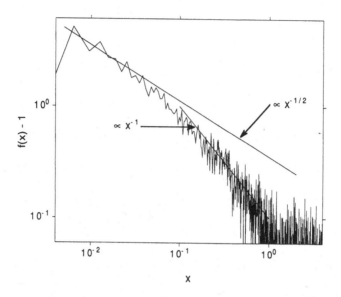

Fig. 3. Scaled staying probability $\xi P_s(\xi_x) - 1$, as a function of the scaled time $x = n/\xi$, for the QKR. Parameter values are $K = 20$ and $\tau/2\pi = 0.05/((\sqrt{5} - 1)/2)$. The left and right straight lines indicate algebraic decay as $x^{-1/2}$ and as x^{-1}, respectively.

way the staying probability approaches a constant for $n \to \infty$. It shows that this approach can be described by $f(x) - 1 \sim 1/x$ for $x \gtrsim 1$.

The scaling theory implies the well-known relation between the classical diffusion constant D and the participation ratio ξ which, in turn, is proportional to the localization length. In the present notation, it reads

$$\xi = cD. \qquad (13)$$

Our numerical data yield the estimate $c = 0.3 \pm 0.05$. The proportionality of ξ and D, eq.(13), is demonstrated by the inset in Fig. 4, which is based on the same three sets of data as Fig. 2. The histogram shown in Fig. 4 represents the distribution of 512 individual values of the inverse participation ratio ξ^{-1}. This distribution exhibits a single pronounced maximum, confirming that the mean inverse participation ratio is a statistically meaningful quantity.

The constant c determines the crossover time n_{qm} at which the asymptote $f(x) = 1$ is reached. We observe that $f(x) - 1 \approx 0.1$ at $x \approx 1$ (see Fig. 3), which implies $n_{qm} \approx \xi$. This result is consistent with the known proportionality [22] obtained for the time evolution of the energy gain. In fact, the time scale n_{qm} has a very general significance — it gives the time beyond which interference effects dominate the dynamics of the QKR (see below).

Summarizing, we state as a crude estimate of the time evolution of the staying

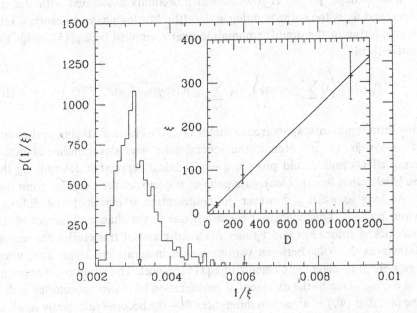

Fig. 4. Probability distribution of individual values of the inverse participation ratio ξ^{-1}. Parameter values are $K = 20$ and $\tau/2\pi = 0.05/((\sqrt{5} - 1)/2)$. The histogram is based on the long-time asymptotes of the staying probability for 512 different initial conditions. The arrow indicates the mean. Inset: Mean values and rms deviations of the measured participation ratios ξ, as a function of the diffusion constant D, for the three cases shown in Fig. 2.

probability,

$$P_s(n) \approx \begin{cases} 1 & n = 0, \\ \sqrt{2c/\pi n \xi} & 1 \lesssim n_d < n \leq \xi, \\ 1/\xi & n > \xi. \end{cases} \quad (14)$$

It becomes exact in the semiclassical limit, which amounts here to $\xi \to \infty$. Eq.(14) is not valid for very short times, due to our neglect of the non-generic first kicks. It also fails to reproduce the crossover from diffusion to quasiperiodicity as a smooth transition.

The above scaling theory, which combines both strictly classical and genuine quantum-mechanical behaviour in a single expression, is most naturally studied from a semiclassical point of view. The staying probability is expressed semiclassically, as a sum over paths, by

$$P_s(n) = \left\langle \left| \sum_r \sqrt{p_r(l; n)} e^{i\Phi_r} \right|^2 \right\rangle. \quad (15)$$

The index r runs over the various classical orbits which start at, and return to,

l after n time steps, $p_r(l;n)$ is the classical probability associated with the rth trajectory, and Φ_r is the corresponding action (the Maslov phase is absorbed into Φ_r). A reordering of the double summation over r, implied by eq.(15), yields the alternative form

$$P_s(n) = \left\langle 2 \sum_r p_r(l;n) \right\rangle + \left\langle \sum_{r \neq r'} \sqrt{p_r(l;n) p_{r'}(l;n)} e^{i(\Phi_r - \Phi_{r'})} \right\rangle. \qquad (16)$$

The first term represents, apart from a factor of 2, the classical staying probability which decays as $1/\sqrt{n}$. Hence, the second term includes genuine quantum coherence effects and should provide a semiclassical expression determining the inverse localization length. Conjugate pairs of trajectories are excluded from this term. As long as $n \lesssim n_{qm}$, however, its contribution is marginal and diffusive behaviour prevails. This behaviour is entirely due to the chaotic dynamics of the classical kicked rotor. For short times n, closed classical trajectories are scarce, the differences $\Phi_r - \Phi_{r'}$ between various action integrals are large, and, upon averaging the non-diagonal elements in eq.(16), cancel. The situation changes as soon as $n \gtrsim n_{qm}$. Due to the exponential proliferation of closed trajectories with n and the fact that $\langle \Phi_r \rangle \sim n^2$, action differences $\Phi_r - \Phi_{r'}$ become sufficiently small to reach the order of unity, and therefore interference terms cease to average out. A direct evaluation of the inverse participation ratio from the non-diagonal term of eq.(16) is beyond our present understanding of the semiclassical approximation. Yet, its genuine quantum-mechanical origin is clear.

3 The local quasienergy spectrum

In their work on the analogy between dynamical localization in the QKR and Anderson localization, Fishman *et al.* [5] adopted a concept well known in the theory of electronic states in disordered solids and introduced the *local spectrum* for periodically driven systems. Its density is defined by

$$P_1(l;\omega) = \sum_\alpha |\langle \alpha | l \rangle|^2 \delta(\omega - \omega_\alpha). \qquad (17)$$

(Here and in the following, delta functions of quasienergies are understood to be 2π-periodic.) The definition (17) assigns a weight $|\langle \alpha | l \rangle|^2$ (where $0 \leq |\langle \alpha | l \rangle|^2 \leq 1$) to each quasienergy ω_α. The summation in eq.(17) comprises the entire Floquet basis, so that the local density is normalized to unity.

By resolving the periodic delta functions in eq.(17) into a Fourier sum and using the defining property of the states $|\alpha\rangle$ to be eigenstates of the Floquet operator U, the local spectral density can alternatively be written as

$$P_1(l;\omega) = \frac{1}{2\pi} \sum_{n=-\infty}^{\infty} e^{-i\omega n} \langle l | U^n | l \rangle, \qquad (18)$$

where $\langle l | U^n | l \rangle$ gives the *amplitude* to be in the initial state $|l\rangle$ after n cycles of the driving force.

Similarily, a two-point density can be defined by

$$P_2(l; \omega, \chi) = \sum_{\alpha \neq \beta} |\langle \alpha | l \rangle|^2 |\langle \beta | l \rangle|^2 \delta(\omega - \omega_\alpha)\delta(\chi - \omega_\beta). \tag{19}$$

Here, each *pair* of quasienergy values ω_α, ω_β is weighted by the *product* of overlaps of the corresponding eigenstates $|\alpha\rangle$, $|\beta\rangle$ with the reference state $|l\rangle$.

Systems like the QKR are statistically homogeneous along the unperturbed basis, i.e., besides fluctuations, the properties of the Floquet eigenstates do not depend on their position in that basis. Therefore, relevant information is extracted from quantities like the complete two-point distribution function defined in eq.(19) by averaging over the reference state $|l\rangle$.

Furthermore, the statistical homogeneity of the quasienergy spectrum around the unit circle suggests that the average over the mean position of quasienergy pairs should also be taken. When this is done, the two-point correlation function depends only on the separation $\eta = \omega - \chi$ and reads

$$P_2^{loc}(\eta) = \frac{1}{2\pi} \sum_{n=-\infty}^{\infty} e^{-i\eta n} \left(\langle |\langle l | U^n | l \rangle|^2 \rangle - \xi^{-1} \right). \tag{20}$$

Angle brackets denote averaging over a sufficiently large ensemble of unperturbed eigenstates $|l\rangle$. Note that

$$\xi^{-1} = \lim_{N \to \infty} \frac{1}{N} \sum_{n=0}^{N-1} \langle |\langle l | U^n | l \rangle|^2 \rangle \tag{21}$$

(see eq.(6)), which ensures regular behaviour of $P_2^{loc}(\eta)$ at $\eta = 0$. Eq.(20) can also be written as

$$P_2^{loc}(\eta) = \frac{1}{2\pi} \sum_{n=-\infty}^{\infty} e^{-i\eta n} \left(P_s(n) - \xi^{-1} \right). \tag{22}$$

Here, $P_s(n)$ is the mean *probability* to stay at a site $|l\rangle$ after n kicks. We conclude that the staying probability $P_s(n)$ is the Fourier transform of the local two-point correlation function P_2^{loc}.

Eqs.(20), (22) are the basic analytical tools used in the subsequent arguments, since they relate *spectral* correlations to *dynamical* staying amplitudes or probabilities, respectively. The function P_2^{loc} has a close relative in random-matrix theory: Multiplying it by the factor ξ^2 defines the function

$$R_2^{loc}(\eta) = \xi^2 P_2^{loc}(\eta). \tag{23}$$

It can be considered as the local analogue of the pair-distribution function $R_2(\eta)$ [2,3], introduced in random-matrix theory as the probability for any two eigenphases to be separated by an angle η, integrated over the positions of all other eigenphases. Accordingly, R_2^{loc} is normalized to the effective number of non-identical quasienergy pairs,

$$\int_0^{2\pi} d\eta\, R_2^{loc}(\eta) = \xi(\xi - 1). \tag{24}$$

In the statistical analysis of spectra it is customary to define the *two-point cluster function* $Y_2(r)$ [1,3]. It is the difference between the pair-distribution function for an uncorrelated spectrum and the actual pair distribution. The argument r corresponds to the difference η measured in units of the mean spacing $2\pi/\xi$. It is important to note that whenever $Y_2(r)$ vanishes, the spectrum does not possess pair correlations. If $Y_2(r)$ is *negative*, the pairs with this value of r have a *higher* probability than in an uncorrelated spectrum, while if it is *positive*, the corresponding probability is *lower*. Since in the present case the effective number of quasienergy values is ξ, we get

$$
\begin{aligned}
Y_2^{loc}(r) &= \frac{1}{\xi}\left(1 - 2\sum_{n=1}^{\infty}(\xi P_s(n) - 1)\cos\left(\frac{2\pi}{\xi}rn\right)\right) \\
&= 1 - 2\pi P_2^{loc}\left(\frac{2\pi}{\xi}r\right). \tag{25}
\end{aligned}
$$

The fact that $\xi P_s(n) - 1 \sim n^{-1}$ for $x \gg 1$ (see Fig. 3) ensures convergence of the Fourier sum in the definition of $Y_2^{loc}(r)$. As was shown in Section 2, the function $\xi P_s(n)$ can be written in the form $f(n/\xi)$, where $f(x)$ is independent of the parameters of the QKR (see eqs.(11), (14) and Fig. 2). Hence, in the limit of large ξ, $Y_2^{loc}(r)$ also approaches a function which no longer depends on ξ. Using numerical data for the staying probability, we can calculate Y_2^{loc} for a typical case of the QKR (see Fig. 5). The most striking feature is that $Y_2^{loc}(r)$ is *negative* at small distances (it must become positive for large arguments, however, to ensure proper normalization — see inset in Fig. 5). This implies that the *weighted* quasienergy values tend to *cluster* at short distances.

This is a surprising result. It appears to be incompatible with the known, intuitive picture of how dynamical localization gradually takes over in the time evolution of the QKR [26]. According to this view, as time n increases, the dynamics is influenced by ever finer quasienergy scales $\Delta\omega$, given by the energy–time uncertainty relation $n\Delta\omega \gtrsim 2\pi$. As soon as the scale of the mean nearest-neighbour separation in the spectrum is reached, the dynamics crosses over from diffusive behaviour to quasiperiodicity. Arguing in the same spirit, we show in what follows that the clustering of quasienergies in the local spectrum is not at variance with this picture, but refines it in essential points.

To begin with, it is important to note that the preparation of a very narrow initial state, compared even with the localized quasienergy states, is a necessary condition for diffusion to occur in the first place; there will be essentially no spreading of the wave packet if the system has been prepared in a state as wide as a typical quasienergy state. In order to understand how a narrow

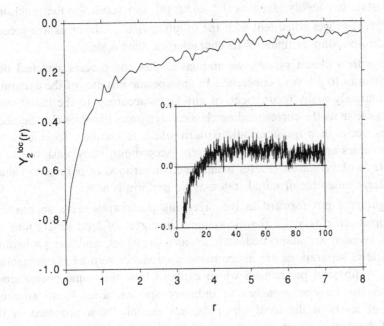

Fig. 5. Local cluster function for the QKR. Parameter values are $K = 20$ and $\tau/2\pi = 0.05/((\sqrt{5} - 1)/2)$. The function shown has been obtained by Fourier transformation, according to eq.(25), of the average quantal time evolution of the staying probability depicted in Fig. 1. The inset shows the same function over an extended r range.

state can arise by superposing eigenstates which decay only on a much larger spatial scale, finer details of their structure have to be taken into account. The superposition of those eigenstates to form a narrow state is enabled only by the typical, erratic fluctuations of the phase, as well as of the amplitude around its smooth exponential envelope. Diffusive spreading can then be interpreted as a gradual *coherent* destruction of the initial constellation of relative phases of the participating quasienergy states as time evolves. They will mutually run out of phase due to the factors $e^{i(\omega_\alpha - \omega_\beta)n}$ generating the dynamics in the quasienergy representation. Two states cease to contribute in a constructive way to the sharp initial state as soon as their accumulated relative phase $n(\omega_\alpha - \omega_\beta)$ has changed by an amount of the order of π. This process, in turn, is determined by the distribution of quasienergy differences $\Delta\omega = \omega_\alpha - \omega_\beta$, *weighted* by the relative contribution of the associated pair of quasienergy states $|\alpha\rangle$, $|\beta\rangle$ to the initial state. Thus it depends on the *local* two-point level correlation P_2^{loc}. As soon as the finest quasienergy scale is reached, the quantum state no longer spreads on average, but merely fluctuates quasiperodically around the superposition of the exponentially decaying envelopes of the participating Floquet states. Since the effective fraction of Floquet states no longer contributing constructively to the initial state is roughly given by $\int_{\pi/n}^{\infty} d\eta \, P_2^{loc}(\eta)$, our result for the functional form of $P_2^{loc}(\eta)$ proves consistent with the intuitive view of the crossover process and the corresponding estimate of its characteristic time scale.

Dealing with a closed system, we emphasize that the process sketched here does not amount to a loss of coherence. In any specific instance of the dynamics, there are infinitely many recurrences, of any given accuracy, to the initial state. However, as soon as the corresponding classical dynamics becomes chaotic, these recurrences occur in a quasiperiodic pattern which is uniquely specific to the parameter values and the initial state chosen. Accordingly, they wash out in an appropriate *incoherent average* over a small neighbourhood of parameter values, or over a large ensemble of initial states, as we perform it here.

The arguments put forward in the preceding paragraphs leave an essential question open. In classically chaotic systems, instances of level separations far below the average are associated with avoided crossings, and the probability that small level separations are encountered approaches zero as the separation decreases. A physical mechanism which explains how this same phenomenon which gives rise to level *repulsion* in unbiased spectra, leads to an apparent *attraction* of levels in the local spectrum, has recently been proposed in the context of spectral features associated with Anderson localization [10–12]. It goes back to an idea originally proposed by Mott [13].

Assume two localized quasienergy states $|a\rangle$, $|b\rangle$ associated with a pair ω_α, ω_β of quasienergy levels somewhat below an avoided crossing (see upper part of Fig. 6), and follow the association of states with levels through the crossing. The eigenstates will remain associated with the eigenvalues in the same way as they would if the levels crossed without interaction, i.e., their actual association is gradually interchanged according to the turn of a mixing angle from 0 to $\pi/2$. Correspondingly, at the crossing point, they in fact consist of the symmetric and antisymmetric superpositions $|\alpha\rangle = (|a\rangle + |b\rangle)/\sqrt{2}$ and $|\beta\rangle = (|a\rangle - |b\rangle)/\sqrt{2}$, respectively, of the original states $|a\rangle$, $|b\rangle$ (lower part of Fig. 6). The states $|\alpha\rangle$ and $|\beta\rangle$, though still orthogonal to one another, will contribute to the local two-point correlation function with an exceedingly high weight, because this function is based on the overlap of probabilities and not of amplitudes. This proves sufficient to overcompensate the decrease of unbiased probability for small separations.

Dynamically, this mechanism is associated with tunneling between $|a\rangle$ and $|b\rangle$ and corresponding Rabi oscillations on an extremely slow time scale $2\pi/|\omega_\alpha - \omega_\beta|$. It is effective not only at the precise parameter value where the quasienergy separation is minimal, but also in its neighbourhood. Its existence can be verified by demonstrating that there are "double-hump states" among the quasienergy eigenstates. Such states have indeed been observed at level crossings in systems related to the QKR [23,24].

We can summarize our arguments by stating that the study of the local

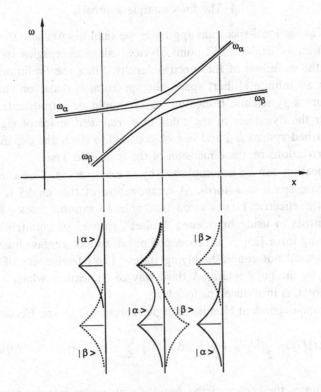

Fig. 6. Schematic representation, for a system with dynamical localization, of two quasi-energies ω_α, ω_β and the corresponding eigenstates $|\alpha\rangle$, $|\beta\rangle$, as functions of a parameter x, in the neighbourhood of an avoided crossing. The exchange of identity of the two eigenstates proceeds, by a continuous turn of a mixing angle, through an intermediate stage where $|\alpha\rangle$ and $|\beta\rangle$ form symmetric and antisymmetric "double-hump states", respectively.

quasienergy spectrum suggests a new feature of the quantum dynamics of classically chaotic, periodically driven systems. The local quasienergy correlations prefer clustering rather than repulsion of quasienergy pairs. This phenomenon is closely connected to the suppression of classical diffusion by Anderson-like localization, and it is most clearly reflected in the probability to stay in the initial state. This latter function displays an interesting scaling property which emphasizes the intimate relation between the quantum localization length and the classical diffusion constant.

Our analysis also shows that among the various parameters which can be used to give a quantitative measure of localization, the participation ratio ξ seems to emerge as the natural choice. It stems from the dynamical analysis of the staying probability and does not rely on an assumption concerning a particular functional dependence (e.g., exponential decay) of the wave function. It is distributed rather sharply about its mean (see Fig. 4), and thus is a useful tool in practice.

4 The finite-sample approach

In contrast to the local-spectrum approach, we shall discuss here the quasienergy spectrum from a "democratic" point of view: all quasienergies are given equal weights in the definition of the spectral density. Since the evolution operator U operates in an infinite Hilbert space, the spectrum is dense on the unit circle, and therefore a systematic truncation of U should be introduced. To this end we consider the dynamics of the rotor in a *truncated* space of eigenvectors of the unperturbed system H_0, and our objective is to study the dependence of the spectral correlations on the dimension of the truncated basis.

The truncation can be accomplished by various means. One can restrict the classical phase space to a torus. A quantization of this model is achieved by restricting the effective Planck's constant $\tau/2\pi$ to rational values [9]. Another method consists in using broadened "pulses", instead of infinitely sharp kicks, for the driving force [23]. The broadened pulses do not contain high frequencies and therefore do not couple highlying states. The effective size of the space is determined by the pulse width. A third way of truncation, which we use in the present chapter, is introduced as follows:

Let the time-dependent Hamiltonian be represented by the *Hermitian* matrix

$$\langle l|H^{(L,l_c)}(t)|l'\rangle = \frac{1}{2}l^2\delta_{l-l'} + k\langle l|\cos\hat{\theta}|l'\rangle \sum_{n=-\infty}^{\infty}\delta(t-n\tau), \quad |l-l_c|,|l'-l_c| \le \hat{L},$$

$$(26)$$

where l_c denotes the center of the angular-momentum interval considered, and the dimension of the basis is

$$L = 2\hat{L} + 1. \tag{27}$$

The Hamiltonian (26) controls the *unitary* evolution in the Hilbert space spanned by the L eigenvectors of the free rotor. If the dimension L by far exceeds the localization length (given, say, by the participation ratio ξ), this evolution approximates that of the unmodified QKR ($L \to \infty$). Therefore, this method of truncation is also the most natural one to be used in numerical simulations of the QKR.

The classical analogue of the quantum system described above is obtained by considering a finite phase-space cylinder whose length is L and whose midpoint on the l axis is l_c. The classical equations of motion are modified by imposing elastic reflection conditions at the boundaries $|l - l_c| = \hat{L}$, which corresponds to the truncation imposed in the quantal treatment, i.e., if at the nth iteration $|l_n - l_c| > \hat{L}$, one enforces $l_n = l_{n-1}$ and $\theta_n = 2\pi - \theta_{n-1}$. This amounts to switching from a trajectory to its conjugate counterpart in the same way as for elastic reflection from a hard wall.

There are several scales on the angular-momentum axis which are of importance for the understanding of the dependence of the spectral fluctuations on the size of the system. Classical dynamics provides us with two characteristic scales:

The first one has to do with an intrinsic symmetry of the classical equations of motion in the unmodified phase space (see eq.(2)). Since the angles are defined $mod\, 2\pi$, a shift of l by an integer multiple of $2\pi/\tau$ does not affect the dynamics. Thus $L_u = 2\pi/\tau = \sqrt{2D}2\pi/K$ determines the size of the "unit cell" in the classical description ($K = \tau k$ and D are the classical stochasticity parameter and diffusion constant, respectively — see Section 2). Another classical scale arises from the fact that the classical evolution is diffusive along the l axis. Hence we can define $L_D = \sqrt{D}$, which gives a measure of the broadening along the l axis of a phase-space distribution during one cycle of the driving force. Both L_D and L_u are proportional to $\sqrt{D}$. Localization provides a quantum-mechanical scale through the mean localization length, for which the mean participation ratio ξ is an adequate measure. Since $\xi \approx 0.3D$ and, in the semiclassical limit, $\xi \gg 1$, we have

$$L_u \lesssim L_D \ll \xi. \tag{28}$$

In what follows we shall discuss the spectral fluctuations for systems with arbitrary length L. We shall see that their statistics depends crucially on the value of L relative to ξ. This ratio is given in terms of the parameter

$$\gamma = D/L \approx 3\xi/L. \tag{29}$$

The following time scales are relevant to the discussion of spectral correlations in the QKR: The shortest one is the time n_d which is inversely proportional to the classical Lyapunov exponent. It determines the time after which the diffusion approximation becomes valid. A cylinder of length L will be covered diffusively during a time $n_L \approx L^2/D$. It is convenient to define

$$n_L = \frac{L^2}{2\pi D}. \tag{30}$$

As long as $\gamma < 1$, we have $n_L > L > \xi$. For $\gamma > 1$ and increasing γ, n_L decreases until it reaches the value $n_L = 1$ at $\gamma \approx \xi^{1/2}$. At this point, $L \approx L_u$. In other words, the classical phase-space cylinder is truncated to the size of the unit cell, and the concept of diffusion along the l axis becomes inadequate. Quantum mechanics introduces another time scale n_{qm} which was discussed in the preceding sections and will be a central issue here as well. This is the time after which the quantal evolution departs from the corresponding classical one due to the fact that quantum-interference effects dominate the dynamics. For systems which display Anderson localization,

$$n_{qm} \approx \min(\xi, L). \tag{31}$$

The Floquet operator for the system (26) is given by

$$U^{(L, l_c)} = e^{-ikC}e^{-i\hat{l}^2}, \quad |l - l_c|, |l' - l_c| \le \hat{L}. \tag{32}$$

The matrix C is defined by $C_{l, l'} = \langle l | \cos\hat{\theta} | l' \rangle$. The eigenvalues $e^{i\omega_\alpha}$, $\alpha = 1, \ldots, L$, of $U^{(L, l_c)}$ are the object of our study.

The spectral density is

$$d^{(L,l_c)}(\omega) = \sum_{\alpha=1}^{L} \delta(\omega - \omega_\alpha) = \frac{1}{2\pi} \sum_{n=-\infty}^{\infty} e^{in\omega} \mathrm{tr}\left((U^{(L,l_c)})^n\right). \tag{33}$$

The diagonal elements of $(U^{(L,l_c)})^n$ are the probability *amplitudes* to stay at the initial state l after n periods of the driving force. Therefore, this relation provides the link between the spectral distribution and the dynamics of the rotor.

A straightforward calculation (see Section 3) gives the the *ensemble-averaged* two-point cluster function

$$Y_2^{(L)}(r) = \frac{1}{L}\left(1 - 2\sum_{n=1}^{\infty} b^{(L)}(n)\cos\left(n\frac{2\pi}{L}r\right)\right), \tag{34}$$

with

$$b^{(L)}(n) = \langle b^{(L,l_c)}(n)\rangle = \frac{1}{L_c}\sum_{l_c=-\hat{L}_c}^{\hat{L}_c}\left(\frac{1}{L}\left|\mathrm{tr}\left((U^{(L,l_c)})^n\right)\right|^2 - 1\right). \tag{35}$$

The averaging interval is $-\hat{L}_c \le l_c \le \hat{L}_c$, $L_c = 2\hat{L}_c + 1$.

The large-n behaviour of $b^{(L)}(n)$ can best be deduced by writing

$$b^{(L)}(n) = \frac{1}{L}\left\langle\left|\sum_{\alpha=1}^{L} e^{in\omega_\alpha}\right|^2\right\rangle - 1. \tag{36}$$

It must fluctuate about the value zero as soon as $2\pi/n$ is smaller than the minimum distance between any pair of phases, and provided the ω_α are not rationally related to 2π. Under these conditions, and because of the averaging over the set of l_c, we may write

$$\lim_{n\to\infty} b^{(L)}(n) = 0. \tag{37}$$

We are not able to give an explicit expression describing the approach to this asymptote. We shall argue, however, that it is already reached for $n > n_{qm}$, the time scale defined in eq.(31) above.

The semiclassical theory will now be used to calculate the coefficients $b^{(L)}(n)$ in the domain $n < n_{qm}$. $\mathrm{Tr}(U^n)$ can be expressed in terms of the periodic orbits of the classical mapping. The resulting expression is analogous to Gutzwiller's trace formula (see, e.g., [25]),

$$\mathrm{Tr}(U^n) \approx \sum_{\substack{p \\ \eta_p \rho_p = n}} \eta_p A_{p,\rho_p} e^{i\rho_p \Phi_p} \tag{38}$$

The summation is over all primitive periodic orbits p of period η_p which, if repeated ρ_p times, will reach a period of length n. The action Φ_p includes the

Maslov phase. The amplitudes A_{p,ρ_p} are given by

$$A_{p,\rho_p} = \frac{1}{\sqrt{|\det(I - M_p^{\rho_p})|}}, \tag{39}$$

where M_p is the matrix of the tangent mapping (the Jacobian), calculated for the primitive periodic orbit p.

Because of intrinsic symmetries of the dynamics, there may exist classical orbits which are physically distinct, but whose contributions to the expression (38) add coherently to the sum. In the present problem, for any periodic orbit $x(j)$, $j = 1, \ldots, n$, there exists a corresponding time-reversed orbit $Tx(j)$, where x denotes the phase-space point (θ, l), and T stands for the time-reversal operator. The two trajectories have the same action and the same prefactor. We now write

$$b^{(L)}(n) \approx \frac{1}{L} \left\langle \left| 2n \sum_p{}' A_p e^{i\Phi_p} \right|^2 \right\rangle - 1, \tag{40}$$

where the sum is over primitive periodic orbits of period n, and the symbol $\sum'$ indicates that only one representative of each set of symmetry-conjugated orbits is included. In writing eq.(40), we ignored self-retracing orbits as well as the contributions from orbits which are composed of repetitions of primitive orbits. It is well known that for hyperbolic mappings of the kind we consider here, this can be justified for sufficiently large n. We can rewrite $b^{(L)}(n)$ as

$$b^{(L)}(n) \approx 2\pi \frac{(2n)^2}{\Omega} \left(\left\langle \sum_p{}' |A_p|^2 \right\rangle + \left\langle \sum_{p \neq p'}{}' A_p A_{p'}^* e^{i(\Phi_p - \Phi_{p'})} \right\rangle \right) - 1, \tag{41}$$

where, for later use we have introduced the phase-space area $\Omega = 2\pi L$. For sufficiently small values of n, the second term (the "non-diagonal contribution") is expected to vanish upon averaging. This is not an exact result, but can be made plausible by the following argument: In chaotic systems, action differences approach each other and get arbitrarily small for long periodic orbits due to the exponential proliferation of the orbits on the one hand, and the fact that actions are quadratic in n on the mean, on the other hand. However, as long as n is not large, action differences are still much larger than $\hbar$ so that upon averaging, the non-diagonal term vanishes. For large values of n, this term dominates the sum, and it is plausible that intricate correlations between the classical actions conspire in such a way that the semiclassical expression for $b^{(L)}(n)$ satisfies eq.(37).

The diagonal contribution can be written as

$$b^{(L)}(n) = \frac{2\pi}{\Omega} (2n)^2 \left\langle \sum_p{}' \frac{1}{|\det(I - M_p)|} \right\rangle. \tag{42}$$

This sum allows a classical interpretation which was first used in this context by

Hannay and Ozorio de Almeida [26] and Berry [27], and was generalized in refs. [22,37,38]. Consider the classical mapping, which can formally be written as

$$x_n = F(x_{n-1}) = \cdots = F^n(x_0), \tag{43}$$

where again, $x = (\theta, l)$. If, instead of following the evolution of phase-space points, we study the evolution of phase-space distributions, we can introduce the classical evolution operator $U_{cl}(x', x; n)$ such that any initial phase-space distribution $\rho(x; 0)$ evolves into

$$\rho(x'; n) = \int U_{cl}(x', x; n)\rho(x; 0)dx. \tag{44}$$

It is easy to check that $U_{cl}(x', x; n)$ is the unitary Frobenius–Perron operator [28],

$$U_{cl}(x', x; n) = \delta(x' - F^n(x)). \tag{45}$$

We here introduce an important concept, the classical probability density to perform periodic motion with period n. It will be denoted by $p_{cl}(n)$. In particular, we confine our attention to periodic motion which starts and ends inside Ω. It is clear from the definition of the Frobenius–Perron operator that

$$p_{cl}(n) = \frac{1}{\Omega} \int_\Omega U_{cl}(x, x; n)dx = \frac{1}{\Omega} 2n \sum_p{}' \frac{1}{|\det(I - M_p)|}. \tag{46}$$

The sum is over periodic orbits of period n. It is obtained by straightforward integration, remembering that the matrix M is the linearization of F^n or, in other words, the Jacobian of the transformation $x_0 \to x_n$. Notice also that in eq.(46), we have a factor $2n$ which takes account of the contributions from the n periodic points on each orbit, and from each pair of symmetry-conjugated orbits.

We can now substitute eq.(46) into eq.(42) and get

$$b^{(L)}(n) \approx (2\pi)2np_{cl}(n) - 1, \tag{47}$$

which is one of the most fundamental results in the present chapter. It shows that the semiclassical expression for the two-point correlation function depends on the classical probability for period motion, and the only remnant of quantum interference comes through the factor $2n$ which is due to the fact that quantum mechanically, one adds amplitudes and then takes the absolute square, whereas classically, one adds probabilities.

The further evaluation of eq.(47) depends on whether $\gamma \sim \xi/L$ is smaller or larger than a critical value γ_c, which is of order unity and whose exact value will be determined below. Therefore, we shall discuss the two cases separately.

$$\gamma < \gamma_c$$

This is the domain where the truncated classical dynamics is diffusive for some time, and the Floquet eigenstates are known to localize (for $\tau/2\pi$ sufficiently irrational). In this domain, the classical probability density to perform periodic motion is given by

$$p_{cl}(n) = \frac{1}{2\pi} \frac{1}{\sqrt{2\pi nD}}, \qquad n_d < n < n_L, \tag{48}$$

where n_d and n_L were defined in the introduction to the present section.

We can substitute this approximation in eq.(47) and use the asymptotic expression (42) to write

$$b^{(L)}(n) \approx \begin{cases} \sqrt{2n/\pi D} - 1 & n_d < n < n_{qm}, \\ 0 & n > n_{qm}. \end{cases} \tag{49}$$

We skipped the first few time steps $1 \leq n \leq n_d$ where the dynamics is not yet generic. Lacking an expression that interpolates $b^{(L)}(n)$ between the two asymptotes in eq.(49), we extrapolate both asymptotes to their intersection point $n = \pi D/2$. This amounts to a renormalization of n_{qm}, consistent with the required proportionality [22] (see eq.(13)) between the diffusion constant and the participation ratio ξ. By substituting eq.(49) into the expression (34) for the two-point cluster function, we obtain

$$Y_2^{(L)}(r) = \frac{1}{L}\left(1 - 2\sum_{n=n_d+1}^{n_{qm}} \left(\sqrt{\frac{2n}{\pi D}} - 1\right) \cos\left(n\frac{2\pi}{L}r\right)\right). \tag{50}$$

We now go to the continuum limit $L, \xi \to \infty$, keeping their ratio γ constant. Taking the lower limit of the summation in eq.(50) to zero, we can rewrite this expression as

$$Y_2(\gamma;r) \approx \frac{1}{\pi}\int_0^{\pi^2\gamma} dv \left(1 - \frac{1}{\pi}\sqrt{v/\gamma}\right)\cos(vr). \tag{51}$$

The last integral can be evaluated numerically. The function $Y_2(\gamma;r)$ is shown in Fig. 7(a) for various values of γ. The fact that it is positive for small r and decays to zero for large r is an expression of the level repulsion, consistent with the numerical results reported in [9]. $Y_2(\gamma;r)$ attains its maximum value $\gamma\pi/3$ at $r = 0$. As γ decreases, this maximum flattens out, which proves that the spectral correlations become Poissonian in the limit $\gamma \to 0$.

$$\gamma > \gamma_c$$

We have shown above that as γ increases beyond γ_c, the finite size of the accessible phase space limits the applicability of the free-diffusion description to shorter and shorter times. At $\gamma \approx \sqrt{\xi}$, classical diffusion becomes completely irrelevant. We shall now show how the change in the duration of the initial, diffusive regime of the dynamics is reflected in the spectral correlations.

We first consider the range $1 < \gamma < \sqrt{\xi}$. This amounts to

$$1 < n_L < L < \xi; \qquad L > L_u. \tag{52}$$

In other words, there exists a range of time $n_d < n < n_L$ for which the classical dynamics can still be described in terms of the diffusion approximation. Hence, the semiclassical theory which yielded eq.(48) also holds for the reduced range $n_d < n < n_L$.

In the range $n > n_L$, the discussion which led to eq.(48) is no longer valid and we should start again from eq.(47). Now most of the classical periodic orbits with period $p > n_L$ are affected by the scattering off the boundaries of the truncated cylinder. In other words, the dynamics in the truncated cylinder is ergodic for $n > n_L$. Thus we can write $p_{cl}(n) = 1/\Omega$, where now $\Omega = 2\pi L$. Substituting into eq.(47), we get

$$b^{(L)}(n) \approx \begin{cases} \sqrt{2n/\pi D} - 1 & n_d < n < n_L, \\ (2n/L) - 1 & n_L < n < L, \\ 0 & n > L, \end{cases} \tag{53}$$

where again we skipped the non-generic regime $1 \le n < n_d$ of the dynamics.

For $\gamma > \sqrt{\xi}$, $n_L < 1$ and hence, classical diffusion is suppressed from the very beginning. The expression for $b^{(L)}(n)$, in this range, is obtained from eq.(53) by leaving out the first option. The resulting function is analogous to the result obtained for chaotic mappings of bounded domains [27], which in turn approximates rather well the asymptotic behaviour of the corresponding result obtained by Dyson [2] for the COE.

The generic part of the two-point cluster function for the present case can be derived easily. We obtain

$$Y_2(\gamma; r) \approx \frac{1}{\pi} \int_0^{\gamma^{-1}} dv \left(1 - \frac{1}{\pi}\sqrt{v/\gamma}\right) \cos(vr) + \frac{1}{\pi} \int_{\gamma^{-1}}^{\pi} dv \left(1 - \frac{1}{\pi}v\right) \cos(vr). \tag{54}$$

For $\gamma = 1/\pi$, the second integral in eq.(54) vanishes and the first term coincides with the expression (51). Therefore, a consistent choice of the critical value γ_c is

$$\gamma_c = \frac{1}{\pi}. \tag{55}$$

For $\gamma \to \infty$, on the other hand, the first integral vanishes. The remaining second term gives

$$Y_2(\infty; r) = \frac{1}{(\pi r)^2}\left(1 - \cos(\pi r)\right). \tag{56}$$

We cannot expect to reproduce Dyson's result rigorously since we lack a semiclassical theory which interpolates between the three time domains in eq.(53). Nevertheless, except for oscillations superposed on $Y_2(\infty; r)$ due to the singularity in the form factor $b^{(L)}(n)$ at $n = L$, there is rough quantitative agreement between eq.(56) and the exact COE result (dotted line in Fig. 7(a)). Fig. 7(a) also shows the function $Y_2(\gamma; r)$ for some values of $\gamma > \gamma_c$.

The two-point cluster function is the input to calculate the other spectral

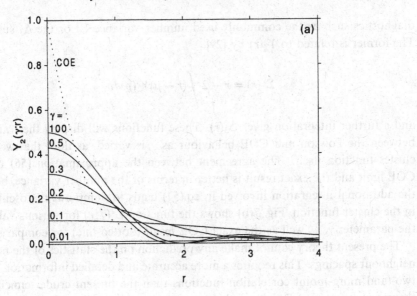

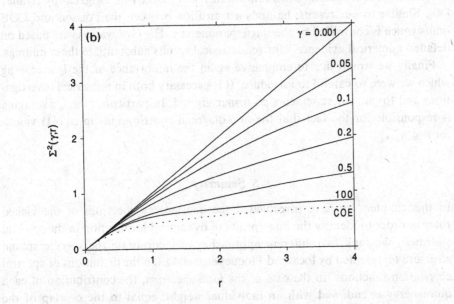

Fig. 7. Two-level cluster function $Y^2(\gamma;r)$ (a) and number variance $\Sigma^2(\gamma;r)$ (b) for the QKR on a truncated Hilbert space, for various values of the parameter $\gamma = D/L$.

diagnostics such as the commonly used number variance Σ^2 or the Δ_3 statistics. The former is related to $Y_2(r)$ by [29]

$$\Sigma^2(r) = r - 2 \int_0^r (r - \rho) Y_2(\rho) d\rho, \tag{57}$$

and a further integration gives $\Delta_3(r)$. These functions will display the transition between the Poisson and COE behaviour as γ is varied, as does the two-point cluster function itself. The agreement between the approximation (56) for the COE limit and the exact result is better in terms of the number variance, because the additional integration involved in eq.(57) tends to damp out the oscillations in the cluster function. Fig. 7(b) shows the function $\Sigma^2(\gamma; r)$ for various values of the parameter γ, as well as the exact COE result (dotted line) for comparison.

The present theory cannot make any prediction on the statistics of the nearest-neighbour spacings. This requires a more accurate and detailed information on the two-(and more-)point correlation functions than the present crude semiclassical theory can provide. This is rather unfortunate, since the dependence of the nearest-neighbour spacing distributions on γ was recently investigated by Izrailev [30]. Similar to our results, he finds a transition between the Poisson and COE limits which is controlled by the *single* parameter γ. His observations are based on detailed numerical evidence. Our semiclassical results substantiate these findings.

Finally we would like to emphasize again the importance of the l_c averaging which we were so careful to introduce. It is necessary both in numerical investigations and to put our statements on firmer ground. In particular, the l_c averaging is responsible for the fact that the non-diagonal contributions in eq.(41) vanish for $n < n_{qm}$.

5 Summary

In this chapter, we have investigated the quasienergy spectrum of the kicked rotor in order to identify the fingerprints of dynamical localization in the spectral statistics. We took two different approaches to incorporate the specific spatial structure represented by localized Floquet eigenstates in the definitions of spectral correlation functions. In the case of the *local spectrum*, the contribution of each quasienergy is endowed with an individual weight, equal to the overlap of the corresponding eigenstate with a reference state in the unperturbed basis. The *unbiased spectrum*, on the other hand, is defined by restricting the Hilbert space to a finite subset of the unperturbed basis, in analogy to a finite sample of a one-dimensional solid. Both approaches reveal features not encountered in any one of the canonical quasienergy ensembles provided by random-matrix theory.

The local spectrum of the QKR is characterized by *positive* correlations on small quasienergy scales, corresponding to an apparent *attraction* of levels. At avoided crossings, which dominate the small quasienergy scales, quasienergies are associated with eigenstates showing two (or more) centers of localization. Such "double-hump states" endow quasienergies at avoided crossings with a sufficiently high weight in the local spectrum such that the level repulsion in the unbiased spectrum is effectively masked.

For the unbiased spectrum, we find that the spectral correlations depend on the parameter γ, the ratio of the localization length to the dimension of the truncated Hilbert space. The spectral fluctuations show a smooth transition between COE behaviour for $\gamma \to \infty$ and Poissonian statistics for $\gamma \to 0$, the limit leading back to the unmodified QKR. This transition is described analytically. Resembling the results for the local spectrum, the decisive dynamical element which determines the character of the spectral correlations turns out to be the initial diffusive regime of the dynamics and the way it leads to the coverage of the accessible phase space. It should be emphasized that for any finite value of γ, the cluster function deviates in an essential way from the predictions of any of the canonical random-matrix ensembles.

An interesting observation emerges from our study of correlations in the unbiased spectrum: We have shown that the generic part of the Fourier coefficients of the two-point cluster function is characterized by two asymptotes. The long-time asymptote is of quantal origin. It sets a natural cutoff to the time correlations and hence to the quasienergy correlations. The short-time asymptote is determined entirely by the classical dynamics. The only information required is the classical probability to perform periodic motion, which depends only on the way phase space is filled by classical trajectories. As long as the accessible phase space is finite and covered uniformly, one obtains a behaviour which is similar to the one predicted by the canonical random-matrix ensembles. This is intuitively clear: The main assumption underlying Dyson's random-matrix theory is that there are no spatial preferences in the ensemble, and the only relevant parameter is the mean level spacing which depends only on the phase-space volume. The classical analogues of these ensembles are characterized by a finite phase space covered ergodically by classical orbits. The analogy fails if phase space is infinite and covered diffusively. This type of classical dynamics has no counterpart in the systems described by the canonical random-matrix ensembles — hence an agreement in the statistics of the spectral fluctuations cannot be expected, either.

The statistical fluctuations of the spectra of Floquet operators which are quantized in a finite Hilbert space were studied by Izrailev, from a different point of view, and emphasizing other aspects of the problem [9,30]. Among other properties, Izrailev *et al.* analyzed the dependence of the localization length on the system length (finite size scaling) and emphasized the relation of the present problem to a banded-random-matrix model. A summary of these results and their relation to properties extracted from the theory of random banded matrices [31] can be found in a recent review [32].

The theory presented here has been extended in various directions. Argaman *et al.* [33] showed that the correlations in single-electron spectra in disordered media (in the metallic regime) can be calculated semiclassically by using a variant of the theory presented above. As a matter of fact, they reproduced the two-point correlation function derived diagrammatically by Al'tshuler and Shklovski [12]. Argaman *et al.* extended their numerical tests to systems in two spatial degrees of freedom. Their results confirm the analogue of eq.(49) for diffusion in two dimensions. Doron *et al.* [34] studied a model composed of chaotic billiards connected aperiodically be narrow ducts. The calculated spectral correlations agree well with the expressions for $Y_2(\gamma, r)$ derived in the present paper for a large range of γ values. Finally, Smilansky *et al.* [35] showed that the connection between $b^{(L)}(n)$ and $p_{cl}(n)$ (see eq.(46)) also applies to systems in which the coverage of phase space is chaotic but, due to dynamical barriers, involves more than one time scale.

Acknowledgements

We have profitted from discussions with R. Graham, J. L. Pichard, J. Imry, F. M. Izrailev and B. Al'tshuler. M. V. Berry made very useful comments on the manuscript. We are most obliged for their help. TD would like to thank the Department of Nuclear Physics at the Weizmann Institute in Rehovot for the warm hospitality extended to him during a postdoctoral stay and acknowledges financial support by *Minerva*.

REFERENCES

[1]   G. M. Zaslavsky, Sov. Phys. Usp. **22**, 788 [Usp. Fiz. Nauk **129**, 211 (1979)]; O. Bohigas and M.-J. Giannoni, in *Chaotic Motion and Random Matrix Theories*, Lecture Notes in Physics **209**, 1, Springer, Berlin (1984); M. V. Berry, Proc. R. Soc. Lond. A **400**, 229 (1985).

[2]   F. J. Dyson, J. Math. Phys. **3**, 140 (1962); *ibid.*, 1191.

[3]   C. E. Porter, *Statistical Theory of Spectral Fluctuations*, Academic Press, New York (1965).

[4]   R. Blümel and U. Smilansky, Phys. Rev. Lett. **60**, 477 (1988); Physica D **36**, 111 (1989).

[5]   S. Fishman, D. R. Grempel, and R. E. Prange, Phys. Rev. Lett. **49**, 509 (1982); Phys. Rev. A **29**, 1639 (1984).

[6]   D. L. Shepelyansky, Phys. Rev. Lett. **56**, 677 (1986).

[7]   M. Feingold, S. Fishman, D. R. Grempel, and R. E. Prange, Phys. Rev. B **31**, 6852 (1985).

[8]   S. A. Molcanov, Commun. Math. Phys. **78**, 429 (1981).

[9]   F. M. Izrailev, Phys. Rev. Lett. **56**, 541 (1986).

[10] L. P. Gor'kov, O. N. Dorokhov, and F. V. Prigara, Zh. Eksp. Teor. Fiz. **84**, 1440 (1983) [Sov. Phys. – JETP **57**, 838 (1983)].

[11] U. Sivan and Y. Imry, Phys. Rev. B **35**, 6074 (1987).

[12] B. L. Al'tshuler and B. I. Shklovski, Zh. Eksp. Teor. Fiz. **91**, 220 (1986) [Sov. Phys. JETP **64**, 127 (1986)].

[13] N. F. Mott, Philos. Mag. **22**, 7 (1970).

[14] T. Dittrich and U. Smilansky, Nonlinearity **4**, 59 (1991).

[15] T. Dittrich and U. Smilansky, *ibid*. 85.

[16] B. V. Chirikov, Phys. Rep. **52**, 263 (1979).

[17] G. Casati, B. V. Chirikov, F. M. Izrailev, and J. Ford, Lecture Notes in Physics **93**, 334, Springer, Berlin (1979).

[18] V. Enß and K. Veselić, Ann. Inst. Henri Poincaré A **XXXIX**, 159 (1983).

[19] G. Casati and I. Guarneri, Commun. Math. Phys. **95**, 121 (1984).

[20] G. Casati, J. Ford, I. Guarneri, and F. Vivaldi, Phys. Rev. A **34**, 1413 (1986).

[21] I. Guarneri, Europhys. Lett. **10**, 95 (1989).

[22] B. V. Chirikov, F. M. Izrailev, and D. L. Shepelyansky, Sov. Sci. Rev. **2C**, 209 (1981).

[23] R. Blümel, S. Fishman, and U. Smilansky, J. Chem. Phys. **84**, 2604 (1986).

[24] D. Cohen, Phys. Rev. A **44**, 2292 (1991).

[25] M. Tabor, Physica **6D**, 195 (1983).

[26] J. H. Hannay and A. M. Ozorio de Almeida, J. Phys. A **17**, 3429 (1984).

[27] M. V. Berry, Proc. R. Soc. Lond. A **400**, 229 (1985).

[28] See, e.g., *Deterministic Chaos* by H.G. Schuster (2nd edition), VCH, Weinheim (1989), pp.29 – 30.

[29] T. A. Brody, J. Flores, J. B. French, P. A. Mello, A. Pandey, and S. S. M. Wong, Rev. Mod. Phys. **53**, 385 (1981).

[30] F. M. Izrailev, Phys. Lett. A **134**, 13 (1988); J. Phys. A **22**, 865 (1989).

[31] Y. V. Fyodorov and A. D. Mirlin, Phys. Rev. Lett. **67**, 2049 (1991); *ibid*. 2405.

[32] F. M. Izrailev, Phys. Rep. **196**, 299 (1990).

[33] N. Argaman, Y. Imry, and U. Smilansky, Phys. Rev. B **47**, 4440 (1993).

[34] E. Doron, U. Smilansky, and T. Dittrich, Physica B **179**, 1 (1992); J. Phys. A **27**, 79 (1994).

[35] U. Smilansky, S. Tomsovic, and O. Bohigas, J. Phys. A **25**, 3261 (1992).

Unbounded quantum diffusion and fractal spectra

T. GEISEL, R. KETZMERICK, AND G. PETSCHEL

Institut für Theoretische Physik und SFB Nichtlineare Dynamik,
Universität Frankfurt, Postfach 11 19 32,
D-60054 Frankfurt/Main, Federal Republic of Germany

Abstract

A few years ago we found the existence of an *unbounded* quantum mechanical diffusion process which strongly contrasted with the dynamical localization of the kicked rotator and other previously studied quantum systems. Here we review how this phenomenon is related to multifractal properties of the spectrum, its level statistics, and to the algebraic decay of correlations. We investigate the influence of classical chaos on uncountable fractal spectra and demonstrate that the concepts of level statistics can be related to multifractal concepts in these cases. First we point out a new class of level statistics where the level spacing distribution follows *inverse* power laws $p(s) \sim s^{-\beta}$ with $1 < \beta < 2$ and $\beta = 1 + D_0$ where D_0 is the fractal dimension of the spectrum. It is characteristic of hierarchical level clustering rather than level repulsion and appears to be universal for systems exhibiting unbounded quantum diffusion where the mean square displacement increases as $t^{2\delta}$ with $\delta = \beta - 1$. A realization of this class with $\beta = 3/2$ is a model of Bloch electrons in a magnetic field (Harper's equation), for which lateral surface superlattices on semiconductor heterojunctions presently serve as experimental realizations. While here diffusion is linear in time ($\delta = 1/2$), in the Fibonacci chain model the spread of wave packets also shows *anomalous* diffusion ($\delta \neq 1/2$) with $0 < \delta < 1$. In the kicked Harper model we study how a Cantor spectrum is affected by the onset of classical chaos. While the spectrum undergoes visible changes, its level spacing distribution is unaffected on *small* scales in contradistinction to level statistics of discrete spectra. In the time domain there is a crossover between two diffusive regimes characterized by a classical and a quantum mechanical diffusion coefficient. We then prove that the temporal autocorrelation function $C(t)$ for quantum systems with Cantor spectra has an algebraic decay $C(t) \sim t^{-\nu}$ where ν equals the generalized dimension D_2 of the spectral measure. In the kicked Harper model we demonstrate that the quantum mechanical decay is unrelated to the existence of classical chaos.

1 Introduction

The preparation of lateral surface superlattices [1] offers new and fascinating possibilities to study the transition from classical mechanics to quantum mechanics. In these systems one can realize periodic potentials for the dynamics of ballistic electrons in the semiclassical regime. In the presence of a magnetic field the classical dynamics is known to show chaotic behaviour [2], whereas the simplest quantum models of the system like the Harper model already have a fractal spectrum [3]. It is thus important to study the dynamical behaviour in these quantum systems and how they are influenced by classical chaos.

For discrete spectra, level statistics is a well established tool for the analysis of complex structures of excited states, e.g. in systems that are chaotic in the classical limit [4]. Depending on symmetry properties of the Hamiltonian one distinguishes three universality classes. Level repulsion causes a power law behaviour $p(s) \sim s^{\beta}$ of the probability density of the nearest neighbor level spacings s, where $\beta = 1, 2, 4$ and $s \to 0$. These universality classes apply to discrete spectra, i.e. countable sets. Here we extend the concept to Cantor spectra, i.e. uncountable sets which show some type of hierarchical level clustering. We point out a new class of level statistics characterized by an *inverse* power law $p(s) \sim s^{-\beta}$ with $\beta = 3/2$. The exponent β is related to the fractal dimension $D_0 = \beta - 1$ of the spectrum. This class appears to be universal for quantum systems with unbounded diffusion in one dimension [3]. We found such a case of unbounded diffusion for Bloch electrons in a magnetic field as described by Harper's equation [5–8]. The level statistics can exhibit also other values of β with $1 < \beta < 2$ and $D_0 = \beta - 1$ which are related to unbounded *anomalous* diffusion in one dimension where the mean square displacement var$(t) \sim t^{2(\beta-1)}$ [9]. This type of diffusion is met in a one-dimensional model of quasi-crystals [10], the so called Fibonacci chain [11–13]. In general we find that for quantum systems with Cantor spectra level statistics is equivalent to a multifractal analysis, i.e. nearest neighbor distributions, number statistics, and Δ_3-statistics can be related to the generalized dimensions D_q of the spectrum.

The classical limits of these models, however, are integrable systems and show no chaotic behaviour. On the other hand, for the dynamics of electrons in crystal lattices it is known that the real classical analog has diffusive chaotic dynamics [2]. In fact, we have shown [14] that the chaotic dynamics close to nonlinear resonances is responsible for a series of magnetoresistance peaks found in lateral surface superlattices on semiconductor heterojunctions [1]. As the ratio of superlattice period to Fermi wavelength decreases an interplay between fractal spectrum and classical chaotic dynamics will occur. While the influence of classical chaos on discrete spectra has been studied intensely in the past, here we can address the question of how a Cantor spectrum is affected by the onset of classical chaos. To this end we assume the kicked Harper model as a classically chaotic modification of Harper's equation. We find that the inverse power law

$p(s) \sim s^{-3/2}$ accounts also for the classically chaotic case. A mere inspection of the spectrum, on the other hand, reveals considerable changes. We explain the discrepancy by a crossover on an energy scale s^* and on a corresponding time scale t^* [15]. On small energy scales $s < s^*$ the spectral statistics shows hierarchical level clustering $p(s) \sim s^{-3/2}$ unaffected by classical chaos. The level clustering is destroyed only on scales $s > s^*$. In the time domain there is a corresponding crossover from a mimicking of the chaotic classical diffusion below t^* to a purely quantum mechanical diffusion above t^*. The onset of unbounded quantum diffusion clearly contrasts the behaviour known as dynamical localization for the kicked rotator [16, 17], where after the crossover time quantum interferences impose a finite bound on the diffusive growth of the variance [18]. The crossover time t^* roughly scales like $1/\hbar$ or $1/\hbar^2$ depending on whether the classical dynamics is regular or chaotic, respectively. In the strong quantum limit we show that the system can be transformed to the integrable Harper system. Thereby we explain why the spectrum has the statistics of the integrable Harper system, even if the classical phase space is strongly chaotic.

The temporal decay of correlation functions plays an important role in classical physics, as it can be used in ergodic theory to define mixing, a somewhat weaker property than chaotic behaviour. As quantum mechanics precludes sensitive dependence on initial conditions, the possibility that quantum systems might exhibit mixing behaviour through their decay of correlations has attracted considerable attention in recent years. The situation is far more complex than in classical physics and some of the investigations have arrived at controversial conclusions [19]. The complication is due to the absence of one-to-one relations between the nature of the decay and the spectral type (absolutely continuous, singular continuous, pure point or any mixture). Only decays faster than any power law can be uniquely related to an absolutely continuous spectrum. Slow power-law decays, however, can be compatible with a singular continuous spectrum, as well as an absolutely continuous spectrum. Therefore, in order to relate the decay to the spectral type, Avron and Simon [20] had to introduce a distinction between "transient" and "recurrent" absolutely continuous spectra. Many problems, however, still remain to be solved. In this situation it is useful to investigate the correlation decay of the above systems, for which the spectral types are known and to develop a new general concept for a *quantitative* determination of the correlation decay.

We therefore analyze the decay of the correlation function $C(t)$ [21] in the unkicked and kicked Harper model for localized, critical, and extended states, as well as in the Fibonacci chains. Numerically we find slow algebraic decays $C(t) \sim t^{-\nu}$ with $0 < \nu \leq 0.84$. This is the first quantitative determination of the correlation decay in these systems and confirms the conjecture of anomalous transport. For the regime of extended states of the Harper model the power-law decay has an exponent $\nu = 0.84 \pm 0.01$ reflecting a recurrent absolutely continuous spectrum, whereas the singular continuous spectrum in the critical case gives rise to an extremely slow decay with $v = 0.14 \pm 0.01$. The singular continuous spectrum of the Fibonacci chains shows variable exponents $0 < v \leq 0.84$ with v approaching 0.84 as $V \to 0$, where the spectrum becomes absolutely continuous. In the kicked Harper model, which is classically chaotic we demonstrate that the decay of the quantum correlation function is unaffected and cannot be interpreted as a signature for the existence of classical chaos. We then show analytically that for Cantor set spectra (singular continuous or absolutely continuous) the correlation function decays algebraically $C(t) \sim t^{-v}$, that v is given by the generalized dimension D_2 of the spectral measure $v = D_2$, and that the Hausdorff dimension D_0 gives an upper bound $D_0 \geq v$. Thus, while there are no one-to-one relations between spectral types and the decay of correlations, there does exist a relation determining the correlation decay from multifractal properties of the spectrum. This relation is simple and of general validity for fractal spectra of all types. In other words, the specific spectral type is rather irrelevant in the present context, whereas the multifractal concept is relevant.

2 Harper's equation

We first study the level statistics of Bloch electrons in a magnetic field B in the framework of the Peierls substitution, which leads to a discrete Schrödinger equation in a quasiperiodic potential (Harper's equation)[3–6]

$$\psi_{n+1} + \psi_{n-1} + \lambda \cos(2\pi n\sigma - \varphi_0)\psi_n = E\psi_n, \qquad (1)$$

where ψ_n is the wavefunction at site n and $\lambda = 2$. The dimensionless parameter $\sigma = a^2 eB/(hc)$ gives the number of flux quanta per unit cell of area a^2 and determines the incommensurability of the system. For comparison we will also consider cases $\lambda \neq 2$, as it is known that $\lambda = 2$ is a critical case [11] separating a regime of extended states ($\lambda < 2$) from a regime of localized states ($\lambda > 2$) for irrational σ [22]. For $\lambda = 2$ the states are neither localized nor extended and called critical. The spectrum is a Cantor set for a dense set of parameter pairs (λ, σ) [23]. Considering it as a perturbation, the incommensurate potential breaks translational symmetry, lifts the twofold degeneracy, and introduces a dense set of gaps into the tight binding band $E(k)$ [7]. (see e.g. Fig. 1 below). For σ a Liouville number and $\lambda > 2$, the spectrum is singular continuous [24].

The levels and their spacings are obtained numerically with the use of transfer matrices $\underline{M}_1(n, E)$ [25]. One can replace Eq. (1) by a matrix equation

$$\begin{pmatrix} \psi_{n+1} \\ \psi_n \end{pmatrix} = \underline{M}_1(n, E) \begin{pmatrix} \psi_n \\ \psi_{n-1} \end{pmatrix} \qquad (2)$$

where

$$\underline{M}_1 = \begin{pmatrix} E - \lambda \cdot \cos(2\pi n\sigma - \varphi_0) & -1 \\ 1 & 0 \end{pmatrix}. \qquad (3)$$

We approximate the irrational incommensurability σ by successive rational

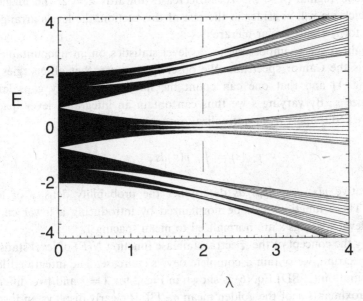

Fig. 1. Allowed energies as a function of the parameter λ for a rational approximant $\sigma = 34/55$ of the golden mean. The magnetic field case ($\lambda = 2$) is at the transition between regimes of extended states ($\lambda < 2$) and localized states ($\lambda > 2$) for incommensurate σ. As λ approaches $\lambda = 2$ from above, the levels arrange in clusters.

convergents of its continued fraction expansion. For $\sigma = p/q$ the potential is periodic with period q. We thus analyze the matrix product

$$\underline{M}_q(E) = \prod_{n=0}^{q-1} \underline{M}_1(n, E), \tag{4}$$

which transfers the states (ψ_0, ψ_{-1}) into the states (ψ_q, ψ_{q-1}). According to the Bloch theorem, $\psi_{n+q} = e^{ikq}\psi_n$ and thus $\underline{M}_q(E)$ has eigenvalues $e^{\pm ikq}$, i.e.

$$\mathrm{Tr}\underline{M}_q(E) = 2cos(kq). \tag{5}$$

This leads to the condition $|\mathrm{Tr}\,\underline{M}_q(E)| \leq 2$, from which one can determine the allowed eigenvalues E of Eq. (1) [7]. The eigenfunctions at sites $n = 0$ and $n = -1$ form the corresponding eigenvectors of $\underline{M}_q$. The eigenfunction at site m is obtained by multiplying with the matrix $\underline{M}_m(E)$.

3 Level statistics and multifractal properties

Figure 1 illustrates the spectral changes, i.e. the allowed energies as a function of λ for an approximant of the golden mean $\sigma_G = (\sqrt{5}-1)/2$. Energy levels indicative of the localized regime on the right hand side ($\lambda > 2$) turn into pronounced bands in the extended regime ($\lambda < 2$). As λ decreases towards $\lambda = 2$, the magnetic field case which we will consider in the rest of this section, the levels arrange in clusters and form a self-similar hierarchy.

At first sight it seems impossible to do level statistics on an uncountable set of levels as is the Cantor spectrum. We observe, however, that all energies are bounded (Fig. 1) and that one can count the number of energy gaps larger than some size s. By varying s we thus can obtain an integrated level spacing distribution (ILSD) apart from normalization

$$p_{int}(s) = \int_s^\infty p(s')ds', \tag{6}$$

whose derivative $p(s) = -dp_{int}/ds$ determines the probability density of level spacings s. These functions can be normalized by introducing a lower cut-off $s_0 > 0$. The level spacings s are normalized to mean spacing 1.

If we apply the concept of the spectral staircase function $N(E)$ of level statistics [4] to this spectrum, we obtain a complete devil's staircase. The integrated level spacing distribution (ILSD) Eq. (6) is shown in Fig. 2 for $\lambda = 2$ and two different rational approximants σ of the golden mean σ_G [3]. It clearly displays an inverse power law

$$p_{int}(s) \sim s^{1-\beta}, \tag{7}$$

and thus the level spacing distribution (LSD) behaves as

$$p(s) \sim s^{-\beta}, \tag{8}$$

where $\beta = 1.5009 \pm 0.0010$. This equation expresses the self-similarity of the structure of gaps. Unfolding the spectrum by a smoothened spectral staircase as in other cases would not change the power law, as the spectral fluctuations remain self-similar on all scales. In Fig. 2 the ILSD levels off at a small value s_0, since for all rational approximants of σ_G the total number of gaps is finite. The cut-off s_0, however, can be shifted to arbitrarily small values for higher approximants. The LSD of Eq. (8) behaves very differently from Poisson, Wigner, and intermediate distributions found in other systems [4].

The increasing probability for smaller spacings indicates what we call hierarchical level-clustering (see also Fig. 1). This property is more pronounced in another quantity [4], the probability density $\mu(x)$ defined by the conditional probability of finding a level in $[x_0 + x, x_0 + x + dx]$, if there is a level at x_0 and no level in $]x_0, x_0 + x[$. For a Poisson distribution one has $\mu(x) = const$ corresponding to independent level positions. For a Wigner distribution, $\mu(x) \sim x$ reflecting the repulsion of levels. In our case we have found $\mu(x) = (\beta - 1)/x$ expressing a preference of clustering other levels in the vicinity of a given one. This property also affects the Δ_3-statistics of the spectrum shown in Fig. 3. We find that $\Delta_3(L)$ closely follows a power law $\Delta_3(L) \sim L^\gamma$ with $\gamma = 1.48 \pm 0.06$, in clear contrast to a

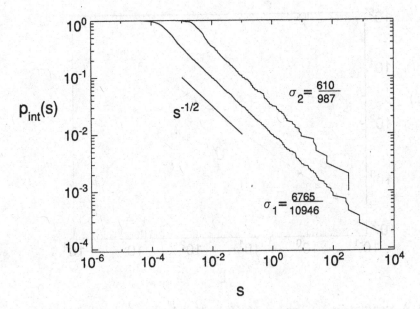

Fig. 2. ILSD $\lambda = 2$ for two approximants of the golden mean displaying an *inverse* power law $p_{int} \sim s^{1-\beta}$ with $\beta = 1.5009 \pm 0.0010$. As is seen the lower cut-off of the scaling region decreases for higher approximants.

Poissonian spectrum ($L/15$) and random matrix theories ($\ln L$) [4]. The spectrum thus is even less rigid than a Poissonian spectrum.

Of course one would like to understand what causes this new class of level statistics. In accordance with random matrix theories we here have degenerate levels that are split by the perturbation. In distinction, however, the degeneracy is not accidental, but systematically twofold (for states k and $-k$). The matrix elements of the perturbation therefore are not random, but are due to the quasiperiodic potential. Besides, neighboring splittings of levels are not independent, but affect each other as levels are nowhere isolated. The exponent β of the LSD can be related to the *fractal dimension* D_0 of the spectrum as follows. Dividing the energy range into boxes of length l we obtain for the number of empty boxes

$$\text{\# of empty boxes} = \frac{1}{l} \int_l^\infty sp(s)ds. \tag{9}$$

Using the normalization of the mean spacing we now write

$$\text{\# of filled boxes} = \frac{1}{l} \int_0^l sp(s)ds \sim \frac{1}{l} \int_0^l ss^{-\beta}ds \sim l^{1-\beta} \tag{10}$$

and thus

$$\beta = D_0 + 1. \tag{11}$$

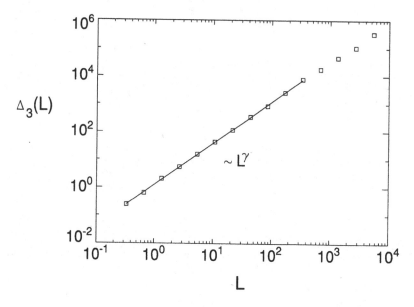

Fig. 3. Δ_3-statistics for $\lambda = 2$ and $\sigma = 6765/10946$. A least-squares fit (straight line) yields $\Delta_3(L) \sim L^\gamma$ with $\gamma = 1.48 \pm 0.06$, in accordance with Eq. (16) ($\gamma = 1 + D_2 = 1.493 \pm 0.002$).

The numerical value of β appears to be largely independent of the incommensurability σ. Exceptions are the Liouville numbers, where we could not find an inverse power-law behaviour, whereas for a variety of irrational values of σ having small numbers in their continued fraction expansion we always found $\beta = 3/2$. The fact that $\lambda = 2$ is the critical point of the delocalization transition suggests that $\beta = 3/2$ is a universal exponent. There are renormalization techniques for Eq. (1) [11, 26, 27], but local scaling properties (e.g. near $E = 0$) are not sufficient to explain the global power law Eq. (8). In fact it was found that the spectrum is a multifractal [28]. We can relate the number statistics [4] $n(L)$, which counts the number of levels in an interval of length L to the *multifractal* scaling properties [29]. The moments of their distribution are

$$\langle n^q(L) \rangle \sim \frac{1}{N} \sum_{i=1}^{N} p_i^q, \tag{12}$$

where $N \sim 1/L$ is the number of boxes of length L and p_i is the probability for a level to be in box i. If D_q are the generalized dimensions, the scaling behaviour of the sum is given by

$$\sum_{i=1}^{N} p_i^q \sim L^{(q-1)D_q}. \tag{13}$$

For the moments of the number statistics we thus obtain

$$\langle n^q(L) \rangle \sim L^{1+(q-1)D_q}. \tag{14}$$

In particular, for $q = 2$ we have $\langle n^2(L) \rangle \sim L^{1+D_2}$ and with the relation [4]

$$\Delta_3(L) = \frac{2}{L^4} \int_0^L (L^3 - 2L^2 r + r^3) \langle n^2(r) \rangle dr, \tag{15}$$

we find

$$\Delta_3(L) \sim L^\gamma \quad \text{with} \quad \gamma = 1 + D_2. \tag{16}$$

A numerical determination of $D_2 = 0.493 \pm 0.002$ yields an improved value of γ consistent with Fig. 3 and shows that γ is different from $3/2$.

4 Unbounded diffusion and level statistics

The global character of the exponent β asks for a global argument for its explanation. We can give a heuristic argument similar in spirit to arguments developed by Allen [30] and Chirikov et al. [17] for localization problems. We consider successive rational approximants $\sigma_i = p_i/q_i$ of the continued fraction expansion of σ. If we want to resolve the spectrum with a finite resolution only, it suffices to confine the potential to a finite interval of length q_i. On this length scale the periodicity of the potential is not manifest and we may assume that a wave packet moves diffusively (with diffusion coefficient D) inside, i.e.

$$\text{var}(t) = \langle x^2(t) \rangle \sim 2Dt. \tag{17}$$

The maximum distance q_i to be travelled defines a longest time scale $\tau \sim q_i^2/(2D)$ and a smallest energy difference between levels $s \sim \hbar/\tau$. The number of states living in the interval is $\sim q_i$ and thus determines the number of states with spacing $\Delta E \geq s$, whence $p_{int}(s) \sim q_i \sim (2D\tau)^{1/2} = (2D\hbar)^{1/2} s^{-1/2}$, i.e.

$$p(s) \sim s^{-3/2}. \tag{18}$$

For a refined energy spectrum consider the next approximant p_{i+1}/q_{i+1}, where again the potential looks quasiperiodic within a period q_{i+1}. Repeating the argument yields the observed LSD Eq. (8) on all scales.

This argument rests on the assumption of diffusion, to be repeated on all length scales. It suggests that the exponent $\beta = 3/2$ is universal for systems showing unbounded quantum diffusion in one dimension. The assumption, however, is not obvious in our case. In particular one might also expect that the diffusive growth is nonlinear in time. This motivated us to analyze the time evolution of a wave packet $\phi(t)$ released at site 0, using the eigenenergies and eigenfunctions obtained as above [3]. Fig. 4 shows the variance

$$\text{var}(t) = \sum_{n=-q_i/2}^{q_i/2} n^2 |\phi_n(t)|^2, \tag{19}$$

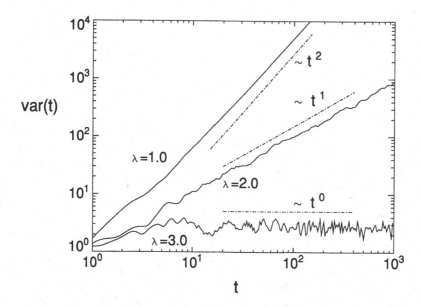

Fig. 4. Time evolution of the variance of a wave packet, Eq. (19). In the case $\lambda = 2$ the spread is purely diffusive (var$(t) \sim 2Dt$) [2].

for three different values of λ in a finite lattice of $q_i = 987$ sites. The quadratic growth for $\lambda = 1$ and the boundedness for $\lambda = 3$ correspond to extended and localized eigenstates, respectively. In the critical case $\lambda = 2$ we find clear-cut diffusion (var$(t) \sim 2Dt$) with $D \approx 1/2$ where all eigenstates are exited by the initial wave packet. Hiramoto and Abe [31] reported a slightly different exponent var$(t) \sim t^{2\delta}$ with $2\delta = 0.97$ obtained from a numerical simulation. The slight deviation from linear diffusion probably is an artifact (see discussion below). We mention that the wave packet reaches the boundary of the lattice of q_i sites in a finite time. We found that this time scales like q_i^2 and thus diffusion never stops for $q_i \to \infty$. In this way the validity of the heuristic argument is verified step by step and in particular the assumption of unlimited diffusion is confirmed.

5 Anomalous diffusion in the fibonacci chain

The validity of the heuristic argument can be verified in more generality in the Fibonacci chain model of a one-dimensional quasicrystal where other scaling exponents can be obtained. Here the potential only takes the two values $+V$ and $-V$ arranged in a Fibonacci sequence and replaces the cosine potential in Eq. (1) [11]. It was shown that the spectrum is a Cantor set with zero Lebesgue-measure for all values of $V > 0$ [32]. The eigenenergies for a finite lattice can be determined efficiently using the trace map [11, 33]. We have applied the same

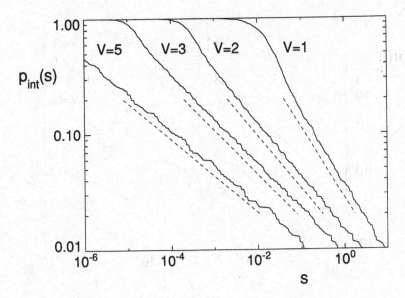

Fig. 5. ILSD for different values of the potential strength V on a finite lattice with 2584 sites displaying an *inverse* power law. For comparison the power laws s^{-D_0} (dashed lines) are plotted, where the Hausdorff dimension D_0 was determined independently from the spectrum.

procedure as in the Harper model to obtain level statistics and dynamics. As Fig. 5 shows we again find inverse power laws for the ILSD. Here the exponent depends on the potential strength V, while Eq. (11) and Eq. (14) still relate the level statistics to the generalized dimensions D_q.

For the spread of a wave packet we find anomalous diffusion with various exponents (see Fig. 6). Hiramoto and Abe [13] have obtained similar results and give an explanation for the values of the exponents based on renormalization group ideas in the limiting cases $V \to 0$ and $V \to \infty$. We will show below that our numerical data do not agree with these explanations even in the limiting cases. Instead we generalize the heuristic argument introduced for the Harper model in the previous section. Using straigthforward modifications this argument relates anomalous diffusion

$$\text{var}(t) \sim t^{2\delta}, \tag{20}$$

to a LSD

$$p(s) \sim s^{-\beta} \quad \text{with} \quad \beta = 1 + \delta. \tag{21}$$

On the other hand, if the scaling exponent β or the fractal dimension $D_0 = \beta - 1$ (Eq. (11)) is given, the variance should grow as

$$\text{var}(t) \sim t^{2D_0}. \tag{22}$$

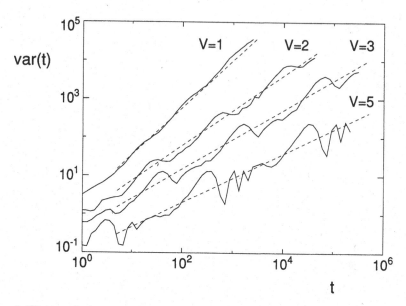

Fig. 6. Time evolution of the variance var(t) of a wave packet for different values of V corresponding to Fig. 5. The spread shows anomalous unbounded diffusion. For comparison the power laws t^{2D_0} (dashed lines) according to Eq. (22) are shown.

This relation is supported by the numerical results shown in Fig. 7. The values for the dynamical exponent δ coincide with the Hausdorff dimension D_0 for all values of V within the numerical error. According to the renormalization group ideas of Hiramoto and Abe [13] the dynamical exponent δ would be determined by the maximum and minimum scaling indices of the spectrum in the limiting cases, but even there the latter do not agree as well with the dynamical exponent δ as D_0 does (Fig. 7). The validity of Eq. (22) was recently confirmed by Artuso, Casati, and Shepelyansky [34] for the kicked Harper model. A mathematical proof, however, has still to be worked out.

As the heuristic argument is confirmed numerically in the one-dimensional case, we now assume that it can be generalized to the multi-dimensional case. Observing that the number of states living on a d-dimensional cube of length q is q^d the generalization of Eq. (22) becomes

$$\text{var}(t) \sim t^{2D_0/d} = t^{2(\beta-1)/d}. \tag{23}$$

6 Cantor spectra in a classically chaotic quantum system

While the influence of classical chaos on discrete spectra and their level statistics was studied intensely in the past, we now want to investigate how a Cantor

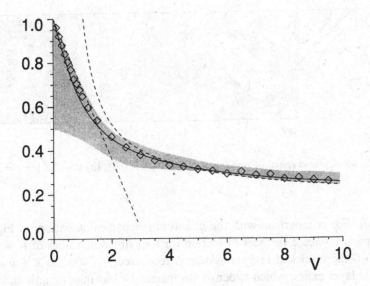

Fig. 7. The dynamical exponent δ (diamonds) of the anomalous diffusion var$(t) \sim t^{2\delta}$ depending on the potential strength V. It coincides with the Hausdorff dimension D_0 (solid line) thus confirming Eq. (22). The gray area gives the range of scaling indices obtained from multifractal $f(\alpha)$ curves for every V. Its maximum value $1 - V/\pi$ (for $V \to 0$) and its minimum value $-3ln(\sigma_g)/2ln(2V)$ (for $V \to \infty$) are indicated by the dashed lines. Even in these limits they do not agree as well with δ as D_0 does.

spectrum, i.e. an uncountable spectrum is affected [15]. As an unperturbed system we assume Harper's equation, which was described above. It is a stationary Schrödinger equation $\hat{H}\psi = E\psi$, which can be formally expressed in terms of a Hamiltonian

$$\hat{H} = 2cos(\hat{p}) + \lambda cos(\hat{x}). \tag{24}$$

Here $\lambda = 2$ and $\hat{p} = -i\hbar\partial/\partial x$ plays the role of a momentum operator with an effective $\hbar = 2\pi\sigma$. This allows one to vary $\hbar$ and to consider the semiclassical limit $\hbar \to 0$ [35]. Note that the classical limit of the Hamiltonian Eq. (24) is integrable. We study the above questions in the kicked Harper model [36, 37] as the simplest nonintegrable perturbation of Eq. (24)

$$\hat{H} = Lcos(\hat{p}) + Kcos(\hat{x}) \cdot \delta_1(t) \tag{25}$$

where $\delta_1(t)$ is a periodic delta function of period one. The corresponding classical equation of motion is an iterated map from one kick to the next

$$p' = p + Ksin(x), \qquad\qquad x' = x - Lsin(p'). \tag{26}$$

For $K = L$ this map was studied as a model for the motion of a charged particle in a plane perpendicular to a magnetic field and subject to an electromagnetic

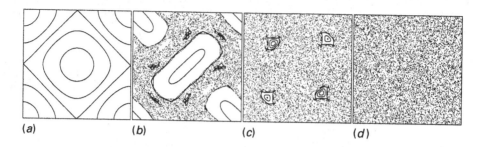

(a)   (b)   (c)   (d)

Fig. 8. Classical phase space of one unit cell for (a) $\kappa = 0.01$, (b) $\kappa = 2.0$, (c) $\kappa = 3.5$, and (d) $\kappa = 5.0$.

wave [38]. For comparison with the quantum mechanical situation, in Fig. 8 we show typical Poincare surfaces of section for four different values of $K = L \equiv \kappa$. For $\kappa \to 0$ the unkicked Harper system is recovered. As soon as $\kappa > 0$ a small stochastic layer exists, which widens with increasing kicking strength and allows for diffusive motion in p and x. In an intermediate regime there is a mixed phase space with regions of regular and stochastic motion. For $\kappa = 5$ no islands are visible and the dynamics is strongly chaotic.

The quantum dynamics of Eq. (25) is described by the time-dependent Schrödinger equation. As the kicks are periodic in time, the Floquet theorem applies and one can use quasienergy eigenstates $\Psi_\omega(x, t)$ [39]. The system is also periodic in x, which requires the Bloch theorem to hold, i.e. $\Psi_\omega(x, t) = e^{i\theta_x x}\psi_\omega(x, t)$ where θ_x is the quasimomentum in x-direction and $\psi_\omega(x + 2\pi, t) = \psi_\omega(x, t)$. The periodicity of $\psi_\omega(x, t)$ now implies that the operator $\hat{p}$ has eigenvalues of the form $\hbar(n + \theta_x)$ with integer n. The time evolution operator acting on periodic functions in x for one period of time is then given by

$$\hat{U} = e^{-i(L/\hbar)cos(\hbar(\hat{n}+\theta_x))}e^{-i(K/\hbar)cos(x)} \tag{27}$$

and determines the quasienergies ω by $\hat{U}\psi_\omega(x, t) = e^{i\omega}\psi_\omega(x, t)$.

We have investigated the ILSD and the time evolution of the variance

$$\mathrm{var}(t) = \hbar^2 \langle n^2(t) \rangle = \hbar^2 \sum_n n^2 |\phi_n(t)|^2 \tag{28}$$

of a wave packet $\phi(t)$ for different values of κ and $\hbar = 2\pi\sigma$ in the kicked Harper model. For the time evolution of a wave packet the operator $\hat{U}$ was iterated using the FFT-method [40] with up to 10^5 momentum eigenstates. We have analyzed the variance for initial Gaussian wave packets with equal width in position and momentum. The spectrum was studied for rational values of $\sigma = r/q$. Here the system is periodic in x and p and another Bloch phase θ_p is introduced. This reduces the time-evolution operator in Eq. (27) to a $q \times q$-matrix [41]. A Fourier transform of the time evolution of a wave packet yields those quasienergies

$\omega \in [-\pi, \pi]$ which belong to eigenstates excited by the wave packet. To obtain all q quasienergies it turns out to be most efficient to start with a random initial wave packet. The quasienergies ω depend on the two phases θ_p and θ_x and form energy bands, as long as σ is rational.

We first study how Hofstadter's butterfly is affected by classical chaos, i.e. the spectrum (of quasienergies in our case) as a function of σ. In Fig. 9 these bands are plotted as functions of σ for four different values of $\kappa/\hbar$. As can be seen from Eq. (27), for $\kappa/\hbar = const.$, the spectrum is symmetric with respect to $\sigma = 1/2$. For $\kappa/\hbar = 1$ (top) the spectrum looks like the original Hofstadter butterfly. With increasing values of $\kappa/\hbar$ we observe that the large gaps become smaller and eventually disappear. For small denominators q of σ the q single bands broaden until they touch each other and form *one* single band from $-\pi$ to π. For σs close to irrational numbers there is a transition from a hierarchical band clustering to a more uniform distribution of bands all over the interval. Although Fig. 9 exhibits visible changes in the butterfly, the asymptotic form of the LSD $p(s) \sim s^{-3/2}$ remains unchanged (see below). We get more insight into this behaviour by studying time-dependent phenomena. In the integrable Harper model we had found an unbounded linear (i.e. diffusive) spread of wave packets [3] (see Sect.4). Subsequently unbounded diffusion was also observed in the classically chaotic kicked Harper model by Lima and Shepelyansky [37]. We now investigate how these observations are related [15]. Figures 10(a) and (b) compare the growth of the variance of a quantum mechanical wave packet and a corresponding classical distribution in phase space for kicking strengths κ corresponding to Figs. 8(c) and (d), respectively. The quantum behaviour mimics the classical diffusion for a finite time as is seen in the insets. There is a crossover time t^*, after which unbounded quantum diffusion due to hierarchical level clustering like in the integrable case dominates. Thus there are a classical and a quantum mechanical diffusive regime with different diffusion coefficients and the transition can be detected. This crossover is to be contrasted to the diffusion-localization crossover known from the kicked rotator [16–18].

For the wave packet of Fig. 10(a) ($\kappa = 3.5, \hbar = 2\pi/(400 + \sigma_g)$), which was centered initially on the outermost torus of the upper right island, the transition occurs after about 20 time steps. In this situation the crossover time t^* scales as $t^* \sim 1/\hbar$. Here the classical diffusion coefficient is reduced due to the fact that many classical trajectories are trapped within the island and do not contribute to diffusion. More typically, however, we find the classical diffusion coefficient to exceed the quantum mechanical one. An example is shown in Fig. 10(b) ($\kappa = 5.0, \hbar = 2\pi/(10 + \sigma_g)$) where the wave packet moves in a classically fully chaotic phase space. In this case the crossover time numerically appears to scale roughly as $\tau \sim 1/\hbar^2$. Thus the fingerprints of classical chaos in the diffusion behaviour of the kicked Harper model are the short time diffusion and its scaling. One expects the crossover in the time domain to correspond to a crossover at some scale s^* in the spectrum. For the integrated level spacing distribution (Fig. 11)

Fig. 9. Quasienergy spectrum versus $\sigma \in [0, 1]$ for $\kappa/\hbar = 1, 3, 6$, and 10 (top to bottom). The quasienergies ω range from $-\pi$ to π (-2 to 2 in the top figure). In the Hofstadter butterfly increasing $\kappa/\hbar$ removes the large energy gaps and broadens the bands. The hierarchical level clustering is removed on large scales. Note that we keep $\kappa/\hbar$ constant as $\sigma = 2\pi\hbar$ is varied.

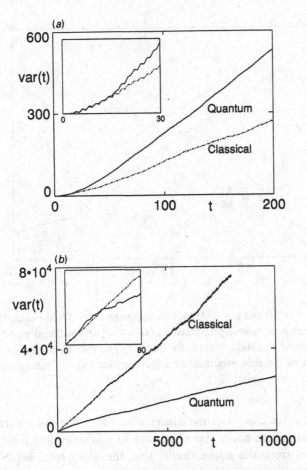

Fig. 10. (*a*) Quantum versus classical time evolution (full and dot-dashed line, respectively) of the variance of a wave packet initially centered on the outermost torus of the upper right island in Fig. 8(*c*) with $\kappa = 3.5$ and $\hbar = 2\pi/(400 + \sigma_g)$. The inset shows that after a crossover time a second diffusion process sets in, i.e. quantum diffusion due to hierarchical level clustering. (*b*) Same as (*a*) for the fully chaotic phase space of Fig. 8(*d*) with $\kappa = 5$ and $\hbar = 2\pi/(10 + \sigma_g)$. Here the quantum diffusion coefficient is smaller than the classical [10].

we find an inverse power law $p_{int}(s) \sim s^{-1/2}$ for small spacings $s < s^*$. On small scales the spectrum thus is unaffected by classical chaos. The figure clearly shows a crossover on larger scales where the hierarchical level clustering is destroyed due to classical chaos. Thus in contrast to the level repulsion occurring on small scales for discrete spectra, the small-scale fractal nature of the Cantor spectrum remains unchanged by classical chaos. If, however, a finite approximation of the Cantor spectrum (for a rational approximant of σ) is used, there are no arbitrarily small scales and level repulsion may show up at the smallest occurring

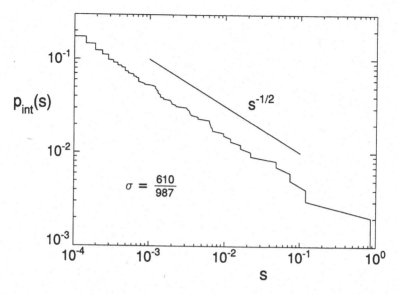

Fig. 11. ILSD for $\kappa = 5.0$ and $\sigma = 233/2474$ an approximant of $1/(10+\sigma_g)$ (corresponding to Fig. 10(*b*)). There is an *inverse* power law $p_{int}(s) \sim s^{-1/2}$ on small scales and a crossover to a regime influenced by classical chaos for larger s. The spectrum was determined by a Fourier transform of the time evolution of a wave packet over 2^{17} time steps yielding a quasienergy resolution of about 10^{-4}.

scale (Fig. 12). In this case, with increasing values of κ one finds a transition of the entire spectrum from an inverse power law to a Wigner distribution.

One concludes from the above results that the asymptotic behaviour ($t \rightarrow \infty, s \rightarrow 0$) of the kicked system is determined by the *integrable* Harper's equation already. In the strong quantum limit ($K/\hbar, L/\hbar \rightarrow 0$) we can explain this by mapping the eigenvalue equation of $\hat{U}$ to the integrable Harper's equation. Denoting by ψ^- and ψ^+ the eigenfunctions before and after the kick, in analogy to Grempel et. al. [40] we define

$$\bar{\psi}_\omega(x, t) = \psi_\omega^+(x, t)/\{1 - i \tan[(K/2\hbar)\cos x]\}. \tag{29}$$

We use the Floquet theorem and write $\bar{\psi}_\omega(x, t) = e^{i\omega t}\bar{u}_\omega(x, t)$ where $\bar{u}(x, t + 1) = \bar{u}(x, t)$. For the Fourier components $\varphi_n = (2\pi)^{-1}\int_0^{2\pi} e^{inx}\bar{u}_\omega(x, t)dx$ we formally obtain a tight-binding equation

$$\sum_{r(\neq 0)} W_r\varphi_{n+r} + T_n\varphi_n = -W_0\varphi_n \tag{30}$$

where the hopping terms W_r are the Fourier coefficients of $W(x) = -\tan[(K/2\hbar)\cos x]$ and

$$T_n = \tan[\frac{\omega}{2} - \frac{L}{2\hbar}\cos(\hbar(n + \theta_x))]. \tag{31}$$

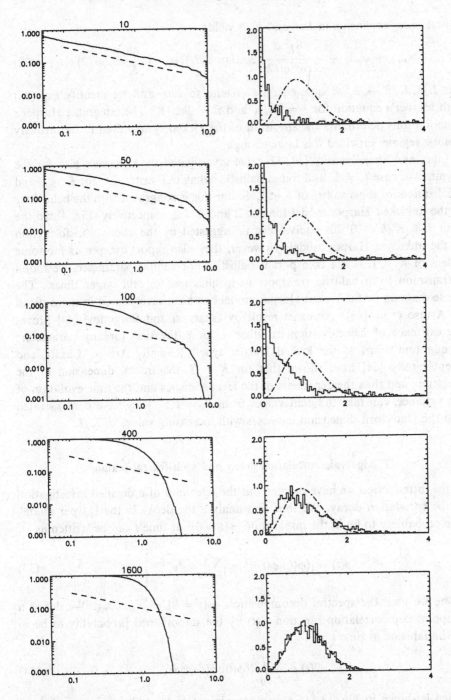

Fig. 12. ILSD (left) and LSD (right) for a *finite approximation* of the Cantor spectrum ($\sigma = 987/1597$) and increasing values of $\kappa = 10, 50, 100, 400, 1600$ displaying a transition from an inverse power law (dashed line) to a Wigner distribution (dash-dotted line). This transition of the entire spectrum is only possible because the fractal scaling does not extend to arbitrarily small scales for rational σ.

A first order expansion in $L/\hbar$ and $K/\hbar$ yields

$$\varphi_{n+1} + \varphi_{n-1} + \frac{2L}{K\cos^2\omega/2} \cdot \cos(\hbar(n + \theta_x))\varphi_n = \frac{4\hbar}{K}\tan(\omega/2)\varphi_n. \quad (32)$$

For $L/\hbar, K/\hbar \to 0$ the quasienergies ω tend to zero and we identify Eq. (32) with Harper's equation for $\lambda = 2L/K$ and $E = 2\hbar\omega/K$. The integrable Harper's equation thus determines the spectrum of the kicked system even in the strongly chaotic regime, provided $\hbar$ is large enough.

Lima and Shepelyansky [37, 42] have investigated the dynamics also for the asymmetric case $K \neq L$, and found ballistic behaviour var$(t) \sim t^2$ for $K > L$ and localization for most values of $K < L$. So far this is congruent with the behaviour of the unkicked Harper model for $\lambda < 2$ and $\lambda > 2$, respectively (Fig. 4). In the limit $L/\hbar, K/\hbar \to 0$ this equivalence is suggested by the above transformation to the unkicked Harper model. However, they also report exceptions for some values of $K < L$ where transport is ballistic. For similar parameters we found a transition from ballistic transport to localization for still larger times. The phase diagram of the kicked Harper model has been investigated in more detail by Artuso et al. [43]. An exact result by Guarneri and Borgonovi [44] proves the existence of delocalization in either x- or p-direction for any parameters in quantum maps of the kicked Harper type. Recently Artuso, Casati, and Shepelyansky [34] have shown that for $K = L$ the fractal dimension of the spectrum, and thus the exponents of the level statistics and the time evolution of the variance, can take different values from $D_0 = 0.5$. They have demonstrated that the Hausdorff dimension increases with increasing values of $K/\hbar$.

7 Algebraic correlation decay and multifractal scaling

In the introduction we have pointed out the relevance of a detailed investigation of the correlation decay. We thus first analyze the decay in the Harper model. The probability to be in the initial state $|\phi(t = 0)\rangle$ at time t can be written as

$$p(t) = |\langle\phi(0)|\phi(t)\rangle|^2 = \left|\sum_k |a_k|^2 e^{-i\omega_k t}\right|^2, \quad (33)$$

where we used the spectral decomposition $\phi_n(t = 0) = \sum_k a_k \psi_{n,k}$. We define a temporal autocorrelation function $C(t)$ by the smoothened probability to be in the initial state at time t

$$C(t) = \frac{1}{t}\int_0^t |\langle\phi(0)|\phi(t')\rangle|^2 dt', \quad (34)$$

which is shown in Fig. 13 for σ an approximant of the golden mean and three different values of λ. All numerical simulations were started with localized initial wave packets. For $\lambda = 2$ we find an extremely slow decay of the correlation function following a power law $C(t) \sim t^{-\nu}$ with $\nu = 0.14 \pm 0.01$. For $\lambda < 2$ the

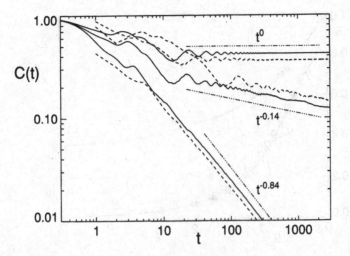

Fig. 13. Correlation function in the Harper model (solid lines) for $\sigma = 1597/2584$ an approximant of the golden mean displaying power laws $C(t) \sim t^{-\nu}$ with $\nu = 0.84 \pm 0.01$, $\nu = 0.14 \pm 0.01$, and $\nu = 0$ for $\lambda = 1$, 2, and 3, respectively. For the kicked Harper model (dashed lines) with σ the golden mean, the same asymptotic behaviour is obtained for $K = 6$ and $L = 3$, 6, and 9 corresponding to the extended, critical, and localized regime, respectively.

correlation decays with $\nu = 0.84 \pm 0.01$, while the extended eigenstates and the absolutely continuous spectrum might have suggested an exponential decay. The slow power-law decay with $\nu = 0.84$ thus must be a manifestation of a recurrent absolutely continuous spectrum [20]. The singular continuous spectrum for $\lambda = 2$ on the other hand, has an even slower decay ($\nu = 0.14$). For $\lambda > 2$ there is no decay, as expected for localized states.

In order to study the influence of classical chaos on the decay of $C(t)$ we numerically investigate the kicked Harper system, which has a classically chaotic phase space [36, 37, 15]. The results for the correlation function Eq. (34) are shown by the dashed lines in Fig. 13 for parameters in the strongly chaotic regime. The asymptotic behaviour of $C(t)$ is the same as for the integrable Harper model and thus the decay of the quantum correlation function does *not* reflect the existence of classical chaos.

In the Fibonacci chain model the decay of the correlation function also shows power-law behaviour and the exponent ν depends on the parameter V. This dependence is shown in Fig. 14 where we find values of ν ranging from 0 to 0.84. Here the spectrum is singular continuous for all $V > 0$ and becomes absolutely continuous for $V = 0$ [32].

Summarizing the above results we find asymptotic power-law decays $C(t) \sim t^{-\nu}$ with $\nu < 1$ for all three quasiperiodic systems. This asymptotic behaviour calls for

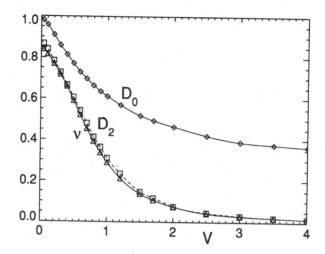

Fig. 14. Power-law exponents ν (triangles) of the correlation function decay for an initially localized wave packet and various potential strengths V in the Fibonacci model. Theory predicts the equality of ν and the generalized dimension D_2 (squares) according to Eq. (40) and gives an upper bound $D_0 \geq \nu$ (see Eq. (41)), where D_0 (diamonds) is the fractal dimension.

a general explanation. We give an analytic derivation of the correlation function $C(t)$ for spectra (singular continuous or absolutely continuous) characterized by generalized dimensions D_q [29]. First we specify the Hamiltonian $\hat{H}$ through its spectral projections $\hat{P}_\omega$ [45], where $\omega \in \text{spec}(H)$. The spectral measure $\mu_\phi(\omega)$ with respect to an initial state vector $|\phi\rangle$ is defined by $\mu_\phi(\omega) = \langle \phi | \hat{P}_\omega | \phi \rangle$. The probability $p(t) = |\langle \phi(0) | \phi(t) \rangle|^2$ generally can be written as [20]

$$p(t) = \left| \int_{-\infty}^{+\infty} e^{-i\omega t} d\mu_\phi(\omega) \right|^2 = 2\pi \, \tilde{\mu}_\phi(t) \, \tilde{\mu}_\phi(t)^* \tag{35}$$

where $\tilde{\mu}_\phi(t)$ denotes the Fourier-Stieltjes transform of $\mu_\phi(\omega)$. Among the generalized dimensions D_q in particular we need D_2, which is defined by the scaling behaviour

$$\gamma(l) = \int d\mu_\phi(\omega) \int_{\omega-l/2}^{\omega+l/2} d\mu_\phi(\omega') \sim l^{D_2} \quad (l \to 0). \tag{36}$$

The function $\gamma(l)$ gives the probability that two eigenfunctions picked (with the according probability) from the spectral decomposition of $|\phi\rangle$ have an energy difference less than l [46]. We now show that $\gamma(l)$ is related to $p(t)$. To this end we introduce the characteristic function

$$\chi_A(x) = \begin{cases} 1 & \text{if } x \in A \\ 0 & \text{if } x \notin A \end{cases} \tag{37}$$

with $A = [-l/2, l/2]$ and rewrite Eq. (36)

$$\gamma(l) = \int d\mu_\phi(\omega) \int \chi_A(\omega - \omega') d\mu_\phi(\omega'). \tag{38}$$

Using the convolution theorem we write the second integral as the inverse Fourier transform of the product of the Fourier transforms of $\mu_\phi(\omega)$ and $\chi_A(\omega)$, yielding

$$\gamma(l) = \frac{1}{\pi} \int_{-\infty}^{+\infty} d\mu_\phi(\omega) \int_{-\infty}^{+\infty} e^{i\omega\tau} \tilde{\mu}_\phi(\tau) \frac{\sin(l\tau/2)}{\tau} d\tau$$

$$= \sqrt{2/\pi} \int_{-\infty}^{+\infty} \tilde{\mu}_\phi(\tau)^* \tilde{\mu}_\phi(\tau) \frac{\sin(l\tau/2)}{\tau} d\tau = \sqrt{2/\pi^3} \int_0^{+\infty} p(\tau) \frac{\sin(l\tau/2)}{\tau} d\tau \tag{39}$$

where we made use of Eq. (35). We thus have a simple relation between the temporal probability $p(t)$ and the spectral probability $\gamma(l)$. Assuming a pure power-law behaviour either for $\gamma(l)$ or for $C(t)$ we derive from Eq. (39) the relation

$$\gamma(l) \sim l^\alpha \quad (l \to 0) \quad \Leftrightarrow \quad C(t) \sim t^{-\alpha} \quad (t \to \infty) \tag{40}$$

which holds for $0 < \alpha < 1$ and where $\alpha = D_2$ according to Eq. (36). A scaling behaviour in $\gamma(l)$ is not necessarily restricted to Cantor spectra, whereas a Cantor spectrum always implies the scaling behaviour of Eq. (40). If the spectrum has continuous *and* pure point components, $C(t)$ will not decay even if $\gamma(l)$ shows scaling behaviour. A similar relation with $C(t)$ replaced by $p(t)$ (i.e. without smoothing) does not hold in general, e.g. for the standard middle third Cantor set $C(t)$ decays like $C(t) \sim t^{-D_2}$ whereas $p(t)$ does not decay to zero.

This result connects the algebraic decay of the correlation function with the multifractal structure of the spectral measure $\mu_\phi(\omega)$. The value of the exponent, i.e. the generalized dimension D_2 can only be computed if the spectrum, the eigenfunctions, and $\phi(t = 0)$ are known. If only the spectrum is known, it is possible to give an upper bound for D_2 as follows. On the one hand, from $\int d\mu_\phi(\omega) = 1$ it follows that $D_2(\mu_\phi(\omega)) \leq D_0(\mu_\phi(\omega))$, where $D_0(\mu_\phi(\omega))$ is the Hausdorff dimension of the spectral measure [29]. On the other hand, $D_0(\mu_\phi(\omega))$ is the fractal dimension of the subset of spec(H) that is excited by the initial wavepacket, whereby $D_0(\mu_\phi(\omega)) \leq D_0(\text{spec}(H))$. Therefore the upper bound $D_0(\text{spec}(H))$ gives a lower bound for the decay of the correlation function, which is consistent with an earlier bound given by Salem [47]

$$C(t) \geq c\, t^{-D_0(\text{spec}(H))} \quad (t \to \infty) \tag{41}$$

where c is an appropriate constant.

To illustrate these analytical results we have determined the dimension D_2 for the Harper model and the Fibonacci model by dividing the energy range into

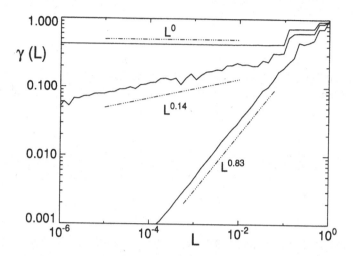

Fig. 15. Probability $\gamma(l)$ versus l (Eq. (42)) for Harper's equation, from which we deduce $D_2 = 0.83 \pm 0.01$, $D_2 = 0.14 \pm 0.01$, and $D_2 = 0$ for $\lambda = 1, 2$, and 3, respectively. The values of D_2 equal those of ν in Fig. 13 within the numerical errors.

boxes B_i of length l and computing the function $\gamma(l)$ as [46]

$$\gamma(l) = \sum_i \left(\sum_{\omega_k \in B_i} |a_k|^2 \right)^2 \sim l^{D_2} \quad (l \to 0). \tag{42}$$

For the Harper model this function is shown in Fig. 15 for three different values of λ corresponding to those of Fig. 13. The values of D_2 agree with those of ν in Fig. 13. In the same way we have computed the dimension D_2 for the Fibonacci model as a function of the potential strength V. This result is shown by the dashed line in Fig. 14 where the values of ν and D_2 also coincide within the numerical errors and D_0 is recognized as an upper bound. Due to the spectral measure of the initial state ϕ involved in Eq. (36) these values of D_2 are very different from the dimensions D_2 of Fig. 3 and of the scaling region in the Fibonacci chain model shown in Fig. 7. This is visualized in Fig. 16, where we have plotted the different $f(\alpha)$-curves for the uniform measure [28] and for the spectral measure of an initially localized wave packet. It can be seen that the curve has widened and the range of scaling-indices α is increased. This comes along with decreasing values of D_q for $q > 0$, e.g. $D_2 = 0.14$ instead of 0.493.

We thus have shown that the decay of the quantum autocorrelation function is determined by the generalized dimension D_2 of the spectral measure. In contrast, the growth of the variance of a wave packet is governed by the fractal dimension, i.e. $\mathrm{var}(t) \sim t^{2D_0}$. As a curiosity we note that in a different context, i.e. in dissipative dynamical systems D_2 already goes by the name of correlation dimension as it describes the spatial correlations of an attractor.

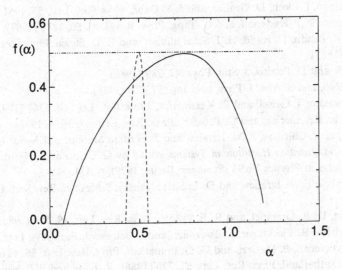

Fig. 16. The $f(\alpha)$-curve for the spectral measure of an initially localized wave packet (solid line) compared with the curve for the uniform measure [28] (dashed line).

Acknowledgements

We acknowledge stimulating discussions with I. Guarneri. This work was supported by the Deutsche Forschungsgemeinschaft.

REFERENCES

[1] D. Weiss, M. L. Roukes, A. Menschig, P. Grambow, K. v. Klitzing, and G. Weimann, Phys. Rev. Lett. **66**, 2790 (1991).

[2] T. Geisel, J. Wagenhuber, P. Niebauer, and G. Obermair, Phys. Rev. Lett. **64**, 1581 (1990); J. Wagenhuber, T. Geisel, P. Niebauer, and G. Obermair, Phys. Rev. B **45**, 4372, (1992).

[3] T. Geisel, R. Ketzmerick, and G. Petschel, Phys. Rev. Lett. **66**, 1651 (1991).

[4] see e.g. O. Bohigas and M. J. Giannoni, in *Mathematical and Computational Methods in Nuclear Physics*, edited by J. S. Dehesa, J. M. G. Gomez, and A. Polls, Lecture Notes in Physics Vo.209 (Springer Berlin, 1984), p. 1; F. Haake, *Quantum Signatures of Chaos*, Springer Series in Synergetics Vol.54 (Springer, Berlin, 1991).

[5] P. G. Harper, Proc. Roy. Soc. Lond. **A68**, 874 (1955).

[6] M. Ya. Azbel', Sov. Phys.–JETP **19**, 634 (1964).

[7] D. R. Hofstadter, Phys. Rev. **B14**, 2239 (1976).

[8] For reviews see e.g.: B. Simon, Adv. Appl. Math. **3**, 463 (1982); J. B. Sokolov, Phys. Rep. **126**, 189 (1985).

[9] T. Geisel, R. Ketzmerick, and G. Petschel, in *Quantum Chaos - Theory and Experiment*, edited by P. Cvitanovic, I. C. Percival, and A. Wirzba (Kluwer, Dordrecht, 1992), p. 43.

[10] D. Shechtman, I. Blech, D. Gratias, and J. V. Cahn, Phys. Rev. Lett. **53**, 1951 (1984).

[11] M. Kohmoto, L. P. Kadanoff, and C. Tang, Phys. Rev. Lett. **50**, 1870 (1983); S. Ostlund, R. Pandit, D. Rand, H. J. Schellnhuber, and E. D. Siggia, Phys. Rev. Lett. **50**, 1873 (1983).

[12] J. M. Luck and D. Petritis, J. Stat. Phys. **42**, 289 (1986).

[13] H. Hiramoto, and S. Abe, J. Phys. Soc. Jap. **57**, 230 (1988).

[14] R. Fleischmann, T. Geisel, and R. Ketzmerick, Phys. Rev. Lett. **68**, 1367 (1992).

[15] T. Geisel, R. Ketzmerick, and G. Petschel, Phys. Rev. Lett. **67**, 3635 (1991).

[16] G. Casati, B. V. Chirikov, F. M. Izrailev, and J. Ford, in *Stochastic Behavior in Classical and Quantum Hamiltonian Systems*, edited by G. Casati and J. Ford, Lecture Notes in Physics Vo.93 (Springer, Berlin, 1979), p. 334.

[17] B. V. Chirikov, F. M. Izrailev, and D. L. Shepelyansky, Soviet Sci. Rev. Sect. **C2**, 209 (1981).

[18] S. Fishman, D. R. Grempel, and R. E. Prange, Phys. Rev. Lett. **49**, 509 (1982).

[19] M. Samuelides, R. Fleckinger, L. Touzillier, and J. Bellissard, Europhys. Lett. **1**, 203 (1986); Y. Pomeau, B. Dorizzi, and G. Grammaticos, Phys. Rev. Lett. **56**, 681 (1986); B. Sutherland, Phys. Rev. Lett. **57**, 770 (1986); R. Badii and P. F. Meier, Phys. Rev. Lett. **58**, 1045 (1987); J. M. Luck, H. Orland, and U. Smilansky, J. Stat. Phys. **53**, 551 (1988); R. Graham, Europhys. Lett. **8**, 717 (1989); T. Geisel, Phys. Rev. A **41**, 2989 (1990); G. Jona-Lasinio, C. Presilla, and F. Capasso, Phys. Rev. Lett. **68**, 2269 (1992).

[20] J. Avron and B. Simon, J. Funct. Anal. **43**, 1 (1981).

[21] R. Ketzmerick, G. Petschel, and T. Geisel, Phys. Rev. Lett. **69**, 695 (1992).

[22] S. Aubry and G. André, in *Proceedings of the Israel Physical Society*, edited by C. G. Kuper (Hilger, Bristol, 1979), Vol.3, p.133.

[23] J. Bellissard and B. Simon, J. Funct. Anal. **48**, 408 (1982).

[24] J. Avron and B. Simon, Bull. Am. Math. Soc. **6**, 81 (1982).

[25] R. E. Borland, in *Mathematical Physics in One Dimension*, edited by E. H. Lieb and D. C. Mattis (Academic Press, New York, 1966), p.929.

[26] S. Ostlund and R. Pandit, Phys. Rev. **B29** 1394, (1984).

[27] M. Wilkinson, J. Phys. **A20**, 4337 (1987).

[28] C. Tang and M. Kohmoto, Phys. Rev. **B34**, 2041 (1986).

[29] T. C. Halsey, M. H. Jensen, L. P. Kadanoff, I. Procaccia, and B. I. Shraiman, Phys. Rev. **A33**, 1141 (1986).

[30] P. B. Allen, J. Phys. **C13**, L667 (1980).

[31] H. Hiramoto, S. Abe, J. Phys. Soc. Jap. **57**, 1365 (1988).

[32] J. Bellissard, B. Iochum, E. Scoppola, and D. Testard, Comm. Math. Phys. **125**, 527 (1989).

[33] M. J. Feigenbaum and B. Hasslacher, Phys. Rev. Lett. **49**, 605 (1982).

[34] R. Artuso, G. Casati, and D. Shepelyansky. Phys. Rev. Lett. **68**, 3826 (1992).

[35] M. Wilkinson, Proc. Roy. Soc. London **A391**, 305 (1984).

[36] P. Leboeuf, J. Kurchan, M. Feingold, and D. P. Arovas, Phys. Rev. Lett. **65**, 3076 (1990).

[37] R. Lima and D. L. Shepelyansky, Phys. Rev. Lett. **67**, 1377 (1991).

[38] G. M. Zaslavskii, M. Yu. Zakharov, R. Z. Sagdeev, D. A. Usikov, and A. A. Chernikov, JETP Letters **44**, 451 (1986).

[39] Ya. B. Zeldovich, Zh. Eksp. Teor. Fiz. **51**, 1492 (1966) [Sov. Phys.– JETP **24**, 1006 (1967)].

[40] D. R. Grempel, R. E. Prange, and S. Fishman, Phys. Rev. **A29**, 1639 (1984).

[41] S.-J. Chang and K.-J. Shi, Phys. Rev. **A34**, 7 (1986).

[42] D. Shepelyansky, in *Quantum Chaos – Quantum Measurement*, edited by P. Cvitanovic, I. Percival, and A. Wirzba (Kluwer, Dordrecht, 1992), p.81.

[43] R. Artuso, F. Borgonovi, I. Guarneri, L. Rebuzzini and G. Casati, Phys. Rev. Lett **69**, 3302 (1992).

[44] I. Guarneri and F. Borgonovi, J. Phys. A **26**, 119 (1993).

[45] M. Reed and B. Simon, *Methods of Mathematical Physics, I. Functional Analysis* (Academic Press, New York, 1978).

[46] P. Grassberger and I. Procaccia, Phys. Rev. **A28**, 2591 (1983).

[47] R. Salem, Ark. Math. **1**, 353 (1950).

Microwave studies in irregularly shaped billiards

H.-J. STÖCKMANN, J. STEIN, M. KOLLMANN

Fachbereich Physik der Philipps-Universität,
D-35032 Marburg, Federal Republic of Germany

Abstract

Until a short time ago, the study of the quantum-mechanical behaviour of classically chaotic systems was an exclusive domain of the theoretical physicist. Model systems amongst others were irregularly shaped billiards, such as the Sinai and the stadium billiard. An alternative experimental approach to study these systems has recently been demonstrated using the fact that the time-independent Schrödinger and wave equations are mathematically equivalent (though, in general, the boundary conditions are different). This chapter concentrates on the presentation of microwave studies of billiards, but experiments with vibrating plates and water surfaces waves are also briefly discussed. Topics are the statistical properties of spectra, level dynamics with respect to geometrical changes of the billiards, and the close connection between quantum-mechanical spectra and classical trajectories.

1 Introduction

One day in February 1808 Ernst F. Chladni who was then on a long stay in Paris received an invitation to the Tuileries to give a demonstration of his famous experiments with vibrating plates in the presence of Napoleon and an illustrious audience. The Emperor was impressed by the performance and offered a prize of 3000 francs for the correct mathematical explanation of the sound figures which was still lacking at that time. The prize was paid in 1816 to the French mathematician Sophie Germain, in spite of the fact that her solution was still incomplete. The correct explanation for circular plates was not given until 1850 by Robert Kirchhoff (all the above details are taken from the book of F. Melde on the life of Chladni [1]). For irregularly shaped plates the solution was not found in the last century, though, interestingly enough, in a 1891 physics handbook Chladni figures on a plate with an irregular boundary were reproduced [2].

In a way the Chladni figures can be considered as precursors of the microwave studies presented in this chapter. The recent interest in these nearly 200 years old experiments arose from a number of theoretical studies of spectra and eigenfunctions of electrons confined to irregularly shaped boxes with infinitely high walls, called billiards in analogy to the corresponding classical systems. Nonintegrable billiards are perhaps conceptually the simplest systems exhibiting classical chaos. In contrast to other chaotic model systems such as coupled anharmonic oscillators or periodically kicked tops here the chaos does not originate in a nonlinear Hamiltonian but in irregular boundary conditions. Bohigas et al. [3] showed that in the Sinai billiard, a rectangle with an excised central circle, the eigenenergy distances obey a Wigner distribution exactly as is observed, for example, in highly excited states of nuclei [4] and molecules [5]. McDonald and Kaufman [6, 7] calculated eigenfunctions of a stadium-shaped billiard and observed a pattern of meandric nodal lines, starting and ending at the boundary, or forming closed loops, but avoiding crossings. Heller [8] discovered that in the stadium there are wavefunctions which are not distributed more or less homogeneously over the total area, but which exhibit regions of extra high amplitude, the so-called scars, close to classical periodic orbits. The appearance of periodic orbits suggested that semiclassical methods, as developed by Gutzwiller [9], should be an appropriate tool for the understanding of billiard wavefunctions. It will be shown in the following that the Gutzwiller theory indeed plays a dominating role in the interpretation of billiard spectra and eigenfunctions.

All the above-mentioned theoretical studies of nonintegrable billiards were hampered by the fact that the irregular boundary conditions make the calculations extremely time consuming. If a large number of spectra is to be calculated, as is necessary, for example, for the study of level dynamics, these limitations become prohibitive. For the same reason only two-dimensional billiards have been studied up to now. This gave an opportunity for experimentalists to contribute to a field which up to now has seemed to be the exclusive domain of theorists.

2 Experimental techniques

The basic idea for the experimental approach to the study of billiards is the equivalence of the time-independent Schrödinger equation

$$\Delta\psi_n = \frac{2mE_n}{\hbar^2}\psi_n, \tag{1}$$

and the time-independent wave equation, the Helmholtz equation,

$$\Delta\psi_n = -k_n^2\psi_n. \tag{2}$$

Substituting billiards by correspondingly shaped resonators, the eigenfrequencies can be determined by resonance absorption. Comparison of eqs. (1) and (2) shows that up to a factor the square of the wavenumber corresponds to the quantum-mechanical eigenenergy, while the field amplitude corresponds to the quantum-mechanical wavefunction. As in both cases the boundary conditions may be different, in general this equivalence is incomplete.

One idea suggests itself immediately: namely to study the vibrations of metal and glass plates. This brings us back to the Chladni figures. In fact, vibrating plates do not obey the Helmholtz equation but the fourth order differential equation [10]

$$(\Delta)^2 \psi_n = k_n^4 \psi_n, \tag{3}$$

but this difference is not of relevance for the following qualitative discussion. Fig. 1 shows a number of Chladni figures for differently shaped glass plates. All the plates were fixed in the center, placed on an overhead projector, and excited by vibrations from a loudspeaker. Typical vibration frequencies were in the range 300–2000 Hz. The upper figure shows a nodal pattern for a circular plate, as was produced by Chladni himself. One observes a regular net of intersecting circles and straight lines which is typical for an integrable system. The photograph in the middle shows a Chladni figure for a rectangular plate. Here the central clamp disturbs the integrability making the system pseudointegrable. For small frequencies the clamp has only a little influence, for higher frequencies it leads to the avoidance of some of the crossings as observed in the figure. The bottom figure shows a Chladni figure for a nonintegrable plate, a quartered Sinai billiard [11]. One observes a pattern of meandric nodal lines, exactly as was seen for the first time in the calculations of McDonald and Kaufman [6]. A detailed account on the behaviour of nodal lines in billiards can be found in chapter 15 of ref. [9].

Another beautiful analog experiment was performed by Blümel and coworkers [12]. They excited water surface waves in small tanks in the shape of a circle, and a stadium. Frequencies were typically between 0 and 50 Hz. All the features known from calculations were found in the experiments, including scars. For several high frequencies above 700 Hz fine ridge-like structures were seen. Similar ridges were produced in calculations by O'Connor et al. [13] by the superposition of plane waves of the same frequency, but with random directions, amplitudes and phases.

Both experiments, Chladni figures and water surface waves, are easily able to demonstrate very well the qualitative difference between integrable and nonintegrable systems. They are, however, not suited for quantitative measurements as the strong damping leads to resonance widths of the order of 1 Hz. This corresponds to rather low quality factors $Q = \nu/\Delta\nu$ of the order of 10–100 with the consequence of an intolerable loss of eigenfrequencies. A better alternative is the use of microwave resonators. Here qualities of 10^3–10^4 are easily accessible, and with superconducting cavities even values of 10^5–10^7 have been obtained [14]. Microwave measuring techniques are highly evolved, and analyzing systems in the frequency range 0–25 GHz are commercially available. The corresponding wavelengths are of the order of mm–cm, the fabrication of the corresponding cavities is easy and cheap. As an example Fig. 2 shows a part of the spectrum of a quartered stadium billiard [15]. Every minimum in the reflected microwave power corresponds to the excitation of an eigenfrequency. The height of the

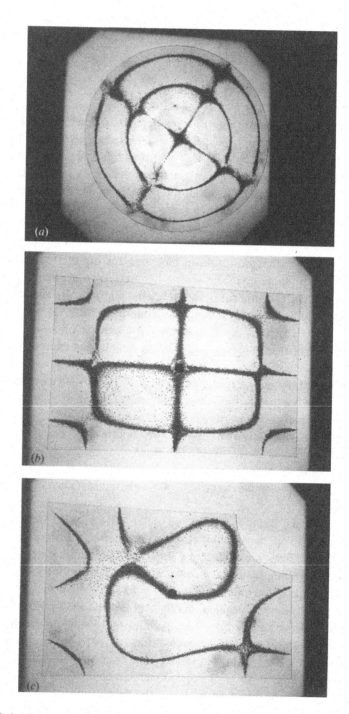

Fig. 1. Chladni figures of glass plates of (*a*) circular (*r*=10.5 cm), (*b*) rectangular (*a*=21 cm, *b*=14 cm), and (*c*) quartered Sinai billiard (*a*=21 cm, *b*=14 cm, *r*=5 cm) shapes [11]. All plates were 1 mm thick. The plates correspond to integrable, pseudointegrable, and nonintegrable systems, respectively (*Photographs: K. H. Kretschmer*)

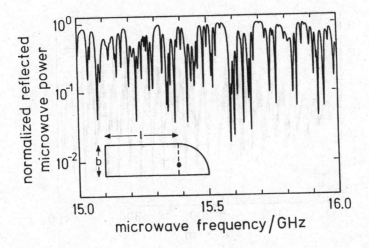

Fig. 2. Part of the eigenfrequency spectrum of a microwave cavity shaped like a quartered stadium billiard (l=36 cm, b=20 cm) [15] (see inset, the dot marks the position of the coupling wire).

resonator was $d = 8$ mm. An elementary calculation shows that for frequencies $v < v_{max} = c/2d$ the billiards are quasi-two-dimensional. The electric field vector is perpendicular to the large faces of the resonator, and the elecromagnetic boundary conditions reduce to a single one, $E = 0$ on the boundary. In this special case the time-independent Schrödinger equation and the wave equation are completely equivalent, including the boundary conditions, where the electric field strength corresponds to the wavefunction. As the depth of a resonance is proportional to the square of the electric field at the position of the coupling wire, a variation of this position thus even allows the experimental determination of quantum-mechanical wavefunctions.

If the coupling wire points exactly to a nodal line, the resonance cannot be excited, and the eigenvalue is lost. This problem can be partially overcome by using several coupling wires. A second cause for losses is the finite width of the resonances limiting the minimum resolvable distance to some MHz for quality factors of some thousands. In microwave cavities the damping is caused by the absorption of microwave energy in the resonator walls. For typical microwave frequencies of about 10 GHz and a good metallic conductor such as brass the microwave penetration depth, limited by the skin effect, is of the order of 1 μm. The loss of eigenvalues can be controlled by comparing the experimental integrated density of states $n_{exp}(E)$ with the value expected from the extended Weyl formula [16],

$$n_{Weyl}(E) = \frac{A}{4\pi} E - \frac{L}{4\pi} \sqrt{E} + \cdots, \qquad (4)$$

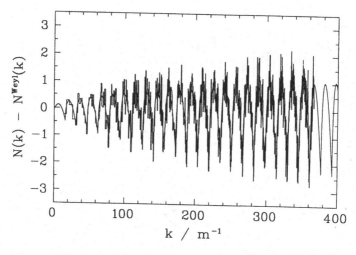

Fig. 3. The difference between the experimental integrated density of states obtained from a superconducting cavity of the shape of a quartered stadium billiard, and the corresponding value calculated from the extended Weyl formula [14] (*by courtesy of A. Richter*).

where A and L are the area and circumference of the billiard, respectively. The terms not written depend on the corners and the curvature of the boundary and are of the order of 1. In the whole paper it has been assumed that $\hbar^2/2m = 1$, i.e. $E = k^2$, where k is the de Broglie wave number of the particle.

In the first experiments of our group [15] the relative loss of eigenfrequencies amounted to about 15 per cent, but recently it was possible to reduce the loss to zero by following the eigenfrequencies as a function of the billiard length (see section 3). The loss of eigenfrequencies is also zero in superconducting cavities. Here, however, rapid changes of the billiard shape and of the position of the coupling wire are no longer possible. In Fig. 3 the difference between the experimental integrated density of states and the value calculated from the Weyl formula is shown as obtained from a superconducting cavity of the shape of a quartered stadium [14]. On average the difference is zero thus demonstrating that no eigenfrequency was lost. The periodic fluctuations reflect the presence of periodic orbits. Here it is predominantly the bouncing ball orbit between the long sides of the stadium which is responsible for the observed modulations. The influence of periodic orbits on the spectra of billiards will be discussed in more detail in section 4.

3 Energy distance distributions and level dynamics

One of the signatures of quantum systems whose classical counterparts are chaotic is the avoidance of level crossings if an external parameter is varied. This leads to characteristic differences between nearest neighbor spacing distributions for integrable and nonintegrable systems, respectively. In Fig. 4(a) the energy-distance histogram for a rectangular-shaped microwave resonator is shown [17]. All frequencies in the range 5–10 GHz were considered. Because of the equivalence $E \sim k^2 \sim v^2$ the distribution of $S_n = v_n^2 - v_{n-1}^2$ is plotted where all s_n were normalized to a mean distance of $< s >= 1$. The distribution is well described by a Poissonian,

$$p(s) = \exp(-s), \tag{5}$$

a behaviour typical of integrable systems [18]. The loss of eigenvalues at small distances is caused by the linewidth limiting the resolution to several MHz (see section 2). Therefore the small but significant deviations found in calculated spectra for distances close to zero [19] are not seen in the experiment. In reality, the system is not integrable but pseudointegrable, as the coupling wire necessary for the irradiation of the microwaves constitutes an unavoidable disturbance. As with the Chladni figures (see Fig. 1(b)) the influence of the wire is nearly negligible at low frequencies but becomes manifest at higher frequencies. Here the distance histogram deviates more and more form a Poissonian, and in the frequency range 15–18 GHz one observes the distribution shown in Fig. 4(b). In ref. [17] this change with frequency was quantitatively described by a random matrix model assuming Poissonian distributed diagonal elements and Gaussian distributed with frequency increasing nondiagonal elements. For the high frequency range shown in Fig. 4(b) the observed distance histogram is the same as for a rectangle with a delta-like singularity [20]. In Fig. 4(c) the energy-distance histogram for the quartered Sinai billiard is shown. The histogram is well described by a Wigner distribution,

$$p(s) = \frac{\pi}{2} s \exp(-\frac{\pi}{4} s^2) \tag{6}$$

being universal for nonintegrable systems. In calculations of billiard eigenvalues this was found for the first time by Bohigas and coworkers [3] (in recent experiments with superconducting cavities significant deviations from the Wigner distribution were observed which could be traced back to the bouncing ball orbit [14]).

Another quantity well suited to discriminate between integrable and nonintegrable systems is the spectral rigidity $\Delta_3(L)$ introduced by Dyson and Mehta [21]. This is defined by

$$\Delta_3(L) = \min_{a,b} \frac{1}{L} \int_{E-L/2}^{E+L/2} [n(\epsilon) - a\epsilon - b]^2 d\epsilon. \tag{7}$$

Fig. 5 shows spectral rigidities for a cavity shaped like a quartered stadium billiard for different lengths l [22]. For small l values the experimental points are close to the straight line corresponding to integrable systems, whereas with increasing l the data more and more approach the curved line expected for

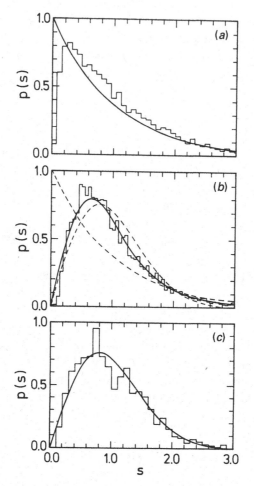

Fig. 4. Energy–distance histogram for a rectangular billiard in the frequency ranges (*a*) 5–10 GHz, (*b*) 15–18 GHz [17], and (*c*) for a Sinai billiard [15]. The full lines correspond to the expected distributions for integrable, pseudointegrable and nonintegrable systems, respectively. The histogram for pseudointegrable systems lies in between the distributions expected for integrable, and non-integrable systems (broken lines in Fig. 4(*b*)).

nonintegrable systems (for a more detailed discussion of $\Delta_3(L)$ see the review articles of Brody and coworkers [23] and Eckhardt [24]). For large L values one expects a saturation behaviour of $\Delta_3(L)$ reflecting the presence of periodic orbits [25]. Exactly the opposite behaviour was found in the above-mentioned experiment using a super conducting cavity of the shape of a quartered stadium [14]. This rather unexpected behaviour is again caused by the bouncing ball orbit which, being marginally stable, changes the rigidity for large L values

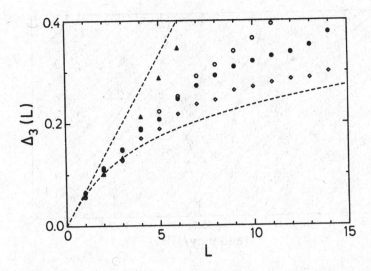

Fig. 5. Spectral rigidity $\Delta_3(L)$ for quartered stadium billiards ($r=20$ cm) of different lengths l (▲ : 0 cm, ○ : 0.8 cm, ● : 1.3 cm, ◇ : 3.5 cm) [22]. The case $l=0$ corresponds to the quarter circle. The straight and the curved lines correspond to the expected L-dependences for integrable, and nonintegrable systems, respectively. The experimental points were obtained by taking the average of $\Delta_3(L)$ over the frequency range 0–18 GHz for $l > 0$ cm and 0–9.6 GHz for $l=0$ cm.

dramatically. For a detailed explanation we refer the reader to the original publication.

It was mentioned at the beginning of this section that the observed distance distributions are a consequence of avoiding the crossings of eigenvalues under change of an external parameter. The simplest possible model Hamiltonian was introduced by Pechukas [26] and is given by

$$\mathcal{H} = \mathcal{H}_0 + \lambda V, \tag{8}$$

where $\mathcal{H}_0$ is integrable, and V is nonintegrable. The coupling strength λ takes the role of a time. As a function of λ, the eigenvalues move on the energy axis just like the particles of a one-dimensional gas with a repulsive interaction. If this gas is treated using statistical mechanics the Wigner distribution results as the thermal equilibrium distribution. In reality there are small deviations from the Wigner distribution, details of this point and the whole level dynamics complex are found in chapter 6 of ref. [27].

For billiards things are a bit different. Here the Hamiltonian is given by the free particle Hamiltonian $\mathcal{H}_0 = -\Delta$, and parameter dependences result from a change of the billiard shape, but not from a change of the Hamiltonian. As an experimental example Fig. 6 shows a part of the spectrum of a quartered Sinai billiard as a function of the length [28,29]. The spectrum is not normalized

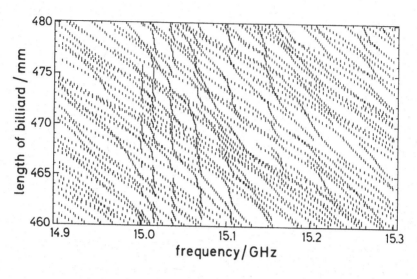

Fig. 6. Level dynamics of a quartered Sinai billiard as a function of length ($b = 20$ cm, $r = 7$ cm) [28,29].

to a constant density of states. This leads to the observed overall decrease of eigenfrequencies with length, which is an immediate consequence of the fact that due to the Weyl formula (4) the density of states is asymptotically proportional to the area of the billiard. The figure shows that regions exhibiting strong dynamics between the eigenvalues change with comparatively calm regions. These inhomogeneities appear in the whole studied frequency range. They were observed already in ref. [15] and associated there with the bouncing ball orbit. The temporary disappearance of some eigenvalues is caused by the passing of a nodal line through the position of the coupling wire (see section 2). The corresponding loss of eigenvalues can therefore be corrected for as the shift of eigenvalues is studied as a function of the length. The same is true for the case in which two near neighbor eigenvalues cannot be separated because of the limit of resolution. In the present example all 720 eigenvalues expected from the extended Weyl formula (4) were recovered in this way.

The analysis of the level dynamics allows the determination not only of the positions, but also of the velocities and accelerations of the eigenvalues. From the study of avoided crossings information on the repulsive "potential" can be obtained. As an example Fig. 7 shows the distribution of the normalized curvatures $a_n = \ddot{x}_n / <\dot{x}_n^2>$ (the x_n are the eigenvalues normalized to constant density, the dot means derivation with respect to the length, and $<\dot{x}_n^2>$ is a local average of the squared velocity) [28]. The accelerations have their maxima near the avoided crossing. Therefore the asymptotic curvature distribution depends only on the short range behaviour of the nearest neighbor distribution, i.e. on

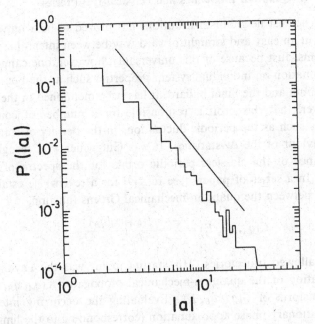

Fig. 7. Double logarithmic plot of the distribution of normalized curvatures $a_n = \ddot{x}_n/ < \dot{x}_n^2 >$ for a quartered Sinai billiard [28]. The straight line has a slope of -3.

the repulsion exponent v. Gaspard et al. [30] derived the expression

$$P(|a|) \simeq |a|^{-v-2} \tag{9}$$

for the asymptotic distribution. As in billiards energy distances are Wigner distributed with a repulsion exponent $v = 1$, an asymptotic behavior $P(|a|) \simeq |a|^{-3}$ is expected. This is exactly what is seen in Fig. 7, and what was also found in a calculated stadium spectrum [31].

The question arises of whether the level dynamic approach of Pechukas can also be applied to billiards. As the parameter dependence results from the shape but not from the Hamiltonian the best procedure to use is not obvious. One possible way is shown in ref. [29], where the billiard was transformed by a conformal mapping to a standard shape, thus shifting the parameter dependence from the shape to the Hamiltonian. This new Hamiltonian was then used to obtain a dynamic equation system following the method of Pechukas. Whether this model is again able to make predictions on nearest neighbor spacing distributions etc. is an open question.

4 Quantum mechanics and classical trajectories

The analysis of nearest neighbor distributions has proved to be a universal tool to discriminate in an easy and straightforward way between integrable and non-integrable systems. Just because of this universality, however, one cannot expect to get any information on individual system properties such as the bouncing ball orbit in the stadium and the Sinai billiard. It has been mentioned in the previous sections, however, that this orbit is responsible for a number of conspicuous spectral features such as the periodic fluctuations in the density of states or the saturation behaviour of the Δ_3-statistics. It was Gutzwiller who recognized the central importance of the classical periodic orbits for the spectra of noninte-grable systems. In a series of papers (see ref. [9] for a review) he established a correspondence between the quantum-mechanical Greens function,

$$G(q_A, q_B, E) = \sum_n \frac{\psi_n^*(q_A)\psi_n(q_B)}{E_n - E}, \qquad (10)$$

and classically allowed trajectories. His starting point was the Feynman path integral formulation of the quantum-mechanical propagator $K(q_A, q_B, t)$, being the Fourier transform of $G(q_A, q_B, E)$. Evaluating the occurring integrals by means of the stationary phase approximation (corresponding to the limit $h \to 0$) he got a semiclassical approximation for the Greens function, containing only classically allowed trajectories,

$$G(q_A, q_B, E) = \frac{1}{i(2\pi i)^{1/2}} \sum_r |\Delta_r|^{1/2} \, \exp[\frac{i}{h} \, S_r(q_A, q_B, E) - im_r\frac{\pi}{2}]. \qquad (11)$$

The sum is over all trajectories starting at q_A and ending at q_B. $S_r(q_A, q_B, E)$ is the classical action, and m_r is the Maslow index counting the number of conjugate points. The prefactor $|\Delta_r|^{1/2}$ takes into account the stability of the trajectory. It can be written as a Jacobian determinant,

$$\Delta_r = |\partial p_A^\alpha(q_A, q_B, E)/\partial q_B^\beta|, \quad \alpha, \, \beta = 1, \, 2, \qquad (12)$$

where the p_A^α are the two components of the momentum at the starting point, and the q_B^β are the components at the end position. Classically, Δ_r corresponds to the density of trajectories at point q_B starting isotropically from point q_A (for a discussion of these questions see ref. [32]).

The density of states can be expressed in terms of the Green's function as

$$\rho(E) = \lim_{\epsilon \to 0} \text{Im} \int G(q, q, E + i\epsilon)d^2q. \qquad (13)$$

Inserting expression (10) for the Green's function and applying again the stationary phase approximation, one ends at the Gutzwiller trace formula for the density of states,

$$\rho(E) = \rho_0(E) + \sum_r \rho_r \, \exp(\frac{i}{\hbar}S_r(E) - im_r\frac{\pi}{2}), \qquad (14)$$

where the sum is now restricted to all *periodic* orbits. If $\rho(E)$ is integrated over E, the first term on the right-hand side yields the extended Weyl formula (4), whereas the second term is responsible for the periodic fluctuations of $n(E)$ (see Fig. 2).

A direct calculation of $\rho(E)$ by summing up the different contributions of the periodic orbits is difficult, as the sum (13) is divergent and must first be made convergent by a skilful rearrangement of terms. That it is indeed possible in this way to reproduce the quantum-mechanical spectrum was recently demonstrated for two different billiards [33, 34] and the anisotropic Kepler problem [34].

The opposite procedure is easier, i.e. to start with the spectrum and to extract the contributions of the different periodic orbits out of it. If the action $S_r(E)$ scales with the energy, this can be done with the help of a Fourier transform. By this Wintgen [35] showed that the spectrum of a hydrogen atom in a strong magnetic field can be understood completely in terms of classical orbits of the electron. The same procedure is also possible for billiards. Here the action is given by

$$S(q_A, q_B, E) = \hbar k q, \tag{15}$$

where q is the length of the trajectory and $k = E^{1/2}$ is the de Broglie wave number of the particle. The Maslov index m_r equals just twice the number of reflections, as every reflection gives rise to a phase jump of π. Inserting these expressions into eq. (13), one sees that the Fourier transform of $\rho(E)$ with respect to the variable k,

$$\hat{\rho}(l) = \int \rho(k^2) \exp(-ikl) \, dk \sim \sum_n \exp(ik_n l), \tag{16}$$

where the sum is over all eigenfrequencies, should show maxima for all l values corresponding to lengths of periodic orbits. This is demonstrated in Fig. 8, where $|\hat{\rho}(l)|^2$ is plotted for a number of Sinai billiards of variable length [22]. Every resonance of $|\hat{\rho}(l)|^2$ can be attributed to a periodic orbit. If the length of the billiard is changed, all orbits but the bouncing ball orbit change their lengths too, leading to shifts of the corresponding resonances. The bouncing ball orbit resonance (and its harmonics corresponding to repeated orbits) is independent of the billiard length as it should be.

The close correspondence between quantum mechanics and classical trajectories is perhaps even more conspicuous, if wavefunctions are considered. Fig. 9 shows three eigenfunctions of the stadium billiard, obtained experimentally by a scan of the position of the coupling wire, as described in section 2 [11]. All the wavefunctions shown exhibit regions of extra high amplitude close to paths of periodic orbits, namely the bouncing ball orbit, the double diamond, and the whispering gallery. These scars were observed for the first time in calculations by Heller [8]. Approaches to explain the structures within the framework of the Gutzwiller theory have been made by Bogomolny [36] and Berry [37].

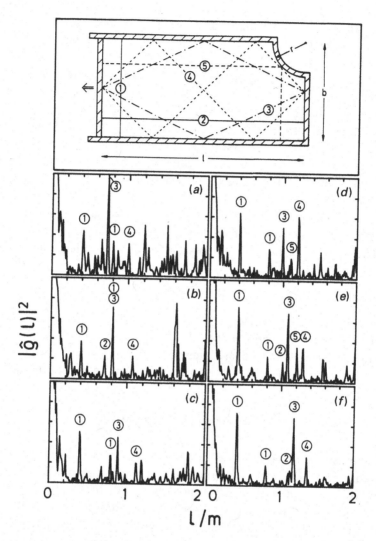

Fig. 8. Absolute square $|\hat{\rho}(l)|^2$ of the Fourier transform of the spectrum of a quartered Sinai billiard ($b=20$ cm, $r=7$ cm) for different billiard lengths $l((a)$: 30 cm, (b): 35 cm, (c): 40 cm, (d): 45 cm, (e): 50 cm, (f): 55 cm) [22]. Every resonance can be associated with a periodic orbit.

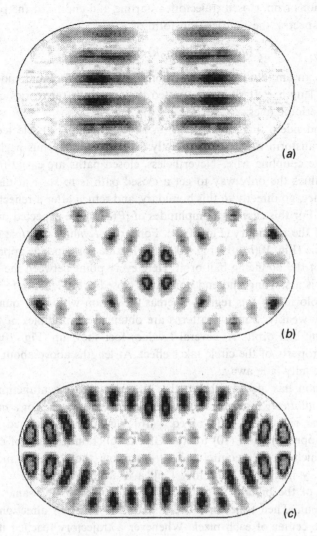

Fig. 9. Experimentally determined eigenfunctions for a stadium billiard [11]. The eigenfunctions show scars corresponding to the bouncing ball (*a*), the double diamond (*b*), and the whispering gallery orbit (*c*).

As the reflected microwave power corresponds to the square of the electric field strength at the wire position, i.e. to the absolute square of the wavefunction, the reflected microwave power measured as a function of position q and of wave number k yields directly the quantity

$$P(q, k) \sim \sum_n \frac{\lambda}{(E - E_n)^2 + \lambda^2} |\psi(q)|^2 \sim \text{Im} G(q, q, E - i\lambda).$$ (17)

Inserting expression (10) for the Greens function one gets $P(q, k)$ as a sum over contributions from closed trajectories starting and ending at the point q. As in the case of spectra, the Fourier transform of $P(q, k)$,

$$\hat{P}(q, l) = \sum P(q, k) \exp(-ikl) \, dk \tag{18}$$

should show high amplitudes at all positions q being starting and end points of trajectories of length l. It should be noted that now *all* trajectories contribute, not only the periodic ones.

The left-hand side of Fig. 10 shows $|\hat{P}(q, l)|$ for a number of selected l values. The Fourier transform was applied directly to the raw data, thus neglecting the influence of the coupling wire. Nevertheless, closed paths are easily recognized. For small l values the only way to get a closed path is to start at distance $l/2$ from a boundary, go directly to this boundary, and return after a reflection to the starting point. For this case high amplitudes of $|\hat{P}(q, l)|$ are expected on a rim of distance $l/2$ of the boundary (Fig. 10(a)). For larger l values multiple reflections become possible (Fig. 10(b)). In Fig. 10(c) the l value corresponds approximately to the length of the bouncing ball orbit. Here every point between the long sides of the billiard is a starting and end point of a closed orbit, leading to a more or less uniform coloring of this region, whereas the region within the quarter circle is colored only weakly. Similar patterns are obtained at multiples of the length of the bouncing ball orbit. At several l values foci show up (Fig. 10(d)). Here the focusing property of the circle takes effect. At lengths above about 1.5 m the structures gradually fade away.

The discussion has already shown that the measured eigenfunctions can be interpreted *qualitatively* in terms of classical closed trajectories. For a *quantitative* comparison the right-hand side of eq. (16) has to be calculated using the semiclassical approximation (10) for the Greens function. Instead of calculating the $|\Delta_r|^{1/2}$, which would require one to determine the so-called monodromy matrix for every individual trajectory [36], a brute force method was applied. First the area of the billiard was subdivided into pixels in the same way as in the measurement. Then trajectories were started in different directions in steps of 1° from the center of each pixel. Whenever a trajectory reached its starting pixel after length l, a corresponding array variable was increased or decreased by 1, depending on the number of reflections. The value calculated for a pixel is proportional to the density of returning trajectories, i.e. to Δ_r (see the remark following eq. (12)). This is just the square of the weight factor of the trajectory in the semiclassical Greens function. The described procedure is therefore able to reproduce expression (17) quantitatively. On the right-hand side of Fig. 10 the absolute values of the obtained array variables are represented in a grey scale. One observes a nearly complete correspondence between the figures on the left-hand side, obtained from the wave-functions, and their classical counterparts, obtained from classical trajectories. This shows that Gutzwillers semiclassical approximation of the Greens function is able to account not only for eigenvalue spectra but also for eigenfunction patterns. In other words, *geometrical optics*

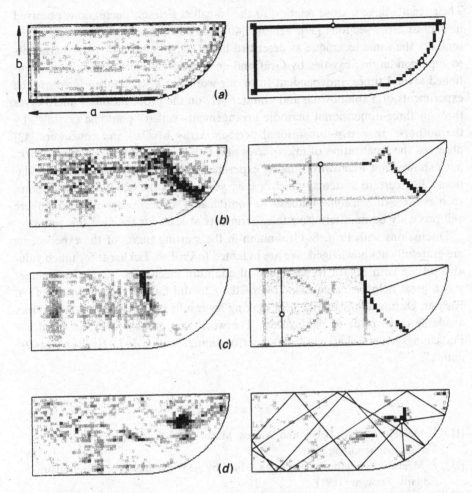

Fig. 10. Absolute values of the Fourier transform $|\hat{P}(q, l)|$ of the position-resolved spectrum of a quartered stadium billiard ($a=18$ cm, $b=13.5$ cm) for different l values ((a): 2.4 cm, (b): 19.5 cm, (c): 27.4 cm, (d): 84.1 cm, left column) and its classical counterparts calculated from classical trajectories (right column) [11]. The degree of blackening corresponds to the number of trajectories of length l starting and ending at the pixel. For details see text.

and the de Broglie relation are sufficient to understand both billiard eigenvalue and eigenfunction properties.

A brief survey closes this review. Stimulated by our first work on the experimental determination of billiard eigenvalues [15] a number of other experiments have been started to study quantum-mechanical systems by means of microwave analogs. First, the work of Doron and coworkers [38], who studied the chaotic reflection of microwaves in an elbow-shaped resonator, has to be mentioned.

Their results have a close relation to the so-called Ericson fluctuations observed in nuclear cross sections [39]. Sridhar [40] studied billiard eigenfunctions by essentially the same technique as described here. The extension of the experiments to superconducting cavities by Gräf and coworkers [14] has already been mentioned several times. Independent from our work but also closely related are the experiments of Yablonovitch and Gmitter [41] on the propagation of microwaves through three-dimensional periodic arrangements called "photonic crystals" by the authors. In a two-dimensional periodic array McCall and coworkers [42] observed the localization of microwaves at a lattice imperfection. All these works have shown that microwave analog experiments can compete well with calculations. For certain systems, which are as yet beyond the scope of calculations, such as three-dimensional billiards or complicated scattering arrangements, there will probably be no alternative to experiments at least for the next few years.

Discussions with Prof. S. Großmann in the starting phase of the experiments are gratefully acknowledged. We are indebted to Prof. B. Eckhardt for much valuable advice with regard to semiclassical quantum mechanics. K. H. Kretschmer was a great help in the construction of the Chladni figures experiment. Prof. A. Richter, Darmstadt, is thanked for making his results on superconducting cavities available to us prior to publication. The work was sponsored by the Deutsche Forschungsgemeinschaft via the Sonderforschungsbereich 185 "Nichtlineare Dynamik".

REFERENCES

[1]   F. Melde: Chladni's Leben und Wirken. Marburg: N. G. Elwertsche Verlagsbuchhandlung (1888).

[2]   F. Melde, in: Handbuch der Physik, edited by A. Winkelmann, p. 732. Breslau: Eduardt Trewent (1891)

[3]   O. Bohigas, M. J. Giannoni, C. Schmit: Phys. Rev. Lett. **52**, 1 (1984).

[4]   C. E. Porter: Statistical Theories of Spectra: fluctuations. New York: Academic (1965).

[5]   Th. Zimmermann, H. Köppel, L. S. Cederbaum, G. Persch, W. Demtröder, Phys. Rev. Lett. **61**, 3 (1988).

[6]   S. W. McDonald, A. N. Kaufman: Phys. Rev. Lett. **42**, 1189 (1979).

[7]   S. W. McDonald, A. N. Kaufman: Phys. Rev. **A37**, 3067 (1988).

[8]   E. J. Heller: Phys. Rev. Lett. **53**, 1515 (1984).

[9]   M. C. Gutzwiller: Chaos in Classical and Quantum Mechanics, Kap. 15. New York: Springer (1990).

[10] L. D. Landau, E. M. Lifschitz: Elastizitätstheorie, Lehrbuch der theoretischen Physik, Band VII. §25. Berlin: Akademie (1965).

[11] J. Stein, H.-J. Stöckmann: Phys. Rev. Lett. **68**, 2867 (1992).

[12] R. Blümel, I. H. Davidson, W. P. Reinhardt, H. Lin, M. Sharnoff: Phys. Rev. **A45**, 2641 (1992).

[13] P. O'Connor, J. Gehlen, E. J. Heller: Phys. Rev. Lett. **58**, 1296 (1987).

[14] H.-D. Gräf, H. L. Harney, H. Lengeler, C. H. Lewenkopf, C. Rangacharyulu, A. Richter, P. Schardt, H. A. Weidenmüller: Phys. Rev. Lett. **69**, 1296 (1992).

[15] H.-J. Stöckmann, J. Stein: Phys. Rev. Lett. **64**, 2215 (1990).

[16] H. P. Baltes, E. R. Hilf: Spectra of Finite Systems. Mannheim: B.-I.-Wissenschafts (1976).

[17] F. Haake, G. Lenz, P. Šeba, J. Stein, H.-J. Stöckmann, K. Życzkowski: Phys. Rev. **A44**, R6161 (1991).

[18] M. V. Berry, M. Tabor: Proc. R. Soc. Lond. **A349**, 101 (1976).

[19] G. Casati, B. V. Chirikov, I. Guarneri: Phys. Rev. Lett. **54**, 1350 (1985).

[20] P. Šeba: Phys. Rev. Lett **64**, 1855 (1990).

[21] F. J. Dyson, M. C. Mehta: J. Math. Phys. **4**, 701 (1963).

[22] J. Stein: Diploma work, Marburg 1990.

[23] T. A. Brody, J. Flores, J. B. French, P. A. Mello, A. Pandey, S. S. M. Wong: Rev. Mod. Phys. **53**, 385 (1981).

[24] B. Eckhardt: Phys. Rep. **163**, 205 (1988).

[25] M. V. Berry: Proc. R. Soc. Lond. **A400**, 229 (1985).

[26] P. Pechukas: Phys. Rev. Lett. **51**, 943 (1983).

[27] F. Haake: Quantum Signatures of Chaos. Heidelberg: Springer (1991).

[28] M. Kollmann: Diploma work, Marburg 1992.

[29] M. Kollmann, J. Stein, U. Stoffregen, H.-J. Stöckmann, B. Eckhardt: Phys. Rev. **E49**, R1 (1994).

[30] P. Gaspard, S. A. Rice, H. J. Mikeska, K. Nakamara: Phys. Rev. **A42**, 4015 (1990).

[31] T. Takami: J. Phys. Soc. Jap. **60**, 2489 (1991).

[32] M. V. Berry, M. E. Mount: Rep. Prog. Phys. **35**, 315 (1972).

[33] M. Sieber, F. Steiner: Phys. Rev. Lett. **67**, 1941 (1991).

[34] G. Tanner, P. Scherer, E. B. Bogomolny, B. Eckhardt, D. Wintgen: Phys. Rev. Lett. **67**, 2410 (1991).

[35] D. Wintgen: Phys. Rev. Lett. **58**, 1589 (1987).

[36] E. B. Bogomolny: Physica **D31**, 169 (1988).

[37] M. V. Berry: Proc. R. Soc. Lond. **A423**, 219 (1989).

[38] E. Doron, U. Smilansky, A. Frenkel: Phys. Rev. Lett. **65**, 3072 (1990).

[39] T. Ericson: Ann. Phys. **23**, 390 (1963).

[40] S. Sridhar: Phys. Rev. Lett. **67**, 785 (1991).

[41] E. Yablonovitch, T. J. Gmitter: Phys. Rev. Lett. **63**, 1950 (1989).

[42] S. L. McCall, P. M. Platzmann, R. Dalichaouch, D. Smith, S. Schultz: Phys. Rev. Lett. **67**, 2017 (1991).

Index

absorption spectra 221
action billiards 537
adiabatic motion 298
Aharonov–Bohm flux 205
Akulin–Dykhne model 107
anisotropic Kepler problem 290
anti-Stokes curve 522
Arnold cat maps 84, 457
atoms in a resonant cavity 439
avoided crossings 93

backscattering 24, 33, 190
Baker's transformation 453, 485
  quantal 455
  symmetric 489
band bending 217
Bargmann transform 515
bending modes 94
billiards 108, 326, 396, 430, 511, 537, 554, 662
Bloch states 30, 39, 41, 78, 79
Bohr–Sommerfeld quantization rule 278, 522
break time 72, 436
broadband noise 252
Brody formula 595

caustics 354, 448
chaos border 11, 239
Chladni figures 663
classical limit 12, 158, 436, 462
coherent state 113, 475, 494
conductance fluctuations 185
continued fractions 39, 127, 138
coupled kicked rotor 148
correlation dimension 655
correlation functions 5, 105, 442
  density–density 106
  pair 80, 602
  temporal auto 652
  two-level 200
correspondence principle 8, 41, 263
crossover time 647
cycle expansions 295, 420

Δ_3 statistics *see* rigidity
delocalization 27, 90
density of states 33, 108, 200, 320, 387, 394, 398, 544, 622
destruction time scale 15

diatomic molecules 88, 92
Dicke model 439
diffusion
  anomalous 41, 643
  classical 11, 13, 24, 26, 62, 100, 108, 148, 527, 560, 608
  quantum 12, 158, 175, 237, 641
dissipative quantum mechanics 164
double hump states 618
doubled Poisson brackets 363
doubled wave function 359
drift states 261
driven square well 144
Dyson Coulomb model 571
Dyson ensemble 124

EBK quantization 293
Ehrenfest theorem 14, 64
entropy 22, 45, 563
  information 195, 564, 596
  Kolmogorov–Sinai 4
ergodicity parameter 20
Ericson fluctuations 306
escape rate 413, 417

Fermi accelerating disk 591
Feynman diagrams 392
Feynman path integral 388
Fibonacci model 642
Floquet operator 29, 41, 88, 112, 158, 228, 558, 608
Fokker Planck equation 24
fractal dimension 639
Fredholm determinant 399, 414
frequency multiplexing 480
fundamental domain 290, 408

generalized dimensions 640
Ginibre statistics 309
Green's function 11, 24, 391, 448

$\hbar$ expansions 399
Hamilton–Jacobi equation 346
harmonic generation 224
Harper equation 636
Harper operator 465, 493
He atoms 288, 302
Helmholtz equation 397, 511, 662